Springer Handbook of Enzymes Volume 30

Dietmar Schomburg and
Ida Schomburg (Eds.)

Springer Handbook of Enzymes

Volume 30
Class 2 · Transferases III
EC 2.3.1.60–2.3.3.15

coedited by Antje Chang

Second Edition

Springer

Professor DIETMAR SCHOMBURG
e-mail: d.schomburg@uni-koeln.de

Dr. IDA SCHOMBURG
e-mail: i.schomburg@uni-koeln.de

Dr. ANTJE CHANG
e-mail: a.chang@uni-koeln.de

University to Cologne
Institute for Biochemistry
Zülpicher Strasse 47
50674 Cologne
Germany

ISBN-10 3-642-43627-7 2nd Edition Springer Berlin Heidelberg New York
ISBN-13 978-3-642-43627-7 2nd Edition Springer Berlin Heidelberg New York

The first edition was published as Volume 11 (ISBN 3-540-60295-X) and Volume 12 (ISBN 3-540-60703-X) of the "Enzyme Handbook".

Springer is a part of Springer Science+Business Media
springer.com

Softcover re-print of the Hardcover 2nd edition 2006

Cover design: Erich Kirchner, Heidelberg
Typesetting: medionet AG, Berlin

Printed on acid-free paper 2/3141m-5 4 3 2 1 0

Preface

Today, as the full information about the genome is becoming available for a rapidly increasing number of organisms and transcriptome and proteome analyses are beginning to provide us with a much wider image of protein regulation and function, it is obvious that there are limitations to our ability to access functional data for the gene products – the proteins and, in particular, for enzymes. Those data are inherently very difficult to collect, interpret and standardize as they are widely distributed among journals from different fields and are often subject to experimental conditions. Nevertheless a systematic collection is essential for our interpretation of genome information and more so for applications of this knowledge in the fields of medicine, agriculture, etc. Progress on enzyme immobilisation, enzyme production, enzyme inhibition, coenzyme regeneration and enzyme engineering has opened up fascinating new fields for the potential application of enzymes in a wide range of different areas.

The development of the enzyme data information system BRENDA was started in 1987 at the German National Research Centre for Biotechnology in Braunschweig (GBF) and is now continuing at the University at Cologne, Institute of Biochemistry. The present book "Springer Handbook of Enzymes" represents the printed version of this data bank. The information system has been developed into a full metabolic database.

The enzymes in this Handbook are arranged according to the Enzyme Commission list of enzymes. Some 4,000 "different" enzymes are covered. Frequently enzymes with very different properties are included under the same EC-number. Although we intend to give a representative overview on the characteristics and variability of each enzyme, the Handbook is not a compendium. The reader will have to go to the primary literature for more detailed information. Naturally it is not possible to cover all the numerous literature references for each enzyme (for some enzymes up to 40,000) if the data representation is to be concise as is intended.

It should be mentioned here that the data have been extracted from the literature and critically evaluated by qualified scientists. On the other hand, the original authors' nomenclature for enzyme forms and subunits is retained. In order to keep the tables concise, redundant information is avoided as far as possible (e.g. if K_m values are measured in the presence of an obvious cosubstrate, only the name of the cosubstrate is given in parentheses as a commentary without reference to its specific role).

The authors are grateful to the following biologists and chemists for invaluable help in the compilation of data: Cornelia Munaretto and Dr. Antje Chang.

Cologne
Spring 2006

Dietmar Schomburg, Ida Schomburg

List of Abbreviations

A	adenine
Ac	acetyl
ADP	adenosine 5'-diphosphate
Ala	alanine
All	allose
Alt	altrose
AMP	adenosine 5'-monophosphate
Ara	arabinose
Arg	arginine
Asn	asparagine
Asp	aspartic acid
ATP	adenosine 5'-triphosphate
Bicine	N,N'-bis(2-hydroxyethyl)glycine
C	cytosine
cal	calorie
CDP	cytidine 5'-diphosphate
CDTA	trans-1,2-diaminocyclohexane-N,N,N,N-tetraacetic acid
CMP	cytidine 5'-monophosphate
CoA	coenzyme A
CTP	cytidine 5'-triphosphate
Cys	cysteine
d	deoxy-
D-	(and L-) prefixes indicating configuration
DFP	diisopropyl fluorophosphate
DNA	deoxyribonucleic acid
DPN	diphosphopyridinium nucleotide (now NAD^+)
DTNB	5,5'-dithiobis(2-nitrobenzoate)
DTT	dithiothreitol (i.e. Cleland's reagent)
EC	number of enzyme in Enzyme Commission's system
E. coli	Escherichia coli
EDTA	ethylene diaminetetraacetate
EGTA	ethylene glycol bis(-aminoethyl ether) tetraacetate
ER	endoplasmic reticulum
Et	ethyl
EXAFS	extended X-ray absorption fine structure
FAD	flavin-adenine dinucleotide
FMN	flavin mononucleotide (riboflavin 5'-monophosphate)
Fru	fructose
Fuc	fucose
G	guanine
Gal	galactose

GDP	guanosine 5'-diphosphate
Glc	glucose
GlcN	glucosamine
GlcNAc	N-acetylglucosamine
Gln	glutamine
Glu	glutamic acid
Gly	glycine
GMP	guanosine 5'-monophosphate
GSH	glutathione
GSSG	oxidized glutathione
GTP	guanosine 5'-triphosphate
Gul	gulose
h	hour
H4	tetrahydro
HEPES	4-(2-hydroxyethyl)-1-piperazineethane sulfonic acid
His	histidine
HPLC	high performance liquid chromatography
Hyl	hydroxylysine
Hyp	hydroxyproline
IAA	iodoacetamide
Ig	immunoglobulin
Ile	isoleucine
Ido	idose
IDP	inosine 5'-diphosphate
IMP	inosine 5'-monophosphate
ITP	inosine 5'-triphosphate
K_m	Michaelis constant
L-	(and D-) prefixes indicating configuration
Leu	leucine
Lys	lysine
Lyx	lyxose
M	mol/l
mM	millimol/l
m-	*meta-*
Man	mannose
MES	2-(N-morpholino)ethane sulfonate
Met	methionine
min	minute
MOPS	3-(N-morpholino)propane sulfonate
Mur	muramic acid
MW	molecular weight
NAD^+	nicotinamide-adenine dinucleotide
NADH	reduced NAD
$NADP^+$	NAD phosphate
NADPH	reduced NADP
NAD(P)H	indicates either NADH or NADPH
NBS	N-bromosuccinimide

NDP	nucleoside 5'-diphosphate
NEM	N-ethylmaleimide
Neu	neuraminic acid
NMN	nicotinamide mononucleotide
NMP	nucleoside 5'-monophosphate
NTP	nucleoside 5'-triphosphate
o-	*ortho-*
Orn	ornithine
p-	*para-*
PBS	phosphate-buffered saline
PCMB	*p*-chloromercuribenzoate
PEP	phosphoenolpyruvate
pH	$-\log10[H^+]$
Ph	phenyl
Phe	phenylalanine
PHMB	*p*-hydroxymercuribenzoate
PIXE	proton-induced X-ray emission
PMSF	phenylmethane-sulfonylfluoride
p-NPP	*p*-nitrophenyl phosphate
Pro	proline
Q_{10}	factor for the change in reaction rate for a 10°C temperature increase
Rha	rhamnose
Rib	ribose
RNA	ribonucleic acid
mRNA	messenger RNA
rRNA	ribosomal RNA
tRNA	transfer RNA
Sar	N-methylglycine (sarcosine)
SDS-PAGE	sodium dodecyl sulfate polyacrylamide gel electrophoresis
Ser	serine
T	thymine
t_H	time for half-completion of reaction
Tal	talose
TDP	thymidine 5'-diphosphate
TEA	triethanolamine
Thr	threonine
TLCK	N^α-*p*-tosyl-L-lysine chloromethyl ketone
T_m	melting temperature
TMP	thymidine 5'-monophosphate
Tos-	tosyl-(*p*-toluenesulfonyl-)
TPN	triphosphopyridinium nucleotide (now $NADP^+$)
Tris	tris(hydroxymethyl)-aminomethane
Trp	tryptophan
TTP	thymidine 5'-triphosphate
Tyr	tyrosine
U	uridine
U/mg	μmol/(mg*min)

UDP	uridine 5'-diphosphate
UMP	uridine 5'-monophosphate
UTP	uridine 5'-triphosphate
Val	valine
Xaa	symbol for an amino acid of unknown constitution in peptide formula
XAS	X-ray absorption spectroscopy
Xyl	xylose

List of Deleted and Transferred Enzymes

Since its foundation in 1956 the Nomenclature Committee of the International Union of Biochemistry and Molecular Biology (NC-IUBMB) has continually revised and updated the list of enzymes. Entries for new enzymes have been added, others have been deleted completely, or transferred to another EC number in the original class or to different EC classes, catalyzing other types of chemical reactions. The old numbers have not been allotted to new enzymes; instead the place has been left vacant or cross-references given to the changes in nomenclature.

Deleted and Transferred Enzymes
For EC class 2.3.1.60–2.3.3.15 these changes are:

Recommended name	Old EC number	Alteration
6'-deoxychalcone synthase	2.3.1.120	deleted, reaction listed is due to EC 2.3.1.74
diacylglycerol acyltransferase	2.3.1.124	deleted, identical to EC 2.3.1.20

Index of Recommended Enzyme Names

Description of Data Fields

All information except the nomenclature of the enzymes (which is based on the recommendations of the Nomenclature Committee of IUBMB (International Union of Biochemistry and Molecular Biology) and IUPAC (International Union of Pure and Applied Chemistry) is extracted from original literature (or reviews for very well characterized enzymes). The quality and reliability of the data depends on the method of determination, and for older literature on the techniques available at that time. This is especially true for the fields *Molecular Weight* and *Subunits.*
The general structure of the fields is: **Information – Organism – Commentary – Literature**
The information can be found in the form of numerical values (temperature, pH, K_m etc.) or as text (cofactors, inhibitors etc.).
Sometimes data are classified as *Additional Information.* Here you may find data that cannot be recalculated to the units required for a field or also general information being valid for all values. For example, for *Inhibitors*, *Additional Information* may contain a list of compounds that are not inhibitory.
The detailed structure and contents of each field is described below. If one of these fields is missing for a particular enzyme, this means that for this field, no data are available.

1 Nomenclature

EC number

The number is as given by the IUBMB, classes of enzymes and subclasses defined according to the reaction catalyzed.

Systematic name

This is the name as given by the IUBMB/IUPAC Nomenclature Committee

Recommended name

This is the name as given by the IUBMB/IUPAC Nomenclature Committee

Synonyms

Synonyms which are found in other databases or in the literature, abbreviations, names of commercially available products. If identical names are frequently used for different enzymes, these will be mentioned here, cross references are given. If another EC number has been included in this entry, it is mentioned here.

CAS registry number

The majority of enzymes have a single chemical abstract (CAS) number. Some have no number at all, some have two or more numbers. Sometimes two enzymes share a common number. When this occurs, it is mentioned in the commentary.

2 Source Organism

For listing organisms their systematic name is preferred. If these are not mentioned in the literature, the names from the respective literature are used. For example if an enzyme from yeast is described without being specified further, yeast will be the entry. This field defines the code numbers for the organisms in which the enzyme with the respective EC number is found. These code numbers (form <_>) are displayed together with each entry in all fields of BRENDA where organism-specific information is given.

3 Reaction and Specificity

Catalyzed reaction

The reaction as defined by the IUBMB. The commentary gives information on the mechanism, the stereochemistry, or on thermodynamic data of the reaction.

Reaction type

According to the enzyme class a type can be attributed. These can be oxidation, reduction, elimination, addition, or a name (e.g. Knorr reaction)

Natural substrates and products

These are substrates and products which are metabolized in vivo. A natural substrate is only given if it is mentioned in the literature. The commentary gives information on the pathways for which this enzyme is important. If the enzyme is induced by a specific compound or growth conditions, this will be included in the commentary. In *Additional information* you will find comments on the metabolic role, sometimes only assumptions can be found in the references or the natural substrates are unknown.
In the listings, each natural substrate (indicated by a bold **S**) is followed by its respective product (indicated by a bold **P**). Products are given with organisms and references included only if the respective authors were able to demonstrate the formation of the specific product. If only the disappearance of the substrate was observed, the product is included without organisms of references. In cases with unclear product formation only a **?** as a dummy is given.

Substrates and products

All natural or synthetic substrates are listed (not in stoichiometric quantities). The commentary gives information on the reversibility of the reaction,

on isomers accepted as substrates and it compares the efficiency of substrates. If a specific substrate is accepted by only one of several isozymes, this will be stated here.
The field *Additional Information* summarizes compounds that are not accepted as substrates or general comments which are valid for all substrates.
In the listings, each substrate (indicated by a bold **S**) is followed by its respective product (indicated by a bold **P**). Products are given with organisms and references included if the respective authors demonstrated the formation of the specific product. If only the disappearance of the substrate was observed, the product will be included without organisms or references. In cases with unclear product formation only a **?** as a dummy is given.

Inhibitors

Compounds found to be inhibitory are listed. The commentary may explain experimental conditions, the concentration yielding a specific degree of inhibition or the inhibition constant. If a substance is activating at a specific concentration but inhibiting at a higher or lower value, the commentary will explain this.

Cofactors, prosthetic groups

This field contains cofactors which participate in the reaction but are not bound to the enzyme, and prosthetic groups being tightly bound. The commentary explains the function or, if known, the stereochemistry, or whether the cofactor can be replaced by a similar compound with higher or lower efficiency.

Activating Compounds

This field lists compounds with a positive effect on the activity. The enzyme may be inactive in the absence of certain compounds or may require activating molecules like sulfhydryl compounds, chelating agents, or lipids. If a substance is activating at a specific concentration but inhibiting at a higher or lower value, the commentary will explain this.

Metals, ions

This field lists all metals or ions that have activating effects. The commentary explains the role each of the cited metal has, being either bound e.g. as Fe-S centers or being required in solution. If an ion plays a dual role, activating at a certain concentration but inhibiting at a higher or lower concentration, this will be given in the commentary.

Turnover number (min^{-1})

The k_{cat} is given in the unit min^{-1}. The commentary lists the names of the substrates, sometimes with information on the reaction conditions or the type of reaction if the enzyme is capable of catalyzing different reactions with a single substrate. For cases where it is impossible to give the turnover number in the defined unit (e.g., substrates without a defined molecular weight, or an undefined amount of protein) this is summarized in *Additional Information*.

Specific activity (U/mg)

The unit is micromol/minute/milligram of protein. The commentary may contain information on specific assay conditions or if another than the natural substrate was used in the assay. Entries in *Additional Information* are included if the units of the activity are missing in the literature or are not calculable to the obligatory unit. Information on literature with a detailed description of the assay method may also be found.

K_m-Value (mM)

The unit is mM. Each value is connected to a substrate name. The commentary gives, if available, information on specific reaction condition, isozymes or presence of activators. The references for values which cannot be expressed in mM (e.g. for macromolecular, not precisely defined substrates) are given in *Additional Information.* In this field we also cite literature with detailed kinetic analyses.

K_i-Value (mM)

The unit of the inhibition constant is mM. Each value is connected to an inhibitor name. The commentary gives, if available, the type of inhibition (e.g. competitive, non-competitive) and the reaction conditions (pH-value and the temperature). Values which cannot be expressed in the requested unit and references for detailed inhibition studies are summerized under *Additional information.*

pH-Optimum

The value is given to one decimal place. The commentary may contain information on specific assay conditions, such as temperature, presence of activators or if this optimum is valid for only one of several isozymes. If the enzyme has a second optimum, this will be mentioned here.

pH-Range

Mostly given as a range e.g. 4.0–7.0 with an added commentary explaining the activity in this range. Sometimes, not a range but a single value indicating the upper or lower limit of enzyme activity is given. In this case, the commentary is obligatory.

Temperature optimum (°C)

Sometimes, if no temperature optimum is found in the literature, the temperature of the assay is given instead. This is always mentioned in the commentary.

Temperature range (°C)

This is the range over which the enzyme is active. The commentary may give the percentage of activity at the outer limits. Also commentaries on specific assay conditions, additives etc.

4 Enzyme Structure

Molecular weight

This field gives the molecular weight of the holoenzyme. For monomeric enzymes it is identical to the value given for subunits. As the accuracy depends on the method of determination this is given in the commentary if provided in the literature. Some enzymes are only active as multienzyme complexes for which the names and/or EC numbers of all participating enzymes are given in the commentary.

Subunits

The tertiary structure of the active species is described. The enzyme can be active as a monomer a dimer, trimer and so on. The stoichiometry of subunit composition is given. Some enzymes can be active in more than one state of complexation with differing effectivities. The analytical method is included.

Posttranslational modifications

The main entries in this field may be proteolytic modification, or side-chain modification, or no modification. The commentary will give details of the modifications e.g.:

- proteolytic modification <1> (<1>, propeptide Name) [1];
- side-chain modification <2> (<2>, N-glycosylated, 12% mannose) [2];
- no modification [3]

5 Isolation / Preparation / Mutation / Application

Source / tissue

For multicellular organisms, the tissue used for isolation of the enzyme or the tissue in which the enzyme is present is given. Cell-lines may also be a source of enzymes.

Localization

The subcellular localization is described. Typical entries are: cytoplasm, nucleus, extracellular, membrane.

Purification

The field consists of an organism and a reference. Only references with a detailed description of the purification procedure are cited.

Renaturation

Commentary on denaturant or renaturation procedure.

Crystallization

The literature is cited which describes the procedure of crystallization, or the X-ray structure.

Cloning

Lists of organisms and references, sometimes a commentary about expression or gene structure.

Engineering

The properties of modified proteins are described.

Application

Actual or possible applications in the fields of pharmacology, medicine, synthesis, analysis, agriculture, nutrition are described.

6 Stability

pH-Stability

This field can either give a range in which the enzyme is stable or a single value. In the latter case the commentary is obligatory and explains the conditions and stability at this value.

Temperature stability

This field can either give a range in which the enzyme is stable or a single value. In the latter case the commentary is obligatory and explains the conditions and stability at this value.

Oxidation stability

Stability in the presence of oxidizing agents, e.g. O_2, H_2O_2, especially important for enzymes which are only active under anaerobic conditions.

Organic solvent stability

The stability in the presence of organic solvents is described.

General stability information

This field summarizes general information on stability, e.g., increased stability of immobilized enzymes, stabilization by SH-reagents, detergents, glycerol or albumins etc.

Storage stability

Storage conditions and reported stability or loss of activity during storage.

References

Authors, Title, Journal, Volume, Pages, Year.

Gentamicin 3'-N-acetyltransferase 2.3.1.60

1 Nomenclature

EC number
2.3.1.60

Systematic name
acetyl-CoA:gentamicin-C $N^{3'}$-acetyltransferase

Recommended name
gentamicin 3'-N-acetyltransferase

Synonyms
acetyltransferase, gentamicin 3-
aminoglycoside acetyltransferase AAC(3)-1
gentamicin acetyltransferase I

CAS registry number
58500-58-6

2 Source Organism

<1> *Escherichia coli* (strain C600 containing R-factor JR88 [1,4,5]; strain R135/C600 [2]; strains JR88 and 72091801 [6]; strain JR88 [7]) [1, 2, 4, 5, 6, 7, 8]
<2> *Pseudomonas aeruginosa* (strain Travers PST [2]; plasmid pPK237 transferred to and expressed in E. coli K12 [3]) [2, 3]

3 Reaction and Specificity

Catalyzed reaction
acetyl-CoA + gentamicin C = CoA + $N^{3'}$-acetylgentamicin C

Reaction type
acyl group transfer

Substrates and products
S acetyl-CoA + N-ethylsisomicin <1> (Reversibility: ? <1> [5]) [5]
P CoA + $N^{3'}$-acetylethylsisomicin
S acetyl-CoA + dactimicin <1> (<1> the percentage of dactimicin affinity relative to gentamicin C_{1a} is very high (74.6%) only with AAC(3)-I, to the most other enzymes tested, in spite of the bifunctional AAC(6')/APH(2"), dactamicin is stable [8]) (Reversibility: ? <1> [8]) [8]

P CoA + $N^{3'}$-acetyldactamicin

S acetyl-CoA + dibekacin <1, 2> (<2> acetylation at 8% the rate of gentamicin C_1 [3]) (Reversibility: ? <1,2> [3,7]) [3, 7]

P CoA + $N^{3'}$-acetyldibekacin

S acetyl-CoA + fortimicin A <1> (Reversibility: ? <1> [5]) [5]

P CoA + $N^{3'}$-acetylfortimicin A

S acetyl-CoA + gentamicin A <1, 2> (<2> acetylation at 6% the rate of gentamicin C_1 [3]) (Reversibility: ? <1,2> [3,5]) [3, 5]

P CoA + $N^{3'}$-acetylgentamicin A

S acetyl-CoA + gentamicin B <1> (Reversibility: ? <1> [5]) [5]

P CoA + $N^{3'}$-acetylgentamicin B

S acetyl-CoA + gentamicin B_1 <1> (Reversibility: ? <1> [5]) [5]

P CoA + $N^{3'}$-acetylgentamicin B1

S acetyl-CoA + gentamicin C <2> (<2> acetylation at 86% the rate of gentamicin C_1 [3]) (Reversibility: ? <2> [3]) [3]

P CoA + $N^{3'}$-acetylgentamicin C

S acetyl-CoA + gentamicin C_1 <1, 2> (<2> best substrate [3]) (Reversibility: ? <1,2> [2,3,5]) [2, 3, 5]

P CoA + $N^{3'}$-acetylgentamicin C1

S acetyl-CoA + gentamicin C_{1a} <1, 2> (<1> best substrate [5]; <1> enzyme modifies the 3-N-position of the deoxystreptamine ring [5]; <2> acetylation at 60% the rate of gentamicin C_1 [3]) (Reversibility: ? <1,2> [1-8]) [1, 3-5, 7, 8]

P CoA + $N^{3'}$-acetylgentamicin C_{1a} <1, 2> [1, 3-5]

S acetyl-CoA + gentamicin C_2 <1, 2> (<2> acetylation at 84% the rate of gentamicin C_1 [3]) (Reversibility: ? <1,2> [3,5]) [3, 5]

P CoA + $N^{3'}$-acetylgentamicin C_2

S acetyl-CoA + gentamicin XK62-2 <1> (Reversibility: ? <1> [5]) [5]

P CoA + $N^{3'}$-acetylgentamicin XK62-2

S acetyl-CoA + kanamycin A <1, 2> (<2> acetylation at 7% the rate of gentamicin C_1 [3]) (Reversibility: ? <1,2> [2,3]) [2, 3]

P CoA + $N^{3'}$-acetylkanamycin A

S acetyl-CoA + kanamycin B <1> (Reversibility: ? <1> [5]) [5]

P CoA + $N^{3'}$-acetylkanamycin B

S acetyl-CoA + neamine <1> (Reversibility: ? <1> [5]) [5]

P CoA + $N^{3'}$-acetylneamine

S acetyl-CoA + nebramine <1> (Reversibility: ? <1> [5]) [5]

P CoA + $N^{3'}$-acetylnebramine

S acetyl-CoA + neomycin B <1> (Reversibility: ? <1> [2,7]) [2, 7]

P CoA + $N^{3'}$-acetylneomycin B

S acetyl-CoA + netilmicin <1> (<1> the much higher threshold value for the modification of netilmicin as compared with the one for the modification of sisomicin is due to the 1-N side chain [7]) (Reversibility: ? <1> [7]) [7]

P CoA + $N^{3'}$-acetylnetilmicin

S acetyl-CoA + paromomycin <1> (<1> not E. coli [2]) (Reversibility: ? <1> [2]) [2]

P CoA + $N^{3'}$-acetylparomycin B

S acetyl-CoA + sisomicin <1, 2> (<1,2> best substrate [2,5,7]; <2> acetylation at 51% the rate of gentamicin C_1 [3]) (Reversibility: ? <1,2> [2-5,7]) [2-5, 7]

P CoA + $N^{3'}$-acetylsisomicin <1, 2> [2]

S acetyl-CoA + tobramycin <1, 2> (<2> acetylation at 6% the rate of gentamicin C_1 [3]) (Reversibility: ? <1,2> [2-5]) [2-5]

P CoA + $N^{3'}$-acetyltobramycin

S Additional information <1, 2> (<1> substrate structure/activity relationship [5]; <1,2> no substrates are amikacin [2,3]; <2> lividomycin A, butirosin A and B [3]) [2, 3, 5]

P ?

Inhibitors

2-deoxystreptamine <1> (<1> weak [5]) [5]
CoA <1> (<1> product inhibition [4]) [4]
EDTA <2> (<2> above 0.5 mM [3]) [3]
N^3-acetyltobramycin <1> (<1> product inhibition [4]) [4]
Zn^{2+} <2> [3]
acetyltobramycin <1> [4]
amikacin <1> (<1> i.e. BB-K 8 [5]) [4, 5]
butyryl-CoA <1> (<1> dead-end inhibition pattern [4]) [4]
garamine A <1> [5]
gentamicin C_{1a} <1> (<1> substrate inhibition [1,4]) [1, 4]
kanamine <1> [5]
kanamycin A <1> [5]
lividomycin B <1> [5]
neomycin <1> (<1> best inhibitor [5]) [4, 5]
paromomycin I <1> [5]
ribostamycin <1> [4, 5]
Additional information <1, 2> (<1> inhibition kinetics [4, 5]; <2> no inhibition by NH_4^+, K^+, Na^+, DTT, N-methylmaleimide [3]) [3-5]

Activating compounds

Additional information <1> (<1> N-methylmaleimide and DTT have no effect on activity [1]) [1]

Metals, ions

Ca^{2+} <2> (<2> 1.0 mM, about 100% activity enhancement [3]) [3]
Mg^{2+} <2> (<2> 1.5 mM, about 100% activity enhancement [3]) [3]
NH_4^+ <2> (<2> slight activation [3]) [3]

Turnover number (min^{-1})

214 <1> (N-acetylgentamicin) [1]

Specific activity (U/mg)

3.8 <1> [1]
34 <2> [3]

K_m-Value (mM)
0.00012 <1> (gentamicin C_{1a}) [4]
0.0003-0.0005 <1> (gentamicin C_{1a}) [1]
0.0013 <1> (acetyl-CoA, <1> cosubstrate gentamicin C_{1a} [4]) [4]
0.0016 <1> (tobramycin) [4]
0.0017 <1> (acetyl-CoA, <1> cosubstrate tobramycin [4]) [4]
0.0039-0.0048 <1> (acetyl-CoA) [1]
0.066 <2> (gentamicin C1) [3]
Additional information <1> (<1> kinetic pattern [4,5]) [4, 5]

K_i-Value (mM)
0.0004 <1> (neomycin) [5]
0.002 <1> (lividomycin) [5]
0.0036 <1> (garamine) [5]
0.0083 <1> (paromomycin) [5]
0.024 <1> (amikacin) [5]
0.026 <1> (kanamycin C) [5]
0.07 <1> (ribostamycin) [5]
0.093 <1> (kanamycin A) [5]
0.16 <1> (gentamicin) [1]
0.208 <1> (kanamine) [5]
2 <1> (2-deoxystreptamine) [5]

pH-Optimum
7-7.6 <2> [3]
7.7-8 <1> [1]

pH-Range
5.5-8.5 <1> [1]

Temperature optimum (°C)
25 <1> (<1> assay at [1,4,5]) [1, 4, 5]
30 <1, 2> (<1,2> assay at [2]) [2]
37 <2> (<2> assay at [3]) [3]

4 Enzyme Structure

Molecular weight
63000 <1> (<1> equilibrium sedimentation centrifugation [1]) [1]
Additional information <2> (<2> amino acid analysis [3]) [3]

Subunits
tetramer <1> (<1> 4 * 17000, SDS-PAGE [1]) [1]

5 Isolation/Preparation/Mutation/Application

Localization
cytoplasm <1> [1]

Purification
<1> (partial [7]) [1, 7]
<2> (plasmid pPK237 transferred to and expressed in E. coli K12, affinity chromatography [3]) [3]

Cloning
<2> (Pseudomonas aeruginosa plasmid pPK237 transferred to and expressed in Escherichia coli K12 [3]) [3]

6 Stability

pH-Stability
4 <1> (<1> irreversible inactivation [1]) [1]
4.5 <1> (<1> no activity, restorable by neutralization [1]) [1]

Temperature stability
40 <2> (<2> above, inactivation within 2 min [3]) [3]

General stability information
<1>, lyophilization, stable to [1]
<2>, dialysis against EDTA with subsequent Mg^{2+}-addition to dialyzed enzyme has no effect on activity [3]
<2>, gel filtration leads to complete inactivation [3]

Storage stability
<1>, 20°C, lyophilized, up to 1 month [1]
<1>, frozen, redissolved ammonium sulfate precipitate, at least 1 month [1]
<1>, refrigerated, lyophilized, up to 1 month [1]
<2>, -20°C, partially purified, several months [3]
<2>, -20°C, purified, several days [3]

References

[1] Williams, J.W.; Northrop, D.B.: Purification and properties of gentamicin acetyltransferase I. Biochemistry, **15**, 125-131 (1976)
[2] Biddlecome, S.; Haas, M.; Davies, J.; Rane, D.F.; Daniels, P.J.L.: Enzymatic modification of aminoglycoside antibiotics: a new 3-N-acetylating enzyme from a Pseudomonas aeruginosa isolate. Antimicrob. Agents Chemother., **9**, 951-955 (1976)
[3] Angelatou, F.; Litsas, S.B.; Kontomichalou, P.: Purification and properties of two gentamicin-modifying enzymes, coded by a single plasmid pPK237 originating from Pseudomonas aeruginosa. J. Antibiot., **35**, 235-244 (1982)

[4] Williams, J.W.; Northrop, D.B.: Kinetic mechanisms of gentamicin acetyltransferase I. Antibiotic-dependent shift from rapid to nonrapid equilibrium random mechanisms. J. Biol. Chem., **253**, 5902-5907 (1978)
[5] Williams, J.W.; Northrop, D.B.: Substrate specificity and structure-activity relationships of gentamicin acetyltransferase I. The dependence of antibiotic resistance upon substrate V_{max}/K_m values. J. Biol. Chem., **253**, 5908-5914 (1978)
[6] Vliegenthart, John S.; Ketelaar-Van Gaalen, Petra A. G.; Van de Klundert, Jos A. M.: Identification of three genes coding for aminoglycoside-modifying enzymes by means of the polymerase chain reaction. J. Antimicrob. Chemother., **25**, 759-765 (1990)
[7] Bongaerts, Ger P. A.; Vliegenthart, John S.: Effect of aminoglycoside concentration on reaction rates of aminoglycoside-modifying enzymes. Antimicrob. Agents Chemother., **32**, 740-746 (1988)
[8] Gomez-Lus, R.; Gomez-Lus, S.; Goni, M. P.; Rivera, M. J.; Martin, C.; Rubio-Calvo, M. C.: Stability of dactimicin to aminoglycoside-modifying enzymes produced by 341 bacterial clinical isolates. Drugs Exp. Clin. Res., **15**, 129-132 (1989)

Dihydrolipoamide S-succinyltransferase 2.3.1.61

1 Nomenclature

EC number
2.3.1.61

Systematic name
enzyme-dihydrolipoyllysine:succinyl-CoA S-succinyltransferase

Recommended name
dihydrolipoyllysine-residue succinyltransferase

Synonyms
OGDC-E2 <32> [14]
dihydrolipoamide S-succinyltransferase
dihydrolipoamide succinyltransferase
dihydrolipoic transsuccinylase
dihydrolipolyl transsuccinylase
dihydrolipoyl transsuccinylase
lipoate succinyltransferase
lipoic transsuccinylase
lipoyl transsuccinylase
succinyl-CoA:dihydrolipoamide S-succinyltransferase
succinyl-CoA:dihydrolipoate S-succinyltransferase

CAS registry number
9032-28-4

2 Source Organism

<1> *Acetobacter xylinum* [7]
<2> *Acinetobacter lwoffii* [4]
<3> *Agrobacterium tumefaciens* [18]
<4> *Ralstonia eutropha* (H16, wild-type, ATCC 17699 [11]) [11, 16]
<5> *Azotobacter vinelandii* [9, 16, 17]
<6> *Azotobacter vinelandii* [11]
<7> *Geobacillus stearothermophilus* [16, 17]
<8> *Bacillus subtilis* [9, 15-17]
<9> *Bacillus subtilis* [11]
<10> *Bos taurus* [1, 14, 17]
<11> *Brucella abortus* (strain S19 [18]) [18]
<12> *Brucella melitensis* (strain 16M) [18]
<13> *Brucella ovis* [15]

<14> *Corynebacterium glutamicum* (synonym Brevibacterium lactofermentum [16]) [16]
<15> *Coxiella burnetii* (nucleotide sequence deposited in the GenBank and DDBJ [15]) [15]
<16> *Escherichia coli* [1-3, 6, 8, 9, 14-18]
<17> *Escherichia coli* (tE2oCD, Protein Data Bank accession number [16]) [16]
<18> *Escherichia coli* [11]
<19> *Escherichia coli* (Strain K12 [18]) [18]
<20> *Enterococcus faecalis* [16, 17]
<21> *Haemophilus influenzae* [16]
<22> *Homo sapiens* (human [10,12-14,16,17]) [10, 12-14, 16, 17]
<23> *Macaca fuscata* (japanese monkey [12]) [12]
<24> *Ochrobactrum anthropi* [18]
<25> *Phyllobacterium rubiacearum* [18]
<26> *Pseudomonas aeruginosa* [16]
<27> *Rattus norvegicus* [10, 12, 13, 17]
<28> *Rattus norvegicus* [11]
<29> *Rhizobium leguminosarum* [18]
<30> *Saccharomyces cerevisiae* (yeast [11]) [11]
<31> *Saccharomyces cerevisiae* (yeast [12,16,17]) [12, 16, 17]
<32> *Sus scrofa* (pig [1,5,14]; porcine [8,14,17]) [1, 5, 8, 14, 17]

3 Reaction and Specificity

Catalyzed reaction

succinyl-CoA + enzyme N^6-(dihydrolipoyl)lysine = CoA + enzyme N^6-(S-succinyldihydrolipoyl)lysine (EC 2.3.1.61 is a component of the multienzyme 2-oxoglutarate dehydrogenase complex (EC 1.2.4.2))

Reaction type

acyl group transfer
oxidative decarboxylation

Natural substrates and products

S succinyl-CoA + dihydrolipoamide <1-32> (<32> plays both a catalytic and a structural role in the 2-oxoglutarate dehydrogenase complex [5]) (Reversibility: ? <1-32> [1-18]) [1-18]
P CoA + S-succinyldihydrolipoamide

Substrates and products

S succinyl-CoA + dihydrolipoamide <1-32> (Reversibility: ? <1-32> [1-18]) [1-18]
P CoA + S-succinyldihydrolipoamide <1-32> [1-18]
S succinyl-CoA + dihydrolipoic acid <32> (Reversibility: ? <32> [5]) [5]
P CoA + succinyldihydrolipoate <32> [5]
S succinyl-CoA + dihydrolipoyllysine <32> (Reversibility: ? <32> [5]) [5]
P CoA + succinyldihydrolipoyllysine <32> [5]

Cofactors/prosthetic groups

lipoyl-protein <16, 32> (<16> contains 1 lipoyl moiety per chain [8]; <16> transsuccinylase core consists of 24 similar polypeptide chains, 12 of these chains contain a covalently bound lipoyl moiety [3]; <32> enzyme contains 8 mol of protein-bound lipoic acid per mol of enzyme [5]) [2, 3, 5, 8]

K_m-Value (mM)

0.06 <2> (succinyl-CoA) [4]
0.1 <32> (succinyl-CoA) [17]
4.2 <32> (dihydrolipoamide) [17]

pH-Optimum

7.2 <32> [5]

Temperature optimum (°C)

30 <32> (<32> at pH 7.2 [5]) [5]

4 Enzyme Structure

Molecular weight

41530 <32> (<32> amino acid sequence [17]) [17]
45000 <12> (<12> SDS-PAGE [18]) [18]
50000 <15> (<15> SDS-PAGE [15]) [15]
900000 <1> (<1> analytical ultracentrifugation [7]) [7]
958000 <32> (<32> analytical ultracentrifugation [5]) [5]
992000-1027000 <32> (<32> different preparations of the enzyme, analytical ultracentrifugation [5]) [5]
1000000 <16, 32> (<16> sucrose density gradient centrifugation, sedimentation equilibrium [3]; <32> sedimentation equlibrium [5]; <32> gel filtration, FPLC [17]) [1, 3, 5, 17]

Subunits

dodecamer <1> (<1> 12 * 75000, E2, gel filtration, SDS-PAGE [7]) [7]
polymer <16> (<16> made up of 24 similar structurally identical polypeptide chains [2,3,8]; <16> consists of 8 morphological subunits arranged in a cube-like structure [1]; <16> multienzyme complex contains 24 copies of the E20 subunit as an octahedral core that forms a truncated cube [9]; <16> 24 * 39000-42000, sedimentation equilibrium analysis [3]; <16> 24 * 47000-51000, SDS-PAGE [3]; <16> 24 * 40000 [6]; <32> 24 * 48000, SDS-PAGE [5,8]; <32> 24 * 41000, sedimentation equilibrium [5]; <32> 24 * 48000, SDS-PAGE [17]) [1-3, 5, 6, 8, 9, 16, 17]
trimer <16> (<16> 3 * 37000, E2o is normally a 24-mer, found as a trimer, when E2o is expressed with a C-terminal [His]6 tag, SDS-PAGE [16]) [16]

5 Isolation/Preparation/Mutation/Application

Source/tissue

Meynert's basal nucleus cell <23> [12]
blood <22> (<22> peripheral blood cells [10]) [10]
brain <32> [17]
cerebral cortex <23> [12]
fibroblast <22> [10, 12]
heart <22, 23, 27, 32> [1, 5, 8, 10, 12-14, 17]
hippocampus <23> [12]
kidney <10, 23> [1, 12]
liver <22> [14]
locus ceruleus <23> [12]
muscle <32> [5]
skeletal muscle <22, 27> [13]
substantia nigra <23> [12]

Localization

mitochondrion <22, 27, 28, 32> [10-14]
plasma membrane <22, 27> [13]

Purification

<1> [7]
<2> [4]
<16> [8, 16]
<22> [13]
<27> [12, 13]
<32> (recombinant enzyme [17]) [5, 14, 17]

Renaturation

<32> (10% recovery of enzyme activity after renaturation of SDS denatured enzyme (3%) [5]) [5]

Crystallization

<16> (space group F432 [2]; crystal structure of the truncated octahedral cubic E2 core reported [17]; trimeric from of the E2o catalytical domain crystallized in space group P3(1)21 [16]) [2, 3, 16, 17]

Cloning

<4> (cloning of the whole 2-oxoglutarase dehydrogenase complex structural genes, heterologous expression in Escherichia coli XL1-Blue and in the kgdA mutant Pseudomonas putida JS 347 [11]) [11]
<5> (molecular cloning and sequence analysis of the gene encoding the succinyltransferase component [16]) [16]
<8> (citM, the structural gene for dihydrolipoamide transsuccinylase, cloned and expressed in Escherichia coli [15]) [15]
<10> (cDNA identification [14]) [14]
<12> (sucB gene encoding E2o cloned, sequenced and expressed in Escherichia coli [18]) [18]

<14> (molecular cloning of the odhA gene [16]) [16]
<15> (sucB gene cloned by immunological screening of a lambda EMBL3 genomic library from strain Nine Mile DNA, expressed in Escherichia coli [15]) [15]
<16> [16]
<22> (isolation of the gene and pseudogene for the dihydrolipoamide succinyltransferase component of the human 2-oxoglutarate dehydrogenase complex from a human genomic DNA library, cloning vector lambda EMBL3 subcloned into plasmid vector pUC18, gene is located on chromosome 14 at q24.2-q24.3, pseudogene is located on chromosome at p31 [10,12]) [10, 12-14, 16, 17]
<27> (cDNA, sequence analysis [10,12,13]) [10, 12-14]
<32> (cDNA encoding porcine E20 cloned, with expression vector pET-11d overexpressed in Escherichia coli [17]; cDNA identification [14]) [14, 17]

Application

medicine <12, 15, 22> (<22> possible relationship between enzyme and familial Alzheimer's disease is discussed [10,12]; <15> enzyme may serve as a potential target antigen for diagnostic assays for Q fever [15]; <12> most pathogenic species for human [18]) [10, 12, 13, 15, 18]

6 Stability

Temperature stability

70 <32> (<32> loss of activity is slow up to 70°C, but rapid beyond 75°C [5]) [5]

General stability information

<32>, lyophilized enzyme loses more than 50% of its original activity [5]
<32>, stable to frequent freezing and thawing [5]

Storage stability

<32>, -18°C, 0.05 M potassium phosphate buffer, pH 7.0, containing 0.05 mM EDTA, stable for over 6 months without significant loss of activity, despite frequent freezing and thawing [5]

References

[1] Reed, L.J.; Cox, D.J.: Multienzyme complexes. The Enzymes, 3rd Ed. (Boyer, P.D., ed.), **1**, 213-240 (1970)
[2] Derosier, D.J.; Oliver, R.M.; Reed, L.J.: Crystallization and preliminary structural analysis of dihydrolipoyl transsuccinylase, the core of the 2-oxoglutarate dehydrogenase complex. Proc. Natl. Acad. Sci. USA, **68**, 1135-1137 (1971)
[3] Pettit, F.H.; Hamilton, L.; Munk, P.; Namihira, G.; Eley, M.H.; Willms, C.R.; Reed, L.J.: α-Keto acid dehydrogenase complexes. XIX. Subunit structure of

the Escherichia coli α-ketoglutarate dehydrogenase complex. J. Biol. Chem., **248**, 5282-5290 (1973)

[4] Hall, E.R.; Weitzman, P.D.J.: A continuous spectrophotometric assay for the transacylase (E2) component of pyruvate and α-oxoglutarate dehydrogenase enzyme complexes. Anal. Biochem., **62**, 286-290 (1974)

[5] Tanaka, N.; Koike, K.; Otsuka, K.I.; Hamada, M.; Ogasahara, K.; Koike, M.: Mammalian α-keto acid dehydrogenase complexes. 8. Properties and subunit composition of the pig heart lipoate succinyltransferase. J. Biol. Chem., **249**, 191-198 (1974)

[6] Angelides, K.J.; Hammes, G.G.: Structural and mechanistic studies of the α-ketoglutarate dehydrogenase multienzyme complex from Escherichia coli. Biochemistry, **18**, 5531-5537 (1979)

[7] De Kok, A.; Kornfeld, S.; Benziman, M.; Milner, Y.: Subunit composition and partial reactions of the 2-oxoglutarate dehydrogenase complex of Acetobacter xylinum. Eur. J. Biochem., **106**, 49-58 (1980)

[8] White, R.H.; Bleile, D.M.; Reed, L.J.: Lipoic acid content of dihydrolipoyl transacylases determined by isotope dilution analysis. Biochem. Biophys. Res. Commun., **94**, 78-84 (1980)

[9] Robien, M.A.; Clore, G.M.; Omichinski, J.G.; Perham, R.N.; Appella, E.; Sakaguchi, K.; Gronenborn, A.M.: Three-dimensional solution structure of the E3-binding domain of the dihydrolipoamide succinyltransferase core from the 2-oxoglutarate dehydrogenase multienzyme complex of Escherichia coli. Biochemistry, **31**, 3463-3471 (1992)

[10] Nakano, K.; Takase, C.; Sakamoto, T.; Nakagawa, S.; Inazawa, J.; Ohta, S.; Matuda, S.: Isolation, characterization and structural organization of the gene and pseudogene for the dihydrolipoamide succinyltransferase component of the human 2-oxoglutarate dehydrogenase complex. Eur. J. Biochem., **224**, 179-189 (1994)

[11] Hein, S.; Steinbuechel, A.: Cloning and characterization of the Alcaligenes eutrophus 2-oxoglutarate dehydrogenase complex. FEMS Microbiol. Lett., **136**, 231-238 (1996)

[12] Takase, C.; Nakano, K.; Ohta, S.; Nakagawa, S.; Matuda, S.: Different distribution of dihydrolipoamide succinyltransferase, dihydrolipoamide acetyltransferase and ATP synthase β-subunit in monkey brain. In Vivo, **10**, 495-502 (1996)

[13] Matuda, S.; Kodama, J.; Goshi, N.; Takase, C.; Nakano, K.; Nakagawa, S.; Ohta, S.: A polypeptide derived from mitochondrial dihydrolipoamide succinyltransferase is located on the plasma membrane in skeletal muscle. Biochem. Biophys. Res. Commun., **241**, 151-156 (1997)

[14] Koike, K.; Ishibashi, H.; Koike, M.: Immunoreactivity of porcine heart dihydrolipoamide acetyl- and succinyl-transferases (PDC-E2, OGDC-E2) with primary biliary cirrhosis sera: characterization of the autoantigenic region and effects of enzymic delipoylation and relipoylation. Hepatology, **27**, 1467-1474 (1998)

[15] Nguyen, S.V.; To, H.; Yamaguchi, T.; Fukushi, H.; Hirai, K.: Characterization of the Coxiella burnetii sucB gene encoding an immunogenic dihydrolipoamide succinyltransferase. Microbiol. Immunol., **43**, 743-749 (1999)

[16] Knapp, J.E.; Carroll, D.; Lawson, J.E.; Ernst, S.R.; Reed, L.J.; Hackert, M.L.: Expression, purification, and structural analysis of the trimeric form of the catalytic domain of the Escherichia coli dihydrolipoamide succinyltransferase. Protein Sci., **9**, 37-48 (2000)
[17] Koike, K.; Suematsu, T.; Ehara, M.: Cloning, overexpression and mutagenesis of cDNA encoding dihydrolipoamide succinyltransferase component of the porcine 2-oxoglutarate dehydrogenase complex. Eur. J. Biochem., **267**, 3005-3016 (2000)
[18] Zygmunt, M.S.; Diaz, M.A.; Teixeira-Gomes, A.P.; Cloeckaert, A.: Cloning, nucleotide sequence, and expression of the Brucella melitensis sucB gene coding for an immunogenic dihydrolipoamide succinyltransferase homologous protein. Infect. Immun., **69**, 6537-6540 (2001)

2-Acylglycerophosphocholine O-acyltransferase 2.3.1.62

1 Nomenclature

EC number
2.3.1.62

Systematic name
acyl-CoA:2-acyl-sn-glycero-3-phosphocholine O-acyltransferase

Recommended name
2-acylglycerophosphocholine O-acyltransferase

Synonyms
2-acyl-glycerol-3-phosphorylcholine acyltransferase <5> [3]
2-acylglycerophosphocholine acyltransferase <1, 2> [5, 6]
acyltransferase <1> [1]

CAS registry number
64295-73-4

2 Source Organism

<1> *Rattus norvegicus* (male Wistar strain [1]) [1, 2, 5]
<2> *Cavia porcellus* [2, 6]
<3> *Bos taurus* [2]
<4> *Sus scrofa* [2]
<5> *Tetrahymena pyriformis* (NT-1 [3]) [3]
<6> *Saccharomyces cerevisiae* (OC-2 [4]) [4]

3 Reaction and Specificity

Catalyzed reaction
acyl-CoA + 2-acyl-sn-glycero-3-phosphocholine = CoA + phosphatidylcholine

Reaction type
acyl group transfer

Natural substrates and products

S acyl-CoA + 2-acyl-sn-glycero-3-phosphocholine <1-6> (<1-6> reaction in phosphatidylcholine metabolism [1-6]) (Reversibility: ? <1-6> [1-6]) [1-6]

P CoA + phosphatidylcholine <1-6> [1-6]

Substrates and products

S acyl-CoA + 2-acyl-sn-glycero-3-phosphocholine <1-6> (Reversibility: ? <1-6> [1-6]) [1-6]

P CoA + phosphatidylcholine <1-6> [1-6]

S arachidonoyl-CoA + 2-arachidonoyl-sn-glycero-3-phosphocholine <2> (Reversibility: ? <2> [6]) [6]

P CoA + 1-arachidonoyl-2-arachidonoyl-sn-glycero-3-phosphocholine <2> [6]

S arachidonoyl-CoA + 2-palmitoyl-sn-glycero-3-phosphocholine <2> (Reversibility: ? <2> [6]) [6]

P CoA + 1-arachidonoyl-2-palmitoyl-sn-glycero-3-phosphocholine <2> [6]

S γ-linolenoyl-CoA + 2-acyl-sn-glycero-3-phosphocholine <5> (Reversibility: ? <5> [3]) [3]

P CoA + 1-γ-linolenoyl-2-acyl-sn-glycero-3-phosphocholine <5> [3]

S linoleoyl-CoA + 2-acyl-sn-glycero-3-phosphocholine <1-5> (Reversibility: ? <1-5> [2, 3]) [2, 3]

P CoA + 1-linoleoyl-2-acyl-sn-glycero-3-phosphocholine <1-5> [2, 3]

S linoleoyl-CoA + 2-arachidonoyl-sn-glycero-3-phosphocholine <2> (Reversibility: ? <2> [6]) [6]

P CoA + 1-linoleoyl-2-arachidonoyl-sn-glycero-3-phosphocholine <2> [6]

S linoleoyl-CoA + 2-palmitoyl-sn-glycero-3-phosphocholine <2> (Reversibility: ? <2> [6]) [6]

P CoA + 1-linoleoyl-2-palmitoyl-sn-glycero-3-phosphocholine <2> [6]

S myristoyl-CoA + 2-acyl-sn-glycero-3-phosphocholine <1-5> (Reversibility: ? <1-5> [2, 3]) [2, 3]

P CoA + 1-myristoyl-2-acyl-sn-glycero-3-phosphocholine <1-5> [2, 3]

S oleoyl-CoA + 2-acyl-sn-glycero-3-phosphocholine <1-6> (Reversibility: ? <1-6> [2-4]) [2-4]

P CoA + 1-oleoyl-2-acyl-sn-glycero-3-phosphocholine <1-6> [2-4]

S oleoyl-CoA + 2-arachidonoyl-sn-glycero-3-phosphocholine <2> (Reversibility: ? <2> [6]) [6]

P CoA + 1-oleoyl-2-arachidonoyl-sn-glycero-3-phosphocholine <2> [6]

S oleoyl-CoA + 2-palmitoyl-sn-glycero-3-phosphocholine <1, 2> (Reversibility: ? <2> [1, 6]) [1, 6]

P CoA + 1-oleoyl-2-palmitoyl-sn-glycero-3-phosphocholine <1, 2> [1, 6]

S palmitoleoyl-CoA + 2-acyl-sn-glycero-3-phosphocholine <5> (Reversibility: ? <5> [3]) [3]

P CoA + 1-palmitoleoyl-2-acyl-sn-glycero-3-phosphocholine <5> [3]

S palmitoyl-CoA + 2-acyl-sn-glycero-3-phosphocholine <1-6> (Reversibility: ? <1-6> [2-4]) [2-4]

P CoA + 1-palmitoyl-2-acyl-sn-glycero-3-phosphocholine <1-6> [2-4]

S palmitoyl-CoA + 2-arachidonoyl-sn-glycero-3-phosphocholine <1, 2> (Reversibility: ? <1, 2> [1, 6]) [1, 6]
P CoA + 1-palmitoyl-2-arachidonoyl-sn-glycero-3-phosphocholine <1, 2> [1, 6]
S palmitoyl-CoA + 2-linolenoyl-sn-glycero-3-phosphocholine <1> (Reversibility: ? <1> [1]) [1]
P CoA + 1-palmitoyl-2-linolenoyl-sn-glycero-3-phosphocholine <1> [1]
S palmitoyl-CoA + 2-linoleoyl-sn-glycero-3-phosphocholine <1> (Reversibility: ? <1> [1]) [1]
P CoA + 1-palmitoyl-2-linoleoyl-sn-glycero-3-phosphocholine <1> [1]
S palmitoyl-CoA + 2-oleoyl-sn-glycero-3-phosphocholine <1> (Reversibility: ? <1> [1]) [1]
P CoA + 1-palmitoyl-2-oleoyl-sn-glycero-3-phosphocholine <1> [1]
S palmitoyl-CoA + 2-palmitoyl-sn-glycero-3-phosphocholine <1, 2> (Reversibility: ? <1, 2> [1, 6]) [1, 6]
P CoA + 1,2-dipalmitoyl-sn-glycero-3-phosphocholine <1, 2> [1, 6]
S palmitoyl-CoA + 2-stearoyl-sn-glycero-3-phosphocholine <1> (Reversibility: ? <1> [1]) [1]
P CoA + 1-palmitoyl-2-stearoyl-sn-glycero-3-phosphocholine <1> [1]
S stearoyl-CoA + 2-acyl-sn-glycero-3-phosphocholine <1-5> (Reversibility: ? <1-5> [2, 3]) [2, 3]
P CoA + 1-stearoyl-2-acyl-sn-glycero-3-phosphocholine <1-5> [2, 3]
S stearoyl-CoA + 2-arachidonoyl-sn-glycero-3-phosphocholine <1, 2> (Reversibility: ? <1, 2> [1, 6]) [1, 6]
P CoA + 1-stearoyl-2-arachidonoyl-sn-glycero-3-phosphocholine <1, 2> [1, 6]
S stearoyl-CoA + 2-linolenoyl-sn-glycero-3-phosphocholine <1> (Reversibility: ? <1> [1]) [1]
P CoA + 1-stearoyl-2-linolenoyl-sn-glycero-3-phosphocholine <1> [1]
S stearoyl-CoA + 2-linoleoyl-sn-glycero-3-phosphocholine <1> (Reversibility: ? <1> [1]) [1]
P CoA + 1-stearoyl-2-linoleoyl-sn-glycero-3-phosphocholine <1> [1]
S stearoyl-CoA + 2-oleoyl-sn-glycero-3-phosphocholine <1> (Reversibility: ? <1> [1]) [1]
P CoA + 1-stearoyl-2-oleoyl-sn-glycero-3-phosphocholine <1> [1]
S stearoyl-CoA + 2-palmitoyl-sn-glycero-3-phosphocholine <1, 2> (Reversibility: ? <1, 2> [1, 6]) [1, 6]
P CoA + 1-stearoyl-2-palmitoyl-sn-glycero-3-phosphocholine <1, 2> [1, 6]
S stearoyl-CoA + 2-stearoyl-sn-glycero-3-phosphocholine <1> (Reversibility: ? <1> [1]) [1]
P CoA + 1,2-distearoyl-sn-glycero-3-phosphocholine <1> [1]

Inhibitors

5,5'-dithiobis-(2-nitrobenzoic acid) <2> (<2> 0.5 mM 76% activity [6]) [6]
N-ethylmaleimide <2> (<2> 0.5 mM 68% activity [6]; <1> not inhibitory [2]) [6]
dithiothreitol <2> (<2> 0.5 mM 47% activity [6]) [6]

p-chloromercuribenzoate <1> (<1> above 2.0 mM [2]) [2]
palmitoyl-CoA <1> (<1> above 0.064 mM [1]; <1> bovine serum albumin or boiled microsomal preparation restores activity [2]) [1, 2]
stearoyl-CoA <1> (<1> above 0.064 mM [1]) [1]
Additional information <1, 2> (<1> not inhibited by iodoacetamide [2]; <2> not inhibited by glutathione or iodoacetate [6]) [2, 6]

Activating compounds
bovine serum albumin <1> (<1> fatty acid deficient, 2 mg per ml [2]) [2]

Metals, ions
Mg^{2+} <2> (<2> enhanced activity by 75% [6]) [6]
Additional information <2> (<2> Ca^{2+} does not affect activity [6]) [6]

Specific activity (U/mg)
Additional information <5> (<5> specific activities of microsomal preparations from cells grown at different temperatures [3]) [3]

5 Isolation/Preparation/Mutation/Application

Source/tissue
brain <1> [2]
heart <2> [6]
intestine <1> [2]
liver <1-4> [1, 2, 5]

Localization
microsome <1-6> [1-6]

6 Stability

Storage stability
<6>, -70°C, membrane preparation, at least 2 months [4]

References

[1] Holub, B.: The suitability of different acyl acceptors as substrates for the acyl-CoA:2-acyl-sn-glycero-3-phosphorylcholine acyltransferase in rat liver microsomes. Biochim. Biophys. Acta, **664**, 221-228 (1981)
[2] Lands, W.E.M.; Hart, P.: Metabolism of glycerolipids. VI. Specificities of acyl-CoA:phospholipid acyltransferases. J. Biol. Chem., **240**, 1905-1911 (1965)
[3] Yoshioka, S.; Kameyama, Y.; Nozawa, Y.: Mechanism for adaptive modification during cold acclimation of phospholipid acyl chain composition in Tetrahymena. II. Activities of 2-acyl-sn-glycerol-3-phosphorylcholine and 2-acyl-sn-glycerol-3-phosphorylethanolamine acyltransferases involving the reacylation. Biochim. Biophys. Acta, **793**, 34-41 (1984)

[4] Yamada, K.; Okuyama, H.; Endo, Y.; Ikezawa, H.: Acyltransferase systems involved in phospholipid metabolism in Saccharomyces cerevisiae. Arch. Biochem. Biophys., **183**, 281-289 (1977)
[5] Van den Bosch, H.; Van Golde, L.M.G.; Slotboom, A.J.; Van Deenen, L.L.M.: The acylation of isomeric monoacyl phosphatidylcholines. Biochim. Biophys. Acta, **152**, 694-703 (1968)
[6] Arthur, G.: Acylation of 2-acyl-glycerophosphocholine in guinea pig heart microsomal fractions. Biochem. J., **261**, 575-580 (1989)

1-Alkylglycerophosphocholine O-acyltransferase 2.3.1.63

1 Nomenclature

EC number
2.3.1.63

Systematic name
acyl-CoA:1-alkyl-sn-glycero-3-phosphocholine O-acyltransferase

Recommended name
1-alkylglycerophosphocholine O-acyltransferase

Synonyms
Additional information (may be identical with EC 2.3.1.23)

CAS registry number
58693-63-3

2 Source Organism

<-1> no activity in *Rattus norvegicus* (liver microsomes [3]) [3]
<1> *Oryctolagus cuniculus* [1, 3]
<2> *Mus musculus* [1-4]
<3> *Homo sapiens* [3]

3 Reaction and Specificity

Catalyzed reaction
acyl-CoA + 1-alkyl-sn-glycero-3-phosphocholine = CoA + 2-acyl-1-alkyl-sn-glycero-3-phosphocholine

Reaction type
acyl group transfer

Substrates and products
S acyl-CoA + 1-alkyl-sn-glycero-3-phosphocholine <1, 2, 3> (<1,2> i.e. lysophospholipid or lysoPAF [1]; <1,2> stereospecific, unsaturated acyl-CoA preferred over saturated species [1,3]) (Reversibility: ? <1, 2, 3> [1, 3]) [1-4]
P CoA + 2-acyl-1-alkyl-sn-glycero-3-phosphocholine <1, 2, 3> (<1,2> i.e. plasmanylcholine [1,3]) [1-4]

S arachidonoyl-CoA + 1-alkyl-sn-glycero-3-phosphocholine <2> (<2> Ehrlich ascites tumor cells [1]; <2> specific for polyunsaturated fatty acids [2]) (Reversibility: ? <2> [1]) [1, 2]
P CoA + 2-arachidonoyl-1-alkyl-sn-glycero-3-phosphocholine <2> [1, 2]
S linolenoyl-CoA + 1-alkyl-sn-glycero-3-phosphocholine <1, 2> (<2> specific for polyunsaturated fatty acids [2]) (Reversibility: ? <1, 2> [1, 2]) [1, 2, 3]
P CoA + 2-linolenoyl-1-alkyl-sn-glycero-3-phosphocholine <1, 2> [1, 2, 3]
S linoleoyl-CoA + 1-alkyl-sn-glycero-3-phosphocholine <1, 2> (<2> preferred acyl-donor, specific for polyunsaturated fatty acids, no activity with palmitoyl-CoA, stearoyl-CoA and oleoyl-CoA [2,4]) (Reversibility: ? <1, 2> [1, 3, 4]) [1, 3, 4]
P CoA + 2-linoleoyl-1-alkyl-sn-glycero-3-phosphocholine <1, 2> [1, 3, 4]

Metals, ions
Additional information <1> (<1> no cation requirement [1]) [1]

Specific activity (U/mg)
0.00046 <1> [1]

pH-Optimum
7-8 <1> [1]

Temperature optimum (°C)
25 <1, 2> (<1,2> assay at [1]) [1]

5 Isolation/Preparation/Mutation/Application

Source/tissue
Ehrlich ascites carcinoma cell <2> [1, 2, 4]
erythrocyte <3> (<3> low activity [3]) [3]
hematopoietic cell <1, 2> [1]
muscle <1, 2> (<1> hind legs and back [3]) [1, 3]

Localization
membrane <3> [3]
microsome <2> (<2> Ehrlich ascites tumor cells [4]) [4]
sarcoplasmic reticulum <1> [1]

References

[1] Choy, P.C.; McMaster, C.R.: 1-Alkyl- and 1-alkenylglycerophosphocholine acyltransferases. Methods Enzymol., **209**, 86-92 (1992)
[2] Waku, K.; Nakazawa, Y.: Regulation of the fatty acid composition of alkyl ether phospholipid in Ehrlich ascites tumor cells. The substrate specificities of 1-O-alkylglycerol 3-phosphate and 1-O-alkylglycero-3-phosphocholine acyltransferases. J. Biochem., **82**, 1779-1784 (1977)

[3] Waku, K.; Nakazawa, Y.: Acyltransferase activity to 1-O-alkyl-glycero-3-phosphorylcholine in sarcoplasmic reticulum. J. Biochem., **68**, 459-466 (1970)
[4] Waku, K.; Nakazawa, Y.: Acyltransferae activity to 1-acyl-, 1-O-alkenyl-, and 1-O-alkyl-glycero-3-phosphorylcholine in Ehrlich ascites tumor cells. J. Biochem., **72**, 495-497 (1972)

Agmatine N^4-coumaroyltransferase 2.3.1.64

1 Nomenclature

EC number
2.3.1.64

Systematic name
4-coumaroyl-CoA:agmatine N^4-coumaroyltransferase

Recommended name
agmatine N^4-coumaroyltransferase

Synonyms
agmatine coumaroyltransferase
coumaroyltransferase, agmatine
p-coumaroyl-CoA-agmatine N-*p*-coumaroyltransferase

CAS registry number
85030-72-4

2 Source Organism

<1> *Hordeum vulgare* (barley, cv. claret [1]; cv. Proctor [3]; isoenzymes 1-3 [4]) [1-4]

3 Reaction and Specificity

Catalyzed reaction
4-coumaroyl-CoA + agmatine = CoA + N-(4-guanidinobutyl)-N-4-hydroxycinnamamide

Reaction type
acyl group transfer

Natural substrates and products

S 4-coumaroyl-CoA + agmatine <1> (<1> reaction in hordatine anabolism [1]) (Reversibility: ? <1> [1]) [1]

P CoA + 4-coumaroylagmatine <1> [1]

Substrates and products

S 4-coumaroyl-CoA + agmatine <1, 1> (<1> no activity with putrescine, spermidine, spermine, cadaverine, arginine, homoarginine [2,3]; <1> no activity with N-carbamoylputrescine, homoagmatine [3]) (Reversibility: ? <1> [1-3]) [1-3]

P CoA + 4-coumaroylagmatine <1> [1-3]
S caffeoyl-CoA + agmatine <1> (Reversibility: ? <1> [2, 3]) [2, 3]
P CoA + caffeoylagmatine <1> [2, 3]
S cinnamoyl-CoA + agmatine <1> (Reversibility: ? <1> [3]) [3]
P CoA + cinnamoylagmatine <1> [3]
S feruloyl-CoA + agmatine <1> (Reversibility: ? <1> [2, 3]) [2, 3]
P CoA + feruloylagmatine <1> [2, 3]
S sinapoyl-CoA + agmatine <1> (Reversibility: ? <1> [3]) [3]
P CoA + sinapoylagmatine <1> [3]

Inhibitors

$CuSO_4$ <1> (<1> 10 mM, 85% inhibition [4]) [4]
$MnCl_2$ <1> (<1> 10 mM, 29% inhibition [4]) [4]
$ZnSO_4$ <1> (<1> 10 mM, 99% inhibition [4]) [4]
dithiothreitol <1> (<1> 30% inhibition [4]) [4]

Activating compounds

1-mercapto-2,3-propanediol <1> (<1> 1 mM is required for optimal activity [4]) [4]
2-mercaptoethanol <1> (<1> 1 mM is required for optimal activity [4]) [4]

Specific activity (U/mg)

0.393 <1> [2]
1.22 <1> [3]
1.78 <1> [4]
11.34 <1> (<1> recombinant enzyme [4]) [4]

K_m-Value (mM)

0.00074 <1> (4-coumaroyl-CoA) [2]
0.001 <1> (4-coumaroyl-CoA, <1> isoenzyme 1 [4]) [4]
0.0013 <1> (feruloyl-CoA) [3]
0.0015 <1> (4-coumaroyl-CoA) [3]
0.0016 <1> (4-coumaroyl-CoA, <1> isoenzyme 3 [4]) [4]
0.0017 <1> (feruloyl-CoA, <1> isoenzyme 3 [4]) [4]
0.0021 <1> (4-coumaroyl-CoA, <1> isoenzyme 2 [4]) [4]
0.0021 <1> (4-coumaroyl-CoA, <1> recombinant enzyme [4]) [4]
0.0033 <1> (caffeoyl-CoA) [3]
0.0033 <1> (cinnamoyl-CoA) [3]
0.0041 <1> (feruloyl-CoA, <1> isoenzyme 2 [4]) [4]
0.0043 <1> (caffeoyl-CoA, <1> isoenzyme 2 [4]) [4]
0.0043 <1> (caffeoyl-CoA, <1> recombinant enzyme [4]) [4]
0.0052 <1> (agmatine, <1> isoenzyme 1 [4]) [4]
0.0052 <1> (agmatine, <1> isoenzyme 3 [4]) [4]
0.0053 <1> (caffeoyl-CoA, <1> isoenzyme 3 [4]) [4]
0.0058 <1> (caffeoyl-CoA, <1> isoenzyme 1 [4]) [4]
0.0061 <1> (feruloyl-CoA, <1> isoenzyme 1 [4]) [4]
0.0065 <1> (agmatine, <1> isoenzyme 2 [4]) [4]

0.0066 <1> (feruloyl-CoA, <1> recombinant enzyme [4]) [4]
0.0081 <1> (agmatine, <1> recombinant enzyme [4]) [4]
0.0095 <1> (agmatine) [3]
0.012 <1> (agmatine) [2]
0.05 <1> (sinapoyl-CoA) [3]

pH-Optimum
7.5 <1> [2, 3, 4]
8.5 <1> [1]

pH-Range
6.9-8.1 <1> (<1> half-maximal activity at pH 6.9 and pH 8.1 [4]) [4]
7-8 <1> (<1> approx. half-maximal activity at pH 7.0 and 8.0 [2,3]) [2, 3]

Temperature optimum (°C)
30 <1> (<1> assay at [1]) [1]
40 <1> [4]

Temperature range (°C)
20-50 <1> [4]

4 Enzyme Structure

Molecular weight
40000 <1> (<1> gel filtration [2]) [2, 3, 4]

Subunits
monomer <1> (<1> 1 * 48000, SDS-PAGE [4]) [4]

5 Isolation/Preparation/Mutation/Application

Source/tissue
seedling <1> (<1> shoots, not seeds [2]; <1> maximal activity 3-4 days after germination at 22°C in the dark [2]) [1-3]

Localization
soluble <1> [3]

Purification
<1> (ammonium sulfate, Biogel A, Sepharose 4B-agmatine [2]; Blue Sepharose, t-butyl-Sepharose, Resource Q, Superose 12 HR [4]) [2, 3, 4]

Cloning
<1> (expression in Escherichia coli [4]) [4]

6 Stability

General stability information

<1>, 2-mercaptoethanol stabilizes [3]

Storage stability

<1>, -40°C, concentrated preparation at least 1 month, negligible loss of activity [2]

<1>, 0°C, 100 mM Tris, pH 7.5, 1 mM EDTA, 10 mM 2-mercaptoethanol, 50 mM KCl [4]

References

[1] Bird, C.R.; Smith, T.A.: The biosynthesis of coumaroylagmatine in barley seedlings. Phytochemistry, **20**, 2345-2346 (1981)

[2] Bird, C.R.; Smith, T.A.: Agmatine coumaroyltransferase (barley seedlings). Methods Enzymol., **94**, 344-347 (1983)

[3] Bird, C.R.; Smith, T.A.: Agmatine coumaroyltransferase from barley seedlings. Phytochemistry, **22**, 2401-2403 (1983)

[4] Burhenne, K.; Kristensen, B.K.; Rasmussen, S.K.: A new class of N-hydroxycinnamoyltransferases. Purification, cloning, and expression of barley agmatine coumaroyltransferase (EC 2.3.1.64). J. Biol. Chem., **278**, 13919-13927 (2003)

Glycine N-choloyltransferase 2.3.1.65

1 Nomenclature

EC number
2.3.1.65

Systematic name
choloyl-CoA:glycine N-choloyltransferase

Recommended name
bile acid-CoA:amino acid N-acyltransferase

Synonyms
BAT
acyltransferase, glycine-taurine N-
amino acid N-choloyltransferase
glycine N-choloyltransferase
glycine-taurine N-acyltransferase
hBAT <2> [9]
mBAT <6> [10]

CAS registry number
74506-32-4

2 Source Organism

<1> *Bos taurus* [1-3]
<2> *Homo sapiens* [4, 9]
<3> *Rattus norvegicus* (male Sprague-Dawley rats [5]; male Wistar rats, 4 weeks old [8]) [5, 8]
<4> *Canis familiaris* [6]
<5> *Gallus gallus* [7]
<6> *Mus musculus* [10]

3 Reaction and Specificity

Catalyzed reaction
choloyl-CoA + glycine = CoA + glycocholate (<1> ping-pong mechanism [2,3]; <2> catalytic mechanism [9])

Reaction type
acyl group transfer

Natural substrates and products

S choloyl-CoA + glycine <1, 2> (<1,2> taurine can replace glycine, involved in conjugation of bile acids [2,4,9]; <1> conjugation of bile acids in the liver with glycine or taurine prior to being secreted into the bile duct [1,3]; <2> cholesterol degradation, glycine conjugates of bile acids occur in adult [4]; <2> conjugation of bile acids with glycine or taurine favors their excretion into bile and uptake from portal blood into the liver, it promotes absorption of fat and fat-soluble vitamins in the acidic environment of the small intestine by lowering the pKa of the bile acids and hence maintaining its solubility [9]) (Reversibility: ? <1, 2> [1-4, 9]) [1-4, 9]

P CoA + glycocholate

S choloyl-CoA + taurine <1-4, 6> (<1,2> taurine can replace glycine, involved in conjugation of bile acids [2,4,9]; <1> conjugation of bile acids in the liver with glycine or taurine prior to being secreted into the bile duct [1,3]; <2> cholesterol degradation, bile acids are secreted as taurine conjugates in fetus and neonatus [4]; <3> taurine conjugates predominate in rat bile at normal hepatocellular concentrations of glycine and taurine [5,8]; <4> conjugation of the CoA adducts of bile acids with taurine [6]; <3> bile acid conjugation [8]; <2> conjugation of bile acids with glycine or taurine favors their excretion into bile and uptake from portal blood into the liver, it promotes absorption of fat and fat-soluble vitamins in the acidic environment of the small intestine by lowering the pKa of the bile acids and hence maintaining its solubility [9]; <6> taurine-specific enzyme, gallbladder bile contains only taurine conjugates of bile acids, bile acid conjugates increases their detergent properties, prevents their precipitation in the acidic milieu of the upper small intestine, in intestines they are responsible for the solubilization and absorption of fats, vitamins and fat-soluble compounds [10]) (Reversibility: ? <1-4, 6> [1-6, 8-10]) [1-6, 8-10]

P CoA + taurocholate

Substrates and products

S chenodeoxycholoyl-CoA + glycine <1-3> (<1> rate of conjugation in decreasing order: deoxycholyl-CoA, cholyl-CoA, lithocholyl-CoA, chenodeoxycholyl-CoA [2]; <2> 91.4% of activity with choloyl-CoA [4]) (Reversibility: ? <1-3> [2, 4, 5]) [2, 4, 5]

P CoA + glycochenodeoxycholate

S chenodeoxycholoyl-CoA + taurine <2, 3> (<2> 101.4% of activity with choloyl-CoA [4]) (Reversibility: ? <2, 3> [4, 5]) [4, 5]

P CoA + taurochenodeoxycholate

S choloyl-CoA + 2-fluoro-β-alanine <2> (<2> 2-fluoro-β-alanine is an excellent substrate like taurine [9]) (Reversibility: ? <2> [9]) [9]

P CoA + 2-fluoro-β-alanylcholate

S choloyl-CoA + D-α-alanine <2> (<2> about 10% of the rate with glycine [4]) (Reversibility: ? <2> [4]) [4]

P CoA + D-α-alanylcholate

S choloyl-CoA + β-alanine <2, 3> (<2> about 10% of the rate with glycine [4]; <2> β-alanine is a poor substrate [9]) (Reversibility: ? <2, 3> [4, 5, 9]) [4, 5, 9]
P CoA + β-alanylcholate
S choloyl-CoA + glycine <1-3> (<1> rate of conjugation in decreasing order: deoxycholyl-CoA, cholyl-CoA, lithocholyl-CoA, chenodeoxycholyl-CoA [2]; <2> ratio of glycine-/taurine-dependent activity is 1.0-1.4 [4]; <3> produces taurine- but little glycine-conjugated bile acid at physiological concentrations of glycine and taurine, dietary manipulations modify the glycine:taurine ratio [8]) (Reversibility: ir <1> [2]; ? <1-3> [1, 3-5, 8, 9]) [1-5, 8, 9]
P CoA + glycocholate <1-3> [1-5, 9]
S choloyl-CoA + taurine <1-4, 6> (<2> ratio of glycine-/taurine-dependent activity is 1.0-1.4 [4]; <3> preferred substrate [5]; <4> synthesizes only taurine but no gylcine conjugates [6]; <3> produces taurine- but little glycine-conjugated bile acid at physiological concentrations of glycine and taurine, dietary manipulations modify the hepatic taurine concentration and glycine:taurine ratio [8]; <2> taurine is an excellent substrate [9]; <6> taurine-specific enzyme, no use of glycine or fluoro-β-alanine as substrates [10]) (Reversibility: ? <1-4, 6> [1-6, 8-10]) [1-6, 8-10]
P CoA + taurocholate <1-3> [1, 2, 4, 5, 9]
S deoxycholoyl-CoA + aminomethanesulfonate <3> (Reversibility: ? <3> [5]) [5]
P CoA + aminomethanesulfonyldeoxycholate
S deoxycholoyl-CoA + glycine <1-3> (<1> rate of conjugation in decreasing order: deoxycholyl-CoA, cholyl-CoA, lithocholyl-CoA, chenodeoxycholyl-CoA [2]; <2> 92.4% of activity with choloyl-CoA [4]) (Reversibility: ? <1-3> [2, 4, 5]) [2, 4, 5]
P CoA + glycodeoxycholate <3> [5]
S deoxycholoyl-CoA + taurine <2, 3> (<2> 81.7% of activity with choloyl-CoA [4]) (Reversibility: ? <2, 3> [4, 5]) [4, 5]
P CoA + taurodeoxycholate <3> [5]
S lithocholoyl-CoA + glycine <1-3> (<1> rate of conjugation in decreasing order: deoxycholyl-CoA, cholyl-CoA, lithocholyl-CoA, chenodeoxycholyl-CoA [2]; <2> 58.4% of activity with choloyl-CoA [4]) (Reversibility: ? <1-3> [2, 4, 5]) [2, 4, 5]
P CoA + glycolithocholate
S lithocholoyl-CoA + taurine <2, 3> (<2> 43% of activity with choloyl-CoA [4]) (Reversibility: ? <2, 3> [4, 5]) [4, 5]
P CoA + taurolithocholate
S ursodeoxycholoyl-CoA + glycine <2> (<2> 24.8% of activity with choloyl-CoA [4]) (Reversibility: ? <2> [4]) [4]
P CoA + glycoursodeoxycholate
S ursodeoxycholoyl-CoA + taurine <2> (<2> 35.3% of activity with choloyl-CoA [4]) (Reversibility: ? <2> [4]) [4]
P CoA + tauroursodeoxycholate

S Additional information <1-4, 6> (<1> in absence of amino acid substrate choloyl-CoA is cleaved with the release of CoA and formation of a covalent cholate-enzyme complex [3]; <2> essential catalytic triad consisting of Cys-235, His-362 and Asp-328 with Cys-235 serving as the probable nucleophile and thus the site of covalent attachment of the bile acid molecule [9]; <6> Cys-234 may be involved in the formation of a bile acid thioester at the active site [10]; <1,3> not: L-α-alanine, L-glutamine, L-ornithine [2,5]; <2> not: L-α-alanine, DL-cysteic acid, 4-aminobutyric acid, cysteine, sulfanilic acid [4]; <1> not: L-serine, β-alanine [2]; <3> not: D-alanine, L-arginine, L-lysine, 2-hydroxyethanesulfonate, 2-aminoethanesulfonate, 1-aminoethanephosphonate and 2-aminoethanephosphonate, 3-aminopropanesulfonate [5]; <3> not as acyl-CoA donors: succinoyl-CoA, palmitoyl-CoA [5]; <1,3> not as acyl-CoA donors: acetyl-CoA, phenylacetyl-CoA and benzoyl-CoA [2,5]; <4> not: choloyl-CoA + glycine [6]; <6> not: glycine, fluoro-β-alanine [10]) [2-6, 9, 10]

P ?

Inhibitors

5,5'-dithiobis(2-nitrobenzoic acid) <2, 3> [5, 9]
N-ethylmaleimide <2> (<2> 10 min, at 0.2 mM: 90% loss of activity, preincubation with cholyl-CoA before NEM-treatment protects [9]) [9]
NaCl <1> (<1> above 50 mM [2]) [2]
chenodeoxycholate <2> [4]
cholate <1, 2> [2, 4]
cholesterol <2> [4]
conjugated bile acids <1> (<1> efficient competitive inhibitors of cholyl-CoA binding [2]) [2]
deoxycholate <2> [4]
glycocholate <1, 2> (<1> competitive to cholyl-CoA, uncompetitive to glycine [2]) [2, 4]
glycodeoxycholate <2> [4]
p-mercuribenzoate <1> (<1> reversible by dithioerythritol, substrates protect, serine and alanine protect [2]) [2]
taurocholate <1, 2> [2, 4]
taurodeoxycholate <2> [4]
ursodeoxycholate <2> [4]
Additional information <1, 2, 6> (<1> not inhibited by N-ethylmaleimide, iodoacetate [2]; <2> not inhibited by 2-mercaptoethanol, $CaCl_2$, EDTA, glutathione [4]; <6> not inhibited by glycine [10]) [2, 4, 10]

Activating compounds

EDTA <2> (<2> 2fold activation, lowers the K_m-value [4]) [4]
L-cysteine <2> (<2> 2fold activation [4]) [4]
pectin <3> (<3> dietary pectin relative to cellulose activates, especially the glycine-dependent activity [8]) [8]
reduced glutathione <2> (<2> 2fold activation [4]) [4]

Turnover number (min^{-1})

Additional information <1> (<1> higher k_{cat} for glycine than for taurine [1,2]) [1, 2]

Specific activity (U/mg)

0.0027 <4> [6]
0.42 <2> (<2> taurine [4]) [4]
0.43 <2> (<2> glycine [4]) [4]
1.99 <3> (<3> taurine [5]) [5]
3.88 <3> (<3> glycine [5]) [5]
35.2 <1> (<1> taurine [1]) [1]
43-59 <1> (<1> glycine [2]) [2]
46.9 <1> (<1> glycine [1]) [1]
Additional information <3> [5]

K_m-Value (mM)

0.0072 <1> (choloyl-CoA) [2]
0.05 <2> (choloyl-CoA, <2> + 60 mM glycine [4]) [4]
0.087 <2> (choloyl-CoA, <2> + 40 mM taurine [4]) [4]
0.6 <2> (taurine) [4]
0.8 <3> (aminomethanesulfonate, <3> + 0.05 mM deoxycholoyl-CoA [5]) [5]
0.8 <3> (taurine, <3> + 0.05 mM deoxycholoyl-CoA [5]) [5]
0.8-2.5 <3> (taurine) [8]
0.83 <3> (taurine, <3> rats fed with cellulose [8]) [8]
1.02 <3> (taurine, <3> rats fed with pectin [8]) [8]
1.9 <6> (taurine, <6> recombinant enzyme, expressed in Escherichia coli [10]) [10]
3.2 <2> (glycine) [4]
6.7 <1> (taurine) [2]
8.8 <1> (glycine) [2]
31 <3> (glycine, <3> + 0.05 mM deoxycholoyl-CoA [5]) [5]
35-40 <3> (glycine) [8]
38 <3> (glycine, <3> rats fed with cellulose [8]) [8]
40 <3> (glycine, <3> rats fed with pectin [8]) [8]
175 <3> (β-alanine) [5]
Additional information <1, 3> [1, 2, 5]

K_i-Value (mM)

0.038 <1> (glycocholate) [2]

pH-Optimum

6.5 <2> (<2> taurine-dependent activity [4]) [4]
7.2 <2> (<2> glycine-dependent activity [4]) [4]
7.8-8 <3> [5]
Additional information <1-3> (<2> pI: 5.85 [4]; <1> pI: 6.6 [1]; <3> pI: 7.2 [5]) [1, 4, 5]

pH-Range

5.6-8.2 <2> (<2> taurine: about half-maximal activity at pH 5.2 and pH 8.2, glycine: 80% of maximal activity at pH 8.2 [4]) [4]
6-9 <3> (<3> about 70% of maximal activity at pH 6.0, about half-maximal activity at pH 9.0, inactivation rate is similar for both taurine and glycine [5]) [5]

Temperature optimum (°C)

30 <1, 3> (<1,3> assay at [2,5]) [2, 5]
37 <2, 3> (<2,3> assay at [4,8,9]) [4, 8, 9]

4 Enzyme Structure

Molecular weight

45700 <4> (<4> gel filtration [6]) [6]
47000 <1> (<1> gel filtration [1]) [1]
63000-65000 <5> [7]
100000 <2> (<2> gel filtration [4]) [4]
118000 <2> (<2> sucrose density gradient centrifugation [4]) [4]
Additional information <1> (<1> amino acid composition [3]) [3]

Subunits

? <2, 6> (<2> x * 50000, SDS-PAGE [9]; <6> x * 46525, calculated from the amino acid sequence, x * 45000, recombinant enzyme, SDS-PAGE [10]) [9, 10]
homodimer <2> (<2> 2 * 52000, SDS-PAGE [4]) [4]
monomer <1> (<1> 1 * 50900, SDS-PAGE [1]) [1]

5 Isolation/Preparation/Mutation/Application

Source/tissue

liver <1-4, 6> [1-6, 8, 10]
Additional information <6> (<6> not in heart, brain, spleen, lung, skeletal muscle, kidney or testis tissue [10]) [10]

Localization

soluble <1, 4> [1, 6]
Additional information <3> (<3> in the post-mitochondrial fraction of liver homogenates [8]) [8]

Purification

<1> (900fold [1]) [1-3]
<2> (about 100fold [4]) [4, 9]
<3> (partial, 200fold [5]) [5]
<4> (partial [6]) [6]
<6> (partial purification of recombinant enzyme, expressed in Escherichia coli DH5α [10]) [10]

Cloning

<2> (gene/full-length cDNA is cloned and expressed in Escherichia coli XL1-Blue or BL21 [9]) [9]
<6> (cDNA encoding enzyme is cloned, sequenced and expressed in Escherichia coli DH5α, the open-reading frame codes for a 420 amino acid protein, Baat gene is mapped to mouse chromosome 4, single gene encodes enzyme [10]) [10]

Engineering

C235S <2> (<2> mutant with lower enzyme activity [9]) [9]
C372A <2> (<2> mutant with low enzyme activity [9]) [9]
D328A <2> (<2> inactive enzyme [9]) [9]
H362A <2> (<2> inactive enzyme [9]) [9]

6 Stability

pH-Stability

6 <3> (<3> phosphate buffer, $t_{1/2}$: 2 h [5]) [5]
9 <3> (<3> Tris-buffer, $t_{1/2}$: 3 h [5]) [5]

Temperature stability

45 <3> (<3> stable below [5]) [5]
50 <1, 3> (<1> 10 min, stable up to [1]; <3> inactivation above [5]) [1, 5]
57 <1> (<1> 10 min, about 70% loss of activity [1]) [1]
60 <1> (<1> inactivation within 10 min [1]) [1]
Additional information <1> (<1> taurocholate stabilizes against temperature denaturation [1]) [1]

General stability information

<1>, freezing inactivates [1]
<1>, high ionic strength stabilizes [1, 2]
<2>, glycerol, 40% v/v, stabilizes during storage [4]
<4>, unstable enzyme [6]

Storage stability

<1>, 0°C, 50 mM phosphate buffer, pH 7.8, with 0.5 M NaCl, 2 mM dithioerythritol, 1 month, 25% loss of activity [1]
<2>, -20°C, in the presence of 40% v/v glycerol [4]
<3>, -70°C, several weeks, stable [5]
<3>, 0° to 20°C, several h, stable [5]

References

[1] Vessey, D.A.: The co-purification and common identity of cholyl CoA:glycine- and cholyl CoA:taurine-N-acyltransferase activities from bovine liver. J. Biol. Chem., **254**, 2059-2063 (1979)

[2] Czuba, B.; Vessey, D.A.: Kinetic characterization of cholyl-CoA glycine-taurine N-acyltransferase from bovine liver. J. Biol. Chem., **255**, 5296-5299 (1980)

[3] Czuba, B.; Vessey, D.A.: Structural characterization of cholylcoenzyme A: glycine-taurine N-acyltransferase and a covalent substrate intermediate. J. Biol. Chem., **261**, 6260-6263 (1986)

[4] Kimura, M.; Okuno, E.; Inada, J.; Ohyama, H.; Kido, R.: Purification and characterization of amino-acid N-choloyltransferase from human liver. Hoppe-Seyler's Z. Physiol. Chem., **364**, 637-645 (1983)

[5] Killenberg, P.G.; Jordan, J.T.: Purification and characterization of bile acid-CoA:amino acid N-acyltransferase from rat liver. J. Biol. Chem., **253**, 1005-1010 (1978)

[6] Czuba, B.; Vessey, D.A.: Identification of a unique mammalian species of cholyl-CoA:amino acid N-acyltransferase. Biochim. Biophys. Acta, **665**, 612-614 (1981)

[7] Poley, J.R.; Dower, J.C.; Owen, J.R.; Owen C.A.; Stickler, G.B.: Bile acids in infants and children. J. Lab. Clin. Med., **63**, 838-846 (1964)

[8] Ide, T.; Kano, S.; Murata, M.; Yanagita, T.; Sugano, M.: Dietary modifications of the biliary bile acid glycine:taurine ratio and activity of hepatic bile acid-CoA:amino acid N-acyltransferase (EC 2.3.1) in the rat. Br. J. Nutr., **72**, 93-100 (1994)

[9] Sfakianos, M.K.; Wilson, L.; Sakalian, M.; Falany, C.N.; Barnes, S.: Conserved residues in the putative catalytic triad of human bile acid coenzyme A:amino acid N-acyltransferase. J. Biol. Chem., **277**, 47270-47275 (2002)

[10] Falany, C.N.; Fortinberry, H.; Leiter, E.H.; Barnes, S.: Cloning, expression, and chromosomal localization of mouse liver bile acid CoA:amino acid N-acyltransferase. J. Lipid Res., **38**, 1139-1148 (1997)

Leucine N-acetyltransferase 2.3.1.66

1 Nomenclature

EC number
2.3.1.66

Systematic name
acetyl-CoA:L-leucine N-acetyltransferase

Recommended name
leucine N-acetyltransferase

Synonyms
acetyltransferase, leucine
leucine acyltransferase

CAS registry number
75496-56-9

2 Source Organism

<1> *Streptomyces roseus* (strain MA839-A1, also present in strain MA839-A1 LN-S [1]) [1]

3 Reaction and Specificity

Catalyzed reaction
acetyl-CoA + L-leucine = CoA + N-acetyl-L-leucine

Reaction type
acyl group transfer

Natural substrates and products
S acetyl-CoA + L-leucine <1> (Reversibility: ? <1> [1]) [1]
P CoA + N-acetyl-L-leucine

Substrates and products
S acetyl-CoA + L-arginine <1> (Reversibility: ? <1> [1]) [1]
P CoA + ?
S acetyl-CoA + L-leucine <1> (Reversibility: ? <1> [1]) [1]
P CoA + N-acetyl-L-leucine <1> [1]
S acetyl-CoA + L-phenylalanine <1> (Reversibility: ? <1> [1]) [1]
P CoA + N-acetyl-L-phenylalanine

S acetyl-CoA + L-valine <1> (Reversibility: ? <1> [1]) [1]
P CoA + N-acetyl-L-valine
S propionyl-CoA + L-leucine <1> (<1> 10% of activity compared to acetyl-CoA [1]) (Reversibility: ? <1> [1]) [1]
P CoA + propionyl-L-leucine
S Additional information <1> (<1> dipeptides, e.g. L-leucyl-L-leucine, are acetylated more rapidly than L-leucine or peptides with more than two amino acids, D-isomers are acetylated more slowly than L-isomers [1]) [1]
P ?

Inhibitors
Mg^{2+} <1> (<1> weak [1]) [1]
Mn^{2+} <1> (<1> weak [1]) [1]
N-ethylmaleimide <1> (<1> strong [1]) [1]
p-chloromercuribenzoate <1> (<1> strong [1]) [1]

Metals, ions
Cu^{2+} <1> (<1> slight stimulation [1]) [1]
Fe^{2+} <1> (<1> stimulates strongly [1]) [1]

K_m-Value (mM)
0.005 <1> (acetyl-CoA) [1]
0.06 <1> (L-leucine) [1]

pH-Optimum
9 <1> (<1> assay at [1]) [1]

pH-Range
7-9 <1> (<1> 10% of maximal activity at pH 7, maximal activity at pH 9 [1]) [1]

Temperature optimum (°C)
27 <1> (<1> assay at [1]) [1]

4 Enzyme Structure

Molecular weight
27000 <1> (<1> gel filtration [1]) [1]

5 Isolation/Preparation/Mutation/Application

Purification
<1> (partial [1]) [1]

6 Stability

Storage stability

<1>, -180°C, 3 months, 70% remaining activity [1]

References

[1] Umezawa, H.: Biosynthesis of leupeptin. III. Isolation and properties of an enzyme synthesizing acetyl-L-leucine. J. Antibiot., **33**, 857-862 (1980)

1-Alkylglycerophosphocholine O-acetyltransferase 2.3.1.67

1 Nomenclature

EC number
2.3.1.67

Systematic name
acetyl-CoA:1-alkyl-sn-glycero-3-phosphocholine 2-O-acetyltransferase

Recommended name
1-alkylglycerophosphocholine O-acetyltransferase

Synonyms
1-alkyl-2-lyso-sn-glycero-3-phosphocholine acetyltransferase
PAF acetyltransferase
acetyl-CoA:1-alkyl-2-lyso-sn-glycero-3-phosphocholine 2-O-acetyltransferase
acetyl-CoA:lyso-PAF acetyltransferase
acetyltransferase, 1-alkyl-2-lysolecithin
acetyltransferase, 1-alkylglycerophosphocholine
acyl-CoA:1-alkyl-sn-glycero-3-phosphocholine acyltransferase
blood platelet-activating factor acetyltransferase
lyso-GPC:acetyl CoA acetyltransferase
lyso-platelet activating factor:acetyl-CoA acetyltransferase
lyso-platelet-activating factor:acetyl-CoA acetyltransferase
lysoPAF:acetyl CoA acetyltransferase
platelet-activating factor acylhydrolase
platelet-activating factor-synthesizing enzyme

CAS registry number
76773-96-1

2 Source Organism

<1> *Rattus norvegicus* [1, 2, 3, 4, 5, 6, 8, 11, 14, 15, 18]
<2> *Canis familiaris* [2]
<3> *Homo sapiens* [7, 11, 12]
<4> *Sus scrofa* [9]
<5> *Oryctolagus cuniculus* [10, 17, 19]
<6> *Mus musculus* [13]
<7> *Bos taurus* [16]

3 Reaction and Specificity

Catalyzed reaction

acetyl-CoA + 1-alkyl-sn-glycero-3-phosphocholine = CoA + 2-acetyl-1-alkyl-sn-glycero-3-phosphocholine

Reaction type

acyl group transfer

Natural substrates and products

S acetyl-CoA + 1-O-alkyl-sn-glycero-3-phosphocholine <1> (<1> final step in remodelling pathway of PAF biosynthesis [1]) (Reversibility: ? <1> [1]) [1]

P CoA + 2-acetyl-1-alkyl-6-phosphocholine <1> [1]

Substrates and products

S acetyl-CoA + 1-acyl-sn-glycero-3-phosphate <5> (Reversibility: ? <5> [10]) [10]

P CoA + 2-acetyl-1-acyl-sn-glycero-3-phosphate <5> [10]

S acetyl-CoA + 1-acyl-sn-glycero-3-phosphocholine <5> (Reversibility: ? <5> [10]) [10]

P CoA + 2-acetyl-1-acyl-sn-glycero-3-phosphocholine <5> [10]

S acetyl-CoA + 1-hexadecenyl-lyso-sn-glycero-3-phosphocholine <3, 6> (<3> 40% higher activity respect to the C_{18} analogue [7]) (Reversibility: ? <3, 6> [7, 13]) [7, 13]

P CoA + 2-acetyl-1-hexadecenyl-sn-glycero-3-phosphocholine <3, 6> [7, 13]

S acetyl-CoA + 1-octadecenyl-lyso-sn-glycero-3-phosphocholine <3, 6> (Reversibility: ? <3, 6> [7, 13]) [7, 13]

P CoA + 2-acetyl-octadecenyl-sn-glycero-3-phosphocholine <3, 6> [7, 13]

S acetyl-CoA + 1-palmitoyl-2-lyso-sn-glycero-3-phosphocholine <1> (<1> 12% of the activity observed in presence of alkyl-lyso-sn-glycero-phosphocholine [2]; <1> rate reduced to 50% [3]; <1> rate of synthesis is 55% of the rate observed with the ether linked substrate [5]) (Reversibility: ? <1> [2, 3, 5]) [2, 3, 5]

P CoA + 2-acetyl-1-palmitoyl-sn-glycero-3-phosphocholine <1> [2, 3, 5]

S acetyl-CoA + acetyl-phosphatidylcholine <3> (Reversibility: ? <3> [7]) [7]

P ?

S acetyl-CoA + acyl-lyso-glycero-3-phosphocholine <1> (<1> worse substrate than alkyl-lyso-glycero-3-phosphocholine [5]) (Reversibility: ? <1> [1, 5]) [1, 5]

P CoA + acetyl-acyl-glycero-3-phosphocholine

S acetyl-CoA + alk-1-enyl-lyso-glycero-3-phosphoethanolamine <1, 6> (Reversibility: ? <1, 6> [1, 13]) [1, 13]

P CoA + acetyl-alk-1-enyl-glycero-3-phosphoethanolamine <1, 6> [1, 13]

S acetyl-CoA + alkyl-lyso-N',N'-dimethylethanamine <1> (Reversibility: ? <1> [1, 3]) [1, 3]

P ?

S acetyl-CoA + alkyl-lyso-glycero-3-phosphoethanolamine <1> (<1> least enzyme activity [3]) (Reversibility: ? <1> [1, 3]) [1, 3]

P CoA + 2-acetyl-alkyl-glycero-3-phosphoethanolamine <1> [1, 3]

S acetyl-CoA + alkyl-lyso-monomethylethanamine <1> (Reversibility: ? <1> [1, 3]) [1, 3]

P ?

S acetyl-CoA + alkyl-sn-glycero-3-phosphocholine <1, 5, 7> (<1> alkyl-lyso-sn-glycero-3-phosphocholine is preferred to its saturated counterpart as substrate [1,3]; <1> the higher the degree of methylation of the nitrogen base, the lower the enzyme activity [3]; <1> the longer the acyl-CoA length, the smaller the values of apparent K_m and V_{max} [3]) (Reversibility: ? <1, 5, 7> [1-3, 5, 6, 8-10, 15, 16, 18, 19]) [1-3, 5, 6, 8-10, 15, 16, 18, 19]

P CoA + 2-acetyl-1-alkyl-sn-glycero-3-phosphocholine <1, 5, 7> [1-3, 5, 6, 8-10, 15, 16, 18, 19]

S acetyl-CoA + hexadecyl-lyso-sn-glycero-3-phosphocholine <3, 5> (Reversibility: ? <3, 5> [1, 3, 5, 11, 17]) [1, 3, 5, 11, 17]

P CoA + 2-acetyl-1-hexadecyl-sn-glycero-3-phosphocholine <3, 5> [1, 3, 5, 11, 17]

S acetyl-CoA + lyso-phosphatidylcholine <3> (Reversibility: ? <3> [7]) [7]

P CoA + 2-acetyl-1-acylphosphatidylcholine <3> [7]

S acetyl-CoA + lyso-phosphatidylethanolamine <3> (Reversibility: ? <3> [7]) [7]

P CoA + 2-acetyl-1-acyl-phosphatidylethanolamine <3> [7]

S acetyl-CoA + octadecyl-lyso-sn-glycero-3-phosphocholine <1, 6> (Reversibility: ? <1, 6> [1, 3, 13, 14]) [1, 3, 13, 14]

P CoA + 2-acetyl-1-octadecyl-sn-glycero-3-phosphocholine <1, 6> [1, 3, 13, 14]

S acetyl-CoA + phosphatidylcholine <3> (Reversibility: ? <3> [7]) [7]

P ?

S oleoyl-CoA + alkyl-lyso-glycero-3-phosphoethanolamine <1> (<1> 10% of the activity observed in presence of acetyl-CoA [2]) (Reversibility: ? <1> [2]) [2]

P CoA + 1-alkyl-2-oleoyl-glycero-3-phosphoethanolamine <1> [2]

S palmitoyl-CoA + alkyl-lyso-glycero-3-phosphoethanolamine <1> (<1> 5% of the activity observed in presence of acetyl-CoA [2]) (Reversibility: ? <1> [2]) [2]

P CoA + acetyl-2-palmitoyl-glycero-3-phosphoethanolamine <1> [2]

S propionyl-CoA + alkyl-lyso-glycero-phosphocholine <1> (Reversibility: ? <1> [1]) [1]

P CoA + 2-propionyl-1-alkyl-lyso-sn-glycero-3-phosphocholine <1> [1]

Inhibitors

1-O-alkyl-2-acetyl-sn-glycero-3-phosphocholine <3> (<3> irreversible inhibition dependent of both 1-O-alkyl-2-acetyl-sn-glycero-3-phosphocholine and enzyme concentration [7]) [7]

1-O-alkyl-sn-glycero-3-phosphocholine <3> (<3> inhibition at high concentrations due to its detergent effect [7]) [1, 3, 5, 7]
1-acyl-sn-glycero-3-phosphate <5> (<5> competitive inhibition for 1-alkyl-sn-glycero-3-phosphate [10]) [10]
1-methoxyphaseolidin <1> (<1> IC50: 0.048 mM [15]) [15]
1-methoxyphaseolin <1> (<1> IC50: 0.057 mM [15]) [15]
4-(1-naphthylvinyl)pyridine <1> (<1> IC50: 0.043 mM [14]) [14]
6,8-diprenylgenistein <1> (<1> IC50: 0.019 mM [15]) [15]
$CaCl_2$ <1> (<1> 15% inhibition [5]) [5]
EDTA <7> (<7> 67.9 and 72.7% inhibition at 0.5 and 1 mM respectively [1,5,16]) [1, 5, 16]
EGTA <1> [1, 5]
Fe^{2+} <7> (<7> 31% inhibition at 0.01 mM [16]) [16]
H_2O_2 <7> (<7> 8% inhibition at 1 mM [16]) [16]
$MgATP^{2-}$ <5> (<5> 32% and 54% inhibition at 0.125 and 1 mM respectively in neuronal nuclear fraction [19]) [19]
$MgCl_2$ <1> (<1> 98% inhibition [5]) [1, 5]
$MnCl_2$ <1> (<1> 50% inhibition [5]) [1, 5]
N-ethylmaleimide <1> [1]
Triton X-100 <1> (<1> 75 and 95% inhibition at 0.2 and 0.6 mM respectively due to detergent [5]) [5]
acetyl-salicylic acid <1, 7> (<7> IC50: 1.52 mM [16]; <1> weak inhibition [11]) [11, 16]
arachidonoyl-CoA <1> [1]
ascorbic acid <7> (<7> 18.3% inhibition at 1 mM [16]) [16]
balcalein <1, 7> (<7> concentration dependent inhibition, IC50: 0.105 mM [16]; <1> IC50: 0.148 mM [11]) [11, 16]
chiyusaponin <1> (<1> IC50: 0.2 mM [11]) [11]
dexamethasone <1> (<1> 2 mg/kg for 3 days in liver and spleen [18]) [18]
diisopropylfluorophosphate <1> (<1> 98% inhibition at 10 mM [5]) [1, 5]
ebselen <7> (<7> suppresses alkyl-2-acetyl-sn-glycero-3-phosphocholine synthesis by 7.7% and 26.6% in presence of glutathione [16]) [16]
gallic acid <7> (<7> 5.3% inhibition at 0.1 mM [16]) [16]
glutathione <7> (<7> suppresses alkyl-2-acetyl-sn-glycero-3-phosphocholine synthesis by 18.9% [16]) [16]
honokiol <1, 3> (<1> IC50: 0.15 mM [11]; <3> IC50: 0.06 mM, reversible inhibition [11]) [11]
indomethacin <1, 7> (<7> IC50: 0.26 mM [16]; <1> weak inhibition at 1mM [11]) [11, 16]
licoricidin <1> (<1> IC50: 0.0077 mM [15]) [15]
luteolin <1> (<1> IC50: 0.045 mM [14]) [14]
lysophosphatidylcholine <3> (<3> inhibition at high concentrations due to its detergent effect [7]) [7]
magnolol <1, 3> (<1> IC50: 0.15 mM [11]; <3> IC50: 0.07 mM, reversible inhibition [11]) [11]
medroxyprogesterone <1> (<1> 50 mg/kg in liver [18]) [18]

myricetin <7> (<7> suppresses alkyl-2-acetyl-sn-glycero-3-phosphocholine synthesis by 17% [16]) [16]
nordihydroguaiaretic acid <1, 3, 7> (<7> concentration dependent inhibition, IC50: 0.075 mM [16]; <1> IC50: 0.29 mM [11]; <3> IC50: 0.06 mM [11]) [11, 16]
oleic acid <5> (<5> 1-acyl lysolipids less inhibited than the 1-alkyl species [10]) [10]
oleoyl-CoA <5> (<5> 16% and 36% inhibition at 0.002 and 0.01 mM respectively in neuronal nuclear fraction [19]; <5> 45-55% inhibition [10]) [1, 10, 19]
p-bromophenacyl bromide <1> (<1> 90% inhibition at 0.1 mM [5]) [1, 5]
p-chloromercuribenzoate <1> [1]
palmitoyl-CoA <1> (<1> 85% inhibition at 10 uM due to detergent effect [5]) [1, 5]
palmitoyllyso-glycero-3-phosphocholine <1> (<1> competitive inhibitor for alkyl-sn-glycero-3-phosphocholine [3]) [1, 3]
phenylmethylsulfonyl fluoride <3> (<3> 30% inhibition at 50 mM [7]) [7]
phosphatidic acid <3> [12]
propanolol <3> [12]
quercetin <7> (<7> suppresses alkyl-2-acetyl-sn-glycero-3-phosphocholine synthesis by 8.8% [16]; <1> IC50: 0.08 mM [14]) [14, 16]
serine/threonine phosphatase <5> (<5> diminishes rates of acetylation for 1-alkyl-sn-glycero-3-phosphocholine and 1-acyl-sn-glycero-3-phosphocholine [10]) [10]
sodium dodecylsulfate <1> (<1> inactivation at 0.1% [2]) [2]
trifnoperazine <1> [8]
urea <1> (<1> inactivation at 8 mM [2]) [2]

Activating compounds

A23187 <3> [7]
AMP-dependent protein kinase <1> (<1> activity increases 2-3 fold in presence of $MgATP^{2-}$, activation is reversible, stimulation is optimal with 30 U/ml protein kinase [6]) [1, 6]
N-formyl-methionyl-leucyl-phenylalanine <3> (<3> stimulation is dependent on the preincubation of cells with cytochalasin B, addition of tumor necrosis factor increases 50% alkyl-2-acetyl-sn-glycero-3-phosphocholine production [12]) [12]
bovine serum albumin <1> (<1> optimal concentration of 1-1.5 mg/ml at short incubation periods and 2-2.5 mg bovine serum albumin/ml during longer incubation periods [1]) [1]
lipopolysaccharide <6> (<6> activity increases in a time-dependent fashion, maximal activation at 0.010 mg/ml lipopolysaccharide [13]) [13]

Metals, ions

Ca^{2+} <1> (<1> optimal activity at 0.0002 mM [1]; <1> activation reversed by addition of EGTA in excess of Ca^{2+}, Ca^{2+} reduces the apparent K_m for acetyl-CoA, maximum effect between 0.0001 and 0.01 mM, the action of Ca^{2+} seems to be independent of the presence of calmodulin or phosphorylation [8]) [1, 8]

Specific activity (U/mg)

0.0002 <1> (<1> kidney cortex [5]) [5]
0.0002 <1> (<1> liver [5]) [5]
0.0004 <1> (<1> liver specific activity is reduced in presence of dexamethasone and medroxyprogesterone [18]) [18]
0.001 <1> (<1> spleen specific activity is reduced in presence of dexamethasone [18]) [18]
0.002 <1> (<1> kidney medulla [5]) [5]
0.0021 <1> (<1> A23187 stimulated cell [14]) [14]
0.0025 <3> (<3> specific activity is reduced to 0.0015 in presence of acetylphosphatidylcholine without preincubation at 37°C for 60 min [7]) [7]
0.0026 <1> (<1> bone marrow [5]) [5]
0.0034 <3> (<3> specific activity is reduced by preincubation at 37°C for 60 min [7]) [7]
0.0039 <1> (<1> lymph nodes [5]) [5]
0.004 <1> (<1> lung [5]) [5]
0.004 <1> (<1> thymus [5]) [5]
0.0045 <3> (<3> specific activity is reduced in presence of phosphatidylcholine by preincubation at 37°C for 60 min [7]) [7]
0.0047 <3> (<3> specific activity is reduced in presence of lyso-phosphatidylethanolamine by preincubation at 37°C for 60 min [7]) [7]
0.007 <3> (<3> specific activity is reduced in presence of alkyl-lyso-sn-glycero-3-phosphocholine by preincubation at 37°C for 60 min [7]) [7]
0.0072 <3> (<3> specific activity is reduced in presence of lyso-phosphatidylcholine by preincubation at 37°C for 60 min [7]) [7]
0.0076 <3> [7]
0.0078 <3> [7]
0.0086 <3> [7]
0.009 <3> (<3> specific activity is reduced in presence of 1-hexadecenyl-lyso-sn-glycero-3-phosphocholine by preincubation at 37°C for 60 min [7]) [7]
0.01 <1> (<1> spleen microsome [5]) [5]
0.0105 <3> [7]
0.317 <1> [2]
Additional information <1, 2, 5> [2, 19]

K_m-Value (mM)

0.009 <5> (1-alkyl-sn-glycero-3-phosphate) [10]
0.011 <5> (1-acyl-sn-glycero-3-phosphate) [10]
0.018 <5> (1-alkyl-sn-glycero-3-phosphocholine) [10]
0.02 <1> (hexanoyl-CoA) [3]
0.046 <1> (butyryl-CoA) [3]
0.067 <1> (acetyl-CoA) [5]
0.11 <4> (acetyl-CoA) [9]
0.128 <1> (propionyl-CoA) [3]
0.137 <1> (acetyl-CoA) [2]

0.146 <5> (acetyl-CoA, <5> microsomal fraction [19]) [19]
0.173 <1> (acetyl-CoA) [3]
0.214 <5> (acetyl-CoA, <5> neuronal nuclear fraction [19]) [19]
0.274 <1> (acetyl-CoA, <1> K_m is induced to 0.118 mM in presence of Ca^{2+} [8]) [8]

pH-Optimum
6.9 <1, 6, 7> (<1,6,7> assay at [1,5,13-16]) [1, 5, 13-16]
7 <3> (<3> assay at [7]) [7]
7.4 <1, 4, 5> (<1,4,5> assay at [6,8,9,17]) [6, 8, 9, 17]
7.5 <1, 3> (<1,3> assay at [12,18]) [12, 18]
8 <1> [2]
8.4 <1, 5> (<5,1> assay at [4,10,19]) [4, 10, 19]

pH-Range
6-9 <5> (<5> neuronal nuclear fraction and microsomal fraction [19]) [19]

Temperature optimum (°C)
23 <1> (<1> assay at [4]) [4]
37 <1, 3-7> (<1, 3-7> assay at [1, 5, 7, 9-11, 13, 14, 16-18]) [1, 5, 7, 9-11, 13, 14, 16-18]

4 Enzyme Structure

Molecular weight
800000 <1> (<1> MW after gel filtration, aggregate of several proteins or protein aggregates coated with lipids [2]) [2]

Subunits
? <1> (<1> x * 29000, SDS-PAGE and affinity labelling experiments [2]) [2]

5 Isolation/Preparation/Mutation/Application

Source/tissue
HL-60 cell <1> (<1> differentiated [1]) [1]
IC-21 cell <6> (<6> peritoneal macrophage [13]) [13]
RBL-2H3 cell <1> (<1> mastocytoma cell, mucosal-type [14,15]) [14, 15]
bone marrow <1> [5]
cerebral cortex <5> [10, 19]
corneal epithelium <7> [16]
endometrial cell line <5> [17]
leukocyte <4> [9, 11, 12]
liver <1, 2> [2, 5, 18]
lung <1> [5, 18]
lymph node <1> [5]
neutrophil <3> (<3> polymorphonuclear, ionophore A23187 stimulates [7]) [7]

renal cortex <1> [5]
renal medulla <1> [5]
spleen <1> (<1> most abundant [1]) [1-6, 8, 11, 18]
thymus <1> [5]
uterus <1> [18]

Localization
membrane <3> [11]
microsome <1, 5, 7> [1-6, 8, 11, 16-19]
nucleus <5> (<5> neuronal nuclear fraction [19]) [10, 19]

Purification
<1> (solubilization in sodium deoxycholate, precipitation by ammonium sulfate, ultragel AcA-22 chromatography, DEAE-sepharose CL-6B chromatography, covalent chromatography with thiopropyl-sepharose 6B [2]) [2]

6 Stability

Temperature stability
37 <3> (<3> 85% loss of activity after 2 h, 35% loss of activity after 2 h in the presence of DTT and 1-alkyl-2-lyso-sn-glycero-3-phosphate, microsomal preparation [7]) [7]

General stability information
<1>, stable to lyophilization [2]
<3>, 1-alkyl-sn-glycero-3-phosphocholine at low concentration and lysophosphatidylcholine stabilizes microsomal preparation [7]
<1, 3>, dithiothreitol, 1 mM stabilizes microsomal preparation [1, 7]

References

[1] Lee, T.C.; Vallari, D.S.; Snyder, F.: 1-Alkyl-2-lyso-sn-glycero-3-phosphocholine acetyltransferase. Methods Enzymol., **209**, 396-401 (1992)

[2] Gomez-Cambronero, J.; Velasco, S.; Sanchez-Crespo, M.; Vivanco, F.; Mato, J.M.: Partial purification and characterization of 1-O-alkyl-2-lyso-sn-glycero-3-phosphocholine:acetyl-CoA acetyltransferase from rat spleen. Biochem. J., **237**, 439-445 (1986)

[3] Lee, T.C.: Biosynthesis of platelet activating factor. Substrate specificity of 1-alkyl-2-lyso-sn-glycero-3-phosphocholine:acetyl-CoA acetyltransferase in rat spleen microsomes. J. Biol. Chem., **260**, 10952-10955 (1985)

[4] Lee, T.C.; Malone, B.; Snyder, F.: A new de novo pathway for the formation of 1-alkyl-2-acetyl-sn-glycerols, precursors of platelet activating factor. Biochemical characterization of 1-alkyl-2-lyso-sn-glycero-3-P:acetyl-CoA acetyltransferase in rat spleen. J. Biol. Chem., **261**, 5373-5377 (1986)

[5] Wykle, R.L.; Malone, B.; Snyder, F.: Enzymatic synthesis of 1-alkyl-2-acetyl-sn-glycero-3-phosphocholine, a hypotensive and platelet-aggregating lipid. J. Biol. Chem., **255**, 10256-10260 (1980)
[6] Gomez-Cambronero, J.; Velasco, S.; Mato, J.M.; Sanchez-Crespo, M.: Modulation of lyso-platelet activating factor: acetyl-CoA acetyltransferase from rat splenic microsomes. The role of cyclic AMP-dependent protein kinase. Biochim. Biophys. Acta, **845**, 516-519 (1985)
[7] Ninio, E.; Joly, F.; Bessou, G.: Biosynthesis of paf-acether. XI. Regulation of acetyltransferase by enzyme-substrate imbalance. Biochim. Biophys. Acta, **963**, 288-294 (1988)
[8] Gomez-Cambronero, J.; Nieto, M.L.; Mato, J.M.; Sanchez-Crespo, M.: Modulation of lyso-platelet-activating factor: acetyl-CoA acetyltransferase from rat splenic microsomes. The role of calcium ions. Biochim. Biophys. Acta, **845**, 511-515 (1985)
[9] Kume, K.; Waga, I.; Shimizu, T.: Microplate chromatography assay for acetyl-CoA:lysoplatelet-activating factor acetyltransferase. Anal. Biochem., **246**, 118-122 (1997)
[10] Roy Baker, R.; Chang, H.y.: Substrate specificities of neuronal nuclear acetyltransferases involved in the synthesis of platelet-activating factor: Differences in the use of 1-alkyl and 1-acyl lysophospholipid acceptors. Biochim. Biophys. Acta, **1390**, 215-224 (1998)
[11] Yamazaki, R.; Sugatani, J.; Fujii, I.; Kuroyanagi, M.; Umehara, K.; Ueno, A.; Suzuki, Y.; Miwa, M.: Development of a novel method for determination of acetyl-CoA:1-alkyl-sn-glycero-3-phosphocholine acetyltransferase activity and its application to screening for acetyltransferase inhibitors. Inhibition by magnolol and honokiol from Magnoliae cortex. Biochem. Pharmacol., **47**, 995-1006 (1994)
[12] Garcia, C.; Montero, M.; Alvarez, J.; Sanchez Crespo, M.: Biosynthesis of platelet-activating factor (PAF) induced by chemotactic peptide is modulated at the lyso-PAF:acetyl-CoA acetyltransferase level by calcium transient and phosphatidic acid. J. Biol. Chem., **268**, 4001-4008 (1993)
[13] Svetlov, S.I.; Liu, H.; Chao, W.; Olson, M.S.: Regulation of platelet-activating factor (PAF) biosynthesis via coenzyme A-independent transacylase in the macrophage cell line IC-21 stimulated with lipopolysaccharide. Biochim. Biophys. Acta, **1346**, 120-130 (1997)
[14] Yanoshita, R.; Chang, H.W.; Son, K.H.; Kudo, I.; Samejina, Y.: Inhibition of lysoPAF acetyltransferase activity by flavonoids. Inflamm. Res., **45**, 546-549 (1996)
[15] Nagumo, S.; Fukuju, A.; Takayama, M.; Nagai, M.; Yanoshita, R.; Samejima, Y.: Inhibition of lysoPAF acetyltransferase activity by components of licorice root. Biol. Pharm. Bull., **22**, 1144-1146 (1999)
[16] Hurst, J.S.; Bazan, H.E.P.: The sensitivity of bovine corneal epithelial lyso-PAF acetyltransferase to cyclooxygenase and lipoxygenase inhibitors is independent of arachidonate metabolites. J. Ocul. Pharmacol. Ther., **13**, 415-426 (1997)
[17] Kudolo, G.B.; Yang, Y.Q.; Chen, D.B.; Jones, M.A.; Harper, M.J.K.: Differential metabolism of exogenous platelet-activating factor by glandular epithe-

lial and stromal cells of rabbit endometrium. J. Reprod. Fertil., **105**, 315-324 (1995)
[18] Ihara, Y.; Frenkel, R.A.; Johnston, J.M.: Hormonal regulation of platelet-activating factor-acetyltransferase activity in rat tissues. Arch. Biochem. Biophys., **304**, 503-507 (1993)
[19] Baker, R.R.; Chang, H.Y.: Alkylglycerophosphate acetyltransferase and lyso platelet activating factor acetyltransferase, two key enzymes in the synthesis of platelet activating factor, are found in neuronal nuclei isolated from cerebral cortex. Biochim. Biophys. Acta, **1302**, 257-263 (1996)

Glutamine N-acyltransferase 2.3.1.68

1 Nomenclature

EC number
2.3.1.68

Systematic name
acyl-CoA:L-glutamine N-acyltransferase

Recommended name
glutamine N-acyltransferase

CAS registry number
9030-00-6

2 Source Organism

<1> *Homo sapiens* [1]
<2> *Macaca mulatta* [1]

3 Reaction and Specificity

Catalyzed reaction
acyl-CoA + L-glutamine = CoA + N-acyl-L-glutamine

Reaction type
acyl group transfer

Substrates and products
S indolacetyl-CoA + L-glutamine <1, 2> (Reversibility: ? <1, 2> [1]) [1]
P CoA + indolacetyl-CoA
S phenylacetyl-CoA + L-glutamine <1, 2> (Reversibility: ? <1, 2> [1]) [1]
P CoA + phenylacetyl-L-glutamine

Inhibitors
benzoyl-CoA <2> (<2>, weak competitive [1]) [1]
butyryl-CoA <2> (<2>, 0.4 mM, in presence of 100 mM L-glutamine, 50% loss of activity [1]) [1]

K_m-Value (mM)
0.035 <2> (phenylacetyl-CopA) [1]

Additional information <2> (<2>, K_m-value for glutamine is only approximate, 0.6 mM, since 150 mM is the highest concentration at which glutamine can be tested [1]) [1]

K_i-Value (mM)

0.17 <2> (benzoyl-CoA, <2>, with phenylacetyl-CoA as substrate [1]) [1]

5 Isolation/Preparation/Mutation/Application

Source/tissue

liver <1, 2> [1]

Localization

mitochondrion <1, 2> [1]

Purification

<1> [1]
<2> [1]

References

[1] Webster, L.T.; Siddiqui, U.A.; Lucas, S.V.; Strong, J.M.; Mieyal, J.J.: Identification of separate acyl-CoA:glycine and acyl-CoA:L-glutamine N-acyltransferase activities in mitochondrial fractions from liver of rhesus monkey and man. J. Biol. Chem., **251**, 3352-3358 (1976)

Monoterpenol O-acetyltransferase 2.3.1.69

1 Nomenclature

EC number
2.3.1.69

Systematic name
acetyl-CoA:monoterpenol O-acetyltransferase

Recommended name
monoterpenol O-acetyltransferase

Synonyms
acetyltransferase, monoterpenol
menthol transacetylase

CAS registry number
78990-59-7

2 Source Organism

<1> *Mentha piperita* (peppermint [1,2]) [1, 2]

3 Reaction and Specificity

Catalyzed reaction
acetyl-CoA + a monoterpenol = CoA + a monoterpenol acetate ester

Reaction type
acyl group transfer

Substrates and products

S (+)-2-terpineol + acetyl-CoA <1> (<1> low activity [2]) (Reversibility: ? <1> [2]) [2]
P CoA + (+)-2-terpineol acetate
S (+)-isomenthol + acetyl-CoA <1> (Reversibility: ? <1> [1, 2]) [1, 2]
P CoA + (+)-isomenthol acetate
S (+)-menthol + acetyl-CoA <1> (Reversibility: ? <1> [2]) [2]
P CoA + (+)-menthol acetate
S (+)-neoisomenthol + acetyl-CoA <1> (Reversibility: ? <1> [1, 2]) [1, 2]
P CoA + (+)-neoisomenthol acetate
S (-)-isomenthol + acetyl-CoA <1> (Reversibility: ? <1> [2]) [2]

P CoA + (-)-isomenthol acetate
S (-)-neoisomenthol + acetyl-CoA <1> (Reversibility: ? <1> [2]) [2]
P CoA + (-)-neoisomenthol acetate
S (-)-neomenthol + acetyl-CoA <1> (Reversibility: ? <1> [2]) [2]
P CoA + (-)-neomenthol acetate
S acetyl-CoA + (+)-neomenthol <1> (Reversibility: ? <1> [1, 2]) [1, 2]
P CoA + neomenthyl acetate <1> [1]
S acetyl-CoA + (-)-menthol <1> (Reversibility: ? <1> [1, 2]) [1, 2]
P CoA + (-)-menthyl acetate <1> [1, 2]
S acetyl-CoA + borneol <1> (Reversibility: ? <1> [2]) [2]
P CoA + borneol acetate
S acetyl-CoA + cyclohexanol <1> (Reversibility: ? <1> [2]) [2]
P CoA + cyclohexanol acetate
S acetyl-CoA + n-decanol <1> (<1> best substrate [2]) (Reversibility: ? <1> [2]) [2]
P CoA + n-decanol acetate
S Additional information <1> (<1> specific for acetyl-CoA, propionyl-CoA reacts with 11% of the activity of acetyl-CoA, butyryl-CoA reacts with 2% of the activity of acetyl-CoA, enzyme has a slight preference for stereomers in which the hydroxyl and isopropyl are cis located [2]) [2]
P ?

Inhibitors

EDTA <1> (<1> slightly inhibitory [2]) [2]
diisopropylfluorophosphate <1> (<1> strong inhibitor, 50% inhibition at 0.1 mM [2]) [2]
p-hydroxymercuribenzoate <1> [2]

Activating compounds

dithioerythritol <1> (<1> required for maximal activity [2]) [2]

K_m-Value (mM)

0.2 <1> ((-)-menthol, <1> flower head enzyme [2]) [2]
0.24 <1> (D-neoisomenthol) [1]
0.3 <1> ((-)-menthol, <1> leaf enzyme [2]) [2]
0.31 <1> (D-isomenthol) [1]
0.32 <1> (L-menthol) [1]
0.35 <1> (D-neomenthol) [1]

pH-Optimum

9 <1> [2]

pH-Range

7.5-10 <1> (<1> pH 7.5: about 50% of activity maximum, increase of activity up to pH 9.5, pH 10: about 40% of activity maximum [2]) [2]

4 Enzyme Structure

Molecular weight
37000 <1> (<1> gel filtration [2]) [2]

5 Isolation/Preparation/Mutation/Application

Source/tissue
flower <1> [2]
leaf <1> (<1> midstem leaf of flowering plants [1]) [1, 2]

Localization
soluble <1> (100000 g supernatant fraction [2]) [2]

Purification
<1> (partial [2]) [2]

6 Stability

General stability information
<1>, dithioerythritol, required for maximal stability [2]

Storage stability
<1>, 0°C, 5 mM dithioerythritol, 15% glycerol, 60% loss of activity after 24 h [2]

References

[1] Martinkus, C.; Croteau, R.: Metabolism of monoterpenes. Evidence for compartmentation of L-menthone metabolism in peppermint (Mentha piperita) leaves. Plant Physiol., **68**, 99-106 (1981)
[2] Croteau, R.; Hooper, C.L.: Metabolism of monoterpenes. Acetylation of (-)-menthol by a soluble enzyme preparation from peppermint (Mentha piperita) leaves. Plant Physiol., **61**, 737-742 (1978)

CDP-Acylglycerol O-arachidonoyltransferase 2.3.1.70

1 Nomenclature

EC number
2.3.1.70

Systematic name
arachidonoyl-CoA:CDP-acylglycerol O-arachidonoyltransferase

Recommended name
CDP-acylglycerol O-arachidonoyltransferase

Synonyms
CDP-acylglycerol O-arachidonyltransferase
arachidonyl-CoA:CDP-acylglycerol O-arachidonyltransferase
arachidonyl-CoA:cytidine diphosphate monoacylglycerol acyltransferase
cytidine diphosphomonoacylglycerol arachidonyltransferase

CAS registry number
70771-22-1

2 Source Organism

<1> *Rattus norvegicus* (rat [1]) [1]

3 Reaction and Specificity

Catalyzed reaction
arachidonoyl-CoA + CDP-acylglycerol = CoA + CDP-diacylglycerol (highly specific for both donor and acceptor)

Reaction type
acyl group transfer

Natural substrates and products
S CDP-monoacylglycerol + arachidonoyl-CoA <1> (<1> high specificity for both substrates [1]) (Reversibility: ? <1> [1]) [1]
P CoA + CDP-diacylglycerol <1> [1]

Substrates and products
S CDP-monoacylglycerol + arachidonoyl-CoA <1> (<1> high specificity for both substrates [1]) (Reversibility: ? <1> [1]) [1]
P CoA + CDP-diacylglycerol <1> [1]

Specific activity (U/mg)
0.073 <1> [1]
0.916 <1> (<1> stabilized by asolectin in the purification procedure [1]) [1]

4 Enzyme Structure

Molecular weight
35000 <1> (<1> major band, SDS-PAGE [1]) [1]
61000 <1> (<1> very minor band, SDS-PAGE [1]) [1]

5 Isolation/Preparation/Mutation/Application

Source/tissue
liver <1> [1]

Localization
microsome <1> [1]

Purification
<1> [1]

6 Stability

General stability information
<1>, stabilization of the enzyme by asolectin in the purification procedure [1]

Storage stability
<1>, -15°C [1]

References

[1] Thomson, W.; MacDonald, G.: Isolation of a specific arachidonoyl coenzyme A: Cytidine diphoshate monoacylglycerolacyltransferase. J. Biol. Chem., **254**, 3311-3314 (1979)

Glycine N-benzoyltransferase 2.3.1.71

1 Nomenclature

EC number
2.3.1.71

Systematic name
benzoyl-CoA:glycine N-benzoyltransferase

Recommended name
glycine N-benzoyltransferase

Synonyms
GNAT
aralkyl-CoA:glycine N-acyltransferase
benzoyl CoA-amino acid N-acyltransferase
benzoyl-CoA:glycine N-acyltransferase
benzoyltransferase, glycine
glycine N-acyltransferase

CAS registry number
71567-07-2

2 Source Organism

<1> *Bos taurus* [1, 2, 4, 5, 6]
<2> *Rattus norvegicus* [3, 8, 9]
<3> *Homo sapiens* [5, 7]

3 Reaction and Specificity

Catalyzed reaction
benzoyl-CoA + glycine = CoA + N-benzoylglycine (<1>, sequential reaction mechanism in which the acyl-CoA substrate adds to the enzyme first, glycine adds before CoA leaves and the peptide product dissociates last [2]; <1>, sequential two-substrate mechanism [5])

Reaction type
acyl group transfer

Natural substrates and products

S Additional information <2> (<2>, enzyme synthesis is induced in response to cholestasis [9]) [9]

P ?

Substrates and products

S 3-methylcrotonyl-CoA + glycine <1> (<1>, 32% of the activity with benzoyl-CoA [1]) (Reversibility: ? <1> [1, 2]) [1, 2]

P CoA + N-3-methylcrotonylglycine

S acetyl-CoA + glycine <1> (<1>, 32% of the activity with benzoyl-CoA [1]) (Reversibility: ? <1> [1, 2]) [1, 2]

P CoA + N-acetylglycine

S benzoyl-CoA + L-asparagine <1> (<1>, weak activity [2]; <1>, 1.88% of the activity with Gly [5]) (Reversibility: ? <1> [1, 2, 5]) [1, 2, 5]

P CoA + N-benzoylasparagine

S benzoyl-CoA + L-glutamine <1> (<1>, weak activity [1]; <1>, 0.52% of the activity with Gly [5]) (Reversibility: ? <1> [1, 5]) [1, 5]

P CoA + N-benzoylglutamine

S benzoyl-CoA + alanine <1, 3> (<1>, 0.23% of the activity with Gly [5]; <3>, 5% of the activity with Gly [5]) (Reversibility: ? <1, 3> [5]) [5]

P CoA + N-benzoylalanine <1> [5]

S benzoyl-CoA + glutaminic acid <1, 3> (<1>, 0.06% of the activity with Gly [5]) (Reversibility: ? <1> [5]) [5]

P CoA + N-benzoylglutamic acid

S benzoyl-CoA + glycine <1, 2, 3> (Reversibility: ? <1, 2, 3> [1, 2, 3, 4, 5, 6, 7, 8, 9]) [1, 2, 3, 4, 5, 6, 7, 8, 9]

P CoA + N-benzoylglycine <1> [1, 2]

S benzoyl-CoA + serine <1, 3> (<1>, 0.03% of the activity with Gly [5]; <3>, 0.88% of the activity with Gly [5]) (Reversibility: ? <1, 3> [5]) [5]

P CoA + N-benzoylserine

S heptanoyl-CoA + glycine <1, 3> (<1>, 3.9% of the activity with benzoyl-CoA [6]; <3>, 9.5% of the activity with benzoyl-CoA [7]) (Reversibility: ? <1, 3> [6, 7]) [6, 7]

P CoA + N-heptanolyglycine <1> [6]

S indoleacetyl-CoA + glycine <3> (Reversibility: ? <3> [7]) [7]

P CoA + N-indoleacetylglycine

S isobutyryl-CoA + glycine <1> (<1>, 25% of the activity with benzoyl-CoA [1]) (Reversibility: ? <1> [1, 2]) [1, 2]

P CoA + N-isobutyrylglycine

S isovaleryl-CoA + glycine <1, 3> (<1>, 103% of the activity with benzoyl-CoA [1]; <1>, 8.4% of the activity with benzoyl-CoA [6]; <3>, 9.1% of the activity with benzoyl-CoA [7]) (Reversibility: ? <1, 3> [1, 2, 6, 7]) [1, 2, 6, 7]

P CoA + N-isovalerylglycine

S methylmalonyl-CoA + glycine <1> (Reversibility: ? <1> [1]) [1]

P CoA + N-methylmalonylglycine

S n-butyryl-CoA + glycine <1, 3> (<1>, 223% of the activity with benzoyl-CoA [1]; <1>, 19% of the activity with benzoyl-CoA [6]; <3>, 28% of the activity with benzoyl-CoA [7]) (Reversibility: ? <1, 3> [1, 2, 6, 7]) [1, 2, 6, 7]
P CoA + N-butyrylglycine
S naphthylacetyl-CoA + glycine <1> (Reversibility: ? <1> [6]) [6]
P CoA + N-naphthylacetylglycine
S phenylacetyl-CoA + glycine <1> (Reversibility: ? <1> [6]) [6]
P CoA + N-phenylacetylglycine
S propionyl-CoA + glycine <1> (<1>, 53% of the activity with benzoyl-CoA [1]) (Reversibility: ? <1> [1, 2]) [1, 2]
P CoA + N-propionylglycine
S salicyl-CoA + glycine <1, 3> (<1>, 15% of the activity with benzoyl-CoA [6]; <3>, 22% of the activity with benzoyl-CoA [7]) (Reversibility: ? <1, 3> [4, 6, 7]) [4, 6, 7]
P CoA + N-salicylglycine
S salicyl-CoA + glycine <1> (<1>, 35% of the activity with benzoyl-CoA [1]) (Reversibility: ? <1> [1, 2]) [1, 2]
P CoA + N-salicylglycine
S tiglyl-CoA + glycine <1> (<1>, 115% of the activity with benzoyl-CoA [1]) (Reversibility: ? <1> [1, 2]) [1, 2]
P CoA + N-tiglylglycine

Inhibitors

3'-dephospho-CoA <1> [4]

CoA <1> (<1>, in absence of KCl, 0.1 mM CoA inhibits activity over 40% irrespective of the concentration of glycine. In presence of KCl, CoA inhibits activity only slightly, less than 10%. In presence of potassium phosphate the inhibition is reduced to less than 2%. 2.5 mM, almost complete inhibition of salt-free enzyme [4]) [4]

K^+ <1> (<1>, 200 mM KCl, 44% inhibition [1]; <1>, K^+ is a competitive inhibitor for benzoyl-coenzyme A, not a competitive inhibitor for salicylyl-CoA, K^+ increases K_m-value for glycine 10fold [4]; <1>, 100 mM, 50% inhibition [6]) [1, 4, 6]

Li^+ <1> (<1>, 110 mM [4]) [4]

Mg^{2+} <1> (<1>, 100 mM $MgCl_2$, 90% inhibition [1]) [1, 2, 4]

Na^+ <1> (<1>, 110 mM [4]) [4]

Ni^{2+} <1> (<1>, 10 mM $NiCl_2$, 98% inhibition [1]) [1, 2]

Rb^+ <1> (<1>, 110 mM [4]) [4]

Zn^{2+} <1> (<1>, 1 mM $ZnCl_2$, 13% inhibition [1]) [1, 2]

benzoylalanine <1> [5]

benzoylasparagine <1> [5]

benzoylglutamic acid <1> [5]

benzoylglycine <1> [5]

benzoylserine <1> [5]

hippuric acid <1> (<1>, competitive with respect to benzoyl-CoA [4]) [4, 6]

indolacetyl-CoA <1> [1]

p-hydroxymercuribenzoate <1> (<1>, 1 mM, 24°C, 90% inhibition after 40 min [6]) [6]
phenylacetyl-CoA <1> [1]
potassium phosphate <1> [4]

Specific activity (U/mg)
1 <1> [4]
24 <1> [1, 2]

K_m-Value (mM)
2 <1> (glycine, <1>, reaction with benzoyl-CoA or salicylyl-CoA [4]) [4]
3 <1> (glycine, <1>, reaction with benzoyl-CoA [1]) [1]
6 <1> (glycine, <1>, reaction with benzoyl-CoA [6]) [6]
6.2 <1> (glycine) [5]
6.4 <3> (glycine) [5]
7.6 <1> (salicyl-CoA, <1>, reaction with salicylyl-CoA [6]) [6]
8 <1> (glycine, <1>, reaction with salicyl-CoA [1]) [1]
13 <3> (benzoyl-CoA, <1>, reaction with Gly [5]) [5]
15.2 <3> (benzoyl-CoA, <1>, reaction with Ala [5]) [5]
20 <1> (benzoyl-CoA) [1]
41 <1> (benzoyl-CoA, <1>, reaction with Ala [5]) [5]
45 <1> (benzoyl-CoA, <1>, reaction with L-Asn or L-Gln [1]) [1]
50 <1> (glycine, <1>, reaction with butyryl-CoA [1]) [1]
79 <1> (glycine, <1>, reaction with butyryl-CoA [6]) [6]
105 <1> (benzyl-CoA, <1>, reaction with Gln [5]) [5]
129 <1> (asparagine) [5]
130 <1> (L-glutamine) [1]
130 <1> (butyryl-CoA) [1]
157 <1> (benzoyl-CoA, <1>, reaction with Asn [5]) [5]
160 <1> (benzoyl-CoA, <1>, reaction with Gly [5]) [5]
170 <1> (L-asparagine) [1]
353 <1> (glutamine) [5]
360 <1> (methylmalonyl-CoA) [1]
997 <3> (alanine) [5]
998 <1> (benzoyl-CoA, <1>, reaction with Glu [5]) [5]
1148 <1> (glutaminic acid) [5]
1573 <1> (alanine) [5]

K_i-Value (mM)
0.03 <1> (hippuric acid) [4]
0.075 <1> (hippuric acid) [6]
0.2 <1> (benzoylglycine, <1>, with variable Gly concentration [5]) [5]
8.2 <1> (benzoylserine, <1>, with variable Gly concentration [5]) [5]
8.6 <1> (benzoylasparagine, <1>, with variable Gly concentration [5]) [5]
11.8 <1> (benzoylglutamic acid, <1>, with variable Gly concentration [5]) [5]
55.2 <1> (benzoylalanine, <1>, with variable Gly concentration [5]) [5]

pH-Optimum
8.4-8.6 <1> [1, 2]

4 Enzyme Structure

Molecular weight

30500 <3> (<3>, gel filtration [7]) [7]
32400 <1> (<1>, sucrose density gradient centrifugation [1]) [1]
33500 <1> (<1>, gel filtration [6]) [6]
33800 <1> (<1>, gel filtration [1]) [1]
34000 <1> (<1>, gel filtration [4]) [4]

Subunits

? <1, 3> (<3>, x * 27000, SDS-PAGE [5]; <1>, x * 36000, SDS-PAGE [5]) [5]
monomer <1> (<1>, 1 * 33000, SDS-PAGE [1,2]; <1>, 1 * 34000, SDS-PAGE [6]) [1, 2, 6]

5 Isolation/Preparation/Mutation/Application

Source/tissue

kidney <1> [6]
liver <1, 2, 3> (<2>, enzyme activity is reduced by 66% in bile duct ligation animals [3]; <2>, enzyme activity is reduced by approximately 60% in hepatic mitochondria from cirrhotic rats [8]) [1, 2, 3, 4, 5, 7, 8, 9]

Localization

cytosol <2> [9]
microsome <2> (<2>, enzyme activity shows significant increase between the 1st and 7th day after common bile duct ligation [9]) [9]
mitochondrion <1, 2, 3> (<2>, enzyme activity shows significant increase between the 1st and 7th day after common bile duct ligation [9]) [1, 2, 3, 4, 6, 7, 8, 9]

Purification

<1> [1, 2, 4]
<3> (partial [7]) [7]

6 Stability

Storage stability

<1>, -70°C, enzyme retains more than half of its activity after several months [1]
<1>, 4°C, 3 weeks, enzyme retains about 80% of its initial activity [1, 2]

References

[1] Nandi, D.L.; Lucas, S.V.; Webster, L.T.: Benzoyl-coenzyme A:glycine N-acyltransferase and phenylacetyl-coenzyme A:glycine N-acyltransferase from bovine liver mitochondria. Purification and characterization. J. Biol. Chem., **254**, 7230-7237 (1979)

[2] Webster, L.T.: Benzoyl-CoA: amino acid and phenylacetyl-CoA: amino acid N-acyltransferases. Methods Enzymol., **77**, 301-308 (1981)

[3] Krahenbuhl, L.; Reichen, J.; Talos, C.; Krahenbuhl, S.: Benzoic acid metabolism reflects hepatic mitochondrial function in rats with long-term extrahepatic cholestasis. Hepatology, **25**, 278-283 (1997)

[4] Kelley, M.; Vessey, D.A.: The effects of ions on the conjugation of xenobiotics by the aralkyl-CoA and arylacetyl-CoA N-acyltransferases from bovine liver mitochondria. J. Biochem. Toxicol., **5**, 125-135 (1990)

[5] Van der Westhuizen, F.H.; Pretorius, P.J.; Erasmus, E.: The utilization of alanine, glutamic acid, and serine as amino acid substrates for glycine N-acyltransferase. J. Biochem. Mol. Toxicol., **14**, 102-109 (2000)

[6] Kelley, M.; Vessey, D.A.: Isolation and characterization of mitochondrial acyl-CoA:glycine N-acyltransferases from kidney. J. Biochem. Toxicol., **8**, 63-69 (1993)

[7] Kelley, M.; Vessey, D.A.: Characterization of the acyl-CoA:amino acid N-acyltransferases from primate liver mitochondria. J. Biochem. Toxicol., **9**, 153-158 (1994)

[8] Krahenbuhl, L.; Ledermann, M.; Lang, C.; Krahenbuhl, S.: Relationship between hepatic mitochondrial functions in vivo and in vitro in rats with carbon tetrachloride-induced liver cirrhosis. J. Hepatol., **33**, 216-223 (2000)

[9] Kim, Y.J.; Kim, Y.H.: Benzoyltransferase and phenylacetyltransferase activities in cholestatic rat liver induced by common bile duct ligation. J. Biochem. Mol. Biol., **32**, 67-71 (1999)

Indoleacetylglucose-inositol O-acyltransferase 2.3.1.72

1 Nomenclature

EC number
2.3.1.72

Systematic name
indole-3-acetyl-β-1-D-glucoside:myo-inositol indoleacetyltransferase

Recommended name
indoleacetylglucose-inositol O-acyltransferase

Synonyms
IAA-myo-inositol synthase
indoleacetic acid-inositol synthase
indoleacetyltransferase, indoleacetylglucose-myo-inositol

CAS registry number
74082-57-8

2 Source Organism

<1> *Zea mays* (sweet corn, var. Seneca Horizon [3]) [1-3]

3 Reaction and Specificity

Catalyzed reaction
indole-3-acetyl-β-1-D-glucoside + myo-inositol = D-glucose + indole-3-acetyl-myo-inositol

Reaction type
acyl group transfer

Natural substrates and products
S indole-3-acetyl-β-1-D-glucoside + myo-inositol <1> (<1> involved in biosynthesis of indole-3-acetyl-β-1,4-glucan [3]) (Reversibility: ? <1> [3]) [3]
P D-glucose + indole-3-acetyl-myo-inositol

Substrates and products
S indole-3-acetyl-β-1-D-glucoside + myo-inositol <1> (<1> i.e. 1-O-indol-3-yl-acetyl-β-D-glucose, high specificity [1,2]; <1> scyllo-inositol or myo-inosose-2 can replace myo-inositol [3]; <1> no substrates are naphthalene 1-acetylglucose, glycerol, propan-2-ol, aspartic acid, galactosyl-myo-inosi-

tol, arabinosyl-myo-inositol [1]; <1> myo-inositol-D-galactopyranose, cyclohexanol, mannitol, glycerol [3]) (Reversibility: ? <1> [1-3]) [1-3]

P D-glucose + indole-3-acetyl-myo-inositol <1> [1, 2]

Inhibitors

2-mercaptoethanol <1> (<1> 74% inhibition at 5 mM [3]) [3]
dithionite <1> (<1> inhibition at 40 mM [3]) [3]
dithiothreitol <1> (<1> 98% inhibition at 5 mM [3]) [3]

Specific activity (U/mg)

13.3 <1> [3]

K_m-Value (mM)

0.03 <1> (1-O-indol-3-yl-acetyl-β-D-glucose) [3]
4.1 <1> (myo-inositol) [3]

pH-Optimum

6.9 <1> [1]
Additional information <1> (<1> pI: 6.1 [3]) [3]

Temperature optimum (°C)

25 <1> (<1> assay at [3]) [3]
37 <1> (<1> assay at [1,2]) [1, 2]

4 Enzyme Structure

Molecular weight

59000 <1> (<1> gel filtration [3]) [3]
100000 <1> (<1> MW below 100000, gel filtration [1]) [1]

5 Isolation/Preparation/Mutation/Application

Source/tissue

kernel <1> (<1> endosperm [3]) [1-3]

Purification

<1> (partial [2]) [2, 3]

6 Stability

General stability information

<1>, lyophilization, stable to [3]

Storage stability

<1>, -20°C, crude liquid endosperm, several years, no loss of activity [3]
<1>, 4°C, more than 2 months [3]

References

[1] Michalczuk, L.; Bandurski, R.S.: Enzymic synthesis of 1-O-indol-3-ylacetyl-β-D-glucose and indol-3-ylacetyl-myo-inositol. Biochem. J., **207**, 273-281 (1982)

[2] Michalczuk, L.; Bandurski, R.S.: UDP-glucose:indoleacetic acid glucosyl transferase and indoleacetyl-glucose:myo-inositol indoleacetyl transferase. Biochem. Biophys. Res. Commun., **93**, 588-592 (1980)

[3] Kesy, J.M.; Bandurski, R.S.: Partial purification and characterization of indol-3-ylacetylglucoe:myo-inositol indol-3-ylacetyltransferase (indoleacetic acid-inositol synthase). Plant Physiol., **94**, 1598-1604 (1990)

Diacylglycerol-sterol O-acyltransferase 2.3.1.73

1 Nomenclature

EC number
2.3.1.73

Systematic name
1,2-diacyl-sn-glycerol:sterol O-acyltransferase

Recommended name
diacylglycerol-sterol O-acyltransferase

Synonyms
1,2-diacyl-sn-glycerol:sterol acyl transferase

CAS registry number
79586-23-5

2 Source Organism

<1> *Spinacia oleracea* [1, 2]

3 Reaction and Specificity

Catalyzed reaction
1,2-diacyl-sn-glycerol + sterol = monoacylglycerol + sterol ester

Reaction type
acyl group transfer

Substrates and products
S 1,2-diacyl-sn-glycerol + sterol <1> (Reversibility: ? <1> [1, 2]) [1, 2]
P monoacylglycerol + sterol ester <1> [1, 2]
S campesterol + 1,2-diacyl-sn-glycerol <1> (Reversibility: ? <1> [1]) [1]
P monoacylglycerol + campesterol-3-O-acyl ester <1> [1]
S cholesterol + 1,2-diacyl-sn-glycerol <1> (Reversibility: ? <1> [1]) [1]
P monoacylglycerol + cholesterol-3-O-acyl ester <1> [1]
S diacylglycerol + 1,2-diacyl-sn-glycerol <1> (Reversibility: ? <1> [1]) [1]
P monoacylglycerol + triacyl-sn-glycerol <1> [1]
S dipalmitoylglycerol + sterol <1> (Reversibility: ? <1> [2]) [2]
P monopalmitoylglycerol + steryl palmitate <1> [2]
S sitosterol + 1,2-diacyl-sn-glycerol <1> (Reversibility: ? <1> [1]) [1]

P monoacylglycerol + sitosterol-3-O-acyl ester <1> [1]

S Additional information <1> (<1> monoacylglycerol and triacylglycerol are not suitable substrates [1]; <1> testing of diacylglycerols with only one type of fatty acid, preference for fatty acid transfer [1]; <1> all reactions observed in a mixed micelle system are transacetylation reactions involving various acceptors, examination of pathways of metabolism of diacylglycerol [2]) [1, 2]

P ?

Inhibitors

digitonin <1> (<1> effective inhibitor [1]) [1]
ethanol <1> (<1> at 4% inhibits [2]) [2]

Activating compounds

bovine serum albumin <1> (<1> stimulates [2]) [2]
phosphatidylcholine <1> (<1> indispensable for activity [1]) [1]
Additional information <1> (<1> it is possible that Triton X-100 inhibits diacylglycerol degradation and that bovine serum albumin prevents this enzymic inhibition by binding the nonmicellar Triton X-100 [2]) [2]

Specific activity (U/mg)

0.0002 <1> (<1> leaf homogenate [1]) [1]
0.0014 <1> (<1> 20000 g pellet from differential centrifugation of the leaf homogenate [1]) [1]
0.0022 <1> (<1> 88000 g pellet from differential centrifugation of the leaf homogenate [1]) [1]

pH-Optimum

6 <1> (<1> MES-NaOH buffer [1]) [1]

5 Isolation/Preparation/Mutation/Application

Source/tissue

leaf <1> [1, 2]

Localization

Additional information <1> (<1> differential centrifugation of the leaf homogenate shows the 20000 g pellet to have the highest amount of enzyme of the fractions tested but the highest specific activity is in the 88000g pellet. The active fractions contain endoplasmic reticulum, ribosomes and other membrane vesicles [1]) [1]

6 Stability

Storage stability

<1>, -15°C, in a desiccator, several weeks [1]

References

[1] Garcia, R.E.; Mudd, J.B.: 1,2-Diacyl-sn-glycerol:sterol acyl transferase from spinach leaves (Spinacia olerecea L.). Methods Enzymol., **71**, 768-772 (1981)
[2] Garcia, R.E.; Mudd, J.B.: Metabolism of monoacylglycerol and diacylglycerol by enzyme preparations from spinach leaves. Arch. Biochem. Biophys., **191**, 487-493 (1978)

Naringenin-chalcone synthase 2.3.1.74

1 Nomenclature

EC number
2.3.1.74

Systematic name
malonyl-CoA:4-coumaroyl-CoA malonyltransferase (cyclizing)

Recommended name
naringenin-chalcone synthase

Synonyms
6'-deoxychalcone synthase
CHS
DOCS
EC 2.3.1.120 (formerly)
chalcone synthase
chalcone synthetase
flavanone synthase
flavanone synthetase
synthase, flavanone

CAS registry number
56803-04-4

2 Source Organism

<1> *Glycyrrhiza echinata* [1]
<2> *Petroselinum hortense* (parsley [2,4,17,19]) [2, 4, 17, 19, 30]
<3> *Glycine max* (soybean [5]) [3, 5, 11]
<4> *Juglans sp.* (walnut, Juglans nigra * Juglans regia [6]) [6]
<5> *Sinapis alba* [5, 7]
<6> *Secale cereale* (rye [8]) [8, 30]
<7> *Spinacia oleracea* (spinach [9]) [9]
<8> *Fagopyrum esculentum* (buckwheat [10,12]) [10, 12]
<9> *Avena sativa* (oat [13]) [13]
<10> *Daucus carota* (carrot, ssp. sativa [14]; carrot, var Kurdagosun [16]) [14-16, 30]
<11> *Phaseolus vulgaris* [18, 30]
<12> *Tulipa sp.* (cv. Apeldoorn [20,21,23]) [20, 21, 23]
<13> *Cosmos sulphureus* [21]

<14> *Haplopappus gracilis* [22]
<15> *Arabidopsis thaliana* [5, 25]
<16> *Stellaria longipes* (genotype alpine and prairie [24]) [24]
<17> *Hordeum vulgare* (barley, inoculated with fungus blumeria graminis [26]) [26]
<18> *Gerbera hybrida* (var. Regina [27]) [27]
<19> *Medicago sativa* (alfalfa [28]) [28]
<20> *Ruta graveolens* [29]
<21> *Silene sp.* [30]
<22> *Verbena sp.* [30]
<23> *Pinus sylvestris* (Scots pine [30]) [30]
<24> *Sinapsis alba* (white mustard [30,34]) [30, 34]
<25> *Scutellaria baicalensis* (labiatae [31]) [31]
<26> *Cassia alata* (ringworm bush [32]) [32]
<27> *Pueraria lobata* [33, 35]
<28> *Rubus idaeus* (cv. Royalty, raspberry [36]) [36]

3 Reaction and Specificity

Catalyzed reaction
3 malonyl-CoA + 4-coumaroyl-CoA = 4 CoA + naringenin chalcone + 3 CO_2

Reaction type
acyl group transfer

Natural substrates and products
S 4-coumaroyl-CoA + malonyl-CoA <1, 2, 8, 9, 10, 14, 15> (<1> key enzyme in flavonoid synthesis [1]; <9> first step in flavonoid biosynthesis in plants [13]) (Reversibility: ? <1, 2, 8, 9, 10, 14, 15> [1, 5, 12, 13, 15, 14, 19, 22]) [1, 5, 12, 13, 15, 14, 19, 22]
P naringenin chalcone + CoA + CO_2 <9> [13, 22]
S caffeoyl-CoA + malonyl-CoA <14> (Reversibility: ? <14> [22]) [22]
P eriodictyol + CoA + CO_2 <14> [22]

Substrates and products
S caffeoyl-CoA + malonyl-CoA <3, 6, 8, 10, 12, 13, 14, 17, 21, 22> (<8> 80% enzyme activity [12]; <17> wild type enzyme uses cinnamoyl-CoA and 4-coumaroyl-CoA at comparable rates whereas feruloyl-CoA is a poor substate, HvCHS2 converts feruloyl-CoA and caffeoyl-CoA at the highest rate whereas cinnamoyl-CoA is a poor substrate [26]; <22> less efficient than 4-coumaroyl-CoA [30]) (Reversibility: ? <3, 6, 8, 10, 12, 13, 14, 17, 21, 22> [8, 11, 12, 15, 20-23, 26, 30]) [8, 11, 12, 15, 20-23, 26, 30]
P eriodictyol + CoA + CO_2 <3, 6, 10, 12> [8, 11, 15, 23]
S cinnamoyl-CoA + malonyl-CoA <3, 17, 23, 24, 26> (<3> 15% of the reaction rate with 4-coumaroyl-CoA [11]; <17> wild type enzyme uses cinnamoyl-CoA and 4-coumaroyl-CoA at comparable rates whereas feruloyl-CoA is a poor substate, HvCHS2 converts feruloyl-CoA and caffeoyl-CoA

at the highest rate whereas cinnamoyl-CoA is a poor substrate [26]) (Reversibility: ? <3, 17, 23, 24, 26> [11, 26, 30, 32]) [11, 26, 30, 32]

P 5,7-dihydroxyflavanone + CoA + CO_2 <3, 23> (<3,23> i.e. pinocembrin [11,30]) [11, 30]

S ferulyl-CoA + malonyl-CoA <3, 8, 12, 13, 14, 17> (<3> reaction gives one major and 3 minor reaction products, none of those is identical with homoeriodictyol [11]; <8> 80% enzyme activity [12]; <17> wild type enzyme uses cinnamoyl-CoA and 4-coumaroyl-CoA at comparable rates whereas feruloyl-CoA is a poor substrate, HvCHS2 converts feruloyl-CoA and caffeoyl-CoA at the highest rate whereas cinnamoyl-CoA is a poor substrate [26]) (Reversibility: ? <3, 8, 12, 13, 14, 17> [11, 12, 20, 21-23, 26]) [11, 12, 20, 21-23, 26]

P homoeriodictyol + CoA + CO_2 <12, 14> (<14> formation of by-products [22]) [22, 23]

S malonyl CoA + 4-coumaroyl-CoA + NADH <1, 2, 3, 24, 27> (Reversibility: ? <1, 2, 3, 24, 27> [1, 3, 19, 34, 33]) [1, 3, 19, 34, 33]

P 6'-deoxychalcone + CoA + CO_2 + NAD^+ <1, 3, 24> (<3,24> the formation of this product requires an additional reductase [3,34]) [1, 3, 33, 34]

S malonyl-CoA + 4-coumaroyl-CoA <1-18, 21, 22, 27, 28> (<17> wild type enzyme uses cinnamoyl-CoA and 4-coumaroyl-CoA at comparable rates whereas feruloyl-CoA is a poor substrate, HvCHS2 converts feruloyl-CoA and caffeoyl-CoA at the highest rate whereas cinnamoyl-CoA is a poor substrate [26]; <18> no formation of naringenin [27]) (Reversibility: ? <1-18, 21, 22, 27, 28> [1, 2, 3, 5, 6-9, 11-13, 15, 16, 18, 20-24, 26, 27, 30, 35, 36]) [1, 2, 3, 5, 6-9, 11-13, 15, 16, 18, 20-24, 26, 27, 30, 35, 36]

P naringenin chalcone + CoA + CO_2 <1-6, 8-12, 14, 16, 27, 28> (<1> in the presence of NADH liquiritigenin i.e. 5-deoxyflavanone is formed as a by-product [1]; <2,4> due to the presence of isomerase the immediate product naringenin chalcone is not detectable, instead 4,2',4',6'-tetrahydroxychalcone is found [2,6]; <5,9> due to the presence of isomerase the immediate product naringenin chalcone is not detectable, instead 4,2',4',6'-tetrahydroxychalcone is found, the latter cyclizes spontaneously to naringenin [7,13]; <10> chalcone is the initial product that spontaneously transforms to naringenin [16]; <2> amount of byproducts varies with with concentration of 2-mercaptoethanol [17]; <27> byproduct 2.7-4.2% reservatrol [35]; <28> PKS1 produces mainly naringenin chalcone as product with small amounts of 4-coumaroyltriacetic acid lactone as byproduct, PKS3 produces mainly 4-coumaroyltriacetic acid lactone and only small amounts of naringenin chalcone [36]) [1, 2, 3, 5, 6-8, 11-13, 15, 16, 17, 18, 20, 22, 23, 24, 35, 36]

S malonyl-CoA + 4-coumaroyl-CoA + NADPH <1, 3> (Reversibility: ? <1, 3> [1, 3]) [1, 3]

P 5'-deoxyflavanone + CoA + CO_2 + $NADP^+$ <1, 3> (<3> the formation of this product requires an additional reductase [3]) [1, 3]

S malonyl-CoA + acetyl-CoA <2, 26> (<2> poor substrate [19]) (Reversibility: ? <2, 26> [19, 32]) [19, 32]

P 4-hydroxy-6-methyl-2-pyrone + CoA + CO_2 <2> (<2> i.e. triacetic acid lactone [19]) [19, 32]

S malonyl-CoA + benzoyl-CoA <2, 18, 25, 26> (<2> poor substrate at pH 8, most efficient substrate at pH 6.5 [19]) (Reversibility: ? <2, 18, 25, 26> [19, 27, 31, 32]) [19, 27, 31, 32]

P phlorobenzophenone + CO_2 + CoA <25> (<25> i.e. 2,4,6-trihydroxy-benzophenone along with byproducts [31]) [31]

S malonyl-CoA + butyryl-CoA <2> (Reversibility: ? <2> [19]) [19]

P phlorobutyrophenone + CO_2 + CoA <2> [19]

S malonyl-CoA + hexanoyl-CoA <2, 26> (Reversibility: ? <2, 26> [19, 32]) [19, 32]

P phlorocaprophenone + CO_2 + CoA <2> [19]

S malonyl-CoA + isovaleryl-CoA <26> (Reversibility: ? <26> [32]) [32]

P ?

S malonyl-CoA + n-butyryl-CoA <26> (Reversibility: ? <26> [32]) [32]

P ?

S malonyl-CoA + octanoyl-CoA <2> (<2> poor substrate [19]) (Reversibility: ? <2> [19]) [19]

P ?

S malonyl-CoA + phenylacetyl-CoA <25> (Reversibility: ? <25> [31]) [31]

P phlorobenzylketone + CoA + Co2 <25> (<25> i.e. 2,4,6-trihydroxyphenylbenzylketone [31]) [31]

S Additional information <25> (<25> enzyme accepts both aromatic and aliphatic CoA esters as starter substrate [31]) [31]

P ?

Inhibitors

2'-hydroxygenistein <11> [18]

2-mercaptoethanol <12, 18> (<12> above 5 mM [23]; <18> 100 mM [27]) [23, 27]

3'-nucleotidase <10> (<10> i.e. EC 3.1.3.6 hydrolyzes the phosphate group in the 3'-position of adenosine, a part of the CoA thioester substrates [14]) [14]

4-coumaroyl-CoA <4, 10> (<4> substrate inhibition above 0.03 mM [6]; <10> above 0.01 mM [15]) [6, 15]

Ca^{2+} <12> (<12> 10-20% decrease at 1 mM [23]) [23]

CoA <2, 6, 12> (<6> 50% inhibition at 0.036 mM [8]; <12> total inhibition at 0.25 mM [23]; <2> 50% inhibition at 0.01 mM [30]) [8, 23, 30]

Cu^{2+} <12> (<12> 50% decrease above 1 mM [23]) [23]

Mg^{2+} <12> (<12> 10-20% decrease at 1 mM [23]) [23]

Tris <10> [16]

Zn^{2+} <12> (<12> 50% decrease above 1 mM [23]) [23]

apigenin <6, 9> (<6> 50% inhibition at 0.009 mM [8]; <9> competitive with respect to 4-coumaroyl-CoA and non competitive with respect to malonyl-CoA [13]) [8, 13, 30]

cerulenin <5, 10, 11> (<5> 50% inhibition at 0.001 mg per assay [7]) [7, 15, 18]

dalbergioidin <11> [18]

diethyl diphosphate <27> (<27> inhibition of wild-type enzyme is pH independent, inhibition of C164S mutant increases with increasing pH between pH 6.2 to pH 7.4 [33]) [33]
dithiothreitol <12> (<12> 60% inhibition [23]) [23]
eriodictyol <6, 10, 14> (<6> 50% inhibition at 0.045 mM [8]) [8, 15, 22]
eriodictyol chalcone <10> [15]
ethylene glycol <14> [22]
glycerol <14> [22]
iodoacetamide <19, 27> (<27> due to labeling of C164 [33]) [28, 33]
iodoacetic acid <19> [28]
isoliquiritigenin <11> [18]
isovitexin 2"-O-arabinoside <6> (<6> 50% inhibition at 0.062 mM [8]) [8]
kievitone <11> (<11> 50% inhibition at 0.001 mM [18]) [18, 30]
luteolin <6> (<6> 50% inhibition at 0.013 mM [8]) [8, 30]
malonyl-3'-dephospho-CoA <10> (<10> 50% inhibition at 0.003 mM [14]) [14]
malonyl-CoA <4, 10> (<10> substrate inhibition above 0.05 mM [6, 15]) [6, 15]
naringenin <2, 6, 11, 12, 14> (<6> 50% inhibition at 0.045 mM [8]; <10> 50% inhibition of naringenin formation and total inhibition of eriodictyol formation [15]; <12> total inhibition at 0.4 mM [23]; <2> 50% inhibition at 0.01 mM [30]; <6> 0.1 mM [30]) [8, 18, 22, 23, 30]
naringenin chalcone <6, 10> (<10> 50% inhibition of naringenin formation and total inhibition of eriodictyol formation [15]; <6> 0.1 mM [30]) [15, 30]
p-chloromercuribenzoate <12> (<12> 50% inhibition at 2.5 mM [23]) [23]
potassium ascorbate <14> [22]

Activating compounds

2-mercaptoethanol <4, 10, 11, 12> (<4,12> activation [6,23]; <10> also leads to increased formation of by-products [15]; <11> 2fold increase at 20 mM [18]) [6, 15, 23, 18]
Dowex <4> (<4> 77% increase of activity [6]) [6]
EDTA <4> (<4> 50 mM, 9% increase of activity [6]) [6]
N_2 <4> (<4> 79% increase of activity [6]) [6]
bovine serum albumin <10, 14> (<14> activation [22]) [15, 22]
chlorogenic acid <10> (<10> increases eriodictyol formation by 20% [15]) [15]
cysteine <4> (<4> dependent on [6]) [6]
dithiothreitol <11> (<11> 4fold increase at 20 mM [18]) [18]
polyethyleneglycol <4> (<4> activity strongly dependent on 1.5% [6]) [6]
polyvinylpyrrolidone <4> (<4> 10% w/v essential [6]) [6]
sucrose <4> (<4> 25% increase of activity [6]) [6]

Metals, ions

$CaCl_2$ <4> (<4> 1 mM, 15% increase of activity [6]) [6]

Turnover number (min^{-1})

1.26 <25> (4-coumaroyl-CoA) [31]
Additional information <19> [28]

Specific activity (U/mg)

0.00083 <14> [22]
0.00176 <11> [18]
0.0064 <6> [8]
0.0085 <3> [11]
0.00918 <9> [13]
0.0161 <2> [4]
0.0188 <10> [16]
0.162 <10> [15]
Additional information <2, 8, 12, 27> [12, 17, 23, 33, 35]

K_m-Value (mM)

0.0006 <6, 8, 10> (4-coumaroyl-CoA) [8, 12, 15]
0.0007 <2> (hexanoyl-CoA) [19]
0.0008 <7> (4-coumaroyl-CoA, <7> isozyme AI [9]) [9]
0.001 <7> (4-coumaroyl-CoA, <7> isozyme AII [9]) [9]
0.001 <17> (feruloyl-CoA) [26]
0.001 <8> (malonyl-CoA) [12]
0.0014-0.0015 <6> (malonyl-CoA) [8]
0.00145 <6> (caffeoyl-CoA) [8]
0.0015 <9> (4-coumaroyl-CoA) [13]
0.0016 <4> (4-coumaroyl-CoA) [6]
0.0016 <12> (caffeoyl-CoA) [23]
0.0017 <12> (4-coumaroyl-CoA) [23]
0.0019 <7> (malonyl-CoA, <7> isozyme AII [9]) [9]
0.002 <17> (4-coumaroyl-CoA) [26]
0.002 <7> (malonyl-CoA, <7> isozyme AI [9]) [9]
0.0022 <2> (4-coumaroyl-CoA) [19]
0.0023 <11> (4-coumaroyl-CoA) [18]
0.0023 <24> (4-coumaroyl-CoA, <24> C169S/Q100E/K180Q triple mutant [34]) [34]
0.0025 <24> (4-coumaroyl-CoA, <24> wild-type enzyme [34]) [34]
0.0025 <12> (feruloyl-CoA) [23]
0.0026 <4> (malonyl-CoA) [6]
0.003 <10> (malonyl-CoA) [15]
0.0057 <10> (4-coumaroyl-CoA) [16]
0.0063 <9> (malonyl-CoA) [13]
0.0077 <10> (caffeoyl-CoA) [15]
0.017 <3> (NADPH) [3]
0.018 <10> (malonyl-CoA) [16]
0.0214 <11> (malonyl-CoA) [18]
0.0361 <25> (4-coumaroyl-CoA) [31]

K_i-Value (mM)

0.0004 <9> (apigenin) [13]
0.0207 <11> (kievitone) [18]

pH-Optimum

6 <3> (<3> caffeoyl-CoA [11]) [11]
6.5 <6, 14> (<6> caffeoyl-CoA [8]; <14> caffeoyl-CoA and feruloyl-CoA [22]) [8, 22]
6.5-7 <14> (<14> caffeoyl-CoA [22]) [22]
6.8 <4, 10> (<4> assay at [6]; <10> caffeoyl-CoA [15]) [6, 15]
7.5 <3> (<3> 4-coumaroyl-CoA [11]) [11]
7.9 <10> (<10> 4-coumaroyl-CoA [15]) [15]
8 <4, 6, 8, 9, 10, 11, 12, 14> (<4, 6, 10, 14> 4-coumaroyl-CoA [6, 8, 16, 22]) [6, 8, 12, 13, 16, 18, 22, 23]

Temperature optimum (°C)

30 <12> [23]
35 <4> [6]
40 <11> [18]
45 <3> [11]

Temperature range (°C)

20-50 <12> (<12> less than 50% of maximal activity above and below [23]) [23]
20-60 <12> (<12> 50% activity at 20°C, 50°C and 60°C [23]) [23]
25-45 <4> (<4> less than 40% of maximal activity above and below [6]) [6]

4 Enzyme Structure

Molecular weight

42800 <28> (<28> calculated from amino acid sequence [36]) [36]
43000 <15, 17, 28> (<15> calculated from amino acid sequence [25]; <17> wild type enzyme migrates at 43000 Da, HvCHS2 migrated at a slightly lower molecular mass [26]; 2<8> SDS-PAGE [36]) [25, 26, 36]
48000 <7> (<7> isozyme AI, gel filtration, differences in MW may be due to different spatial conformations of AI and AII [9]) [9]
55000 <12> (<12> gel filtration, gradient centrifugation [23]) [23]
62000 <7> (<7> isozyme AII, gel filtration, differences in MW may be due to different spatial conformations of AI and AII [9]) [9]
69000 <15> (<15> fusion protein with glutathione S-transferase, SDS-PAGE [25]) [25]
75000 <3> (<3> gel filtration [11]) [11]
77000 <2, 11> (<2> sedimentation equilibrium centrifugation [4]; <11> gel filtration [18]) [4, 18]
78000 <7> (<7> isozyme AII, non-denaturing gel electrophoresis, differences in MW may be due to different spatial conformations of AI and AII [9]) [9]
80000-85000 <10> (<10> gel filtration, PAGE under non-denaturing conditions [16]) [16]
83000 <8> (<8> gel filtration [12]) [12]
88000 <7> (<7> isozyme AI, non-denaturing gel electrophoresis, differences in MW may be due to different spatial conformations of AI and AII [9]) [9]

Additional information <15> (<15> comparison of amino acid sequence with resveratrol synthase, EC 2.3.1.95 [5]) [5]

Subunits

dimer <2, 3, 6, 7, 8, 10> (<2> 2 * 40000-45000, SDS-PAGE [4]; <6> 1 * 43000 + 1 * 44000, SDS-PAGE [8]; <7> 1 * 78000 + 1 * 88000, non-denaturating PAGE, 1 * 48000 + 1 * 62000, gel filtration [9]; <3> 2 * 40000, SDS-PAGE [11]; <8> 2 * 41000, SDS-PAGE [12]; <10> 2 * 46000, SDS-PAGE [16]) [4, 8, 9, 11, 12, 16]

5 Isolation/Preparation/Mutation/Application

Source/tissue

anther <12> (<12> 2% activity in pollen, 98% activity in tapetum [20]) [20, 21, 23]
bark <4> [6]
cell suspension culture <1-3, 10, 11, 14> [1, 3, 4, 11, 14-19, 22]
hypocotyl <8> [10, 12]
leaf <6, 7, 9, 25, 26> (<6> seedling [8]) [8, 9, 13, 31, 32]
petal <13> [21]
root <26> [32]
seedling <15> [25]
stem <26> [32]

Localization

cytosol <2> [2]
endoplasmic reticulum membrane <8> [10]
Additional information <7> (<7> no activity in chloroplasts [9]) [9]

Purification

<2> [17]
<2> (homogeneity [2,4]) [2, 4]
<3> [5, 11]
<6> (450fold [8]) [8]
<7> [9]
<8> (homogeneity [12]) [12]
<9> (500fold [13]) [13]
<10> (partial [15]; homogeneity [16]) [15, 16]
<11> (31fold [18]) [18]
<12> (partial [23]) [21, 23]
<14> (62fold [22]) [22]
<15> (near homogeneity [25]) [25]
<19> (homogeneity [28]) [28]
<25> [31]
<27> (11fold [25]) [35]
<28> (PKS3 mutant [36]) [36]

Crystallization

<19> [28]

Cloning

<5> (expression in Escherichia coli, site directed mutagenesis, sequence alignment with EC 2.3.1.95 [7]) [7]
<15> (fusion protein with glutathione S-transferase [25]) [5, 25]
<17> (wild type enzyme and HvCHS2 with different substrate requirements [26]) [26]
<18> (GCHS2 and GCHS26 with different enzymatic and structural properties [27]) [27]
<19> [28]
<20> [29]
<24> [34]
<25> [31]
<26> (isoforms [32]) [32]
<27> (wild type and mutants [33]) [33, 35]
<28> (3 genes PKS1, PKS2 and PKS3 [26]) [36]

Engineering

A133S <20> (<20> fully functional enzyme [29]) [29]
C135A <5> (<5> 77% enzyme activity [7]) [7]
C164A <19> (<19> reduced activity [28]) [28]
C164S <27> (<27> inhibition by diethyl diphosphate is pH dependent [33]) [33]
C169A <5> (<5> no enzyme activity [7]) [7]
C169S <24> (<24> no enzyme activity [34]) [34]
C169S/Q100E <24> (<24> 28% of wild-type activity [34]) [34]
C169S/Q100E/K180Q <24> (<24> 15% of wild-type activity [34]) [34]
C195A <5> (<5> increased enzyme activity [7]) [7]
C347A <5> (<5> 44% enzyme activity [7]) [7]
C65A <5> (<5> 15% enzyme activity [7]) [7]
C89A <5> (<5> 71% enzyme activity [7]) [7]
H303Q <19> (<19> reduced activity [28]) [28]
K180Q <24> (<24> no enzyme activity [34]) [34]
Q100E <24> (<24> 14% of wild-type activity [34]) [34]
S132T <20> (<20> fully functional enzyme [29]) [29]
S158C <24> (<24> 90% of wild-type activity [34]) [34]
V265F <20> (<20> reduced activity [29]) [29]
V265F/S132T/A133S <20> (<20> triple mutant, reduced specific activity [29]) [29]

6 Stability

Temperature stability

0 <10> (<10> half-life 3 h in 1.4 mM 2-mercaptoethanol, increase of stability with 14 mM 2-mercaptoethanol [15]) [15]

4 <7> (<7> half-life 4-5 d, partially purified enzyme [9]) [9]
25 <12> (<12> 50% inactivation after 90 min [23]) [23]
100 <12> (<12> boiling for 10 min leads to complete inactivity [20]) [20]

Oxidation stability

<2>, exclusion of oxygen during purification gives improved yield and purity of product [17]
<4>, sensitive to oxygen [6]

General stability information

<4>, 0.1% bovine serum albumin stabilizes [6]
<7>, 5% activity after freezing for 1 h without presence of glycerol [9]
<10>, withdrawal of 2-mercaptoethanol after ammonium sulfate precipitation leads to increased stability during further purification [15]

Storage stability

<2>, -70°C, 0.04 mM potassium phosphate buffer, pH 8.0, 14 mM 2-mercaptoethanol, 20% v/v glycerol [2]
<2>, -70°C, no loss of activity in 0.1 M imidazole-HCl buffer, pH 6.8, 20 mM sodium ascorbate, 10% v/v glycerol for several months [17]
<4>, -70°C, 0.1% bovine serum albumin, at least 3 weeks [6]
<6>, -20°C, 4 mg/ml bovine serum albumin [8]
<7>, -20°C, 0.4 M phosphate buffer, pH 6.5, 20% v/v glycerol, 4 weeks [9]
<8>, -20°C, 10% glycerol, 6 months, 20% loss of activity [12]
<9>, -20°C, 3 mg protein/ml [13]
<10>, -70°C, decrease of activity in solutions below 1 mg/ml [15]
<12>, -20°C, 30% initial loss of activity, remaining activity stable for 14 d [23]
<14>, -20°C, 5 mg/ml bovine serum albumin, 12 d [22]

References

[1] Ayabe, S.; Udagawa, A.; Furuya, T.: NAD(P)H-dependent 6-deoxychalcone synthase activity in Glycyrrhiza echinata cells induced by yeast extract. Arch. Biochem. Biophys., **261**, 458-462 (1988)
[2] Heller, W.; Hahlbrock, K.: Highly purified flavanone synthase from parsley catalyzes the formation of naringenin chalcone. Arch. Biochem. Biophys., **200**, 617-619 (1980)
[3] Welle, R.; Grisebach, H.: Isolation of a novel NADPH-dependent reductase which coacts with chalcone synthase in the biosynthesis of 6'-deoxychalcone. FEBS Lett., **236**, 221-225 (1988)
[4] Kreuzaler, F.; Ragg, H.; Heller, W.; Tesch, R.; Witt, I.; Hammer, D.; Hahlbrock, K.: Flavanone synthase from Petroselinum hortense. Molecular weight, subunit composition, size of messenger RNA, and absence of pantetheinyl residue. Eur. J. Biochem., **99**, 89-96 (1979)

[5] Schröder, J.; Schröder, G.: Stilbene and chalcone synthases: Related enzymes with key functions in plant-specific pathways. Z. Naturforsch. C, **45c**, 1-8 (1990)
[6] Claudot, A.C.; Drouet, A.: Preparation and assay of chalcone synthase from walnut tree tissue. Phytochemistry, **31**, 3377-3380 (1992)
[7] Lanz, T.; Tropf, S.; Marner, F.J.; Schröder, J.; Schröder, G.: The role of cysteines in polyketide synthases. Site-directed mutagenesis of resveratrol and chalcone synthases, two key enzymes in different plant-specific pathways. J. Biol. Chem., **266**, 9971-9976 (1991)
[8] Peters, A.; Schneider-Poetsch, H.A.W.; Schwarz, H.; Weissenböck, G.: Biochemical and immunological characterization of chalcone synthase from rye leaves. J. Plant Physiol., **133**, 178-182 (1988)
[9] Beerhues, L.; Wiermann, R.: Chalcone synthases from spinach (Spinacia oleracea L.). I. Purification, peptide patterns, and immunological properties of different forms. Planta, **173**, 532-543 (1988)
[10] Hrazdina, G.; Zobel, A.M.; Hoch, H.C.: Biochemical, immunological, and immunocytochemical evidence for the association of chalcone synthase with endoplasmic reticulum membranes. Proc. Natl. Acad. Sci. USA, **84**, 8966-8970 (1987)
[11] Welle, R.; Grisebach, H.: Purification and properties of chalcone synthase from cell suspension cultures of soybean. Z. Naturforsch. C, **42c**, 1200-1206 (1987)
[12] Hrazdina, G.; Lifson, E.; Weeden, N.F.: Isolation and characterization of buckwheat (Fagopyrum esculentum M.) chalcone synthase and its polyclonal antibodies. Arch. Biochem. Biophys., **247**, 414-419 (1986)
[13] Knogge, W.; Schmelzer, E.; Weissenböck, G.: The role of chalcone synthase in the regulation of flavonoid biosynthesis in developing oat primary leaves. Arch. Biochem. Biophys., **250**, 364-372 (1986)
[14] Hinderer, W.; Seitz, H.U.: In vitro inhibition of carrot chalcone synthase by 3'-nucleotidase: the role of the 3-phosphate group of malonyl-coenzyme A in flavonoid biosynthesis. Arch. Biochem. Biophys., **246**, 217-224 (1986)
[15] Hinderer, W.; Seitz, H.U.: Chalcone synthase from cell suspension cultures of Daucus carota L. Arch. Biochem. Biophys., **240**, 265-272 (1985)
[16] Ozeki, Y.; Sakano, K.; Komamine, A.; Tanaka, Y.; Noguchi, H.; Sankawa, U.; Suzuki, T.: Purification and some properties of chalcone synthase from a carrot suspension culture induced for anthocyanin synthesis and preparation of its specific antiserum. J. Biochem., **98**, 9-17 (1985)
[17] Britsch, L.; Grisebach, H.: Improved preparation and assay of chalcone synthase. Phytochemistry, **24**, 1975-1976 (1985)
[18] Whitehead, I.M.; Dixon, R.A.: Chalcone synthase from cell suspension cultures of phaseolus vulgaris L.. Biochim. Biophys. Acta, **747**, 298-303 (1983)
[19] Schuz, R.; Heller, W.; Hahlbrock, K.: Substrate specificity of chalcone synthase from Petroselinum hortense. Formation of phloroglucinol derivatives from aliphatic substrates. J. Biol. Chem., **258**, 6730-6734 (1983)
[20] Sutfeld, R.; Wiermann, R.: Chalcone synthesis with enzyme extracts from tulip anther tapetum using a biphasic enzyme assay. Arch. Biochem. Biophys., **201**, 64-72 (1980)

[21] Sutfeld, R.; Wiermann, R.: Purification of chalcone synthase from tulip anthers and comparison with the synthase from cosmos petals. Z. Naturforsch. C, **36c**, 30-34 (1981)
[22] Saleh, N.A.M.; Fritsch, H.; Kreuzaler, F.; Grisebach, H.: Flavanone synthase from cell suspension cultures of Haplopappus gracilis and comparison with the synthase from parsely. Phytochemistry, **17**, 183-186 (1978)
[23] Sutfeld, R.; Kehrel, B.; Wiermann, R.: Characterization and development and localization of Flavanone synthase in tulip anthers. Z. Naturforsch. C, **33c**, 841-846 (1978)
[24] Alokam, S.; Li, Y.; Li, W.; Chinnappa, C.C.; Reid, D.M.: Photoregulation of phenylalanine ammonia-lyase (PAL) and chalcone synthase (CHS) in the accumulation of anthocyanin in alpine and prairie ecotypes of Stellaria longipes under varied R/FR. Physiol. Plant., **116**, 531-538 (2002)
[25] Cain, C.C.; Saslowsky, D.E.; Walker, R.A.; Shirley, B.W.: Expression of chalcone synthase and chalcone isomerase proteins in Arabidopsis seedlings. Plant Mol. Biol., **35**, 377-381 (1997)
[26] Christensen, A.B.; Gregersen, P.L.; Schroder, J.; Collinge, D.B.: A chalcone synthase with an unusual substrate preference is expressed in barley leaves in response to UV light and pathogen attack. Plant Mol. Biol., **37**, 849-857 (1998)
[27] Helariutta, Y.; Kotilainen, M.; Elomaa, P.; Kalkkinen, N.; Bremer, K.; Teeri, T.H.; Albert, V.A.: Duplication and functional divergence in the chalcone synthase gene family of Asteraceae: Evolution with substrate change and catalytic simplification. Proc. Natl. Acad. Sci. USA, **93**, 9033-9038 (1996)
[28] Jez, J.M.; Noel, J.P.: Mechanism of chalcone synthase: pKa of the catalytic cysteine and the role of the conserved histidine in a plant polyketide synthase. J. Biol. Chem., **275**, 39640-39646 (2000)
[29] Lukacin, R.; Schreiner, S.; Matern, U.: Transformation of acridone synthase to chalcone synthase. FEBS Lett., **508**, 413-417 (2001)
[30] Martin, C.R.: Structure, function, and regulation of the chalcone synthase. Int. Rev. Cytol., **147**, 233-284 (1993)
[31] Morita, H.; Takahashi, Y.; Noguchi, H.; Abe, I.: Enzymatic Formation of Unnatural Aromatic Polyketides by Chalcone Synthase. Biochem. Biophys. Res. Commun., **279**, 190-195 (2000)
[32] Samappito, S.; Page, J.; Schmidt, J.; De-Eknamkul, W.; Kutchan, T.M.: Molecular characterization of root-specific chalcone synthases from Cassia alata. Planta, **216**, 64-71 (2002)
[33] Suh, D.Y.; Kagami, J.; Fukuma, K.; Sankawa, U.: Evidence for catalytic cysteine-histidine dyad in chalcone synthase. Biochem. Biophys. Res. Commun., **275**, 725-730 (2000)
[34] Tropf, S.; Kaercher, B.; Schroeder, G.; Schroeder, J.: Reaction mechanisms of homodimeric plant polyketide synthases (stilbene and chalcone synthase). A single active site for the condensing reaction is sufficient for synthesis of stilbenes, chalcones, and 6'-deoxychalcones. J. Biol. Chem., **270**, 7922-7928 (1995)

[35] Yamaguchi, T.; Kurosaki, F.; Suh, D.Y.; Sankawa, U.; Nishioka, M.; Akiyama, T.; Shibuya, M.; Ebizuka, Y.: Cross-reaction of chalcone synthase and stilbene synthase overexpressed in Escherichia coli. FEBS Lett., **460**, 457-461 (1999)

[36] Zheng, D.; Schroder, G.; Schroder, J.; Hrazdina, G.: Molecular and biochemical characterization of three aromatic polyketide synthase genes from Rubus idaeus. Plant Mol. Biol., **46**, 1-15 (2001)

Long-chain-alcohol O-fatty-acyltransferase 2.3.1.75

1 Nomenclature

EC number
2.3.1.75

Systematic name
acyl-CoA:long-chain-alcohol O-acyltransferase

Recommended name
long-chain-alcohol O-fatty-acyltransferase

Synonyms
acyl-CoA:alcohol transacylase <1> [1, 2]
acyltransferase, long-chain alcohol
fatty acyl-coenzyme A:fatty alcohol acyltransferase <1> [3]
wax synthase
wax-ester synthase

CAS registry number
64060-40-8

2 Source Organism

<1> *Simmondsia chinensis* [1-3]

3 Reaction and Specificity

Catalyzed reaction
acyl-CoA + a long-chain alcohol = CoA + a long-chain ester (Transfers saturated or unsaturated acyl residues of chain-length C_{18} to C_{20} to long-chain alcohols, forming waxes. The best acceptor is cis-icos-11-en-1-ol)

Reaction type
acyl group transfer

Natural substrates and products
S cis-11-eicosenol + acyl-CoA <1> (Reversibility: ? <1> [1]) [1]
P cis-11-eicosenyl acetate + CoA

Substrates and products
S eicosanoyl-CoA + cis-11-eicosenol <1> (<1> cis-11-eicosenol is the best acceptor [1]) (Reversibility: ? <1> [1]) [1]

P cis-11-eicosenyl eicosanoate + CoA
S eicosanoyl-CoA + cis-13-eicosenol <1> (Reversibility: ? <1> [1]) [1]
P cis-13-eicosenyl eicosanoate + CoA
S eicosanoyl-CoA + cis-9-octadecenol <1> (Reversibility: ? <1> [1]) [1]
P cis-9-octadecenyl eicosanoate + CoA
S eicosanoyl-CoA + decanol <1> (<1> low activity [1]) (Reversibility: ? <1> [1]) [1]
P decanyl eicosanoate + CoA
S eicosanoyl-CoA + docosanol <1> (<1> low activity [1]) (Reversibility: ? <1> [1]) [1]
P docosanyl eicosanoate + CoA
S eicosanoyl-CoA + dodecanol <1> (Reversibility: ? <1> [1]) [1]
P dodecanyl eicosanoate + CoA
S eicosanoyl-CoA + eicosanol <1> (<1> low activity [1]) (Reversibility: ? <1> [1]) [1]
P eicosanyl eicosanoate + CoA
S eicosanoyl-CoA + hexadecanol <1> (Reversibility: ? <1> [1]) [1]
P hexadecanyl eicosanoate + CoA
S eicosanoyl-CoA + octadecanol <1> (<1> low activity [1]) (Reversibility: ? <1> [1]) [1]
P octadecanyl eicosanoate + CoA
S eicosanoyl-CoA + tetradecanol <1> (Reversibility: ? <1> [1]) [1]
P tetradecanyl eicosanoate + CoA
S eicosenoyl-CoA + docosenol <1> (Reversibility: ir <1> [2]) [2]
P docosenyl eicosenoate + CoA <1> [2]
S eicosenoyl-CoA + octadecenol <1> (<1> best substrates [3]) (Reversibility: ? <1> [3]) [3]
P octadecenyl eicosenoate + CoA
S octadecenoyl-CoA + dodecanol <1> (Reversibility: ir <1> [2]) [2]
P dodecanyl octadecanoate + CoA
S Additional information <1> (<1> substrate specificity, overview [3]; <1> eicosanoyl-CoA can be replaced by stearoyl-CoA or cis-11-eicosenoyl-CoA with equal reactivity [1]) [1, 3]
P ?

Inhibitors

bovine serum albumin <1> (<1> above 5 mg/ml [1]) [1]
polyvinylpyrrolidone <1> (<1> slight inhibition [1]) [1]

Cofactors/prosthetic groups

ATP <1> (<1> activation [1]) [1]
CTP <1> (<1> activation [1]) [1]
UTP <1> (<1> activation [1]) [1]

Activating compounds

GTP <1> (<1> activation [1]) [1]
bovine serum albumin <1> (<1> defatted, 2fold activation in crude enzyme extract at concentration below 5 mg/ml, inhibition above [1]) [1]

Metals, ions

Additional information <1> (<1> no influence of NaCl or KCl up to 20 mM [1]) [1]

pH-Optimum

8 <1> [2]
8-8.1 <1> [1]
Additional information <1> (<1> pI: 9.8 [3]) [3]

Temperature optimum (°C)

30 <1> (<1> assay at [1]) [1]
30-40 <1> [2]

4 Enzyme Structure

Molecular weight

40000 <1> (<1> gel filtration [3]) [3]
Additional information <1> (<1> amino acid sequence analysis and alignment [3]) [3]

Subunits

? <1> (<1> x * 30000, SDS-PAGE [3]) [3]

5 Isolation/Preparation/Mutation/Application

Source/tissue

cotyledon <1> (<1> fat pad [1]) [1, 2]
embryo <1> (<1> developing [3]) [3]

Localization

microsome <1> [3]

Purification

<1> (partially [3]) [3]

Cloning

<1> (cloning of cDNA, DNA sequence analysis, construction of plasmid to express the cDNA in transgenic Arabidopsis thaliana plants [3]) [3]

Engineering

Additional information <1> (<1> cloned cDNA is expressed in transgenic Arabidopsis thaliana plants, in combination with cDNAs encoding the jojoba fatty acyl-CoA reductase and a β-ketoacyl-CoA synthase from Lunaria annua under a seed specific promotor, highly enhanced wax content in transgenic seeds [3]) [3]

Application

agriculture <1> (<1> transgenic plants expressing the enzyme allow the production of long-chain liquid waxes at reasonable cost for use in commercial applications [3]) [3]

6 Stability

Temperature stability

25 <1> (<1> reduced stability [1]) [1]
50 <1> (<1> 10 min, complete inactivation [1]) [1]
Additional information <1> (<1> no improvement of heat-stability by polyvinylpyrrolidone [1]) [1]

General stability information

<1>, no improvement of heat-stability by polyvinylpyrrolidone [1]

Storage stability

<1>, 0°C, fairly stable [1]
<1>, 4°C, 2-3 h stable [1]

References

[1] Wu, X.Y.; Moreau, R.A.; Stumpf, P.K.: Studies of biosynthesis of waxes by developing jojoba seed: III. Biosynthesis of wax esters from acetyl-CoA and long-chain alcohols. Lipids, **16**, 897-902 (1981)
[2] Garver, W.S.; Kemp, J.D.; Kuehn, G.D.: A high-performance liquid chromatography-based radiometric assay for acyl-CoA:alcohol transacylase from jojoba. Anal. Biochem., **207**, 335-340 (1992)
[3] Lardizabal, K.D.; Metz, J.G.; Sakamoto, T.; Hutton, W.C.; Pollard, M.R.; Lassner, M.W.: Purification of a jojoba embryo wax synthase, cloning of its cDNA, and production of high levels of wax in seeds of transgenic Arabidopsis. Plant Physiol., **122**, 645-655 (2000)

Retinol O-fatty-acyltransferase 2.3.1.76

1 Nomenclature

EC number
2.3.1.76

Systematic name
acyl-CoA:retinol O-acyltransferase

Recommended name
retinol O-fatty-acyltransferase

Synonyms
retinol fatty-acyltransferase

CAS registry number
81295-48-9

2 Source Organism

<1> *Rattus norvegicus* (Sprague-Dawley [1,5,6,7]) [1-3, 5-8]
<2> *Felis silvestris* (cat [1]) [1]
<3> *Bos taurus* (bovine calf or adult [4]) [4]
<4> *Homo sapiens* (cell cultures [9]) [9]
<5> *Salvelinus fontinalis* (fish, brook trout [10]) [10]

3 Reaction and Specificity

Catalyzed reaction
acyl-CoA + retinol = CoA + retinyl ester

Reaction type
acyl group transfer

Natural substrates and products
S acyl-CoA + retinol <1-5> (Reversibility: ? <1-5> [1-10]) [1-10]
P CoA + retinyl ester <1-5> [1-10]

Substrates and products
S lauroyl-CoA + retinol <1> (Reversibility: ? <1> [1]) [1]
P CoA + retinyllaurate <1> [1]
S octanoyl-CoA + retinol <1> (Reversibility: ? <1> [1]) [1]
P CoA + retinyloctanoate <1> [1]

S oleolyl-CoA + retinol <1, 3-4> (Reversibility: ? <1, 3-4> [1, 4, 5, 9]) [1, 4, 5, 9]
P CoA + retinyloleolate <1, 3-4> [1, 4, 5, 9]
S palmitoyl-CoA + retinol <1-5> (<1> better effect with long chain fatty acids [1]; <1> polyunsaturated derivatives are poor substrates [2]; <1> predominant fatty acid among tissue retinyl esters [3]) (Reversibility: ? <1-5> [1-10]) [1-10]
P CoA + retinylpalmitate <1-5> [1-10]
S stearoyl-CoA + retinol <1> (Reversibility: ? <1> [1, 4, 5]) [1, 4, 5]
P CoA + retinylstearate <1> [1, 4, 5]

Inhibitors

2,3,7,8-tetrachlorodibenzo-*p*-dioxin <1> (<1> decreased activity [7]) [7]
5,5'-dithiobis(2-nitrobenzoic acid) <1> (<1> activity inhibited above 2 mM [5]) [5]
cellular retinol-binding protein <1> (<1> specific inhibition of retinol fatty-acyltransferase by capturing substrate [6]) [6]
p-chloromercuribenzoate <1> (<1> activity abolished at 5 mM [1]) [1]
p-chloromercuriphenylsulfonic acid <1> (<1> activity abolished at 6 mM [5]) [5]
p-hydroxymercuribenzoate <1> (<1> activity inhibited at 2 and 6 mM [5]) [5]
progesterone <1> (<1> activity inhibited above 0.05 mM [1]) [1]
sodium deoxycholate <1> (<1> activity inhibited at 0.5% [1]) [1]
sodium taurocholate <1> (<1> activity abolished at 8 mM [5]) [5]
sodium taurolate <1> (<1> activity inhibited at 0.5% [1]) [1]
Additional information <1, 5> (<1> not inhibited by cholesterol [1]; <1> not inhibited by PMSF [3,6]; <1> not inhibited by EDTA 2 mM [5]; <5> not inhibited by 3,3',4,4'-tetrachlorobiphenyl [10]) [1, 10]

Activating compounds

bovine serum albumin <1> (<1> 0.02 mM enhances activity [1]) [1, 5]
dithiothreitol <1> (<1> 5 mM enhances activity [1]) [1, 5]
Additional information <4> (<4> not enhanced by retinoic acids [9]) [9]

Specific activity (U/mg)

Additional information <1-5> (<1> description of assay method [2]; <1-5> activity determined from microsomes [1-10]) [1-10]

K_m-Value (mM)

0.016 <5> (retinol) [10]

pH-Optimum

7-8 <1> (<1> maximum esterification [1]) [1]
7.8 <5> (<5> maximal activity [10]) [10]

pH-Range

3-11 <5> (<5> decreasing activity below pH 5 [10]) [10]
6-10 <1> (<1> maximum esterification between pH 7-8 [1]) [1]

Temperature optimum (°C)

12 <5> (<5> assay at [10]) [10]
37 <1> (<1> assay at [1,2]) [1, 2]

5 Isolation/Preparation/Mutation/Application

Source/tissue

carcinoma cell <4> (<4> squamous SCC12b and SCC13b lines are common type of skin cancer that has reduced capacity to esterify retinol relative to normal keratinocytes [9]) [9]
keratinocyte <4> (<4> D033 line [9]) [9]
kidney <1> [7]
liver <1, 2, 5> [1-3, 7-8, 10]
neural retina <3> [4]
small intestine <1> [6]
testis <1> [5]

Localization

microsome <1-5> (<1-5> membrane fraction [1-10]) [1-10]

Purification

<1-5> (partial purification by differential centrifugation [1-10]) [1-10]

Application

medicine <1, 3-5> (<1, 3, 5> retinyl esterification: storage of vitamin A [4-6, 8, 10]; <4> keratinocyte growth and differentiation [9]) [4-6, 8-10]

6 Stability

pH-Stability

6-11 <5> (<5> activity is decreased below [10]) [10]
7.2 <1> (<1> phosphate buffer [7]) [7]
7.4 <1-2> (<1,2> phosphate buffer [1-3,5,8]) [1-3, 5, 8]
7.8 <5> (<5> Tris-HCl buffer [10]) [10]
8 <3> (<3> Tris acetate [4]) [4]

Temperature stability

70 <1> (<1> activity abolished in 20 min [5]) [5]

Storage stability

<1>, -70°C, phosphate buffer, pH 7.2, DTT, 12 months [6]
<1>, -70°C, phosphate buffer, pH 7.4, DTT, 2 months [1, 2, 3]
<3>, -80°C, Tris-acetate, pH 7, DTT, 6 months [4]

References

[1] Ross, A.C.: Retinol esterification by rat liver microsomes. Evidence for a fatty acyl coenzyme A:retinol acyltransferase. J. Biol. Chem., **257**, 2453-2459 (1982)
[2] Ball, M.D.: Acyl coenzyme A-dependent retinol esterification. Methods Enzymol., **189**, 446-449 (1990)
[3] Ross, A.C.: Measurement of acyl coenzyme A-dependent esterification of retinol. Methods Enzymol., **189**, 442-445 (1990)
[4] Saari, J.C.; Bredberg, D.L.: Acyl-CoA:retinol acyltransferase and lecithin:retinol acyltransferase activities of bovine retinal pigment epithelial microsomes. Methods Enzymol., **190**, 156-163 (1990)
[5] Chaudhary, L.R.; Nelson, E.C.: Some properties and subcellular distribution of acyl-coenzyme A:retinol acyltransferase activity in rat testes. Biochim. Biophys. Acta, **917**, 24-32 (1987)
[6] MacDonald, P.N.; Ong, D.E.: Assay of lecithin-retinol acyltransferase. Methods Enzymol., **189**, 450-459 (1990)
[7] Nilsson, C.B.; Hanber, A.; Trossvik, C.; Haekansson, H.: 2,3,7,8-Tetrachlorodibenzo-*p*-dioxin affects retinol esterification in rat hepatic stellate cells and kidney. Environ. Toxicol. Pharmacol., **2**, 17-23 (1996)
[8] Ross, A.C.; Kempner, E.S.: Radiation inactivation analysis of acyl-CoA:retinol acyltransferase and lecithin:retinol acyltransferase in rat liver. J. Lipid Res., **34**, 1201-1207 (1993)
[9] Jurukovski, V.; Simon, M.: Reduced lecithin:retinol acyl transferase activity in cultured squamous cell carcinoma lines results in increased substrate-driven retinoic acid synthesis. Biochim. Biophys. Acta, **1436**, 479-490 (1999)
[10] Ndayibagira, A.; Spear, P.A.: Esterification and hydrolysis of vitamin A in the liver of brook trout (Salvelinus fontinalis) and the influence of a coplanar polychlorinated biphenyl. Comp. Biochem. Physiol. C, **122C**, 317-325 (1999)

Triacylglycerol-sterol O-acyltransferase 2.3.1.77

1 Nomenclature

EC number
2.3.1.77

Systematic name
triacylglycerol:3β-hydroxysterol O-acyltransferase

Recommended name
triacylglycerol-sterol O-acyltransferase

Synonyms
acyltransferase, triacylglycerol-sterol
triacylglycerol:sterol acyltransferase

CAS registry number
80487-96-3

2 Source Organism

<1> *Sinapis alba* (white mustard [1,2]) [1, 2]

3 Reaction and Specificity

Catalyzed reaction
triacylglycerol + a 3β-hydroxysterol = diacylglycerol + a 3β-hydroxysterol ester

Reaction type
acyl group transfer

Natural substrates and products

S Additional information <1> (<1> the acyltransferase in Sinapis alba roots exhibits group specificity for natural sterols, which are 3β-hydroxysterols containing a planar ring system and differing only in number or position of double bonds or in the structure of side chain [2]) [2]

P ?

Substrates and products

S triacylglycerol + cholest-5-en-3β-ol <1> (<1> triacylglycerol from olive oil [2]) (Reversibility: ? <1> [2]) [2]

P diacylglycerols + cholesteryl ester <1> [2]

S trilinoleoylglycerol + cholest-5-en-3β-ol <1> (Reversibility: ? <1> [2]) [2]
P dilinoleoylglycerol + cholesteryl linoleate
S trioleoylglycerol + cholest-5-en-3β-ol <1> (Reversibility: ? <1> [2]) [2]
P dioleoylglycerol + cholesteryl oleate
S tripalmitoylglycerol + 24R-methylcholest-5-en-3β-ol <1> (<1> i.e. campestero [2]l) (Reversibility: ? <1> [2]) [2]
P ?
S tripalmitoylglycerol + 5α-androstan-3β-ol <1> (<1> i.e. androstanol [2]) (Reversibility: ? <1> [2]) [2]
P ?
S tripalmitoylglycerol + 5α-cholestan-3β-ol <1> (<1> i.e. cholestanol, highest rate of esterification [2]) (Reversibility: ? <1> [2]) [2]
P ?
S tripalmitoylglycerol + 5β-cholestan-3β-ol <1> (<1> i.e. coprostanol [2]) (Reversibility: ? <1> [2]) [2]
P ?
S tripalmitoylglycerol + cholest-5-en-3β-ol <1> (<1> i.e.cholesterol [1]) (Reversibility: ? <1> [1]) [1]
P dipalmitoylglycerol + cholesteryl palmitate <1> [1]
S tripalmitoylglycerol + stigmast-5-en-3β-ol <1> (<1> i.e. sitosterol [2]) (Reversibility: ? <1> [2]) [2]
P ?
S tripalmitoylglycerol + stigmasta-5,22-dien-3β-ol <1> (<1> i.e. stigmasterol [2]) (Reversibility: ? <1> [2]) [2]
P ?
S tristearoylglycerol + cholest-5-en-3β-ol <1> (Reversibility: ? <1> [2]) [2]
P distearoylglycerol + cholesteryl stearate
S Additional information <1> (<1> enzyme has an absolute requirement for a β-configuration of the hydroxyl group at C-3, rate of esterification is higher for sterols containing a planar ring system, triacylglycerols with unsaturated fatty acids are used more efficiently than fatty acids saturated with equal chain length, triacylglycerols with fatty acids C6-C22 can be used as acyl donors [2]) [2]
P ?

Inhibitors

Additional information <1> (<1> reagents containing free sulfhydryl groups e.g. DTT have no effect [2]) [2]

Metals, ions

Additional information <1> (<1> divalent metal ions have no stimulatory effect [2]) [2]

Specific activity (U/mg)

Additional information <1> (<1> specific activity in seedlings after various days of germination [1]) [1]

pH-Optimum

5.8 <1> (<1> assay at [1,2]) [1, 2]

pH-Range
5-6.8 <1> [2]

Temperature optimum (°C)
52 <1> [2]

4 Enzyme Structure

Molecular weight
18000 <1> (<1> gel filtration [2]) [2]
Additional information <1> (<1> two peaks, one more than 100000 Da, the other 18000 Da: the fraction with higher molecular weight may consist of the aggregated form of the enzyme with 18000 Da [2]) [2]

5 Isolation/Preparation/Mutation/Application

Source/tissue
cotyledon <1> [1]
root <1> (<1> of seedling [2]) [2]
seed <1> (<1> germinating [1]) [1]

Localization
cell membrane <1> [2]

Purification
<1> (partial [1,2]) [1, 2]

6 Stability

General stability information
<1>, purification has to be carried out at 0-4°C [2]

Storage stability
<1>, 4°C, solubilized enzyme preparation, 24 h, 10% loss of activity [2]

References

[1] Kalinowska, M.; Wojciechowski, Z.A.: Sterol conjugate interconversions during germination of white mustard (Sinapis alba). Phytochemistry, **23**, 2485-2488 (1984)

[2] Zimowski, J.; Wojciechowski, Z.A.: Partial purification and specificity of triacylglycerol: sterol acyltransferase from Sinapsis alba. Phytochemistry, **20**, 1799-1803 (1981)

Heparan-α-glucosaminide N-acetyltransferase 2.3.1.78

1 Nomenclature

EC number
2.3.1.78

Systematic name
acetyl-CoA:heparan-α-D-glucosaminide N-acetyltransferase

Recommended name
heparan-α-glucosaminide N-acetyltransferase

Synonyms
acetyl-CoA:α-glucosaminide N-acetyltransferase
acetyl-coenzyme A-α-glucosaminide N-acetyltransferase
acetyl-coenzyme:α-D-2-amino-glucosamine transferase <1> [10]
acetyltransferase, α-glucosaminide
Additional information (not identical with EC 2.3.1.3 or EC 2.3.1.4)

CAS registry number
79955-83-2

2 Source Organism

<1> *Homo sapiens* [1, 6-10]
<2> *Rattus norvegicus* [2-5]

3 Reaction and Specificity

Catalyzed reaction
acetyl-CoA + heparan sulfate α-D-glucosaminide = CoA + heparan sulfate N-acetyl-α-D-glucosaminide (<2> transmembranal reaction mechanism [4,5]; <2> di-iso ping-pong mechanism [3,5])

Reaction type
acyl group transfer

Natural substrates and products
S acetyl-CoA + heparan sulfate α-D-glucosaminide <1, 2> (<1> initial step in heparan sulfate degradation [1]) (Reversibility: ? <1, 2> [1, 3-8, 10]) [1, 3-8, 10]
P CoA + heparan sulfate N-acetyl-α-D-glucosaminide <1> [1, 3-5]

S Additional information <1, 2> (<2> enzyme is acetylated at the cytoplasmic side of the lysosome and the acetyl group is then transferred to the inside where it is used to acetylate heparan sulfate [4,5]; <1> enzyme deficiency leads to Sanfilippo syndrome type C, i.e. mucopolysaccharidosis III C [1,8,10]) [1, 4, 5, 8, 10]
P ?

Substrates and products
S acetyl-CoA + 4-methylumbelliferyl-α-D-glucosaminide <1> (Reversibility: ? <1> [8]) [8]
P CoA + 4-methylumbelliferyl-N-acetyl-α-D-glucosaminide <1> [8]
S acetyl-CoA + 4-methylumbelliferyl-β-D-glucosaminide <1> (Reversibility: ? <1> [8]) [8]
P CoA + 4-methylumbelliferyl-N-acetyl-β-D-glucosaminide <1> [8]
S acetyl-CoA + β-D-glucosamine <1> (Reversibility: ? <1> [7, 9]) [7, 9]
P CoA + N-acetyl-β-D-glucosamine <1> [7, 9]
S acetyl-CoA + glucosamine-L-iduronic acid-D-glucosamine <1> (<1> reduced with $NaBH_4$ [6]) (Reversibility: ? <1> [6]) [6]
P ?
S acetyl-CoA + heparan sulfate α-D-glucosaminide <1, 2> (<1,2> brings about the acetylation of glucosamine groups of heparan sulfate and heparin from which the sulfate has been removed [1-5]) (Reversibility: ? <1, 2> [1-5]) [1-5]
P CoA + heparan sulfate N-acetyl-α-D-glucosaminide <1, 2> [1-5]
S acetyl-CoA + heparin α-D-glucosaminide <1> (<1> also acetylates di- and tetrasaccharide fragments of heparin [7]) (Reversibility: ? <1> [7]) [7]
P CoA + heparin N-acetyl-D-glucosamine

Inhibitors
2-mercaptoethanol <2> (<2> membrane-bound enzyme is insensitive, solubilized enzyme is inactivated [3]) [3]
Hg^{2+} <2> (<2> complete inhibition [2]) [2]
N-acetyl-β-D-glucosamine <1, 2> (<1> competitive against glucosamine [7]; <1,2> noncompetitive against acetyl-CoA [3,7]) [3, 5, 7]
N-bromosuccinimide <2> (<2> binds a histidine residue at the active site [4]) [4]
N-laurylsarcosine <2> (<2> 1% [3]) [3]
Zn^{2+} <2> [2]
acetyl-CoA <2> [5]
α-D-glucosamine 6-phosphate <2> (<2> weak [3]) [3]
α-methyl-D-glucose <2> (<2> weak [3]) [3]
cellobiose <2> (<2> weak [3]) [3]
chitin <2> [3]
chitosan <2> [3]
coenzyme A <1, 2> (<1,2> noncompetitive [3,7]) [3, 5, 7]
desulfo-CoA <1> (<1> competitive against acetyl-CoA [7]) [7]

diethyl dicarbonate <2> (<2> reversible by incubation with hydroxylamine, binds a histidine residue at the active site [4]) [4]
dithiothreitol <2> (<2> solubilized protein [3]) [3]
maltose <2> (<2> weak [3]) [3]
p-chloromercuribenzoate <2> (<2> binding site is not the active site [4]) [3-5]
sucrose <2> (<2> weak [3]) [3]
zwittergent <2> (<2> 1% [3]) [3]
Additional information <2> (<2> iodoacetamide and N-ethylmaleimide are not inhibitory [3-5]) [3-5]

Activating compounds

Triton X-100 <1, 2> (<1> 1.5 to 2.5fold stimulation at 0.25% w/v [8]; <2> 2 to 3fold increase of activity [3]) [3, 8]
methyl methanethiolsulfonate <2> (<2> activates the membrane-bound enzyme 1.5fold [3]) [3]
taurodeoxycholate <2> (<2> 2 to 3fold increase of activity [3]) [3]

Specific activity (U/mg)

0.000017 <1> (<1> leukocyte, substrate 4-methylumbelliferyl-α-D-glucosaminide + acetyl-CoA [8]) [8]
0.000022 <1> (<1> leukocyte, substrate 4-methylumbelliferyl-β-D-glucosaminide + acetyl-CoA [8]) [8]
0.000058 <1> (<1> fibroblast, substrate 4-methylumbelliferyl-β-D-glucosaminide + acetyl-CoA [8]) [8]
0.000059 <1> (<1> fibroblast, substrate 4-methylumbelliferyl-α-D-glucosaminide + acetyl-CoA [8]) [8]
0.000673 <1> (<1> lysosomal fraction [7]) [7]
0.008 <2> [2]
Additional information <1> (<1> Sanfilippo syndrome type C patients in comparison [8]) [1, 8]

K_m-Value (mM)

0.0025 <1> (acetyl-CoA, <1> pH 7.0 [7]) [7]
0.003 <1> (β-D-glucosamine, <1> pH 7.0 [7]) [7]
0.054 <1> (acetyl-CoA, <1> pH 5.0 [7]) [7]
0.29-0.3 <2> (α-D-glucosamine) [3, 5]
0.37 <1> (β-D-glucosamine, <1> pH 5.0 [7]) [7]
0.55 <2> (acetyl-CoA) [3, 5]

K_i-Value (mM)

0.032 <1> (coenzyme A, <1> versus acetyl-CoA [7]) [7]
0.082 <1> (coenzyme A, <1> versus β-D-glucosamine [7]) [7]
0.21 <1> (desulfo-CoA, <1> versus β-D-glucosamine [7]) [7]
0.22 <1> (desulfo-CoA, <1> versus acetyl-CoA [7]) [7]
3 <2> (coenzyme A, <2> versus acetyl-CoA [5]) [5]
3-7 <2> (coenzyme A) [3]
3.8 <1> (N-acetyl-β-D-glucosamine, <1> versus acetyl-CoA [7]) [7]
5.6 <1> (N-acetyl-β-D-glucosamine, <1> versus glucosamine [7]) [7]

7 <2> (coenzyme A, <2> versus N-acetyl-α-D-glucosamine [5]) [5]
15 <2> (N-acetyl-α-D-glucosamine, <2> versus glusosamine, no detergents added [5]) [5]
15-17 <2> (N-acetyl-α-D-glucosamine) [3]
16.5 <2> (N-acetyl-α-D-glucosamine, <2> versus acetyl-CoA, no detergents added [5]) [5]

pH-Optimum

5.5 <1, 2> (<1> about [1]; <2> above, broad maximum [3]) [1, 3]
5.7 <1> [8]
6 <2> [5]
6-8 <2> [2]
6.5 <1> (<1> tetrasaccharide substrate [7]) [7]
6.5-8 <1> (<1> mono-and disaccharide substrates [7]) [7]
Additional information <1, 2> (<2> transmembrane transfer [5]) [5, 7]

pH-Range

5-5.5 <2> (<2> pH 5.0: about 30% of activity maximum, pH 5.5: maximum activity above [5]) [5]
5-7 <1> (<1> half-maximal activity at pH 5.0 and pH 7.0 [1]) [1]

Temperature optimum (°C)

37 <1, 2> (<1,2> assay at [1,5,8]) [1, 5, 8]
45 <2> (<2> membrane-bound form [3]) [3]

Temperature range (°C)

45-55 <2> [3]

5 Isolation/Preparation/Mutation/Application

Source/tissue

leukocyte <1> (<1> deficiency on leukocytes from patients with Sanfillippo syndrome type C [8]) [8]
liver <2> [2-5]
placenta <1> [7, 9]
skin fibroblast <1> (<1> deficiency on fibroblasts from patients with Sanfillippo syndrome type C [1,6,8]) [1, 6, 8]

Localization

lysosome <1, 2> (<1,2> enzyme is acetylated at the cytoplasmic side of the lysosome and the acetyl group is then transferred to the inside where it is used to acetylate heparan sulfate [4,5,9]; <1> integral membrane protein [7,9]; <2> membrane bound, about one fourth of the activity [2]) [2, 4, 5, 7-9]
microsome <2> (<2> membrane-bound, greater part of activity is recovered in microsomes [2]) [2]
Additional information <1> (<1> enzyme is associated to detergent-resistent microdomains of lysosomal membrane [9]) [9]

Purification

<1> (partial [7]) [7]

Application

medicine <1> (<1> overview, design and synthesis of substrate and internal standard conjugates as substrates in enzyme assay for confirmation of enzyme deficiency using a combination of affinity chromatography and electrospray ionization mass spectrometry [10]) [10]

6 Stability

Temperature stability

37 <2> (<2> 5 min, solubilized enzyme, 50% loss of activity [3]) [3]

General stability information

<2>, solubilized enzyme, not stable [3]

Storage stability

<2>, -70°C, membrane-bound enzyme, 12 months stable [3]
<2>, 4°C, solubilized enzyme, 25% loss of activity after 7 days [3]

References

[1] Klein, U.; Kresse, H.; Von Figura, K.: Sanfilippo syndrome type C: deficiency of acetyl-CoA:α-glucosaminide N-acetyltransferase in skin fibroblasts. Proc. Natl. Acad. Sci. USA, **75**, 5185-5189 (1978)

[2] Pohlmann, R.; Klein, U.; Fromme, H.G.; Von Figura, K.: Localisation of acetyl-CoA:α-glucosaminide N-acetyltransferase in microsomes and lysosomes of rat liver. Hoppe-Seyler's Z. Physiol. Chem., **362**, 1199-1207 (1981)

[3] Bame, K.J.; Rome, L.H.: Acetyl-CoA:α-glucosaminide N-acetyltransferase from rat liver. Methods Enzymol., **138**, 607-611 (1987)

[4] Bame, K.J.; Rome, L.H.: Acetyl-coenzyme A:α-glucosaminide N-acetyltransferase. Evidence for an active site histidine residue. J. Biol. Chem., **261**, 10127-10132 (1986)

[5] Bame, K.J.; Rome, L.H.: Acetyl coenzyme A:α-glucosaminide N-acetyltransferase. Evidence for a transmembrane acetylation mechanism. J. Biol. Chem., **260**, 11293-11299 (1985)

[6] Reynertson, R.; Campbell, P.; Ford, J.D.; Jacobsson, I.; Roden, L.; Thompson, J.N.: New oligosaccharides from heparin and heparan sulfate and their use as substrates for heparin-degrading enzymes. J. Biol. Chem., **258**, 7449-7459 (1983)

[7] Meikle, P.J.; Whittle, A.M.; Hopwood, J.J.: Human acetyl-coenzyme A:α-glucosaminide N-acetyltransferase. Kinetic characterization and mechanistic interpretation. Biochem. J., **308**, 327-333 (1995)

[8] Voznyi, Y.V.; Karpova, E.A.; Dudukina, T.V.; Tsvetkova, I.V.; Boer, A.M.; Janse, H.C.; van Diggelen, O.P.: A fluorometric enzyme assay for the diag-

nosis of Sanfilippo disease C (MPS III C). J. Inher. Metab. Dis., **16**, 465-472 (1993)
[9] Taute, A.; Watzig, K.; Simons, B.; Lohaus, C.; Meyer, H.E.; Hasilik, A.: Presence of detergent-resistant microdomains in lysosomal membranes. Biochem. Biophys. Res. Commun., **298**, 5-9 (2002)
[10] Gerber, S.A.; Turecek, F.; Gelb, M.H.: Design and synthesis of substrate and internal standard conjugates for profiling enzyme activity in the Sanfilippo syndrome by affinity chromatography/electrospray ionization mass spectrometry. Bioconjugate Chem., **12**, 603-615 (2001)

Maltose O-acetyltransferase 2.3.1.79

1 Nomenclature

EC number
2.3.1.79

Systematic name
acetyl-CoA:maltose O-acetyltransferase

Recommended name
maltose O-acetyltransferase

Synonyms
MAT <1> [3]
maa gene product <1> [3]
maltose transacetylase

CAS registry number
81295-47-8

2 Source Organism

<1> *Escherichia coli* (strain K12 [1,2]) [1-3]
<2> *Bacillus subtilis* [3]

3 Reaction and Specificity

Catalyzed reaction
acetyl-CoA + maltose = CoA + acetyl-maltose (not identical with EC 2.3.1.18 galactoside O-acetyltransferase)

Reaction type
acyl group transfer

Natural substrates and products
S acetyl-CoA + maltose <1, 2> (Reversibility: ? <1, 2> [1-3]) [1-3]
P CoA + acetyl-maltose

Substrates and products
S acetyl-CoA + 3-O-methyl-D-glucopyranoside <1> (Reversibility: ? <1> [3]) [3]
P CoA + ?
S acetyl-CoA + D-fructose <1> (Reversibility: ? <1> [2, 3]) [2, 3]

P CoA + acetyl-D-fructose
S acetyl-CoA + D-galactose <1> (Reversibility: ? <1> [2, 3]) [2, 3]
P CoA + acetyl-D-galactose
S acetyl-CoA + D-glucopyranose 6-phosphate <1> (Reversibility: ? <1> [3]) [3]
P CoA + ?
S acetyl-CoA + D-glucosamine <1> (Reversibility: ? <1> [3]) [3]
P CoA + acetyl-D-glucosamine
S acetyl-CoA + D-glucose <1> (Reversibility: ? <1> [2, 3]) [2, 3]
P CoA + acetyl-D-glucose
S acetyl-CoA + D-mannose <1> (Reversibility: ? <1> [2, 3]) [2, 3]
P CoA + acetyl-D-mannose
S acetyl-CoA + isomaltose <1> (Reversibility: ? <1> [3]) [3]
P CoA + acetyl-isomaltose
S acetyl-CoA + isopropyl-β-D-thiogalactoside <1> (Reversibility: ? <1> [3]) [3]
P CoA + ?
S acetyl-CoA + isopropyl-β-D-thioglucopyranoside <1> (Reversibility: ? <1> [3]) [3]
P CoA + ?
S acetyl-CoA + lactose <1> (Reversibility: ? <1> [3]) [3]
P CoA + acetyl-lactose
S acetyl-CoA + maltodextrine <1> (Reversibility: ? <1> [1]) [1]
P CoA + ?
S acetyl-CoA + maltoheptaose <1> (Reversibility: ? <1> [3]) [3]
P CoA + acetyl-maltoheptaose
S acetyl-CoA + maltohexaose <1> (Reversibility: ? <1> [3]) [3]
P CoA + acetyl-maltohexaose
S acetyl-CoA + maltooligosaccharide <1> (Reversibility: ? <1> [2]) [2]
P CoA + ?
S acetyl-CoA + maltopentaose <1> (Reversibility: ? <1> [3]) [3]
P CoA + acetyl-maltopentaose
S acetyl-CoA + maltose <1, 2> (Reversibility: ? <1, 2> [1-3]) [1-3]
P CoA + acetyl-maltose
S acetyl-CoA + maltotetraose <1> (Reversibility: ? <1> [3]) [3]
P CoA + acetyl-maltotetraose
S acetyl-CoA + maltotriose <1> (Reversibility: ? <1> [1, 2]) [1, 2]
P CoA + acetyl-maltotriose
S acetyl-CoA + methyl α-D-glucopyranoside <1> (Reversibility: ? <1> [3]) [3]
P CoA + ?
S acetyl-CoA + methyl β-D-glucopyranoside <1> (Reversibility: ? <1> [3]) [3]
P CoA + ?
S acetyl-CoA + methyl β-D-glucopyranoside <1> (Reversibility: ? <1> [3]) [3]
P CoA + ?

S acetyl-CoA + n-decyl β-D-glucopyranoside <1> (Reversibility: ? <1> [3]) [3]
P CoA + ?
S acetyl-CoA + n-decyl β-D-maltoside <1> (Reversibility: ? <1> [3]) [3]
P CoA + ?
S acetyl-CoA + n-dodecyl β-D-maltoside <1> (Reversibility: ? <1> [3]) [3]
P CoA + ?
S acetyl-CoA + n-heptyl β-D-glucopyranoside <1> (Reversibility: ? <1> [3]) [3]
P CoA + ?
S acetyl-CoA + n-heptyl-β-D-thioglucopyranoside <1> (Reversibility: ? <1> [3]) [3]
P CoA + ?
S acetyl-CoA + n-hexyl β-D-glucopyranoside <1> (Reversibility: ? <1> [3]) [3]
P CoA + ?
S acetyl-CoA + n-nonyl β-D-glucopyranoside <1> (Reversibility: ? <1> [3]) [3]
P CoA + ?
S acetyl-CoA + n-octyl α-glucopyranoside <1> (Reversibility: ? <1> [3]) [3]
P CoA + ?
S acetyl-CoA + n-octyl β-D-glucopyranoside <1> (Reversibility: ? <1> [3]) [3]
P CoA + ?
S acetyl-CoA + n-octyl-β-D-thioglucopyranoside <1> (Reversibility: ? <1> [3]) [3]
P CoA + ?
S acetyl-CoA + nigerose <1> (Reversibility: ? <1> [3]) [3]
P CoA + acetyl-nigerose
S acetyl-CoA + *p*-nitrophenyl α-D-glucopyranoside <1> (Reversibility: ? <1> [3]) [3]
P CoA + ?
S acetyl-CoA + *p*-nitrophenyl α-D-maltoside <1> (Reversibility: ? <1> [3]) [3]
P CoA + ?
S acetyl-CoA + *p*-nitrophenyl β-D-glucopyranoside <1> (Reversibility: ? <1> [3]) [3]
P CoA + ?
S acetyl-CoA + *p*-nitrophenyl β-D-maltoside <1> (Reversibility: ? <1> [3]) [3]
P CoA + ?
S acetyl-CoA + sucrose <1> (Reversibility: ? <1> [3]) [3]
P CoA + acetyl-sucrose
S acetyl-CoA + thiomaltose <1> (Reversibility: ? <1> [1]) [1]
P CoA + acetyl-thiomaltose
S acetyl-CoA + triethanolamine <1> (Reversibility: ? <1> [2]) [2]
P CoA + acetyl-triethanolamine

S Additional information <1> (<1> rhamnose, arabinose, lactose, isopropyl-1-thio-β-D-galactoside and trehalose are not substrates [2]; <1> fucose and myo-inositol are not substrates, glucose-1-phosphate, UDP-glucose or ADP-glucose are very poor substrates or nonsubstrates, C6 glucose derivatives are not substrates [3]) [2, 3]

P ?

Inhibitors

CoA-SH <1> (<1> product inhibition [2]) [2]

Specific activity (U/mg)

24.8 <1> [2]

K_m-Value (mM)

0.018 <1> (acetyl-CoA) [2]
40 <1> (triethanolamine) [2]
62 <1> (D-glucose) [2]
90 <1> (maltose) [2]

pH-Optimum

7.8 <1> [3]

pH-Range

4-9 <1> [3]

4 Enzyme Structure

Molecular weight

40000 <1> (<1> gel filtration [2]) [2]

Subunits

dimer <1> (<1> 2 * 20000, SDS-PAGE [2]) [2]

5 Isolation/Preparation/Mutation/Application

Purification

<1> [2]

Crystallization

<1> (crystals belong to space group C222(1), a 64.7 A, b 106.6 A and c 176.0 A [3]) [3]

Cloning

<1> (cloning of the structural gene mac [2]; cloning of the maa gene and overexpression of MAT [3]) [2, 3]
<2> [3]

Application

nutrition <1> (<1> biotechnologically attractive for the modification of starch and maltooligosaccharides [3]) [3]

6 Stability

Temperature stability

40 <1> (<1> stable for at least 4 h [3]) [3]

Storage stability

<1>, -70°C, enzyme stored frozen in presence of 0.1% 2-mercaptoethanol and 40% glycerol without appreciable loss of activity [2]

References

[1] Freundlieb, S.; Boos, W.: Maltose transacetylase of Escherichia coli: A preliminary report. Ann. Microbiol., **133 A,** 181-189 (1982)
[2] Brand, B.; Boos, W.: Maltose transacetylase of Escherichia coli. Mapping and cloning of its structural, gene, mac, and characterization of the enzyme as a dimer of identical polypeptides with a molecular weight of 20,000. J. Biol. Chem., **266,** 14113-14118 (1991)
[3] Lo Leggio, L.; Dal Degan, F.; Poulsen, P.; Moeller Andersen, S.; Larsen, S.: The structure and specificity of Escherichia coli maltose acetyltransferase give new insight into the LacA family of acyltransferases. Biochemistry, **42,** 5225-5235 (2003)

Cysteine-S-conjugate N-acetyltransferase 2.3.1.80

1 Nomenclature

EC number
2.3.1.80

Systematic name
acetyl-CoA:S-substituted L-cysteine N-acetyltransferase

Recommended name
cysteine-S-conjugate N-acetyltransferase

Synonyms
acetyltransferase, cysteine S-conjugate N-

CAS registry number
81725-80-6

2 Source Organism

<1> *Rattus norvegicus* [1]
<2> *Sus scrofa* [2-4]

3 Reaction and Specificity

Catalyzed reaction
acetyl-CoA + an S-substituted L-cysteine = CoA + an S-substituted N-acetyl-L-cysteine

Reaction type
acyl group transfer

Natural substrates and products
S acetyl-CoA + an S-substituted L-cysteine <2> (<2> final reaction of mercapturic biosynthesis [3]) [3]
P CoA + an S-substituted N-acetyl-L-cysteine

Substrates and products
S acetyl-CoA + O-benzyl-L-serine <1> (Reversibility: ? <1> [1]) [1]
P CoA + N-acetyl-O-benzyl-L-cysteine
S acetyl-CoA + S-4-nitrobenzyl-L-cysteine <2> (Reversibility: <2> [2, 4]) [2, 4]
P CoA + ?

S acetyl-CoA + S-benzyl-L-cysteine <1> (<1> best substrate [1]) (Reversibility: ? <1> [1]) [1]
P CoA + N-acetyl-S-benzyl-L-cysteine
S acetyl-CoA + S-butyl-L-cysteine <1> (Reversibility: ? <1> [1]) [1]
P CoA + N-acetyl-S-butyl-L-cysteine
S acetyl-CoA + S-ethyl-L-cysteine <1> (Reversibility: ? <1> [1]) [1]
P CoA + N-acetyl-S-ethyl-L-cysteine
S acetyl-CoA + S-propyl-L-cysteine <1> (Reversibility: ? <1> [1]) [1]
P CoA + N-acetyl-S-propyl-L-cysteine
S Additional information <1, 2> (<1> no substrate: aniline, *o*-aminobenzoate, D-cysteine, L-methionine, L-leucine, L-phenylalanine, L-glutamate [1]; <2> comparison of haloalkene-derived cysteine S-conjugates [3]) [1, 3]
P ?

Inhibitors

N-ethylmaleimide <2> [2]
p-chloromercuribenzoate <2> [2]
probenecid <2> (<2> less pronounced inhibition [2]) [2]
Additional information <2> (<2> not inhibitory: PMSF [2]) [2]

Specific activity (U/mg)

0.394 <1> [1]

K_m-Value (mM)

0.026 <1> (S-acetyl-CoA) [1]
0.063 <1> (S-butyl-L-cysteine) [1]
0.14 <1> (S-benzyl-L-cysteine) [1]
0.176 <2> (S-4-nitrobenzyl-L-cysteine) [2]
0.67 <1> (S-propyl-L-cysteine) [1]
2.6 <1> (O-benzyl-L-serine) [1]
7.1 <1> (S-ethyl-L-cysteine) [1]
Additional information <1, 2> (<1> substitution of oxygen for sulfur as in O-benzyl-L-serine, greatly increases the apparent K_m, with only slight decrease in maximal velocity [1]; <2> radioactive assay [3]; <2> nonradioactive assay and separation by HPLC, kinetics [4]) [1, 3, 4]

pH-Optimum

6.5 <2> [2]
6.8-7 <1> [1]

Temperature optimum (°C)

37 <1> (<1> assay at [1]) [1]

4 Enzyme Structure

Molecular weight

32000 <2> (<2> gel filtration [2]) [2]

Subunits

monomer <2> (<2> 1 * 34000, SDS-PAGE [2]) [2]

5 Isolation/Preparation/Mutation/Application

Source/tissue

kidney <1, 2> [1-4]

Localization

microsome <1> [1]

Purification

<1> (partial [1]) [1]

<2> (amino acid analysis [2]) [2]

6 Stability

Storage stability

<2>, -20°C, stable for several months [2]

<2>, 4°C, glycine/deoxyBIGCHAP buffer, pH 7.0, considerable inactivation [2]

References

[1] Duffel, M.W.; Jakoby, W.B.: Cysteine S-conjugate N-acetyltransferase. Methods Enzymol., **113**, 516-521 (1985)

[2] Aigner, A.; Jaeger, M.; Pasternack, R.; Weber, P.; Wienke, D.; Wolf, S.: Purification and characterization of cysteine-S-conjugate N-acetyltransferase from pig kidney. Biochem. J., **317**, 213-218 (1996)

[3] Kraus, T.; Uttamsingh, V.; Anders, M.W.; Wolf, S.: Porcine kidney microsomal cysteine S-conjugate N-acetyltransferase-catalyzed N-acetylation of haloalkene-derived cysteine S-conjugates. Drug Metab. Dispos., **28**, 440-445 (2000)

[4] Aigner, A.; Jaeger, M.; Weber, P.; Wolf, S.: A nonradioactive assay for microsomal cysteine-S-conjugate N-acetyltransferase activity by high-pressure liquid chromatography. Anal. Biochem., **223**, 227-231 (1994)

Aminoglycoside $N^{3'}$-acetyltransferase 2.3.1.81

1 Nomenclature

EC number
2.3.1.81

Systematic name
acetyl-CoA:2-deoxystreptamine-antibiotic $N^{3'}$-acetyltransferase

Recommended name
aminoglycoside $N^{3'}$-acetyltransferase

Synonyms
3'-aminoglycoside acetyltransferase
3-N-acetyltransferase [1, 3]
3-N-aminoglycoside acetyltransferase [5]
aminoglycoside 3-N-acetyltransferase <4, 6> [2, 4]
Additional information (not identical to EC 2.3.1.60 [1]) [1]

CAS registry number
60120-42-5

2 Source Organism

<1> *Escherichia coli* (strains K802N, W677, p75-24, S3035, JR225, JR226, JR227 [1]) [1]
<2> *Klebsiella pneumoniae* [1]
<3> *Arizona sp.* [1]
<4> *Serratia marcescens* [2]
<5> *Streptomyces griseus* (SS-1198 and SS-1198PR mutant [3]) [3]
<6> *Streptomyces griseus* [4]
<7> *Pseudomonas aeruginosa* (phenotype of VA-182/00 [5]) [5]

3 Reaction and Specificity

Catalyzed reaction
acetyl-CoA + a 2-deoxystreptamine antibiotic = CoA + $N^{3'}$-acetyl-2-deoxystreptamine antibiotic

Reaction type

acyl group transfer

Substrates and products

- **S** acetyl-CoA + amikacin <7> (<7> wild-type and recombinant enzyme [5]) (Reversibility: ? <7> [5]) [5]
- **P** CoA + N$^{3'}$-acetylamikacin <7> [5]
- **S** acetyl-CoA + apramycin <1-3> (<1-3> a wide range of antibiotics containing the 2-deoxystreptamine ring can act as acceptors, including gentamicin, kanamycin, tobramycin, neomycin [1]) (Reversibility: ? <1-3> [1]) [1]
- **P** CoA + N$^{3'}$-acetylapramycin <1-3> [1]
- **S** acetyl-CoA + gentamicin <4> (Reversibility: ? <4> [2]) [2]
- **P** CoA + N$^{3'}$-acetylgentamicin <4> [2]
- **S** acetyl-CoA + gentamicin <7> (<7> wild-type and recombinant enzyme [5]) (Reversibility: ? <7> [5]) [5]
- **P** CoA + N$^{3'}$-acetylgentamicin <7> [5]
- **S** acetyl-CoA + isepamycin <7> (<7> wild-type and recombinant enzyme [5]) (Reversibility: ? <7> [5]) [5]
- **P** CoA + N$^{3'}$-acetylisepamycin <7> [5]
- **S** acetyl-CoA + kanamycin <5> (Reversibility: ? <5> [3]) [3]
- **P** CoA + N$^{3'}$-acetyl kanamycin <5> [3]
- **S** acetyl-CoA + kanamycin <7> (<7> wild-type and recombinant enzyme [5]) [5]
- **P** CoA + N$^{3'}$-acetylkanamycin <7> [5]
- **S** acetyl-CoA + neomycin <7> (<7> wild-type and recombinant enzyme [5]) (Reversibility: ? <7> [5]) [5]
- **P** CoA + N$^{3'}$-acetylneomycin <7> [5]
- **S** acetyl-CoA + netilmicin <7> (<7> wild-type and recombinant enzyme [5]) (Reversibility: ? <7> [5]) [5]
- **P** CoA + N$^{3'}$-acetylnetilmicin <7> [5]
- **S** acetyl-CoA + sisomicin <7> (<7> wild-type and recombinant enzyme [5]) (Reversibility: ? <7> [5]) [5]
- **P** CoA + N$^{3'}$-acetylsisomicin <7> [5]
- **S** acetyl-CoA + spectinomycin <7> (<7> wild-type and recombinant enzyme [5]) (Reversibility: ? <7> [5]) [5]
- **P** CoA + N$^{3'}$-acetylspectinomycin <7> [5]
- **S** acetyl-CoA + streptomycin <7> (<7> wild-type and recombinant enzyme [5]) (Reversibility: ? <7> [5]) [5]
- **P** CoA + N$^{3'}$-acetylstreptomycin <7> [5]
- **S** acetyl-CoA + tobramycin <7> (<7> wild-type and recombinant enzyme [5]) (Reversibility: ? <7> [5]) [5]
- **P** CoA + N$^{3'}$-acetyltobramycin <7> [5]

4 Enzyme Structure

Molecular weight

18630 <4> (<4> predicted from nucleotide sequence: 18649 Da, found by mass spectrometry [2]) [2]

Posttranslational modification

side-chain modification <4> (<4> cleavage of the hexahistidine tag by tobacco etch virus protease [2]) [2]

5 Isolation/Preparation/Mutation/Application

Localization

soluble <1-7> [1-5]

Crystallization

<4> (X-ray structure of a recombinant enzyme, expressed in its selenomethionine substituted form, cocrystallization with 5 mM CoA at 4°C and pH: 7.8, crystal cryoprotection achieved with 17% glycerol and 8% 2R,3R-butanediol, active center found by X-ray analysis and model building [2]) [2]

Cloning

<2> (expression in Escherichia coli K802N by mating [1]) [1]
<3> (expression in Escherichia coli K802N by mating [1]) [1]
<4> (wild-type and truncated enzymes expressed in Escherichia coli [2]) [2]
<5> (plasmid transfer pIJ702 to Streptomyces lividans TK21 [3]) [3]
<7> (expression in Escherichia coli DH5α [5]) [5]

Application

medicine <1-3, 5> (<1-3> studies on the novel aminoglycoside antibiotic apramycin [1]; <5> mutation in the gene promoter affects the transcription level and enhances resistance to antibiotics [3]) [1, 3]

References

[1] Davies, J.; O'Connor, S.: Enzymatic modification of aminoglycoside antibiotics: 3-N-acetyltransferase with broad specificity that determines resistance to the novel aminoglycoside apramycin. Antimicrob. Agents Chemother., **14**, 69-72 (1978)

[2] Wolf, E.; Vassilev, A.; Makino, Y.; Sali, A.; Nakatani, Y.; Burley, S.K.: Crystal structure of a GCN5-related N-acetyltransferase: Serratia marcescens aminoglycoside 3-N-acetyltransferase. Cell, **94**, 439-449 (1998)

[3] Ishikawa, J.; Sunada, A.; Oyama, R.; Hotta, K.: Identification and characterization of the point mutation which affects the transcription level of the chromosomal 3-N-acetyltransferase gene of Streptomyces griseus SS-1198. Antimicrob. Agents Chemother., **44**, 437-440 (2000)

[4] Tsuchizaki, N.; Hamadal, M.; Hotta, K.: Rapid characterization by colony direct PCR of distribution specificity in Streptomyces of kan gene encoding a specific aminoglycoside-3-N-acetyltransferase. Actinomycetologica, **15**, 23-29 (2001)
[5] Riccio, M.L.; Docquier, J.D.; DellÁmico, E.; Luzzaro, F.; Amicosante, G.; Rossolini, G.M.: Novel 3-N-aminoglycoside acetyltransferase gene, aac(3)-Ic, from a Pseudomonas aeruginosa integron. Antimicrob. Agents Chemother., **47**, 1746-1748 (2003)

Aminoglycoside $N^{6'}$-acetyltransferase 2.3.1.82

1 Nomenclature

EC number
2.3.1.82

Systematic name
acetyl-CoA:kanamycin-B $N^{6'}$-acetyltransferase

Recommended name
aminoglycoside $N^{6'}$-acetyltransferase

Synonyms
AAC(6')
acetyltransferase, aminoglycoside 6'-N-
acetyltransferase, kanamycin
aminoglycoside 6'-N-acetyltransferase
aminoglycoside-6'-N-acetyltransferase
aminoglycoside-6'-acetyltransferase
aminoglycoside-6-N-acetyltransferase

CAS registry number
56467-65-3

2 Source Organism

<1> *Escherichia coli* (strains W677 carrying either R factor R-5 or R-79 [1]; CS2R2 [2]; NR79 [4]; DH5α, AAC(6')-Ib protein and AAC(6')-IIa protein [6]) [1, 2, 4, 6]
<2> *Moraxella sp.* (strains 2513 and 731 [2]) [2]
<3> *Pseudomonas aeruginosa* (strains GN315 and 141 [4]) [4]
<4> *Serratia marcescens* (strain VU12944/77 [4]) [4]
<5> *Staphylococcus aureus* (harbouring plasmid RPAL [5]) [5]
<6> *Streptomyces kanamyceticus* [3, 4]
<7> *Enterococcus faecium* (6'-N-acetyltransferase type li, transformation of E. coli methionine auxotroph B834(DE3)/pLys S with a pPLaac-1 overexpression plasmid containing the aac(6')-li gene [7]) [7]
<8> *Staphylococcus epidermidis* (RYC 13036 [8]) [8]
<9> *Pseudomonas fluorescens* (strain BM2687, enzyme Ib' [9]) [9]
<10> *Enterobacter cloacae* (strain EC1562 and EC1563, enzyme variant with Ser119 [10]) [10]
<11> *Citrobacter freundii* (strain CFr564, enzyme variant with Ser119 [10]) [10]

<12> *Actinomyces sp.* (strain 8 [11]) [11]
<13> *Klebsiella pneumoniae* [12, 14]
<14> *Enterococcus faecium* [13]
<15> *Salmonella enterica* (wild-type enzyme and mutant enzymes C109A and C109A/C70A [15]) [15, 16]
<16> *Klebsiella pneumoniae* (plasmid pJHC-MW1 mutant AAC6'-IbG175A 6'-N-acetyltransferase gene [17]) [17]
<17> *Klebsiella pneumoniae* (plasmid pJHC-MW1 mutant AAC6'-IbQ174A 6'-N-acetyltransferase gene [17]) [17]
<18> *Klebsiella pneumoniae* (plasmid pJHC-MW1 mutant AAC6'-IbR173A 6'-N-acetyltransferase gene [17]) [17]
<19> *Klebsiella pneumoniae* (plasmid pJHC-MW1 mutant AAC6'-IbE172A 6'-N-acetyltransferase gene [17]) [17]
<20> *Klebsiella pneumoniae* (plasmid pJHC-MW1 mutant AAC6'-IbF171A 6'-N-acetyltransferase gene [17]) [17]
<21> *Klebsiella pneumoniae* (plasmid pJHC-MW1 mutant AAC6'-IbG170A 6'-N-acetyltransferase gene [17]) [17]
<22> *Klebsiella pneumoniae* (plasmid pJHC-MW1 mutant AAC6'-IbK168A 6'-N-acetyltransferase gene [17]) [17]
<23> *Klebsiella pneumoniae* (plasmid pJHC-MW1 mutant AAC6'-IbE167A 6'-N-acetyltransferase gene [17]) [17]
<24> *Klebsiella pneumoniae* (plasmid pJHC-MW1 mutant AAC6'-IbY166A 6'-N-acetyltransferase gene [17]) [17]
<25> *Klebsiella pneumoniae* (plasmid pJHC-MW1 mutant AAC6'-IbC165A 6'-N-acetyltransferase gene [17]) [17]
<26> *Klebsiella pneumoniae* (plasmid pJHC-MW1 mutant AAC6'-IbR164A 6'-N-acetyltransferase gene [17]) [17]
<27> *Klebsiella pneumoniae* [18, 19]
<28> *Citrobacter diversus* [20]
<29> *Enterobacter aerogenes* [21]
<30> *Serratia marcescens* [22]
<31> *Klebsiella pneumoniae* (plasmid pJHC-MW1 mutant AAC6'-IbP155A 6'-N-acetyltransferase gene [17]) [17]
<32> *Klebsiella pneumoniae* (plasmid pJHC-MW1 mutant AAC6'-IbS156A 6'-N-acetyltransferase gene [17]) [17]
<33> *Klebsiella pneumoniae* (plasmid pJHC-MW1 mutant AAC6'-IbP157A 6'-N-acetyltransferase gene [17]) [17]
<34> *Klebsiella pneumoniae* (plasmid pJHC-MW1 mutant AAC6'-IbS158A 6'-N-acetyltransferase gene [17]) [17]
<35> *Klebsiella pneumoniae* (plasmid pJHC-MW1 mutant AAC6'-IbN159A 6'-N-acetyltransferase gene [17]) [17]
<36> *Klebsiella pneumoniae* (plasmid pJHC-MW1 mutant AAC6'-IbL160A 6'-N-acetyltransferase gene [17]) [17]
<37> *Klebsiella pneumoniae* (plasmid pJHC-MW1 mutant AAC6'-IbR161A 6'-N-acetyltransferase gene [17]) [17]
<38> *Klebsiella pneumoniae* (plasmid pJHC-MW1 mutant AAC6'-IbI163A 6'-N-acetyltransferase gene [17]) [17]

3 Reaction and Specificity

Catalyzed reaction

acetyl-CoA + kanamycin-B = CoA + $N^{6'}$-acetylkanamycin-B (<5>, mechanism [5])

Reaction type

acyl group transfer

Natural substrates and products

S Additional information <1-38> (<1-38>, involved in resistance to antibiotics [1-22]; <1>, constitutive enzyme [1]) [1-22]

P ?

Substrates and products

S acetyl-CoA + 2'-N-ethylnetilmicin <1, 9> (Reversibility: ? <1,9> [6,9]) [6, 9]

P CoA + $N^{6'}$-acetyl-2'-N-ethylnetilmicin

S acetyl-CoA + amikacin <1, 2, 3, 4, 6, 8, 9, 10, 11, 13, 14, 15> (<1>, the AAC(6')-Ib protein is unable to efficiently modify gentamicin C1, 1.7% relative to sisomicin, however it is capable of modifying amikacin, 65.5% relative to sisomycin. The mutant enzyme AAC(6')-Ib L119S shows a 2.8fold increase in acetylation of gentamicin C_1, but a 8.7fold reduction in the ability to modify amikacin. The AAC(6')-IIa protein modifies gentamicin C_1 at 10.1% relative to sisomicin, however it shows low activity towards amikacin, 4.1% relative to sisomycin. The mutation AAC(6')-IIa S119L results in a 4.8fold reduction in the acetylation of gentamicin C_1, but causes an 2fold increase in the ability to modify amikacin [6]; <9>, gentamicin complex: C_1 41%, C_{1a} 26% and C_2 33%. The aac(6')-Ib' gene from strain BM2687 and the aac(6')-Ib gene from strain BM2656 show total identity with the exception of a C to T transition that results in a Ser to Leu substitution at position 83 of the deduced polypeptide. The enzyme encoded by aac(6')-Ib' shows resistance to gentamicin but not to amikacin. The enzyme encoded by aac(6')-Ib shows resistance to amikacin but not to gentamicin [9]; <10, 11>, the enzyme variant of enzyme type Ib has at position 119 a Ser instead of the Leu, conferring resistance to amikacin [10]; <13>, wild-type enzyme shows activity, mutant enzymes show no activity or reduced activity [12]) (Reversibility: ? <1, 2, 3, 4, 6, 8, 9, 10, 11, 13, 14, 15> [2, 4, 6, 8, 9, 10, 12, 13, 16]) [2, 4, 6, 8, 9, 10, 12, 13, 16]

P CoA + $N^{6'}$-acetylamikacin

S acetyl-CoA + butirosin <14> (Reversibility: ? <14> [13]) [13]

P CoA + $N^{6'}$-acetylbutirosin

S acetyl-CoA + dibekacin <1, 3, 4, 6, 12, 14, 15> (Reversibility: ? <1, 3, 4, 6, 12, 14, 15> [4, 11, 13, 16]) [4, 11, 13, 16]

P CoA + $N^{6'}$-acetyldibekacin

S acetyl-CoA + gentamicin <8> (Reversibility: ? <8> [8]) [8]

P CoA + $N^{6'}$-acetylgentamicin

S acetyl-CoA + gentamicin B <14> (Reversibility: ? <14> [13]) [13]
P CoA + $N^{6'}$-acetylgentamicin B
S acetyl-CoA + gentamicin C <8> (Reversibility: ? <8> [8]) [8]
P CoA + $N^{6'}$-acetylgentamicin C
S acetyl-CoA + gentamicin C_1 <1> (<1>, the AAC(6')-Ib protein is unable to efficiently modify gentamicin C_1, 1.7% relative to sisomicin, however it is capable of modifying amikacin, 65.5% relative to sisomycin. The mutant enzyme AAC(6')-Ib L119S shows a 2.8fold increase in acetylation of gentamicin C_1, but a 8.7fold reduction in the ability to modify amikacin. The AAC(6')-IIa protein modifies gentamicin C_1 at 10.1% relative to sisomicin, however it shows low activity towards amikacin, 4.1% relative to sisomycin. The mutation AAC(6')-IIa S119L results in a 4.8fold reduction in the acetylation of gentamicin C_1, but causes an 2fold increase in the ability to modify amikacin [6]) (Reversibility: ? <1> [6]) [6]
P CoA + $N^{6'}$-acetylgentamicin C_1
S acetyl-CoA + gentamicin C_{1a} <1-6, 9> (<1>, poor substrate, AAC-(6')-II type [6]; <9>, gentamicin complex: C_1 41%, C_{1a} 26% and C_2 33%. The aac(6')-Ib' gene from strain BM2687 and the aac(6')-Ib gene from strain BM2656 show total identity with the exception of a C to T transition that results in a Ser to Leu substitution at position 83 of the deduced polypeptide. The enzyme encoded by aac(6')-Ib' shows resistance to gentamicin but not to amikacin. The enzyme encoded by aac(6')-Ib shows resistance to amikacin but not to gentamicin [9]) (Reversibility: ? <1-6,9> [1-6]) [1-6, 9]
P CoA + $N^{6'}$-acetylgentamicin C_{1a} <1, 6> [1, 3]
S acetyl-CoA + gentamicin C_2 <1> (<1>, poor substrate, not gentamicin A and C_1 [1]) (Reversibility: ? <1> [1]) [1]
P CoA + $N^{6'}$-acetylgentgamicin C_2
S acetyl-CoA + histone <7> (<7>, a mixture of calf histones enriched in H3 and H4 [7]) (Reversibility: ? <7> [7]) [7]
P CoA + ?
S acetyl-CoA + hybrimycin A_1 <1> (Reversibility: ? <1> [1]) [1]
P CoA + $N^{6'}$-acetylhybrimycin A_1
S acetyl-CoA + hybrimycin A_2 <1> (Reversibility: ? <1> [1]) [1]
P CoA + $N^{6'}$-acetylhybrimycin A_2
S acetyl-CoA + hybrimycin A3 <1> (Reversibility: ? <1> [1]) [1]
P CoA + $N^{6'}$-acetylhybrimycin A_3
S acetyl-CoA + hybrimycin B1 <1> (Reversibility: ? <1> [1]) [1]
P CoA + $N^{6'}$-acetylhybrimycin B_1
S acetyl-CoA + hybrimycin B2 <1> (Reversibility: ? <1> [1]) [1]
P CoA + $N^{6'}$-acetylhybrimycin B_2
S acetyl-CoA + hybrimycin B3 <1> (Reversibility: ? <1> [1]) [1]
P CoA + $N^{6'}$-acetylhybrimycin B_3
S acetyl-CoA + istamycin-B <12> (Reversibility: ? <12> [11]) [11]
P CoA + $N^{6'}$-acetylistamycin-B

S acetyl-CoA + kanamycin <12, 13> (<13>, wild-type enzyme shows activity, mutant enzymes show no activity or reduced activity [12]) (Reversibility: ? <12,13> [11,12]) [11, 12]
P CoA + $N^{6'}$-acetylkanamycin
S acetyl-CoA + kanamycin A <1-4, 6, 15> (Reversibility: ? <1-4,6,15> [1,2,4,16]) [1, 2, 4, 16]
P CoA + $N^{6'}$-acetylkanamycin A <1, 2, 15> [1, 2, 16]
S acetyl-CoA + kanamycin B <1, 2, 5, 14, 15> (Reversibility: ? <1,2,5,14,15> [1,2,5,13,16]) [1, 2, 5, 13, 16]
P CoA + $N^{6'}$-acetylkanamycin B <1, 2> [1, 2]
S acetyl-CoA + micromomicin <12> (Reversibility: ? <12> [11]) [11]
P CoA + $N^{6'}$-acetylmicromomicin
S acetyl-CoA + myelin basic protein <7> (Reversibility: ? <7> [7]) [7]
P CoA + ?
S acetyl-CoA + neamine <1, 2, 5, 14> (Reversibility: ? <1, 2, 5, 14> [1, 2, 5, 13]) [1, 2, 5, 13]
P CoA + 6'-N-acetylneamine <5> [5]
S acetyl-CoA + nebramycin factor 4 <1> (<1>, no activity with factor 2 [1]) (Reversibility: ? <1> [1]) [1]
P CoA + $N^{6'}$-acetylnebramycin factor 4
S acetyl-CoA + nebramycin factor 6 <1> (<1>, no activity with factor 2 [1]) (Reversibility: ? <1> [1]) [1]
P CoA + $N^{6'}$-acetylnebramycin factor 6
S acetyl-CoA + neomycin A <15> (Reversibility: ? <15> [16]) [16]
P CoA + $N^{6'}$-neomycin A
S acetyl-CoA + neomycin B <1> (Reversibility: ? <1> [1]) [1]
P CoA + $N^{6'}$-acetylneomycin B <1> [1]
S acetyl-CoA + neomycin C <1, 2, 14, 15> (Reversibility: ? <1, 2, 14, 15> [1,2,13,16]) [1, 2, 13, 16]
P CoA + $N^{6'}$-acetylneomycin C
S acetyl-CoA + netilmicin <8, 9, 13, 14, 15> (<13>, wild-type enzyme shows activity, mutant enzymes show no activity or reduced activity [12]) (Reversibility: ? <8,9,13,14,15> [8,9,12,13,16]) [8, 9, 12, 13, 16]
P CoA + $N^{6'}$-acetylnetilmicin
S acetyl-CoA + poly-L-Lys <7> (Reversibility: ? <7> [7]) [7]
P CoA + ?
S acetyl-CoA + ribostamycin <14, 15> (Reversibility: ? <14,15> [13,16]) [13, 16]
P CoA + $N^{6'}$-acetylribostamycin
S acetyl-CoA + sisomicin <1, 2, 12, 14, 15> (Reversibility: ? <1, 2, 12, 14, 15> [2,6,11,13,16]) [2, 6, 11, 13, 16]
P CoA + $N^{6'}$-acetylsisomicin <1, 2> [2]

S acetyl-CoA + tobramycin <1, 2, 4-6, 8, 9, 13, 14, 15> (<13>, wild-type enzyme shows activity, mutant enzymes show no activity or reduced activity [12]) (Reversibility: ? <1,2,4-6,8,9,13,14,15> [2,4-6,8,9,12,13,16]) [2, 4-6, 8, 9, 12, 13, 16]
P CoA + $N^{6'}$-acetyltobramycin
S butyryl-CoA + tobramycin <15> (Reversibility: ? <15> [16]) [16]
P CoA + $N^{6'}$-butyryltobramycin
S malonyl-CoA + tobramycin <15> (Reversibility: ? <15> [16]) [16]
P CoA + $N^{6'}$-malonyltobramycin
S propionyl-CoA + tobramycin <15> (<15>, tobramycin exhibits a rapid, tobramycin-independent rate of hydrolysis that is linearly proportional to enzyme [16]) (Reversibility: ? <15> [16]) [16]
P CoA + $N^{6'}$-propionyltobramycin
S Additional information <1, 7, 14, 15, 16, 26, 31-38> (<1>, the smallest antibiotic moiety required for recognition as a substrate by the acetylating enzyme is a 6-amino-6-deoxy-hexose glycosidically linked to a streptamine or deoxystreptamine ring [1]; <1>, a single amino acid, Leu119 in AAC(6')-Ib and S119 in AAC(6')-IIa, is largely responsible for determining the specificity of the AAC(6')-Ib and AAC(6')-IIa proteins. Changing this amino acid in either the AAC(6')-Ib or the AAC(6')-IIa protein results in a dramatic change in substrate specificity [6]; <7>, no activity with aminoglycoside kinase APH(3')-IIIa, yeast homoserine dehydrogenase, spermine and serotonin [7]; <14>, acetyl-CoA and propionyl-CoA are comparable substrates, but butyryl-CoA is not [13]; <15>, C70 is directly involved in aminoglycoside binding [15]; <16-26,31-38>, acetylation reaction occurs through a direct mechanism rather than a ping-pong mechanism that includes a transient transfer of the acetyl group to a cysteine residue [17]) [1, 6, 7, 13, 15, 17]
P ?

Inhibitors

6'-N-acetylneamine <5> (<5>, product inhibition [5]) [5]
AMP <5> [5]
CoA <5> (<5>, product inhibition [5]) [5]
$N^{6'}$-acetylgentamicin C_{1a} <1> [1]
$N^{6'}$-acetylneomycin B <1> [1]
amikacin <8> (<8>, substrate inhibition [8]) [8]
butyryl-CoA <14> [13]
dibekacin <14> [13]
dithio-CoA <15> [16]
gentamicin <8> (<8>, substrate inhibition [8]) [8]
gentamicin A <1> [1]
gentamicin C_1 <2, 5> (<5>, competitive inhibitor of neamine, non-competitive inhibitor of acetyl-CoA [5]) [2, 5]
gentamine A <1> [1]

iodoacetamide <15> (<15>, inactivation in a biphasic manner, half of the activity is lost rapidly and the other half more slowly, tobramycin but not acetyl-CoA protects [16]) [16]
kanamycin A <14> [13]
kanamycin B <14> [13]
lividomycin <15> [16]
netilmicin <14> [13]
netilmicin <8> (<8>, substrate inhibition [8]) [8]
paromamine <1> [1]
paromomycin <1, 5> (<5>, inhibits reaction with neamine [5]) [1, 5]
poly-L-Lys <7> (<7>, substrate inhibition [7]) [7]
sisomycin <14> [13]
tobramycin <8, 14> (<8>, substrate inhibition [8]) [8, 13]
Additional information <5> (<5>, no inhibition by GTP [5]) [5]

Metals, ions
Mg^{2+} <1> (<1>, required [1]) [1]

Turnover number (min^{-1})
0.114 <7> (poly-L-Lys, <7>, pH 7.5, 37°C [7]) [7]
6.36 <14> (amikacin, <14>, pH 6.0, 37°C [13]) [13]
12.3 <14> (neomycin C, <14>, pH 6.0, 37°C [13]) [13]
19.02 <14> (propionyl-CoA, <14>, pH 6.0, 37°C [13]) [13]
20.28 <14> (ribostamycin, <14>, pH 6.0, 37°C [13]) [13]
22.2 <14> (sisomicin, <14>, pH 6.0, 37°C [13]) [13]
23.28 <14> (gentamicin B, <14>, pH 6.0, 37°C [13]) [13]
24.18 <14> (acetyl-CoA, <14>, pH 6.0, 37°C [13]) [13]
25.14 <14> (neamine, <14>, pH 6.0, 37°C [13]) [13]
27.96 <14> (butirosin, <14>, pH 6.0, 37°C [13]) [13]
30.73 <14> (netilmicin, <14>, pH 6.0, 37°C [13]) [13]
37.92 <14> (dibekacin, <14>, pH 6.0, 37°C [13]) [13]
48.96 <14> (kanamycin A, <14>, pH 6.0, 37°C [13]) [13]
64.8 <14> (kanamycin B, <14>, pH 6.0, 37°C [13]) [13]
66.6 <14> (tobramycin, <14>, pH 6.0, 37°C [13]) [13]

Specific activity (U/mg)
0.304 <5> [5]
Additional information <15> (<15>, equilibrium binding and the directly determined thermodynamic parameters for different aminoglycosides and acyl-CoA derivatives to the wild-type enzyme and two mutant enzymes c109A and C109A/C70A using fluorescence spectroscopy and isothermal titration calorimetry [15]) [15]

K_m-Value (mM)
0.001 <3, 4> (kanamycin A, <3>, pH 6.0 [4]; <4>, pH 7.5 [4]) [4]
0.004 <1, 2> (tobramycin) [2]
0.005 <3> (kanamycin A, <3>, pH 6.0 [4]) [4]
0.00531 <14> (neomycin C, <14>, pH 6.0, 37°C [13]) [13]

0.00582 <14> (neamine, <14>, pH 6.0, 37°C [13]) [13]
0.006 <4> (kanamycin A, <4>, pH 7.5 [4]) [4]
0.006 <15> (netilmicin, <15>, mutant enzyme C109A [16]) [16]
0.00638 <14> (netilmicin, <14>, pH 6.0, 37°C [13]) [13]
0.008 <2> (kanamycin) [2]
0.008 <15> (netilmicin, <15>, wild-type enzyme [16]) [16]
0.00908 <14> (ribostamycin, <14>, pH 6.0, 37°C [13]) [13]
0.01 <15> (acetyl-CoA, <15>, reaction with tobramycin [16]) [16]
0.0117 <14> (sisomicin, <14>, pH 6.0, 37°C [13]) [13]
0.012 <15> (sisomicin, <15>, wild-type enzyme [16]) [16]
0.0131 <14> (amikacin, <14>, pH 6.0, 37°C [13]) [13]
0.014 <14> (butirosin, <14>, pH 6.0, 37°C [13]) [13]
0.014 <15> (neomycin C, <15>, wild-type enzyme [16]) [16]
0.017 <15> (acetyl-CoA, <15>, reaction with tobramycin, mutant enzyme C109A [16]) [16]
0.017 <15> (kanamycin B, <15>, mutant enzyme C109A [16]) [16]
0.0189 <14> (kanamycin B, <14>, pH 6.0, 37°C [13]) [13]
0.0195 <14> (propionyl-CoA, <14>, pH 6.0, 37°C [13]) [13]
0.0199 <14> (kanamycin A, <14>, pH 6.0, 37°C [13]) [13]
0.022 <1, 14> (tobramycin, <14>, pH 6.0, 37°C [13]) [2, 13]
0.0223 <14> (gentamicin B, <14>, pH 6.0, 37°C [13]) [13]
0.0235 <14> (acetyl-CoA, <14>, pH 6.0, 37°C [13]) [13]
0.026 <15> (tobramycin, <15>, mutant enzyme C109A [16]) [16]
0.026 <8> (tobramycin, <8>, pH 7.5, 37°C [8]) [8]
0.03 <15> (dibekacin, <15>, wild-type enzyme [16]) [16]
0.036 <8> (amikacin, <8>, pH 7.5, 37°C [8]) [8]
0.036 <14> (dibekacin, <14>, pH 6.0, 37°C [13]) [13]
0.038 <7> (poly-L-Lys, <7>, pH 7.5, 37°C [7]) [7]
0.05 <15> (amikacin, <15>, wild-type enzyme [16]) [16]
0.05 <15> (ribostamycin, <15>, wild-type enzyme [16]) [16]
0.057 <1> (amikacin) [2]
0.066 <15> (tobramycin, <15>, wild-type enzyme [16]) [16]
0.07 <15> (acetyl-CoA, <15>, reaction with tobramycin, mutant enzyme C109A/C70A [16]) [16]
0.075 <15> (amikacin, <15>, mutant enzyme C109A [16]) [16]
0.077 <8> (gentamicin, <8>, pH 7.5, 37°C [8]) [8]
0.077 <8> (tobramycin, <8>, pH 6.5, 37°C [8]) [8]
0.079 <15> (kanamycin B, <15>, wild-type enzyme [16]) [16]
0.105 <15> (kanamycin A, <15>, wild-type enzyme [16]) [16]
0.216 <8> (tobramicin, <8>, pH 5.5, 37°C [8]) [8]
0.25 <15> (neomycin C, <15>, mutant enzyme C109A/C70A [16]) [16]
0.259 <8> (netilmicin, <8>, pH 7.5, 37°C [8]) [8]
0.27 <15> (propionyl-CoA, <15>, reaction with tobramycin, wild-type enzyme [16]) [16]
0.297 <8> (gentamicin, <8>, pH 6.5, 37°C [8]) [8]
0.364 <8> (amikacin, <8>, pH 6.5, 37°C [8]) [8]
0.46 <15> (netilmicin, <15>, mutant enzyme C109A/C70A [16]) [16]

0.53 <15> (neomycin A) [16]
0.54 <15> (tobramycin, <15>, mutant enzyme C109A/C70A [16]) [16]
0.571 <8> (amikacin, <8>, pH 5.5, 37°C [8]) [8]
0.653 <8> (gentamicin, <8>, pH 5.5, 37°C [8]) [8]
0.82 <15> (malonyl-CoA, <15>, reaction with tobramycin, wild-type enzyme [16]) [16]
0.9 <15> (kanamycin B, <15>, mutant enzyme C109A/C70A [16]) [16]
2.44 <8> (netilmicin, <8>, pH 6.5, 37°C [8]) [8]
9.3 <15> (butyryl-CoA, <15>, reaction with tobramycin, wild-type enzyme [16]) [16]

K_i-Value (mM)

0.00059 <8> (tobramycin, <8>, pH 7.5, 37°C, substrate inhibition [8]) [8]
0.0023 <8> (tobramycin, <8>, pH 6.5, 37°C, substrate inhibition [8]) [8]
0.0036 <5> (CoA, <5>, pH 6.9, 37°C [5]) [5]
0.00499 <8> (amikacin, <8>, pH 7.5, 37°C, substrate inhibition [8]) [8]
0.00647 <8> (tobramycin, <8>, pH 5.5, 37°C, substrate inhibition [8]) [8]
0.0087 <8> (amikacin, <8>, pH 6.5, 37°C, substrate inhibition [8]) [8]
0.02206 <8> (amikacin, <8>, pH 5.5, 37°C, substrate inhibition [8]) [8]
0.02256 <8> (gentamicin, <8>, pH 5.5, 37°C, substrate inhibition [8]) [8]
0.0482 <8> (gentamicin, <8>, pH 6.5, 37°C, substrate inhibition [8]) [8]
0.059 <14> (tobramycin, <14>, pH 6.0, 37°C [13]) [13]
0.063 <14> (sisomicin, <14>, pH 6.0, 37°C [13]) [13]
0.06446 <8> (netilmicin, <8>, pH 5.5, 37°C, substrate inhibition [8]) [8]
0.072 <14> (dibekacin, <14>, pH 6.0, 37°C [13]) [13]
0.075 <8> (netilmicin, <8>, pH 6.5, 37°C, substrate inhibition [8]) [8]
0.079 <8> (gentamicin, <8>, pH 7.5, 37°C, substrate inhibition [8]) [8]
0.117 <14> (kanamycin B, <14>, pH 6.0, 37°C [13]) [13]
0.142 <14> (netilmicin, <14>, pH 6.0, 37°C [13]) [13]
0.168 <8> (netilmicin, <8>, pH 7.5, 37°C, substrate inhibition [8]) [8]
0.196 <14> (kanamycin A, <14>, pH 6.0, 37°C [13]) [13]
1.78 <7> (poly-L-Lys, <7>, pH 7.5, 37°C [7]) [7]
4 <5> (gentamicin C_1, <5>, pH 6.9, 37°C, inhibition of the reaction with neamine [5]) [5]
6 <5> (AMP, <5>, pH 6.9, 37°C, inhibition of the reaction with neamine [5]) [5]
7.7 <5> (paromomycin, <5>, pH 6.9, 37°C, inhibition of the reaction with neamine [5]) [5]

pH-Optimum

5.3 <2> (<2>, reaction with tobramycin, gentamicin C_{1a} or neomycin [2]) [2]
5.8 <1> (<1>, reaction with kanamycin, neomycin, hybrimycin, nebramycin factor 4 and 6 [1]) [1]
6 <1, 3> [4]
6.5 <2, 3, 8> (<2>, reaction with kanamycin, sisomicin, neamine, amikacin [2]; <8>, reaction with tobramycin, amikacin or gentamicin [8]) [2, 4, 8]
7 <6> [4]
7.4 <6> [3]

7.5 <4> [4]
7.5 <8> (<8>, activity with netilmicin [8]) [8]
7.6 <1> [1]

pH-Range

5.3-7.6 <1> (<1>, pH 5.3: about 50% of maximal activity, pH 7.6: about 20% of maximal activity, kanamycin A [1]) [1]

4 Enzyme Structure

Molecular weight

35000 <15> (<15>, gel filtration [16]) [16]
55000 <5> (<5>, sucrose density gradient centrifugation [5]) [5]

Subunits

? <3, 4, 6> (<4>, x * 55000, strain GN315, SDS-PAGE [4]; <4>, x * 57000, SDS-PAGE [4]; <3>, x * 59000, strain 141, SDS-PAGE [4]; <6>, x * 75000, SDS-PAGE [4]) [4]
dimer <15> (<15>, 2 * 17000, SDS-PAGE [16]; <15>, 2 * 17173, calculation from nucleotide sequence [16]; <15>, 2 * 17184, electrospray ionization mass spectrometry [16]) [16]

5 Isolation/Preparation/Mutation/Application

Source/tissue

aerial mycelium <12> [11]

Purification

<1> (partial [1]) [1, 4]
<2> [2]
<3> (strain GN315 [4]) [4]
<4> [4]
<5> (unseparable from 2-O-aminoglycoside phosphotransferase by affinity chromatography or sucrose density gradient ultracentrifugation [5]) [5]
<6> [4]
<14> [13]
<15> [15]

Crystallization

<7> (hanging drop vapor diffusion technique, crystal structure of the 6'-N-acetyltransferase type li in complex with acetyl-CoA determined at 2.7 A resolution [7]) [7]

Cloning

<9> (aac(6')-Ib' gene [9]) [9]
<12> (expression in Streptomyces lividans TK21/pANT-S2 [11]) [11]
<13> (expression of mutant enzymes F171L and Y80C [14]) [14]

<14> (expression in Escherichia coli [13]) [13]
<15> [16]
<28> [20]
<1, 3> (Pseudomonas aeruginosa 141, gene transfer via plasmid pBP30 to Escherichia coli Hb101 [4]; isogenic plasmids pSCH4663 containing aac(6')-Ib gene and pSCHB4105 containing aac(6')-IIa gene, transferred to Escherichia coli DH5α [6]) [4, 6]

Engineering

C109A <15> (<15>, mutation neither abolishes activity nor alters the biphasic inactivation by iodoacetamide [16]) [16]
C109A/C70A <15> (<15>, mutant enzyme is not inactivated by iodoacetamide. Double mutant exhibits large increases in K_m-values for both acetyl-CoA and aminoglycoside substrates [15]) [15]
C165A <25> (<25>, mutant enzyme shows levels of resistance to both antibiotics no more than threefold lower than that for the wild type [17]) [17]
E167A <23> (<23>, highly reduced ability to confer resistance to kanamycin and amikacin [17]) [17]
E172A <19> (<19>, mutant enzyme shows levels of resistance to both antibiotics no more than threefold lower than that for the wild type [17]) [17]
F171G <13> (<13>, the mutant enzyme is unable to confer resistance against amikacin, kanamycin, netilmicin and tobramicin [12]) [12]
F171I <13> (<13>, the mutant enzyme shows reduced resistance against amikacin, kanamycin, netilmicin and tobramicin [12]) [12]
F171K <13> (<13>, the mutant enzyme is unable to confer resistance against amikacin, kanamycin, netilmicin and tobramicin [12]) [12]
F171L <13, 14> (<13>, the mutant enzyme shows reduced resistance against amikacin, kanamycin, netilmicin and tobramicin [12]; <13>, the mutant enzyme of enzyme variant Ib shows lower specific activity than the wild-type enzyme when either kanamycin or its semisynthetic derivative amikacin is used as substrate. Alteration of substrate specificity at 42°C. The acetylating activity for kanamycin is higher at 42°C than at 30°C, the ability to use amikacin as substrate is reduced at 42°C. E. coli cells expressing the mutant enzyme are resistant the amikacin at 37°C but susceptible at 42°C [14]) [12-14]
F171M <13> (<13>, the mutant enzyme is able to confer detectable resistance against amikacin, kanamycin, netilmicin and tobramicin, although the levels are considerably lower than those conferred by the wild-type enzyme [12]) [12]
F171N <13> (<13>, the mutant enzyme is unable to confer resistance against amikacin, kanamycin, netilmicin and tobramicin [12]) [12]
F171S <13> (<13>, the mutant enzyme is unable to confer resistance against amikacin, kanamycin, netilmicin and tobramicin [12]) [12]
F171W <13> (<13>, the mutant enzyme is able to confer detectable resistance against amikacin, kanamycin, netilmicin and tobramicin, although the levels are considerably lower than those conferred by the wild-type enzyme [12]) [12]

G170A <21> (<21>, mutant enzyme confers high-level resistance to kanamycin but loses the ability to confer resistance to amikacin [17]) [17]
G175A <16> (<16>, mutant enzyme shows levels of resistance to both antibiotics no more than threefold lower than that for the wild type [17]) [17]
I163A <38> (<38>, the percentage of loss of resistance to one of the two antibiotics, kanamycin and amikacin, is no more than twice the percentage of loss of resistance to the other antibiotic [17]) [17]
K168A <22> (<22>, the percentage of loss of resistance to one of the two antibiotics, kanamycin and amikacin, is no more than twice the percentage of loss of resistance to the other antibiotic [17]) [17]
L119S <1> (<1>, AAC(6')-Ib L119S. The AAC(6')-Ib protein is unable to efficiently modify gentamicin C_1, 1.7% relative to sisomicin, however it is capable of modifying amikacin, 65.5% relative to sisomycin. The mutation results in a 2.8fold increase in acetylation of gentamicin C_1, but causes an 8.7fold reduction in the ability to modify amikacin [6]) [6]
L160A <26> (<26>, the percentage of loss of resistance to one of the two antibiotics, kanamycin and amikacin, is no more than twice the percentage of loss of resistance to the other antibiotic [17]) [17]
N159A <35> (<35>, mutant enzyme conders high-level resistance to kanamycin but loses the ability to confer resistance to amikacin [17]) [17]
P155A <31> (<31>, the percentage of loss of resistance to one of the two antibiotics, kanamycin and amikacin, is no more than twice the percentage of loss of resistance to the other antibiotic [17]) [17]
P157A <33> (<33>, the percentage of loss of resistance to one of the two antibiotics, kanamycin and amikacin, is no more than twice the percentage of loss of resistance to the other antibiotic [17]) [17]
Q174A <17> (<17>, mutant enzyme shows levels of resistance to both antibiotics no more than threefold lower than that for the wild type [17]) [17]
R161A <27> (<27>, mutant enzyme shows levels of resistance to both antibiotics no more than threefold lower than that for the wild type [19]) [19]
R164A <26> (<26>, mutant enzyme shows levels of resistance to both antibiotics no more than threefold lower than that for the wild type [17]) [17]
R173A <18> (<18>, mutant enzyme shows levels of resistance to both antibiotics no more than threefold lower than that for the wild type [17]) [17]
S119L <1> (<1>, AAC(6')-IIa S119L. The AAC(6')-IIa protein modifies gentamicin C_1 at 10.1% relative to sisomicin, however it shows low activity towards amikacin, 4.1% relative to sisomycin. The mutation results in a 4.8fold reduction in the acetylation of gentamicin C_1, but causes an 2fold increase in the ability to modify amikacin [6]) [6]
S156A <32> (<32>, mutant enzyme shows levels of resistance to both antibiotics no more than threefold lower than that for the wild type [17]) [17]
S158A <24> (<24>, the MICs of amikacin and kanamycin are higher than those for the wilde-type enzyme [17]) [17]
S158A <34> (<34>, mutant enzyme shows levels of resistance to both antibiotics no more than threefold lower than that for the wild type [17]) [17]
S83L <9> (<9>, the aac(6')-Ib' gene from strain BM2687 and the aac(6')-Ib gene from strain BM2656 show total identity with the exception of a C to T

transition that results in a Ser to Leu substitution at position 83 of the deduced polypeptide. The enzyme encoded by aac(6')-Ib' shows resistance to gentamicin but not to amikacin. The enzyme encoded by aac(6')-Ib shows resistance to amikacin but not to gentamicin [9]) [9]
Y166A <24> (<24>, mutant enzyme confers high-level resistance to kanamycin but loses the ability to confer resistance to amikacin [17]) [17]
Y80C <13> (<13>, the mutant enzyme shows only marginal levels of activity when either amikacin, kanamycin, tobramycin or netilmicin is used as substrate [14]) [14]

6 Stability

Temperature stability

37 <1> (<1>, pH 7.4, 10 mM $MgCl_2$, 0.6 mM 2-mercaptoethanol, about 35% loss of activity after 10 min, acetyl-CoA prevents, kanamycin protects slightly [1]) [1]
42 <1> (<1>, pH 7.4, 10 mM $MgCl_2$, 0.6 mM 2-mercaptoethanol, about 70% loss of activity after 10 min, acetyl-CoA prevents, kanamycin protects slightly [1]) [1]
Additional information <5> (<5>, GTP, CoA, gentamicin C, neamine protect against thermal inactivation [5]) [5]

General stability information

<1>, Mg^{2+} stabilizes [1]
<1>, dialysis in the absence of Mg^{2+} causes irreversible loss of activity [1]
<5>, GTP, CoA, gentamicin C, neamine protect against thermal inactivation [5]

Storage stability

<1>, -20°C, several months [1]
<2>, -30°C, stable [2]

References

[1] Benveniste, R.; Davies, J.: Enzymatic acetylation of aminoglycoside antibiotics by Escherichia coli carrying an R factor. Biochemistry, **10**, 1787-1796 (1971)
[2] Le Goffic, F.; Martel, A.: Resistance to aminosides induced by an isoenzyme, kanamycin acetyltransferase. Biochimie, **56**, 893-897 (1974)
[3] Benveniste, R.; Davies, J.: Aminoglycoside antibiotic-inactivating enzymes in actinomycetes similar to those present in clinical isolates of antibiotic-resistant bacteria. Proc. Natl. Acad. Sci. USA, **70**, 2276-2280 (1973)
[4] Meyer, J.F.; Wiedemann, B.: Characterization of aminoglycoside 6-N-acetyltransferases [AAC(6)] from gram-negative bacteria and Streptomyces kanamyceticus. J. Antimicrob. Chemother., **15**, 271-282 (1985)

[5] Martel, A.; Masson, M.; Moreau, N.; Le Goffic, F.: Kinetic studies of aminoglycoside acetyltransferase and phosphotransferase from Staphylococcus aureus RPAL. Relationship between the two activities. Eur. J. Biochem., **133**, 515-521 (1983)
[6] Rather, P.N.; Munayyer, H.; Mann, P.A.; Hare, R.S.; Miller, G.H.; Shaw, K.J.: Genetic analysis of bacterial acetyltransferases: identification of amino acids determining the specificities of the aminoglycoside 6-N-acetyltransferase Ib and IIa proteins. J. Bacteriol., **174**, 3196-3203 (1992)
[7] Wybenga-Groot, L.E.; Draker, K.-a.; Wright, G.D.; Berghuis, A.M.: Crystal structure of an aminoglycoside 6'-N-acetyltransferase: defining the GCN5-related N-acetyltransferase superfamily fold. Structure, **7**, 497-507 (1999)
[8] Culebras, E.; Martinez, J.L.; Baquero, F.; Perez-Diaz, J.C.: pH Modulation of aminoglycoside resistance in Staphylococcus epidermidis harboring 6'-N-aminoglycoside acetyltransferase. J. Antimicrob. Chemother., **37**, 881-889 (1996)
[9] Lambert, T.; Ploy, M.C.; Courvalin, P.: A spontaneous point mutation in the aac(6')-Ib' gene results in altered substrate specificity of aminoglycoside 6'-N-acetyltransferase of a Pseudomonas fluorescens strain. FEMS Microbiol. Lett., **115**, 297-304 (1994)
[10] Casin, I.; Bordon, F.; Bertin, P.; Coutrot, A.; Podglajen, I.; Brasseur, R.; Collatz, E.: Aminoglycoside 6'-N-acetyltransferase variants of the Ib type with altered substrate profile in clinical isolates of Enterobacter cloacae and Citrobacter freundii. Antimicrob. Agents Chemother., **42**, 209-215 (1998)
[11] Zhu, C.B.; Sunada, A.; Ishikawa, J.; Ikeda, Y.; Kondo, S.; Hotta, K.: Role of aminoglycoside 6'-acetyltransferase in a novel multiple aminoglycoside resistance of an actinomycete strain No. 8: inactivation of aminoglycosides with 6'-amino group except arbekacin and neomycin. J. Antibiot., **52**, 889-894 (1999)
[12] Chavideh, R.; Sholly, S.; Panaite, D.; Tolmasky, M.E.: Effects of F171 mutations in the 6'-N-acetyltransferase type Ib [AAC(6')-Ib] enzyme on susceptibility to aminoglycosides. Antimicrob. Agents Chemother., **43**, 2811-2812 (1999)
[13] Wright, G.D.; Ladak, P.: Overexpression and characterization of the chromosomal aminoglycoside 6'-N-acetyltransferase from Enterococcus faecium. Antimicrob. Agents Chemother., **41**, 956-960 (1997)
[14] Panaite, D.M.; Tolmasky, M.E.: Characterization of mutants of the 6'-N-acetyltransferase encoded by the multiresistance transposon Tn1331: effect of Phe171-to-Leu171 and Tyr80-to-Cys80 substitutions. Plasmid, **39**, 123-133 (1998)
[15] Hegde, S.S.; Dam, T.K.; Brewer, C.F.; Blanchard, J.S.: Thermodynamics of aminoglycoside and acyl-Coenzyme A binding to the Salmonella enterica AAC(6')-Iy aminoglycoside N-acetyltransferase. Biochemistry, **41**, 7519-7527 (2002)
[16] Magnet, S.; Lambert, T.; Courvalin, P.; Blanchard, J.S.: Kinetic and mutagenic characterization of the chromosomally encoded Salmonella enterica AAC(6')-Iy aminoglycoside N-acetyltransferase. Biochemistry, **40**, 3700-3709 (2001)

[17] Shmara, A.; Weinsetel, N.; Dery, K.J.; Chavideh, R.; Tolmasky, M.E.: Systematic analysis of a conserved region of the aminoglycoside 6'-N-acetyltransferase type Ib. Antimicrob. Agents Chemother., **45**, 3287-3292 (2001)

[18] Nobuta K., Tolmasky M.E., Crosa L.M., Crosa J.H.: Sequencing and expression of the 6'-N-acetyltransferase gene of transposon Tn1331 from Klebsiella pneumoniae. J. Bacteriol., **170**, 3769-3773 (1988)

[19] Tolmasky M.E.: Sequencing and expression of aadA, bla, and tnpR from the multiresistance transposon Tn1331. Plasmid, **24**, 218-226 (1990)

[20] Tenover F.C., Filpula D., Phillips K.L., Plorde J.J.: Cloning and sequencing of a gene encoding an aminoglycoside 6'-N-acetyltransferase from an R factor of Citrobacter diversus. J. Bacteriol., **170**, 471-473 (1988)

[21] Bunny K.L., Hall R.M., Stokes H.W.: New mobile gene cassettes containing an aminoglycoside resistance gene, aacA7, and a chloramphenicol resistance gene, catB3, in an integron in pBWH301. Antimicrob. Agents Chemother., **39**, 686-693 (1995)

[22] Van Nhieu G.T., Collatz E.: Primary structure of an aminoglycoside 6'-N-acetyltransferase AAC(6')-4, fused in vivo with the signal peptide of the Tn3-encoded β-lactamase. J. Bacteriol., **169**, 5708-5714 (1987)

Phosphatidylcholine-dolichol O-acyltransferase 2.3.1.83

1 Nomenclature

EC number
2.3.1.83

Systematic name
3-sn-phosphatidylcholine:dolichol O-acyltransferase

Recommended name
phosphatidylcholine-dolichol O-acyltransferase

CAS registry number
9012-30-0

2 Source Organism

<1> *Rattus norvegicus* (female, Sprague-Dawley strain [1]) [1]

3 Reaction and Specificity

Catalyzed reaction
3-sn-phosphatidylcholine + dolichol = 1-acyl-sn-glycero-3-phosphocholine + acyldolichol

Reaction type
acyl group transfer

Substrates and products
S 3-sn-phosphatidylcholine + dolichol <1> (<1> dolichols with 16-22 isopren units, no substrates are phosphatidylethanolamine, lysophosphatidylcholine, sphingomyelin and palmitoyl-CoA [1]) (Reversibility: ? <1> [1]) [1]
P 1-acyl-sn-glycero-3-phosphocholine + acyldolichol <1> [1]

Inhibitors
Triton X-100 <1> (<1> high concentration, stimulation at low concentration [1]) [1]
Additional information <1> (<1> not inhibited by retinol [1]) [1]

Cofactors/prosthetic groups
Additional information <1> (<1> no requirement for CoA, ATP and Mg^{2+}, separately or together [1]) [1]

Activating compounds

Triton X-100 <1> (<1> 1%, several fold activation, inhibition at 3% [1]) [1]

Metals, ions

Additional information <1> (<1> no Mg^{2+} required [1]) [1]

K_m-Value (mM)

0.3 <1> (dolichol) [1]

pH-Optimum

6 <1> [1]

pH-Range

4.8-7.5 <1> (<1> approx. half-maximal activity at pH 4.8 and 7.5 [1]) [1]

Temperature optimum (°C)

38 <1> (<1> assay at [1]) [1]

5 Isolation/Preparation/Mutation/Application

Source/tissue

brain <1> [1]
intestinal mucosa <1> [1]
kidney <1> [1]
liver <1> (predominant [1]) [1]
pancreas <1> [1]
spleen <1> [1]

Localization

microsome <1> (<1> predominant, subcellular distribution [1]) [1]

6 Stability

pH-Stability

6 <1> (<1> and below, inactivation [1]) [1]

Temperature stability

Additional information <1> (<1> inactivation at room temperature and above [1]) [1]

References

[1] Keenan, R.W.; Kruczek, M.E.: The esterification of dolichol by rat liver microsomes. Biochemistry, **15**, 1586-1591 (1976)

Alcohol O-acetyltransferase 2.3.1.84

1 Nomenclature

EC number
2.3.1.84

Systematic name
acetyl-CoA:alcohol O-acetyltransferase

Recommended name
alcohol O-acetyltransferase

Synonyms
acetyltransferase, alcohol
alcohol acetyltransferase

CAS registry number
80237-89-4

2 Source Organism

<-1> no activity in *Cucumus melo var. rochet* [13]
<1> *Saccharomyces uvarum* (brewery Lager yeast [1]) [1, 6]
<2> *Saccharomyces cerevisiae* (sake yeast strain Kyokai No. 7 [2,5]; alcohol acetyltransferases 1 and 2 [11]; wine yeast strain VIN13 [12]) [2, 5, 6, 8, 10, 11, 12]
<3> *Hansenula anomala* [3]
<4> *Musa sapientum* (banana [4]) [4]
<5> *Cladosporium cladosporioides* [5]
<6> *Lycopersicon esculentum* [7]
<7> *Fragaria x ananassa* (strawberry [9]) [9]
<8> *Hansenula mrakii* [10]
<9> *Saccharomyces pastorianus* (bottom-fermenting lager yeast, alcohol acetyltransferases 1 and 2 [11]) [11]
<10> *Cucumis melo* (var. arava, aromatic melon [13]) [13]

3 Reaction and Specificity

Catalyzed reaction
acetyl-CoA + an alcohol = CoA + an acetyl ester

Reaction type
acyl group transfer

Natural substrates and products

- **S** isoamylalcohol + acetyl-CoA <2> (<2> alcohol acetyltransferase 2 is probably the key enzyme in a steroid detoxification mechanism, alcohol acetyltransferase 1 accounts for 80% of isoamyl acetate production and 30% of ethyl acetate production and might be involved in fatty acid metabolism [11]) (Reversibility: ? <2> [11]) [11]
- **P** isoamyl acetate + CoA <2> [11]

Substrates and products

- **S** 1-octen-3-ol + acetyl-CoA <10> (Reversibility: ? <10> [13]) [13]
- **P** 1-octen-3-acetate + CoA <10> [13]
- **S** 3-methylthio-1-propanol + acetyl-CoA <10> (Reversibility: ? <10> [13]) [13]
- **P** 3-methylthio-1-acetate + CoA <10> [13]
- **S** benzyl alcohol + acetyl-CoA <10> (Reversibility: ? <10> [13]) [13]
- **P** benzyl acetate + CoA <10> [13]
- **S** butanol + acetyl-CoA <1, 10> (Reversibility: ? <1, 10> [1, 13]) [1, 13]
- **P** butyl acetate + CoA <1, 10> [1, 13]
- **S** ethanol + acetyl-CoA <1> (Reversibility: ? <1> [1]) [1]
- **P** ethyl acetate + CoA <1> [1]
- **S** ethanol + butyryl-CoA <1> (Reversibility: ? <1> [1]) [1]
- **P** ethyl butyrate + CoA <1> [1]
- **S** hexanol + acetyl-CoA <1, 10> (Reversibility: ? <1, 10> [1, 13]) [1, 13]
- **P** hexyl acetate + CoA <1, 10> [1, 13]
- **S** isoamylalcohol + acetyl-CoA <1, 2, 4, 9, 10> (<2> slightly higher activity with straight-chain alcohols than with branched-chain alcohols, activity with straight-chain alcohols in descending order: pentanol, butanol, propanol, very low activity with ethanol, activity with branched-chain alcohols in descending order of carbon number: C_5, C_4, C_3 [5]) (Reversibility: ? <1, 2, 4, 9, 10> [1, 2, 4, 11, 13]) [1, 2, 4, 5, 11, 13]
- **P** isoamyl acetate + CoA <1, 2, 4, 9, 10> [1, 2, 4, 5, 11, 13]
- **S** isobutanol + acetyl-CoA <3> (Reversibility: ? <3> [3]) [3]
- **P** isobutyl acetate + CoA <3> [3]
- **S** methanol + acetyl-CoA <1> (Reversibility: ? <1> [1]) [1]
- **P** methyl acetate + CoA <1> [1]
- **S** octanol + acetyl-CoA <10> (Reversibility: ? <10> [13]) [13]
- **P** octyl acetate + CoA <10> [13]
- **S** pentanol + acetyl-CoA <1> (Reversibility: ? <1> [1]) [1]
- **P** penthyl acetate + CoA <1> [1]
- **S** phenylethyl alcohol + acetyl-CoA <10> (Reversibility: ? <10> [13]) [13]
- **P** phenylethyl acetate + CoA <10> [13]
- **S** propanol + acetyl-CoA <1> (<1> low activity [1]) (Reversibility: ? <1> [1]) [1]
- **P** propyl acetate + CoA <1> [1]

Inhibitors

2,4,6-trinitrobenzenesulfonic acid <1, 2> (<2> 54% inhibition [2]) [1, 2]
5,5'-dithiobis(2-nitrobenzoate) <1, 2> (<1> 99% inhibition [1]; <2> complete inhibition [2]; <2> complete inhibition [5]) [1, 2, 5]
Ag^{+} <1, 2> (<1> 1 mM, 92% inhibition [1]; <2> complete inhibition [2]) [1, 2]
Cd^{2+} <1, 2> (<1> 1 mM, complete inhibition [1]; <2> 1 mM, 95% inhibition [2]) [1, 2, 5]
Co^{2+} <1> (<1> 1 mM, 83% inhibition [1]) [1]
Cu^{2+} <1, 2> (<1> 1 mM, 99% inhibition [1]; <2> 1 mM, complete inhibition [2]) [1, 2, 5]
Hg^{2+} <1, 2> (<1> 1 mM, 98% inhibition [1]; <2> 1 mM, complete inhibition [2]) [1, 2, 5]
Ni^{2+} <1> (<1> 1 mM, 68% inhibition [1]) [1]
Pb^{2+} <1> (<1> 1 mM, complete inhibition [1]) [1]
Tween 60 <1> (<1> 2 mM, 77% inhibition [1]) [1]
Tween 80 <1, 1> (<1> 2 mM, 89% inhibition [1]) [1]
Zn^{2+} <1, 2> (<1> 1 mM, 97% inhibition [1]; <2> 1 mM, 89% inhibition [2]) [1, 2, 5]
ergosterol <1> (<1> 2 mM, 73% inhibition [1]) [1]
iodoacetate <1> [1]
linoleic acid <1, 2> (<1> 2 mM, 97% inhibition [1]; <2> 2 mM, 86% inhibition [5]) [1, 5]
linolenic acid <1, 2> (<1> 2 mM, 97% inhibition [1]; <2> 2 mM, 68% inhibition [5]) [1, 5]
monoolein <1> (<1> 2 mM, 76% inhibition [1]) [1]
monostearin <1> (<1> 2 mM, 58% inhibition [1]) [1]
oleoyl-CoA <1, 1> (<1> 2 mM, 87% inhibition [1]) [1]
p-chloromercuribenzoate <1, 2> (<1> 98% inhibition [1]; <2> 96% inhibition [2]; <2> complete inhibition [5]) [1, 2, 5]
phosphatidylinositol <1, 1> (<1> 2 mM, 97% inhibition [1]) [1]
phosphatidylserine <1, 1> (<1> 2 mM, 88% inhibition [1]) [1]

Metals, ions

Mg^{2+} <2> (<2> 1 mM, 10% activation [2]) [2]

Specific activity (U/mg)

0.00729 <1> [1]
5.47 <2> [5]
190.4 <2> [2]
Additional information <4, 10> (<4,10> activity increases during ripening [4,13]) [4, 13]

K_m-Value (mM)

0.025 <2> (acetyl-CoA, <2> strain NCYC 366 [5]; <2> isoamyl acetate synthesis [11]) [5, 11]

0.025 <2> (isoamyl alcohol) [11]
0.045 <2> (acetyl-CoA, <2> ethyl acetate synthesis [11]) [11]
0.05 <4> (acetyl-CoA) [4]
0.19 <2> (acetyl-CoA) [5]
0.4 <4> (isoamylalcohol) [4]
29.8 <2> (isoamyl alcohol) [5]

pH-Optimum
5.9 <5> [5]
7 <3> [3]
7-8 <1, 2> [1, 2]
8 <2, 6> [5, 7]
8.5 <4> [4]

pH-Range
5-8.5 <1> (<1> approx. 40% of maximal activity at pH 5.0, 80% of maximal activity at pH 8.5 [1]) [1]

Temperature optimum (°C)
25 <2> [5]
30 <1-4> [1-4]
30-40 <6> [7]

Temperature range (°C)
10-40 <1> (<1> 25% of maximal activity at 10°C, 50% of maximal activity at 40°C [1]) [1]

4 Enzyme Structure

Molecular weight
17000 <6> (<6> gel filtration [7]) [7]
40000 <4> (<4> gel filtration [4]) [4]
50000 <2> (<2> gel filtration [2]) [2]
57000 <2> (<2> alcohol acetyltransferase I [11]) [11]
220000 <1> (<1> gel filtration [1]) [1]
270000 <2> (<2> gel filtration [5]) [5]

Subunits
? <2, 5, 9> (<2> x * 60000, SDS-PAGE [5]; <5> x * 22000, SDS-PAGE [5]; <2> x * 61059, deduced from nucleotide sequence [6]; <2> x * 61100, alcohol acetyltransferase I [11]; <9> x * 63200, alcohol acetyltransferase I [11]; <2> x * 61900, alcohol acetyltransferase II [11]) [5, 6, 11]
monomer <2> (<2> 1 * 50000, SDS-PAGE [2]) [2]

5 Isolation/Preparation/Mutation/Application

Source/tissue

fruit <4, 10> [4, 13]
pulp <6, 7> [7, 9]

Localization

cytosol <2, 7> (<2> alcohol acetyltransferase 1 is located within the cellular vacuoles [12]) [9, 12]
membrane <2> (<2> membrane associated [2]; <1> cell membrane bound [1]) [1, 2]
soluble <4> [4]

Purification

<1> (Sepharose 6B, DEAE-Sephadex A50, partially purified [1]) [1]
<2> (ammonium sulfate, Biofine HIC-PH, DEAE-5PW, Mono-P [2]; PBE 94 column, DEAE-cellulose, Toypearl HW60, hydroxyapatite, octyl Sepharose [5]) [2, 5, 11]
<3> (ammonium sulfate, Sephadex G-150, DEAE-Sephadex A-50 [3]) [3]
<4> [4]
<6> (partially purified [7]) [7]

Cloning

<1> (cloning of Lg-ATF1 gene encoding alcohol acetyltransferase, expression in Saccharomyces cerevisiae [6]) [6]
<2> (cloning of ATF1 gene encoding alcohol acetyltransferase, expression in Saccharomyces cerevisiae [6]) [6, 10]
<2> (cloning of the ATF1 gene encoding alcohol acetyltransferase 1 from wine yeast [12]) [12]
<9> (cloning of ATF1 and ATF2 genes encoding alcohol acetyltransferases [11]) [11]

6 Stability

pH-Stability

6.5-7 <2> [2]
7.5-8 <2> (<2> no loss of activity after 22 h at 4°C [5]) [5]
7.5-9 <4> [4]

Temperature stability

10 <2> (<2> up to [2]; <2> loss of 20% activity [5]) [2, 5]
20 <2> (<2> rapid inactivation above [5]) [5]
30 <2> (<2> rapid inactivation [2]; <2> 55% loss of activity after 30 min [10]) [2, 10]

Organic solvent stability

ethanol <2> (<2> 15%, stable [2]) [2]

General stability information

<4>, very labile at pH values lower than pH 7.0 [4]

Storage stability

<2>, -80°C, 15% glycerol, no loss of activity [2]

References

[1] Yoshioka, K.; Hashimoto, N.: Ester formation by alcohol acetyltransferase from brewer's yeast. Agric. Biol. Chem., **45**, 2183-2190 (1981)

[2] Akita, O.; Suzuki, S.; Obata, T.: Purification and some properties of alcohol acetyltransferase from sake yeast. Agric. Biol. Chem., **54**, 1485-1490 (1990)

[3] Kang, H.S.; Kim, J.H.; Shin, H.K.: Partial purifiction and catalytic properties of alcohol acetyltransferase from Hansenula anomala. Sanop Misaengmul Hakhoechi, **14**, 69-73 (1986)

[4] Harada, M.; Ueda, Y.; Iwata, T.: Purification and some properties of alcohol acetyltransferase from banana fruit. Plant Cell Physiol., **26**, 1067-1074 (1985)

[5] Minetoki, T.; Bogaki, T.; Iwamatsu, A.; Fujii, T.; Hamachi, M.: The purification, properties and internal peptide sequences of alcohol acetyltransferase isolated from Saccharomyces cerevisiae Kyokai No. 7. Biosci. Biotechnol. Biochem., **57**, 2094-2098 (1993)

[6] Fujii, T.; Nagasawa, N.; Iwamatsu, A.; Bogaki, T.; Tamai, Y.; Hamachi, M.: Molecular cloning, sequence analysis, and expression of the yeast alcohol acetyltransferase gene. Appl. Environ. Microbiol., **60**, 2786-2792 (1994)

[7] Ueda, Y.; Nishimura, K.; Chachin, K.: Alcohol acetyltransferase in tomato (Lycopersicon esculentum Mill). Appl. Biol. Sci., **3**, 25-32 (1997)

[8] Del Carmen Plata, M.; Mauricio, J.C.; Millan, C.; Ortega, J.M.: In vitro specific activity of alcohol acetyltransferase and esterase in two flor yeast strains during biological aging of sherry wines. J. Ferment. Bioeng., **85**, 369-374 (1998)

[9] Noichinda, S.; Ueda, Y.; Imahori, Y.; Chachin, K.: Subcellular localization of alcohol acetyltransferase in strawberry fruit. Food Sci. Technol. Res., **5**, 239-242 (1999)

[10] Fukuda, K.; Yamamoto, N.; Kiyokawa, Y.; Yanagiuchi, T.; Wakai, Y.; Kitamoto, K.; Inoue, Y.; Kimura, A.: Balance of activities of alcohol acetyltransferase and esterase in Saccharomyces cerevisiae is important for production of isoamyl acetate. Appl. Environ. Microbiol., **64**, 4076-4078 (1998)

[11] Mason, A.B.; Dufour, J.P.: Alcohol acetyltransferases and the significance of ester synthesis in yeast. Yeast, **16**, 1287-1298 (2000)

[12] Lilly, M.; Lambrechts, M.G.; Pretorius, I.S.: Effect of increased yeast alcohol acetyltransferase activity on flavor profiles of wine and distillates. Appl. Environ. Microbiol., **66**, 744-753 (2000)

[13] Shalit, M.; Katzir, N.; Tadmor, Y.; Larkov, O.; Burger, Y.; Shalekhet, F.; Lastochkin, E.; Ravid, U.; Amar, O.; Edelstein, M.; Karchi, Z.; Lewinsohn, E.: Acetyl-CoA:Alcohol acetyltransferase activity and aroma formation in ripening melon fruits. J. Agric. Food Chem., **49**, 794-799 (2001)

Fatty-acid synthase 2.3.1.85

1 Nomenclature

EC number
2.3.1.85

Systematic name
acyl-CoA:malonyl-CoA C-acyltransferase (decarboxylating, oxoacyl- and enoyl-reducing and thioester-hydrolysing)

Recommended name
fatty-acid synthase

Synonyms
FAS
yeast fatty acid synthase
Additional information (the animal enzyme is a multi-functional protein catalyzing the reactions of EC 2.3.1.38, EC 2.3.1.39, EC 2.3.1.41, EC 1.1.1.100, EC 4.2.1.61, EC 1.3.1.10, EC 3.1.2.14)

CAS registry number
9045-77-6

2 Source Organism

<1> *Columba sp.* (pigeon [1,5-8,14]) [1, 5-8, 14]
<2> *Gallus gallus* [7, 8, 10, 12, 13, 15, 20, 22, 26, 28]
<3> *Anser sp.* (goose [7]) [7]
<4> *Rattus norvegicus* (Sprague-Dawley [3]; female Wistar rat [16]; male Sprague-Dawley [29]; wild-type and mutant [31,32]) [2, 3, 7-9, 14, 16, 19, 21, 27, 29, 31, 32]
<5> *Mus musculus* (male FVB-N [30]) [7, 30]
<6> *Oryctolagus cuniculus* (female [4]) [4, 7]
<7> *Capra hircus* (goat, female [11]) [11]
<8> *Homo sapiens* [7, 17, 18, 25]
<9> *Canis familiaris* [7]
<10> *Bos taurus* [7]
<11> *Brevibacterium ammoniagenes* [23, 24]

3 Reaction and Specificity

Catalyzed reaction

acetyl-CoA + n malonyl-CoA + 2n NADPH + 2n H^+ = long-chain fatty acid + (n+1) CoA + n CO_2 + 2n $NADP^+$ (<1-6,8-10> reaction mechanism [7,8,13,14]; <1-10> structure [7,11-13] <1-6,8-10> structure and regulation [7])

Reaction type

acyl group transfer
decarboxylation
redox reaction
thioester hydrolysis

Natural substrates and products

S acetyl-CoA + malonyl-CoA + NADPH <1-6, 8-10> (<1-6,8-10> multifunctional enzyme, involved in animal fat synthesis [7,8]; <2> participates in energy metabolism in vivo which is related to adiposis and cancer [26]) (Reversibility: ? <1-6, 8-10> [7, 8, 26]) [7, 8, 26]

P palmitate + CoA + CO_2 + $NADP^+$

Substrates and products

S acetyl-CoA + malonyl-CoA + NADPH <1-11> (<1> specific for malonyl-CoA, acetyl-CoA can be replaced by propionyl-CoA or butyryl-CoA [8]; <7> decanoyl-CoA, acetyl-CoA, butyryl-CoA and malonyl-CoA are bound to the same active site [11]; <8> acetoacetyl-CoA can substitute for acetyl-CoA [18]) (Reversibility: ? <1-11> [4, 7, 8, 11-13, 17, 18, 19, 21-23, 26, 28]) [4, 7, 8, 11-13, 17, 18, 19, 21-23, 26, 28]

P palmitate + CoA + CO_2 + $NADP^+$ <1-6, 8-10, 11> (<4> palmitate and stearate in equal amounts [2]; <1,2,4> in the absence of NADPH the product is triacetic acid lactone [8]; <1-6,8-10> C_{20} and C_{22} fatty acids in the absence of thioesterase activity [7]; <3> methylmalonyl-CoA instead of malonyl-CoA yields branched fatty acid, e.g. 2,4,6,8-tetramethyldecanoic acid [7]; <4,6> lactating mammary gland thioesterase II of fatty acid synthetase: products are medium chain fatty acids from C_8 to C_{12} [7]; <8> small amounts of stearate and myristate are also produced [18]; <4> products are fatty acid chains from C_{14} to C_{20} [19]; <11> FAS-A mainly synthesizes the C_{18} fatty acids oleate and stearate with only traces of palmitate, the major product of FAS-B is pamitate [23]; <11> synthesis of saturated and unsaturated fatty acids, FAS-B cannot synthesize oleic acid [24]) [2, 7, 8, 13, 14, 17, 18, 19, 23, 24]

S Additional information <1, 4> (<4> synthesizes equal amounts of C_{14} and C_{16} fatty acids [2]; <1> carries acetyl- and malonyl-CoA transacylase, condensing enzyme, β-ketoacyl reductase, β-hydroxyacyl dehydrase, enoylacyl reductase, palmitoyl-CoA thioesterase activities on each multifunctional subunit [5]; <1,4> fatty acid synthetases of vertebrates and yeast are stable enzyme complexes of multifunctional polypeptide chains, the fatty acid synthetases of plants and E. coli consist of non-associated individual enzymes [14]) [2, 5, 14]

P ?

Inhibitors

1,3-dibromo-2-propanone <1-6, 8-10> (<2> covalently cross-links subunits, inactivates β-ketoacyl synthetase- and overall-fatty acid synthase reaction [10]; <2,6> acetyl-CoA, not malonyl-CoA, protects against inactivation, both protect against cross-linking of the subunits [4,10]) [4, 7, 10]
2,3-butanedione <3> [7]
2,3-trans-octenoyl-CoA <4> (<4> competitively inhibits malonyl-transferase reaction [9]) [9]
3-oxooctanoyl-CoA <4> (<4> competitively inhibits malonyl-transferase reaction [9]) [9]
CoA <4> (<4> competitive inhibition [9]) [9]
N-ethylmaleimide <4> (<4> inhibition of elongation process and malonyl transfer at 10 mM [9]) [9]
PMSF <1-6, 8-10> [7]
S-(4-bromo-2,3-dioxobutyl)-CoA <1> (<1> irreversible, acetyl transacetylase and β-ketoacyl synthase activity, 4 mol inhibitor per mol enzyme complex, mechanism, dithiothreitol protects [5]; <1> irreversible, 50% inhibition after 10 s at 0.02 mM, 6.5 s at 0.06 mM and 4.5 s at 0.09 mM, specific for acetyl transacetylase and β-ketoacyl synthase activity, 4 mol inhibitor per mol enzyme complex, acetyl-CoA, malonyl-CoA, cysteine and pantetheine protect [6]) [5, 6]
Zn^{2+} <2> (<2> 80% loss of activity at 0.008 mM, interacts with SH groups, substrates of the reaction protect, malonyl-CoA being the most effective, addition of dithiothreitol leads to a recovery of 70% enzyme activity [26]) [26]
aryl-acyl-β-alanyl $NADP^+$ <2> (<2> inhibits β-ketoacyl-reductase [12]) [12]
cerulenin <2> (<2> acetyl-CoA protects [13]) [13]
chloroacetyl-CoA <6> [4]
crotonyl-CoA <1> (<1> dehydrase activity [6]) [6]
diisopropylfluorophosphate <1-6, 8-10> [7]
iodoacetamide <1-6, 8-10> (<1> kinetics, 50% inactivation after 5 min at 1 mM and after 0.5 min at 20 mM [6]; <4> inhibition of elongation processs [9]; <2> inhibits β-ketoacyl synthetase activity, acetyl-CoA but not malonyl CoA protects [10]; <1-6,8-10> β-ketoacyl synthetase activity, acetyl-CoA but not malonyl-CoA protects [7]) [6, 7, 9, 10]
long-chain acyl-CoA <4> (<4> malonyl-transferase reaction [9]) [9]
malonyl pantetheine <4> (<4> malonyl-transferase reaction [9]) [9]
malonyl-CoA <4> (<4> competitive, malonyl-transferase reaction [9]) [9]
octanoyl-CoA <4> (<4> competitively inhibits malonyl-transferase reaction [9]) [9]
phenylglyoxal <3> [7]
pyridoxal 5'-phosphate <1-6, 8-10> (<1-6,8-10> enoyl reductase activity, NADPH protects [7]) [7]

sodium dodecylsulfate <2> (<2> causes conformational changes at higher concentrations [22]) [22]
urea <2> (<2> enzyme forms inactive aggregates in the presence of urea, cyclodextrins prevent aggregation [20]; <2> non-competitive inhibitor for NADPH, competitive inhibitor for acetyl-CoA and malonyl-CoA, complete inactivation occurs at lower concentration than obvious conformational changes, aggregation occurs at 3-4 M [28]) [20, 28]

Cofactors/prosthetic groups

4'-phosphopantetheine <1-6, 8-10> (<1-6,8-10> requirement, 1 mol associated with 1 mol subunit [3,7]; <4> no covalently bound subunit component [3]; <6> present in substoichiometric amounts [4]; <1,6> location of active site [4-6]) [1, 3-9, 13]
NADH <1> (<1> requirement, 10% as effective as NADPH [8]) [8]
NADPH <1-6, 8-10> (<1-6,8-10> requirement, high specificity [7,12]) [7, 12, 8, 18, 22, 28]
Additional information <1-6, 8-10> (<1-6,8-10> no requirement for FMN [7,8]) [7, 8]

Turnover number (min^{-1})

0.012 <4> (malonyl-CoA, <4> R606K mutant [32]) [32]
0.1 <4> (malonyl-CoA, <4> R606A mutant [32]) [32]
1.2 <4> (malonyl-CoA, <4> wild-type enzyme [32]) [32]
1.9 <4> (acetyl-CoA, <4> wild-type enzyme [32]) [32]
3.4 <4> (acetyl-CoA, <4> R606K mutant [32]) [32]
14.2 <4> (acetyl-CoA, <4> R606A mutant [32]) [32]

Specific activity (U/mg)

0.015 <4> (<4> for malonyl-CoA, R606A mutant [32]) [32]
0.078-0.1 <1> (<1> palmitate [8]) [8]
0.14-0.16 <1> [5]
0.154 <4> (<4> for malonyl-CoA, R606K mutant [32]) [32]
0.31 <11> (<11> FAS-A [23]) [23]
0.37 <11> (<11> FAS-B [23]) [23]
0.6 <8> (<8> NADPH [18]) [18]
1.12-1.4 <1> (<1> NADPH, 30°C [8]) [8]
1.13-2.77 <4> (<4> various tagged wild-type enzyme [19]) [19]
1.2 <2> (<2> NADPH [10,12]) [10, 12]
1.2-1.6 <2> (<2> NADPH [13]) [13]
1.3 <2> (<2> NADPH [15]) [15]
1.61 <4> (<4> for malonyl-CoA, wild-type enzyme [32]) [32]
2.049 <4> (<4> wild-type enzyme, activity in mutants is less than 0.5% [27]) [27]
2.6 <2> (<2> NADPH [12]) [12]
2.68 <4> (<4> for acetyl-CoA, wild-type enzyme [32]) [32]
4.53 <4> (<4> for acetyl-CoA, R606K mutant [32]) [32]
17.8 <4> (<4> for acetyl-CoA, R606A mutant [32]) [32]

Additional information <2, 4> (<4> specific activity in liver and fat cells [3]; <2> specific activity of recombinant domain I and its subdomains [15]; <4> specific activities after diet with sesamin [29]; <4,8> specific activities of subdomains [18,31]; <5> increase after exposure of animals to cold temperatures for 48 h [30]) [3, 15, 18, 27, 29, 30, 31]

K_m-Value (mM)

0.0013 <4> (malonyl-CoA, <4> R606K mutant [32]) [32]
0.0018 <4> (acetyl-CoA, <4> R606A mutant [32]) [32]
0.0019 <4> (malonyl-CoA, <4> wild-type enzyme [32]) [32]
0.0029 <4> (acetyl-CoA, <4> R606K mutant [32]) [32]
0.0039 <4> (acetyl-CoA, <4> wild-type enzyme [32]) [32]
0.0041 <2> (NADPH) [12]
0.0043 <2> (acetyl-CoA) [22]
0.005 <2> (malonyl-CoA) [22]
0.00796 <2> (NADPH) [22]
0.008 <8> (acetyl-CoA) [18]
0.0162 <4> (malonyl-CoA, <4> R606A mutant [32]) [32]
0.02 <8> (malonyl-CoA) [18]
0.025 <8> (NADPH) [18]
0.096 <4> (malonyl-CoA) [21]
0.1 <4> (malonyl-CoA, <4> alcohol-fed animals [21]) [21]
Additional information <1> (<1> K_m-values for various partial activities of fatty acid synthetase [8,9,12]) [8, 9, 12]

K_i-Value (mM)

0.02 <4> (CoA) [9]
0.14 <2> (sodium dodecylsulfate) [22]
1.04 <4> (malonyl pantetheine) [9]
210 <2> (urea, <2> substrate NADPH [28]) [28]
570 <2> (urea, <2> substrate malonyl-CoA [28]) [28]
580 <2> (urea, <2> substrate acetyl-CoA [28]) [28]

pH-Optimum

6.5-6.7 <8> [18]
6.8 <1, 2, 4> (<1,2,4> assay at [8]) [8]

Temperature optimum (°C)

25 <1, 2, 4> (<1,2,4> assay at [3,8,10,12]) [3, 8, 10, 12]
30 <1, 2, 4> (<1,2,4> assay at [8]) [8]

4 Enzyme Structure

Molecular weight

236000 <8> (<8> SDS-PAGE [25]) [25]
265000 <8> (<8> SDS-PAGE [18]) [18]
270000 <2> (<2> subunit, SDS-PAGE [26]) [26]

272000 <8> (<8> SDS-PAGE [17]) [17]
272500 <8> (<8> calculated from nucleotide sequence [18]) [18]
310000 <4> (<4> SDS-PAGE [2]) [2]
324900 <11> (<11> calculated from nucleotide sequence [24]) [24]
400000 <2> (<2> value above 400000 Da, polymer, SDS-PAGE [26]) [26]
425000 <4> (<4> sedimentation equilibrium method [3]) [3]
434000 <4> (<4> ultracentrifugation in 0.5-0.8 M phosphate buffer [3]) [3]
450000 <1> [8]
480000 <2> (<2> SDS-PAGE [10]) [10]
Additional information <1, 2, 4, 7> (<1,2,6> molecular weight of fragments after proteolytic cleavage [1,4,12,13,15]; <4> sedimentation coefficients of centrifugation of active enzyme [3]; <4> amino acid composition [3]; <7> active-site mapping of isolated acetyl/malonyl transferase activity domain [11]) [1, 3, 11, 12, 13]

Subunits

dimer <2, 4, 6> (<4> 2 identical subunits, SDS-PAGE [2]; <2> 2 * 220000, SDS-PAGE [10]; <1> 2 * 240000, SDS-PAGE [6]; <4> 2 * 244000, ultracentrifugation in 6 M guanidinium chloride [3]; <6> 2 * 250000, SDS-PAGE [4]) [2-4, 10]
hexamer <11> (<11> homohexamer [24]) [24]
Additional information <1-6, 8-10, 11> (<6> α_2-dimer of multifunctional subunits: each subunit contains up to seven active centres required for palmitate synthesis [4,7]; <2> α_2-dimer of multifunctional subunits: each subunit contains up to seven active centres required for palmitate synthesis and an acyl carrier site, arranged in three domains [13]; <4> α_2-dimer of multifunctional subunits: each subunit contains up to seven active centres required for palmitate synthesis and an acyl carrier site, arranged in three domains but dimer is the only catalytically active form of fatty acid synthase [3,7]; <2> condensing enzyme activity requires both subunits [7,10]; <4> fatty acid synthetases of vertebrates and yeast are stable enzyme complexes of multifunctional polypeptide chains, the fatty acid synthetases of plants and E. coli consist of non-associated individual enzymes [14]; <11> identical subunits, SDS-PAGE [23]) [3, 4, 7, 10, 13, 14, 23]

5 Isolation/Preparation/Mutation/Application

Source/tissue

adipose tissue <4, 5> (<4> epididymal fat pads [3]; <5> brown adipose tissue [30]) [3, 7, 30]
brain <8> [17, 18]
erythrocyte <4> [2]
liver <1-6, 8-10> (<4> cultured hepatocytes [16]; <8> hepatoma cell line HepG2 [18]) [1, 3, 5-10, 12-14, 16, 17, 18, 21, 26, 28, 29]
lung <8> (<8> fetal epithelial cells [25]) [18, 25]

mammary gland <4, 6, 7> (<6> lactating [4,7]) [4, 7, 8, 11]
uropygial gland <3> [7]

Localization
cytosol <4> [21]
microsome <4> [9, 29]
soluble <1, 2, 4> [8]

Purification
<1> (partial [1]; homogeneity [8]) [1, 8]
<2> (homogeneity [8]; 2 β-ketoacyl reductase containing fragments after sequential proteolysis [12]; domain I with acetyl and malonyl transacetylase activity and β-hydroxyacyl dehydratase activity after expression in E. coli [15]) [10, 12, 13, 15, 28]
<4> (680fold, erythrocyte enzyme immunologically not related to liver enzyme [2]; homogeneity [8]; isolation of modified dimers containing independently mutated subunits [19]; 95% pure, transacyclase domain [32]) [2, 3, 8, 9, 19, 27, 32]
<6> [4]
<7> [11]
<8> (fusion protein with maltose-binding protein [17]; near homogeneity [18]) [17, 18]
<11> (purification of 2 structurally related but functionally differentiated fatty acid synthases: FAS-A and FAS-B, homogeneity [23]; FAS-A and FAS-B [24]) [23, 24]

Cloning
<2> (domain I with acetyl and malonyl transacetylase activity and β-hydroxyacyl dehydratase activity [15]) [15]
<3> (gene sequences [7]) [7]
<4> (transacyclase domain [32]) [27, 31, 32]
<8> (full length cDNA, cDNA encoding domain I and cDNA encoding domains II and III [17]) [17]
<11> [23, 24]

Engineering
C161A <4> (<4> no β-ketoacyl synthase activity [31]) [31]
C161T <4> (<4> defective in β-ketoacyl synthase, no overall fatty acid synthase activity [27]) [27]
K326A <4> (<4> defective in β-ketoacyl synthase, no overall fatty acid synthase activity [27]) [27]
K326A <4> (<4> no β-ketoacyl synthase activity [31]) [31]
R606A <4> (<4> reduced malonyl/acetyltransferase activity, increased transacylase activity, 16000fold increased selectivity for acetyl-CoA, 8.5fold increase of K_m for malonyl-CoA [32]) [32]
R606K <4> (<4> reduced malonyl/acetyltransferase activity, increased transacylase activity, 16fold increased selectivity for acetyl-CoA [32]) [32]
S215A <4> (<4> defective in acyl carrier protein [27]) [27]
S518A <4> (<4> no malonyl/acetyltransferase activity [31]) [31]

6 Stability

pH-Stability

8.4 <1> (<1> inactivation, decreased by 0.2 M KCl or 0.02 mM NADP(H) [8]) [8]

Temperature stability

25 <1> (<1> 50% loss of activity in dilute, mildly alkaline solution within 3.5 h, high ionic strength together with DTT reactivates [8]) [8]

Organic solvent stability

acetonitrile <6> (<6> up to 10% v/v, stable to [4]) [4]

General stability information

<1>, dithiothreitol stabilizes [8]
<1>, foaming leads to surface inactivation [8]
<1>, low ionic strength leads to dissociation of native enzyme and inactivation [8]
<4>, phosphate buffer, 0.5 M, reassociates enzyme subunits [3]
<1-6, 8-10>, dissociation of native enzyme leads to loss of activity [7]

Storage stability

<1>, -20°C, at least a month in the presence of DTT [8]
<1>, -20°C, under nitrogen at least 1-2 months [8]
<1>, 4°C, 90% loss of activity in 8 days with 40% dissociation of native enzyme, incubation at 38°C with 10 mM DTT restores activity [8]
<1-6, 8-10>, 0°C, inactivation after 12 h, reactivation after 2 h at 25°C in the presence of NADPH, not acetyl-CoA or NADH [7]

References

[1] Puri, R.N.; Porter, J.W.: Isolation and partial purification of elastase-released peptide domains that contain the partial activities of pigeon liver fatty acid synthetase. Biochem. Biophys. Res. Commun., **107**, 1212-1218 (1982)
[2] Jenik, R.A.; Porter, J.W.: Red blood cell fatty acid synthetase. Nonidentity with the enzyme from liver. Int. J. Biochem., **10**, 609-617 (1979)
[3] Stoops, J.K.; Ross, P.; Arslanian, M.J.; Aune, K.C.; Wakil, S.J.: Physicochemical studies of the rat liver and adipose fatty acid synthetases. J. Biol. Chem., **254**, 7418-7426 (1979)
[4] McCarthy, A.D.; Hardie, D.G.: The multifunctional polypeptide chains of rabbit-mammary fatty-acid synthase. Stoichiometry of active sites and active-site mapping using limited proteolysis. Eur. J. Biochem., **130**, 185-193 (1983)
[5] Katiyar, S.S.; Pan, D.; Porter, J.W.: Role of cysteine and 4-phosphopantetheine in the inactivation of pigeon liver fatty acid synthetase by S-(4-

bromo-2,3-dioxobutyl)-coenzyme A. Biochem. Biophys. Res. Commun., **104**, 517-522 (1982)
[6] Katiyar, S.S.; Pan, D.; Porter, J.W.: Inactivation of 3-oxoacyl synthetase activity of pigeon liver fatty acid synthetase by S-(4-bromo-2,3-dioxobutyl)-coenzyme A. Eur. J. Biochem., **130**, 177-184 (1983)
[7] Wakil, S.J.; Stoops, J.K.; Joshi, V.C.: Fatty acid synthesis and its regulation. Annu. Rev. Biochem., **52**, 537-579 (1983)
[8] Muesing, R.A.; Porter, J.W.: Fatty acid synthase from pigeon liver. Methods Enzymol., **35**, 45-49 (1975)
[9] Podack, E.R.; Saathoff, G.; Seubert, W.: On the mechanism and control of the malonyl-CoA-dependent chain elongation of fatty acids. The malonyl-transfer reaction. Eur. J. Biochem., **50**, 237-243 (1974)
[10] Stoops, J.K.; Wakil, S.J.: Animal fatty acid synthetase. A novel arrangement of the β-ketoacyl synthetase sites comprising domains of the two subunits. J. Biol. Chem., **256**, 5128-5133 (1981)
[11] Mikkelsen, J.; Hoejrup, P.; Hansen, H.F.; Hansen, J.K.; Knudsen, J.: Evidence that the medium-chain acyltransferase of lactating-goat mammary-gland fatty acid synthetase is identical with the acetyl/malonyltransferase. Biochem. J., **227**, 981-985 (1985)
[12] Wong, H.; Mattick, J.S.; Wakil, S.J.: The architecture of the animal fatty acid synthetase. III. Isolation and characterization of β-ketoacyl reductase. J. Biol. Chem., **258**, 15305-15311 (1983)
[13] Tsukamoto, Y.; Wong, H.; Mattick, J.S.; Wakil, S.J.: The architecture of the animal fatty acid synthetase complex. IV. Mapping of active centers and model for the mechanism of action. J. Biol. Chem., **258**, 15312-15322 (1983)
[14] Vagelos, R.P.: Acyl group transfer (acyl carrier protein). The Enzymes, 3rd Ed. (Boyer, P.D., ed.), **8**, 155-199 (1973)
[15] Chirala, S.S.; Huang, W.Y.; Jayakumar, A.; Sakai, K.; Wakil, S.J.: Animal fatty acid synthase: Functional mapping and cloning and expression of the domain I constituent activities. Proc. Natl. Acad. Sci. USA, **94**, 5588-5593 (1997)
[16] Foretz, M.; Foufelle, F.; Ferre, P.: Polyunsaturated fatty acids inhibit fatty acid synthase and spot-14-protein gene expression in cultured rat hepatocytes by a peroxidative mechanism. Biochem. J., **341**, 371-376 (1999)
[17] Jayakumar, A.; Huang, W.Y.; Raetz, B.; Chirala, S.S.; Wakil, S.J.: Cloning and expression of the multifunctional human fatty acid synthase and its subdomains in Escherichia coli. Proc. Natl. Acad. Sci. USA, **93**, 14509-14514 (1996)
[18] Jayakumar, A.; Tai, M.H.; Huang, W.Y.; Al-Feel, W.; Hsu, M.; Abu-Elheiga, L.; Chirala, S.S.; Wakil, S.J.: Human fatty acid synthase: properties and molecular cloning. Proc. Natl. Acad. Sci. USA, **92**, 8695-8699 (1995)
[19] Joshi, A.K.; Rangan, V.S.; Smith, S.: Differential affinity labelling of the two subunits of the homodimeric animal fatty acid synthase allows isolation of heterodimers consisting of subunits that have been independently modified. J. Biol. Chem., **273**, 4937-4943 (1998)

[20] Park, Y.D.; Wu, B.N.; Tian, W.X.; Zhou, H.M.: Effects of osmolytes on unfolding of chicken liver fatty acid synthase. Biochemistry (Moscow), **67**, 914-917 (2002)
[21] Simpson, K.J.; Venkatesan, S.; Peters, T.J.: Fatty acid synthesis by rat liver after chronic ethanol feeding with a low-fat diet. Clin. Sci., **87**, 441-446 (1994)
[22] Shi, Y.; Luo, W.; Tian, W.X.; Zhang, T.; Zhou, H.M.: Inactivation and conformational changes of fatty acid synthase from chicken liver during unfolding by sodium dodecyl sulfate. Int. J. Biochem. Cell Biol., **30**, 1319-1330 (1998)
[23] Stuible, H.P.; Meurer, G.; Schweizer, E.: Heterologous expression and biochemical characterization of two functionally different type I fatty acid synthases from Brevibacterium ammoniagenes. FEBS Lett., **247**, 268-273 (1997)
[24] Stuible, H.P.; Wagner, C.; Andreou, J.; Guter, G,; Haselmann, J.; Schweizer, E.: Identification and functional differentiation of two type I fatty acid synthases in Brevibacterium ammoniagenes. J. Bacteriol., **178**, 4787-4793 (1996)
[25] Wagle, S.; Bui, A.; Ballard, P.L.; Shuman, H.; Gonzales, J.; Gonzales, L.W.: Hormonal regulation and cellular localization of fatty acid synthase in human fetal lung. Am. J. Physiol., **277**, L381-L390 (1999)
[26] Wang, F.; Wang, X.; Liu, Y.; Tian, W.X.; Zhou, H.M.: Inhibitive effect of zinc ion on fatty acid synthase from chicken liver. Int. J. Biochem. Cell Biol., **35**, 391-400 (2003)
[27] Witkowski, A.; Joshi, S.; Smith, S.: Fatty acid synthase: in vitro complementation of inactive mutants. Biochemistry, **35**, 10569-10575 (1996)
[28] Wu, B.N.; Park, Y.D.; Tian, W.X.; Zhou, H.M.: Unfolding and inactivation of fatty acid synthase from chicken liver during urea denaturation. Biochim. Biophys. Acta, **1549**, 112-121 (2001)
[29] Ide, T.; Ashakumary, L.; Takahashi, Y.; Kushiro, M.; Fukuda, N.; Sugano, M.: Sesamin, a sesame lignan, decreases fatty acid synthesis in rat liver accompanying the down-regulation of sterol regulatory element binding protein-1. Biochim. Biophys. Acta, **1534**, 1-13 (2001)
[30] Yu, X.X.; Lewin, D.A.; Forrest, W.; Adams, S.H.: Cold elicits the simultaneous induction of fatty acid synthesis and β-oxidation in murine brown adipose tissue: prediction from differential gene expression and confirmation in vivo. FASEB J., **16**, 155-168 (2002)
[31] Rangan, V.S.; Joshi, A.K.; Smith, S.: Fatty acid synthase dimers containing catalytically active β-ketoacyl synthase or malonyl/acetyltransferase domains in only one subunit can support fatty acid synthesis at the acyl carrier protein domains of both subunits. J. Biol. Chem., **273**, 34949-34953 (1998)
[32] Rangan, V.S.; Smith, S.: Alteration of the substrate specificity of the malonyl-CoA/acetyl-CoA:acyl carrier protein S-acyltransferase domain of the multifunctional fatty acid synthase by mutation of a single arginine residue. J. Biol. Chem., **272**, 11975-11978 (1997)

Fatty-acyl-CoA synthase 2.3.1.86

1 Nomenclature

EC number
2.3.1.86

Systematic name
acyl-CoA:malonyl-CoA C-acyltransferase (decarboxylating, oxoacyl- and enoyl-reducing)

Recommended name
fatty-acyl-CoA synthase

Synonyms
yeast fatty acid synthase
Additional information (the yeast enzyme is a multi-functional protein catalyzing the reactions of EC 2.3.1.38, EC 2.3.1.39, EC 2.3.1.41, EC 1.1.1.100, EC 4.2.1.61, EC 1.3.1.9)

CAS registry number
9045-77-6

2 Source Organism

<1> *Saccharomyces cerevisiae* (strain X-2180-A [1]; baker's yeast [2,6]; wild-type haploid strain X2180-1A and 52 fas-mutant strains [1]; mutants lacking endogenous de novo fatty acid synthesis [15]; brewer's yeast [3]; strain Fleishmann, wild type and protease-negative pep4-mutant [8]) [1, 2, 3, 4, 5, 6, 7-11, 13, 15, 19]
<2> *Brevibacterium ammoniagenes* [20, 21]
<3> *Escherichia coli* [13]
<4> *Mycobacterium smegmatis* [7]
<5> *Mycobacterium phlei* [12, 13]
<6> *Photobacterium profundum* (strain SS9 [14]) [14]
<7> *Euglena gracilis* (var. bacillaris [16]) [16, 13]
<8> *Pisum sativum* (pea [17,18]) [17, 18]
<9> *Spinacia oleracea* (spinach [18]) [18]

3 Reaction and Specificity

Catalyzed reaction

acetyl-CoA + n malonyl-CoA + 2n NADH + 2n NADPH + 4n H^+ = long-chain-acyl-CoA + n CoA + n CO_2 + 2n NAD^+ + 2n $NADP^+$ (<1,4> reaction mechanism [6,7,9]; <1> structure [7-10]; <1> structure and regulation [7])

Reaction type

acyl group transfer
decarboxylation
redox reaction

Natural substrates and products

S acetyl-CoA + malonyl-CoA + NADH + NADPH <1, 4> (<1,4> multifunctional enzyme involved in yeast fat metabolism [1-8]) (Reversibility: ? <1, 4> [1-8]) [1-8]

P ?

Substrates and products

S acetyl-CoA + malonyl-CoA + NADH + NADPH <1, 4, 8, 9, 2> (<1> S-acetylpantetheine and S-malonylpantetheine and saturated acetyl-CoA derivatives can replace acetyl-CoA and malonyl-CoA [6]; <1> mutants require acyl-CoA primers of 10 or more carbon atoms, maximal activity with 12-14 carbon atoms [15]; <5> primers instead of acetyl-CoA: propionyl-CoA, butyryl-CoA or hexanoyl-CoA [12]; <1> stable multifunctional enzyme complex: carries acetyl-CoA and malonyl-CoA transacylase, β-ketoacyl reductase, β-hydroxyacyl dehydrase activities on β-subunits and condensing enzyme, i.e. β-ketoacyl synthetase, enoylacyl reductase activities and acyl-carrier-protein components on α-subunits [7]; <1> intermediates are never released into the medium [10]) (Reversibility: ? <1, 2, 4, 8, 9> [1-7, 10, 15, 18, 20, 21]) [1-7, 10, 15, 18, 20, 21]

P palmitoyl-CoA + CoA + CO_2 + NAD^+ + $NADP^+$ <1, 2> (<1> palmitoyl-CoA and steraoyl-CoA are main products, myristoyl-CoA is produced in small amounts [6]; <2> FAS-A mainly synthesizes the C_{18} fatty acids oleate and stearate with only traces of palmitate, the major product of FAS-B is pamitate [20]; <2> synthesis of saturated and unsaturated fatty acids, FAS-B cannot synthesize oleic acid [21]) [6, 20, 21]

S Additional information <1, 3> (<1,3> fatty acid synthetases of vertebrates and yeast are stable enzyme complexes of multifunctional polypeptide chains, the fatty acid synthetases of plants and E. coli consist of non-associated individual enzymes [13]) [13]

P ?

Inhibitors

1,3-dibromo-2-propanone <1> (<1> complete inhibition at 0.005 mM after 1 min, cross-links α-, not β-subunits, inhibits only β-ketoacyl synthetase reaction, acetyl-CoA prevents, malonyl-CoA prevents only slightly [9]) [7, 9]

5,5'-dithio-bis(2-nitrobenzoic acid) <1> (<1> covalent binding to palmitoyl residues, malonyl-CoA protects [10]) [10, 11]
N-ethylmaleimide <1, 7> (<1> no inhibition of acetyl transferase activity [2,6]; <1> pH-dependent [6]; <7> complete inhibition of type II fatty acid synthase at 1 mM [16]) [2, 6, 16]
cerulenin <7> (<7> complete inhibition of type II fatty acid synthase at 0.1 mM [16]) [16]
iodoacetamide <1, 7> (<1> β-ketoacyl synthetase [2]; <1> no inhibition of acetyl transferase activity [2,6]; <1> pH-independent between 5.0 and 9.0 [6]; <1> irreversible [9]; <7> complete inhibition of type II fatty acid synthase at 1 mM [16]) [2, 6, 9, 16]
methylamine tungstad <1> (<1> inactivation within 24 h [8]) [8]
p-chloromercuribenzoate <7> (<7> complete inhibition of type II fatty acid synthase at 1 mM [16]) [16]
thiolactomycin <8> (<8> antibiotic and it's analogues, 50% inhibition of the overall activity at 0.17 mM, effect of antibiotic on compounds of the enzyme [17]) [17]

Cofactors/prosthetic groups

4'-phosphopantetheine <1> (<1> requirement, pantetheinate-free mutants have no β-ketoacyl synthetase activity [1]; <1> 4.0-5.0 mol per mol enzyme [7]; <1> 4.0-5.0 mol per mol enzyme or about 1 mol per 2 subunits [7,8]; <1> α-subunit bears prosthetic group [7]) [1, 7, 8]
FMN <1> (<1> 4 mol FMN per mol of enzyme complex [6]; <1> requirement, enoyl reductase activity, associated with β-subunit, 6 mol per mol synthetase [7]) [6, 7]
NADH <1, 5> (<1> requirement, 25% as efficient as NADPH [6]) [6, 12]
NADPH <1, 4, 5> (<1,4,5> requirement [1-8,12]) [1-8, 12, 13, 15]

Activating compounds

3-O-methylmannose <5> (<5> 0.1 mM, lowers K_m 4fold, activation by relieving product inhibition by binding long-chain acyl-CoA [12]) [12]
6-O-methylglucose <5> (<5> 0.1 mM, lowers K_m 4fold, activation by relieving product inhibition by binding long-chain acyl-CoA [12]) [12]
ATP <1> (<1> stimulatory at suboptimal but not at saturating substrate concentrations [15]) [15]
Triton X-100 <8, 9> (<8,9> leads to increasing turnover rate of acyl-CoA [18]) [18]
cysteine <1> (<1> activation [6]; <1> not [5]) [6]
glutathione <1> (<1> activation [6]) [6]
Additional information <1> (<1> no activation by DTT [5]) [5]

Specific activity (U/mg)

0.062 <1> (<1> substrates: acetoacetyl-cysteamine, malonyl-CoA and NADPH [5]) [5]
0.31 <2> (<2> FAS-A [20]) [20]
0.37 <2> (<2> FAS-B [20]) [20]
0.375 <5> [12]

0.53 <1> [1]
0.8-0.83 <1> (<1> malonyl-CoA [5]) [5]
1.25-1.75 <1> (<1> substrate: C_2-unit [3]) [3]
1.5-2.5 <1> [6]
1.5-4 <1> [8]
1.6 <1> (<1> substrates: acetoacetyl-CoA, malonyl-CoA and NADPH H [5]) [5]
2.5-3.5 <1> (<1> NADPH [3]) [3]
3 <1> [4, 9]
Additional information <1> (<1> data of partial reactions of fatty acid synthetase, different fas-mutants of Saccharomyces cerevisiae [1]; <1> comparison of partial activities of α- and β-subunits with that of the native complex [4]; <1> specific activities of components [5]) [1, 4, 5]

K_m-Value (mM)

0.0096 <5> (malonyl-CoA, <5> in presence of polysaccharide [12]) [12]
0.04 <5> (malonyl-CoA) [12]
0.05 <1> (lauroyl-CoA) [15]
0.067 <1> (NADPH) [6]
0.09 <5> (acetyl-CoA, <5> in presence of polysaccharide [12]) [12]
0.13 <1> (palmitoyl-CoA) [15]
0.33 <1> (octanoyl-CoA) [15]
0.4 <1> (myristoyl-CoA) [15]
0.8 <5> (acetyl-CoA) [12]
0.83 <1> (decanoyl-CoA) [15]
Additional information <1> (<1> specific activities for acetyl-CoA derivatives [6]; <1> kinetics of transacylase activity [10]) [6, 10]

pH-Optimum

6.5-7 <1> [6]

Temperature optimum (°C)

22 <1> (<1> β-ketoacyl synthetase activity, assay at [1]) [1]
25 <1> (<1> assay at [4,9,10]) [4, 9, 10]
37 <1> (<1> assay at [5,6]) [5, 6]

4 Enzyme Structure

Molecular weight

324900 <2> (<2> calculated from nucleotide sequence [21]) [21]
Additional information <1, 4> (<1,7> MW of components [1,9,16]; <1> $\alpha_6\beta_6$-complex of multifunctional subunits [4,7,8]; <1> amino acid composition [3,5,8]; <1> amino acid sequence of acyl-carrier-protein [7]) [1, 3-5, 7, 8, 9]
1390000 <5> (<5> sedimentation velocity data [12]) [12]
2300000 <1> (<1> sedimentation equilibrium [6]; <1> analytical ultracentrifugation [3]) [3, 6]
2370000 <1> (<1> sedimentation equilibrium method [8]) [8]
2400000 <1> (<1> sedimentation velocity [4,7,8]) [4, 7, 8]

Subunits

dodecamer <1> (<1> α_6,β_6, 6 * 185000 + 6 * 180000 , SDS-PAGE [7]; <1> α_6,β_6, 6 * 213000 + 6 * 203000, Tris-glycine-SDS-PAGE [4,7,8]; <1> active enzyme centrifugation [8]) [4, 7, 8]
hexamer <2> (<2> homohexamer [21]) [21]
Additional information <1, 7, 2, 3> (<1> multifunctional subunits [9]; <1,3> fatty acid synthetases of vertebrates and yeast are stable enzyme complexes of multifunctional polypeptide chains, the fatty acid synthetases of plants and E. coli consist of non-associated individual enzymes [13]; <7> two de novo fatty acid synthases, a true multienzyme complex in the cytosol and a plastid-localized type II fatty acid synthase composed of discrete enzymes and acyl carrier protein [16]; <2> identical subunits, SDS-PAGE [20]) [3, 4, 7-9, 13, 16, 20]

5 Isolation/Preparation/Mutation/Application

Source/tissue

leaf <8> [17]

Localization

chloroplast <8, 9> [17, 18]
soluble <3> [13]

Purification

<1> (mutants [1]; 900fold [5]; isolation of α- and β-subunits by acylation [3]; isolation of α- and β-subunits [4,7]; isolation of α- and β-subunits by citraconic and dimethylmaleic anhydride modification [7]; isolation of α- and β-subunits by citraconic and dimethylmaleic anhydride modification or as 3,4,5,6-tetrahydrophthalate derivatives [4]; isolation of peptides after proteolysis [8]; isolation and sequencing of active-site peptides with transacylase activity [11]) [1-8, 10, 11, 19]
<2> (purification of 2 structurally related but functionally differentiated fatty acid synthases: FAS-A and FAS-B, homogeneity [20]; FAS-A and FAS-B [21]) [20, 21]
<3> (purification of acyl carrier protein [13]) [13]
<5> (homogeneity [12]) [12]
<7> (partial purification of components [16]) [16]

Renaturation

<1> (hydrolysis under mild acidic condition leads to unmodified subunits, which can be reconstituted to form a complex displaying about 60% of the original activity [3]) [3]
<5> (low concentration of phosphate buffer causes dissociation into inactive species, reaggregation and reactivation can be partially achieved by dialysis against 0.5 M phosphate buffer [12]) [12]

Cloning

<1> (transformation technique, plasmid YEpFAS2 transformed to and expressed in Escherichia coli maxi-cells [7]; mutants [19]) [7, 19, 15]
<2> [20, 21]
<6> [14]

Engineering

C1305A <1> (<1> peripheral SH-group defective [19]) [19]
S180G <1> (<1> central SH-group defective [19]) [19]
S5421Q <1> (<1> malonyl/palmitoyl transferase defective [19]) [19]
S819Q <1> (<1> acyltransferase defective [19]) [19]
T181G <1> (<1> central SH-group defective [19]) [19]

6 Stability

Oxidation stability

<1>, under 100 atm nitrogen, at 4°C 1-20 h stable, 40% loss of activity after 40 h [5]
<1>, under 100 atm oxygen, at 4°C 1-2 h stable, 48% and 90% loss of activity after 20 h and 40 h, respectively, DTT does not restore activity [5]

General stability information

<1>, 2-mercaptoethanol, inactivation during storage [6]
<1>, PMSF ensures isolation of native enzyme [8]
<1>, acylation of free protein amino groups leads to reversible dissociation into subunits [3]
<1>, low ionic strength leads to inactivation [3]
<5>, low ionic strength, 0.005 M, leads to dissociation into subunits, partially reactivated by increasing the ionic strength to 0.5 M [12]

Storage stability

<1>, -15°C, stable in 90% ammonium sulfate solution containing 0.1 M potassium phosphate, pH 6.5, and 1 mM DTT [3, 6]
<1>, -20°C, 1 month [8]
<1>, 4°C, precipitate in 50% ammonium sulfate solution, several days [6]

References

[1] Schweizer, E.; Kniep, B.; Castorph, H.; Holzner, U.: Pantetheine-free mutants of the yeast fatty-acid-synthetase complex. Eur. J. Biochem., **39**, 353-362 (1973)
[2] Ziegenhorn, J.; Niedermeier, R.; Nussler, C.; Lynen, F.: Study of the acetyltransferase component of fatty acid synthetase of yeast. Eur. J. Biochem., **30**, 285-300 (1972)

[3] Wieland, F.; Renner, L.; Verfurth, C.; Lynen, F.: Studies on the multi-enzyme complex of yeast fatty-acid synthetase. Reversible dissociation and isolation of two polypeptide chains. Eur. J. Biochem., **94**, 189-197 (1979)
[4] Stoops, J.K.; Wakil, S.J.: The isolation of the two subunits of yeast fatty acid synthetase. Biochem. Biophys. Res. Commun., **84**, 225-231 (1978)
[5] Yein, F.; Brown, O.R.: Comparative inactivation of yeast fatty acid synthetase component enzymes by 100 atmospheres of oxygen. Biochim. Biophys. Acta, **486**, 421-428 (1977)
[6] Lynen, F.: Yeast fatty acid synthase. Methods Enzymol., **14**, 17-33 (1969)
[7] Wakil, S.J.; Stoops, J.K.; Joshi, V.C.: Fatty acid synthesis and its regulation. Annu. Rev. Biochem., **52**, 537-579 (1983)
[8] Stoops, J.K.; Wakil, S.J.: Studies on the yeast fatty acid synthetase. Subunit composition and structural organization of a large multifunctional enzyme complex. J. Biol. Chem., **253**, 4464-4475 (1978)
[9] Stoops, J.K.; Wakil, S.J.: Yeast fatty acid synthetase: structure-function relationship and nature of the β-ketoacyl synthetase site. Proc. Natl. Acad. Sci. USA, **77**, 4544-4548 (1980)
[10] Engeser, H.; Hubner, K.; Straub, J.; Lynen, F.: Identity of malonyl and palmitoyl transferase of fatty acid synthetase from yeast. Functional interrelationships between the acyl transferases. Eur. J. Biochem., **101**, 407-412 (1979)
[11] Engeser, H.; Hubner, K.; Straub, J.; Lynen, F.: Identity of malonyl and palmitoyl transferase of fatty acid synthetase from yeast. 2. A comparison of active-site peptides. Eur. J. Biochem., **101**, 413-422 (1979)
[12] Vance, D.E.; Mitsuhashi, O.; Bloch, K.: Purification and properties of the fatty acid synthetase from Mycobacterium phlei. J. Biol. Chem., **248**, 2303-2309 (1973)
[13] Vagelos, R.P.: Acyl group transfer (acyl carrier protein). The Enzymes, 3rd Ed. (Boyer, P.D., ed.), **8**, 155-199 (1973)
[14] Allen, E.E.; Bartlett, D.H.: Structure and regulation of the omega-3 polyunsaturated fatty acid synthase genes from the deep-sea bacterium Photobacterium profundum strain SS9. Microbiology, **148**, 1903-1913 (2002)
[15] Dittrich, F.; Zajonc, D.; Huhne, K.; Hoja, U.; Ekici, A.; Greiner, E.; Klein, H.; Hofmann, J.; Bessoule, J.J.; Sperling, P.; Schweizer, E.: Fatty acid elongation in yeast. Biochemical characteristics of the enzyme system and isolation of elongation-defective mutants. Eur. J. Biochem., **252**, 477-485 (1998)
[16] Worsham, L.M.S.; Williams, S.G.; Ernst-Fonberg, M.L.: Early catalytic steps of Euglena gracilis chloroplast type II fatty acid synthase. Biochim. Biophys. Acta, **1170**, 62-71 (1993)
[17] Jones, L.A.; Herbert, D.; Rutter, A.J.; Dancer, J.E.; Harwood, J.L.: Novel inhibitors of the condensing enzymes of the type II fatty acid synthase of pea (pisum sativum). Biochem. J., **347**, 205-209 (2000)
[18] Roughan, P.G.: Stromal concentrations of coenzyme A and its esters are insufficient to account for rates of chloroplast fatty acid synthesis: evidence for substrate channeling within the chloroplast fatty acid synthase. Biochem. J., **327**, 267-273 (1997)

[19] Schuster, H.; Rautenstrauss, B.; Mittag, M.; Stratmann, D.; Schweizer, E.: Substrate and product binding sites of yeast fatty acid synthase. Stoichiometry and binding kinetics of wild-type and in vitro mutated enzymes. Eur. J. Biochem., **228**, 417-424 (1995)
[20] Stuible, H.P.; Meurer, G.; Schweizer, E.: Heterologous expression and biochemical characterization of two functionally different type I fatty acid synthases from Brevibacterium ammoniagenes. FEBS Lett., **247**, 268-273 (1997)
[21] Stuible, H.P.; Wagner, C.; Andreou, J.; Guter, G,; Haselmann, J.; Schweizer, E.: Identification and functional differentiation of two type I fatty acid synthases in Brevibacterium ammoniagenes. J. Bacteriol., **178**, 4787-4793 (1996)

Aralkylamine N-acetyltransferase 2.3.1.87

1 Nomenclature

EC number
2.3.1.87

Systematic name
acetyl-CoA:2-arylethylamine N-acetyltransferase

Recommended name
aralkylamine N-acetyltransferase

Synonyms
AA-NAT
AANAT
N-acetyltransferase
acetyl-CoA:aralkylamine N-acetyltransferase
acetyltransferase, arylalkylamine N-
arylalkylamine N-acetyltransferase
serotonin acetylase
serotonin acetyltransferase
Additional information (cf. EC 2.3.1.5)

CAS registry number
92941-56-5

2 Source Organism

<1> *Bos taurus* [6]
<2> *Ovis aries* (glutathione S-transferase fusion protein, fused and cleaved form [20]) [1, 2, 4, 11, 15, 20, 22, 23, 24]
<3> *Rattus norvegicus* (regulation by circadian clock at posttranscriptional level [16]) [1, 2, 4, 6, 8-10, 16]
<4> *Gallus gallus* (embryo, light-dark regulation of enzyme [19]) [5, 7, 19]
<5> *Macrobrachium rosenbergii* (giant freshwater prawn [3]) [3]
<6> *Sparus aurata* (light regulation of circadian rhythm in melatonin secretion [12]) [12]
<7> *Esox lucius* (light regulation of circadian rhythm in melatonin secretion [12]) [12]
<8> *Oncorhynchus mykiss* (light regulation of circadian rhythm in melatonin secretion [12]) [12]
<9> *Macaca mulatta* [13]
<10> *Onchocerca volvulus* (human filarial parasite [14]) [14]

<11> *Drosophila melanogaster* (AANAT2 [17]) [17]
<12> *Rana perezi* (regulated by a daily photocycle [18]) [18]
<13> *Rattus norvegicus* [21]
<14> *Homo sapiens* (recombinant enzyme [24]) [24]
<15> *Periplaneta americana* (cockroach, two isoforms [25]) [25, 26]

3 Reaction and Specificity

Catalyzed reaction

acetyl-CoA + a 2 arylethylamine = CoA + an N-acetyl-2-arylethylamine (<11> mechanism [17]; <2> ordered bi-bi sequential mechanism [20]; <2> mechanism [23])

Reaction type

acyl group transfer

Natural substrates and products

S acetyl-CoA + serotonin <2-4> (<4> key enzyme of circadian rhythm of melatonin synthesis [5]; <2,3> initial reaction in melatonin synthesis from serotonin [1]) [1, 5]
P CoA + N-acetylserotonin

Substrates and products

S acetyl-CoA + 3-indolebutylamine <2> (<2> 60-fold less efficiently than serotonin [23]) (Reversibility: ? <2> [23]) [23]
P CoA + N-acetyl-(3-indol-3-yl-butyl)-amine
S acetyl-CoA + 3-indolepropylamine <2> (<2> 20-fold less efficiently than serotonin [23]) (Reversibility: ? <2> [23]) [23]
P CoA + N-acetyl-(3-indol-3-yl-propyl)-amine
S acetyl-CoA + 3-methoxytryptamine <3, 14> (Reversibility: ? <3, 14> [9, 24]) [9, 24]
P CoA + N-acetyl-3-methoxytryptamine
S acetyl-CoA + 5-hydroxytryptamine <10, 15> (Reversibility: ? <10, 15> [14, 26]) [14, 26]
P CoA + N-acetyl-5-hydroxytryptamine
S acetyl-CoA + 5-methoxytryptamine <2-4> (Reversibility: ? <2-4> [2, 5]) [2, 5]
P CoA + N-acetyl-5-methoxytryptamine
S acetyl-CoA + 6-fluorotryptamine <2, 3> (Reversibility: ? <2, 3> [2]) [2]
P CoA + N-acetyl-6-fluorotryptamine
S acetyl-CoA + N^{ω}-methyltryptamine <2> (<2> less efficiently than serotonin [23]) (Reversibility: ? <2> [23]) [23]
P CoA + ?
S acetyl-CoA + α-methyltryptamine <2> (<2> racemic, 9:1 stereoselectivity for R-enantiomer, less efficiently than serotonin [23]) (Reversibility: ? <2> [23]) [23]
P CoA + N-acetyl-α-methyltryptamine

S acetyl-CoA + β-phenylethylamine <2-4, 10, 14> (<3> also substrate: phenylethylamine derivatives without a β-hydroxy group [9]) (Reversibility: ? <2-4, 10, 14> [1, 2, 5, 9, 14, 24]) [1, 2, 5, 14, 24]
P CoA + N-(2-phenylethyl)-acetaminde
S acetyl-CoA + octopamine <10, 15> (Reversibility: ? <10, 15> [14, 26]) [14, 26]
P CoA + N-acetyloctopamine
S acetyl-CoA + *p*-phenetidine <12> (Reversibility: ? <12> [18]) [18]
P CoA + N-(4-ethoxyphenyl)-aecetamide
S acetyl-CoA + serotonin <2, 3, 14, 15> (i.e. 5-hydroxytryptamine) (Reversibility: ? <2, 3, 14, 15> [1, 2, 9, 24, 26]) [1, 2, 9, 24, 26]
P CoA + N-acetylserotonin <2, 3> [1]
S acetyl-CoA + tryptamine <1-8, 10, 11, 14, 15> (Reversibility: ? <1-8, 10, 11, 14, 15> [1-3, 5, 6, 9, 11, 12, 14, 17, 24, 26]) [1-3, 5, 6, 9, 11, 12, 14, 17, 24, 26]
P CoA + N-acetyltryptamine <2, 3> [1]
S acetyl-CoA + tryptophol <2> (<2> structural analogue to tryptamine [20]) (Reversibility: ? <2> [20]) [20]
P CoA + N-acetyltryptophol
S acetyl-CoA + tyramine <10, 14, 15> (<14> very poor substrate [24]) (Reversibility: ? <10, 14, 15> [14, 24, 26]) [14, 24, 26]
P CoA + N-acetyltyramine
S Additional information <2, 3, 14> (<2,3> arylamines, such as aniline or *p*-phenetidine are very poor substrates [1,2]; <3> no substrates are phenylethanolamine derivatives with a β-hydroxy group [9]; <14> further substrates: selected synthetic amines [24]) [1, 2, 9, 24]
P ?

Inhibitors

N-ethylmaleimide <13> (<13> irreversible, acetyl-CoA protects [21]) [21]
Zn^{2+} <2, 3> (<2,3> at low concentration [1]) [1]
α-trifluoromethyltryptamine <2> (<2> modest, competitive [23]) [23]
desulfo-CoA <2> (<2> dead end inhibitor analog, competitive versus CoA [20]) [20]
disulfides <2, 3> (<2,3> in vivo and in vitro, reversible by dithiothreitol [1]) [1]
glutathione <13> (<13> reversible, acetyl-CoA protects [21]) [21]
melatonin <14> (<14> IC50 0.16 mM [24]) [24]
oxygen <13> (<13> reversible, acetyl-CoA protects [21]) [21]
p-chloro-mercuribenzoate <10> [14]
peptides containing a disulfide bond <2, 3> [1]
serotonin <12> (<12> strong, mixed kinetics [18]) [18]
tryptophol <2> (<2> dead end inhibitor analog, competitive versus tryptamine [20]) [20]
Additional information <14> (<14> inhibition values of peptide inhibitors [24]) [24]

Metals, ions

Additional information <2, 3> (<2,3> activation by salts as a function of ionic strength [1]) [1]

Specific activity (U/mg)

0.000195 <2> [1]
Additional information <4> [5]

K_m-Value (mM)

0.0015 <10> (tyramine) [14]
0.002 <10> (octopamine) [14]
0.002 <10> (tryptamine) [14]
0.003 <12> (acetyl-CoA) [18]
0.009 <10> (β-phenylethylamine) [14]
0.011 <12> (*p*-phenetidine) [18]
0.024 <10> (5-hydroxytryptamine) [14]
0.0287 <1> (acetyl-CoA, <1> cosubstrate tryptamine [6]) [6]
0.0315 <1> (tryptamine) [6]
0.05 <3> (acetyl-CoA, <3> cosubstrate tryptamine, pineal gland) [10]
0.11 <2> (tryptamine) [24]
0.125 <3> (acetyl-CoA, cosubstrate tryptamine, liver) [10]
0.125 <2> (serotonin) [24]
0.173 <14> (β-phenylethylamine) [24]
0.18 <2> (acetyl-CoA) [24]
0.2 <14> (5-methoxytryptamine) [24]
0.24 <3> (tryptamine, <3> liver [10]) [10]
0.53 <3> (tryptamine, <3> pineal gland [10]) [10]
0.55 <14> (acetyl-CoA) [24]
0.6 <3> (tryptamine) [9]
0.91 <14> (tryptamine) [24]
1.23 <14> (serotonin) [24]
2 <3> (serotonin) [9]
Additional information <2, 3, 13> (<2,3> apparent K_m-values [2]; <13> formation and cleavage of a disulfide bond produce active/inactive states of enzyme [21]) [2, 21]

pH-Optimum

6 <15> [26]
6.5 <5> [3]
8.5 <10> [14]

Temperature optimum (°C)

25 <5> [3]

4 Enzyme Structure

Molecular weight

26000 <3> (<3> liver, gel filtration [10]) [10]
30000 <15> (<15> gel filtration [26]) [26]
30000 <3> (<3> gel filtration with sodium citrate [8]) [8]
37000 <10> (<10> gel filtration [14]) [14]
39000 <3> (<3> pineal gland, gel filtration [10]) [10]
Additional information <2, 3> (<2> multiple forms, from 10000 via 39000 to 100000 Da [1]; <3> two molecular forms: 10000 and 95000, HPLC size exclusion chromatography with ammonium acetate [8]) [1, 8]

Subunits

? <3, 7, 11> (<3> x * 11000, SDS-PAGE [8]; <7> x * 26000, SDS-PAGE [12]; <11> x * 29200, SDS-PAGE [17]) [8, 12, 17]
dimer <3> (<3> 2 * 12000, liver, gel filtration in the presence of cysteamine [10]) [10]
monomer <15> (<15> 1 * 28000, SDS-PAGE [25,26]) [25, 26]
tetramer <3> (<3> 4 * 10000, pineal gland, gel filtration in the presence of cysteamine [10]) [10]

5 Isolation/Preparation/Mutation/Application

Source/tissue

brain <3> (<3> suprachiasmatic nucleus [16]) [16]
kidney <4> [5]
liver <3> (<3> non-inducible [10]) [9, 10]
midgut <15> (<15> female [25]) [25]
optic lobe <5> [3]
pineal gland <1-4, 7, 9> (<3> inducible [10]; <9> 4fold increase at night [13]) [1, 2, 4-10, 12, 13]
pituitary gland <2> (<2> mainly pars tuberalis, much less than in pineal gland, probably different regulation [11]) [11]
retina <4, 9> (<9> 4-fold increase at night [13]; <4> cultured photoreceptor cells [19]) [13, 19]
testis <15> (testicular acessory gland) [25, 26]

Localization

cytosol <3> [10]
soluble <12> [18]

Purification

<2> (affinity chromatography on Sepharose CoA [1]) [1, 2, 4]
<3> (partial [2]) [1, 2, 8, 10]
<10> (partial [14]) [14]
<11> [17]
<15> [26]

Cloning

<11> (AANAT2 [17]) [17]

Engineering

C177A <13> (<13> fully active, not sensitive to oxidation or N-ethylmaleimide [21]) [21]
C61A <13> (<13> fully active, not sensitive to oxidation or N-ethylmaleimide [21]) [21]
H120Q <2> (<2> crystallographic studies, role in enzymic reaction [22]) [22]
H122Q <2> (<2> crystallographic studies, role in enzymic reaction [22]) [22]
H122Q/H120Q <2> (<2> crystallographic studies, role in enzymic reaction [22]) [22]
Y168F <2> (<2> crystallographic studies, role in enzymic reaction [22]) [22]

6 Stability

Temperature stability

25 <5> (<5> stable for 3 h [3]) [3]

General stability information

<3>, cysteamine stabilizes [10]
<2, 3>, ATP stabilizes [1]
<2, 3>, acetyl-CoA stabilizes [1, 2]
<2, 3>, freezing markedly decreases activity [1]
<2, 3>, polyanions stabilize [1]

Storage stability

<2>, 4°C, 25% loss of activity within 48 h [2]
<3>, 4°C, $t_{1/2}$: 24 h [2]
<2, 3>, 0-2°C, up to 72 h [1]

References

[1] Namboodiri, M.A.A.; Dubbels, R.; Klein, D.C.: Arylalkylamine N-acetyltransferase from mammalian pineal gland. Methods Enzymol., **142**, 583-590 (1987)
[2] Voisin, P.; Namboodiri, M.A.A.; Klein, D.C.: Arylamine N-acetyltransferase and arylalkylamine N-acetyltransferase in the mammalian pineal gland. J. Biol. Chem., **259**, 10913-10918 (1984)
[3] Withyachumnarnkul, B.; Pongsa-Asawapaiboon, A.; Poolsanguan, B.: Characteristics of the enzyme N-acetyltransferase in the optic lobe of the giant freshwater prawn, Macrobrrachium rosenbergii. Comp. Biochem. Physiol. B Comp. Biochem., **104**, 449-454 (1993)

[4] Namboodiri, M.A.A.; Brownstein, M.J.; Voisin, P.; Weller, J.L.; Klein, D.C.: A simple and rapid method for the purification of ovine pineal arylalkylamine N-acetyltransferase. J. Neurochem., **48**, 580-585 (1987)
[5] Ohtomi, M.; Sasaki, M.; Deguchi, T.: Two arylamine N-acetyltransferases from chicken pineal gland as identified by cDNA cloning. Eur. J. Biochem., **185**, 253-261 (1989)
[6] Fajardo, N.; Abreu, P.; Alonso, R.: Determination of kinetic properties of serotonin-N-acetyltransferase in bovine pineal gland using HPLC with fluorimetric detection. J. Pineal Res., **13**, 80-84 (1992)
[7] Rodriguez-Cabello, J.C.; Agapito, M.T.; Garcia-Herrero, I.; Recio, J.M.: Effects of EGTA and calmodulin, neutral thiol proteinases and protein kinase C inhibitors on loss of chicken pineal serotonin N-acetyltransferase activity. J. Comp. Physiol. B, **159**, 583-588 (1989)
[8] Namboodiri, M.A.A.; Brownstein, M.J.; Weller, J.L.; Voisin, P.; Klein, D.C.: Multiple forms of arylalkylamine N-acetyltransferases in the rat pineal gland: purification of one molecular form. J. Pineal Res., **4**, 235-246 (1987)
[9] Deguchi, T.: Characteristics of serotonin-acetyl coenzyme A N-acetyltransferase in pineal gland of rat. J. Neurochem., **24**, 1083-1085 (1975)
[10] Morrissey, J.J.; Edwards, S.B.; Lovenberg, W.: Comparison of rat pineal gland and rat liver serotonin-N-acetyltransferase. Biochem. Biophys. Res. Commun., **77**, 118-123 (1977)
[11] Fleming, J.V.; Barrett, P.; Coon, S.L.; Klein, D.C.; Morgan, P.J.: Ovine arylalkylamine N-acetyltransferase in the pineal and pituitary glands: differences in function and regulation. Endocrinology, **140**, 972-978 (1999)
[12] Falcon, J.; Galarneau, K.M.; Weller, J.L.; Ron, B.; Chen, G.; Coon, S.L.; Klein, D.C.: Regulation of arylalkylamine N-acetyltransferase-2 (AANAT2, EC 2.3.1.87) in the fish pineal organ: evidence for a role of proteasomal proteolysis. Endocrinology, **142**, 1804-1813 (2001)
[13] Coon, S.L.; Del Olmo, E.; Young, W.S.; Klein, D.C.: Melatonin synthesis enzymes in Macaca mulatta: focus on arylalkylamine N-acetyltransferase (EC 2.3.1.87). J. Clin. Endocrinol. Metab., **87**, 4699-4706 (2002)
[14] Aisien, S.O.; Hellmund, C.; Walter, R.D.: Characterization of the arylalkylamine N-acetyltransferase in Onchocerca volvulus. Parasitol. Res., **82**, 369-371 (1996)
[15] Obsil, T.; Ghirlando, R.; Klein, D.C.; Ganguly, S.; Dyda, F.: Crystal structure of the 14-3-3ζ:serotonin N-acetyltransferase complex: a role for scaffolding in enzyme regulation. Cell, **105**, 257-267 (2001)
[16] Hamada, T.; Ootomi, M.; Horikawa, K.; Niki, T.; Wakamatu, H.; Ishida, N.: The expression of the melatonin synthesis enzyme: arylalkylamine N-acetyltransferase in the suprachiasmatic nucleus of rat brain. Biochem. Biophys. Res. Commun., **258**, 772-777 (1999)
[17] Amherd, R.; Hintermann, E.; Walz, D.; Affolter, M.; Meyer, U.A.: Purification, cloning, and characterization of a second arylalkylamine N-acetyltransferase from Drosophila melanogaster. DNA Cell Biol., **19**, 697-705 (2000)
[18] Alonso-Gomez, A.L.; Valenciano, A.I.; Alonso-Bedate, M.; Delgado, M.J.: Differential characteristics and regulation of arylamine and arylalkylamine

N-acetyltransferases in the frog retina (Rana perezi). Neurochem. Int., **26**, 223-231 (1995)
[19] Ivanova, T.N.; Michael Iuvone, P.: Melatonin synthesis in retina: circadian regulation of arylalkylamine N-acetyltransferase activity in cultured photoreceptor cells of embryonic chicken retina. Brain Res., **973**, 56-63 (2003)
[20] De Angelis, J.; Gastel, J.; Klein, D.C.; Cole, P.A.: Kinetic analysis of the catalytic mechanism of serotonin N-acetyltransferase (EC 2.3.1.87). J. Biol. Chem., **273**, 3045-3050 (1998)
[21] Tsuboi, S.; Kotani, Y.; Ogawa, K.i.; Hatanaka, T.; Yatsushiro, S.; Otsuka, M.; Moriyama, Y.: An intramolecular disulfide bridge as a catalytic switch for serotonin N-acetyltransferase. J. Biol. Chem., **277**, 44229-44235 (2002)
[22] Scheibner, K.A.; De Angelis, J.; Burley, S.K.; Cole, P.A.: Investigation of the roles of catalytic residues in serotonin N-acetyltransferase. J. Biol. Chem., **277**, 18118-18126 (2002)
[23] Khalil, E.M.; De Angelis, J.; Cole, P.A.: Indoleamine analogs as probes of the substrate selectivity and catalytic mechanism of serotonin N-acetyltransferase. J. Biol. Chem., **273**, 30321-30327 (1998)
[24] Ferry, G.; Loynel, A.; Kucharczyk, N.; Bertin, S.; Rodriguez, M.; Delagrange, P.; Galizzi, J.P.; Jacoby, E.; Volland, J.P.; Lesieur, D.; Renard, P.; Canet, E.; Fauchere, J.L.; Boutin, J.A.: Substrate specificity and inhibition studies of human serotonin N-acetyltransferase. J. Biol. Chem., **275**, 8794-8805 (2000)
[25] Ichihara, N.; Okada, M.; Takeda, M.: Characterization and purification of polymorphic arylalkylamine N-acetyltransferase from the American cockroach, Periplaneta americana. Insect Biochem. Mol. Biol., **32**, 15-22 (2001)
[26] Ichihara, N.; Okada, M.; Nakagawa, H.; Takeda, M.: Purification and characterization of arylalkylamine N-acetyltransferase from cockroach testicular organs. Insect Biochem. Mol. Biol., **27**, 241-246 (1997)

Peptide α-N-acetyltransferase 2.3.1.88

1 Nomenclature

EC number
2.3.1.88

Systematic name
acetyl-CoA:peptide N^{α}-acetyltransferase

Recommended name
peptide α-N-acetyltransferase

Synonyms
NAT
N^{α}-acetyltransferase
Xat-1 <26> [17]
acetyltransferase, peptide N-terminal
amino-terminal amino acid-acetylating enzyme
arylamine N-acetyltransferase <13> [18]
β-endorphin acetyltransferase
peptide acetyltransferase
protein N-terminal acetyltransferase

CAS registry number
83452-29-3

2 Source Organism

<1> *Aeropyrum pernix* [16]
<2> *Arabidopsis thaliana* [16]
<3> *Arabidopsis thaliana* [17]
<4> *Bos taurus* (calf [1,7,8]; bovine, cow [2,4,8]) [1, 2, 4, 7, 8]
<5> *Caenorhabditis elegans* [16]
<6> *Caenorhabditis elegans* [17]
<7> *Drosophila melanogaster* [2, 16]
<8> *Drosophila melanogaster* [17]
<9> *Escherichia coli* [8]
<10> *Gallus gallus* (chicken, hen [1,2,7-11,13,14]) [1, 2, 7-11, 13, 14]
<11> *Homo sapiens* [17]
<12> *Homo sapiens* (human [17]) [17]
<13> *Homo sapiens* (human [16,18]) [16, 18]
<14> *Methanobacterium thermoautotrophicum* [16]

<15> *Mus musculus* (mouse, acetyltransferase Tubedown-1 [17]) [17]
<16> *Mus musculus* (mouse [16,17]) [16, 17]
<17> *Mus musculus* (mouse, L-cells, Ehrlich ascites cells [7,12]) [7, 12]
<18> *Neurospora crassa* [7]
<19> *Oryctolagus cuniculus* (rabbit [2,4,6,7]) [2, 4, 6, 7]
<20> *Pyrococcus abyssi* [16]
<21> *Rattus norvegicus* (rat [1-4,7,8,13,14]; Sprague-Dawley, albino [3]) [1-4, 7, 8, 13, 14]
<22> *Saccharomyces cerevisiae* [8, 12-18]
<23> *Schizosaccharomyces pombe* [17]
<24> *Sus scrofa* (pig [2]) [2]
<25> *Triticum aestivum* [5, 7, 8]
<26> *Xenopus laevis* [17]

3 Reaction and Specificity

Catalyzed reaction

acetyl-CoA + peptide = CoA + N^{α}-acetylpeptide (acetylates N-terminal alanine, serine, methionine and glutamate residues in a number of peptides and proteins, including β-endorphin, corticotropins and melanotropin. cf. EC 2.3.1.108 tubulin N-acetyltransferase)

Reaction type

acetylation
acyl group transfer

Natural substrates and products

S acetyl-CoA + peptide <1-26> (<21> physiological significance of β-endorphin acetyltransferase is to modulate the activity of endorphin secreted from opiomelanotropinergic cells and neurons [3]) (Reversibility: ? <1-26> [1-18]) [1-18]
P CoA + N^{α}-acetylpeptide

Substrates and products

S acetyl-CoA + ADH I-(1-24) <22> (Reversibility: ? <22> [12]) [12]
P CoA + ?
S acetyl-CoA + ATPase inhibitor (1-24)(yeast, mitochondrial) <22> (Reversibility: ? <22> [12]) [12]
P CoA + ?
S acetyl-CoA + [D-Ser1,Lys17,18]-adrenocorticotropic hormone 1-18-NH_2 <10> (<10> corticotropin, synthetic peptide [1]) (Reversibility: ? <10> [1]) [1]
P CoA + ?
S acetyl-CoA + [Gly1] adrenocorticotropic hormone 1-18-NH_2 <10> (<10> corticotropin, synthetic peptide [9]) (Reversibility: ? <10> [9]) [9]
P CoA + ?

S acetyl-CoA + [N^6-PTC-Lys11,15,16,21]-adrenocorticotropic hormone 2-24 <10> (<10> corticotropin, synthetic peptide [1]) (Reversibility: ? <10> [1]) [1]
P CoA + ?
S acetyl-CoA + [Phe2] ACTH(1-24) <22> (Reversibility: ? <22> [8]) [8]
P CoA + ?
S acetyl-CoA + acetyl-β-endorphin <21> (Reversibility: ? <21> [3]) [3]
P CoA + ?
S acetyl-CoA + adrenocorticotropic hormone 1-10 <21> (<21> adrenocorticotropic hormone, synthetic peptide [4]) (Reversibility: ? <21> [4]) [4]
P CoA + ?
S acetyl-CoA + adrenocorticotropic hormone 1-13-NH_2 <4> (<10> adrenocorticotropic hormone, synthetic peptide [2]) (Reversibility: ? <4> [2]) [2]
P CoA + αMSH
S acetyl-CoA + adrenocorticotropic hormone 1-18-NH_2 <10> (<10> corticotropin, synthetic peptide [9]) (Reversibility: ? <10> [9]) [9]
P CoA + ?
S acetyl-CoA + adrenocorticotropic hormone 1-24 <10, 21, 22> (<10,21,22> corticotropin, synthetic peptide [1,4,8,12]) (Reversibility: ? <10, 21, 22> [1, 4, 8, 12]) [1, 4, 8, 12]
P CoA + ?
S acetyl-CoA + adrenocorticotropic hormone 1-39 <4, 10, 21> (<10> corticotropin, synthetic peptide, porcine [1]) (Reversibility: ? <4, 10, 21> [1, 2, 4]) [1, 2, 4]
P CoA + ?
S acetyl-CoA + adrenocorticotropic hormone 1-8 <10, 21> (<10,21> ACTH, synthetic peptide [1,14]) (Reversibility: ? <10, 21> [1, 14]) [1, 4, 14]
P CoA + ?
S acetyl-CoA + adrenocorticotropic hormone 4-10 <21> (<21> adrenocorticotropic hormone, synthetic peptide [4]) (Reversibility: ? <21> [4]) [4]
P CoA + ?
S acetyl-CoA + adrenocorticotropic hormone 5-24 <10> (<10> corticotropin, synthetic peptide [1]) (Reversibility: ? <10> [1]) [1]
P CoA + ?
S acetyl-CoA + α-endorphin (1-16) <21> (Reversibility: ? <21> [3]) [3]
P CoA + ?
S acetyl-CoA + α-melanotropin (1-27) <4, 21> (<4,21> α-melanocyte-stimulating hormone, α-melanophore-stimulating hormone [2]) (Reversibility: ? <4, 21> [2, 4]) [2, 4]
P CoA + α-N,O-diacetyl-α-melanophore-stimulating hormone
S acetyl-CoA + β-endorphin (1-27) <4, 21> (Reversibility: ? <4, 21> [2-4]) [2-4]
P CoA + α-N-acetyl-β-endorphin
S acetyl-CoA + β-endorphin (1-31) <4> (Reversibility: ? <4> [2]) [2]
P CoA + α-N-acetyl-β-endorphin

S acetyl-CoA + camel β-endorphin (1-31) <21> (Reversibility: ? <21> [3]) [3]
P CoA + ?
S acetyl-CoA + cytochrome c oxidase (1-24)(yeast, mitochondrial subunit VI) <22> (Reversibility: ? <22> [12]) [12]
P CoA + ?
S acetyl-CoA + human β-endorphin (1-31) <21> (Reversibility: ? <21> [3]) [3]
P CoA + ?
S acetyl-CoA + human superoxide dismutase (1-24) <22> (Reversibility: ? <22> [8]) [8]
P CoA + ?
S acetyl-CoA + peptide <1-26> (Reversibility: ? <1-26> [1-18]) [1-18]
P CoA + N^{α}-acetylpeptide
S acetyl-CoA + yeast alcohol dehydrogenase I (1-24) <22> (Reversibility: ? <22> [8, 12]) [8, 12]
P CoA + ?
S acetyl-CoA + yeast alcohol dehydrogenase II (1-24) <22> (Reversibility: ? <22> [8]) [8]
P CoA + ?
S Additional information <4, 21, 22> (<4> adrenocorticotropic hormone (18-39), lysine-rich histones, adrenocorticotropic hormone (1-2)[Ser-Tyr], adrenocorticotropic hormone (11-24), α-N,O-diacetyl-αMSH and α-N-acetylated β-endorphin-related molecules are not acetylated [2]; <21> methionine enkephalin, leucine enkephalin, Des-Tyr-β-endorphin and Arg-β-endorphin are not substrates [3]; <21> enzyme will not use adrenocorticotropic hormone (2-10), adrenocorticotropic hormone (3-10) or adrenocorticotropic hormone (1-8) as substrates [4]; <22> does not catalyze the N^{α}-acetylation of adrenocorticotropic hormone (11-24), adrenocorticotropic hormone (7-38), adrenocorticotropic hormone (18-39), human β-endorphin, yeast superoxide dismutase(1-24) [8]) [2-4, 8]
P ?

Inhibitors

Ca^{2+} <10> [9, 13]
Cl^- <21> [13]
ClO_4^- <21> [13]
Cu^{2+} <22> [8, 13]
Fe^{2+} <10> [9, 13]
I^- <21> [13]
Mg^{2+} <10> [9, 13]
Mn^{2+} <10> [9, 13]
N-bromosuccinimide <22> [8, 13]
N-ethylmaleimide <10> [13]
NH_4^+ <21> (<21> partially inhibitory [13]) [13]
SCN^- <21> [13]

Ser-Tyr <4> (<4> inhibits NH_2-terminal acetylation of adrenocorticotropic hormone (1-13)NH_2 [2]) [2]
Zn^{2+} <10, 22> [8, 9, 13]
acetate <21> [13]
acetyl-β-endorphin(1-27) <4> (<4> competitive inhibition [2]) [2]
adrenocorticotropic hormone (11-24) <4> (inhibits NH_2-terminal acetylation of adrenocorticotropic hormone (1-13)NH_2 [2]) [2]
α-N,O-diacetyl-αMSH <4> (<4> inhibits NH_2-terminal acetylation of adrenocorticotropic hormone (1-13)NH_2 competitive [2]) [2]
diethyldicarbonate <22> [8]
iodoacetamide <10> [13]
iodoacetic acid <10, 21> [13]
p-chloromercuribenzoate <10> [13]
sulfate <21> [13]

Activating compounds
2-mercaptoethanol <10> [9]
Cl^- <25> [8]
cysteine <10> [9]
dithiothreitol <10> [9]
glutathione <10> [9]

Specific activity (U/mg)
0.000001 <21> (<21> serum [4]) [4]
0.0000102 <21> (<21> heart [4]) [4]
0.0000142 <21> (<21> kidney [4]) [4]
0.0000166 <21> (<21> liver [4]) [4]
0.0000169 <21> (<21> lens [4]) [4]
0.0000221 <21> (<21> lung [4]) [4]
0.000029 <21> (<21> muscle [4]) [4]
0.0000378 <21> (<21> brain [4]) [4]
0.0000383 <21> (<21> pituitary lobe, posterior-intermediate [4]) [4]
0.0000654 <21> (<21> pituitary lobe, anterior [4]) [4]
0.000093 <21> (<21> pituitary [4]) [4]
0.00027 <21> [4]

K_m-Value (mM)
0.0022 <21> (acetylCoA) [4]
0.0042 <21> (adrenocorticotropic hormone (1-24)) [4]
0.008 <21> (acetylCoA) [2]
0.011 <21> (adrenocorticotropic hormone (1-18)) [2]
0.017 <21> (adrenocorticotropic hormone (1-24)) [2]
0.024 <21> (adrenocorticotropic hormone (1-16)NH_2) [2]
0.035 <4> (adrenocorticotropic hormone (1-13)NH_2) [2]
0.037 <4> (adrenocorticotropic hormone (4-10)NH_2) [4]
0.043 <21> (β-endorphin(1-27)) [2]
0.05 <21> (adrenocorticotropic hormone (1-13)NH_2) [2]
0.05 <21> (adrenocorticotropic hormone (1-39)) [2]

0.065 <21> (β-endorphin(1-31)) [2]
0.096 <21> (adrenocorticotropic hormone (1-10)) [4]
0.16 <21> (adrenocorticotropic hormone (1-13)NH_2) [2]
0.2 <21> (adrenocorticotropic hormone (1-24)) [2]
0.276 <21> (adrenocorticotropic hormone (1-10)) [2]
0.286 <21> (adrenocorticotropic hormone (1-39)) [2]

K_i-Value (mM)
0.1 <21> (acetyl-β-endorphin(1-27)) [2]
0.106 <21> (adrenocorticotropic hormone (1-24)) [2]
0.113 <21> (adrenocorticotropic hormone (1-2)[Ser-Tyr]) [2]
0.253 <21> (α-N,O-diacetyl-αMSH) [2]

pH-Optimum
6.5-7.5 <25> [8]
6.8 <21> [3]
7 <4> [2]
7.2 <10> [1, 8]
7.8 <10> [9, 13]
8 <21> [13]
9 <22> [8, 13]

pH-Range
5-8 <21> [3]
5-11 <22> [8]
6-8.5 <21> [4]
6.2-9.8 <21> [13]

Temperature optimum (°C)
30-42 <22> [8]

Temperature range (°C)
5-55 <22> [8]

4 Enzyme Structure

Molecular weight
170000 <4> [8]
180000 <22> (<22> gel filtration [8,13]) [8, 13]
190000 <21> (<21> gel filtration [13]) [13]
200000 <21> [8]
240000 <10> (<10> gel filtration [9]) [9]
241000 <10> (<10> holoenzyme, calculated on the basis of the subunits [10]) [10]
250000 <10> (<10> gel filtration [1]) [1, 8, 13]

Subunits

dimer <21, 26> (<21> 2 * 95000, homodimer, SDS-PAGE [13]; <22> 2 * 95000, SDS-PAGE [8,13]; <22> 2 * 98575, cDNA sequence [13]; <26> 2 * 98800, SDS-PAGE [17]) [8, 13, 17]
trimer <10> (<10> 2 * 79000 + 1 * 83000, 2 protein subunits, 1 RNA subunit, heterotrimer, SDS-PAGE, formamide-PAGE [9,10,13]) [9, 10, 13]

5 Isolation/Preparation/Mutation/Application

Source/tissue

brain <12, 21> [3, 4, 8, 17]
breast <10> [11]
egg <26> [17]
embryo <3, 6, 8, 12, 16, 26> [2, 17]
germ <25> [7, 8]
heart <21> [4]
hypothalamus <21> [3]
kidney <21> [4]
lens <4, 21> [1, 4, 7, 8]
liver <21> [1, 2, 4, 7, 8, 13, 14]
lung <21> [4]
muscle <21> [4]
oviduct <10> [1, 2, 7-10, 13, 14]
pituitary gland <4, 21> (<21> intermediate lobe of pituitary gland, highest activity in neurointermediate lobe of pituitary, little activity in the anterior lobe of pituitary [2,3]) [1-4, 7, 8, 13]
pronephros <12> [17]
reticulocyte <19> [2, 4, 6, 7]
serum <21> [4]
somite <12> [17]

Localization

Golgi apparatus <4, 21> [2, 13]
cytoplasm <21> [13, 14]
cytosol <4, 21> [2]
endoplasmic reticulum <4, 17, 19, 21> (<19,21> rough endoplasmic reticulum [2]) [2, 5, 12, 13]
polysome <21> [4, 13, 14]
ribosome <25> [5]
vesicular fraction <12> [17]

Purification

<10> (partial [1,8]) [1, 8, 9, 13]
<19> (partially [4]) [4]
<21> (partially [8]) [13]
<22> [8, 12, 14]

Cloning

<13> (human NAT1 and NAT2 genes located on chromosome 8p22 [18]) [18]
<22> (full-length cDNA clone [13]; 3 N-terminal acetyltransferases, Ard1p/Nat1p3, Nat3p and Mak3p [15]; N-terminal acteyltransferases, NatA, NatB and NatC encoded by orthologous genes [16]) [13-17]
<26> (cDNA isolation, Xat-1 recombinant protein in vitro translated in rabbit reticulocyte lysate [17]) [17]

6 Stability

pH-Stability

6-8 <21> (<21> about 25% maximal activity at pH 6.0 and 8.0 [3]) [3]
7.6 <21> (<21> pronounced decline in reaction rate at pH values above [4]) [4]

Temperature stability

65 <22> (<22> irreversible denaturation after 1 min [8]) [8]

General stability information

<10>, enzyme is completely inactivated by lyophilization [1]
<10>, loses activity by digestion with bovine pancreatic RNase A, Staphylococcus aureus nuclease or proteinase K [10]
<21>, greatly stablized by inclusion of EDTA and 0.01% deoxycholate in the isolation buffer [13]

Storage stability

<10>, -20°C, enzyme is completely inactivated by freezing [1]
<10>, 0-4°C, 10% of the activity is lost after storage for a week [1]
<21>, -20°C, stored as Percoll fraction, activity declines by about 50% in 5 days [2]
<21>, -70°C, 10-20% glycerol, activity is stable for at least 5 months [2]
<21>, -70°C, stored as Percoll fraction, activity declines by about 50% in 5 days [2]
<21>, 4°C, stored as Percoll fraction, activity declines by about 50% in 5 days [2]
<22>, 0°C, sensitive to freezing, more than 90% loss of activity per freeze-thaw cycle [8]
<22>, 4°C, stable when stored in 20 mM HEPES buffer, pH 7.4, 0.5 mM dithiothreitol, 10% v,v glycerol, 0.02% NaN_3 containing 0.2 M KCl [8]

References

[1] Tsunasawa, S.; Kamitani, K.; Narita, K.: Partial purification and properties of the amino-terminal amino acid-acetylating enzyme from hen's oviduct. J. Biochem., **87**, 645-650 (1980)

[2] Glembotski, C.C.: Characterization of the peptide acetyltransferase activity in bovine and rat intermediate pituitaries responsible for the acetylation of β-endorphin and α-melanotropin. J. Biol. Chem., **257**, 10501-10509 (1982)
[3] O'Donohue, T.L.: Identification of endorphin acetyltransferase in rat brain and pituitary gland. J. Biol. Chem., **258**, 2163-2167 (1983)
[4] Dixon, J.E.; Woodford, T.A.: Rat pituitary N α-acetyltransferase. Methods Enzymol., **106**, 170-179 (1984)
[5] Kido, H.; Vita, A.; Horecker, B.L.: Amino-terminal protein transacetylase from wheat germ. Methods Enzymol., **106**, 193-197 (1984)
[6] Redman, K.L.; Rubenstein, P.A.: Actin amino-terminal acetylation and processing in a rabbit reticulocyte lysate. Methods Enzymol., **106**, 179-192 (1984)
[7] Tsunasawa, S.; Sakiyama, F.: Amino-terminal acetylation of proteins: An overview. Methods Enzymol., **106**, 165-170 (1984)
[8] Lee, F.J.S.; Lin, L.W.; Smith, J.A.: Purification and characterization of an N α-acetyltransferase from Saccharomyces cerevisiae. J. Biol. Chem., **263**, 14948-14955 (1988)
[9] Kamitani, K.; Narita, K.; Fumio, S.: Purification and characterization of hen oviduct N α-acetyltransferase. J. Biol. Chem., **264**, 13188-13193 (1989)
[10] Kamitani, K.; Sakiyama, F.: Hen oviduct N α-acetyltransferase is a ribonucleoprotein having 7 S RNA. J. Biol. Chem., **264**, 13194-13198 (1989)
[11] Karam, L.R.; Simic, M.G.: Formation of *ortho*-tyrosine by radiation and organic solvents in chicken tissue. J. Biol. Chem., **265**, 11581-11585 (1990)
[12] Lee, F.J.S.; Lin, L.W.; Smith, J.A.: Model peptides reveal specificity of N α-acetyltransferase from Saccharomyces cerevisiae. J. Biol. Chem., **265**, 11576-11580 (1990)
[13] Yamada, R.; Bradshaw, R.A.: Rat liver polysome N α-acetyltransferase: isolation and characterization. Biochemistry, **30**, 1010-1016 (1991)
[14] Yamada, R.; Bradshaw, R.A.: Rat liver polysome N α-acetyltransferase: substrate specificity. Biochemistry, **30**, 1017-1021 (1991)
[15] Polevoda, B.; Norbeck, J.; Takakura, H.; Blomberg, A.; Sherman, F.: Identification and specificities of N-terminal acetyltransferases from Saccharomyces cerevisiae. EMBO J., **18**, 6155-6168 (1999)
[16] Polevoda, B.; Sherman, F.: N^{α}-terminal acetylation of eukaryotic proteins. J. Biol. Chem., **275**, 36479-36482 (2000)
[17] Choi, S.C.; Chang, J.Y.; Han, J.K.: A novel Xenopus acetyltransferase with a dynamic expression in early development. Biochem. Biophys. Res. Commun., **285**, 1338-1343 (2001)
[18] Sekine, A.; Saito, S.; Iida, A.; Mitsunobu, Y.; Higuchi, S.; Harigae, S.; Nakamura, Y.: Identification of single-nucleotide polymorphisms (SNPs) of human N-acetyltransferase genes NAT1, NAT2, AANAT, ARD1, and L1CAM in the Japanese population. Hum. Genet., **46**, 314-319 (2001)

Tetrahydrodipicolinate N-acetyltransferase 2.3.1.89

1 Nomenclature

EC number
2.3.1.89

Systematic name
acetyl-CoA:(S)-2,3,4,5-tetrahydrodipicolinate N^2-acetyltransferase

Recommended name
tetrahydrodipicolinate N-acetyltransferase

Synonyms
acetyltransferase, tetrahydrodipicolinate
tetrahydrodipicolinate acetylase
tetrahydrodipicolinate:acetyl-CoA acetyltransferase

CAS registry number
83588-91-4

2 Source Organism

<1> *Bacillus megaterium* (CII 19 [1]) [1]

3 Reaction and Specificity

Catalyzed reaction
acetyl-CoA + (S)-2,3,4,5-tetrahydrodipicolinate = CoA + L-2-acetamido-6-oxoheptanedioate

Reaction type
acyl group transfer

Natural substrates and products
S Additional information <1> (<1> enzyme is involved in the biosynthesis of lysine [1]) [1]
P ?

Substrates and products
S acetyl-CoA + L-tetrahydrodipicolinate <1> (<1> succinyl-CoA is ineffective [1]) (Reversibility: ? <1> [1]) [1]
P CoA + L-N-acetyl-2-amino-6-oxopimelate <1> [1]

Specific activity (U/mg)
0.011 <1> [1]

pH-Optimum
8 <1> (<1> assay at [1]) [1]

Temperature optimum (°C)
30 <1> (<1> assay at [1]) [1]

5 Isolation/Preparation/Mutation/Application

Purification
<1> [1]

References

[1] Chatterjee, S.P.; White, P.J.: Activities and regulation of the enzymes of lysine biosynthesis in a lysine-excreting strain of Bacillus megaterium. J. Gen. Microbiol., **128**, 1073-1081 (1982)

β-Glucogallin O-galloyltransferase 2.3.1.90

1 Nomenclature

EC number
2.3.1.90

Systematic name
1-O-galloyl-β-D-glucose:1-O-galloyl-β-D-glucose O-galloyltransferase

Recommended name
β-glucogallin O-galloyltransferase

Synonyms
β-glucogallin (β-glucogallin donor)
galloyltransferase

CAS registry number
87502-55-4

2 Source Organism

<1> *Quercus robur* (oak [1-3]) [1-3]

3 Reaction and Specificity

Catalyzed reaction
2 1-O-galloyl-β-D-glucose = D-glucose + 1-O,6-O-digalloyl-β-D-glucose

Reaction type
acyl group transfer

Natural substrates and products

- **S** 1-O-galloyl-β-D-glucose + 1-O-galloyl-β-D-glucose <1> (<1> first intermediate in the gallotannin biosynthesis in higher plants, serving as acyl donor and acceptor [1,3]) (Reversibility: ? <1> [1-3]) [1-3]
- **P** 1,6-di-O-galloyl-β-D-glucose + D-glucose <1> [1-3]

Substrates and products

- **S** 1-O-4-hydroxybenzoyl-β-D-glucose + 1-O-4-hydroxybenzoyl-β-D-glucose <1> (Reversibility: ? <1> [1]) [1]
- **P** 1,2-di-O-benzoyl-β-D-glucose + D-glucose <1> [1]
- **S** 1-O-galloyl-β-D-glucose + 1,6-di-O-galloyl-β-D-glucose <1> (Reversibility: ? <1> [2]) [2]

P 1,2,6-tri-O-galloyl-β-D-glucose + D-glucose <1> [2]
S 1-O-galloyl-β-D-glucose + 1-O-galloyl-β-D-glucose <1> (<1> no reaction with galloyl-CoA [2]; <1> no reaction with 1-O-syringoylglucose [1]) (Reversibility: ? <1> [1-3]) [1-3]
P 1,6-di-O-galloyl-β-D-glucose + D-glucose <1> [1-3]
S 1-O-protocatechuoyl-β-D-glucose + 1-O-protocatechuoyl-β-D-glucose <1> (Reversibility: ? <1> [1]) [1]
P 1,6-di-O-protocatechuoyl-β-D-glucose + D-glucose <1> [1]
S 1-O-veratroyl-β-D-glucose + 1-O-veratroyl-β-D-glucose <1> (Reversibility: ? <1> [1]) [1]
P 1,2-di-O-veratroyl-β-D-glucose + D-glucose <1> [1]

Inhibitors
2-mercaptoethanol <1> (<1> above 50 mM [3]) [3]
dithioerythritol <1> (<1> above 20 mM [3]) [3]
Additional information <1> (<1> no inhibition by metal ions [3]) [3]

Cofactors/prosthetic groups
Additional information <1> (<1> no cofactor requirement [2]) [2]

Metals, ions
Additional information <1> (<1> no metal ion requirement [3]) [3]

Specific activity (U/mg)
0.0743 <1> [3]

K_m-Value (mM)
2.5 <1> (1-O-4-hydroxybenzoyl-β-D-glucose) [1]
44.4 <1> (1-O-protocatechuoyl-β-D-glucose) [1]
50 <1> (1-O-veratroyl-β-D-glucose) [1]
66.7 <1> (1-O-galloyl-β-D-glucose) [1]
Additional information <1> (<1> the activity of the enzyme is 58% for 1-O-protocatechouylglucose, 8% for 1-O-*p*-hydroxybenzoylglucose and 7% for 1-O-veratroylglucose relative to galloylglucose [1]; <1> normal Michaelis-Menten kinetics [3]; <1> substrate saturation at 10 mM [2]; <1> substrate saturation not reached at 20 mM [3]) [1-3]

pH-Optimum
6-6.5 <1> [2, 3]

pH-Range
4.2-8 <1> (<1> no activity below pH 4.2 and above pH 8.0 [2]) [2]
5-7.2 <1> (<1> about half-maximal activity at pH 5.0 and pH 7.2 [2]) [2]
5.5-7.5 <1> (<1> about half-maximal activity at pH 5.5 and pH 7.5 [3]) [3]

Temperature optimum (°C)
30 <1> [3]
Additional information <1> (<1> 25% residual activity at 0°C [3]) [3]

4 Enzyme Structure

Molecular weight

400000 <1> (<1> gel filtration [3]) [3]

5 Isolation/Preparation/Mutation/Application

Source/tissue

leaf <1> (<1> 2-3 months old [2,3]) [2, 3]

Purification

<1> (partial purification with ammonium sulfate precipitation, DEAE-cellulose chromatography and gel filtration [1-3]) [1-3]

6 Stability

pH-Stability

4.5-6.5 <1> (<1> highest stability at 30°C [3]) [3]

Temperature stability

30 <1> (<1> highest stability at pH 4.5-6.5 [3]) [3]
55-60 <1> (<1> denaturation [3]) [3]

General stability information

<1>, dilution destabilizes [3]
<1>, glycerol, 10%, stabilizes [3]

Storage stability

<1>, -20°C, stable in the presence of 10% glycerol [3]
<1>, 2-4°C, $t_{1/2}$: 2 days, dilute preparation [3]
<1>, 2-4°C, $t_{1/2}$: 7 days, concentrated preparation [3]

References

[1] Gross, G.G.; Denzel, K.; Schilling, G.: Enzymic synthesis of di-O-phenylcarboxyl-β-D-glucose esters by an acyltransferase from oak leaves. Z. Naturforsch. C, **45**, 37-41 (1990)

[2] Gross, G.G.: Synthesis of mono-, di- and trigalloyl-β-D-glucose by β-glucogallin-dependent galloyltransferases from oak leaves. Z. Naturforsch. C, **38**, 519-523 (1983)

[3] Schmidt, S.W.; Denzel, K.; Schilling, G.; Gross, G.G.: Enzymatic synthesis of 1,6-digalloylglucose from β-glucogallin by β-glucogallin:β-glucogallin 6-0-galloyltransferase from oak leaves. Z. Naturforsch. C, **42**, 87-92 (1987)

Sinapoylglucose-choline O-sinapoyltransferase 2.3.1.91

1 Nomenclature

EC number
2.3.1.91

Systematic name
1-O-(4-hydroxy-3,5-dimethoxycinnamoyl)-β-D-glucose:choline 1-O-(4-hydroxy-3,5-dimethoxycinnamoyl)transferase

Recommended name
sinapoylglucose-choline O-sinapoyltransferase

Synonyms
SCT <1, 2> [1, 2]
sinapine synthase
sinapoylglucose-choline sinapoyltransferase
sinapoyltransferase, sinapoylglucose-choline

CAS registry number
85205-00-1

2 Source Organism

<1> *Raphanus sativus* (var. sativus cv. Saxa [1,2]) [1, 2, 4]
<2> *Sinapis alba* [1, 4]
<3> *Brassica napus* (cv. Westar [3,4]) [3, 4]
<4> *Berteroa incana* (low activity [4]) [4]
<5> *Biscutella lyrata* [4]
<6> *Brassica sp.* (overview [4]) [4]
<7> *Camelina sp.* (overview [4]) [4]
<8> *Capsella bursa-pastoris* (low activity [4]) [4]
<9> *Cheiranthus sp.* (overview [4]) [4]
<10> *Isatis tinctoria* [4]
<11> *Eruca sativa* (low activity [4]) [4]
<12> *Descurainia sophia* (low activity [4]) [4]
<13> *Lepidium sp.* (overview [4]) [4]
<14> *Malcomia africana* [4]
<15> *Matthiola incana* [4]
<16> *Neslia paniculata* [4]
<17> *Raphanus raphanistrum* [4]
<18> *Rapistrum rugosum* [4]

<19> *Rorippa sp.* (overview [4]) [4]
<20> *Sinapis arvensis* [4]
<21> *Sisymbrium sp.* (overview [4]) [4]
<22> *Turritis glabra* (low activity [4]) [4]

3 Reaction and Specificity

Catalyzed reaction

1-O-sinapoyl-β-D-glucose + choline = D-glucose + sinapoylcholine

Reaction type

acyl group transfer

Natural substrates and products

S 1-O-sinapoyl-β-D-glucose + choline <1-22> (<1,2> involved in sinapine biosynthesis [1]) (Reversibility: ? <1-22> [1, 2, 4]) [1, 2, 4]
P D-glucose + sinapoylcholine <1-22> [4]

Substrates and products

S 1,2-di-O-sinapoyl-β-D-glucose + choline <1, 2> (<1,2> best donor substrate [1]) (Reversibility: ? <1, 2> [1]) [1]
P 2-O-sinapoyl-β-D-glucose + sinapoylcholine
S 1-O-sinapoyl-β-D-glucose + choline <1-22> (<3> specific for 1-O-sinapoyl-β-glucose [3]; <1,2> absolute acceptor specificity towards choline [1,2]; <3> no acceptors are betaine or CDP-choline [3]; <1> ethanolamine, myo-inositol, citric, L-malic, L-tartaric, quinic or shikimic acid are no acceptors [2]; <1,2> no donors are 1-benzoylglucose, 1-galloylglucose, 6-sinapoylglucose, 3-sinapoylfructose [1]) (Reversibility: ir <3> [3]; ? <1-22> [1, 2, 4]) [1-4]
P D-glucose + sinapoylcholine <1-22> (<1-22> i.e. sinapine [1-4]) [1-4]
S 1-feruloyl-β-D-glucose + choline <1-3> (<1,3> 40% as active as 1-sinapoylglucose [2,3]) (Reversibility: ? <1-3> [1-3]) [1-3]
P D-glucose + feruloylcholine
S 1-*p*-coumaroyl-β-D-glucose + choline <1-3> (<1> poor substrate [1]; <3> 25% [3]; <1> 13% [2]) (Reversibility: ? <1-3> [1-3]) [1-3]
P D-glucose + *p*-coumaroylcholine
S caffeoyl-β-D-glucose + choline <3> (<3> poor substrate [3]) (Reversibility: ? <3> [3]) [3]
P D-glucose + caffeoylcholine

Inhibitors

Ca^{2+} <1, 2> (<1,2> strong, above 5 mM [1]) [1]
Co^{2+} <1, 2> (<1,2> strong, above 5 mM [1]) [1]
Cu^{2+} <3> [3]
EDTA <1, 2> [1]
Hg^{2+} <3> [3]
Mn^{2+} <1, 2> (<1,2> strong, above 5 mM [1]) [1]
PMSF <3> [3]

SDS <1, 2> [1]
dithioerythritol <1, 2> [1]
dithiothreitol <1, 2> [1]
Additional information <3> (<3> no inhibition by IAA, NEM, tosylphenylalanine chloromethylketone, diethylchlorocarbamate, *p*-nitrophenylchloroformate [3]) [3]

Metals, ions

Additional information <1-3> (<1-3> no requirement for divalent cations [1,3]) [1, 3]

Specific activity (U/mg)

0.258 <2> (<2> partially purified enzyme [1]) [1]
0.486 <1> (<1> partially purified enzyme [1]) [1]
9.6 <3> (<3> purified enzyme [3]) [3]
Additional information <1-22> (<1> Tris buffer is unusable for enzyme assay due to formation of side products [2]; <1-22> distribution [4]) [2, 4]

K_m-Value (mM)

0.11 <1> (1,2-di-O-sinapoyl-β-D-glucose) [1]
0.2 <2> (1,2-di-O-sinapoyl-β-D-glucose) [1]
0.3 <1> (1-O-sinapoyl-β-D-glucose) [2]
0.48 <1> (1-O-sinapoyl-β-D-glucose) [1]
0.71 <2> (1-O-sinapoyl-β-D-glucose) [1]
5.3 <1> (choline) [1]
6.5 <2> (choline) [1]
7.64 <1> (choline) [2]

pH-Optimum

6.6-7.2 <3> [3]
7 <1> (<1> about [2]) [2]
7.2 <1> [1]
7.6 <2> [1]
Additional information <3> (<3> pI: 6.1 [3]) [3]

pH-Range

5.3-9 <1> (<1> about half-maximal activity at pH 5.3 and pH 9.0 [2]) [2]
5.5-10 <1, 2> [1]

Temperature optimum (°C)

45 <1, 2> [1]

4 Enzyme Structure

Molecular weight

60000 <1, 2> (<1,2> gel filtration [1]) [1]
65000 <3> (<3> gel filtration [3]) [3]
Additional information <3> (<3> amino acid composition and sequence [3]) [3]

Subunits

dimer <3> (<3> 2 * 28000, SDS-PAGE [3]) [3]

5 Isolation/Preparation/Mutation/Application

Source/tissue

seed <1-22> (<1-3> immature dark-green [1-5]; <1,2> 20% activity in mature seeds compared to immature seeds [1]) [1-4]

Purification

<1> (partially [1]) [1]
<2> (partially [1]) [1]
<3> [3]

6 Stability

Temperature stability

50 <3> (<3> 15 min stable [3]) [3]

General stability information

<3>, choline chloride stabilizes [3]
<3>, repeated freeze-thawing, stable to [3]

Storage stability

<3>, -70°C, at least 2 months [3]
<3>, 4°C, 2 weeks [3]
<1, 2>, -20°C, crude extract, 1 week, loss of 20% activity [1, 2]
<1, 2>, -20°C, partially purified, 2 weeks upon repeated freezing and thawing [1]

References

[1] Gräwe, W.; Strack, D.: Partial purification and some properties of 1-sinapoylglucose:choline sinapoyltransferase (sinapine synthase) from seeds of Raphanus sativus L. and Sinapis alba L.. Z. Naturforsch. C, **41**, 28-33 (1986)
[2] Strack, D.; Knogge, W.; Dahlbender, B.: Enzymatic synthesis of sinapine from 1-O-sinapoyl-β-D-glucose and choline by a cell-free system from developing seeds of red radish (Raphanus sativus L. var. sativus). Z. Naturforsch. C, **38**, 21-27 (1982)
[3] Vogt, T.; Aebershold, R.; Ellis, B.: Purification and characterization of sinapine synthase from seeds of Brassica napus. Arch. Biochem. Biophys., **300**, 622-628 (1993)
[4] Regenbrecht, J.; Strack, D.: Distribution of 1-sinapoylglucose:choline sinapoyltransferase activity in the Brassicaceae. Phytochemistry, **24**, 407-410 (1985)

Sinapoylglucose-malate O-sinapoyltransferase

2.3.1.92

1 Nomenclature

EC number
2.3.1.92

Systematic name
1-O-sinapoyl-β-D-glucose:(S)-malate O-sinapoyltransferase

Recommended name
sinapoylglucose-malate O-sinapoyltransferase

Synonyms
1-sinapoylglucose-L-malate sinapoyltransferase
SMT <1, 2, 4> [1, 2, 4, 5]
sinapoylglucose:malate sinapoyltransferase

CAS registry number
76095-65-3

2 Source Organism

<1> *Brassica rapa* (rape, ssp. oleifera [4]) [4]
<2> *Arabidopsis thaliana* (Heyn. ecotype Columbia [1,2]; Heyn. ecotype Landsberg erecta [2]) [1, 2, 4]
<3> *Raphanus sativus* (red radish, var. sativus [3,6,7]; two isoforms [3]; cv. Saxa Nova [3]; cv. Saxa [6]) [3, 6, 7, 8]
<4> *Brassica napus* (rapeseed, ssp. napus [5]) [5]

3 Reaction and Specificity

Catalyzed reaction
1-O-sinapoyl-β-D-glucose + (S)-malate = D-glucose + sinapoyl-(S)-malate (<3> mechanism [3])

Reaction type
acyl group transfer

Natural substrates and products

S 1-O-sinapoyl-β-D-glucose + L-malate <3, 4> (<4> key step in overall conversion of seed constituent sinapine to sinapoylmalate [5]; <3> is involved in phenylpropanoid depside formation [8]) (Reversibility: ? <3, 4> [5, 8]) [5, 8]

P D-glucose + sinapoyl-L-malate

Substrates and products

S 1,2-di-O-sinapoyl-β-D-glucose + L-malate <3> (<3> at 26% the rate of the reaction with 1-O-sinapoylglucose [3]) (Reversibility: ? <3> [3]) [3]

P ?

S 1-O-(4-coumaroyl)-β-D-glucose + L-malate <3> (<3> poor substrate [3]) (Reversibility: ? <3> [3]) [3]

P D-glucose + 4-coumaroyl-L-malate

S 1-O-caffeoyl-β-D-glucose + L-malate <3> (<3> at 45% the rate of the reaction with 1-O-sinapoylglucose [3]) (Reversibility: ? <3> [3]) [3]

P D-glucose + caffeoyl-L-malate

S 1-O-feruloyl-β-D-glucose + L-malate <3> (<3> at 85% the rate of the reaction with 1-O-sinapoylglucose [3]) (Reversibility: ? <3> [3]) [3]

P D-glucose + feruloyl-L-malate

S 1-O-sinapoyl-β-D-glucose + L-malate <1, 2, 3, 4> (<3> no donor is sinapic acid [8]) (Reversibility: ir <3> [3]; ? <1, 2, 4> [1, 4, 5, 8]) [1, 2, 3, 4, 5, 7, 8]

P D-glucose + sinapoyl-L-malate <1, 2, 3, 4> [3, 4, 5, 8]

Inhibitors

D-malate <3> (<3> non-competitive [3]) [3]

L-malate <3> (<3> above 0.05 M [3]; <3> above 0.14 M [7]) [3, 7]

Activating compounds

Additional information <3> (<3> no activation by dithioerythritol and 2-mercaptoethanol [3,7]) [3, 7]

Metals, ions

Additional information <3> (<3> no requirement for divalent cations [3,7]) [3, 7]

Specific activity (U/mg)

0.528 <3> [3]

Additional information <1, 2, 4> [2, 4, 5]

K_m-Value (mM)

0.46 <3> (1-O-sinapoyl-β-D-glucose) [7]

54 <3> (L-malate) [7]

Additional information <3> (<3> kinetic data [3]) [3]

K_i-Value (mM)

215 <3> (D-malate) [3]

pH-Optimum

6 <3> [3]

6.3 <3> (<3> various buffer systems [7]) [7]

pH-Range

5.5-6.5 <3> (<3> about half-maximal activity at pH 5.5 and 6.5 [3]) [3]

5.7-6.8 <3> (<3> about half-maximal activity at pH 5.7 and 6.8, various buffer systems [7]) [7]

Temperature optimum (°C)

30 <4> (<4> assay at [5]) [5]

4 Enzyme Structure

Molecular weight

47190 <2> (<2> expressed in Escherichia coli, MALDI-TOF [1]) [1]
52000 <3> (<3> gel filtration [3]) [3]
52000-55000 <2> (<2> expressed in wild-type and in Nicotiana tabacum, SDS-PAGE [1]) [1]

Subunits

monomer <3> (<3> 1 * 51000, SDS-PAGE [3]) [3]

Posttranslational modification

glycoprotein <2> (<2> protein expressed in Escherichia coli has an 8 kDa lower molecular weight than wild-type [1]) [1]

5 Isolation/Preparation/Mutation/Application

Source/tissue

cotyledon <1, 3, 4> (<3> maximal activity in 10-14 days old cotyledons [7]) [3, 4, 5, 7, 8]
flower <2> (<2> traces [1]) [1]
leaf <2> (<2> rosette leaf more abundant than in cauline leaf [1]) [1, 2]
mesophyll <2> [1]
plant epidermis <2> [1]
root <2> (<2> traces [1]) [1]
seed <1, 2, 4> [2, 4, 5]
seedling <1, 2, 4> (<2> only in older seedlings [1]) [1, 2, 4, 5]
stem <2> (<2> flowering stem [1]) [1]

Localization

vacuole <2, 3> (<2> of mesophyll and epidermal cells [1]) [1, 6]

Purification

<2> [1]
<3> (partial [7]; 2 isozymes on isoelectric focusing of purified enzyme, 4 isoforms on isoelectric focusing of crude extract, in varying proportions depending on light conditions during growth [3]) [3, 7]

Cloning

<2> (expressed in Escherichia coli BL21DE3 [1,2]) [1, 2]
<2> (expressed in Nicotiana tabacum cv. Xanthi [1]) [1]

6 Stability

Temperature stability

30 <3> (<3> 14 h stable [7]) [7]

General stability information

<3>, 1 mg/ml bovine serum albumin stabilizes [3]

<3>, L-malate stabilizes during purification [3]

<3>, stable to repeated freeze-thawing [7]

<3>, thawing after freezing at -20°C leads to 50% loss of activity, bovine serum albumin stabilizes [3]

<3>, unstable to dilution [3]

Storage stability

<3>, -20°C, 3 weeks [7]

<3>, 0°C, 3 weeks [7]

References

[1] Hause, B.; Meyer, K.; Viitanen, P.V.; Chapple, C.; Strack, D.: Immunolocalization of 1-O-sinapoylglucose:malate sinapoyltransferase in Arabidopsis thaliana. Planta, **215**, 26-32 (2002)

[2] Lehfeldt, C.; Shirley, A.M.; Meyer, K.; Ruegger, M.O.; Cusumano, J.C.; Viitanen, P.V.; Strack, D.; Chapple, C.: Cloning of the SNG1 gene of Arabidopsis reveals a role for a serine carboxypeptidase-like protein as an acyltransferase in secondary metabolism. Plant Cell, **12**, 1295-1306 (2000)

[3] Gräwe, W.; Bachhuber, P.; Mock, H.P.; Strack, D.: Purification and characterization of sinapoylglucose:malate sinapoyltransferase from Raphanus sativus L.. Planta, **187**, 236-241 (1992)

[4] Mock, H.P.; Vogt, T.; Strack, D.: Sinapoylglucose: malate sinapoyltransferase activity in Arabidopsis thaliana and Brassica rapa. Z. Naturforsch. C, **47c**, 680-682 (1992)

[5] Strack, D.; Ellis, B.E.; Gräwe, W.; Heilemann, J.: Sinapoylglucose: malate sinapoyltransferase activity in seeds and seedlings of rape. Planta, **180**, 217-219 (1990)

[6] Sharma, V.; Strack, D.: Vacuolar localization of 1-sinapoylglucose:L-malate sinapoyltransferase in protoplasts from cotyledons of Raphanus sativus. Planta, **163**, 563-568 (1985)

[7] Strack, D.: Development of 1-O-sinapoyl-β-D-glucose:L-malate sinapoyltransferase activity in cotyledons of red radish (Raphanus sativus L. var. sativus). Planta, **155**, 31-36 (1982)

[8] Tkotz, N.; Strack, D.: Enzymatic synthesis of sinapoyl-L-malate from 1-sinapoylglucose and L-malate by a protein preparation from Raphanus sativus cotyledons. Z. Naturforsch. C, **35c**, 835-837 (1980)

13-Hydroxylupinine O-tigloyltransferase 2.3.1.93

1 Nomenclature

EC number
2.3.1.93

Systematic name
(E)-2-methylcrotonoyl-CoA:13-hydroxylupinine O-2-methylcrotonoyltransferase

Recommended name
13-hydroxylupinine O-tigloyltransferase

Synonyms
13-hydroxylupanine O-tigloyltransferase
13-hydroxylupanine acyltransferase
tigloyl-CoA:13-hydroxylupanine O-tigloyltransferase
tigloyl:13α-hydroxylupanine O-tigloyltransferase <3> (<3> HLTase [2]) [2]
tigloyl:13α-hydroxymultiflorine O-tigloyltransferase <3> (<3> HMTase [2]) [2]
tigloyltransferase, 13-hydroxylupanine

CAS registry number
85341-00-0

2 Source Organism

<-5> no activity in *Conium masculatum* [1]
<-4> no activity in *Cytisus scoparius* [1]
<-3> no activity in *Vicia faba* [1]
<-2> no activity in *Pisum sativum* [1]
<-1> no activity in *Lupinus luteus* [2]
<1> *Lupinus albus* (bitter and sweet variant [2]) [1, 2]
<2> *Lupinus polyphyllus* [1]
<3> *Lupinus termis* (2 isoforms A and B [2]) [2]
<4> *Lupinus polyphyllus x Lupinus arboreus* [2]
<5> *Lupinus hirsutus* [2]
<6> *Cytisus scoparius* [2]

3 Reaction and Specificity

Catalyzed reaction

(E)-2-methylcrotonoyl-CoA + 13-hydroxylupinine = CoA + 13-(2-methylcrotonoyl)oxylupinine

Reaction type

acyl group transfer

Natural substrates and products

S tigloyl-CoA + 13-α-hydroxymultiflorine <3> (<3> isoform A and B [2]; <3> tigloyl-CoA is (E)-2-methylcrotonoyl-CoA [2]) (Reversibility: ? <3> [2]) [2]

P CoA + 13-tigloyloxymultiflorine

S tigloyl-CoA + 13-hydroxylupinine <1-3> (<3> isoform A and B [2]; <1-3> tigloyl-CoA is (E)-2-methylcrotonoyl-CoA [1,2]; <1-3> enzyme is involved in the biosynthesis of quinolizidine alkaloid esters [1,2]) (Reversibility: ? <1-3> [1, 2]) [1, 2]

P CoA + 13-tigloyloxylupinine <1-3> [1, 2]

Substrates and products

S 3-methylbutyryl-CoA + 13-hydroxylupinine <1> (<1> 97% of activity compared to tigloyl-CoA [1]) (Reversibility: ir <1> [1]) [1]

P CoA + 13-(3-methylbutyryl)oxylupinine <1> [1]

S benzoyl-CoA + 13-hydroxylupinine <1> (<1> 111% of activity compared to tigloyl-CoA [1]) (Reversibility: ir <1> [1]) [1]

P CoA + 13-benzoyloxylupinine <1> [1]

S butyryl-CoA + 13-hydroxylupinine <1> (<1> 27% of activity compared to tigloyl-CoA [1]) (Reversibility: ir <1> [1]) [1]

P CoA + 13-butyryloxylupinine <1> [1]

S propionyl-CoA + 13-hydroxylupinine <1> (<1> 25% of activity compared to tigloyl-CoA [1]) (Reversibility: ir <1> [1]) [1]

P CoA + 13-propionyloxylupinine <1> [1]

S tigloyl-CoA + 13-α-hydroxymultiflorine <1, 3, 5> (<3> isoform A and B [2]; <1,3,5> tigloyl-CoA is (E)-2-methylcrotonoyl-CoA [2]) (Reversibility: ? <1, 3, 5> [2]) [2]

P CoA + 13-tigloyloxymultiflorine

S tigloyl-CoA + 13-hydroxylupinine <1-4, 6> (<3> isoform A and B [2]; <1-4, 6> tigloyl-CoA is (E)-2-methylcrotonoyl-CoA [1,2]) (Reversibility: ir <1> [1]; ? <2-4, 6> [1, 2]) [1, 2]

P CoA + 13-tigloyloxylupinine <1-4, 6> [1, 2]

S valeryl-CoA + 13-hydroxylupinine <1> (<1> 73% of activity compared to tigloyl-CoA [1]) (Reversibility: ir <1> [1]) [1]

P CoA + 13-valeryloxylupinine <1> [1]

S Additional information <1, 3> (<3> substrate specificity, overview [2]; <3> baptifoline, epilupinine and lupinine are no substrates [2]; <1> lupinine, 4-hydroxylupinine, and cholesterol and acetyl-CoA are no substrates [1]) [1, 2]

P ?

Inhibitors

(+)-epilupinine <3> [2]
(+)-lupanine <3> [2]
CoA <3> (<3> competitive [2]) [2]
$HgCl_2$ <1> (<1> strong [1]) [1]
N-ethylmaleimide <1, 3> (<1> strong [1]) [1, 2]
diethyldithiocarbamate <1> [1]
iodoacetamide <1> (<1> slight inhibition [1]) [1]
p-chloromercuriphenyl sulfonic acid <3> [2]
p-hydroxymercuribenzoate <1> (<1> strong [1]) [1]
Additional information <1> (<1> EDTA is no inhibitor [1]) [1]

Activating compounds

dithiothreitol <1, 3> (<3> 140% activity at 1 mM [2]; <1> activates about 50-80% [1]) [1, 2]

Metals, ions

Additional information <1> (<1> cations, e.g. Fe^{2+}, Ca^{2+}, Mg^{2+}, NH_4Cl, Fe^{3+} have no significant influence [1]) [1]

Specific activity (U/mg)

0.12 <3> (<3> isoform A [2]) [2]
2.18 <3> (<3> isoform B [2]) [2]

K_m-Value (mM)

0.018 <1> (13-hydroxylupanine) [1]
0.021 <3> (13-α-hydroxymultiflorine) [2]
0.027 <3> (13-hydroxylupanine) [2]
0.14 <1> (tigloyl-CoA) [1]
0.46 <3> (tigloyl-CoA, <3> with 13-α-hydroxymultiflorine [2]) [2]
0.52 <3> (tigloyl-CoA, <3> with 13-α-hydroxylupanine [2]) [2]

K_i-Value (mM)

0.5 <3> (CoA) [2]
1.7 <3> ((+)-lupanine) [2]
3.1 <3> ((+)-epilupinine) [2]

pH-Optimum

7-8 <1> [1]
Additional information <3> (<3> isoform 1 pI: 7.8, isoform 2 pI: 7.6 [2]) [2]

pH-Range

6.5-8 <1> (<1> half-maximal activity at pH 6.5, maximal activity at pH 8.0 [1]) [1]

Temperature optimum (°C)

30 <1> [1]

4 Enzyme Structure

Molecular weight

50000 <3> (<3> isoform A and B, gel filtration [2]) [2]

Subunits

monomer <3> (<3> 1 * 50000, isoform A and B, SDS-PAGE [2]) [2]

5 Isolation/Preparation/Mutation/Application

Source/tissue

leaf <2> (<2> leaflet [1]) [1]
petiole <2> [1]
root <1> [1]
seedling <1, 3> (<1> high activity in stem and root [1]) [1, 2]
stem <2> [1]
Additional information <-1, 2> (<-1> not in leaves and cell suspension culture [1]; <2> not in root and cell suspension culture [1]) [1]

Purification

<1> (partial [1]) [1]
<3> [2]

6 Stability

Temperature stability

70 <3> (<3> 5 min, complete inactivation [2]) [2]

Oxidation stability

<1>, O_2-sensitive, dithioerythritol stabilizes [1]

References

[1] Wink, M.; Hartmann, T.: Enzymatic synthesis of quinolizidine alkaloid esters: a tigloyl-CoA:13-hydroxylupanine O-tigloyl transferase from Lupinus albus L.. Planta, **156**, 560-565 (1982)
[2] Suzuki, H.; Murakoshi, I.; Saito, K.: A novel O-tigloyltransferase for alkaloid biosynthesis in plants. Purification, characterization, and distribution in Lupinus plants. J. Biol. Chem., **269**, 15853-15860 (1994)

Erythronolide synthase 2.3.1.94

1 Nomenclature

EC number
2.3.1.94

Systematic name
malonyl-CoA:propionyl-CoA malonyltransferase (cyclizing)

Recommended name
erythronolide synthase

Synonyms
erythronolide condensing enzyme
synthase, erythronolide
Additional information (cf. EC 2.3.1.74)

CAS registry number
87683-77-0

2 Source Organism

<1> *Streptomyces erythraeus* [1]

3 Reaction and Specificity

Catalyzed reaction
6 malonyl-CoA + propionyl-CoA = 7 CoA + 6-deoxyerythronolide b

Reaction type
acyl group transfer

Natural substrates and products
S methylmalonyl-CoA + propionyl-CoA <1> (<1> formation of a key intermediate in the biosynthesis of erythromycin [1]) (Reversibility: ? <1> [1]) [1]
P 6-deoxyerythronolide b + CoA <1> [1]

Substrates and products
S methylmalonyl-CoA + propionyl-CoA <1> (Reversibility: ? <1> [1]) [1]
P 6-deoxyerythronolide b + CoA <1> [1]

Inhibitors

tetrahydrocerulenin <1> [1]

4 Enzyme Structure

Molecular weight

75000 <1> (<1> gel filtration [1]) [1]

Subunits

? <1> (<1> x * 37000, enzyme may be a component of a multifunctional protein or may exist as a dimer SDS-PAGE [1]) [1]

6 Stability

General stability information

<1>, susceptible to proteolysis [1]

References

[1] Roberts, G.; Leadly, P.F.: Use of [^{3}H]tetrahydrocerulenin to assay condensing enzyme activity in Streptomyces erythreus. Biochem. Soc. Trans., **12**, 642-643 (1984)

Trihydroxystilbene synthase 2.3.1.95

1 Nomenclature

EC number
2.3.1.95

Systematic name
malonyl-CoA:4-coumaroyl-CoA malonyltransferase (cyclizing)

Recommended name
trihydroxystilbene synthase

Synonyms
resveratrol synthase
stilbene synthase
synthase, resveratrol

CAS registry number
128449-70-7

2 Source Organism

<1> *Arachis hypogaea* (peanut [1-3]; L. var. Jinpoong [5,9]) [1-5, 9]
<2> *Vitis vinifera* (grape, L. cv. Corvina [6]) [6]
<3> *Arachis hypogaea* [7]
<4> *Rheum tataricum* (tatar rhubarb [8]) [8]

3 Reaction and Specificity

Catalyzed reaction
3 malonyl-CoA + 4-coumaroyl-CoA = 4 CoA + 3,4',5-trihydroxystilbene + 4 CO_2 (<1> proposed mechanism [1])

Reaction type
acyl group transfer

Natural substrates and products
S 4-Coumaroyl-CoA + malonyl-CoA <1> (<1> key enzyme of stilbene synthesis [2]) (Reversibility: ? <1> [2]) [2]
P 3,4',5-trihydroxystilbene + CoA + CO_2 <1> [2]

Substrates and products

S 3-coumaroyl-CoA + malonyl-CoA <1> (<1> 9% of activity with 4-coumaroyl-CoA [1]) (Reversibility: ? <1> [1]) [1]

P 3,3',5-trihydroxystilbene + CoA + CO_2 <1> [1]

S 4-coumaroyl-CoA + malonyl-CoA <1, 2, 3, 4> (<1> malonyl-CoA cannot be replaced by related compounds e.g. acetyl-CoA [1]) (Reversibility: ? <1, 2, 3, 4> [1, 2, 6, 7, 8]) [1, 2, 4, 6, 7, 8]

P 3,4',5-trihydroxystilbene + CoA + CO_2 <1, 2, 3, 4> (<1> trivial name resveratrol [1]; <1> 79% resveratrol, side products: 10% *p*-coumaroyltriacetic acid lactone, 8.6% bisnoryangonin and 1.7% naringenin [4]; <4> side products: smaller amounts of bisnoryangonin-type and *p*-coumaroyltriacetic acid lactone-type pyrones [8]) [1, 2, 4, 6, 7, 8]

S 4-fluorocinnamoyl-CoA + malonyl-CoA <3> (Reversibility: ? <3> [7]) [7]

P 4'-fluoro-trans-3,5-dihydroxystyrylfuran + CoA + CO_2 <3> (<3> side products: 49% bis-noryangonin and its derivatives, 8.2% 4-coumaroyltriacetic acid and 0.7% chalcone and its derivatives [7]) [7]

S acetyl-CoA + malonyl-CoA <4> (Reversibility: ? <4> [8]) [8]

P 6-acetonyl-4-hydroxy-2-pyrone + 6-methyl-4-hydroxy-2-pyrone + CoA + CO_2 <4> [8]

S benzoyl-CoA + malonyl-CoA <4> (Reversibility: ? <4> [8]) [8]

P bisnoryangonin-type and *p*-coumaroyltriacetic acid lactone-type pyrones + CoA + Co2 <4> [8]

S caffeoyl-CoA + malonyl-CoA <1> (<1> 8% of activity with 4-coumaroyl-CoA [1]) (Reversibility: ? <1> [1]) [1]

P 3,3',4',5-tetrahydroxystilbene + CoA + CO_2 <1> [1]

S cinnamoyl-CoA + malonyl-CoA <1, 4> (<1> 10% of activity with 4-coumaroyl-CoA [1]) (Reversibility: ? <1, 4> [1, 8]) [1, 8]

P 3,5-dihydroxystilbene + CoA + CO_2 <1, 4> (<4> side products: bisnoryangonin-type and *p*-coumaroyltriacetic acid lactone-type pyrones and small amounts of pinocembrin chalcone [8]) [1, 8]

S dihydro-4-coumaroyl-CoA + malonyl-CoA <1> (<1> 13% of activity with 4-coumaroyl-CoA [1]) (Reversibility: ? <1> [1]) [1]

P 3,4',5-trihydroxybibenzyl + CoA + CO_2 <1> [1]

S feruloyl-CoA + malonyl-CoA <1> (<1> 6% of activity with 4-coumaroyl-CoA [1]) (Reversibility: ? <1> [1]) [1]

P 3,4',5-trihydroxy-3'-methoxystilbene + CoA + CO_2 <1> [1]

S isovaleryl-CoA + malonyl-CoA <4> (Reversibility: ? <4> [8]) [8]

P bisnoryangonin-type and *p*-coumaroyltriacetic acid lactone-type pyrones + CoA + CO_2 <4> [8]

S n-butyryl-CoA + malonyl-CoA <4> (Reversibility: ? <4> [8]) [8]

P bisnoryangonin-type and *p*-coumaroyltriacetic acid lactone-type pyrones + CoA + CO_2 <4> [8]

S n-hexanoyl-CoA + malonyl-CoA <4> (Reversibility: ? <4> [8]) [8]

P bisnoryangonin-type and *p*-coumaroyltriacetic acid lactone-type pyrones + CoA + CO_2 <4> [8]

S trans-3-(3-thienyl)acryloyl-CoA + malonyl-CoA <3> (Reversibility: ? <3> [7]) [7]

P trans-3,5-dihydroxystyrylthiophene + CoA + CO_2 <3> (<3> side products: 59% bis-noryangonin and its derivatives, 18.4% 4-coumaroyltriacetic acid and 0.5% chalcone and its derivatives [7]) [7]

S trans-3-furanacryloyl-CoA + malonyl-CoA <3> (Reversibility: ? <3> [7]) [7]

P trans-3,5-dihydroxystyrylfuran + CoA + CO_2 <3> (<3> side products: 32% bis-noryangonin and its derivatives, 15% 4-coumaroyltriacetic acid and 0.6% chalcone and its derivatives [7]) [7]

Inhibitors

mercuribenzoate <1> (<1> inactivation, reactivation by excess of dithiothreitol [1]) [1]

Turnover number (min^{-1})

0.32 <3> (trans-3-furanacryloyl-CoA) [7]
0.65 <3> (trans-3-(3-thienyl)acryloyl-CoA) [7]
0.68 <3> (4-fluorocinnamoyl-CoA) [7]
1.2 <3> (4-coumaroyl-CoA) [7]

Specific activity (U/mg)

0.0035 <1> [1]

K_m-Value (mM)

0.00086 <1> (malonyl-CoA, <1> P375G mutant enzyme [4]) [4]
0.002 <1> (4-coumaroyl-CoA) [1]
0.002 <1> (malonyl -CoA) [4]
0.0066 <3> (trans-3-furanacryloyl-CoA) [7]
0.0075 <3> (trans-3-(3-thienyl)acryloyl-CoA) [7]
0.0084 <3> (4-fluorocinnamoyl-CoA) [7]
0.01 <1> (malonyl-CoA) [1]
0.011 <3> (4-coumaroyl-CoA) [7]
0.016 <1> (*p*-coumaroyl-CoA, <1> P375G mutant enzyme [4]) [4]
0.022 <1> (malonyl-CoA, <1> malonyl-CoA decarboxylation, P375G mutant enzyme [4]) [4]
0.022 <1> (*p*-coumaroyl-CoA) [4]
0.031 <1> (malonyl-CoA, <1> malonyl-CoA decarboxylation [4]) [4]

pH-Optimum

7.5-8.5 <1> [1]

4 Enzyme Structure

Molecular weight

90000 <1> (<1> gel filtration, sucrose density gradient centrifugation [1]) [1]

Subunits

dimer <1> (<1> 2 * 45000, SDS-PAGE [1]) [1]

5 Isolation/Preparation/Mutation/Application

Source/tissue

berry <2> (<2> UV-light greatly stimulates resveratrol synthase mRNA accumulation in berry skin, a weak accumulation is observed during wilting i.e. the post-harvest drying process [6]) [6]
cell suspension culture <1> [1]
leaf <1> (<1> mRNA expression is induced by wounding and stress hormones such as ethylene, jasmonic acid and salicylic acid [5]) [5, 9]
pod <1> (<1> high mRNA level [5]) [5, 9]
rhizome <4> [8]
root <1> (<1> high mRNA levels, expression is induced by yeast extract and UV [5]) [5, 9]

Purification

<1> (ammonium sulfate, DE52 cellulose, hydroxylapatite [1]; recombinant resveratrol synthase, Ni^{2+}-iminodiacetic acid-Sepharose [4]) [1, 4]
<3> (recombinant resveratrol synthase, Ni-affinity chromatography [7]) [7]
<4> (recombinant resveratrol synthase, Co-affinity chromatography [8]) [8]

Cloning

<1> (sequence alignment with naringenin-chalcone synthase, EC 2.3.1.74 [3]; expression of wild-type, P375G and G374L mutant enzyme in Escherichia coli [4]) [3, 4]
<3> (expression Escherichia coli [7]) [7]
<4> (expression in Escherichia coli [8]) [8]

Engineering

G374L <1> (<1> no condensing activity [4]) [4]
P375G <1> (<1> product profile of *p*-coumaroyl condensing reaction changed to: 42% *p*-coumaroyltriacetic acid, 27% resveratrol, 24% bisnoryangonin and 7.3% naringenin [4]) [4]

6 Stability

Storage stability

<1>, -20°C, 1 mM dithiothreitol, 10% sucrose, 3 weeks, 30% loss of activity [1]

References

[1] Schöppner, A.; Kindl, H.: Purification and properties of a stilbene synthase from induced cell suspension cultures of peanut. J. Biol. Chem., **259**, 6806-6811 (1984)

[2] Schröder, J.; Schröder, G.: Stilbene and chalcone synthases: related enzymes with key functions in plant-specific pathways. Z. Naturforsch. C, **45c**, 1-8 (1990)
[3] Lanz, T.; Tropf, S.; Marner, F.J.; Schröder, J.; Schröder, G.: The role of cysteines in polyketide synthases. Site-directed mutagenesis of resveratrol and chalcone synthases, two key enzymes in different plant-specific pathways. J. Biol. Chem., **266**, 9971-9976 (1991)
[4] Suh, D.Y.; Fukuma, K.; Kagami, J.; Yamazaki, Y.; Shibuya, M.; Ebizuka, Y.; Sankawa, U.: Identification of amino acid residues important in the cyclization reactions of chalcone and stilbene synthases. Biochem. J., **350**, 229-235 (2000)
[5] Chung, I.M.; Park, M.R.; Rehman, S.; Yun, S.J.: Tissue specific and inducible expression of resveratrol synthase gene in peanut plants. Mol. Cell., **12**, 353-359 (2001)
[6] Versari, A.; Parpinello, G.P.; Tornielli, G.B.; Ferrarini, R.; Giulivo, C.: Stilbene compounds and stilbene synthase expression during ripening, wilting, and UV treatment in grape cv. Corvina. J. Agric. Food Chem., **49**, 5531-5536 (2001)
[7] Morita, H.; Noguchi, H.; Schroder, J.; Abe, I.: Novel polyketides synthesized with a higher plant stilbene synthase. Eur. J. Biochem., **268**, 3759-3766 (2001)
[8] Samappito, S.; Page, J.E.; Schmidt, J.; De-Eknamkul, W.; Kutchan, T.M.: Aromatic and pyrone polyketides synthesized by a stilbene synthase from Rheum tataricum. Phytochemistry, **62**, 313-323 (2003)
[9] Chung, I.M.; Park, M.R.; Chun, J.C.; Yun, S.J.: Resveratrol accumulation and resveratrol synthase gene expression in response to abiotic stresses and hormones in peanut plants. Plant Sci., **164**, 103-109 (2003)

Glycoprotein N-palmitoyltransferase 2.3.1.96

1 Nomenclature

EC number
2.3.1.96

Systematic name
palmitoyl-CoA:glycoprotein N-palmitoyltransferase

Recommended name
glycoprotein N-palmitoyltransferase

Synonyms
acyltransferase, mucus glycoprotein
mucus glycoprotein fatty acyltransferase

CAS registry number
97162-74-8

2 Source Organism

<1> *Homo sapiens* [1]
<2> *Rattus norvegicus* [2, 3]

3 Reaction and Specificity

Catalyzed reaction
palmitoyl-CoA + glycoprotein = CoA + N-palmitoylglycoprotein

Reaction type
acyl group transfer

Natural substrates and products
S palmitoyl-CoA + glycoprotein <1, 2> (Reversibility: ? <1, 2> [1, 2]) [1, 2]
P CoA + N-palmitoylglycoprotein

Substrates and products
S palmitoyl-CoA + glycoprotein <1, 2> (<2> gastric mucin [2]; <2> mucus glycoprotein [2]) (Reversibility: ? <1, 2> [1, 2]) [1, 2]
P CoA + N-palmitoylglycoprotein <1, 2> [1, 2]

Inhibitors
EDTA <1> (<1> inhibits at 10 mM [1]) [1]
$MgCl_2$ <1, 2> (<1> inhibits at 10 mM [1]; <2> stimulates [2]) [1, 2]

$MnCl_2$ <2> [2]
NaF <1, 2> (<1> 25 mM, optimal concentration for activation, inhibition at higher concentration [1]; <2> stimulates [2]) [1, 2]
sucralfate <2> (<2> antiulcer drug, 40% increase [3]) [3]

Activating compounds
1,4-dithiothreitol <1, 2> (<1,2> activates [1,2]) [1, 2]
Triton X-100 <1, 2> (<1> required for activity [1]; <2> stimulates [2]) [1, 2]
sofalcone <2> (<2> antiulcer drug, enhances acylation [3]) [3]

Metals, ions
$MgCl_2$ <1, 2> (<2> stimulates [2]; <1> inhibits at 10 mM [1]) [1, 2]
NaF <1> (<1> required for activity, 25 mM, inhibition at higher concentration [1]) [1]

Specific activity (U/mg)
Additional information <1, 2> (<1> specific activity in various subcellular fractions of gastric mucosa [1]; <2> specific activity in subcellular fractions of salivary glands [2]) [1, 2]

K_m-Value (mM)
0.00056 <2> (mucin, <2> gastric mucin [3]) [3]
0.00059 <2> (glycoprotein, <2> mucus glycoprotein, submandibular enzyme [2]) [2]
0.033 <2> (palmitoyl-CoA, <2> submandibular enzyme [2]) [2]

K_i-Value (mM)
0.00091 <2> (sucralfate, <2> competitive inhibition [3]) [3]

pH-Optimum
7.4 <1, 2> [1, 2]

pH-Range
6.8-8 <1> (<1> pH 6.8: 40-50% of activity maximum, pH 8.0: about 50% of activity maximum, fundic and antral mucosa enzyme [1]) [1]

5 Isolation/Preparation/Mutation/Application

Source/tissue
gastric mucosa <2> [3]
stomach <1> [1]
sublingual gland <2> [2]
submandibular gland <2> [2]

Localization
microsome <1, 2> [1, 2]
Additional information <2> (<2> Golgi-rich membrane fraction [2]) [2]

6 Stability

Storage stability

<1>, -20°C, solubilized enzyme [1]

References

[1] Liau, Y.H.; Slomiany, B.L.; Slomiany, A.; Piasek, A.; Palmer, D.; Rosenthal, W.S.: Identification of mucus glycoprotein fatty acyltransferase activity in human gastric mucosa. Digestion, **32**, 57-62 (1985)

[2] Slomiany, B.L.; Liau, Y.H.; Carter, S.R.; Zielenski, J.; Slomiany, A.: Enzymic acylation of mucus glycoprotein with palmitic acid in rat submandibular salivary gland. Arch. Oral Biol., **31**, 463-468 (1968)

[3] Slomiany, B.L.; Liau, Y.H.; Mizuta, K.; Slomiany, A.: Fatty acid acylation of mucin by gastric mucosa: effects of sofalcone and sucralfate. Biochem. Pharmacol., **36**, 3273-3276 (1987)

Glycylpeptide N-tetradecanoyltransferase 2.3.1.97

1 Nomenclature

EC number
2.3.1.97

Systematic name
tetradecanoyl-CoA:glycylpeptide N-tetradecanoyltransferase

Recommended name
glycylpeptide N-tetradecanoyltransferase

Synonyms
NMT <1-12> [1-6, 8-16, 18-21, 23, 24, 26, 28, 30-37, 39, 41-44]
Nmt1p <1> [7, 27, 29, 38, 40, 45]
myristoyl-CoA-protein N-myristoyltransferase
myristoyl-coenzyme A:protein N-myristoyl transferase
myristoylating enzymes
myristoyltransferase, protein N-
peptide N-myristoyltransferase
protein N-myristoyltransferase
Additional information <1> (<1> enzyme belongs to the superfamily of GCN5-related N-acetyltransferases [38,40]) [38, 40]

CAS registry number
110071-61-9

2 Source Organism

<-1> no activity in *Escherichia coli* [14, 16]
<1> *Saccharomyces cerevisiae* (recombinant Nmt1p from Escherichia coli [27]; strain BJ405 [21]; gene nmt1 [6,7,12,14,16,29,40]) [1-3, 6, 7, 9-14, 16, 18, 21, 23, 24, 27, 29, 38, 40, 42, 45]
<2> *Triticum aestivum* [1, 3]
<3> *Bos taurus* (recombinant type I and II enzymes from retina and liver [41]; at least 4 isoforms in the brain [26]) [4, 5, 8, 26, 28, 31, 34, 41]
<4> *Candida albicans* (strain ATCC 32354 [6]; single copy gene nmt1 [6]) [6, 43]
<5> *Mus musculus* [1, 10, 15, 19, 23]
<6> *Rattus norvegicus* (2 different splice variants [37]) [1, 8, 10, 17, 20, 22, 24, 36, 37]
<7> *Homo sapiens* (2 different splice variants [37]) [25, 30, 32, 33, 36, 37, 44]

<8> *Homo sapiens* (isoform NMT-1 [35]) [35]
<9> *Homo sapiens* (isoform NMT-2 [35]) [35]
<10> *Mus musculus* (isoform NMT-1 [35]) [35]
<11> *Mus musculus* (isoform NMT-2 [35]) [35]
<12> *Arabidopsis thaliana* (gene nmtl, 2 gene copies on different chromosomes [39]) [39]

3 Reaction and Specificity

Catalyzed reaction

tetradecanoyl-CoA + glycylpeptide = CoA + N-tetradecanoylglycylpeptide (The enzyme from yeast is highly specific for tetradecanoyl-CoA, and highly specific for N-terminal glycine in oligopeptides containing serine in the 5-position. The enzyme from mammalian heart transfers acyl groups to a specific 51 kDa acceptor protein; <1> kinetic analysis [38,40]; <7> Glu290, Val-291 and His293 within conserved region EEVEH, are essential for catalysis [36]; <3> PEST regions for recognition by clapains, putative regulatory enzymes [34]; <1> interactions between enzyme and acyl-CoA and peptide substrates, formation of a high affinity reaction intermediate [16]; <1> cooperativity between acyl-CoA and peptide binding sites [18]; <1> peptide binding site [7,27,29,45]; <1> myristoyl-CoA binding site [29,45]; <1> active site [38]; <1> binding of myristoyl-CoA to the enzyme occurs through at least a 2-step process, X-ray data structure analysis of a binary complex between enzyme and inhibitor S-(2-oxo)-pentadecyl-CoA and ternary with peptide substrate [38,40,45]; <1> requires alanine at position 5 of substrate peptides [2]; <1> ordered bi bi mechanism [27,29,38,40]; <1> ordered mechanism [7]; <1> detailed mechanism [45]; <1> mechanism [1,12,16])

Reaction type

acyl group transfer
addition <3,7> [28,30]
amide bond formation

Natural substrates and products

S myristoyl-CoA + glycylpeptide <1, 3-5, 12> (<3> enzyme is possibly regulated by calpains in vivo [34]; <12> essential for growth [39]; <1> enzyme is essential for vegetative growth [27,29]; <5> diverse protein substrates, protein biosynthesis [19]; <5> specific for Gly at N-terminus [19]) (Reversibility: ir <1> [14]; ? <1, 3-5, 12> [6, 19, 27, 29, 34, 39]) [6, 14, 19, 27, 29, 34, 39]
P N-myristoylglycylpeptide + CoA <1, 5, 12> [14, 19, 27, 29, 39]
S myristoyl-CoA + p60src-derived peptide <3> (<3> retina enzyme [41]) (Reversibility: ? <3> [41]) [41]
P N-myristoylated p60src-derived peptide + CoA <3> [41]
S myristoyl-CoA + protein <1, 12> (<1> natural protein substrates are: catalytic subunit of cAMP-dependent protein kinase, p60src, phosphatases, e.g. calcineurin B, transmembrane signalling proteins, e.g. α-subunits of

heterotrimeric G proteins, gag polyprotein precursors of several retroviruses, capsid proteins of some parvoviruses and picornaviruses [14]; <1> proteins involved in metabolic regulation, e.g. catalytic subunit of protein kinase A and G protein α subunit [13]; <1> overview proteins [7]) (Reversibility: ir <1> [14]; ? <1, 12> [1, 39]) [7, 13, 14, 39]

P ?

S myristoyl-CoA + vinculin <1> (Reversibility: ? <1> [9]) [9]

P ?

Substrates and products

S 11-(ethylthio)-undecanoyl-CoA + glycylpeptide <2> (Reversibility: ? <2> [3]) [3]

P 11-(ethylthio)-undecanoyl-glycylpeptide + CoA

S 11-oxatetradecanoyl-CoA + $G_{0\alpha}$-hexapeptide <1> (Reversibility: ? <1> [13]) [13]

P N-11-oxatetradecanoyl-$G_{0\alpha}$-hexapeptide + CoA

S 11-phenylundecanoyl-CoA + glycylpeptide <1> (Reversibility: ? <1> [18]) [18]

P 11-phenylundecanoyl-glycylpeptide + CoA

S 13-oxatetradecanoyl-CoA + $G_{0\alpha}$-hexapeptide <1> (Reversibility: ? <1> [13]) [13]

P N-13-oxatetradecanoyl-$G_{0\alpha}$-hexapeptide + CoA

S 6-oxatetradecanoyl-CoA + $G_{0\alpha}$-hexapeptide <1> (Reversibility: ? <1> [13]) [13]

P N-6-oxatetradecanoyl-$G_{0\alpha}$-hexapeptide + CoA

S dodecanoyl-CoA + Gly-Asn-Ala-Ala-Ala-Ala-Arg-Arg <1> (Reversibility: ? <1> [16]) [16]

P N-dodecanoyl-Gly-Asn-Ala-Ala-Ala-Ala-Arg-Arg + CoA

S lauroyl-CoA + glycylpeptide <6, 12> (<12> low activity [39]) (Reversibility: ? <6, 12> [22, 39]) [22, 39]

P N-lauroylglycylpeptide + CoA

S myristoleoyl-CoA + glycylpeptide <6> (Reversibility: ? <6> [22]) [22]

P N-myristoleoylglycylpeptide + CoA

S myristoyl-CoA + Arabidopsis thaliana protein CDPK6-derived peptide <12> (Reversibility: ? <12> [39]) [39]

P N-myristoylated Arabidopsis thaliana protein CDPK6-derived peptide + CoA

S myristoyl-CoA + Fen kinase-derived peptide <12> (<12> plant protein peptide [39]) (Reversibility: ? <12> [39]) [39]

P N-myristoylated Fen kinase-derived peptide + CoA

S myristoyl-CoA + Gly-(diaminobutyrate-γ-NH_2-4-azido, 3-iodosalicylamide)-Ala-Ala-Ser-Ala-Arg-Arg <1> (Reversibility: ? <1> [2]) [2]

P N-myristoyl-Gly-(diaminobutyrate-γ-NH_2-4-azido, 3-iodosalicylamide)-Ala-Ala-Ser-Ala-Arg-Arg + CoA

S myristoyl-CoA + Gly-Ala-Ala-Pro-Ser-Lys-Ile-Val <1> (Reversibility: ? <1> [40]) [40]

P N-myristoyl-Gly-Ala-Ala-Pro-Ser-Lys-Ile-Val + CoA <1> [40]

S myristoyl-CoA + Gly-Ala-Arg-Ala-Ala-Ala-Ala-Arg-Arg,Gly-Tyr-Ala-Ala-Ser-Ala-Arg-Arg <1> (Reversibility: ? <1> [2]) [2]
P N-myristoyl-Gly-Ala-Arg-Ala-Ala-Ala-Ala-Arg-Arg,Gly-Tyr-Ala-Ala-Ser-Ala-Arg-Arg + CoA
S myristoyl-CoA + Gly-Ala-Arg-Ala-Ser-Val-Leu-Ser <4> (Reversibility: ? <4> [6]) [6]
P N-myristoyl-Gly-Ala-Arg-Ala-Ser-Val-Leu-Ser + CoA
S myristoyl-CoA + Gly-Ala-Gln-Phe-Ser-Lys-Thr-Ala-Arg-Arg <7> (<7> i.e. myristolated alanine-rich C-kinase substrate MARCKS [44]) (Reversibility: ? <7> [44]) [44]
P N-myristoyl-Gly-Ala-Gln-Phe-Ser-Lys-Thr-Ala-Arg-Arg-derived peptide + CoA
S myristoyl-CoA + Gly-Asn-Ala-Ala-Ala-Ala <1> (Reversibility: ? <1> [2]) [2]
P N-myristoyl-Gly-Asn-Ala-Ala-Ala-Ala + CoA
S myristoyl-CoA + Gly-Asn-Ala-Ala-Ala-Ala-Arg-Arg <1, 4-6, 8, 9> (<8,9> cAMP-dependent protein kinase-derived peptide [35]) (Reversibility: ? <1, 4-6, 8, 9> [2, 6, 7, 10, 11, 13, 16, 19, 23, 35]) [2, 6, 7, 10, 11, 13, 16, 19, 23, 35]
P N-myristoyl-Gly-Asn-Ala-Ala-Ala-Ala-Arg-Arg + CoA <5> [19]
S myristoyl-CoA + Gly-Asn-Ala-Ala-Ala-Arg-Arg <5> (Reversibility: ? <5> [10]) [10]
P N-myristoyl-Gly-Asn-Ala-Ala-Ala-Arg-Arg + CoA
S myristoyl-CoA + Gly-Asn-Ala-Ala-Ser-Ala-Arg-Arg <1, 2, 4> (Reversibility: ? <1, 2, 4> [3, 6, 13]) [3, 6, 13]
P N-myristoyl-Gly-Asn-Ala-Ala-Ser-Ala-Arg-Arg + CoA
S myristoyl-CoA + Gly-Asn-Ala-Ala-Ser-Arg-Arg <1> (Reversibility: ? <1> [7]) [7]
P N-myristoyl-Gly-Asn-Ala-Ala-Ser-Arg-Arg + CoA
S myristoyl-CoA + Gly-Asn-Ala-Ala-Ser-Tyr-Arg-Arg <1> (Reversibility: ? <1> [12, 21]) [12, 21]
P N-myristoyl-Gly-Asn-Ala-Ala-Ser-Tyr-Arg-Arg + CoA <1> [12]
S myristoyl-CoA + Gly-Asn-Ala-Pro-Ala-Ala-Arg-Arg <1> (Reversibility: ? <1> [2]) [2]
P N-myristoyl-Gly-Asn-Ala-Pro-Ala-Ala-Arg-Arg + CoA
S myristoyl-CoA + Gly-Asn-Phe-Ala-Ala-Ala-Arg-Arg <1> (Reversibility: ? <1> [2]) [2]
P N-myristoyl-Gly-Asn-Phe-Ala-Ala-Ala-Arg-Arg + CoA
S myristoyl-CoA + Gly-Gln-Thr-Val-Thr-Thr-Pro-Leu <5> (Reversibility: ? <5> [10]) [10]
P N-myristoyl-Gly-Gln-Thr-Val-Thr-Thr-Pro-Leu + CoA
S myristoyl-CoA + Gly-Leu-Tyr-Ala-Ser-Lys-Leu-Ser <1> (Reversibility: ? <1> [7]) [7]
P N-myristoyl-Gly-Leu-Tyr-Ala-Ser-Lys-Leu-Ser + CoA
S myristoyl-CoA + Gly-Ser-Ser-Lys-Pro-Lys-Asp-Lys-Asp-Pro <5> (Reversibility: ? <5> [10]) [10]
P N-myristoyl-Gly-Ser-Ser-Lys-Pro-Lys-Asp-Lys-Asp-Pro + CoA

S myristoyl-CoA + Gly-Ser-Ser-Lys-Ser-Lys-Pro-Lys <1> (Reversibility: ? <1> [2]) [2]

P N-myristoyl-Gly-Ser-Ser-Lys-Ser-Lys-Pro-Lys + CoA

S myristoyl-CoA + Gly-Ser-Ser-Lys-Ser-Lys-Pro-Lys-Arg <3, 7> (<3,7> pp60src-derived peptide [30,34,44]) (Reversibility: ? <3, 7> [30, 34, 44]) [30, 34, 44]

P N-myristoyl-Gly-Ser-Ser-Lys-Ser-Lys-Pro-Lys-Arg + CoA <7> [30]

S myristoyl-CoA + Gly-Ser-Ser-Lys-Ser-Lys-Pro-Lys-Asp-Pro-Ser-Gln-Arg-Arg-Arg <3> (<3> pp60src-derived peptide [28]) (Reversibility: ? <3> [28]) [28]

P N-myristoyl-Gly-Ser-Ser-Lys-Ser-Lys-Pro-Lys-Asp-Pro-Ser-Gln-Arg-Arg-Arg + CoA

S myristoyl-CoA + M2 gene segment of reovirus type 3-derived peptide <3, 7> (<7> i.e. Gly-Asn-Ala-Ser-Ser-Ile-Lys-Lys-Lys [44]) (Reversibility: ? <3, 7> [34, 44]) [34, 44]

P N-myristoylated M2 gene segment of reovirus type 3-derived peptide + CoA

S myristoyl-CoA + Pr55gag-precursor-derived octapeptide <1> (<1> peptide substrate is derived from human immunodeficiency virus [27]) (Reversibility: ? <1> [27]) [27]

P N-myristoylated Pr55gag-precursor-derived peptide + CoA <1> [27]

S myristoyl-CoA + cAMP-dependent protein kinase-derived peptide <3, 6-9> (<7> i.e. Gly-Asn-Ala-Ala-Ala-Ala-Lys-Lys-Arg-Arg [44]) (Reversibility: ? <3, 6-9> [17, 20, 26, 30, 31, 34, 35, 44]) [17, 20, 26, 30, 31, 34, 35, 44]

P N-myristoylated cAMP-dependent protein kinase-derived peptide + CoA <3, 7> [26, 30]

S myristoyl-CoA + glycylpeptide <1-12> (<1-3, 5, 6, 12> substrate specificity [1, 2, 11, 13, 21-24, 27, 34, 39]; <1, 12> specific for myristoyl-CoA [1, 7, 14, 18, 39]; <1, 3, 6, 7> specific for Gly at N-terminus [1, 2, 14, 17, 18, 23, 24, 26-30, 45]; <1> myristoyl-CoA can be replaced by CoA-derivatives of hydroxytetradecanoic acids, azidoaromatic analogues of myristic acid, ω-nitrocarboxylic acids, halogen- and haloaromatic analogues of myristic acid and, with low activity, by dicarboxylic fatty acids [27]; <1> myristoyl-peptides can compete with myristoyl-CoA for the binding site [2]; <1> specific for uncharged amino acid at position 2 of peptide chain, no peptides with bulky hydrophobic side chains at position 2, broad spectrum of amino acids in position 3, overview [1]; <1> biotinylated peptides [42]; <1, 4, 5> overview peptides and proteins [2, 6, 13, 18, 23]; <6> cAMP-dependent protein kinase-derived peptide [17]; <1, 6> p60src-derived peptides [1, 14, 17, 22]; <1> myristoyl-CoA can be replaced by oxygen or sulfur mono-substituted analogs [11, 21]; <1> myristoyl-CoA can be replaced by oxygen or sulfur di-substituted analogs [11, 18]; <1> myristoyl-CoA can be replaced by phenyl-, furyl-, thienyl-, acylohexyl-substituted fatty acids [11]; <1> myristoyl-CoA can be replaced by unsaturated fatty acids [11, 21]; <1-3, 12> altered fatty acid chain length [3, 21, 39, 41]; <1> stereochemical requirements [14, 27]) (Reversibility: ir <1> [14]; ? <1-12> [1-13, 15-43]) [1-43]

P N-myristoylglycylpeptide + CoA <1, 3, 5-7, 12> [12, 19, 22, 26, 27, 30, 39-41]
S myristoyl-CoA + p60src-derived peptide <3, 6-9, 12> (<8,9> decapeptide, best substrate [35]; <7> hexadecapeptide from N-terminus [32]) (Reversibility: ? <3, 6-9, 12> [17, 20, 22, 26, 31-33, 35, 39]) [17, 20, 22, 26, 31-33, 35, 39]
P N-myristoylated p60src-derived peptide + CoA <3, 6, 7> [22, 26, 32]
S myristoyl-CoA + tumor necrosis factor-derived peptide <8, 9> (Reversibility: ? <8, 9> [35]) [35]
P N-myristoylated tumor necrosis factor-derived peptide + CoA
S n-decanoyl-CoA + glycylpeptide <6, 12> (<12> low activity [39]) (Reversibility: ? <6, 12> [22, 39]) [22, 39]
P N-decanoylglycylpeptide + CoA
S palmitoyl-CoA + Gly-Asn-Ala-Ala-Ala-Ala-Arg-Arg <1> (Reversibility: ? <1> [16]) [16]
P N-palmitoyl-Gly-Asn-Ala-Ala-Ala-Ala-Arg-Arg + CoA
S palmitoyl-CoA + glycylpeptide <1, 3, 12> (<1,3,12> poor substrate [1,34,39]; <4,6> no activity [6,17,22]) (Reversibility: ? <1, 3, 12> [1, 16, 34, 39]) [1, 16, 34, 39]
P N-palmitoylglycylpeptide + CoA
S stearoyl-CoA + glycylpeptide <12> (<12> poor substrate [39]) (Reversibility: ? <12> [39]) [39]
P N-stearoylglycylpeptide + CoA

Inhibitors

1,10-phenanthroline <1> (<1> 0.1 mM, 30% inhibition [2]) [2]
1-O-acetyl-2-fluorotetradecane <5> [15]
1-bromo-2-fluorotetradecane <5> [15]
2-bromomyristic acid <5> [15]
2-bromomyristoyl-CoA <5> [15]
2-fluoromyristic acid <5> [15]
2-fluoromyristoyl-CoA <5> [15]
2-fluorotetradecan-1-ol <5> [15]
2-hydroxymyristic acid <5> [15]
2-hydroxymyristoyl-CoA <5> [15]
Ala-Leu-Tyr-Ala-Ser-Lys-Leu-Ser <1> (<1> competitve against peptide substrate Gly-Leu-Tyr-Ala-Ser-Lys-Leu-Ser [7]) [7]
CoA <1> (<1> noncompetitive [12]) [12]
D-Ala-Leu-Ala-Ala-Ala-Ala-Arg-Arg <1> [23]
D-Ala-Phe-Ala-Ala-Ala-Ala-Arg-Arg <1> [23]
D-Ala-Tyr-Ala-Ala-Ala-Ala-Arg-Arg <1> [23]
D-Ala-Val-Ala-Ala-Ala-Ala-Arg-Arg <1> [23]
L-histidinol <7> (<7> noncompetitive, reversible by L-histidine [36]) [36]
N-myristoylglycylpeptides <6> [22]
S-(2-oxo)-pentadecyl-CoA <1, 5> (<1> competitive [29]; <5> strong [15]) [7, 12, 15, 29, 38, 45]
S-(3-oxohexadecyl)-CoA <1> [9]

SC-58272 <1, 4> (<1> peptidomimetic derived from the N-terminal sequence of the substrate ADP-ribosylation factor-2, i.e. Arf2p [29]) [29, 43]
benzyl-{3-[2-(2-ethoxymethyl-benzofuran-5-ylmethoxymethyl)-3-methyl-benzofuran-4-yloxy]-propyl}-amine <4> [43]
diethyldicarbonate <1> [2]
ethyl 4-[2-hydroxy-3-(isopropylamino)propoxy]-3-methyl-1-benzofuran-2-carboxylate <4> [43]
factor NIP71 <3, 7, 12> (<7> strong inhibition in a concentration dependent manner, noncompetitive [30]; <7,12> bovine brain inhibitor protein [30,39]; <3> competitive against activation factor NAF45 [26]) [26, 30, 39]
histamine <7> (<7> noncompetitive [36]) [36]
inhibitor protein from bovine brain <3> (<3> located in the membrane fraction, heat-stable, monomeric, 71 kDa [4]) [4]
m-calpain <3> (<3> enzyme is inactivated by cleavage, protease is specific for PEST regions, i.e. regions rich in proline, glutamic acid, serine and threonine, calpain-inhibitor calpastatin protects [34]) [34]
myristoyl-carba(dethia)-CoA <1> [9]
octapeptide inhibitors with hydrophobic amino acid in position 2 <1> [23]
p-hydroxymercuribenzoate <1> [2]
palmitoyl-CoA <6> [22]
serinal bisulfite <1> [42]
Additional information <1, 4-7> (<4> inhibitor binding site [43]; <4,7> inhibition mechanism [36,43]; <1,6> inhibitory peptides [22,23]; <5> no inhibition by 10 mM EDTA [19]; <1> overview: inhibition by glycylpeptides with varying amino acids at positions 2 to 8 [1]) [1, 19, 22, 23, 36, 43]
{3-[2-(1H-imidazol-2-yl)-3-methyl-benzofuran-4-yloxy]-propyl}-pyridin-3-ylmethyl-amine <4> [43]
{3-[2-(2,3-difluoro-phenoxymethyl)-3-methyl-benzofuran-4-yloxy]-propyl}-pyridin-3-ylmethyl-amine <4> [43]
{3-[2-(2-ethoxymethyl-benzofuran-5-yloxymethyl)-3-methyl-benzofuran-4-yloxy]-propyl}-pyridin-3-ylmethyl-amine <4> [43]

Activating compounds

EDTA <1, 12> (<12> 18% activation at 1 mM [39]; <1> slight activation [2]) [2, 39]
L-histidine <7> (<7> activation in a concentration dependent manner [36]) [36]
N-myristoyltransferase activator NAF45 <3> (<3> increase of K_m for peptide substrates [26]; <3> absolutely required for activity of brain enzyme isoforms, maximal 3-4fold activation [26]; <3> from brain, 45 kDa protein [26]) [26]
SDS <7> (<7> 3.2fold activation at 1.73 mM [30]) [30]
Tris <7> (<7> activates about 2.5fold at 225 mM compared to 40 mM [30]) [30]
Triton 770 <5> (<5> 2% w/v, activation by pretreatment [10]) [10]
Triton X-100 <12> (<12> 25% activation at 0.1% [39]) [39]
acetonitrile <7> (<7> 2.5fold activation at 5% v/v [30]) [30]

deoxycholate <5, 6> (<5,6> 2% w/v, activation by pretreatment [10]) [10]
dimethyl ammonium chloride <7> (<7> 6.5fold increase in activity with Gly-Ser-Ser-Lys-Ser-Lys-Pro-Lys-Arg as substrate, 2.5fold increase with Gly-Asn-Ala-Ala-Ala-Ala-Lys-Lys-Arg-Arg [44]) [44]
dimethylsulfoxide <7> (<7> activation mechanism [44]; <7> activation requires a serine in the peptide substrate [44]; <7> 8.5fold increase in activity with Gly-Ala-Gln-Phe-Ser-Lys-Thr-Ala-Arg-Arg, 10fold with Gly-Ser-Ser-Lys-Ser-Lys-Pro-Lys-Arg and 7fold increase with Gly-Asn-Ala-Ser-Ser-Ile-Lys-Lys-Lys [44]) [44]
ethanol <7> (<7> 5fold activation at 10% v/v [30]) [30]

Metals, ions

Additional information <1> (<1> no requirement for divalent cations [2]) [2]

Turnover number (min^{-1})

828 <1> (myristoyl-CoA, <1> chemical transformation step [38]) [38]

Specific activity (U/mg)

0.0000019 <6> (<6> microsomes from brain [10]) [10]
0.0000021 <5> (<5> microsomes from leukemia cell line L1210 [10]) [10]
0.0000029 <6> [20]
0.0000055 <7> (<7> cytosol fraction, HeLa cells [32]) [32]
0.0000076 <5> (<5> microsomes from liver [10]) [10]
0.000024 <4> (<4> purified native enzyme [6]) [6]
0.000035 <8> (<8> recombinant enzyme from COS-7 cells [35]) [35]
0.000039 <9> (<9> recombinant enzyme from COS-7 cells [35]) [35]
0.000146 <1> (<1> microsomes [10]) [10]
0.000149 <4> (<4> purified recombinant enzyme [6]) [6]
0.00032 <2> (<2> partially purified enzyme [3]) [3]
0.00076 <6> (<6> partially purified enzyme [24]) [24]
0.00115 <12> (<12> purified recombinant enzyme [39]) [39]
0.0072 <1> (<1> partially purified enzyme [23]) [23]
0.048 <3> (<3> purified recombinant protein [31]) [31]
0.096 <3> (<3> partially puified enzyme [5]) [5]
0.1 <3> (<3> purified enzyme [8]) [8]
0.15 <1> (<1> purified recombinant enzyme [12,16]) [12, 16]
0.248 <7> (<7> purified wild-type enzyme [30]) [30]
Additional information <1, 3-5> (<1> assay development, ELISA with biotinylated peptide substrates [42]; <3> activity of purified inhibitor protein from brain [4]; <1,5> assay method [7,19]) [4, 6, 7, 19, 21, 23, 42]

K_m-Value (mM)

0.000025 <1> (Gly-Asn-Ala-Ala-Ser-Arg-Arg) [7]
0.00007 <1> (Gly-Leu-Tyr-Ala-Ser-Lys-Leu-Ser) [7]
0.0001 <1> (Gly-Asn-Ala-Ala-Ser-Ala-Arg-Arg) [13]
0.0002 <2> (Gly-Asn-Ala-Ala-Ser-Ala-Arg-Arg) [3]
0.00023 <1> (Gly-Asn-Ala-Ala-Ser-Ala-Arg-Arg) [6]
0.0003 <2> (myristoyl-CoA) [3]
0.0004 <1> (myristoyl-CoA) [1]

0.0006 <1> (myristoyl-CoA, <1> with Gly-Asn-Ala-Ala-Ser-Tyr-Arg-Arg [21]) [21]
0.0007 <2> (11-(ethylthio)-undecanoyl-CoA) [3]
0.0009 <1> (Gly-Ala-Ala-Pro-Ser-Lys-Ile-Val) [40]
0.001 <1> (Gly-Asn-Ala-Ala-Ser-Ala-Arg-Arg) [3]
0.0014 <1> (myristoyl-CoA) [40]
0.0044 <1> (Gly-Ala-Arg-Ala-Ser-Val-Leu-Ser) [6]
0.0058 <3> (myristoyl-CoA, <3> recombinant enzyme [34]) [34]
0.006 <4> (Gly-Asn-Ala-Ala-Ser-Ala-Arg-Arg) [6]
0.01 <1> (Gly-Asn-Ala-Ala-Ser-Tyr-Arg-Arg) [21]
0.016 <1> (Gly-(diaminobutyrate-γ-NH_2^-4-azido, 3-iodosalicylamide)-Ala-Ala-Ser-Ala-Arg-Arg) [2]
0.016 <1> (Gly-Asn-Ala-Ala-Ala-Ala-Arg-Arg) [2, 7]
0.016 <3> (p60src-derived peptide, <3> recombinant enzyme [34]) [34]
0.017 <4> (Gly-Ala-Arg-Ala-Ser-Val-Leu-Ser) [6]
0.02 <6> (p60src-derived peptide) [22]
0.0205 <12> (Arabidopsis thaliana protein CDPK6-derived peptide, <12> recombinant enzyme [39]) [39]
0.03 <1> (Gly-Asn-Ala-Ala-Ala-Ala-Arg-Arg) [6]
0.033 <5> (Gly-Ser-Ser-Lys-Pro-Lys-Asp-Lys-Asp-Pro) [10]
0.04 <1> (Gly-Ser-Ser-Lys-Ser-Lys-Pro-Lys) [2]
0.04 <3> (p60src-derived peptide, <3> recombinant enzyme [31]) [31]
0.05 <3> (M2 gene segment of reovirus type 3-derived peptide, <3> recombinant enzyme [34]) [34]
0.06 <1> (Gly-Asn-Ala-Ala-Ala-Ala-Arg-Arg) [2, 13]
0.06 <1> (Gly-Asn-Phe-Ala-Ala-Ala-Arg-Arg) [2]
0.06 <6> (p60src-derived peptide) [17]
0.1 <3, 6> (cAMP-dependent protein kinase-derived peptide, <3> recombinant enzyme [34]) [17, 34]
0.117 <6> (cAMP-dependent protein kinase-derived peptide) [22]
0.17 <5> (Gly-Gln-Thr-Val-Thr-Thr-Pro-Leu) [10]
0.195 <7> (L-histidine) [36]
0.2 <3, 7> (cAMP-dependent protein kinase-derived peptide, <3> recombinant enzyme [31]) [31, 36]
0.3 <1> (Gly-Ala-Arg-Ala-Ala-Ala-Ala-Arg-Arg,Gly-Tyr-Ala-Ala-Ser-Ala-Arg-Arg) [2]
0.3 <1> (Gly-Asn-Ala-Pro-Ala-Ala-Arg-Arg) [2]
0.33 <5> (Gly-Asn-Ala-Ala-Ala-Arg-Arg) [10]
0.6 <4> (Gly-Asn-Ala-Ala-Ala-Ala-Arg-Arg) [6]
1.3 <1> (Gly-Asn-Ala-Ala-Ala-Ala) [2]
Additional information <1-3, 5, 6, 12> (<12> K_m for several protein-derived peptides [39]; <1,2> overview: K_m values for glycylpeptides with varying amino acids sequence, influence of amino acid residue type and substitutions on affinity [1,3,13]; <1> overview fatty acid analogs with various substituents [11,21]) [1-3, 10, 11, 13, 21, 24, 26, 39]

K_i-Value (mM)

0.000005 <1> (S-(2-oxo)-pentadecyl-CoA, <1> wild-type, 37°C [29]) [29]
0.000009 <1> (SC-58272, <1> His-tagged wild-type, 37°C [29]) [29]
0.000014 <1> (SC-58272, <1> wild-type, 37°C [29]) [29]
0.000024 <1> (S-(2-oxo)-pentadecyl-CoA) [9]
0.000055 <1> (SC-58272, <1> His-tagged mutant C217R, 37°C [29]) [29]
0.000082 <1> (SC-58272, <1> His-tagged mutant A202T, 37°C [29]) [29]
0.00025 <1> (S-(3-oxohexadecyl)-CoA) [9]
0.0003 <1> (myristoyl-carba(dethia)-CoA) [9]
0.001-0.0014 <1> (S-(2-oxo)-pentadecyl-CoA) [12]
0.045 <5> (2-hydroxymyristoyl-CoA) [15]
0.06 <1> (D-Ala-Leu-Ala-Ala-Ala-Ala-Arg-Arg) [23]
0.1 <5> (2-bromomyristic acid) [15]
0.16 <1> (D-Ala-Tyr-Ala-Ala-Ala-Ala-Arg-Arg) [23]
0.2 <5> (2-fluoromyristoyl-CoA) [15]
0.2 <5> (2-hydroxymyristic acid) [15]
0.275 <1> (CoA) [12]
0.45 <5> (2-bromomyristoyl-CoA) [15]
Additional information <1, 4> (<1> K_i of several mutants for S-(2-oxo)-pentadecyl-CoA at 24 and 37°C [29]; <1> overview: K_i values for glycylpeptides with varying amino acids at positions 2 to 5 [1]) [1, 29, 43]

pH-Optimum

7-7.5 <6> [22]
7.4 <1, 7> (<1,7> assay at [9,44]) [9, 44]
7.5 <3, 6, 8-11> (<3,6,8-11> assay at [4,5,8,35]) [4, 5, 8, 35]
7.5-8 <1> [2]
7.6 <3> (<3> assay at [28]) [28]
7.7 <1> (<1> assay at [40]) [40]
7.8 <5, 12> (<5> assay at [15,19]) [15, 19, 39]
Additional information <1> (<1> pI of recombinant enzyme: 8.15 [16]) [16]

pH-Range

5.5-8.5 <12> [39]
7-9 <1> (<1> about half-maximal activity above pH 9.0 and below pH 7.0, inactive at pH 6.0 [2]) [2]

Temperature optimum (°C)

25 <8-11> (<8-11> assay at [35]) [35]
30 <1, 3, 5, 7> (<1,3,5,7> assay at [4,5,9,19,32,36,44]) [4, 5, 9, 19, 32, 36, 44]
37 <3> (<3> assay at [28]) [28]

4 Enzyme Structure

Molecular weight

46000 <4> (<4> mass spectroscopy [43]) [43]
50000 <3> (<3> gel filtration [31]) [31]

51880 <4> (<4> calculation from gene sequence [6]) [6]
53000 <1> (<1> amino acid sequence determination [16]) [16]
55000 <1, 3, 6> (<1,3,6> gel filtration [2,8]) [1, 2, 8]
60000 <3> (<3> gel filtration [8]) [8]
60000-66000 <3> (<3> gel filtration, SDS-PAGE [5]) [5]
126000-390000 <3> (<3> high molecular weight aggregates, gel filtration [28]) [28]
Additional information <1, 3, 4, 6-12> (<3> enzyme forms multiple high molecular weight aggregates of fully active monomers [28]; <8-11> amino acid sequence alignment and comparison, isoforms 1 and 2 [35]; <1, 3, 4, 7, 12> amino acid sequence, alignment [6, 34, 37, 39]; <3> partial amino acid sequence [5]) [5, 6, 28, 34, 35, 37, 39]

Subunits

monomer <1, 3, 4, 7, 12> (<4> 1 * 45000, SDS-PAGE [43]; <6,7> 1 * 62000-63000, splice variant 1, SDS-PAGE [37]; <6> 1 * 48000, splice variant 2, SDS-PAGE [37]; <7> 1 * 48000, splice variant 2 from in vitro translation, also recombinant enzyme from Escherichia coli, SDS-PAGE [37]; <8> 1 * 49000-68000, several forms of isoform 1, SDS-PAGE [35]; <9> 1 * 65000, isoform 2, SDS-PAGE [35]; <3,12> 1 * 50000, recombinant enzyme [34,39]; <7> 1 * 62000, native from, SDS-PAGE [33]; <7> 1 * 60000, recombinant h28NMT, SDS-PAGE [33]; <7> 1 * 48000, recombinant h211NMT, SDS-PAGE [33]; <7> x * 63000, SDS-PAGE [32]; <3> 1 * 50000, tag cleaved off by enterokinase, SDS-PAGE [31]; <7> 1 * 49000, tag cleaved off by enterokinase, SDS-PAGE [30]; <1> 1 * 53000, amino acid sequence determination, SDS-PAGE [16]; <1,6> 1 * 55000, SDS-PAGE [1,2,8]; <3> 1 * 60000, SDS-PAGE [8,28]; <3> 1 * 60000, SDS-PAGE [5]; <3> 1 * 47000-49000, active form after storage at 4°C, SDS-PAGE [28]; <3> 1 * 43000, active low MW form generated by proteolysis during storage, SDS-PAGE [5]) [1, 5, 8, 16, 28, 30-35, 37, 39, 43]
Additional information <3, 7> (<7> 2 different splice variants with MW 48000 and 63000 Da [37]; <3> 2 interconvertable forms of 60000 and 47000 Da, both catalytically active, formation of multimeric complexes, molecular weight reduction is not only due to proteolysis but has a possible regulatory role in vivo [28]) [28, 37]

Posttranslational modification

no glycoprotein <3> [28]

5 Isolation/Preparation/Mutation/Application

Source/tissue

BC3H1 cell <5> [23]
HEK-293 cell <7> [37]
HeLa cell <7> [32, 37]
L-1210 cell <5> (<5> leukemic cell line [10]) [10]
brain <3, 5, 6> (<6> distribution in the brain [17]) [4, 5, 8, 10, 15, 17, 19, 20, 22, 26, 28, 37]

cardiac muscle <3> (<3> cardiac, low activity [34]) [34]
flower <12> [39]
germ <2> [1, 3]
gut <6> (<6> low activity [20]) [20]
heart <3, 6> (<6> low activity [20,22]) [20, 22, 41]
intestine <6> (<6> small [22]) [22]
kidney <6> (<6> low activity [20]) [20, 22]
leaf <12> [39]
liver <3, 5-7> (<6> low activity [20,22]) [1, 10, 20, 22, 24, 33, 41]
lung <6> (<6> low activity [20]) [20, 22]
lymphocyte <7> [37]
reticulocyte <2> [3]
retina <3> [41]
root <12> [39]
silique <12> [39]
skeletal muscle <6> (<6> low activity [22]) [22]
spleen <3> [31]
stem <12> [39]
stomach <6> [37]
Additional information <6> (<6> 2 different splicing variants occur in rat tissues, the 62 kDa variant is abundant in most tissues, the 42 kDa variant occurs in brain and stomach [37]) [37]

Localization

cytosol <3, 5-7, 12> (<3> major part [34]; <7> major part [32]) [4, 5, 17, 19, 20, 22, 32, 34, 39]
membrane <1, 5-7> (<7> minor part [32]; <1> peripheral membrane protein [1,32]) [1, 19, 22, 32]
microsome <5, 6> [10]
mitochondrion <3, 12> (<3> occasionally [34]) [34, 39]
myofibril <3> (<3> occasionally [34]) [34]
ribosome <12> (<12> major part [39]) [39]

Purification

<1> (recombinant wild-type and mutants from Escherichia coli [40]; His-tagged recombinant wild-type and mutants from Escherichia coli [29]; recombinant from Escherichia coli [12,16,38,42]; partial [9,21,23]) [1, 2, 9, 12, 16, 21, 23, 29, 38, 40, 42]
<2> (partial [3]) [3]
<3> (cardiac muscle enzyme, recombinant from Escherichia coli [34]; spleen enzyme, recombinant from Escherichia coli [31]; multiple forms from brain [28]; partial [5]; isoform from brain bound to NAF45 [26]) [4, 5, 8, 26, 28, 31, 34]
<4> (native enzyme, and recombinant enzyme from Escherichia coli [6]) [6]
<5> (partially [19]) [19]
<6> (partially [24]) [8, 24]
<7> (recombinant His-tagged wild-type from Escherichia coli [36]; recombinant h28NMT from Escherichia coli [33]; recombinant His-tagged enzyme

from Escherichia coli [32]; recombinant from Escherichia coli [30,37,44]) [30, 32, 33, 36, 37, 44]
<12> (recombinant His-tagged enzyme from Escherichia coli [39]) [39]

Renaturation

<3> (reconstitution of activity by rebuilding enzyme-NAF45 complex [26]) [26]

Crystallization

<1> (hanging drops over a solution of 0.1 M ammonium acetate, 0.1 M sodium cacodylate, pH 6.4, 20°C, 20% polyethylene glycol 4000, protein 25 mg/ml, a few days, structure determination and analysis by X-ray diffraction of binary and ternary complexes of enzyme, peptide substrate or myristoyl-CoA and inhibitor [45]) [45]
<4> (ternary complex formation incubation of protein, 60 mg/ml, with 5fold molar of myristoyl-CoA and inhibitor SC-58272 for a few hours, hanging drop at 4°C, from 0.2 M ammonium acetate, 50 mM HEPES, pH 7.5, 10-12% polyethylene glycol 3350, 2-3 weeks, crystal structure determination by X-ray diffraction analysis [43]) [43]

Cloning

<1> (expression of wild-type and mutants in Escherichia coli [40]; expression of His-tagged wild-type and mutants in Escherichia coli [29]; expression of mutant nmtl-181 in Escherichia coli [14]; expression in Escherichia coli JM101, coexpression of each of the 4 homologous rat α subunits of the signal-transducing, heterotrimeric G proteins to determine in vivo interaction of the 2 enzymes, structural analysis [13]; expression in Escherichia coli [12,14,16,38,42]) [12-14, 16, 29, 38, 41, 42]
<3> (cloning of cardiac muscle enzyme, expression in Escherichia coli, DNA and amino acid sequence determination [34]; construction of expression plasmid encoding the enzyme, lacking the first 8 amino acid residues, fused to a enterokinase cleavage site and a polyhistidine tag, expression in Escherichia coli DH5α, cDNA from spleen [31]) [31, 34]
<4> (expression in Escherichia coli, DNA sequence determination [6]) [6]
<7> (expression in Escherichia coli [44]; gene cloned from cell culture and lymphocytes, in vitro translation of 2 different splice variants, DNA and amino acid sequence determination [37]; expression of wild-type and mutants from Escherichia coli DH5α [36]; cloning of cDNA from liver mRNA, construction of 2 expression plasmids with ATG at position 28 and 211, respectively, termed h28NMT and h211NMT, functional expression in Escherichia coli and h28NMT in HEK 293 cells [33]; expression as His-tagged protein in Escherichia coli [32]; construction of expression plasmid encoding the enzyme, lacking the first 9 amino acid residues, fused to a enterokinase cleavage site and a polyhistidine tag, overexpression in Escherichia coli DH5α [30]) [25, 30, 32, 33, 36, 37, 44]
<8> (overexpression in COS-7 cell, isoform 1 and 2 [35]) [35]
<9> (overexpression in COS-7 cell, isoform 1 and 2 [35]) [35]
<10> (overexpression in COS-7 cell, isoform 1 [35]) [35]

<11> (overexpression in COS-7 cell, isoform 2 [35]) [35]
<12> (functional expression in Escherichia coli as His-tagged protein, DNA sequence analysis [39]) [39]

Engineering

A202T <1> (<1> site-directed mutagenesis, alterations of both peptide and myristoyl binding sites, 3 to 6fold increased K_i for S-(2-oxo)-pentadecyl-CoA and 6 to 9fold increased K_i for SC-58272 [29]) [29]
C217R <1> (<1> site-directed mutagenesis, 3 to 6fold increase in K_i for inhibitor SC-58272, selective alteration of the enzymes peptide binding site [29]) [29]
D417V <1> (<1> site-directed mutagenesis, reduced activity, temperature-sensitive [29]) [29]
D451K <1> (<1> site-directed mutagenesis, no functional complementation of enzyme deficient nmt1-181 mutant at 36°C [14]) [14]
D451N <1> (<1> site-directed mutagenesis, no functional complementation of enzyme deficient nmt1-181 mutant at 36°C [14]) [14]
E167K <1> (<1> site-directed mutagenesis, coexpression of wild-type with Asp451, functional complementation of enzyme deficient nmt1-181 mutant at 36°C [14]) [14]
E167Q <1> (<1> site-directed mutagenesis, kinetics similar to wild-type [40]) [40]
E289G <7> (<7> site-directed mutagenesis, 37.8% remaining activity [36]) [36]
E289G/E290G <7> (<7> site-directed mutagenesis, no remaining activity [36]) [36]
E289G/E290G/E292G <7> (<7> site-directed mutagenesis, no remaining activity [36]) [36]
E289G/E292G <7> (<7> site-directed mutagenesis, 7% remaining activity [36]) [36]
E290G <7> (<7> site-directed mutagenesis, no remaining activity [36]) [36]
E292G <7> (<7> site-directed mutagenesis, 77.6% remaining activity [36]) [36]
E292H <7> (<7> site-directed mutagenesis, 76% remaining activity [36]) [36]
E293K <1> (<1> site-directed mutagenesis, coexpression of wild-type with Asp451, functional complementation of enzyme deficient nmt1-181 mutant at 36°C [14]) [14]
F170A/L171A <1> (<1> site-directed mutagenesis, increased K_i for S-(2-oxo)-pentadecyl-CoA, increased K_m for peptide substrate, altered enzyme conformation which modifies myristoyl-CoA polarization during catalytic reaction [40]) [40]
F413S <1> (<1> site-directed mutagenesis, reduced activity, temperature-sensitive [29]) [29]
H293G <7> (<7> site-directed mutagenesis, no remaining activity [36]) [36]
H293N <7> (<7> site-directed mutagenesis, no remaining activity [36]) [36]

K389I <1> (<1> site-directed mutagenesis, reduced activity, temperature-sensitive [29]) [29]
L171S <1> (<1> site-directed mutagenesis, reduced activity, temperature-sensitive [29]) [29]
L408S <1> (<1> site-directed mutagenesis, reduced activity, temperature-sensitive [29]) [29]
L420S <1> (<1> site-directed mutagenesis, reduced activity, temperature-sensitive [29]) [29]
N102T <1> (<1> site-directed mutagenesis, reduced activity, temperature-sensitive [29]) [29]
N169L <1> (<1> site-directed mutagenesis, slightly increased K_m for peptide substrate, altered kinetics [40]) [40]
N169L/T205A <1> (<1> site-directed mutagenesis, increased K_m for peptide substrate, altered kinetics [40]) [40]
N426I <1> (<1> site-directed mutagenesis, mutation of myristoyl-binding site [29]) [29]
S328P <1> (<1> site-directed mutagenesis, highly reduced activity [29]) [29]
T205A <1> (<1> site-directed mutagenesis, altered kinetics [40]) [40]
V291G <7> (<7> site-directed mutagenesis, no remaining activity [36]) [36]
V395D <1> (<1> site-directed mutagenesis, reduced activity, temperature-sensitive [29]) [29]
Additional information <1, 4, 12> (<4> diverse mutants, evaluation of inhibition mechanism of inhibitor SC-58272 and compounds 1-5 [43]; <1> C-terminal deletion mutants M454 and L455 produce a 300-400fold reduction in the chemical transformation rate, shift of the rate-limite of the process steps [40]; <12> transgenic plants expressing the enzyme under control of the cauliflower mosaic virus 35S promotor, down-regulation by anti-sense expression leads to a growth reduced or lethal phenotype [39]; <1> disruption or deletion of nmt1-gene causes recessive lethality [14]; <1> temperature sensitive mutant strain nmt1-181, exchange of Gly to Asp, 10fold increased K_m for myristoyl-CoA at 36°C [6,14]) [6, 14, 39, 40, 43]

Application

medicine <4> (<4> target for antiviral and antifungal therapy [6,43]) [6, 43]

6 Stability

Temperature stability

100 <6> (<6> inactivation [22]) [22]

Oxidation stability

<7>, acetonitrile, stable in 5% v/v [30]
<7>, dimethylsulfoxide, stable up to 40% [44]
<7>, ethanol, stable in 10% v/v [30]

Storage stability

<1>, -60°C, 12 mg protein/ml [23]

<3>, -15°C, conversion of 60000 Da form into 47000 Da form, loss of 30% activity within 10 days [28]

<3>, 4°C, conversion of 60000 Da form into 47000 Da form, activity is stable over several months [28]

<5>, -100°C, 20 mM Tris/HCl buffer, pH 7.1, 1 mM EDTA, 1 mM DTT, 25% glycerol [19]

References

[1] Towler, D.A.; Gordon, J.I.; Adams, S.P.; Glaser, L.: The biology and enzymology of eukaryotic protein acylation. Annu. Rev. Biochem., **57**, 69-99 (1988)

[2] Towler, D.E.; Adams, S.P.; Eubanks, S.R.; Towery, D.S.; Jackson-Machelski, E.; Glaser, L.; Gordon, J.I.: Purification and characterization of yeast myristoyl CoA:protein N-myristoyltransferase [published erratum appears in Proc Natl Acad Sci U S A 1987 Nov;84(21):7523]. Proc. Natl. Acad. Sci. USA, **84**, 2708-2712 (1987)

[3] Heuckeroth, R.O.; Towler, D.A.; Adams, S.P.; Glaser, L.; Gordon, J.I.: 11-(Ethylthio)undecanoic acid. A myristic acid analogue of altered hydrophobicity which is functional for peptide N-myristoylation with wheat germ and yeast acyltransferase. J. Biol. Chem., **263**, 2127-2133 (1988)

[4] King, M.J.; Sharma, R.K.: Identification, purification and characterization of a membrane-associated N-myristoyltransferase inhibitor protein from bovine brain. Biochem. J., **291**, 635-639 (1993)

[5] McIlhinney, R.A.J.; McGlone, K.; Willis, A.C.: Purification and partial sequencing of myristoyl-CoA:protein N-myristoyltransferase from bovine brain. Biochem. J., **290**, 405-410 (1993)

[6] Wiegand, R.C.; Carr, C.; Minnerly, J.C.; Pauley, A.M.; Carron, C.P.; Langner, C.A.; Duronio, R.J.; Gordon, J.I.: The Candida albicans myristoyl-CoA:protein N-myristoyltransferase gene. Isolation and expression in Saccharomyces cerevisiae and Escherichia coli. J. Biol. Chem., **267**, 8591-8598 (1992)

[7] Rudnick, D.A.; Rocque, W.J.; McWherter, C.A.; Toth, M.V.; Jackson-Machelski, E.; Gordon, J.I.: Use of photoactivatable peptide substrates of Saccharomyces cerevisiae myristoyl-CoA:protein N-myristoyltransferase (Nmt1p) to characterize a myristoyl-CoA-Nmt1p-peptide ternary complex and to provide evidence for an ordered reaction mechanism. Proc. Natl. Acad. Sci. USA, **90**, 1087-1091 (1993)

[8] McIlhenney, R.A.J.; McGlone, K.: Characterization of the myristoyl CoA:-glycylpeptide N-myristoyl transferase from rat and bovine brain. Biochem. Soc. Trans., **20**, 341S (1992)

[9] Wagner, A.P.; Retey, J.: Synthesis of myristoyl-carba(dethia)-coenzyme A and S-(3-oxohexadecyl)-coenzyme A, two potent inhibitors of myristoyl-CoA:protein N-myristoyltransferase. Eur. J. Biochem., **195**, 699-705 (1991)

[10] Boutin, J.A.; Clarenc, J.P.; Ferry, G.; Ernould, A.P.; Remond, G.; Vincent, M.; Atassi, G.: N-myristoyl-transferase activity in cancer cells. Solubilization,

specificity and enzymatic inhibition of a N-myristoyl transferase from L1210 microsomes. Eur. J. Biochem., **201**, 257-263 (1991)
[11] Kishore, N.S.; Lu, T.; Knoll, L.J.; Katoh, A.; Rudnick, D.A.; Mehta, P.P.; Devadas, B.; Huhn, M.; Atwood, J.L.; Adams, S.P.; Gokel, G.W.; Gordon, J.I.: The substrate specificity of Saccharomyces cerevisiae myristoyl-CoA:protein N-myristoyltransferase. Analysis of myristic acid analogs containing oxygen, sulfur, double bonds, triple bonds, and/or an aromatic residue. J. Biol. Chem., **266**, 8835-8855 (1991)
[12] Rudnick, D.A.; McWherter, C.A.; Rocaque, W.J.; Lennon, P.J.; Getman, D.P.; Gordon, J.I.: Kinetic and structural evidence for a sequential ordered Bi Bi mechanism of catalysis by Saccharomyces cerevisiae myristoyl-CoA:protein N-myristoyltransferase. J. Biol. Chem., **266**, 9732-9739 (1991)
[13] Duronio, R.J.; Rudnick, D.A.; Adams, S.P.; Towler, D.A.; Gordon, J.I.: Analyzing the substrate specificity of Saccharomyces cerevisiae myristoyl-CoA: protein N-myristoyltransferase by co-expressing it with mammalian G protein α subunits in Escherichia coli. J. Biol. Chem., **266**, 10498-10504 (1991)
[14] Gordon, J.I.; Duronio, R.J.; Rudnick, D.A.; Adams, S.P.; Gokel, G.W.: Protein N-myristoylation. J. Biol. Chem., **266**, 8647-8650 (1991)
[15] Paige, L.A.; Zheng, G.Q.; DeFrees, S.A.; Cassady, J.M.; Geahlen, R.L.: Metabolic activation of 2-substituted derivatives of myristic acid to form potent inhibitors of myristoyl CoA:protein N-myristoyltransferase. Biochemistry, **29**, 10566-10573 (1990)
[16] Rudnick, D.A.; McWherter, C.A.; Adams, S.A.; Ropson, I.J.; Duronio, R.J.; Gordon, J.I.: Structural and functional studies of Saccharomyces cerevisiae myristoyl-CoA:protein N-myristoyltransferase produced in Escherichia coli. Evidence for an acyl-enzyme intermediate. J. Biol. Chem., **265**, 13370-13378 (1990)
[17] McIlhinney, R.A.J.; McGlone, K.: Characterisation of a myristoyl CoA:glycylpeptide N-myristoyl transferase activity in rat brain: subcellular and regional distribution. J. Neurochem., **54**, 110-117 (1990)
[18] Heuckeroth, R.O.; Jackson-Machelski, E.; Adams, S.P.; Kishore, N.S.; Huhn, M.; Katoh, A.; Lu, T.; Gokel, G.W.; Gordon, J.I.: Novel fatty acyl substrates for myristoyl-CoA:protein N-myristoyl-transferase. J. Lipid Res., **31**, 1121-1129 (1990)
[19] Paige, L.A.; Chafin, D.R.; Cassady, J.M.; Geahlen, R.L.: Detection of myristoyl CoA:protein N-myristoyltransferase activity by ion-exchange chromatography. Anal. Biochem., **181**, 254-258 (1989)
[20] McIlhinney, R.A.J.; McGlone, K.: Characterization, regional and subcellular distribution of an N-myristoyl-CoA:glycyltransferase activity in rat brain. Biochem. Soc. Trans., **17**, 888-889 (1989)
[21] Heuckeroth, R.O.; Glaser, L.; Gordon, J.I.: Heteroatom-substituted fatty acid analogs as substrates for N-myristoyltransferase: an approach for studying both the enzymology and function of protein acylation. Proc. Natl. Acad. Sci. USA, **85**, 8795-8799 (1988)
[22] Glover, C.J.; Goddard, C.; Felsted, R.L.: N-myristoylation of p60src. Identification of a myristoyl-CoA:glycylpeptide N-myristoyltransferase in rat tissues. Biochem. J., **250**, 485-491 (1988)

[23] Towler, D.A.; Eubanks, S.R.; Towery, D.S.; Adams, S.P.; Glaser, L.: Amino-terminal processing of proteins by N-myristoylation. Substrate specificity of N-myristoyl transferase. J. Biol. Chem., **262**, 1030-1036 (1987)

[24] Towler, D.A.; Adams, S.P.; Eubanks, S.R.; Towery, D.S.; Jackson-Machelski, E.; Glaser, L.; Gordon, J.I.: Myristoyl CoA: protein N-myristoyltransferase activities from rat liver and yeast possess overlapping yet distinct peptide substrate specificities. J. Biol. Chem., **263**, 1784-1790 (1988)

[25] Duronio, R.J.; Read, S.I.; Gordon, J.I.: Mutations of human myristoyl-CoA: protein N-myristoyltransferase cause temperature-sensitive myristic acid auxotrophy in Saccharomyces cerevisiae. Proc. Natl. Acad. Sci. USA, **89**, 4129-4133 (1992)

[26] King, M.J.; Sharma, R.K.: Differential activation of bovine brain N-myristoyltransferase(s) by a cytosolic activator. Biochem. Biophys. Res. Commun., **212**, 580-588 (1995)

[27] Lu, T.; Li, Q.; Katoh, A.; Hernandez, J.; Duffin, K.; Jackson-Machelski, E.; Knoll, L.J.; Gokel, G.W.; Gordon, J.I.: The substrate specificity of Saccharomyces cerevisiae myristoyl-CoA:protein N-myristoyltransferase. Polar probes of the enzyme's myristoyl-CoA recognition site. J. Biol. Chem., **269**, 5346-5357 (1994)

[28] Glover, C.J.; Felsted, R.L.: Identification and characterization of multiple forms of bovine brain N-myristoyltransferase. J. Biol. Chem., **270**, 23226-23233 (1995)

[29] Zhang, L.; Jackson-Machelski, E.; Gordon, J.I.: Biochemical studies of Saccharomyces cerevisiae myristoyl-coenzyme A:protein N-myristoyltransferase mutants. J. Biol. Chem., **271**, 33131-33140 (1996)

[30] Raju, R.V.S.; Datla, R.S.S.; Sharma, R.K.: Expression of human N-myristoyltransferase in Escherichia coli. Comparison with N-myristoyltransferases expressed in different tissues. Mol. Cell. Biochem., **155**, 69-76 (1996)

[31] Raju, R.V.S.; Datla, R.S.S.; Kakkarl, R.; Sharma, R.K.: Recombinant bovine spleen myristoyl CoA:protein N-myristoyltransferase. Mol. Cell. Biochem., **189**, 91-97 (1998)

[32] McIlhinney, R.A.J.; McGlone, K.: Immunocytochemical characterization and subcellular localization of human myristoyl-CoA:protein N-myristoyltransferase in HeLa cells. Exp. Cell Res., **223**, 348-356 (1996)

[33] Young, K.; Egerton, M.; Camble, R.; White, A.; McIlhinney, R.A.J.: Immunochemical characterization of human N-myristoyltransferase: evidence for more than one form of the enzyme. Biochem. Soc. Trans., **25**, S631 (1997)

[34] Raju, R.V.S.; Kakkar, R.; Datla, R.S.S.; Radhi, J.; Sharma, R.K.: Myristoyl-CoA:protein N-myristoyltransferase from bovine cardiac muscle: molecular cloning, kinetic analysis, and in vitro proteolytic cleavage by *m*-calpain. Exp. Cell Res., **241**, 23-35 (1998)

[35] Giang, D.K.; Cravatt, B.F.: A second mammalian N-myristoyltransferase. J. Biol. Chem., **273**, 6595-6598 (1998)

[36] Raju, R.V.S.; Datla, R.S.S.; Warrington, R.C.; Sharma, R.K.: Effects of L-histidine and its structural analogs on human N-myristoyltransferase activity and importance of EEVEH amino acid sequence for enzyme activity. Biochemistry, **37**, 14928-14936 (1998)

[37] McIlhinney, R.A.; Young, K.; Egerton, M.; Camble, R.; White, A.; Soloviev, M.: Characterization of human and rat brain myristoyl-CoA:protein N-myristoyltransferase: evidence for an alternative splice variant of the enzyme. Biochem. J., **333**, 491-495 (1998)

[38] Farazi, T.A.; Manchester, J.K.; Gordon, J.I.: Transient-state kinetic analysis of Saccharomyces cerevisiae myristoylCoA:protein N-myristoyltransferase reveals that a step after chemical transformation is rate limiting. Biochemistry, **39**, 15807-15816 (2000)

[39] Qi, Q.; Rajala, R.V.S.; Anderson, W.; Jiang, C.; Rozwadowski, K.; Selvaraj, G.; Sharma, R.; Datla, R.: Molecular cloning, genomic organization, and biochemical characterization of myristoyl-CoA:protein N-myristoyltransferase from Arabidopsis thaliana. J. Biol. Chem., **275**, 9673-9683 (2000)

[40] Farazi, T.A.; Manchester, J.K.; Waksman, G.; Gordon, J.I.: Pre-steady-state kinetic studies of Saccharomyces cerevisiae myristoylCoA:protein N-myristoyltransferase mutants identify residues involved in catalysis. Biochemistry, **40**, 9177-9186 (2001)

[41] Rundle, D.R.; Rajala, R.V.; Anderson, R.E.: Characterization of Type I and Type II myristoyl-CoA:protein N-myristoyltransferases with the Acyl-CoAs found on heterogeneously acylated retinal proteins. Exp. Eye Res., **75**, 87-97 (2002)

[42] Takamune, N.; Hamada, H.; Sugawara, H.; Misumi, S.; Shoji, S.: Development of an enzyme-linked immunosorbent assay for measurement of activity of myristoyl-coenzyme A:protein N-myristoyltransferase. Anal. Biochem., **309**, 137-142 (2002)

[43] Sogabe, S.; Masubuchi, M.; Sakata, K.; Fukami, T.A.; Morikami, K.; Shiratori, Y.; Ebiike, H.; Kawasaki, K.; Aoki, Y.; Shimma, N.; D'Arcy, A.; Winkler, F.K.; Banner, D.W.; Ohtsuka, T.: Crystal structures of Candida albicans N-myristoyltransferase with two distinct inhibitors. Chem. Biol., **9**, 1119-1128 (2002)

[44] Pasha, M.K.; Dimmock, J.R.; Hollenberg, M.D.; Sharma, R.K.: Enhanced activity of human N-myristoyltransferase by dimethyl sulfoxide and related solvents in the presence of serine/threonine-containing peptide substrates. Biochem. Pharmacol., **64**, 1461-1467 (2002)

[45] Farazi, T.A.; Waksman, G.; Gordon, J.I.: Structures of Saccharomyces cerevisiae N-myristoyltransferase with bound myristoylCoA and peptide provide insights about substrate recognition and catalysis. Biochemistry, **40**, 6335-6343 (2001)

Chlorogenate-glucarate O-hydroxycinnamoyltransferase 2.3.1.98

1 Nomenclature

EC number
2.3.1.98

Systematic name
chlorogenate:glucarate O-(hydroxycinnamoyl)transferase

Recommended name
chlorogenate-glucarate O-hydroxycinnamoyltransferase

Synonyms
caffeoyltransferase, chlorogenate-glucarate
chlorogenate:glucarate caffeoyltransferase
chlorogenic acid:glucaric acid O-caffeoyltransferase

CAS registry number
126124-92-3

2 Source Organism

<1> *Lycopersicon esculentum* (tomato [1]) [1]

3 Reaction and Specificity

Catalyzed reaction
chlorogenate + glucarate = quinate + 2-O-caffeoylglucarate

Reaction type
acyl group transfer

Natural substrates and products
S 5-O-caffeoylquinic acid + galactaric acid <1> (Reversibility: ? <1> [1]) [1]
P quinate + caffeoylgalactaric acid
S 5-O-caffeoylquinic acid + glucaric acid <1> (Reversibility: ? <1> [1]) [1]
P quinate + caffeoylglucaric acid

Substrates and products
S 5-O-caffeoylquinic acid + galactaric acid <1> (Reversibility: ? <1> [1]) [1]
P quinate + caffeoylgalactaric acid <1> [1]
S 5-O-caffeoylquinic acid + glucaric acid <1> (<1> i.e. chlorogenic acid [1]) (Reversibility: ? <1> [1]) [1]
P quinate + caffeoylglucaric acid <1> [1]

Inhibitors
Additional information <1> (<1> *p*-chloromercuribenzoate is no inhibitor) [1]

Metals, ions
Ca^{2+} <1> (<1> 150% of activity compared to the activity without addition of divalent ions [1]) [1]
Mg^{2+} <1> (<1> 125% of activity compared to the activity without addition of divalent ions [1]) [1]
Additional information <1> (<1> EDTA has no effect [1]) [1]

Specific activity (U/mg)
Additional information <1> [1]

K_m-Value (mM)
0.4 <1> (glucaric acid) [1]
1.7 <1> (galactaric acid) [1]
20 <1> (chlorogenic acid) [1]

pH-Optimum
5.7 <1> [1]

pH-Range
4.2-6.2 <1> (<1> 50% of maximal activity at pH 4.2 and pH 6.2 [1]) [1]

Temperature optimum (°C)
38 <1> [1]

4 Enzyme Structure

Molecular weight
40000 <1> (<1> SDS-PAGE, gel filtration [1]) [1]

5 Isolation/Preparation/Mutation/Application

Source/tissue
cotyledon <1> [1]

Purification
<1> (2400fold [1]) [1]

6 Stability

General stability information
<1>, freezing and thawing causes 20-30% loss of activity [1]
<1>, purification has to be carried out at 4°C [1]

Storage stability

<1>, -20°C, several months, no apparent loss of activity with crude extract or purified enzyme [1]

References

[1] Strack, D.; Gross, W.: Properties and activity changes of chlorogenic acid: glucaric acid caffeoyltransferase from tomato (Lycopersicon esculentum). Plant Physiol., **92**, 41-47 (1990)

Quinate O-hydroxycinnamoyltransferase 2.3.1.99

1 Nomenclature

EC number
2.3.1.99

Systematic name
feruloyl-CoA:quinate O-(hydroxycinnamoyl)transferase

Recommended name
quinate O-hydroxycinnamoyltransferase

Synonyms
CQT
HQT
hydroxycinnamoyl CoA:quinate O-hydroxycinnamyltransferase
hydroxycinnamoyl CoA:quinate hydroxycinnamyl transferase
hydroxycinnamoyl coenzyme A-quinate transferase
hydroxycinnamoyltransferase, quinate

CAS registry number
60321-02-0

2 Source Organism

<1> *Secale cereale* (rye, var. Kustro [1]) [1]
<2> *Malus sylvestris* (apple, cv. Golden Delicious [2]) [2]
<3> *Lycopersicon esculentum* (tomato, var. Moneymaker Spezialzucht [5]; var. Eurocross BB [3]) [3, 5]
<4> *Solanum tuberosum* (potato, var. Homeguard [4]) [4]
<5> *Cichorium endivia* [6, 9]
<6> *Equisetum arvense* [7]
<7> *Eribotrya japonica* (loquat [8]) [8]

3 Reaction and Specificity

Catalyzed reaction
feruloyl-CoA + quinate = CoA + O-feruloylquinate

Reaction type
acyl group transfer

Natural substrates and products

S Additional information <3, 4, 7> (<3,7> enzyme involves in the biosynthesis of chlorogenic acid [3,8]; <4> enzyme plays a role in the endergonic processes involved in the synthetic reactions leading to esterification of quinic acid with hydroxycinnamoyl residues [4]) [3, 4, 8]

P ?

Substrates and products

S 4-coumaroyl-CoA + quinate <1-5, 7> (<2-4> best substrate [2-4]; <1> at 39% the rate of the reaction with feruloyl-CoA [1]; <3> coumaric acid, caffeic acid cannot replace coumaroyl-CoA [3]; <2> cinnamoyl-CoA cannot replace coumaroyl-CoA [2]; <2> no substrates are citrate, malate, L-(+)-tartrate, UDP-glucose, myo-inositol, tyramine, agmatine [2]; <2,3> not: glucose [2,3]; <3> poor substrates of the reverse reaction are the 3'- or 4'-coumaroylquinates [3]) (Reversibility: r <3, 4> [3, 4]; ? <1, 2, 5, 7> [1, 2, 6, 8]) [1-4, 6, 8]

P CoA + 5'-O-(4-coumaroyl)quinate <2, 3>

S caffeoyl-CoA + quinate <1-7> (<1> condensation at 10% the rate of the reaction with feruloyl-CoA [1]) (Reversibility: r <3, 4, 7> [3, 4, 8]; ? <1, 2, 5, 6> [1, 2, 5, 6, 7]) [1-8]

P CoA + 5'-O-caffeoylquinate <1, 3, 4, 6, 7> (<1,3,4,6,7> i.e. chlorogenic acid [1,3-5,7,8]) [1, 3-5, 7, 8]

S feruloyl-CoA + quinate <1-5> (<1> best substrate [1]; <3> condensation at 10% the rate of the reaction with coumaroyl-CoA [3]) (Reversibility: r <3, 4> [3, 4]; ? <1, 2, 5> [1, 2, 6]) [1-4, 6]

P CoA + O-feruloylquinate <1> [1]

S *p*-coumaroyl-CoA + shikimate <1, 2, 4> (<2,3> poor substrate [2,3]; <4> not [4]) (Reversibility: ? <1, 2, 4> [1, 2, 4]) [1, 2, 4]

P CoA + 5'-O-*p*-coumaroylshikimate

S sinapoyl-CoA + quinate <1, 2> (<1> at 21% the rate of the reaction with feruloyl-CoA, [1]; <2> not [2]) (Reversibility: ? <1> [1]) [1, 2]

P CoA + 5'-O-sinapoylquinate

Inhibitors

4,4(diisothiocyano-2,2)-stilbene disulfonic acid <5> [6]

Ca^{2+} <5> (<5> inhibition at 5 mM [9]) [9]

DTT <5> (<5> concentration above 5 mM results in a marked inhibition of activity [9]) [9]

HEPES <3> [3]

Hg^{2+} <4> [4]

Mg^{2+} <5> (<5> inhibition at 5 mM [9]) [9]

N-[2-acetamido]-2-iminodiacetic acid <3> (<3> i.e. N-[carbamoylmethyl]-iminodiacetic acid or ADA [3]) [3]

Tris/HCl <2-4> (<2> reversible [2]; <4> above 10 mM [4]) [2-4]

Zn^{2+} <4> [4]

caffeic acid <4> [4]

cinnamic acid <4> [4]

coumaric acid <4> [4]

diethyldicarbonate <5> (<5> reversion after treatment with hydroxylamine [6]) [6]
ferulic acid <4> [4]
hydroxylamine <5> [6]
p-chloromercuribenzenesulfonic acid <5> [6]
quinic acid <4> [4]
sodium diphosphate <2> [2]
tricine <2> [2]
Additional information <3-5> (<4> very slight inhibition by Mg^{2+}, Be^{2+}, Ca^{2+}, Cd^{2+} [4]; <3> no inhibition by bovine serum albumin, Mg^{2+} or dithioerythritol [3]; <4> not: shikimate [4]; <5> no effect: 2,3-butanedione, phenylmethylsulfonylfluoride and N-methylmaleimide [6]) [3, 4, 6]

Activating compounds
EDTA <1, 3> (<1,3> activation [1,3]) [1, 3]
bovine serum albumin <3, 4> (<4> activation [4]; <3> not [3]) [3, 4]

Specific activity (U/mg)
0.0042 <1> [1]
0.37 <2> [2]
16.74 <4> [4]
Additional information <3, 6> (<6> change of enzyme activity during a 1-year growth period [7]; <7> change of enzyme activity during growth and ripening [8];) [3, 7, 8]

K_m-Value (mM)
0.003 <3> (coumaroyl-CoA) [3]
0.008 <2> (coumaroyl-CoA) [2]
0.009 <2> (feruloyl-CoA) [2]
0.01 <2> (caffeoyl-CoA) [2]
0.021 <4> (CoA) [4]
0.029 <3> (caffeoyl-CoA) [3]
0.05 <1> (feruloyl-CoA) [1]
0.06 <3> (chlorogenic acid) [3]
0.073 <4> (caffeoyl-CoA) [4]
0.084 <3> (CoA) [3]
0.092 <3> (coumaroylquinate) [3]
0.151 <4> (chlorogenic acid) [4]
0.43 <3> (quinate, <3> with coumaroyl-CoA [3]) [3]
0.63 <3> (quinate, <3> with caffeoyl-CoA [3]) [3]
1.3 <2> (quinate) [2]
4.3 <1> (quinate) [1]
10 <5> (coumaroyl-CoA) [9]
10 <2> (shikimate) [2]
11.4 <5> (feruloyl-CoA) [9]
14.3 <5> (caffeoyl-CoA) [9]

K_i-Value (mM)
0.074 <4> (Zn^{2+}) [4]
0.7 <4> (coumaric acid) [4]
0.82 <4> (caffeic acid) [4]
0.94 <4> (cinnamic acid) [4]
1.1 <4> (quinic acid) [4]
1.17 <4> (ferulic acid) [4]

pH-Optimum
6.3 <4> [4]
6.5 <2, 5> (<2,5> 50 mM potassium phosphate buffer [2,9]) [2, 9]
7 <3> [3]
7.5 <1> (<1> 100 mM potassium phosphate buffer [1]) [1]
Additional information <2, 5> (<2> pI: 5.4 [2]; <5> pI: 5.1 [9]) [2, 9]

pH-Range
5.6-7.9 <2> (<2> about half-maximal activity at pH 5.6 and 7.9 [2]) [2]
6.9-8 <1> (<1> about 80% of maximal activity at pH 6.9 and 8.0 [1]) [1]

Temperature optimum (°C)
45 <2> [2]

4 Enzyme Structure

Molecular weight
40000 <2, 5> (<2,5> gel filtration [2,9]) [2, 9]
41500 <4> (<4> gel filtration [4]) [4]

Subunits
trimer <4> (<4> 3 * 14900, SDS-PAGE [4]) [4]

5 Isolation/Preparation/Mutation/Application

Source/tissue
fruit <2, 3, 7> [2, 3, 5, 8]
leaf <1, 5> (<1> primary [1]) [1, 6, 9]
tuber <4> [4]

Purification
<1> (partial [1]) [1]
<2> (partial [2]) [2]
<3> [3]
<4> [4]
<5> (partial [6,9]) [6, 9]

6 Stability

Temperature stability

50 <4> (<4> 65% loss of activity after 5 min, phosphate buffer protects: 12% loss after 5 min [4]) [4]
80 <1> (<1> denaturation after 5 min [1]) [1]

Storage stability

<1>, -20°C, at least 6 months [1]
<2>, -20°C, with 10% glycerol at least 30 days [2]
<3>, -20°C, at least 2 months [3]
<4>, -20°C, 33% loss of activity after 10 months [4]
<5>, -20°C, for several months, no apparent loss of activity [9]

References

[1] Strack, D.; Keller, H.; Weissenböck, G.: Enzymatic synthesis of hydroxycinnamic acid esters of sugar acids and hydroaromatic acids by protein preparations from rye (Secale cereale) primary leaves. J. Plant Physiol., **131**, 61-73 (1987)

[2] Lotfy, L.; Fleuriet, A.; Macheix, J.J.: Partial purification and characterization of hydroxycinnamoyl CoA:transferase from apple and date fruits. Phytochemistry, **31**, 767-772 (1992)

[3] Rhodes, M.J.C.; Wooltorton, L.S.C.: The enzymic conversion of hydroxycinnamic acids to *p*-coumarylquinic and chlorogenic acids in tomato fruits. Phytochemistry, **15**, 947-951 (1976)

[4] Rhodes, M.J.C.; Wooltorton, L.S.C.; Lourencq, E.J.: Purification and properties of hydroxycinnamoyl CoA quinate hydroxycinnamoyl transferase from potatoes. Phytochemistry, **18**, 1125-1129 (1979)

[5] Strack, D.; Gross, W.; Wray, V.; Grotjahn, L.: Enzymic synthesis of caffeoylglucaric acid from chlorogenic acid and glucaric acid by a protein preparation from tomato cotyledons. Plant Physiol., **83**, 475-478 (1987)

[6] Lotfy, S.: Inactivation and kinetic characterization of hydroxycinnamoyl-CoA:hydroaromatic acid O-hydroxycinnamoyltransferases from Cichorium endivia and Phoenix dactylifera. Plant Physiol. Biochem., **33**, 423-431 (1995)

[7] Hohlfeld, M.; Veit, M.; Strack, D.: Hydroxycinnamoyltransferases involved in the accumulation of caffeic acid esters in gametophytes and sporophytes of Equisetum arvense. Plant Physiol., **111**, 1153-1159 (1996)

[8] Ding, C.K.; Chachin, K.; Ueda, Y.; Imahori, Y.; Wang, C.Y.: Metabolism of phenolic compounds during loquat fruit development. J. Agric. Food Chem., **49**, 2883-2888 (2001)

[9] Lotfy, S.; Fleuriet, A.; Macheix, J.J.: Hydroxycinnamoyl-CoA: transferases in higher plants. II. Characterization in Cichorium endivia and Raphanus sativus and comparison with other plants. Plant Physiol. Biochem., **32**, 355-363 (1994)

[Myelin-proteolipid] O-palmitoyltransferase 2.3.1.100

1 Nomenclature

EC number
2.3.1.100

Systematic name
palmitoyl-CoA:[myelin-proteolipid] O-palmitoyltransferase

Recommended name
[myelin-proteolipid] O-palmitoyltransferase

Synonyms
acyl-protein synthetase
myelin PLP acyltransferase
synthetase, acyl protein

CAS registry number
82657-98-5

2 Source Organism

<1> *Rattus norvegicus* [1]

3 Reaction and Specificity

Catalyzed reaction
palmitoyl-CoA + [myelin proteolipid] = CoA + O-palmitoyl-[myelin proteolipid]

Reaction type
acyl group transfer

Natural substrates and products
- **S** palmitoyl-CoA + myelin proteolipid protein <1> (Reversibility: ? <1> [1]) [1]
- **P** CoA + myelin proteolipid O-palmitoylprotein
- **S** Additional information <1> (<1> enzyme is probably involved in the acylation of endogenous myelin proteolipid protein [1]) [1]
- **P** ?

Substrates and products

- **S** myristoyl-CoA + myelin proteolipid protein <1> (<1> low activity [1]) (Reversibility: ? <1> [1]) [1]
- **P** CoA + myelin proteolipid O-myristoylprotein
- **S** oleoyl-CoA + myelin proteolipid protein <1> (Reversibility: ? <1> [1]) [1]
- **P** CoA + myelin proteolipid O-oleoylprotein
- **S** palmitoyl-CoA + myelin proteolipid protein <1> (Reversibility: ? <1> [1]) [1]
- **P** CoA + myelin proteolipid O-palmitoylprotein <1> [1]
- **S** stearoyl-CoA + myelin proteolipid protein <1> (Reversibility: ? <1> [1]) [1]
- **P** CoA + myelin proteolipid O-stearoylprotein

Inhibitors

Tween 20 <1> (<1> inhibits slightly at 0.5% [1]) [1]
sodium dodecyl sulfate <1> (<1> at concentration 0.5%: 92% loss of activity [1]) [1]

Activating compounds

Nonidet P-40 <1> (<1> stimulates at 0.05% [1]) [1]
Triton X-100 <1> (<1> 0.1%, stimulates [1]) [1]
Tween 20 <1> (<1> 0.05%, slight stimulation [1]) [1]
dithiothreitol <1> (<1> 1 mM, slightly stimulates [1]) [1]
octyl glucoside <1> (<1> slight stimulation [1]) [1]
sodium deoxycholate <1> (<1> stimulates at 0.5% [1]) [1]

Metals, ions

$MgCl_2$ <1> (<1> 2 mM, stimulates [1]) [1]
$MnCl_2$ <1> (<1> stimulates [1]) [1]
NaF <1> (<1> slight stimulation [1]) [1]

K_m-Value (mM)

0.041 <1> (palmitoyl-CoA) [1]
0.046 <1> (oleoyl-CoA) [1]
0.05 <1> (stearoyl-CoA) [1]
0.219 <1> (myristoyl-CoA) [1]

pH-Optimum

7.5 <1> [1]

pH-Range

6-8 <1> (<1> 25% of maximal activity at pH 6, 90% of maximal activity at pH 8 [1]) [1]

Temperature optimum (°C)

37 <1> (<1> assay at [1]) [1]

Temperature range (°C)

10-50 <1> (<1> 45% of maximal acticity at 10°C, 75% of maximal activity at 50°C [1]) [1]

5 Isolation/Preparation/Mutation/Application

Source/tissue

brain myelin <1> (<1> labeled product also found in vitro in brain tissue slices [1]) [1]

Purification

<1> (partial [1]) [1]

6 Stability

Temperature stability

75 <1> (<1> 5 min, complete loss of activity [1]) [1]

Storage stability

<1>, freezing and thawing causes marked decrease in activity [1]

References

[1] Bizzozero, O.A.; McGarry, J.F.; Lees, M.B.: Acylation of endogenous myelin proteolipid protein with different acyl-CoAs. J. Biol. Chem., **262**, 2138-2145 (1987)

Formylmethanofuran-tetrahydromethanopterin N-formyltransferase

2.3.1.101

1 Nomenclature

EC number
2.3.1.101

Systematic name
formylmethanofuran:5,6,7,8-tetrahydromethanopterin 5-formyltransferase

Recommended name
formylmethanofuran-tetrahydromethanopterin N-formyltransferase

Synonyms
FTR
N-formylmethanofuran(CHO-MFR):tetrahydromethanopterin(H4MPT) formyltransferase
formylmethanofuran:tetrahydromethanopterin formyltransferase
formyltransferase, formylmethanofuran-tetrahydromethanopterin
formyltransferase/hydrolase complex

CAS registry number
105669-83-8

2 Source Organism

<1> *Methanopyrus kandleri* [4, 10, 11, 14]
<2> *Methanobacterium thermoautotrophicum* [1, 2, 6, 7]
<3> *Archaeoglobus fulgidus* (strain VC-19, DSM 4304 [3]) [3]
<4> *Methanosarcina barkeri* (strain Fusaro, DSM 804 [7]) [6, 7]
<5> *Methanobacterium thermoautotrophicum* [5]
<6> *Methanopyrus kandleri* (strain DSM 6324 [8]) [8]
<7> *Methanosarcina barkeri* (strain Fusaro, DSM 804 [9]) [9]
<8> *Methylobacterium extorquens* (strain AM 1 [12]) [12, 13]

3 Reaction and Specificity

Catalyzed reaction
formylmethanofuran + 5,6,7,8-tetrahydromethanopterin = methanofuran + N^5-formyl-5,6,7,8-tetrahydromethanopterin (<1,2,3,4> ternary complex type mechanism [1,3,4,6]; <8> pathway via enzyme complex [12,13]; <8> enzyme complex renamed formyltransferase/hydrolase complex [13])

Reaction type

acyl group transfer

Natural substrates and products

S formylmethanofuran + 5,6,7,8-tetrahydromethanopterin <1, 2, 4, 7, 8> (<2,4> involved in the formation of methane from CO_2 [1,2,6,9]) (Reversibility: r <1, 2, 7, 8> [1, 9, 10, 12, 13]; ? <1, 4> [2, 6, 11]) [1, 2, 6, 9-13]

P methanofuran + N^5-formyl-5,6,7,8-tetrahydromethanopterin <1, 2, 4, 7, 8> [1, 2, 6, 9-13]

Substrates and products

S N-furfurylformamide + 5,6,7,8-tetrahydromethanopterin <1, 2, 4> (<1, 2, 4> pseudo-substrate [4,7]) (Reversibility: ? <1, 2, 4> [4, 7]) [4, 7]

P 2-(aminomethyl)furane + N^5-formyl-5,6,7,8-tetrahydromethanopterin

S formylmethanofuran + 5,6,7,8-tetrahydromethanopterin <1-8> (<2,4> i.e. 4-[N-(4,5,7-tricarboxyheptanoyl-γ-L-glutamyl-γ-L-glutamyl)-*p*-(β-aminoethyl)phenoxymethyl]-2-(aminomethyl)furane [2,7]; <2> equilibrium favors transfer of the formyl group to tetrahydromethanopterin at physiological pH [1]; <1> specific for tetrahydromethanopterin [4]) (Reversibility: r <1, 2, 6-8> [1, 8-10, 12-14]; ? <3-5> [2-7, 11]) [1-14]

P methanofuran + N^5-formyl-5,6,7,8-tetrahydromethanopterin <1-8> [1-14]

S Additional information <2, 4, 8> (<2,4> not: N-methylformamide, formamide, formate [7]; <8> formyltransferase/hydrolase complex also catalyzes the hydrolysis of formyl-methanofuran to formate [13]) [7, 13]

P ?

Metals, ions

$(NH_4)_2SO_4$ <1, 3, 8> (<3> can substitute for K_2HPO_4 in stimulation [3]; <8> 1M increases the activity 20fold [12]) [3, 4, 12]

K_2HPO_4 <1, 3, 8> (<3> 1.5 M stimulates 3.8fold [3]; <8> 1.5 M stimulates 10fold [12]; <1,3> no activity in absence, assay mixture contains 2 M [3,4]) [3, 4, 12]

NaCl <8> (<8> 1M increased the activity 20fold [12]) [12]

salts <1> (<1> absolutely dependent on the presence of phosphate or sulfate salts for activity, efficiency of activation in decreasing order: K_2HPO_4, $(NH_4)_2SO_4$, K_2SO_4, Na_2SO_4, Na_2HPO_4, no activation: NaCl, KCl, NH_4Cl [4]; <6> optimal salt concentration for activity: 1.5-2.0 M [8]; <1> stimulation of activity by salt 1000fold [14]) [4, 8, 14]

Additional information <8> (<8> not: Fe, Mo, V, Wo [12]) [12]

Turnover number (min^{-1})

118400 <4> (5,6,7,8-tetrahydromethanopterin) [6]

165900 <2> (5,6,7,8-tetrahydromethanopterin) [6]

Specific activity (U/mg)

539 <2> [1]

610 <4> [6]

800 <1> [14]

1415 <3> [3]

Additional information <1, 6-8> [4, 8, 9, 12]

K_m-Value (mM)

0.017 <3> (5,6,7,8-tetrahydromethanopterin) [3]
0.03 <8> (5,6,7,8-tetrahydromethanopterin) [12]
0.032 <3> (formylmethanofuran) [3]
0.04 <4> (5,6,7,8-tetrahydromethanopterin) [6]
0.04 <4, 6> (formylmethanofuran, <6> recombinant enzyme [8]) [6, 8]
0.05 <1, 2, 6, 8> (formylmethanofuran, <6> wild-type enzyme [8]) [4, 6, 8, 12, 14]
0.06 <2, 4> (formylmethanofuran) [6, 7]
0.1 <1, 6> (5,6,7,8-tetrahydromethanopterin, <6> wild-type enzyme [8]) [4, 8, 14]
0.107 <6> (5,6,7,8-tetrahydromethanopterin, <6> recombinant enzyme [8]) [8]
20 <1> (N-furfurylformamide) [4]
55 <4> (N-furfurylformamide) [7]
70 <2> (N-furfurylformamide) [7]

pH-Optimum

5 <3> [3]
6.5 <1> [4]
7 <4, 8> [6, 12]

Temperature optimum (°C)

60 <2> (<2> assay at [1]) [1]
65 <4, 7> [6, 9]
70 <3> [3]
90 <1, 6> (<6> recombinant and wild-type enzyme [8]) [4, 8, 14]

4 Enzyme Structure

Molecular weight

32000 <4> (<4> gel filtration [6]) [6]
125000 <3> (<3> PAGE [3]) [3]
310000 <8> (<8> enzyme complex with three other polypeptides that shows sequence identities with the subunits of formylmethanofuran-dehydrogenase and are required for stability of enzyme, the subunits are present in 2:2:2:2 stoichiometry [12]) [12]

Subunits

? <8> (<8> enzyme is the 32000 Da subunit of enzyme complex [12]) [12]
homotetramer <1> (<1> composed of two dimers, each subunit is subdivided into two tightly associated lobes both consisting of a predominantly antiparallel β sheet flanked by α helices forming an α/β sandwich structure, amino acid composition [11]) [11]
monomer <1, 2, 4, 6, 7> (<1,6> 1 * 35000, SDS-PAGE [4,8]; <2,4,7> 1 * 32000 SDS-PAGE [6,9]) [4, 6, 8, 9]

tetramer <1, 2, 3> (<2> 4 * 41000, SDS-PAGE before and after treatment with dimethylsuberimidate [1]; <3> 4 * 30000 SDS-PAGE [3]; <1> 4 * 35000 [14]) [1, 3, 14]

5 Isolation/Preparation/Mutation/Application

Purification

<1> (in a anaerobic chamber [14]) [4, 14]
<2> [1, 6]
<3> [3]
<6> (recombinant enzyme [8]) [8]
<8> (purification of enzyme complex [12]) [12]
<4, 7> (recombinant enzyme [9]) [6, 9]

Crystallization

<1> (X-ray diffraction studies of forms M, P and S [10]; crystal form P is grown at a salt concentration of 0.3 M $(NH_4)_2SO_4$, pH 7.0 [11]) [10, 11]

Cloning

<1, 5, 6, 8> (expression in Escherichia coli [5,8,9,12,14]) [5, 8, 9, 12, 14]

6 Stability

Temperature stability

65 <4> (<4> stable for more than 4 h [6]) [6]
70 <7> (<7> up to, in presence of 1.5 M K_2HPO_4, hyperthermophilic enzyme [9]) [9]
80 <3> (<3> complete thermostability in presence of 0.5 M K_2HPO_4 [3]) [3]
90 <1, 3, 6> (<1,3,6> stable up to, in presence of salts [3,4,8,14]) [3, 4, 8, 14]
130 <1> (<1> up to, in the presence of high lyotropic salt concentrations, mechanism of salt-dependent thermoadaption [11]) [11]

Oxidation stability

<1>, inactivated slowly under oxic conditions, purification in anaerobic chamber [14]
<3>, relatively insensitive towards inactivation by molecular oxygen [3]
<4>, stable in presence of molecular oxygen [6]
<8>, if a single chromatographic step was performed 90% loss of activity, in 24 h 100% loss of activity [12]

Organic solvent stability

Additional information <1, 3> (<1> enzyme completely soluble in 80% ammonium sulfate [4,14]; <3> enzyme soluble in 60% ammonium sulfate [3]) [4, 3, 14]

General stability information

<1>, presence of salts, 1.5 M, required for optimal stabilization, order of efficiency in protecting the enzyme from heat inactivation at 90°C: K_2HPO_4, $(NH_4)_2SO_4$, KCl, NH_4Cl, NaCl, Na_2SO_4, Na_2HPO_4 [4]

<3>, salts protect against heat inactivation, order of efficiency: K_2HPO_4, $(NH_4)_2SO_4$, KCl, NH_4Cl, Na_2SO_4, Na_2HPO_4 [3]

Storage stability

<4>, 4°C, 50 mM Tricine/KOH, pH 8, stable for 1 week [6]

References

[1] Donnelly, M.I.; Wolfe, R.S.: The role of formylmethanofuran:tetrahydromethanopterin formyltransferase in methanogenesis from carbon dioxide. J. Biol. Chem., **261**, 16652-16659 (1986)

[2] Leigh, J.A.; Rinehart, K.L.; Wolfe, R.S.: Structure of methanofuran, the carbon dioxide reduction factor of Methanobacterium thermoautotrophicum. J. Am. Chem. Soc., **106**, 3636-3640 (1984)

[3] Schwörer, B.; Breitung, J.; Klein, A.R.; Stetter, K.O.; Thauer, R.K.: Formylmethanofuran:tetrahydromethanopterin formyltransferase and N^5,N^{10}-methylenetetrahydromethanopterin dehydrogenase from the sulfate-reducing Archaeoglobus fulgidus: similarities with the enzymes from methanogenic Archaea. Arch. Microbiol., **159**, 225-232 (1993)

[4] Breitung, J.; Börner, G.; Scholz, S.; Linder, D.; Stetter, K.O.; Thauer, R.K.: Salt dependence, kinetic properties and catalytic mechanism of N-formylmethanofuran:tetrahydromethanopterin formyltransferase from the extreme thermophile Methanopyrus kandleri. Eur. J. Biochem., **210**, 971-981 (1992)

[5] DiMarco, A.A.; Sment, K.A.; Konisky, J.; Wolfe, R.S.: The formylmethanofuran: tetrahydromethanopterin formyltransferase from Methanobacterium thermoautotrophicum δ H. Nucleotide sequence and functional expression of the cloned gene. J. Biol. Chem., **265**, 472-476 (1990)

[6] Breitung, J.; Thauer, R.K.: Formylmethanofuran: tetrahydromethanopterin formyltransferase from Methanosarcina barkeri. Identification of N^5-formyltetrahydromethanopterin as the product. FEBS Lett., **275**, 226-230 (1990)

[7] Breitung, J.; Börner, G.; Karrasch, M.; Berkessel, A.; Thauer, R.K.: N-Furfurylformamide as a pseudo-substrate for formylmethanofuran converting enzymes from methanogenic bacteria. FEBS Lett., **268**, 257-260 (1990)

[8] Shima, S.; Weiss, D.S.; Thauer, R.K.: Formylmethanofuran:tetrahydromethanopterin formyltransferase (Ftr) from the hyperthermophilic Methanopyrus kandleri. Cloning, sequencing and functional expression of the ftr gene and one-step purification of the enzyme overproduced in Escherichia coli. Eur. J. Biochem., **230**, 906-913 (1995)

[9] Kunow, J.; Shima, S.; Vorholt, J.A.; Thauer, R.K.: Primary structure and properties of the formyltransferase from the mesophilic Methanosarcina

barkeri. Comparison with the enzymes from thermophilic and hyperthermophilic methanogens. Arch. Microbiol., **165**, 97-105 (1996)
[10] Shima, S.; Thauer, R.K.; Michel, H.; Ermler, U.: Crystallization and preliminary X-ray diffraction studies of formylmethanofuran:tetrahydromethanopterin formyltransferase from Methanopyrus kandleri. Proteins Struct. Funct. Genet., **26**, 118-120 (1996)
[11] Ermler, U.; Merckel, M.C.; Thauer, R.K.; Shima, S.: Formylmethanofuran: tetrahydromethanopterin formyltransferase from Methanopyrus kandleri - new insights into salt-dependence and thermostability. Structure, **5**, 635-646 (1997)
[12] Pomper, B.K.; Vorholt, J.A.: Characterization of the formyltransferase from Methylobacterium extorquens AM1. Eur. J. Biochem., **268**, 4769-4775 (2001)
[13] Pomper, B.K.; Saurel, O.; Milon, A.; Vorholt, J.A.: Generation of formate by the formyltransferase/hydrolase complex (Fhc) from Methylobacterium extorquens AM1. FEBS Lett., **523**, 133-137 (2002)
[14] Shima, S.; Thauer, R.K.: Tetrahydromethanopterin-specific enzymes from Methanopyrus kandleri. Methods Enzymol., **331**, 317-353 (2001)

N^6-Hydroxylysine O-acetyltransferase 2.3.1.102

1 Nomenclature

EC number
2.3.1.102

Systematic name
acetyl-CoA:N^6-hydroxy-L-lysine 6-acetyltransferase

Recommended name
N^6-hydroxylysine O-acetyltransferase

Synonyms
acetyltransferase, nε-hydroxylysine
N^6-hydroxylysine acetylase
N^6-hydroxylysine:acetyl CoA nε-transacetylase

CAS registry number
101077-53-6

2 Source Organism

<1> *Escherichia coli* (294, with plasmid pABN11 [1]) [1, 2]

3 Reaction and Specificity

Catalyzed reaction
acetyl-CoA + N^6-hydroxy-L-lysine = CoA + N^6-acetyl-N^6-hydroxy-L-lysine

Reaction type
acyl group transfer

Natural substrates and products
- **S** acetyl-CoA + L-N^6-hydroxylysine <1> (Reversibility: ? <1> [2]) [2]
- **P** CoA + N^6-acetyl-N^6-hydroxylysine <1> [2]
- **S** Additional information <1> (<1> enzyme is involved in the biosynthesis of aerobactin in E. coli [1]) [1]
- **P** ?

Substrates and products
- **S** acetyl-CoA + (5-hydroxyamine)pentan-1-amine <1> (Reversibility: ? <1> [1]) [1]
- **P** CoA + N-(5-aminopentyl)-N-hydroxyacetamide

S acetyl-CoA + DL-N^6-hydroxylysine <1> (Reversibility: ? <1> [1]) [1]
P CoA + N^6-acetyl-N^6-hydroxylysine
S acetyl-CoA + L-N^5-hydroxyornithine <1> (Reversibility: ? <1> [1]) [1]
P CoA + N^5-acetyl-N^5-hydroxyornithine
S acetyl-CoA + L-N^6-hydroxylysine <1> (<1> preferred substrate [1]) (Reversibility: ? <1> [1, 2]) [1, 2]
P CoA + N^6-acetyl-N^6-hydroxylysine <1> [1, 2]
S acetyl-CoA + N-methylhydroxylamine <1> (Reversibility: ? <1> [1]) [1]
P CoA + N-acetyl-N-methylhydroxylamine
S acetyl-CoA + hydroxylamine <1> (Reversibility: ? <1> [1]) [1]
P CoA + N-acetylhydroxylamine
S Additional information <1> (<1> propionyl-CoA has 5% of activity compared to acetyl-CoA [1]) [1]
P ?

Inhibitors

Coomassie brilliant blue <1> (<1> mixed inhibitor [1]) [1]

Turnover number (min^{-1})

0.00231 <1> (N-methylhydroxylamine) [1]
0.00286 <1> (N^6-hydroxylysine) [1]
0.00478 <1> (1-amino-5(hydroxyamino)pentane) [1]
0.00671 <1> (N^6-hydroxyornithine) [1]
0.1063 <1> (hydroxylamine) [1]

Specific activity (U/mg)

26 <1> (<1> N^6-hydroxylysine [1]) [1]

K_m-Value (mM)

0.043 <1> (acetyl-CoA, <1> + L-N^6-hydroxylysine [1]) [1]
0.055 <1> (L-N^6-hydroxylysine) [1]
0.079 <1> (acetyl-CoA, <1> + L-N^5-hydroxyornithine [1]) [1]
0.099 <1> (acetyl-CoA, <1> + 1-amino-5-(hydroxyamino)pentane [1]) [1]
0.101 <1> (acetyl-CoA, <1> + N-methylhydroxylamine [1]) [1]
0.317 <1> (L-N^5-hydroxyornithine) [1]
0.33 <1> (1-amino-5-(hydroxyamino)pentane) [1]
0.396 <1> (N-methylhydroxylamine) [1]
1.81 <1> (acetyl-CoA, <1> + hydroxylamine [1]) [1]
329.4 <1> (hydroxylamine) [1]

K_i-Value (mM)

0.00071 <1> (Coomassie brilliant blue) [1]

pH-Optimum

7 <1> [2]

Temperature optimum (°C)

25 <1> (<1> assay at [2]) [2]
40 <1> (<1> sharp decline above [1]) [1]

4 Enzyme Structure

Molecular weight

33000 <1> (<1> SDS-PAGE, calculated from DNA sequence [2]) [2]
150000-200000 <1> (<1> gel filtration, native enzyme [1]) [1]

Subunits

? <1> (<1> x * 33000, SDS-PAGE [1]) [1]

5 Isolation/Preparation/Mutation/Application

Purification

<1> (with plasmid pABN11 [1]) [1, 2]

Cloning

<1> [2]

6 Stability

Storage stability

<1>, 20°C, stable for several days [1]
<1>, 4°C, stable for several months [1]

References

[1] Coy, M.; Paw, B.H.; Bindereif, A.; Neilands, J.B.: Isolation and properties of N^ε-hydroxylysine:acetyl coenzyme A N^ε-transacetylase from Escherichia coli pABN11. Biochemistry, **25**, 2485-2489 (1986)
[2] Lorenzo de, V.; Bindereif, A.; Paw, B.H.; Neilands, J.B.: Aerobactin biosynthesis and transport genes of plasmid ColV-K30 in Escherichia coli K-12. J. Bacteriol., **165**, 570-578 (1986)

Sinapoylglucose-sinapoylglucose O-sinapoyltransferase 2.3.1.103

1 Nomenclature

EC number

2.3.1.103

Systematic name

1-O-(4-hydroxy-3,5-dimethoxycinnamoyl)-β-D-glucoside:1-O-(4-hydroxy-3,5-dimethoxycinnamoyl-β-D-glucoside 1-O-sinapoyltransferase

Recommended name

sinapoylglucose-sinapoylglucose O-sinapoyltransferase

Synonyms

1-(hydroxycinnamoyl)-glucose:1-(hydroxycinnamoyl)-glucose hydroxycinnamoyltransferase
1-O-(4-hydroxy-3,5-dimethoxycinnamoyl)-β-D-glucoside:1-O-(4-hydroxy-3,5-dimethoxycinnamoyl)-β-D-glucoside 1-O-sinapoyltransferase
CGT
hydroxycinnamoyltransferase, hydroxycinnamoylglucose-hydroxycinnamoylglucose

CAS registry number

103537-11-7

2 Source Organism

<1> *Raphanus sativus* (L. var. sativus cv. Saxa [1]) [1]

3 Reaction and Specificity

Catalyzed reaction

2 1-O-sinapoyl β-D-glucoside = D-glucose + 1,2-bis-O-sinapoyl β-D-glucoside

Reaction type

acyl group transfer
sinapoyl group transfer

Natural substrates and products

S 1-O-sinapoyl-β-D-glucopyranose + 1-O-sinapoyl-β-D-glucopyranose <1> (Reversibility: ir <1> [1]) [1]
P 1,2-di-O-sinapoyl-β-D-glucopyranose + glucose

Substrates and products

S 1-(*p*-coumaroyl)-glucose + 1-(*p*-coumaroyl)-glucose <1> (Reversibility: ir <1> [1]) [1]

P 1,2-di-O-*p*-coumaroyl-β-glucopyranose + glucose

S 1-O-sinapoyl-β-D-glucopyranose + 1-O-sinapoyl-β-D-glucopyranose <1> (Reversibility: ir <1> [1]) [1]

P 1,2-di-O-sinapoyl-β-D-glucopyranose + glucose <1> [1]

S 1-feruloyl-β-D-glucose + 1-feruloyl-β-D-glucose <1> (Reversibility: ir <1> [1]) [1]

P 1,2-di-O-feruloyl-β-glucopyranose + glucose

S Additional information <1> (<1> strict specificity of acyl transfer to the C_2 hydroxyl group of the acceptor glucose molecule [1]) [1]

P ?

Inhibitors

2-mercaptoethanol <1> (<1> 1 mM, total inhibition [1]) [1]
Ca^{2+} <1> (<1> divalent cations at concentrations of more than 5 mM are inhibitory, 10 mM, 45% remaining activity [1]) [1]
Co^{2+} <1> (<1> divalent cations at concentrations of more than 5 mM are inhibitory, 10 mM, total inhibition [1]) [1]
Mg^{2+} <1> (<1> divalent cations at concentrations of more than 5 mM are inhibitory, 10 mM, 65% remaining activity [1]) [1]
dithiothreitol <1> (<1> 1 mM, total inhibition [1]) [1]

Activating compounds

Additional information <1> (<1> no requirement for divalent cations or thiols [1]) [1]

K_m-Value (mM)

0.4 <1> (1-feruloyl-β-D-glucose) [1]
0.42 <1> (1-sinapoyl-β-D-glucose) [1]
0.58 <1> (1-(*p*-coumaroyl)-glucose) [1]

pH-Optimum

8 <1> (<1> HEPES-buffer [1]) [1]

pH-Range

7-9 <1> (<1> 50% of maximal activity at pH 7.0 and pH 9.0 [1]) [1]

Temperature optimum (°C)

42 <1> [1]

4 Enzyme Structure

Molecular weight

55000 <1> (<1> gel filtration, SDS-PAGE [1]) [1]

Subunits

monomer <1> (<1> 1 * 55000, SDS-PAGE [1]) [1]

5 Isolation/Preparation/Mutation/Application

Source/tissue

cotyledon <1> (<1> dark grown seedlings show higher activity than light grown seedlings [1]) [1]

Purification

<1> (partial [1]) [1]

6 Stability

Storage stability

<1>, -20°C, presence of 3 mg/ml bovine serum albumin, at least 1 year [1]

References

[1] Dahlbender, B.; Strack, D.: Pruification and properties of 1-(hydroxycinnamoyl)-glucose:1-(hydroxycinnamoyl)-glucose hydroxycinnamoyl-transferase from radish seedlings. Phytochemistry, **25**, 1043-1046 (1986)

1-Alkenylglycerophosphocholine O-acyltransferase 2.3.1.104

1 Nomenclature

EC number
2.3.1.104

Systematic name
acyl-CoA:1-alkenylglycerophosphocholine O-acyltransferase

Recommended name
1-alkenylglycerophosphocholine O-acyltransferase

Synonyms
acyltransferase, alkenylglycerophosphocholine
Additional information (not identical with EC 2.3.1.121)

CAS registry number
102925-32-6

2 Source Organism

<1> *Oryctolagus cuniculus* (rabbit [1]) [1]
<2> *Cavia porcellus* (guinea pig [1,2]) [1, 2]
<3> *Sus scrofa* [1]
<4> *Canis familiaris* [1]
<5> *Mesocricetus auratus* (hamster [1]) [1]

3 Reaction and Specificity

Catalyzed reaction
acyl-CoA + 1-alkenylglycerophosphocholine = CoA + 1-alkenyl-2-acylglycerophosphocholine

Reaction type
acyl group transfer

Natural substrates and products
S arachidonoyl-CoA + 1-alkenylglycerophosphocholine <2> (Reversibility: ? <2> [1]) [1]
P CoA + 1-alkenyl-2-arachidonoylglycerophosphocholine
S arachidoyl-CoA + 1-alkenylglycerophosphocholine <2> (Reversibility: ? <2> [1]) [1]

P CoA + 1-alkenyl-2-arachidoylglycerophosphocholine
S linoleoyl-CoA + 1-alkenylglycerophosphocholine <2> (Reversibility: ? <2> [1, 2]) [1, 2]
P CoA + 1-alkenyl-2-linoleoylglycerophosphocholine
S oleoyl-CoA + 1-alkenylglycerophosphocholine <2> (Reversibility: ? <2> [1]) [1]
P CoA + 1-alkenyl-2-oleoylglycerophosphocholine <2> [1]
S palmitoyl-CoA + 1-alkenylglycerophosphocholine <2> (Reversibility: ? <2> [1]) [1]
P CoA + 1-alkenyl-2-palmitoylglycerophosphocholine
S stearoyl-CoA + 1-alkenylglycerophosphocholine <2> (Reversibility: ? <2> [1]) [1]
P CoA + 1-alkenyl-2-stearoylglycerophosphocholine

Substrates and products

S acyl-CoA + 1-alkenylglycerophosphocholine <2> (<2> unsaturated acyl-CoA preferred [1,2]) (Reversibility: ? <2> [1, 2]) [1, 2]
P CoA + 1-alkenyl-2-acylglycerophosphocholine <2> [1, 2]
S arachidonoyl-CoA + 1-alkenylglycerophosphocholine <2> (Reversibility: ? <2> [1]) [1]
P CoA + 1-alkenyl-2-arachidonoylglycerophosphocholine
S arachidoyl-CoA + 1-alkenylglycerophosphocholine <2> (<2> poor substrate [1]) (Reversibility: ? <2> [1]) [1]
P CoA + 1-alkenyl-2-arachidoylglycerophosphocholine
S linoleoyl-CoA + 1-alkenylglycerophosphocholine <2> (<2> best substrate [1,2]) (Reversibility: ? <2> [1, 2]) [1, 2]
P CoA + 1-alkenyl-2-linoleoylglycerophosphocholine
S oleoyl-CoA + 1-alkenylglycerophosphocholine <2> (Reversibility: ? <2> [1]) [1]
P CoA + 1-alkenyl-2-oleoylglycerophosphocholine <2> [1]
S palmitoyl-CoA + 1-alkenylglycerophosphocholine <2> (<2> poor substrate [1]) (Reversibility: ? <2> [1]) [1]
P CoA + 1-alkenyl-2-palmitoylglycerophosphocholine
S stearoyl-CoA + 1-alkenylglycerophosphocholine <2> (<2> poor substrate [1]) (Reversibility: ? <2> [1]) [1]
P CoA + 1-alkenyl-2-stearoylglycerophosphocholine

Inhibitors

1-acylglycerophosphocholine <2> (<2> non-competitive [1]) [1, 2]
acyl-CoA <2> (<2> high concentrations of acyl-CoA derivatives [1]) [1]
detergent <2> (<2> overview [1]) [1]

Metals, ions

Ca^{2+} <2> (<2> 50% activation at 10 mM [1]) [1]
Mg^{2+} <2> (<2> 30% activation at 10 mM [1]) [1]

Specific activity (U/mg)

0.000039 <1> [1]
0.00007 <5> [1]

0.00011 <4> [1]
0.00013 <3> [1]
0.00019 <2> [1]

pH-Optimum
6.5 <2> [2]

Temperature optimum (°C)
25 <2> (<2> assay at [1]) [1]

5 Isolation/Preparation/Mutation/Application

Source/tissue
heart <2> (<2> most activity [1,2]) [1, 2]

Localization
microsome <2> (<2> predominant [1]) [1, 2]
mitochondrion <2> (<2> only 9% of activity are located in mitochondria [1]) [1, 2]

6 Stability

Temperature stability
55 <2> (<2> inactivation after 1 min [1]) [1]

References

[1] Arthur, G.; Choy, P.C.: Acylation of 1-alkenyl-glycerophosphocholine and 1-acyl-glycerophosphocholine in guinea pig heart. Biochem. J., **236**, 481-487 (1986)
[2] Choy, P.C.; McMaster, C.R.: 1-Alkyl- and 1-alkenylglycerophosphocholine acyltransferases. Methods Enzymol., **209**, 86-92 (1992)

Alkylglycerophosphate 2-O-acetyltransferase 2.3.1.105

1 Nomenclature

EC number
2.3.1.105

Systematic name
acetyl-CoA:1-alkyl-sn-glycero-3-phosphate 2-O-acetyltransferase

Recommended name
alkylglycerophosphate 2-O-acetyltransferase

Synonyms
alkyllyso-GP:acetyl-CoA acetyltransferase

CAS registry number
76773-96-1 (not distinguishable from EC 2.3.1.67 in Chemical Abstracts)

2 Source Organism

<1> *Rattus norvegicus* [1]
<2> *Oryctolagus cuniculus* [2, 3]

3 Reaction and Specificity

Catalyzed reaction
acetyl-CoA + 1-alkyl-sn-glycero-3-phosphate = CoA + 1-alkyl-2-acetyl-sn-glycero-3-phosphate

Reaction type
acyl group transfer

Natural substrates and products
S acetyl-CoA + 1-alkyl-sn-glycero-3-phosphate <1, 2> (<1> involved in biosynthesis of thrombocyte activating factor in animal tissue [1]) (Reversibility: ? <1, 2> [1-3]) [1-3]
P CoA + 1-alkyl-2-acetyl-sn-glycero-3-phosphate <1, 2> [1-3]

Substrates and products
S acetyl-CoA + 1-alkyl-sn-glycero-3-phosphate <1, 2> (Reversibility: ? <1, 2> [1-3]) [1-3]
P CoA + 1-alkyl-2-acetyl-sn-glycero-3-phosphate <1, 2> [1-3]

Inhibitors

ADP <2> (<2> 29% decline in subcellular fraction P3A activity, 43% decline in subcellular fraction P3B activity and 7% decline in subcellular fraction P3D activity [2]) [2]
ATP <2> (<2> little effect on the enzyme activity for the subcellular fraction P3A, 55-64% decline in the subcellular fraction P3B [2]) [2]
$CaCl_2$ <2> (<2> 1 mM $CaCl_2$ inhibits 73% and 24% activity of subcellular fractions P3B and P3D respectively [2]) [2]
EDTA <2> (<2> inhibits 30% fraction P3D, does not inhibit subcellular fraction P3A [2]) [2]
$MgADP^-$ <2> (<2> inhibitory effects less severe than MgATP [2]) [2]
MgAMP <2> (<2> less inhibitory effect than MgADP [2]) [2]
$MgATP^{2-}$ <2> (<2> extreme inhibition in all microsomal fractions [2]) [2]
$MgCl_2$ <2> (<2> at 2 mM: 30, 28 and 58% inhibition of subcellular fractions P3A, P3B and P3D activities respectively [2]) [2]
alkaline phosphatase <2> (<2> 92-95% activity lost for subcellular fractions P3A and P3B [2]) [2]
dithiothreitol <2> [3]
sodium vanadate <1> (<1> inhibits at concentrations above 0.025 mM [1]) [1]
Additional information <2> (<2> ATP, GTP, UTP, CTP show small or no inhibition effects for fraction P3A [2]; <2> not inhibited by 1-alkyl-sn-glycero-3-phosphocholine, 1-alkyl-sn-glycero-3-phosphocholine [3]) [2, 3]

Activating compounds

1-alkyl-sn-glycero-3-phosphocholine <2> (<2> small increase of activity [3]) [3]
EDTA <2> (<2> 47.8% activity in absence of both EDTA and mercaptoethanol [2]) [2]
2-mercaptoethanol <2> (<2> 47.8% activity in absence of both EDTA and mercaptoethanol [2]) [2]
Additional information <2> (<2> activity increases in 2.4 fold by preincubation with ATP, $MgCl_2$ and high speed supernatant fractions [2]) [2]

Metals, ions

Na^+ <1, 2> (<1> 50 nM NaF increases the synthesis of 1-alkyl-2-acetyl-sn-glycero-3-phosphate [1]; <2> activity appears maximal in the presence of NaF [3]) [1, 3]

K_m-Value (mM)

0.11 <2> (acetyl-CoA, <2> apparent K_m for subcellular fraction R [3]) [3]
0.137 <2> (acetyl-CoA, <2> apparent K_m for subcellular fraction P3 [3]) [3]
0.226 <1> (acetyl-CoA, <1> apparent K_m at pH 8.4 and 23°C [1]) [1]

pH-Optimum

8.4 <1> [1]

pH-Range

8-8.8 <1> [1]
8-9 <2> (<2> subcellular fractions R and P3 [3]) [3]
8.4-9 <2> (<2> subcellular fractions P [3]) [3]

5 Isolation/Preparation/Mutation/Application

Source/tissue

brain <1> [1]
cerebral cortex <2> [2, 3]
heart <1> [1]
liver <1> [1]
lung <1> [1]
nerve <2> (<2> brain [3]) [3]
renal cortex <1> [1]
renal medulla <1> [1]
spleen <1> (<1> highest activity [1]) [1]

Localization

cell body <2> (<2> brain [3]) [3]
microsome <1, 2> [1-3]

Purification

<2> (isolation of three subcellular fractions: P3A, P3B and P3D, using NaF, EDTA and mercaptoethanol to preserve the activity [2]; isolation of four subcellular fractions: S, R and P3 from cerebral cortices and P from nerve cell bodies [3]) [2, 3]

References

[1] Lee, T.C.; Malone, B.; Snyder, F.: A new de novo pathway for the formation of 1-alkyl-2-acetyl-sn-glycerols, precursors of platelet activating factor. Biochemical characterization of 1-alkyl-2-lyso-sn-glycero-3-P:acetyl-CoA acetyltransferase in rat spleen. J. Biol. Chem., **261**, 5373-5377 (1986)

[2] Baker, R.R.; Chang, H.Y.: MgATP inhibits the synthesis of 1-alkyl-2-acetyl-sn-glycero-3-phosphate by microsomal acetyltransferase of immature rabbit cerebral cortex. Biochim. Biophys. Acta, **1213**, 27-33 (1994)

[3] Baker, R.R.; Chang, H.Y.: The potential for platelet-activating factor synthesis in brain: properties of cholinephosphotransferase and 1-alkyl-sn-glycero-3-phosphate acetyltransferase in microsomal fractions of immature rabbit cerebral cortex. Biochim. Biophys. Acta, **1170**, 157-164 (1993)

Tartronate O-hydroxycinnamoyltransferase 2.3.1.106

1 Nomenclature

EC number
2.3.1.106

Systematic name
sinapoyl-CoA:2-hydroxymalonate O-(hydroxycinnamoyl)transferase

Recommended name
tartronate O-hydroxycinnamoyltransferase

Synonyms
hydroxycinnamoyl-coenzyme-A:tartronate hydroxycinnamoyltransferase
hydroxycinnamoyltransferase, tartronate
tartronate sinapoyltransferase

CAS registry number
102484-57-1

2 Source Organism

<1> *Phaseolus radiatus* (mung bean [1]) [1]

3 Reaction and Specificity

Catalyzed reaction
sinapoyl-CoA + 2-hydroxymalonate = CoA + sinapoyltartronate

Reaction type
acyl group transfer

Substrates and products
S caffeoyl-CoA + tartronic acid <1> (Reversibility: r <1> [1]) [1]
P CoA + caffeoyltartronic acid <1> [1]
S feruloyl-CoA + tartronic acid <1> (Reversibility: r <1> [1]) [1]
P CoA + feruloyltartronic acid <1> [1]
S hydroxycinnamoyl-CoA + tartronic acid <1> (<1> tartronic acid is hydroxymalonic acid [1]) (Reversibility: r <1> [1]) [1]
P CoA + hydroxycinnamoyltartronic acid <1> [1]
S *p*-coumaroyl-CoA + tartronic acid <1> (Reversibility: r <1> [1]) [1]
P CoA + *p*-coumaroyltartronic acid <1> [1]

S Additional information <1> (<1> 1-(*p*-coumaroyl)-glucose does not act as acyl donor [1]) [1]
P ?

Specific activity (U/mg)
0.000402 <1> (<1> reverse reaction [1]) [1]
0.000414 <1> (<1> forward reaction [1]) [1]

K_m-Value (mM)
0.011 <1> (*p*-coumaroyltartronic acid) [1]
0.02 <1> (caffeoyltartronic acid) [1]
0.061 <1> (CoA, <1> + *p*-coumaroyltartronic acid [1]) [1]
0.063 <1> (CoA, <1> + caffeoyltartronic acid [1]) [1]
0.065 <1> (feruloyltartronic acid) [1]

pH-Optimum
7 <1> (<1> in presence of EDTA and a thiol compound, e.g. dithioerythritol [1]) [1]

Temperature optimum (°C)
30 <1> [1]

5 Isolation/Preparation/Mutation/Application

Source/tissue
leaf <1> [1]
root <1> [1]

Purification
<1> (partial, from leaf [1]) [1]

References

[1] Strack, D.; Ruhoff, R.; Gräwe, W.: Hydroxycinnamoyl-coenzyme-A:tartronate hydroxycinnamoyltransferase in protein preparations from mung bean. Phytochemistry, **25**, 833-837 (1986)

Deacetylvindoline O-acetyltransferase 2.3.1.107

1 Nomenclature

EC number
2.3.1.107

Systematic name
acetyl-CoA:deacetylvindoline 4-O-acetyltransferase

Recommended name
deacetylvindoline O-acetyltransferase

Synonyms
17-O-deacetylvindoline-17-O-acetyltransferase
DAT
acetyl-CoA:17-O-deacetylvindoline 17-O-acetyltransferase
acetyl-CoA-17-O-deacetylvindoline 17-O-acetyltransferase
acetylcoenzyme A-deacetylvindoline 4-O-acetyltransferase
acetylcoenzyme A:deacetylvindoline 4-O-acetyltransferase
acetylcoenzyme A:deacetylvindoline O-acetyltransferase
acetyltransferase, deacetylvindoline
deacetylvindoline acetyltransferase
deacetylvindoline-4-O-acetyltransferase

CAS registry number
100630-41-9

2 Source Organism

<1> *Catharanthus roseus* [1, 2, 3, 4, 5, 6, 7, 8]

3 Reaction and Specificity

Catalyzed reaction
acetyl-CoA + deacetylvindoline = CoA + vindoline

Reaction type
acyl group transfer

Natural substrates and products
S acetyl-CoA + 17-O-deacetylvindoline <1> (<1>, the enzyme catalyzes the final step in the biosynthesis of vindoline [1,2,3,4,7,8]) (Reversibility: r <1> [1]; ? <1> [2, 3, 4, 5, 6, 7, 8]) [1, 2, 3, 4, 7, 8]
P CoA + vindoline <1> [1]

Substrates and products

- **S** acetyl-CoA + 17-O-deacetylvindoline <1> (Reversibility: r <1> [1]; ? <1> [2, 3, 4, 5, 6, 7, 8]) [1, 2, 3, 4]
- **P** CoA + vindoline
- **S** propionyl-CoA + 17-O-deacetylvindoline <1> (Reversibility: ? <1> [5]) [5]
- **P** CoA + 17-O-deacetyl-17-O-propionylvindoline

Inhibitors

CoA <1> (<1>, 0.037 mM, 50% inhibition [2]; <1>, competitive [4]) [2, 4]
secologanin <1> (<1>, 0.5 mM, 25% inhibition [2]) [2]
tabersonine <1> (<1>, 0.045 mM, 50% inhibition [2]) [2]
tryptamine <1> (<1>, 0.5 mM, 28% inhibition [2]) [2]
vindoline <1> (<1>, 0.5 mM, 40% inhibition [2]; <1>, no product inhibition up to 2 mM [4]) [2]

Activating compounds

dithiothreitol <1> (<1>, essential to maintain enzyme activity [5]) [5]

Metals, ions

K^+ <1> (<1>, 10 mM, activity is enhanced by 18% [1]) [1]

Specific activity (U/mg)

Additional information <1> [2, 4]

K_m-Value (mM)

0.0007 <1> (vindoline) [5]
0.0013 <1> (deacetylvindoline) [2]
0.005 <1> (acetyl-CoA, <1>, in HEPES in in Tris-HCl buffer [4]) [4]
0.0054 <1> (acetyl-CoA) [5]
0.0065 <1> (acetyl-CoA) [2]
0.0095 <1> (acetyl-CoA) [6]
0.025 <1> (acetyl-CoA, <1>, in potassium phosphate buffer [4]) [4]
0.03 <1> (deacetylvindoline) [6]

K_i-Value (mM)

0.008 <1> (CoA) [4]

pH-Optimum

7.5-9 <1> [2]
8-9 <1> [5]

pH-Range

5.5-11 <1> (<1>, activity increases from pH 5.5 to 7 and decreases rapidly between pH 9 and 11 [2]) [2]
6.5-9.5 <1> (<1>, about 50% of maximal activity, pH 9.5: about 55% of maximal activity [5]) [5]

4 Enzyme Structure

Molecular weight

45000 <1> (<1>, gel filtration [5,8]) [5, 8]

Subunits

? <1> (<1>, x * 20000 + x * 26000, SDS-PAGE [4]; <1>, x * 54000, SDS-PAGE [8]) [4, 8]

5 Isolation/Preparation/Mutation/Application

Source/tissue

cotyledon <1> (<1>, of light-treated etiolated seedlings [8]) [8]
flower bud <1> (<1>, iodioblasts and laticifer of the bud [6]) [6]
idioblast <1> (<1>, the enzyme is only expressed in specialized iodioblasts and laticifer cells within light exposed tissues like leaves and stems [6]; iodioblasts and laticifer of stem [7]) [6, 7]
laticifer <1> (<1>, the enzyme is only expressed in specialized iodioblasts and laticifer cells within light exposed tissues like leaves and stems [6]) [6, 7]
leaf <1> (<1>, highest activity [1]; <1>, activity is highest in the youngest leaves from the uppermost apical region. The second, third and fourth leaves from the top retain 70, 50, and 25%, respectively, of the former [5];<1>, the enzyme is only expressed in specialized idioblasts and laticifer cells within light exposed tissues like leaves and stems [6]; <1>, idioblasts and laticifer of the leaf [7]) [1, 2, 3, 4, 5, 6, 7, 8]
stem <1> (<1>, the enzyme is only expressed in specialized iodioblasts and laticifer cells within light exposed tissues like leaves and stems [6]) [6, 7]
Additional information <1> (<1>, not detectable in cell suspension culture [1]) [1]

Localization

cytoplasm <1> [3]

Purification

<1> (partial [6]) [2, 4, 5, 6, 8]

Cloning

<1> (enzyme is expressed as recombinant His-tagged protein in Escherichia coli [6]; expression in Escherichia coli [8]) [6, 8]

6 Stability

Storage stability

<1>, 4°C, stable for several months [5]

References

[1] Fahn, W.; Gundlach, H.; Deus-Neumann, B.; Stöckigt, J.: Late enzymes of vindoline biosynthesis, acetyl-CoA:17-O-deacetylvindoline 17-O-acetyltransferase. Plant Cell Rep., **4**, 333-336 (1985)
[2] Power, R.; Kurz, W.G.W.; de Luca, V.: Purification and characterization of acetylcoenzyme A: deacetylvindoline 4-O-acetyltransferase from Catharanthus roseus. Arch. Biochem. Biophys., **279**, 370-376 (1990)
[3] De Luca, V.; Cutler, A.J.: Subcellular localization of enzymes involved in indole alkaloid biosynthesis in Catharanthus roseus. Plant Physiol., **85**, 1009-1102 (1987)
[4] Fahn, W.; Stöckigt, J.: Purification of acetyl-CoA:17-O-deacetylvindoline 17-O-acetyltransferase from Catharanthus roseus leaves. Plant Cell Rep., **8**, 613-616 (1990)
[5] De Luca, V.; Balsevich, J.; Kurz, W.G.W.: Acetyl coenzyme A:deacetylvindoline O-acetyltransferase, a novel enzyme from Catharanthus. J. Plant Physiol., **121**, 417-428 (1985)
[6] Laflamme, P.; St-Pierre, B.; De Luca, V.: Molecular and biochemical analysis of a Madagascar periwinkle root-specific minovincinine-19-hydroxy-O-acetyltransferase. Plant Physiol., **125**, 189-198 (2001)
[7] St-Pierre, B.; Vazquez-Flota, F.A.; De Luca, V.: Multicellular compartmentation of Catharanthus roseus alkaloid biosynthesis predicts intercellular translocation of a pathway intermediate. Plant Cell, **11**, 887-900 (1999)
[8] St-Pierre, B.; Laflamme, P.; Alarco, A.M.; De Luca, V.: The terminal O-acetyltransferase involved in vindoline biosynthesis defines a new class of proteins responsible for coenzyme A-dependent acyl transfer. Plant J., **14**, 703-713 (1998)

α-Tubulin N-acetyltransferase 2.3.1.108

1 Nomenclature

EC number
2.3.1.108

Systematic name
acetyl-CoA:α-tubulin-L-lysine N^6-acetyltransferase

Recommended name
α-tubulin N-acetyltransferase

Synonyms
TAT <2> [2]
acetyl-CoA:α-tubulin-L-lysine N^ε-acetyltransferase
α-tubulin acetylase
α-tubulin acetyltransferase

CAS registry number
99889-90-4

2 Source Organism

<1> *Chlamydomonas reinhardtii* (strain 21 gr, vegetative cells [4]) [1, 3, 4]
<2> *Bos taurus* [1, 2]

3 Reaction and Specificity

Catalyzed reaction
acetyl-CoA + [α-tubulin]-L-lysine = CoA + [α-tubulin]-N^6-acetyl-L-lysine

Reaction type
acyl group transfer

Natural substrates and products
S acetyl-CoA + α-tubulin L-lysine <1, 2> (<1> post-translational modification of α-tubulin [1]) (Reversibility: ? <1, 2> [1, 2]) [1, 2]
P CoA + α-tubulin N^6-acetyl-L-lysine

Substrates and products
S acetyl-CoA + α-tubulin L-lysine <1, 2> (<1> α-tubulin from Chlamydomonas is acetylated [1,3,4]; <1> α-tubulin from mouse is acetylated [1];

<2> acetylation of the lysine residue at position 40 [2]; <1,2> α-tubulin from calf is acetylated [2,4]) (Reversibility: ? <1, 2> [1, 2, 3, 4]) [1, 2, 3, 4]

P CoA + α-tubulin N^6-acetyl-L-lysine <1, 2> [1, 2]

Inhibitors

Ca^{2+} <1> (<1> at 1 mM, almost complete inhibition [3]; <1> inhibition is due to the binding of calcium to the tubulin dimer rather than to the enzyme itself [3]) [1, 3]
CoA <1> (<1> competitive inhibitor [3]) [3]
colchicine <1> (<1> at 0.001 mM is inhibitory to microtubule acetylation [3]) [3]

K_m-Value (mM)

0.002 <1> (acetyl-CoA) [3]
0.003 <2> (acetyl-CoA) [2]

K_i-Value (mM)

0.008 <1> (CoA, <1> competitive inhibitor [3]) [3]

4 Enzyme Structure

Molecular weight

62000 <2> (<2> most prominent polypeptide, SDS-PAGE [1,2]) [1, 2]
130000 <1> (<1> gel filtration [1,3]) [1, 3]

Subunits

dimer <1> (<1> 2 * 67000, SDS-PAGE [1,3]) [1, 3]

5 Isolation/Preparation/Mutation/Application

Source/tissue

brain <2> [1, 3]
neuron <1> [1]
retina <2> [1, 2]

Localization

flagellum <1> [1, 3, 4]
microtubule <1, 2> (<1> 2fold preference for polymerized versus soluble tubulin [1]; <1> enzyme acetylated both dimers and polymers of tubulin [3]) [1, 2, 3]

Purification

<1> [3]
<2> (partial [1]) [1]

References

[1] MacRae, T.H.: Tubulin post-translational modifications. Eur. J. Biochem., **244**, 265-278 (1997)
[2] Lloyd, R.A.; Gentleman, S.; Chader, G.J.: Assay of tubulin acetyltransferase activity in subcellular tissue fractions. Anal. Biochem., **216**, 42-46 (1994)
[3] Maruta, H.; Greer, K.; Rosenbaum, J.L.: The acetylation of α-tubulin and its relationship to the assembly and disassembly of microtubules. J. Cell Biol., **103**, 571-579 (1986)
[4] Greer, K.; Maruta, H.; L'Hernault, S.W.; Rosenbaum, J.L.: α-Tubulin acetylase activity in isolated Chlamydomonas flagella. J. Cell Biol., **101**, 2081-2084 (1985)

Arginine N-succinyltransferase 2.3.1.109

1 Nomenclature

EC number
2.3.1.109

Systematic name
succinyl-CoA:L-arginine N^2-succinyltransferase

Recommended name
arginine N-succinyltransferase

Synonyms
arginine succinyl transferase

CAS registry number
99676-48-9

2 Source Organism

<1> *Aeromonas formicans* [5]
<2> *Escherichia coli* (strain K12 [6]) [6]
<3> *Escherichia coli* (strain K12, Kohara clone 327 [5]) [5]
<4> *Klebsiella aerogenes* [5, 6]
<5> *Pseudomonas aeruginosa* [2, 4]
<6> *Pseudomonas aeruginosa* [5]
<7> *Pseudomonas cepacia* (NCTC 10743 [1]) [1]
<8> *Pseudomonas putida* (IRC204 [3]) [3]
<9> *Salmonella typhimurium* [5]

3 Reaction and Specificity

Catalyzed reaction
succinyl-CoA + L-arginine = CoA + N^2-succinyl-L-arginine (also acts on L-ornithine)

Reaction type
acyl group transfer

Natural substrates and products

S succinyl-CoA + L-arginine <1-9> (<1,3,4,6-9> arginine succinyltransferase pathway [1-6]; <5> catabolism of arginine [2]) (Reversibility: ? <1-9> [1-6]) [1-6]
P CoA + N^2-succinyl-L-arginine
S succinyl-CoA + L-ornithine <5> (<5> ornithine catabolic pathway [2]) (Reversibility: ? <5> [2]) [2, 1-3]
P CoA + N^2-succinyl-L-ornithine

Substrates and products

S succinyl-CoA + L-arginine <1-9> (Reversibility: ? <1-9> [1-6]) [1-6]
P CoA + N^2-succinyl-L-arginine
S succinyl-CoA + L-homoarginine <5> (Reversibility: ? <5> [4]) [4]
P CoA + N^2-succinyl-L-homoarginine
S succinyl-CoA + L-ornithine <5> (Reversibility: ? <5> [2, 4]) [2, 4]
P CoA + N^2-succinyl-L-ornithine
S Additional information <5, 8> (<5> D-arginine, D-ornithine, L-lysine, L-citrulline, 2,4-aminobutyrate, 5-aminovalerate, 2-aminovalerate, agmatine or the polyamines propanediamine, putrescine, cadaverine and spermidine are not substrates [4]; <8> devoid of ornithine succinyltransferase activity [3]) [3, 4]
P ?

Inhibitors

D-arginine <5> (<5> competitive inhibition [4]) [4]
agmatine <5> [4]
spermidine <5> [4]

Activating compounds

L-arginine <5> (<5> allosteric activator of the ornithine succinyltransferase activity [4]) [4]
L-lysine <5> (<5> activator of the ornithine succinyltransferase activity [4]) [4]

Specific activity (U/mg)

0.007-0.37 <7> (<7> depending on growth substrate [1]) [1]

K_m-Value (mM)

0.04 <5> (L-ornithine) [4]
0.04 <7> (succinyl-CoA) [1]
0.5 <5> (L-arginine) [4]
5 <7> (L-arginine) [1]

K_i-Value (mM)

0.9 <5> (spermidine) [4]
2.6 <5> (D-arginine) [4]
3.4 <5> (agmatine) [4]

pH-Optimum

7.6 <7> [1]
8.5 <5> (<5> L-homoarginine as substrate [4]) [4]
8.7 <5> (<5> L-arginine as substrate [4]) [4]
9 <5> (<5> L-ornithine as substrate [4]) [4]

4 Enzyme Structure

Molecular weight

140000 <5> (<5> PAGE [4]) [4]
150000 <5> (<5> gel filtration [4]) [4]

Subunits

tetramer <5> (<5> 2 * 35000 + 2 * 37000, heterotetramer, SDS-PAGE [4]) [4]

5 Isolation/Preparation/Mutation/Application

Purification

<5> (partial [2]) [2, 4]
<7> (partial [1]) [1]

Cloning

<3> (plasmid pLC3-11, enzyme overproduced in Escherichia coli [6]) [5, 6]
<5> [2]
<6> [5]

6 Stability

Storage stability

<5>, -20°C, retains full activity with both substrates for several weeks, loses 10-40% of its activity following repeated freezing and thawing [4]
<5>, 4°C, 10 mM potassium phosphate, pH 7.5, stable for 48 h [4]

References

[1] Vander Wauven, C.; Stalon, V.: Occurence of succinyl derivatives in the catabolism of arginine in Pseudomonas cepacia. J. Bacteriol., **11**, 882-886 (1985)
[2] Vander Wauven, C.; Jann, A.; Haas, D.; Leisinger, T.; Stalon, V.: N^2-Succinylornithine in ornithine catabolism of Pseudomonas aeruginosa. Arch. Microbiol., **150**, 400-404 (1988)
[3] Tricot, C.; Stalon, V.; Legrain, C.: Isolation and characterization of Pseudomonas putida mutants affected in arginine, ornithine and citrulline catabolism: function of the arginine oxidase and arginine succinyltransferase pathways. J. Gen. Microbiol., **137**, 2911-2918 (1991)

[4] Tricot, C.; Vander Wauven, C.; Wattiez, R.; Falmagne, P.; Stalon, V.: Purification and properties of a succinyltransferase from Pseudomonas aeruginosa specific for both arginine and ornithine. Eur. J. Biochem., **224**, 853-861 (1994)
[5] Itoh, Y.: Cloning and characterization of the aru genes encoding enzymes of the catabolic arginine succinyltransferase pathway in Pseudomonas aeruginosa. J. Bacteriol., **179**, 7280-7290 (1997)
[6] Schneider, B.L.; Kiupakis, A.K.; Reitzer, L.J.: Arginine catabolism and the arginine succinyltransferase pathway in Escherichia coli. J. Bacteriol., **180**, 4278-4286 (1998)

Tyramine N-feruloyltransferase 2.3.1.110

1 Nomenclature

EC number
2.3.1.110

Systematic name
feruloyl-CoA:tyramine N-(hydroxycinnamoyl)transferase

Recommended name
tyramine N-feruloyltransferase

Synonyms
HTH
feruloyl-CoA tyramine N-feruloyl-CoA transferase
hydroxycinnamoyl-CoA:tyramine N-(hydroxycinnamoyl)transferase
synthase, feruloyltyramine
tyramine N-feruloyl-CoA transferase
tyramine feruloyltransferase

CAS registry number
95567-96-7

2 Source Organism

<1> *Nicotiana tabacum* (infected with tobacco mosaic virus [1,2]; cv. Bottom special and Xanthi nc [4]; cv. Xanthi [7]) [1, 2, 4, 6, 7]
<2> *Capsicum annuum* (hot pepper [3]) [3]
<3> *Zea mays* (maize, cv. Snwodent 108 [5]) [5]
<4> *Solanum tuberosum* (potato [6,8]) [6, 8]
<5> *Papaver somniferum* (opium poppy [9]) [9]

3 Reaction and Specificity

Catalyzed reaction
feruloyl-CoA + tyramine = CoA + N-feruloyltyramine (<1> mechanism [7])

Reaction type
acyl group transfer

Natural substrates and products

S feruloyl-CoA + tyramine <1, 2, 3, 4, 5> (Reversibility: ? <1, 2, 3, 4, 5> [1, 2, 3, 4, 5, 7, 8, 9]) [1, 2, 3, 4, 5, 7, 8, 9]
P CoA + feruloyltyramine <1, 4, 5> [1, 2, 7, 8, 9]

Substrates and products

S 4-coumaroyl-CoA + octopamine <4> (Reversibility: ? <4> [8]) [8]
P CoA + 4-coumaroyloctopamine <4> [8]
S 4-coumaroyl-CoA + tyramine <1-5> (Reversibility: ? <1-5> [1, 2, 3, 5-9]) [1, 2, 3, 5-9]
P CoA + 4-coumaroyltyramine <1, 4> (<1> i.e. 4-hydroxycinnamoyltyramine [1,2]) [1, 2, 8]
S caffeoyl-CoA + tyramine <1-4> (Reversibility: ? <1-4> [3, 4, 5, 7, 8]) [3, 4, 5, 7, 8]
P CoA + caffeoyltyramine
S cinnamoyl-CoA + tyramine <1, 2, 4, 5> (Reversibility: ? <1, 2, 4, 5> [1, 2, 3, 7, 8, 9]) [1, 2, 3, 7, 8, 9]
P CoA + cinnamoyltyramine
S feruloyl-4'-phosphopantetheine + tyramine <1> (Reversibility: ? <1> [1]) [1]
P 4'-phosphopantetheine + feruloyltyramine
S feruloyl-CoA + (4-hydroxyphenyl)propylamine <1> (Reversibility: ? <1> [7]) [7]
P CoA + feruloyl-(4-hydroxyphenyl)propylamine
S feruloyl-CoA + 3'-methoxyoctopamine <4> (Reversibility: ? <4> [8]) [8]
P CoA + feruloyl-3'-methoxyoctopamine <4> [8]
S feruloyl-CoA + 3-methoxytyramine <1> (Reversibility: ? <1> [1, 7]) [1, 7]
P CoA + feruloyl-3-methoxytyramine
S feruloyl-CoA + N-methyltyramine <1> (Reversibility: ? <1> [1]) [1]
P CoA + N-feruloyl-N-methyltyramine
S feruloyl-CoA + β-phenylethylamine <1> (Reversibility: ? <1> [7]) [7]
P CoA + feruloyl-β-phenylethylamine
S feruloyl-CoA + dopamine <1, 3, 5> (Reversibility: ? <1, 3, 5> [1, 2, 5, 7, 9]) [1, 2, 5, 7, 9]
P CoA + feruloyldopamine
S feruloyl-CoA + homotyramine <1> (Reversibility: ? <1> [1]) [1]
P CoA + feruloylhomotyramine
S feruloyl-CoA + noradrenaline <1, 4> (<1> synonym: norepinephrine [1]) (Reversibility: ? <1, 4> [1, 7, 8]) [1, 7, 8]
P CoA + feruloylnoradrenaline
S feruloyl-CoA + octopamine <1, 2, 4> (Reversibility: ? <1, 2, 4> [1, 3, 7, 8]) [1, 3, 7, 8]
P CoA + feruloyloctopamine <4> [8]
S feruloyl-CoA + phenethylamine <3> (Reversibility: ? <3> [5]) [5]
P CoA + feruloylphenethylamine
S feruloyl-CoA + synephrine <1> (Reversibility: ? <1> [1]) [1]

P CoA + feruloylsynephrine
S feruloyl-CoA + tryptamine <3> (Reversibility: ? <3> [5]) [5]
P CoA + feruloyltryptamine
S feruloyl-CoA + tyramine <1-5> (<1> low specificity for cinnamoyl-CoA derivates and hydroxyphenethylamines [7]) (Reversibility: ? <1-5> [1-5, 7, 8, 9]) [1-5, 7, 8, 9]
P CoA + feruloyltyramine <1, 4, 5> (<1> mainly the trans-form is produced, only traces of the cis-form can be found [1]) [1, 2, 7, 8, 9]
S hydroxycinnamoyl-CoA + tyramine <1> (Reversibility: ? <1> [4]) [4]
P CoA + hydroxycinnamoyltyramine
S isoferuloyl-CoA + tyramine <1> (Reversibility: ? <1> [7]) [7]
P CoA + isoferuloyltyramine
S sinapoyl-CoA + tyramine <1-5> (Reversibility: ? <1-5> [1, 2, 3, 5, 7, 8, 9]) [1, 2, 3, 5, 7, 8, 9]
P CoA + sinapoyltyramine
S Additional information <1> (<1> enzyme has a wide specificity for phenylethylamines and cinnamoyl-CoA thioesters but differs in the affinity for the substrate, thus indicating different isoenzymes [1,2]) [1, 2]
P ?

Inhibitors

(2-hydroxyphenyl)amino sulfinyl acetic acid 1,1-dimethyl ester <1> (<1> i.e. OH-PAS, rapid inhibition, feruloyl-CoA prevents partially [4]) [4]
$CuSO_4$ <5> (<5> almost complete inactivation at 1 mM [9]) [9]
D-tyrosine methyl ester <1> [7]
DL-tyrosine methyl ester <1> [7]
$FeSO_4$ <5> (<5> almost complete inactivation at 1 mM [9]) [9]
L-phenylalanine β-naphthylamide <1, 4> [6]
L-tyrosine 7-amido-4-methylcoumarin <1, 4> [6]
L-tyrosine benzyl ester <1, 4> (<1> irreversible binding to tyramine binding site [7]) [6, 7]
L-tyrosine β-naphthylamide <1, 4> (<1,4> competitive with respect to tyramine, reversible in the presence of high concentrations of tyramine [6]) [6]
L-tyrosine ethyl ester <1> [7]
L-tyrosine methyl ester <1> [7]
L-tyrosine-tert-butyl ester <1> [7]
L-tyrosineamide <1> [7]
L-tyrosinol <1> [7]
N-ethylmaleimide <1> (<1> at 1 mM, 92% inhibition [2]; 50% inactiviation after 10 min at 2 mM [4]) [2, 4]
N-methyltyramine <1> [7]
NaCl <1> (<1> 20% activity at 0.5 M [7]) [7]
$ZnCl_2$ <5> (<5> almost complete inactivation at 1 mM [9]) [9]
ammonium sulfate <1> (<1> 20% activity at 0.5 M [7]) [7]
diethyldicarbonate <1> (<1> 90% loss of activity after 5 min at 0.5 mM, preincubation with feruloyl-CoA protects [4]) [4]
p-chloromercuribenzoate <1> (<1> 1 mM, strong [2]) [2]

tryptamine <3> (<3> 0.1 mM inhibits activity with tyramine [5]) [5]
tyramine <3> (<3> 0.1 mM inhibits activity with tryptamine [5]) [5]

Activating compounds

ethanol <1> (<1> addition of 14% leads to 2fold increase of enzyme activity [7]) [7]

Metals, ions

Additional information <1> (<1> $MgCl_2$ has no influence [1]) [1]

Specific activity (U/mg)

0.0156 <1> [2]
0.132 <3> [5]
4.2 <1> [7]
7.76 <5> [9]

K_m-Value (mM)

0.0006 <1> (feruloyl-CoA) [7]
0.001 <1> (sinapoyl-CoA) [7]
0.002 <1> (4-coumaroyl-CoA) [7]
0.002 <5> (cinnamoyl-CoA) [9]
0.002 <1> (isoferuloyl-CoA) [7]
0.0037 <3> (4-coumaroyl-CoA, <3> with tyramine as acceptor [5]) [5]
0.0043 <1> (tyramine) [7]
0.0047 <1> (cinnamoyl-CoA) [7]
0.0048 <3> (feruloyl-CoA, <3> with tyramine as acceptor [5]) [5]
0.0049 <1> (feruloyl-CoA, <1> + tyramine [2]) [2]
0.0062 <1> (feruloyl-CoA, <1> + tyramine [1]) [1]
0.0067 <1> (*p*-coumaroyl-CoA, <1> + tyramine [1]) [1]
0.0083 <3> (sinapoyl-CoA, <3> with tyramine as acceptor [5]) [5]
0.01 <1> (cinnamoyl-CoA, <1> + tyramine [1]) [1]
0.0113 <1> (octopamine) [7]
0.0125 <1> (feruloyl-4'-phosphopantetheine) [1]
0.017 <5> (4-coumaroyl-CoA) [9]
0.02 <1> (tyramine, <1> + feruloyl-CoA [1]) [1]
0.021 <5> (sinapoyl-CoA) [9]
0.025 <1> (tyramine, <1> + feruloyl-CoA, dopamine + feruloyl-CoA [2]) [2]
0.04 <4> (tyramine, <4> with feruloyl-CoA [8]) [8]
0.045-0.07 <1> (tyramine, <1> + *p*-coumaroyl-CoA, depending on the enzyme fraction used [2]) [2]
0.05 <1> (sinapoyl-CoA, <1> + tyramine [1]) [1]
0.0588-0.071 <1> (tyramine, <1> + feruloyl-CoA, depending on the enzyme fraction used [2]) [2]
0.059 <3> (tryptamine, <3> with feruloyl-CoA as acceptor [5]) [5]
0.062 <5> (feruloyl-CoA) [9]
0.076 <5> (tyramine) [9]
0.078 <5> (dopamine) [9]
0.082 <3> (caffeoyl-CoA, <3> with tyramine as acceptor [5]) [5]

0.1 <4> (feruloyl-CoA) [8]
0.13 <3> (tyramine, <3> with feruloyl-CoA as acceptor [5]) [5]
0.17 <1> (4-hydroxyphenylpropylamine) [7]
0.202 <1> (dopamine) [7]
0.34 <3> (dopamine, <3> with feruloyl-CoA as acceptor [5]) [5]
0.57 <3> (phenethylamine, <3> with feruloyl-CoA as acceptor [5]) [5]
1.57 <1> (noradrenaline) [7]
1.585 <1> (3-methoxytyramine) [7]
4.276 <1> (β-phenylethylamine) [7]

K_i-Value (mM)
0.0003 <4> (L-tyrosine β-naphthylamide) [6]
0.00042 <4> (L-tyrosine 7-amido-4-methylcoumarin) [6]
0.00066 <1> (L-tyrosine β-naphthylamide) [6]
0.00072 <1> (L-tyrosine 7-amido-4-methylcoumarin) [6]
0.003 <1> (L-tyrosine benzyl ester) [6, 7]
0.0034 <4> (L-tyrosine benzyl ester) [6]
0.018 <1> (L-tyrosine-tert-butyl ester) [7]
0.02 <1> (L-tyrosine ethyl ester) [7]
0.026 <1> (L-tyrosine methyl ester) [7]
0.046 <1> (L-tyrosinol) [7]
0.055 <1> (DL-tyrosine methyl ester) [7]
0.091 <4> (L-phenylalanine β-naphthylamide) [6]
0.112 <1> (L-phenylalanine β-naphthylamide) [6]
0.58 <1> (D-tyrosine methyl ester) [7]
2 <1> (L-tyrosineamide) [7]
2.2 <1> (N-methyltyramine) [7]

pH-Optimum
7.5 <1> (<1> feruloyl-CoA [1]) [1]
7.5-8 <5> [9]
7.6-8 <1> (<1> optimum varies slightly between different fractions [2]) [2]
8 <1> (<1> sinapoyl-CoA [1]) [1]
8.5 <1> (<1> cinnamoyl-CoA [1]) [1]
9 <1> (<1> *p*-coumaroyl-CoA [1]) [1]

pH-Range
6.5-9 <1> (<1> formation of feruloyltyramine, half-maximal activity at pH 6.5 and pH 9.0 [1]) [1]
6.5-9.5 <5> (<5> half-maximal activity at pH 6.5 and pH 9.5 [9]) [9]

Temperature optimum (°C)
30 <1> (<1> assay at [1,2]) [1, 2]

4 Enzyme Structure

Molecular weight

24000 <1> (<1> SDS-PAGE [7]) [7]
26000 <1> (<1> calculated from amino acid sequence [4]) [4]
28220 <2> (<2> calculated from amino acid sequence [3]) [3]
28400 <4> (<4> calculated from amino acid sequence [8]) [8]
30000 <4> (<4> SDS-PAGE [8]) [8]
40000 <3> (<3> gel filtration [5]) [5]
45000 <1> (<1> gel filtration [1]) [1]
48000 <1> (<1> gel filtration [7]) [7]
50000 <5> (<5> gel filtration [9]) [9]
55000 <1> (<1> gel filtration [4]) [4]

Subunits

dimer <1> (<1> 2 * 26000 [4]; 2 * 24000, SDS-PAGE [7]; <5> 2 * 25000, SDS-PAGE [9]) [4, 7, 9]

5 Isolation/Preparation/Mutation/Application

Source/tissue

anther <5> [9]
flower <2> [3]
fruit <2> [3]
leaf <1, 2, 3> (<3> increase of enzyme activity after wounding of leaves [5]) [1, 2, 3, 5]
root <2, 5> (<5> xylem [9]) [3, 9]
seed <5> [9]
stem <2> [3]
xylem <5> [9]

Localization

soluble <1> [1]

Purification

<1> (partial [1]; 46fold, separation of 3 fractions with different specificities [2]; partial [4]; homogeneity [7]) [1, 2, 4, 6, 7]
<2> [3]
<3> (partial [5]) [5]
<4> [6, 8]
<5> (988fold [9]) [9]

Cloning

<1> [4]
<2> [3]
<4> [8]

6 Stability

Oxidation stability

<1>, O_2-sensitive [2]
<1>, O_2-sensitive, mercaptoethanol stabilizes, crude extract [1]

General stability information

<1>, 10% glycerol in liquid N_2 stabilizes [2]
<1>, freezing does not alter the properties of the enzyme [2]
<1>, no loss of activity during dialysis for 16 h at 4°C [1]

Storage stability

<1>, 4°C, 10% loss of activity within 7 days, concentrated enzyme [1]
<1>, 4°C, as ammonium sulfate paste, 10 mM mercaptoethanol, 24 h, 25% loss of activity [2]
<1>, 4°C, as ammonium sulfate paste, 5 mM DTT, 10% loss of activity within 24 h [2]
<5>, -20°C, no loss of activity for 2 months in the presence of 2-mercaptoethanol [9]
<5>, -80°C, no loss of activity for 4 months in the presence of 2-mercaptoethanol [9]
<5>, 4°C, no loss of activity for 14 days in the presence of 2-mercaptoethanol [9]

References

[1] Negrel, J.; Martin, C.: The biosynthesis of feruloyltyramine in Nicotiana tabacum. Phytochemistry, **23**, 2797-2801 (1984)
[2] Fleurence, J.; Negrel, J.: Partial purification of tyramine feruloyl transferase from TMV inoculated tobacco leaves. Phytochemistry, **28**, 733-736 (1989)
[3] Back, K.; Jang, S.M.; Lee, B.C.; Schmidt, A.; Strack, D.; Kim, K.M.: Cloning and characterization of a hydroxycinnamoyl-CoA:tyramine N-(hydroxycinnamoyl)transferase induced in response to UV-C and wounding from Capsicum annuum. Plant Cell Physiol., **42**, 475-481 (2001)
[4] Farmer, M.J.; Czermic, P.; Michael, A.; Negrel, J.: Identification and characterization of cDNA clones encoding hydroxycinnamoyl-CoA:tyramine N-hydroxycinnamoyltransferase from tobacco. Eur. J. Biochem., **263**, 686-694 (1999)
[5] Ishihara, A.; Kawata, N.; Matsukawa, T.; Iwamura, H.: Induction of N-hydroxycinnamoyltyramine synthesis and tyramine N-hydroxycinnamoyltransferase (THT) activity by wounding in maize leaves. Biosci. Biotechnol. Biochem., **64**, 1025-1031 (2000)
[6] Negrel, J.; Javelle, F.: L-Tyrosine β-naphthylamide is a potent competitive inhibitor of tyramine N-(hydroxycinnamoyl)transferase in vitro. Phytochemistry, **56**, 523-527 (2001)

[7] Negrel, J.; Javelle, F.: Purification, characterization and partial amino acid sequencing of hydroxycinnamoyl-CoA:tyramine N-(hydroxycinnamoyl)-transferase from tobacco cell-suspension cultures. Eur. J. Biochem., **247**, 1127-1135 (1997)
[8] Schmidt, A.; Grimm, R.; Schmidt, J.; Scheel, D.; Strack, D.; Rosahl, S.: Cloning and expression of a potato cDNA encoding hydroxycinnamoyl-CoA:tyramine N-(hydroxycinnamoyl)transferase. J. Biol. Chem., **274**, 4273-4280 (1999)
[9] Yu, M.; Facchini, P.J.: Purification, characterization, and immunolocalization of hydroxycinnamoyl-CoA: tyramine N-(hydroxycinnamoyl)transferase from opium poppy. Planta, **209**, 33-44 (1999)

Mycocerosate synthase 2.3.1.111

1 Nomenclature

EC number
2.3.1.111

Systematic name
acyl-CoA:methylmalonyl-CoA C-acyltransferase (decarboxylating, oxoacyl- and enoyl-reducing)

Recommended name
mycocerosate synthase

Synonyms
mycocerosic acid synthase

CAS registry number
95229-19-9

2 Source Organism

<1> *Mycobacterium tuberculosis* (var. bovis Bacillus Calmette-Guerin [2-5]) [1-5]
<2> *Mycobacterium smegmatis* [1]

3 Reaction and Specificity

Catalyzed reaction
acyl-CoA + n methylmalonyl-CoA + 2n NADPH + 2n H^+ = multi-methyl-branched acyl-CoA + n CoA + n CO_2 + 2n $NADP^+$

Reaction type
acyl group transfer
decarboxylation

Natural substrates and products
S acyl-CoA + methylmalonyl-CoA + NADPH <1> (<1> fatty acid biosynthesis [1-5]) (Reversibility: ? <1> [1-5]) [1-5]
P tetramethyl-branched mycocerosic acid + CoA + CO_2 + $NADP^+$

Substrates and products
S acyl-CoA + methylmalonyl-CoA + NADPH <1> (<1> multifunctional enzyme, specific for methylmalonyl-CoA, cannot be replaced by malonyl-

CoA, n-acyl-CoAs of C-6 to C-20 are substrates [1]) (Reversibility: ? <1> [1-5]) [1-5]

P tetramethyl-branched mycocerosic acid + CoA + CO_2 + $NADP^+$

S eicosanoyl-CoA + methylmalonyl-CoA + NADPH <1> (Reversibility: ? <1> [1]) [1]

P 2,4,6,8-tetramethyloctacosanoic acid + CoA + CO_2 + $NADP^+$

S hexanoyl-CoA + methylmalonyl-CoA + NADPH <1> (<1> best substrate [1]) (Reversibility: ? <1> [1]) [1]

P ?

Inhibitors

CoA <1> (<1> 1 mM, weak inhibition [1]) [1]
NADPH <1> (<1> above 0.5 mM, weak inhibition [1]) [1]
bovine serum albumin <1> [1]

Cofactors/prosthetic groups

NADPH <1> [1]

Specific activity (U/mg)

0.014 <1> [1]

pH-Optimum

7.5 <1> [1]

4 Enzyme Structure

Molecular weight

490000 <1> (<1> gel filtration [1]) [1]

Subunits

dimer <1, 2> (<1> 2 * 238000, SDS-PAGE [1]; <2> 2 * 238000, amino acid composition [1]) [1]

5 Isolation/Preparation/Mutation/Application

Purification

<1> [1, 3-5]

Cloning

<1> [2, 3]

6 Stability

Organic solvent stability

glycerol <1> (<1> stable in 10% glycerol [1]) [1]

General stability information

<1>, stabilized by glycerol [1]

Storage stability

<1>, -20°C, glycerol-buffer, stable for at least 1 month [1]

References

[1] Rainwater, D.L.; Kolattokudy, P.E.: Fatty acid biosynthesis in Mycobacterium tuberculosis var. bovis Bacillus Calmette-Guerin. Purification and characterization of a novel fatty acid synthase, mycocerosic acid synthase, which elongates n-fatty acyl-CoA with methylmalonyl-CoA. J. Biol. Chem., **260**, 616-623 (1985)

[2] Mathur, M.; Kolattokudy, P.E.: Molecular cloning and sequencing of the gene for mycocerosic acid synthase, a novel fatty acid elongating multifunctional enzyme, from Mycobacterium tuberculosis var. bovis Bacillus Calmette-Guerin. J. Biol. Chem., **267**, 19388-19395 (1992)

[3] Fitzmaurice, A.M.; Kolattukudy, P.E.: An acyl-CoA synthase (acoas) gene adjacent to the mycocerosic acid synthase (mas) locus is necessary for mycocerosyl lipid synthesis in Mycobacterium tuberculosis var. bovis BCG. J. Biol. Chem., **273**, 8033-8039 (1998)

[4] Fernandes, N.D.; Kolattukudy, P.E.: A newly identified methyl-branched chain fatty acid synthesizing enzyme from Mycobacterium tuberculosis var. bovis BCG. J. Biol. Chem., **273**, 2823-2828 (1998)

[5] Li, M.S.; Monahan, I.M.; Waddell, S.J.; Mangan, J.A.; Martin, S.L.; Everett, M.J.; Butcher, P.D.: cDNA-RNA subtractive hybridization reveals increased expression of mycocerosic acid synthase in intracellular Mycobacterium bovis BCG. Microbiology, **147**, 2293-2305 (2001)

D-Tryptophan N-malonyltransferase 2.3.1.112

1 Nomenclature

EC number
2.3.1.112

Systematic name
malonyl-CoA:D-tryptophan N-malonyltransferase

Recommended name
D-tryptophan N-malonyltransferase

Synonyms
malonyltransferase, D-tryptophan

CAS registry number
94490-01-4

2 Source Organism

<1> *Arachis hypogaea* (peanut, cv. fastigiata (Junagarh-11 or cultivar Shulamit)) [1]

3 Reaction and Specificity

Catalyzed reaction
malonyl-CoA + D-tryptophan = CoA + N^2-malonyl-D-tryptophan

Reaction type
acyl group transfer

Natural substrates and products
S malonyl-CoA + D-tryptophan <1> (Reversibility: ? <1> [1]) [1]
P CoA + N^2-malonyl-D-tryptophan

Substrates and products
S malonyl-CoA + D-tryptophan <1> (Reversibility: ? <1> [1]) [1]
P CoA + N^2-malonyl-D-tryptophan
S malonyl-CoA + 1-aminocyclopropan-1-carboxylic acid <1> (<1> condensation at 50% the rate of the condensation with D-tryptophan [1]) (Reversibility: ? <1> [1]) [1]
P CoA + malonyl-1-aminocyclopropan-1-carboxylic acid

S Additional information <1> (<1> no substrates are L-tryptophan, anthranilate and 3,4-dichloroaniline [1]) [1]

P ?

Activating compounds

bovine serum albumin <1> (<1> 2 mg/ml, activation [1]) [1]

Specific activity (U/mg)

0.0002 <1> [1]

K_m-Value (mM)

0.005 <1> (malonyl-CoA) [1]
0.025 <1> (D-tryptophan) [1]

pH-Optimum

8.8 <1> (<1> broad [1]) [1]

pH-Range

7.5-9 <1> (<1> most active [1]) [1]

Temperature optimum (°C)

30 <1> (<1> assay at [1]) [1]

4 Enzyme Structure

Molecular weight

40000 <1> (<1> gel filtration [1]) [1]

Subunits

monomer <1> (<1> 1 * 38000, SDS-PAGE [1]) [1]

5 Isolation/Preparation/Mutation/Application

Source/tissue

seedling <1> [1]

Purification

<1> (partial [1]) [1]

6 Stability

Storage stability

<1>, -20°C, inactivation after 3 months [1]
<1>, 4°C, considerable loss of activity after several weeks [1]
<1>, glycerol, 10%, stabilizes during storage [1]

References

[1] Matern, U.; Feser, C.; Heller, W.: N-malonyltransferases from peanut. Arch. Biochem. Biophys., **235**, 218-227 (1984)

Anthranilate N-malonyltransferase 2.3.1.113

1 Nomenclature

EC number
2.3.1.113

Systematic name
malonyl-CoA:anthranilate N-malonyltransferase

Recommended name
anthranilate N-malonyltransferase

Synonyms
malonyltransferase, anthranilate

CAS registry number
94489-98-2

2 Source Organism

<1> *Arachis hypogaea* (peanut, cv. fastigiata ("Junagarh-11" or cultivar Shulamit)) [1]

3 Reaction and Specificity

Catalyzed reaction
malonyl-CoA + anthranilate = CoA + N-malonylanthranilate

Reaction type
acyl group transfer

Substrates and products
S malonyl-CoA + anthranilate <1> (Reversibility: ? <1> [1]) [1]
P CoA + N-malonylanthranilate <1> [1]
S Additional information <1> (<1> no substrates are D-tryptophan and 3,4-dichloroaniline [1]) [1]
P ?

Activating compounds
bovine serum albumin <1> (<1> 2 mg/ml, activation [1]) [1]

Specific activity (U/mg)
0.00091 <1> [1]

K_m-Value (mM)

0.004 <1> (malonyl-CoA) [1]
0.012 <1> (anthranilate) [1]

pH-Optimum

8.8 <1> (<1> broad [1]) [1]

pH-Range

7.5-9 <1> (<1> most active [1]) [1]

Temperature optimum (°C)

30 <1> (<1> assay at [1]) [1]

4 Enzyme Structure

Molecular weight

50000 <1> (<1> gel filtration [1]) [1]

Subunits

monomer <1> (<1> 1 * 51000, SDS-PAGE [1]) [1]

5 Isolation/Preparation/Mutation/Application

Source/tissue

seedling <1> [1]

Purification

<1> (partial [1]) [1]

6 Stability

Storage stability

<1>, -20°C, inactivtion after 3 months [1]
<1>, 4°C, considerable loss of activity after several weeks [1]
<1>, glycerol, 10%, stabilizes during storage [1]

References

[1] Matern, U.; Feser, C.; Heller, W.: N-malonyltransferases from peanut. Arch. Biochem. Biophys., **235**, 218-227 (1984)

3,4-Dichloroaniline N-malonyltransferase 2.3.1.114

1 Nomenclature

EC number
2.3.1.114

Systematic name
malonyl-CoA:3,4-dichloroaniline N-malonyltransferase

Recommended name
3,4-dichloroaniline N-malonyltransferase

CAS registry number
94489-99-3

2 Source Organism

<-1> no activity in *Triticum aestivum* [3]
<1> *Arachis hypogaea fastigiata* (peanut [1]) [1]
<2> *Glycine max* [2, 3]

3 Reaction and Specificity

Catalyzed reaction
malonyl-CoA + 3,4-dichloroaniline = CoA + N-(3,4-dichlorophenyl)-malonamate

Reaction type
acyl group transfer

Substrates and products
- **S** malonyl-CoA + 2,3,4,5-tetrachloroaniline <2> (<2> 29% of activity with 3,4-dichloroaniline [2]) (Reversibility: ? <2> [2]) [2]
- **P** CoA + N-(2,3,4,5-tetrachlorophenyl)-malonamate <2> [2]
- **S** malonyl-CoA + 2,3,4-trichloroaniline <2> (<2> 50% of activity with 3,4-dichloroaniline [2]) (Reversibility: ? <2> [2]) [2]
- **P** CoA + N-(2,3,4-trichlorophenyl)-malonamate <2> [2]
- **S** malonyl-CoA + 2,4,5-trichloroaniline <2> (<2> 22% of activity with 3,4-dichloroaniline [2]) (Reversibility: ? <2> [2]) [2]
- **P** CoA + N-(4-dichlorophenyl)-malonamate <2> [2]
- **S** malonyl-CoA + 2,4-dichloroaniline <2> (<2> 53% of activity with 3,4-dichloroaniline [2]) (Reversibility: ? <2> [2]) [2]

P CoA + N-(2,4-dichlorophenyl)-malonamate <2> [2]
S malonyl-CoA + 2,5-dichloroaniline <2> (<2> 53% of activity with 3,4-dichloroaniline [2]) (Reversibility: ? <2> [2]) [2]
P CoA + N-(2,5-dichlorophenyl)-malonamate <2> [2]
S malonyl-CoA + 2-chloroaniline <2> (<2> 61% of activity with 3,4-dichloroaniline [2]) (Reversibility: ? <2> [2]) [2]
P CoA + N-(2-chlorophenyl)-malonamate <2> [2]
S malonyl-CoA + 3,4,5-trichloroaniline <2> (<2> 30% of activity with 3,4-dichloroaniline [2]) (Reversibility: ? <2> [2]) [2]
P CoA + N-(3,4,5-dichlorophenyl)-malonamate <2> [2]
S malonyl-CoA + 3,4,5-trichloroaniline <2> (<2> 72% of activity with 3,4-dichloroaniline [2]) (Reversibility: ? <2> [2]) [2]
P CoA + N-(3,4,5-trichlorophenyl)-malonamate <2> [2]
S malonyl-CoA + 3,4-dichloroaniline <1, 2> (<2> preferred substrate [2]) (Reversibility: ? <1, 2> [1, 2]) [1, 2, 3]
P CoA + N-(3,4-dichlorophenyl)-malonamate <1, 2> [1, 2, 3]
S malonyl-CoA + 3,5-dichloroaniline <2> (<2> 48% of activity with 3,4-dichloroaniline [2]) (Reversibility: ? <2> [2]) [2]
P CoA + N-(3,5-dichlorophenyl)-malonamate <2> [2]
S malonyl-CoA + 3-chloroaniline <2> (<2> 67% of activity with 3,4-dichloroaniline [2]) (Reversibility: ? <2> [2]) [2]
P CoA + N-(3-chlorophenyl)-malonamate <2> [2]
S malonyl-CoA + 4-aminophenol <2> (<2> 20% of activity with 3,4-dichloroaniline [2]) (Reversibility: ? <2> [2]) [2]
P CoA + N-(phenyl)-malonamate <2> [2]
S malonyl-CoA + 4-bromoaniline <2> (<2> 75% of activity with 3,4-dichloroaniline [2]) (Reversibility: ? <2> [2]) [2]
P CoA + N-(4-bromophenyl)-malonamate <2> [2]
S malonyl-CoA + 4-dichloroaniline <2> (<2> 87% of activity with 3,4-dichloroaniline [2]) (Reversibility: ? <2> [2]) [2]
P CoA + N-(4-dichlorophenyl)-malonamate <2> [2]
S malonyl-CoA + 4-nitroaniline <2> (<2> 43% of activity with 3,4-dichloroaniline [2]) (Reversibility: ? <2> [2]) [2]
P CoA + N-(4-nitrophenyl)-malonamate <2> [2]
S malonyl-CoA + 7-amino-4-methylcoumarin <2> (<2> 70% of activity with 3,4-dichloroaniline [2]) (Reversibility: ? <2> [2]) [2]
P N-(4-methylcoumaryl)-7-malonamate <2> [2]

Inhibitors

4-chloromercuribenzoate <2> (<2> 0.1 mM, complete inhibition, activity can be restored by incubation with 2 mM dithiothreitol for 30 min at 0°C [2]) [2]

Specific activity (U/mg)

0.143 <2> [2]

K_m-Value (mM)

0.1 <2> (3,4-dichloroaniline) [2]

pH-Optimum

6.3-7.2 <2> [2]

4 Enzyme Structure

Molecular weight

45000 <1> (<1> gel filtration [1]) [1]
48000 <2> (<2> gel filtration [2]) [2]

Subunits

monomer <2> (<2> 1 * 47000, SDS-PAGE [2]) [2]

5 Isolation/Preparation/Mutation/Application

Source/tissue

hypocotyl <1> (<1> of young seedlings [1]) [1]
leaf <1> [1]
seedling <1> [1]

Purification

<1> (ammonium sulfate, DEAE-cellulose, Blue Sepharose, Acrylex P-100, Matrix gel green [1]) [1]
<2> (ammonium sulfate, ion-exchange, Sephadex G-100, hydrophobic interaction, gel filtration [2]) [2]

References

[1] Matern, U.; Feser, C.; Heller, W.: N-Malonyltransferases from peanut. Arch. Biochem. Biophys., **235**, 218-227 (1984)
[2] Sandermann, Jr.H.; Schmitt, r.; Eckey, H.; Bauknecht, T.: Plant biochemistry of xenobiotics: isolation and properties of soybean O- and N-glucosyl and O- and N-malonyltransferases for chlorinated phenols and anilines. Arch. Biochem. Biophys., **287**, 341-350 (1991)
[3] Schmidt, B.; Rivero, C.; Thiede, B.: 3,4-Dichloroaniline N-glucosyl- and N-malonyltransferase activities in cell cultures and plants of soybean and wheat. Phytochemistry, **39**, 81-84 (2000)

Isoflavone-7-O-β-glucoside 6"-O-malonyltransferase 2.3.1.115

1 Nomenclature

EC number
2.3.1.115

Systematic name
malonyl-CoA:isoflavone-7-O-β-D-glucoside 6"-O-malonyltransferase

Recommended name
isoflavone-7-O-β-glucoside 6"-O-malonyltransferase

Synonyms
MAT-7
flavone/flavonol 7-O-β-D-glucoside malonyltransferase
malonyl-CoA:flavone/flavonol 7-O-glucoside malonyltransferase
malonyl-coenzyme A:flavone/flavonol-7-O-glycoside malonyltransferase
malonyl-coenzyme A:isoflavone 7-O-glucoside-6"-malonyltransferase
malonyltransferase, flavone (flavonol) 7-O-glycoside
malonyltransferase, isoflavone 7-O-glucoside 6"-O-

CAS registry number
78413-09-9
93585-97-8

2 Source Organism

<1> *Petroselinum hortense* (parsley [1,2]) [1, 2]
<2> *Cicer arietinum* (chick pea [3]) [3]

3 Reaction and Specificity

Catalyzed reaction
malonyl-CoA + biochanin A 7-O-β-D-glucoside = CoA + biochanin A 7-O-(6-O-malonyl-β-glucoside)

Reaction type
acyl group transfer

Natural substrates and products
S apigenin 7-O-apiosylglucoside + malonyl-CoA <1, 2> (<2> i.e. apiin [3]) (Reversibility: ? <1, 2> [2, 3]) [2, 3]
P apigenin 7-O-apiosylglucoside -6"-O-malonylester+ malonyl-CoA <2> [3]

S Additional information <1> (<1> last step in biosynthesis of flavonoid glycosides [2]) [2]
P ?

Substrates and products

S apigenin 7-O-glucoside + malonyl-CoA <1, 2> (<1> 88% of the activity with apiin [2]) (Reversibility: ? <1, 2> [1-3]) [1-3]
P CoA + apigenin 7-O-glucoside 6"-O-malonylester
S apiin + malonyl-CoA <1> (<1> i.e. apigenin 7-O-apiosylglucoside, best substrate [2]) (Reversibility: ? <1> [2]) [2]
P apigenin 7-O-apiosylglucoside -6"-O-malonylester + malonyl-CoA
S diosmetin 7-O-glucoside + malonyl-CoA <1> (<1> 72% of the activity with apiin [2]) (Reversibility: ? <1> [2]) [2]
P CoA + diosmetin 7-O-glucoside 6"-O-malonylester
S isorhamnetin 3-O-glucoside + malonyl-CoA <1> (<1> 11% of the activity with apiin [2]) (Reversibility: ? <1> [2]) [2]
P CoA + isorhamnetin 3-O-glucoside 6"-O-malonylester
S isorhamnetin 7-O-glucoside + malonyl-CoA <1> (<1> 12% of the activity with apiin [2]) (Reversibility: ? <1> [2]) [2]
P CoA + isorhamnetin 7-O-glucoside 6"-O-malonylester
S kaempferol 3-O-glucoside + malonyl-CoA <1> (<1> 15% of the activity with apiin [2]) (Reversibility: ? <1> [2]) [2]
P CoA + kaempferol 3-O-glucoside 6"-O-malonylester
S luteolin 7-O-glucoside + malonyl-CoA <1, 2> (<1> 30% of the activity with apiin [2]) (Reversibility: ? <1, 2> [1-3]) [2, 3]
P CoA + luteolin 7-O-glucoside 6"-O-malonylester
S malonyl CoA + genistein 7-O-glucoside <2> (Reversibility: ? <2> [3]) [3]
P CoA + genistein 7-O-glucoside 6"-O-malonylester
S malonyl-CoA + 2',4,4'-trihydroxychalcone 4-O-glucoside <2> (Reversibility: ? <2> [3]) [3]
P CoA + 2',4,4'-trihydroxychalcone 4-O-glucoside 6"-O-malonylester
S malonyl-CoA + biochanin A 7-O-glucoside <2> (Reversibility: ? <2> [3]) [3]
P CoA + biochanin A 7-O-glucoside-6"-O-malonylester <2> [3]
S malonyl-CoA + daidzein 7-O-glucoside <2> (Reversibility: ? <2> [3]) [3]
P CoA + daidzein 7-O-glucoside 6"-O-malonylester
S malonyl-CoA + formononetin 7-O-glucoside <2> (Reversibility: ? <2> [3]) [3]
P CoA + formononetin 7-O-glucoside-6"-O-malonylester <2> [3]
S malonyl-CoA + kaempferol 7-O-glucoside <2> (Reversibility: ? <2> [3]) [3]
P CoA + kaempferol 7-O-glucoside 6"-O-malonylester
S malonyl-CoA + maackiain 3-O-glucoside <2> (Reversibility: ? <2> [3]) [3]
P CoA + maackiain 3-O-glucoside 6"-O-malonylester
S malonyl-CoA + orobol 7-O-glucoside <2> (Reversibility: ? <2> [3]) [3]
P CoA + orobol 7-O-glucoside 6"-O-malonylester

S malonyl-CoA + pratensein 7-O-glucoside <2> (Reversibility: ? <2> [3]) [3]
P CoA + pratensein 7-O-glucoside 6”-O-malonylester
S naringenin 7-O-glucoside + malonyl-CoA <1> (<1> 16% of the activity with apiin [2]) (Reversibility: ? <1> [2]) [2]
P CoA + naringenin 7-O-glucoside 6”-O-malonylester
S quercetagetin 7-O-glucoside + malonyl-CoA <2> (Reversibility: ? <2> [3]) [3]
P CoA + quercetagetin 7-O-glucoside 6”-O-malonylester
S quercetin 3-O-glucoside + malonyl-CoA <1> (<1> 8% of the activity with apiin [2]) (Reversibility: ? <1> [2]) [2]
P CoA + quercetin 3-O-glucoside 6”-O-malonylester
S quercetin 7-O-glucoside + malonyl-CoA <2> (Reversibility: ? <2> [3]) [3]
P CoA + quercetin 7-O-glucoside 6”-O-malonylester
S Additional information <1, 2> (<2> succinyl-CoA can replace malonyl-CoA with lower activity [3]; <2> isoflavones with the glucosyl substituent in position 4’ don’t act as acceptors [3]; <1> 4-hydroxy-2,5-dichlorophenoxyacetic acid glucoside and 4-methylumbelliferyl glucoside are not accepted as substrate [1]) [1, 3]
P ?

Specific activity (U/mg)

7.515 <2> [3]
Additional information <1> [1, 2]

K_m-Value (mM)

0.006 <1> (malonyl-CoA) [2]
0.01 <1> (apiin) [2]
0.014 <1> (apigenin 7-O-glucoside) [2]
0.02 <1> (luteolin 7-O-glucoside) [2]
0.024 <2> (formononetin) [3]
0.036 <2> (biochanin A) [3]
0.048 <2> (malonyl-CoA) [3]
0.054 <2> (maackiain 3-O-glucoside) [3]
0.065 <2> (daidzein 7-O-glucoside) [3]
0.067 <2> (kaempferol 7-O-glucoside) [3]
0.088 <2> (genistein 7-O-glucoside) [3]
0.098 <2> (pratensein 7-O-glucoside) [3]
0.101 <2> (succinyl-CoA) [3]
0.11 <2> (2’,4,4’-trihydroxychalcone 4’-O-glucoside) [3]
0.16 <2> (apigenin 7-O-glucoside) [3]
0.335 <2> (quercetagetin 7-O-glucoside) [3]
0.381 <2> (quercetin 7-O-glucoside) [3]
0.46 <2> (orobol 7-O-glucoside) [3]
0.47 <2> (luteolin 7-O-glucoside) [3]

pH-Optimum

8 <1, 2> (<1> most active in Tris buffer [2]) [2, 3]

pH-Range

6.5-9.5 <2> (<2> about 50% of activity maximum at pH 6.5 and 9.5 [3]) [3]

Temperature optimum (°C)

45-50 <2> [3]

4 Enzyme Structure

Molecular weight

50000 <1> (<1> SDS-PAGE [1]; <1> gel filtration [2]) [1, 2]
112000 <2> (<2> gel filtration [3]) [3]

Subunits

? <1> (<1> x * 50000, SDS-PAGE [1]) [1]

Posttranslational modification

glycoprotein <1> [2]

5 Isolation/Preparation/Mutation/Application

Source/tissue

cell culture <1> [1, 2]
root <2> (<2> 4 days old [3]) [3]

Localization

soluble <1> [2]

Purification

<1> (partial [1]) [1, 2]
<2> [3]

6 Stability

pH-Stability

Additional information <1> (<1> acidic pH leads to rapid denaturation [1]) [1]

Temperature stability

60 <2> (<2> above 2 min, destroyed [3]) [3]

General stability information

<2>, most stable in Tris buffer at pH 8, unstable to repeated freezing and thawing, dithioerythritol stabilizes [3]

Storage stability

<1>, -70°C, partially purified enzyme is stable for several months, extensively purified enzyme loses 50% of activity within 2-3 days [2]
<2>, -20°C, 10% glycerol, stable for 4 months [3]

References

[1] Matern, U.; Feser, C.; Hammer, D.: Further characterization and regulation of malonyl-coenzyme A:flavonoid glucoside malonyltransferases from parsley cell suspension cultures. Arch. Biochem. Biophys., **226**, 206-217 (1983)
[2] Matern, U.; Potts, J.R.M.; Hahlbrock, K.: Two flavonoid-specific malonyltransferases from cell suspension cultures of Petroselinum hortense: partial purification and some properties of malonyl-coenzyme A:flavone/flavonol-7-O-glycoside malonyltransferase and malonyl-coenzyme A:flavonol-3-O-glucoside malonyltransferase. Arch. Biochem. Biophys., **208**, 233-241 (1981)
[3] Koester, J.; Bussmann, R.; Barz, W.: Malonyl-coenzyme A:isoflavone 7-O-glucoside-6-O-malonyltransferase from roots of chick pea (Cicer arietinum L.). Arch. Biochem. Biophys., **234**, 513-521 (1984)

Flavonol-3-O-β-glucoside O-malonyltransferase 2.3.1.116

1 Nomenclature

EC number
2.3.1.116

Systematic name
malonyl-CoA:flavonol-3-O-β-D-glucoside 6"-O-malonyltransferase

Recommended name
flavonol-3-O-β-glucoside O-malonyltransferase

Synonyms
MAT-3
malonyl-coenzyme A:flavonol-3-O-glucoside malonyltransferase
malonyltransferase, flavonol 3-O-glucoside

CAS registry number
78413-11-3

2 Source Organism

<1> *Petroselinum hortense* (parsley [1,2]) [1, 2]

3 Reaction and Specificity

Catalyzed reaction
malonyl-CoA + flavonol 3-O-β-D-glucoside = CoA + flavonol 3-O-(6-malonyl-β-D-glucoside)

Reaction type
acyl group transfer

Natural substrates and products
S malonyl-CoA + kaempferol 3-O-glucoside <1> (Reversibility: ? <1> [2]) [2]
P CoA kaempferol 3-O-glucoside malonylester

Substrates and products
S malonyl-CoA + isorhamnetin 3-O-glucoside <1> (<1> 78% of the activity with kaempferol 3-O-glucoside [2]) (Reversibility: ? <1> [2]) [2]
P CoA + isorhamnetin 3-O-glucoside malonylester <1> (<1> probably acylated at position 6 of glucose [2]) [2]

S malonyl-CoA + kaempferol 3-O-glucoside <1> (Reversibility: ? <1> [2]) [2]

P CoA kaempferol 3-O-glucoside malonylester <1> (<1> probably acylated at position 6 of glucose [2]) [2]

S malonyl-CoA + quercetin 3-O-glucoside <1> (<1> 60% of the activity with kaempferol 3-O-glucoside [2]) (Reversibility: ? <1> [2]) [2]

P CoA + quercetin 3-O-glucoside malonylester <1> (<1> probably acylated at position 6 of glucose [2]) [2]

Specific activity (U/mg)

Additional information <1> [1, 2]

K_m-Value (mM)

0.004 <1> (kaempferol 3-O-glucoside) [2]
0.005 <1> (malonyl-CoA) [2]
0.006 <1> (isorhamnetin 3-O-glucoside) [2]

pH-Optimum

8 <1> (<1> most active in Tris buffer containing bovine serum albumin [2]) [2]

4 Enzyme Structure

Molecular weight

50000 <1> (<1> SDS-PAGE [1]; <1> gel filtration [2]) [1, 2]

Subunits

monomer <1> (<1> 1 * 50000, SDS-PAGE [1]) [1]

Posttranslational modification

glycoprotein <1> [2]

5 Isolation/Preparation/Mutation/Application

Source/tissue

cell culture <1> [1, 2]

Localization

soluble <1> [2]

Purification

<1> (partial [1]) [1, 2]

6 Stability

pH-Stability

Additional information <1> (<1> acidic pH leads to rapid denaturation [1]) [1]

Storage stability

<1>, -70°C, partially purified enzyme is stable for several months, extensively purified enzyme loses 50% of activity within 2-3 days [2]

References

[1] Matern, U.; Feser, C.; Hammer, D.: Further characterization and regulation of malonyl-coenzyme A:flavonoid glucoside malonyltransferases from parsley cell suspension cultures. Arch. Biochem. Biophys., **226**, 206-217 (1983)

[2] Matern, U.; Potts, J.R.M.; Hahlbrock, K.: Two flavonoid-specific malonyltransferases from cell suspension cultures of Petroselinum hortense: partial purification and some properties of malonyl-coenzyme A:flavone/flavonol-7-O-glycoside malonyltransferase and malonyl-coenzyme A:flavonol-3-O-glucoside malonyltransferase. Arch. Biochem. Biophys., **208**, 233-241 (1981)

2,3,4,5-Tetrahydropyridine-2,6-dicarboxylate N-succinyltransferase 2.3.1.117

1 Nomenclature

EC number
2.3.1.117

Systematic name
succinyl-CoA:2,3,4,5-tetrahydropyridine-2,6-dicarboxylate N-succinyltransferase

Recommended name
2,3,4,5-tetrahydropyridine-2,6-dicarboxylate N-succinyltransferase

Synonyms
succinyl-CoA:tetrahydrodipicolinate N-succinyltransferase
succinyltransferase, tetrahydrodipicolinate
tetrahydrodipicolinate N-succinyltransferase
tetrahydrodipicolinate succinylase
tetrahydrodipicolinate succinyltransferase

CAS registry number
88086-34-4

2 Source Organism

<1> *Escherichia coli* (enzyme is a member of the hexapeptide acyltransferase superfamily [6]) [1, 2, 6]
<2> *Mycobacterium bovis* (strain BCG [3]) [3, 4, 6, 7]
<3> *Corynebacterium glutamicum* [5]

3 Reaction and Specificity

Catalyzed reaction
succinyl-CoA + (S)-2,3,4,5-tetrahydropyridine-2,6-dicarboxylate + H_2O = CoA + N-succinyl-L-2-amino-6-oxoheptanedioate

Reaction type
acyl group transfer

Natural substrates and products
S succinyl-CoA + tetrahydrodipicolinate <1, 2, 3> (<1,2> involved in the biosynthesis of lysine in bacteria, blue-green algae and higher plants

[1,6]; <3> involved in the four-step succinylase branch of the L-lysine biosynthetic pathway [5]) (Reversibility: r <1> [1]; ? <2, 3> [3, 4, 5]) [1-6]
P CoA + N-succinyl-2-amino-6-keto-L-pimelate <1, 2, 3> [1-6]

Substrates and products

S succinyl-CoA + 3,4-dihydro-2H-1,4-thiazine-3,5-dicarboxylic acid <1> (Reversibility: ? <1> [2]) [2]
P ?
S succinyl-CoA + 6-amino-2-hydroxypimelic acid <1> (Reversibility: ? <1> [2]) [2]
P CoA + N-succinyl-6-amino-2-hydroxypimelate <1> [2]
S succinyl-CoA + L-2-aminopimelate <1> (<1> tetrahydrodipicolinate analogue [2]) (Reversibility: ? <1> [2]) [2]
P CoA + N-succinyl-L-aminopimelate <1> [2]
S succinyl-CoA + L-6-amino-L-2-hydroxypimelic acid <1> (<1> mixture of isomers [2]) (Reversibility: ? <1> [2]) [2]
P CoA + N-succinyl-L-6-amino-L-2-hydroxypimelate <1> [2]
S succinyl-CoA + tetrahydrodipicolinate <1-3> (<1> product formation is strongly preferred [1]) (Reversibility: r <1> [1]; ? <2, 3> [3, 5]) [1-5]
P CoA + N-succinyl-2-amino-6-keto-L-pimelate <1-3> [1-5]

Inhibitors

2,6-dihydroxypimelic acid <1> (<1> 0.54 mM, 66% inhibition [2]) [2]
2-hydroxytetrahydropyran-2,6-dicarboxylic acid <1> (<1> competitive inhibition [2]) [2]
2-oxopimelic acid <1> (<1> 5 mM, 57% inhibition [2]) [2]
4-oxo-(2E,5E)-heptadien-1,7-dioic acid <1> (<1> 0.5 mM, 32% inhibition [2]) [2]
6-aminocaproic acid <1> (<1> 5 mM, 35% inhibition [2]) [2]
$CaCl_2$ <1> (<1> 1 mM, 3% inhibition [1]) [1]
$CoCl_2$ <1> (<1> 1 mM, 95% inhibition [1]) [1]
$CuSO_4$ <1> (<1> 1 mM, complete loss of activity [1]) [1]
D-2-aminopimelate <1> (<1> competitive vs. L-2-aminopimelate and tetrahydropicolinate [2]) [2]
D-2-hydroxypimelic acid <1> (<1> 5 mM, 93% inhibition [2]) [2]
DL-2-amino-5-thiapimelic acid <1> (<1> 5 mM, 75% inhibition [2]) [2]
L-2-hydroxypimelic acid <1> (<1> 5 mM, 88% inhibition [2]) [2]
$MgCl_2$ <1> (<1> 1 mM, 20% inhibition [1]) [1]
$MnCl_2$ <1> (<1> 1 mM, 9% inhibition [1]) [1]
N-ethylmaleimide <1> (<1> 16 mM, 42% inhibition [1]) [1]
$ZnCl_2$ <1> (<1> 1 mM, 88% inhibition [1]) [1]
p-chloromercuriphenylsulfonate <1> (<1> 0.16 mM, 97% inhibition [1]) [1]
pimelic acid <1> (<1> 5 mM, 43% inhibition [2]) [2]
Additional information <1> (<1> not inhibited by EDTA and Fe^{2+} [1]) [1]

Specific activity (U/mg)

36 <1> [1]

K_m-Value (mM)

0.015 <1> (succinyl-CoA) [1]
0.02 <1> (tetrahydrodipicolinate) [2]
0.022 <1> (tetrahydrodipicolinate) [1]
1 <1> (L-2-aminopimelate) [2]
2 <1> (3,4-dihydro-2H-1,4-thiazine-3,5-dicarboxylic acid) [2]

K_i-Value (mM)

0.000058 <1> (2-hydroxytetrahydropyran-2,6-dicarboxylic acid) [2]
0.31 <1> (D-2-aminopimelate, <1> vs. L-2-aminopimelate [2]) [2]
0.76 <1> (D-2-aminopimelate, <1> vs. tetrahydropicolinate [2]) [2]

pH-Optimum

8.2 <1> [1]

pH-Range

7-9 <1> (<1> 50% of maximal activity at pH 7, 75% of maximal activity at pH 9.0 [1]) [1]

Temperature optimum (°C)

37 <1> (<1> assay at [1]) [1]

4 Enzyme Structure

Molecular weight

68000 <1> (<1> gel filtration [1]) [1]
72000 <1> (<1> sucrose density gradient centrifugation [1]) [1]

Subunits

dimer <1> (<1> 2 * 31000, SDS-PAGE [1]) [1]

5 Isolation/Preparation/Mutation/Application

Purification

<1> ($MnCl_2$, ammonium sulfate, DEAE-cellulose, acid treatment, hydroxyapatite, phenyl-Sepharose, chromatofocusing [1]) [1]
<2> (recombinant enzyme [3]) [3]

Crystallization

<2> (crystallized from solutions of 16% poly (ethylene glycol) 4000, 200 mM ammonium sulfate, 100 mM HEPES, pH 7.5, and 10% 2-propanol, crystals belong to space group P21, X-ray structure refined to 2.2 A resolution [3]; crystal structure in complex with L-2-aminopimelate/coenzyme A and L-2-amino-6-oxopimelate/coenzyme A at 2.0 A resolution, hanging drop vapor diffusion from solutions of 10-13% poly(ethylene glycol) 4000, 94 mM MES, pH 6.4, 94 mM ammonium sulfate, and 4.7% 2-propanol in the presence of 16 mM (D,L)-2-aminopimelate and 2.5 mM CoA [4]; crystal structure of the enzyme in ternary complexes with pimelate/succinyl-CoA and L-2-aminopi-

melate with the nonreactive cofactor analog succinamide-CoA, 2.3 and 2.0 A resolution, crystals are prepared by cocrystallization using the hanging drop vapor diffusion method, drops are formed by mixing 0.005 ml 27 mg/ml enzyme with an equal volume of 17% poly(ethylene glycol) 4000, 188 mM ammonium sulfate, 94 mM MES, pH 6.4, 4.7% 2-propanaol, 20 mM pimelate, and 5 mM succinyl-CoA or 5 mM succinamide-CoA [7]) [3, 4, 6, 7]

Cloning

<2> (overexpressed in Escherichia coli [3]) [3]

6 Stability

General stability information

<1>, purification has to be carried out at 0-4°C [1]

Storage stability

<1>, 4°C, 0.001 M 2-mercaptoethanol, 80% remaining activity after 1 month [1]

<1>, 4°C, no 2-mercaptoethanol, 4 days, 80% loss of activity [1]

References

[1] Simms, S.A.; Voige, W.H.; Gilvarg, C.: Purification and characterization of succinyl-CoA: tetrahydrodipicolinate N-succinyltransferase from Escherichia coli. J. Biol. Chem., **259**, 2734-2741 (1984)

[2] Berges, D.A.; DeWolf, W.E.; Dunn, G.L.; Newmann, D.J.; Schmidt, S.J.; Taggart, J.J.; Gilvarg, C.: Studies on the active site of succinyl-CoA:tetrahydrodipicolinate N-succinyltransferase. Characterization using analogs of tetrahydrodipicolinate. J. Biol. Chem., **261**, 6160-6167 (1986)

[3] Beaman, T.W.; Binder, D.A.; Blanchard, J.S.; Roderick, S.L.: Three-dimensional structure of tetrahydrodipicolinate N-succinyltransferase. Biochemistry, **36**, 489-494 (1997)

[4] Beaman, T.W.; Blanchard, J.S.; Roderick, S.L.: The conformational change and active site structure of tetrahydrodipicolinate N-succinyltransferase. Biochemistry, **37**, 10363-10369 (1998)

[5] Shaw-Reid, C.A.; McCormick, M.M.; Sinskey, A.J.; Stephanopoulos, G.: Flux through the tetrahydrodipicolinate succinylase pathway is dispensable for L-lysine production in Corynebacterium glutamicum. Appl. Microbiol. Biotechnol., **51**, 325-333 (1999)

[6] Born, T.L.; Blanchard, J.S.: Structure/function studies on enzymes in the diaminopimelate pathway of bacterial cell wall biosynthesis. Curr. Opin. Chem. Biol., **3**, 607-613 (1999)

[7] Beaman, T.W.; Vogel, K.W.; Drueckhammer, D.G.; Blanchard, J.S.; Roderick, S.L.: Acyl group specificity at the active site of tetrahydrodipicolinate N-succinyltransferase. Protein Sci., **11**, 974-979 (2002)

N-Hydroxyarylamine O-acetyltransferase 2.3.1.118

1 Nomenclature

EC number
2.3.1.118

Systematic name
acetyl-CoA:N-hydroxyarylamine O-acetyltransferase

Recommended name
N-hydroxyarylamine O-acetyltransferase

Synonyms
N-hydroxy-2-aminofluorene-O-acetyltransferase
acetyltransferase, N-hydroxyarylamine O-
arylamine N-acetyltransferase
arylhydroxamate N,O-acetyltransferase

CAS registry number
100984-92-7

2 Source Organism

<1> *Mesocricetus auratus* (Syrian golden hamster [1]) [1, 5]
<2> *Rattus norvegicus* (Sprague-Dawley strain [4,10]) [4, 10]
<3> *Oryctolagus cuniculus* (fast and slow acetylating phenotypes, may be identical with EC 2.3.1.5 [6]) [5, 6]
<4> *Mus musculus* (strain C57BL/67 [5]) [5]
<5> *Homo sapiens* [5, 8]
<6> *Salmonella typhimurium* (strain TA98 [2-4,11]; TA 1538 [3]; derived from strain LT2 [3]; strain LT2 [9]; strain TA100 [11]) [2-5, 9, 11]
<7> *Escherichia coli* [7]

3 Reaction and Specificity

Catalyzed reaction
acetyl-CoA + an N-hydroxyarylamine = CoA + an N-acetoxyarylamine

Reaction type
acyl group transfer

Natural substrates and products

S acetyl-CoA + an N-hydroxyarylamine <1, 4-6> (<1,4-6> involved in mutagenic, metabolic activation of carcinogenic arylamines to form DNA-binding species [1,2,5]; <1,4-6> the enzyme from liver, but not from bacteria, can also catalyse acetylation of arylamines and N,O-acetylation of arylhydroxamates [1,2,5]; <6> unknown endogenous substrate in bacteria [1,2]) (Reversibility: ? <1, 4-6> [1, 2, 5]) [1, 2, 5]

P CoA + an N-acetoxyarylamine <1, 4-6> [1, 2, 5]

S acetyl-CoA + *o*-aminobenzoic acid <7> (<7> *o*-aminobenzoic acid is intermediate in the synthesis of tryptophane in E. coli [7]) (Reversibility: ? <7> [7]) [7]

P CoA + N-acetyl-*o*-aminobenzoic acid <7> [7]

Substrates and products

S ? + N-hydroxy-2-acetylaminofluorene <5> (<5> N,O-acetyltransferase activity [8]) (Reversibility: ? <5> [8]) [8]

P ?

S acetyl-CoA + 2-aminofluorene <5> (<5> N-acetyltransferase activity [8]) (Reversibility: ? <5> [8]) [8]

P CoA + 2-acetyloxyaminofluorene <5> [8]

S acetyl-CoA + 2-hydroxy-amino-6-methyldipyrido[1,2-a: 3',2'-d]imidazole <1, 2, 4-6> (<6> N-hydroxy-2-acetylaminofluorene cannot replace acetyl-CoA [2]) (Reversibility: ? <1, 2, 4-6> [1-5]) [1-5]

P CoA + N-acetoxy-amino-6-methyldipyrido[1,2-a: 3',2'-d]imidazole <1, 2, 6> [1-5]

S acetyl-CoA + 3-hydroxyamino-1-methyl-pyrido[4,3-b]-indole <1, 6> (Reversibility: ? <1, 6> [1, 2]) [1, 2]

P CoA + N-acetoxy-amino-1-methyl-pyrido[4,3-b]-indole <1, 6> [1, 2]

S acetyl-CoA + 4-aminoveratrole <6> (Reversibility: ? <6> [9]) [9]

P CoA + N-acetyl-4-aminoveratrole <6> [9]

S acetyl-CoA + 4-anisidine <6> (Reversibility: ? <6> [9]) [9]

P CoA + N-acetyl-4-anisidine <6> [9]

S acetyl-CoA + 4-iodoaniline <6> (Reversibility: ? <6> [9]) [9]

P CoA + N-acetyl-4-iodoaniline <6> [9]

S acetyl-CoA + N-hydroxy-2-aminofluorene <1, 3, 5, 6> (Reversibility: ? <1, 3, 5, 6> [1, 2, 6, 8]) [1, 2, 6, 8]

P CoA + N-acetoxyaminofluorene <1, 3, 5> [1, 6, 8]

S acetyl-CoA + N-hydroxy-3,2'-dimethyl-4-aminobiphenyl <1> (Reversibility: ? <1> [5]) [5]

P CoA + N-acetoxy-3,2'-dimethyl-4-aminobiphenyl <1> [5]

S acetyl-CoA + N-hydroxy-4-aminobiphenyl <1, 5> (Reversibility: ? <1, 5> [5, 8]) [5, 8]

P CoA + N-acetoxyaminobiphenyl <1, 5> [5, 8]

S acetyl-CoA + aniline <7> (<7> N-acetyltransferase activity, ping-pong bi bi mechanism [7]) (Reversibility: ? <7> [7]) [7]

P CoA + N-acetylaniline <7> [7]

S acetyl-CoA + isoniazid <7> (<7> N-acetyltransferase activity, ping-pong bi bi mechanism [7]) (Reversibility: ? <7> [7]) [7]

P CoA + N-acetyl-isoniazid <7> [7]

S acetyl-CoA + *m*-aminophenol <7> (<7> N-acetyltransferase activity, ping-pong bi bi mechanism [7]) (Reversibility: ? <7> [7]) [7]

P CoA + N-acetyl-*m*-aminophenol <7> [7]

S acetyl-CoA + *o*-aminobenzoic acid <7> (<7> N-acetyltransferase activity, ping-pong bi bi mechanism [7]) (Reversibility: ? <7> [7]) [7]

P CoA + N-acetyl-*o*-aminobenzoic acid <7> [7]

S acetyl-CoA + *o*-aminophenol <7> (<7> N-acetyltransferase activity, ping-pong bi bi mechanism [7]) (Reversibility: ? <7> [7]) [7]

P CoA + N-acetyl-*o*-aminophenol <7> [7]

S acetyl-CoA + *o*-anisidine <7> (<7> N-acetyltransferase activity, ping-pong bi bi mechanism [7]) (Reversibility: ? <7> [7]) [7]

P CoA + N-acetyl-*o*-anisidine <7> [7]

S acetyl-CoA + *p*-aminobenzoic acid <7> (<7> N-acetyltransferase activity, ping-pong bi bi mechanism [7]) (Reversibility: ? <7> [7]) [7]

P CoA + N-acetyl-*p*-aminobenzoic acid <7> [7]

S acetyl-CoA + *p*-aminophenol <7> (<7> N-acetyltransferase activity, ping-pong bi bi mechanism [7]) (Reversibility: ? <7> [7]) [7]

P CoA + N-acetyl-*p*-aminophenol <7> [7]

S acetyl-CoA + *p*-aminotoluene <7> (<7> N-acetyltransferase activity, ping-pong bi bi mechanism [7]) (Reversibility: ? <7> [7]) [7]

P CoA + N-acetyl-*p*-aminotoluene <7> [7]

S acetyl-CoA + *p*-anisidine <7> (<7> N-acetyltransferase activity, ping-pong bi bi mechanism [7]) (Reversibility: ? <7> [7]) [7]

P CoA + N-acetyl-*p*-anisidine <7> [7]

S butyryl-CoA + 2-hydroxy-amino-6-methyldipyrido[1,2-a: 3',2'-d]imidazole <6> (<6> 11% of the activity compared to acetyl-CoA as acyl donor [2]) (Reversibility: ? <6> [2]) [2]

P CoA + N-butyryloxyamino-6-methyldipyrido[1,2-a: 3',2'-d]imidazole <6> [2]

S hexanoyl-CoA + 2-hydroxy-amino-6-methyldipyrido[1,2-a: 3',2'-d]imidazole <6> (<6> 1.3% of the activity compared to acetyl-CoA as acyl donor [2]) (Reversibility: ? <6> [2]) [2]

P CoA + N-hexanoyloxyamino-6-methyldipyrido[1,2-a: 3',2'-d]imidazole <6> [2]

S malonyl-CoA + 2-hydroxy-amino-6-methyldipyrido[1,2-a: 3',2'-d]imidazole <6> (<6> 19% of the activity compared to acetyl-CoA as acyl donor [2]) (Reversibility: ? <6> [2]) [2]

P CoA + N-malonyloxyamino-6-methyldipyrido[1,2-a: 3',2'-d]imidazole <6> [2]

S propionyl-CoA + 2-hydroxyamino-6-methyldipyrido[1,2-a: 3',2'-d]imidazole <6> (<6> 89% of the activity compared to acetyl-CoA as acyl donor [2]) (Reversibility: ? <6> [2]) [2]

P CoA + N-propionyloxyamino-6-methyldipyrido[1,2-a: 3',2'-d]imidazole <6> [2]

S Additional information <6> (<6> the N-acetoxyarylamine products form covalent adducts with cellular macromolecules, e.g. tRNA [2]) [2]
P ?

Inhibitors
1-naphthol <7> [7]
1-nitro-2-naphthol <1, 6> [1, 2, 4]
2,6 dichloro-4-nitrophenol <7> [7]
2-amino-6-methyldipyrido[1,2-a: 3',2'-d]imidazole <6> [2]
2-aminofluorene <1> [1]
2-naphthol <7> [7]
N-ethylmaleimide <6, 7> [2, 7]
acetylsalicylic acid <7> [7]
coenzyme A <6> [2]
harmaline <7> [7]
harmine <7> [7]
iodoacetamide <1, 6, 7> [1, 2, 7]
p-cloromercuribenzoate <6> [2]
paraoxon <5> (<5> no significant effect [8]) [8]
pentachlorophenol <1, 6> [1, 2, 4]
salicylic acid <7> [7]
thiolactomycin <2, 6> [2, 4]
Additional information <6, 7> (<6> no inhibition by paraoxon [1,2]; <6> no inhibition by 2,6-dichloro-4-nitrophenol, chloramphenicol, hydroxylamine [2]; <7> inhibition by salicylic acid and N-ethylmaleimide is non competitive with acetyl-CoA and competitive with *o*-aminobenzoic acid [7]; <1> thiolactomycin does not inhibit enzyme from Syrian golden hamster [1]) [1, 2, 7]

Activating compounds
dithiothreitol <6> [4]

Specific activity (U/mg)
0.001 <6> (<6> N-acetylation of *p*-aminobenzoic acid [3]) [3]
0.00196 <1> (<1> kidney [5]) [5]
0.0024 <1> (<1> intestine [5]) [5]
0.0042 <1> (<1> lung [5]) [5]
0.0058 <1> (<1> liver [5]) [5]
0.034 <3> [6]
0.06 <7> (<7> *p*-aminobenzoic acid [7]) [7]
0.09 <7> (<7> aniline [7]) [7]
0.1 <7> (<7> *o*-anisidine [7]) [7]
0.126 <6> (<6> 2-hydroxyamino-6-methyldipyrido[1,2-a: 3',2'-d]imidazole as substrate [2]) [2]
0.17 <7> (<7> isoniazid [7]) [7]
0.25 <7> (<7> *p*-aminotoluene [7]) [7]
0.28 <7> (<7> *m*-aminophenol [7]) [7]
0.3 <7> (<7> *o*-aminobenzoic acid [7]) [7]
0.33 <7> (<7> *p*-aminophenol [7]) [7]

0.47 <7> (<7> *p*-anisidine [7]) [7]
0.517 <6> (<6> N-acetylation of 2-aminofluorene [3]) [3]
0.66 <6> [9]
0.67 <7> (<7> *o*-aminophenol [7]) [7]
0.82 <1> (<1> 2-hydroxyamino-6-methyldipyrido[1,2-a: 3',2'-d]imidazole as substrate [1]) [1]
1.31 <6> (<6> 2-hydroxyamino-6-methyldipyrido[1,2-a: 3',2'-d]imidazole as substrate [3]) [3]
3.9 <2> (<2> N,O-acetyltransferase assay [10]) [10]
7.8 <2> (<2> O-acetyltransferase assay [10]) [10]
8.72 <6> (<6> N-acetylation of isoniazide [3]) [3]
74 <2> (<2> N-acetyltransferase assay [10]) [10]
Additional information <1, 5> (<1> specific activities in different tissues for three hamster phenotypes [5]; <5> activity in different tissues might be related to carcinogenesis [5]; <5> radiolabel assay and Eadie-Hofstee plot [8]) [5, 8]

K_m-Value (mM)

0.0033 <6> (acetyl-CoA, <6> reaction with O-acetyl transfer to 2-hydroxyamino-6-methyldipyrido[1,2-α: 3',2'-d]imidazole as substrate [2]; <6> fluorometric method [4]) [2, 4]
0.01 <6> (acetyl-CoA, <6> reaction with 2-hydroxy-amino-6-methyldipyrido[1,2-α: 3',2'-d]imidazole as substrate [3]) [3]
0.27 <1> (N-hydroxy-2-acetylaminofluorene) [1]
0.36 <7> (*p*-aminobenzoic acid) [7]
0.37 <1> (acetyl-CoA, <1> reaction of O-acetyl transfer to 2-hydroxyamino-6-methyldipyrido[1,2-α: 3',2'-d]imidazole [1]) [1]
0.41 <7> (*o*-anisidine) [7]
0.48 <7> (*o*-aminobenzoic acid) [7]
0.54 <7> (*p*-aminophenol) [7]
0.55 <7> (aniline) [7]
0.59 <7> (isoniazid) [7]
0.6 <6> (4-aminoveratrole) [9]
0.63 <7> (*p*-aminotoluene) [7]
0.83 <7> (*p*-anisidine) [7]
1.1 <6> (4-anisidine) [9]
1.3 <6> (4-iodoaniline) [9]
1.71 <7> (*m*-aminophenol) [7]
1.94 <7> (*o*-aminophenol) [7]

pH-Optimum

7 <6> (<6> assay at [4]) [4]
7.4 <5> (<5> O-acetyltransferase assay [8]) [8]
7.4 <7> (<7> maximum activity [7]) [7]

Temperature optimum (°C)

37 <5-7> (<5> N-acetyltransferase assay [8]; <6> assay at [3,4]; <7> maximum activity [7]) [3, 4, 7, 8]

4 Enzyme Structure

Molecular weight

32000 <2> (<2> SDS PAGE [10]) [10]
32180 <6> (<6> calculated from nucleotide sequence [3]) [3]
33000 <1> (<1> gel filtration [1,5]) [1, 5]
33500 <5> (<5> SDS PAGE [8]) [8]
33570 <6> (<6> electrospray mass spectrometry [9]) [9]
48000 <6> (<6> gel filtration [2]) [2]
60000 <7> (<7> gel filtration [7]) [7]

Subunits

? <3> (<3> x * 33000 SDS-PAGE [6]) [6]
dimer <7> (<7> 2 * 34000, SDS-PAGE and DNA sequence [7]) [7]
monomer <1, 6> (<1> 1 * 33000 [1]) [1, 3]

5 Isolation/Preparation/Mutation/Application

Source/tissue

bladder <1, 4, 5> (<1,4,5> low level of activity [5]) [5]
colon <5> (<5> different phenotypes [5]) [5, 8]
intestine <1, 4, 6> [5]
kidney <1, 4, 5> [5]
liver <1-5> (<1> high polymorphism, two isozymes with different chromatographic properties and kinetic constants [5]; <5> different phenotypes [5]) [1, 4-6, 10]
lung <1, 4, 6> [5]

Localization

cytosol <1, 2, 4-6> [1, 3, 5, 10]

Purification

<1> (monomeric isozyme [5]; copurifies with arylamine acetyltransferase and arylhydroxamic acid N,O-acetyltransferase [1]; purification by ammonium sulfate followed by DEAE cellulose, gel filtration on GCL-2000 and HPKB-hydroxyapatite chromatography [1]) [1, 5]
<2> (ammonium sulfate precipitation, ion exchange chromatography, gel filtration, immunoaffinity chromatograpy with protein A Sepharose columns [10]; partial, using DEAE Cellulose and Sephacryl S-200 chromatography [4]) [4, 10]
<3> (copurifies and may be identical with N-acetyltransferase EC 2.3.1.5 [6]) [6]
<4> [5]
<6> (partial, by ammonium sulfate precipitation, DEAE-cellulose and Sephadex-G150 column chromatography [2]; nucleotide sequence [3]; single step immobilized metal ion chromatography [9]; streptomycin, ammonium sulfate precipitation, DEAE Cellulose, Sephadex G-150 chromatograpy [2]) [2, 3, 9]
<7> (nickel affinity column chromatography, SDS-PAGE [7]) [7]

Crystallization

<6> (sodium potassium tartrate and X-ray diffraction [9]) [9]

Cloning

<5> (16 recombinant NAT2 allozymes expressed in Escherichia coli JM105 expression system [8]) [8]

<6> (cloned and expression of wild type and two mutant enzymes in Escherichia coli XL-1-Blue maxi cells, plasmid pYG122 [3]; construction of 2 oat-null mutants specifically disrupted by replacing the oat gene with the chloramphenicol resistance gene using the preligation method [11]) [3, 11]

<7> (expression as a hystidine tagged fusion protein [7]) [7]

Engineering

A69C <6> (<6> reduces enzyme activity without affecting stability and mobility on SDS-PAGE [3]) [3]

A69R <6> (<6> reduces enzyme activity without affecting stability and mobility on SDS-PAGE [3]) [3]

6 Stability

Temperature stability

18-40 <6> (<6> stable for 2 hours [9]) [9]

General stability information

<1>, glycerol 50% v/v, stabilizes during storage [1]

Storage stability

<1>, -80°C, glycerol 50% v/v, stabilizes during storage [1]

<6>, -80°C, at least 3 months [2]

<7>, -20°C, 50 mM Tris-HCl, pH: 7.8, 10% glycerol, 1 month [7]

References

[1] Saito, K.; Shinohara, A.; Kamataki, T.; Kato, R.: N-Hydroxyarylamine O-acetyltransferase in hamster liver: identity with arylhydroxamic acid N,O-acetyltransferase and arylamine N-acetyltransferase. J. Biochem., **99**, 1689-1697 (1986)

[2] Saito, K.; Shinohara, A.; Kamataki, T.; Kato, R.: Metabolic activation of mutagenic N-hydroxyarylamines by O-acetyltransferase in Salmonella typhimurium TA98. Arch. Biochem. Biophys., **239**, 286-295 (1985)

[3] Watanabe, M.; Sofuni, T.; Nohmi, T.: Involvement of Cys69 residue in the catalytic mechanism of N-hydroxyarylamine O-acetyltransferase of Salmonella typhimurium. Sequence similarity at the amino acid level suggests a common catalytic mechanism of acetyltransferase for S. typhimurium and higher organisms. J. Biol. Chem., **267**, 8429-8436 (1992)

[4] Saito, K.; Shinohara, A.; Kamataki, T.; Kato, R.: A new assay for N-hydroxyarylamine O-acetyltransferase: reduction of N-hydroxyarylamines through N-acetoxyarylamines. Anal. Biochem., **152**, 226-231 (1986)
[5] Hein, D.W.: Acetylator genotype and arylamine-induced carcinogenesis. Biochim. Biophys. Acta, **948**, 37-66 (1988)
[6] Glowinski, I.B.; Weber, W.W.; Fysh, J.M.; Vaught, J.B.; King, C.M.: Evidence that arylhydroxamic acid N,O-acyltransferase and the genetically polymorphic N-acetyltransferase are properties of the same enzyme in rabbit liver. J. Biol. Chem., **255**, 7883-7890 (1980)
[7] Yamamura, E.; Sayama, M.; Kakikawa, M.; Mori, M.; Taketo, A.; Kodaira, K.: Purification and biochemical properties of an N-hydroxyarylamine O-acetyltransferase from Escherichia coli. Biochim. Biophys. Acta, **1475**, 10-16 (2000)
[8] Hein, D.; Doll, M.; Rustan, T.; Ferguson, R.: Metabolic activation of N-hydroxyarylamines and N-hydroxyarylamides by 16 recombinant human NAT2 allozymes: effects of 7 specific NAT2 nucleic acid substitutions. Cancer Res., **35**, 3531-3536 (1995)
[9] Sinclair, J.; Delgoda, R.; Noble, M.; Jarmin, S.; Goh, N.; Sim, E.: Purification, characterization and crystallization of an N-hydroxyarylamine O-acetyltransferase from Salmonella typhimurium. Protein Expression Purif., **12**, 371-380 (1998)
[10] Land, S.; Zukowski, K.; Lee, M.; Wang, Ch.; King, Ch.: Purification and characterization of a rat hepatic acetyltransferase that can metabolize aromatic amine derivates. Carcinogenesis, **14**, 1441-1449 (1993)
[11] Espinosa-Aguirre, J.; Yamada, M.; Maiso, K.; Watanabe, M.; Sofuni, T.; Nukuri, T.: New O-acetyltransferase deficient ames Salmonella strains generated by specific gene disruption. Mutat. Res., **439**, 159-169 (1999)

Icosanoyl-CoA synthase 2.3.1.119

1 Nomenclature

EC number
2.3.1.119

Systematic name
stearoyl-CoA:malonyl-CoA C-acyltransferase (decarboxylating, oxoacyl- and enoyl-reducing)

Recommended name
icosanoyl-CoA synthase

Synonyms
C_{18}-CoA elongase
acyl-CoA elongase
stearoyl-CoA elongase
synthase, icosanoyl coenzyme A

CAS registry number
141256-55-5

2 Source Organism

<1> *Allium porrum* [1, 2]

3 Reaction and Specificity

Catalyzed reaction
stearoyl-CoA + malonyl-CoA + 2 NAD(P)H + 2 H^+ = icosanoyl-CoA + CO_2 + 2 $NAD(P)^+$ + H_2O

Reaction type
Acyl group transfer

Substrates and products
S stearoyl-CoA + malonyl-CoA + NADPH + NADH <1> [1, 2]
P icosanoyl-CoA + CO_2 + $NAD(P)^+$ + H_2O

Inhibitors
deoxycholate <1> [1]

Cofactors/prosthetic groups

NADH <1> [1, 2]
NADPH <1> [1, 2]

Activating compounds

lipids <1> (requirement, in presence of a mixture of phosphatidylethanolamine, phosphatidylcholine and phosphatidylserine the C_{18}-CoA elongase activity is increased more than 6-fold [1]) [1]

Metals, ions

Mg^{2+} <1> (<1> stimulation [1]) [1]

K_m-Value (mM)

0.0017 <1> (stearoyl-CoA) [1]

pH-Optimum

6.8 <1> (<1> assay at [2]) [2]
7 <1> (<1> assay at [1]) [1]

Temperature optimum (°C)

30 <1> (<1> assay at [1,2]) [1, 2]

4 Enzyme Structure

Molecular weight

300000 <1> (<1> gel filtration [1]) [1]
350000 <1> (<1> gel filtration [2]) [2]

Subunits

Additional information <1> (<1> three major protein bands of MW 56000, 61000, 65000, can be detected, they are probably all constitutive enzymes of the acyl-CoA elongase, SDS-PAGE [1]) [1]

5 Isolation/Preparation/Mutation/Application

Source/tissue

epidermis <1> [1, 2]

Localization

microsome <1> (<1> membrane-associated, probably [1]) [1]

Purification

<1> (partial [1,2]) [1, 2]

6 Stability

General stability information

<1>, Triton X-100, 0.02%, stabilizes [1]

Storage stability

<1>, -20°C, 2% or 20% glycerol, stable for at least 500 h [1]

<1>, 4°C, 24 h, most activity lost [1]

References

[1] Bessoule, J.-J.; Lessire, R.; Cassagne, C.: Partial purification of the acyl-CoA elongase of Allium porrum leaves. Arch. Biochem. Biophys., **268**, 475-484 (1989)

[2] Lessire, R.; Bessoule, J.-J.; Cassagne, C.: Solubilization of C_{18}-CoA and C_{20}-CoA elongases from Allium porrum L. epidermal cell microsomes. FEBS Lett., **187**, 314-320 (1985)

6'-Deoxychalcone synthase 2.3.1.120

1 Nomenclature

EC number
2.3.1.120 (deleted, reaction listed is due to EC 2.3.1.74)

Recommended name
6'-deoxychalcone synthase

1-Alkenylglycerophosphoethanolamine O-acyltransferase 2.3.1.121

1 Nomenclature

EC number
2.3.1.121

Systematic name
acyl-CoA:1-alkenylglycerophosphoethanolamine O-acyltransferase

Recommended name
1-alkenylglycerophosphoethanolamine O-acyltransferase

Synonyms
acyltransferase, alkenylglycerophosphoethanolamine
Additional information (<1> not identical with EC 2.3.1.104 [1]) [1]

CAS registry number
112445-17-7

2 Source Organism

<1> *Cavia porcellus* (guinea pig [1]) [1]

3 Reaction and Specificity

Catalyzed reaction
acyl-CoA + 1-alkenylglycerophosphoethanolamine = CoA + 1-alkenyl-2-acylglycerophosphoethanolamine

Reaction type
acyl group transfer

Natural substrates and products
S acyl-CoA + 1-alkenylglycerophosphoethanolamine <1> (<1> involved in glycerophospholipid metabolism [1]) (Reversibility: ? <1> [1]) [1]
P CoA + 1-alkenyl-2-acylglycerophosphoethanolamine

Substrates and products
S acyl-CoA + 1-alkenylglycerophosphoethanolamine <1> (Reversibility: ? <1> [1]) [1]
P CoA + 1-alkenyl-2-acylglycerophosphoethanolamine <1> [1]
S arachidonoyl-CoA + 1-alkenylglycerophosphoethanolamine <1> (Reversibility: ? <1> [1]) [1]

P CoA + 1-alkenyl-2-arachidonoylglycerophosphoethanolamine
S linoleoyl-CoA + 1-alkenylglycerophosphoethanolamine <1> (<1> best substrate [1]) (Reversibility: ? <1> [1]) [1]
P CoA + 1-alkenyl-2-linoleoylglycerophosphoethanolamine
S oleoyl-CoA + 1-alkenylglycerophosphoethanolamine <1> (<1> poor substrate [1]) (Reversibility: ? <1> [1]) [1]
P CoA + 1-alkenyl-2-oleoylglycerophosphoethanolamine
S palmitoyl-CoA + 1-alkenylglycerophosphoethanolamine <1> (Reversibility: ? <1> [1]) [1]
P CoA + 1-alkenyl-2-palmitoylglycerophosphoethanolamine
S stearoyl-CoA + 1-alkenylglycerophosphoethanolamine <1> (<1> poor substrate [1]) (Reversibility: ? <1> [1]) [1]
P CoA + 1-alkenyl-2-stearoylglycerophosphoethanolamine

Inhibitors

Ca^{2+} <1> (<1> 50% inhibition at 5 mM [1]) [1]
N-ethylmaleimide <1> (<1> inhibition at 0.5 mM [1]) [1]
iodoacetate <1> (<1> inhibition at 0.5 mM [1]) [1]

Activating compounds

dithiothreitol <1> (<1> activation [1]) [1]
glutathione <1> (<1> activation [1]) [1]

pH-Optimum

8-9 <1> (<1> microsomal preparation [1]) [1]

Temperature optimum (°C)

25 <1> (<1> microsomal preparation [1]) [1]

5 Isolation/Preparation/Mutation/Application

Source/tissue

heart <1> [1]

Localization

microsome <1> [1]

6 Stability

Temperature stability

55 <1> (<1> 56% loss of activity after 1 min [1]) [1]

References

[1] Arthur, G.; Page, L.; Choy, P.C.: Acylation of 1-alkenylglycerophosphoethanolamine and 1-acylglycerophosphoethanolamine in guinea-pig heart microsomes. Biochim. Biophys. Acta, **921**, 259-265 (1987)

Trehalose O-mycolyltransferase 2.3.1.122

1 Nomenclature

EC number
2.3.1.122

Systematic name
α,α'-trehalose-6-mycolate:α,α'-trehalose-6-mycolate 6'-mycolyltransferase

Recommended name
trehalose O-mycolyltransferase

Synonyms
α,α'-trehalose 6-monomycolate:α,α'-trehalose mycolyltransferase
mycolyltransferase, trehalose 6-monomycolate-trehalose

CAS registry number
111694-11-2

2 Source Organism

<1> *Mycobacterium smegmatis* [1]
<2> *Mycobacterium tuberculosis* [2, 3]

3 Reaction and Specificity

Catalyzed reaction
2 α,α'-trehalose 6-mycolate = α,α'-trehalose + α,α'-trehalose 6,6'-bismycolate (<2> reaction catalyzed by antigen 85B complex involved in cell wall biosynthesis. Previously unsuspected role of 85B protein in pathogenesis of tuberculosis [2]; <2> reaction catalyzed by 3 members of the Ag85 complex in vitro, namely FbpA, FbpB and FbpC2 [3])

Reaction type
acyl group transfer

Natural substrates and products
S trehalose 6-monomycolate + trehalose <1> (Reversibility: r <1> [1]) [1]
P trehalose + trehalose 6-monomycolate <1> [1]
S trehalose 6-monomycolate + trehalose 6-monomycolate <1> (Reversibility: r <1> [1]) [1]
P trehalose 6,6'-dimycolate + trehalose <1> [1]

Substrates and products

S trehalose 6-monomycolate + trehalose <1> (Reversibility: r <1> [1]) [1]
P trehalose + trehalose 6-monomycolate <1> [1]
S trehalose 6-monomycolate + trehalose 6-monomycolate <1> (Reversibility: r <1> [1]) [1]
P trehalose 6,6'-dimycolate + trehalose <1> [1]
S trehalose 6-monopalmitate + trehalose <1> (Reversibility: r <1> [1]) [1]
P trehalose + trehalose 6-monopalmitate <1> [1]
S Additional information <1> (<1> exchange of mycolyl group, highly specific for α,α'-trehalose as the mycolate acceptor, less specific for the acyl donor group [1]) [1]
P ?

Inhibitors

6-azido-6-deoxytrehalose <2> (<2> inhibits all three members of Ag85 complex in vitro [3]) [3]
ACP <1> (<1> 0.00056 mg/0.6 ml, 68% reduction of activity [1]) [1]
$MgCl_2$ <1> (<1> 16 mM, 91% inhibition [1]) [1]
Tween 80 <1> (<1> strong, might disrupt necessary structure of the TM vesicle/micelle [1]) [1]
iodoacetamide <1> (<1> 16.7 mM, 53% decrease of activity [1]) [1]

Cofactors/prosthetic groups

Additional information <1> (<1> no requirement of cofactors such as ATP or CoA [1]) [1]

Specific activity (U/mg)

0.00779 <1> [1]

K_m-Value (mM)

0.0083 <1> (trehalose) [1]

pH-Optimum

6.5-7 <1> [1]

pH-Range

5-9 <1> (<1> 45% of maximal activity at pH 5.0, 30% of maximal activity at pH 9.0 [1]) [1]

Temperature optimum (°C)

37 <1> [1]

Temperature range (°C)

25-55 <1> (<1> 75% of maximal activity at 25°C, 12% at 55°C [1]) [1]

4 Enzyme Structure

Molecular weight

25000 <1> (<1> SDS-PAGE, gel filtration [1]) [1]

30000 <2> (<2> FbpA: SDS-PAGE, Western Blot, aminoacid sequence. FbpB: aminoacid sequence. FbpC2: SDS-PAGE, Western Blot, aminoacid sequence [3]) [3]

Subunits

monomer <1, 2> (<1> 1 * 25000, SDS-PAGE [1]; <2> FbpA: 1 * 30000, SDS-PAGE. FbpC2: 1 * 30000, SDS-PAGE [3]) [1, 3]

5 Isolation/Preparation/Mutation/Application

Purification

<1> (ammonium sulfate fractionation, gel filtration and Sephadex column chromatography [1]) [1]

<2> (recombinant proteins purified by affinity chromatography on Ni^+ charged His-trap columns [3]) [3]

Crystallization

<2> [3]

Cloning

<2> (expression of FbpA, FbpB and FbpC2 in Escherichia coli, FbpB poorly expressed due to differential efficiency of mRNA translation [3]) [3]

6 Stability

General stability information

<1>, buffer of ionic strength of 0.1 M and above stabilizes [1]

<1>, dithiothreitol, 1 mM, stabilizes [1]

<1>, purification has to be carried out at 4°C [1]

References

[1] Sathyamoorthy, N.; Takayama, K.: Purification and characterization of a novel mycolic acid exchange enzyme from Mycobacterium smegmatis. J. Biol. Chem., **262**, 13417-13423 (1987)

[2] Wilkinson, R. J.; DesJardin, L. E.; Islam, N.; Gibson, B. M.; Kanost, R. A.; Wilkinson, K. A.; Poelman, D.; EisenachK. D.; Toossi, Z.: An increase in expression of a tuberculosis mycolyl transferase gene (fbpB) gene occurs early after infection of human monocytes. Mol. Microbiol., **39**, 813-821 (2001)

[3] Kreme, L.; Maughan, W. N.; Wilson, R. A.; Dover, L. G.; Besra, G. S.: The M. tuberculosis antigen 85 complex and mycolyltransferase activity. Lett. Appl. Microbiol., **34**, 233-237 (2002)

Dolichol O-acyltransferase 2.3.1.123

1 Nomenclature

EC number
2.3.1.123

Systematic name
palmitoyl-CoA:dolichol O-palmitoyltransferase

Recommended name
dolichol O-acyltransferase

Synonyms
acyl-CoA:dolichol acyltransferase
acyltransferase, dolichol

CAS registry number
111839-04-4

2 Source Organism

<1> *Rattus norvegicus* (male [1]) [1]

3 Reaction and Specificity

Catalyzed reaction
palmitoyl-CoA + dolichol = CoA + dolichyl palmitate

Reaction type
acyl group transfer

Natural substrates and products
S acetyl-CoA + dolichol <1> (Reversibility: ? <1> [1]) [1]
P CoA + dolichyl acetate
S arachidonoyl-CoA + dolichol <1> (Reversibility: ? <1> [1]) [1]
P CoA + dolichyl arachidonate
S myristoyl-CoA + dolichol <1> (Reversibility: ? <1> [1]) [1]
P CoA + dolichyl myristate
S oleoyl-CoA + dolichol <1> (Reversibility: ? <1> [1]) [1]
P CoA + dolichyl oleate
S palmitoyl-CoA + dolichol <1> (Reversibility: ? <1> [1]) [1]
P CoA + dolichyl palmitate <1> [1]

S stearoyl-CoA + dolichol <1> (Reversibility: ? <1> [1]) [1]
P CoA + dolichyl stearate
S Additional information <1> (<1> esterfication may play a role in targeting the lipid from the endoplasmic reticulum to lysosomes [1]) [1]
P ?

Substrates and products

S acetyl-CoA + dolichol <1> (<1> 30% of activity compared to palmitoyl-CoA [1]) (Reversibility: ? <1> [1]) [1]
P CoA + dolichyl acetate
S arachidonoyl-CoA + dolichol <1> (<1> 30% of activity compared to palmitoyl-CoA [1]) (Reversibility: ? <1> [1]) [1]
P CoA + dolichyl arachidonate
S myristoyl-CoA + dolichol <1> (<1> 40% of activity compared to palmitoyl-CoA [1]) (Reversibility: ? <1> [1]) [1]
P CoA + dolichyl myristate
S oleoyl-CoA + dolichol <1> (<1> 40% of activity compared to palmitoyl-CoA [1]) (Reversibility: ? <1> [1]) [1]
P CoA + dolichyl oleate
S palmitoyl-CoA + dolichol <1> (Reversibility: ? <1> [1]) [1]
P CoA + dolichyl palmitate <1> [1]
S stearoyl-CoA + dolichol <1> (<1> 40% of activity compared to palmitoyl-CoA [1]) (Reversibility: ? <1> [1]) [1]
P CoA + dolichyl stearate
S Additional information <1> (<1> α-saturated dolichols are acylated more rapidly than α-unsaturated analogues [1]) [1]
P ?

Inhibitors

Ca^{2+} <1> (<1> 20 mM, 50-60% inhibition [1]) [1]
Mn^{2+} <1> (<1> 20 mM, 50-60% inhibition [1]) [1]
β-octylglycopyranoside <1> (<1> strong, almost complete inactivation at 0.04% [1]) [1]
deoxycholate <1> (<1> strong, almost complete inactivation at 0.04% [1]) [1]

Activating compounds

Triton X-100 <1> (<1> activates at low concentrations by 50% [1]) [1]
albumin <1> (<1> activates [1]) [1]

Metals, ions

Additional information <1> (<1> KCl and $MgCl_2$ have no influence [1]) [1]

pH-Optimum

6.4 <1> [1]

pH-Range

5-7 <1> (<1> 20% of maximal activity at pH 5.0, 40% of maximal activity at pH 7.0 [1]) [1]

Temperature optimum (°C)

37 <1> (<1> assay at [1]) [1]

5 Isolation/Preparation/Mutation/Application

Source/tissue

brain <1> (<1> slight activity [1]) [1]
heart <1> (<1> slight activity [1]) [1]
kidney <1> (<1> slight activity [1]) [1]
liver <1> [1]
lung <1> (<1> slight activity [1]) [1]
muscle <1> (<1> slight activity [1]) [1]
small intestine <1> (<1> slight activity [1]) [1]
spleen <1> (<1> slight activity [1]) [1]
thymus <1> (<1> slight activity [1]) [1]

Localization

microsome <1> (<1> liver, esterification takes place on cytoplasmic side of membranes of the endoplasmic reticulum and is subsequently distributed to different organelles [1]) [1]

Purification

<1> (partial [1]) [1]

6 Stability

General stability information

<1>, Triton X-100 stabilizes [1]

Storage stability

<1>, -20°C, microsomes are stable for several months [1]
<1>, 4°C, microsomes in 0.25 M sucrose, after 5 days 50% remaining activity [1]

References

[1] Tollbom, Í.; Valtersson, C.; Chojnacki, T.; Dallner, G.: Esterification of dolichol in rat liver. J. Biol. Chem., **263**, 1347-1352 (1988)

Diacylglycerol acyltransferase 2.3.1.124

1 Nomenclature

EC number
2.3.1.124 (deleted, identical to EC 2.3.1.20)

Recommended name
diacylglycerol acyltransferase

1-Alkyl-2-acetylglycerol O-acyltransferase 2.3.1.125

1 Nomenclature

EC number
2.3.1.125

Systematic name
acyl-CoA:1-O-alkyl-2-acetyl-sn-glycerol O-acyltransferase

Recommended name
1-alkyl-2-acetylglycerol O-acyltransferase

Synonyms
1-hexadecyl-2-acetylglycerol acyltransferase
acyltransferase, 1-hexadecyl-2-acetylglycerol

CAS registry number
114704-90-4

2 Source Organism

<1> *Homo sapiens* [1]

3 Reaction and Specificity

Catalyzed reaction
acyl-CoA + 1-O-alkyl-2-acetyl-sn-glycerol = CoA + 1-O-alkyl-2-acetyl-3-acyl-sn-glycerol

Reaction type
acyl group transfer

Natural substrates and products
S Additional information <1> (<1> related to platelet-activating factor metabolism [1]) [1]
P ?

Substrates and products
S linoleoyl-CoA + 1-O-hexadecyl-2-acetyl-sn-glycerol <1> (<1> best substrate, not acetyl-CoA [1]) (Reversibility: ? <1> [1]) [1]
P CoA + 1-O-hexadecyl-2-acetyl-3-linoleoyl-sn-glycerol <1> [1]

Inhibitors

1-O-hexadecyl-2-oleoyl-sn-glycerol <1> (<1> competitive inhibitor of 1-hexadecyl-2-acetyl-3-acyl-sn-glycerol synthesis [1]) [1]
1-oleoyl-2-acetyl-sn-glycerol <1> (<1> mixed-type inhibition [1]) [1]
Ca^{2+} <1> (<1> 6% inhibition at 2 mM [1]) [1]
Cu^{2+} <1> (<1> 96% inhibition at 2 mM [1]) [1]
Mn^{2+} <1> (<1> 35% inhibition at 2 mM [1]) [1]
Zn^{2+} <1> (<1> 93% inhibition at 2 mM [1]) [1]
dithiothreitol <1> [1]

Activating compounds

bovine serum albumin <1> (<1> activation, molar ratio substrate linoleoyl-CoA/albumin: 2:4:1 [1]) [1]

Metals, ions

Additional information <1> (<1> no metal ion requirement [1]) [1]

K_m-Value (mM)

0.0085 <1> (linoleoyl-CoA) [1]

K_i-Value (mM)

0.032 <1> (1-O-hexadecyl-2-oleoyl-sn-glycerol) [1]

pH-Optimum

6.8 <1> [1]

Temperature optimum (°C)

37 <1> (<1> assay at [1]) [1]

5 Isolation/Preparation/Mutation/Application

Source/tissue

HL-60 cell <1> [1]

Localization

microsome <1> [1]

6 Stability

Temperature stability

45 <1> (<1> stable below [1]) [1]
58 <1> (<1> 50% activity after 15 min [1]) [1]
65 <1> (<1> inactivation [1]) [1]

References

[1] Kawasaki, T.; Snyder, F.: Synthesis of a novel acetylated neutral lipid related to platelet-activating factor by acyl-CoA:1-O-alkyl-2-acetyl-sn-glycerol acyltransferase in HL-60 cells [published erratum appears in J Biol Chem 1988 Jul 25;263(21):10539]. J. Biol. Chem., **263**, 2593-2596 (1988)

Isocitrate O-dihydroxycinnamoyltransferase 2.3.1.126

1 Nomenclature

EC number
2.3.1.126

Systematic name
caffeoyl-CoA:isocitrate 3-O-(3,4-dihydroxycinnamoyl)transferase

Recommended name
isocitrate O-dihydroxycinnamoyltransferase

Synonyms
isocitrate hydroxycinnamoyltransferase

CAS registry number
112352-88-2

2 Source Organism

<1> *Amaranthus chlorostachys* (Willd. [1]) [1]
<2> *Amaranthus cruentus* (L. cv. Oeschberg [1]) [1]
<3> *Amaranthus hybridus* (L., ssp. paniculatus (L.) Hejny [1]) [1]
<4> *Amaranthus paniculatus* (L. [1]) [1]
<5> *Amaranthus tricolor* [1]

3 Reaction and Specificity

Catalyzed reaction
caffeoyl-CoA + isocitrate = CoA + 2-caffeoylisocitrate (Feruoyl-CoA and 4-coumaroyl-CoA can also act as donors)

Reaction type
acyl group transfer

Natural substrates and products
S caffeoyl-CoA + isocitrate <1-5> (Reversibility: r <1-5> [1]) [1]
P CoA + 2-caffeoylisocitrate

Substrates and products
S caffeoyl-CoA + isocitric acid <2> (Reversibility: r <2> [1]) [1]
P CoA + (E)-caffeoylisocitric acid <2> [1]
S feruloyl-CoA + isocitric acid <2> (Reversibility: r <2> [1]) [1]

P CoA + feruloylisocitric acid <2> [1]
S hydroxycinnamoyl-CoA + isocitric acid <2> (<2> strict specificity for isocitric acid [1]) (Reversibility: r <2> [1]) [1]
P CoA + hydroxycinnamoylisocitric acid <2> [1]
S sinapoyl-CoA + isocitric acid <2> (<2> minor activity compared to caffeoyl-CoA [1]) (Reversibility: r <2> [1]) [1]
P CoA + sinapoylisocitric acid <2> [1]
S Additional information <2> (<2> citric, malic tartaric and quinic acids are not accepted as substrates [1]) [1]
P ?

K_m-Value (mM)
0.045 <2> (caffeoyl-CoA) [1]
0.055 <2> (*p*-coumaroyl-CoA) [1]
0.163 <2> (feruloyl-CoA) [1]
0.208 <2> (isocitric acid, <2> second substrate caffeoyl-CoA [1]) [1]
0.257 <2> (isocitric acid, <2> second substrate *p*-coumaroyl-CoA [1]) [1]
0.412 <2> (isocitric acid, <2> second substrate feruloyl-CoA [1]) [1]

5 Isolation/Preparation/Mutation/Application

Source/tissue
cotyledon <2> [1]
leaf <2> [1]
seedling <2> [1]

Purification
<2> (partial [1]) [1]

6 Stability

Storage stability
<2>, -20°C, protein preparation shows no loss of enzymatic activity when stored for several weeks [1]

References

[1] Strack, D.; Leicht, P.; Bokern, M.; Wray, V.; Grotjahn, L.: Hydroxycinnamic acid esters of isocitric acid: Accumulation and enzymatic synthesis in Amaranthus cruentus. Phytochemistry, **26**, 2919-2923 (1987)

Ornithine N-benzoyltransferase 2.3.1.127

1 Nomenclature

EC number
2.3.1.127

Systematic name
benzoyl-CoA:L-ornithine N-benzoyltransferase

Recommended name
ornithine N-benzoyltransferase

Synonyms
ornithine N-acyltransferase
ornithine N-benzoyltransferase

CAS registry number
111693-97-1

2 Source Organism

<1> *Anser sp.* (goose [1]) [1]
<2> *Cairina moschata* (muscovy duck [1]) [1]
<3> *Coturnix coturnix japonica* (japanese quail [1]) [1]
<4> *Gallus gallus* (chicken [1]) [1]
<5> *Meleagris gallopavo* (turkey [1]) [1]

3 Reaction and Specificity

Catalyzed reaction
2 benzoyl-CoA + L-ornithine = 2 CoA + N^2,N^5-dibenzoyl-L-ornithine

Reaction type
acyl group transfer

Natural substrates and products
S benzoyl-CoA + L-ornithine <3> (Reversibility: ? <3> [1]) [1]
P CoA + N^2,N^5-dibenzoyl-L-ornithine <3> (<3> i.e. ornithuric acid [1]) [1]

Substrates and products
S benzoyl-CoA + L-ornithine <3> (Reversibility: ? <3> [1]) [1]
P CoA + N^2,N^5-dibenzoyl-L-ornithine <3> (<3> i.e. ornithuric acid [1]) [1]
S benzoyl-CoA + arginine <3> (Reversibility: ? <3> [1]) [1]

P CoA + ornithuric acid

S benzoyl-CoA + lysine <3> (<3> weak acyl acceptor in vitro [1]) (Reversibility: ? <3> [1]) [1]

P ?

S Additional information <3> (<3> glycine, taurine, glutamine and glutamic acid are not substrates [1]) [1]

P ?

Inhibitors

isoleucine <3> (<3> markedly inhibition when arginine but not when ornithine is the acyl acceptor [1]) [1]

Specific activity (U/mg)

0.0045 <3> [1]

5 Isolation/Preparation/Mutation/Application

Source/tissue

kidney <1-5> [1]
liver <3> (<3> only low levels of activity [1]) [1]

Localization

mitochondrion <3> (<3> renal and hepatic [1]) [1]

Purification

<3> (partial [1]) [1]

References

[1] Seymour, M.A.; Millburn, P.; Tait, G.H.: Renal biosynthesis of ornithuric acid in quail. Biochem. Soc. Trans., **15**, 1108-1109 (1987)

Ribosomal-protein-alanine N-acetyltransferase 2.3.1.128

1 Nomenclature

EC number
2.3.1.128

Systematic name
acetyl-CoA:ribosomal-protein-L-alanine N-acetyltransferase

Recommended name
ribosomal-protein-alanine N-acetyltransferase

Synonyms
ribosomal protein S18 acetyltransferase

CAS registry number
113383-52-1

2 Source Organism

<1> *Escherichia coli* (strain K12 [1]) [1]

3 Reaction and Specificity

Catalyzed reaction
acetyl-CoA + ribosomal-protein L-alanine = CoA + ribosomal-protein N-acetyl-L-alanine (A group of enzymes in Escherichia coli that acetylate the N-terminal alanine residues of specific ribosomal proteins. cf. EC 2.3.1.88, peptide α-N-acetyltransferase)

Reaction type
acyl group transfer

Natural substrates and products
- **S** acetyl-CoA + ribosomal-protein L-alanine <1> (Reversibility: ? <1> [1]) [1]
- **P** CoA + ribosomal-protein N-acetyl-L-alanine <1> [1]

Substrates and products
- **S** acetyl-CoA + ribosomal-protein L-alanine <1> (Reversibility: ? <1> [1]) [1]
- **P** CoA + ribosomal-protein N-acetyl-L-alanine <1> [1]

4 Enzyme Structure

Molecular weight

18230 <1> (<1> calculated from amino acid residues [1]) [1]

5 Isolation/Preparation/Mutation/Application

Cloning

<1> (gene rimI encoding enzyme acetylating ribosomal protein S18 cloned into a mini-F plasmid pRE432 and sequenced [1]) [1]

References

[1] Yoshikawa, A.; Isono, S.; Sheback, A.; Isono, K.: Cloning and nucleotide sequencing of the genes rimI and rimJ which encode enzymes acetylating ribosomal proteins S18 and S5 of Escherichia coli K12. Mol. Gen. Genet., **209**, 481-488 (1987)

Acyl-[acyl-carrier-protein]-UDP-N-acetylglucosamine O-acyltransferase 2.3.1.129

1 Nomenclature

EC number
2.3.1.129

Systematic name
(R)-3-hydroxytetradecanoyl-[acyl-carrier-protein]:UDP-N-acetylglucosamine 3-O-(3-hydroxytetradecanoyl)transferase

Recommended name
acyl-[acyl-carrier-protein]-UDP-N-acetylglucosamine O-acyltransferase

Synonyms
UDP-N-acetylglucosamine acyltransferase
acyltransferase, uridine diphosphoacetylglucosamine
uridine diphosphoacetylglucosamine acyltransferase

CAS registry number
105843-69-4

2 Source Organism

<1> *Escherichia coli* (K12 wild-type [1,3]; strains JB1104 [1,3,5]; R477, R477/pTIH1, MI6383 [1]; MC1061/pSR1 (overproducing strain harbouring plasmid pSR1) [2-4]; W3106 [3]; SM105 [5,8]; SM 101, SM 108 [8]) [1-11]
<2> *Acinetobacter calcoaceticus* [5]
<3> *Enterobacter aerogenes* [5]
<4> *Citrobacter freundii* [5]
<5> *Klebsiella oxytoca* [5]
<6> *Proteus mirabilis* [5]
<7> *Pseudomonas aeruginosa* [5, 11, 12]
<8> *Rhodobacter sphaeroides* [5]
<9> *Serratia marcescens* [5]
<10> *Neisseria meningitidis* [10]
<11> *Bordetella bronchiseptica* [13]
<12> *Bordetella parapertussis* [13]
<13> *Bordetella pertussis* [13]
<14> *Chlamydia trachomatis* [14]
<15> *Helicobacter pylori* [15]

3 Reaction and Specificity

Catalyzed reaction

(R)-3-hydroxytetradecanoyl-[acyl-carrier protein] + UDP-N-acetylglucosamine = [acyl-carrier protein] + UDP-3-O-(3-hydroxytetradecanoyl)-N-acetylglucosamine

Reaction type

acyl group transfer

Natural substrates and products

S (R)-3-hydroxytetradecanoyl-[acyl-carrier-protein] + UDP-N-acetylglucosamine <1-11, 13> (<1> essential for cell growth of gram-negative bacteria [2]; <1> initial reaction in lipid A biosynthesis [2,3]; <1> involved with EC 2.4.1.182 and 2.7.1.130 in the biosynthesis of the phosphorylated glycolipid and outer membrane component, Lipid A [2,7]) (Reversibility: r <1> [4, 6, 9]; ? <2-11, 13> [5, 10, 13]) [2, 5, 10, 13]

P [acyl-carrier-protein] + UDP-3-O-(3-hydroxytetradecanoyl)-N-acetylglucosamine

Substrates and products

S (R)-3-hydroxytetradecanoyl-[acyl-carrier-protein] + UDP-N-butyrylglucosamine <1> (<1> transacylation at 8% the rate of the reaction with UDP-N-acetylglucosamine [2]) (Reversibility: ? <1> [2]) [2]

P [acyl-carrier-protein] + UDP-3-O-(3-hydroxytetradecanoyl)-N-butyrylglucosamine

S (R)-3-hydroxytetradecanoyl-[acyl-carrier-protein] + UDP-N-propionylglucosamine <1> (<1> transacylation at 22% the rate of the reaction with UDP-N-acetylglucosamine [2]) (Reversibility: ? <1> [2]) [2]

P [acyl-carrier-protein] + UDP-3-O-(3-hydroxytetradecanoyl)-N-propionylglucosamine

S (R,S)-3-hydroxydodecanoyl-[acyl-carrier-protein] + UDP-N-acetylglucosamine <1, 7, 8, 10-13> (<1> i.e. 3-hydroxylauroyl-[acyl-carrier-protein], poor substrate [2]; <2,7> main activity in crude extracts [5]) (Reversibility: ? <1, 7, 8, 10-13> [5, 10, 11, 13]) [5, 10, 11, 13]

P [acyl-carrier-protein] + UDP-3-O-(3-hydroxydodecanoyl)-N-acetylglucosamine

S (R,S)-3-hydroxyhexadecanoyl-[acyl-carrier-protein] + UDP-N-acetylglucosamine <11, 13> (Reversibility: ? <11, 13> [13]) [13]

P [acyl-carrier-protein] + UDP-3-O-(3-hydroxyhexadecanoyl)-N-acetylglucosamine

S (R,S)-3-hydroxylauroyl-[acyl-carrier-protein] + UDP-N-acetylglucosamine <1, 7, 10> (Reversibility: ? <1, 7, 10> [10, 12]) [10, 12]

P [acyl-carrier-protein] + UDP-3-O-(3-hydroxylauroyl)-N-acetylglucosamine

S (R,S)-3-hydroxymyristoyl-[acyl-carrier-protein] + UDP-N-acetylglucosamine <1, 10, 14> (Reversibility: ? <1, 10, 14> [10, 14]) [10, 14]

P [acyl-carrier-protein] + UDP-3-O-(3-hydroxymyristoyl)-N-acetylglucosamine
S (R,S)-3-hydroxytetradecanoyl-[acyl-carrier-protein] + UDP-N-acetylglucosamine <1-11, 13> (<1> high specificity with respect to acyl-donor, equilibrium constant favors thioester acyl carrier substrate [4]; <1> no substrates are UDP-glucosamine, UDP-N-(R)-3-hydroxymyristoylglucosamine, ADP-N-acetylglucosamine, GDP-N-acetylglucosamine, CDP-N-acetylglucosamine, (R)-3-hydroxytetradecanoyl-CoA, or palmitoyl-CoA, myristoyl-[acyl-carrier-protein], and palmitoyl-[acyl-carrier-protein] [1,2,4]) (Reversibility: r <1> [4, 6, 9]; ? <2-11, 13> [5, 10, 13]) [1-9, 10-13]
P [acyl-carrier-protein] + UDP-3-O-(3-hydroxytetradecanoyl)-N-acetylglucosamine
S (S)-3-hydroxytetradecanoyl-[acyl-carrier-protein] + UDP-N-acetylglucosamine <1> (<1> transacylation at 7% the rate of (R)-enantiomer [2]) (Reversibility: ? <1> [2]) [2]
P [acyl-carrier-protein] + UDP-3-O-(3-hydroxytetradecanoyl)-N-acetylglucosamine
S decanoyl-[acyl-carrier-protein] + UDP-N-acetylglucosamine <7> (Reversibility: ? <7> [12]) [12]
P [acyl-carrier-protein] + UDP-3-O-decanoyl-N-acetylglucosamine
S myristoyl-[acyl-carrier-protein] + UDP-N-acetylglucosamine <14> (<14> higher specific activity than to (R,S)-3-hydroxymyristoyl-[acyl-carrier-protein] [14]) (Reversibility: ? <14> [14]) [14]
P [acyl-carrier-protein] + UDP-3-O-myristoyl-N-acetylglucosamine

Inhibitors

KCl <1> (<1> 0.5 M [4]) [4]
LiCl <1> (<1> 0.5 M [4]) [4]
NaCl <1> (<1> 0.5 M [4]) [4]
Triton X-100 <1> [3]
diethyldicarbonate <1> [9]
myristoyl-[acyl-carrier-protein] <1> [4]
phenylglyoxal <1> [9]
pyridoxal 5'-phosphate/sodium borohydride <1> [9]
Additional information <1> (<1> no inhibition by N-ethylmaleimide, UDP-2,3-diacyl-N-glucosamine, lipid X, lipid IVA, 3-deoxy-D-mannooctulosonate, octyl-β-D-glucoside [2-4]) [2-4]

Metals, ions

KCl <1> (<1> slight activation at 0.25 M [4]) [4]
LiCl <1> (<1> slight activation at 0.25 M [4]) [4]
NaCl <1> (<1> slight activation at 0.25 M [4]) [4]

Turnover number (min^{-1})

430 <1> (UDP-N-acetylglucosamine, <1> wild-type [9]) [9]
590 <1> (UDP-N-acetylglucosamine, <1> K76R [9]) [9]
630 <1> (UDP-N-acetylglucosamine, <1> K76A [9]) [9]

Specific activity (U/mg)

0.00000002 <7> (<7> (R,S)-3-hydroxytetradecanoyl-[acyl-carrier-protein] as substrate, pTO15 [11]) [11]
0.00000014 <7> (<7> (R,S)-3-hydroxydecanoyl-[acyl-carrier-protein] as substrate, pTO14 [11]) [11]
0.00000028 <7> (<7> (R,S)-3-hydroxydodecanoyl-[acyl-carrier-protein] as substrate, pTO15 [11]) [11]
0.00000034 <7> (<7> (R,S)-3-hydroxydodecanoyl-[acyl-carrier-protein] as substrate, pTO14 [11]) [11]
0.00000072 <13> (<13> (R,S)-3-hydroxyhexadecanoyl-[acyl-carrier-protein] [13]) [13]
0.000001 <7> (<7> (R,S)-3-hydroxydodecanoyl-[acyl-carrier-protein] as substrate, pTO202 [11]) [11]
0.000001 <7> (<7> (R,S)-3-hydroxytetradecanoyl-[acyl-carrier-protein] as substrate, pTO9 [11]) [11]
0.0000011 <14> (<14> (R,S)-3-hydroxymyristoyl-[acyl-carrier-protein] as substrate [14]) [14]
0.0000015 <7> (<7> decanoyl-[acyl-carrier-protein] as substrate [12]) [12]
0.000003 <7> (<7> (R,S)-3-hydroxydecanoyl-[acyl-carrier-protein] as substrate, pTO201 [11]) [11]
0.0000032 <7> (<7> (R,S)-3-hydroxylauroyl-[acyl-carrier-protein] as substrate [12]) [12]
0.000006 <1> (<1> (R,S)-3-hydroxydecanoyl-[acyl-carrier-protein] as substrate, pTO10 and pTO105 [11]) [11]
0.000007 <1> (<1> (R,S)-3-hydroxydodecanoyl-[acyl-carrier-protein] as substrate, pTO10 [11]) [11]
0.00001 <7> (<7> (R,S)-3-hydroxytetradecanoyl-[acyl-carrier-protein] as substrate, pTO202 [11]) [11]
0.000011 <7> (<7> (R,S)-3-hydroxydecanoyl-[acyl-carrier-protein] as substrate, pTO15 [11]) [11]
0.000014 <7> (<7> (R,S)-3-hydroxytetradecanoyl-[acyl-carrier-protein] as substrate, pTO14 [11]) [11]
0.000015 <12> (<12> (R,S)-3-hydroxydodecanoyl-[acyl-carrier-protein] [13]) [13]
0.000019 <14> (<14> (R,S)-3-myristoyl-[acyl-carrier-protein] as substrate [14]) [14]
0.000022 <1> (<1> (R,S)-3-hydroxydecanoyl-[acyl-carrier-protein] as substrate, pTO105 [11]) [11]
0.000023 <1> (<1> (R,S)-3-hydroxydecanoyl-[acyl-carrier-protein] as substrate, pTO103 [11]) [11]
0.000024 <7> (<7> (R,S)-3-hydroxydodecanoyl-[acyl-carrier-protein] as substrate, pTO201 [11]) [11]
0.00003 <1> (<1> SM 101 [8]) [8]
0.000038 <7> (<7> (R,S)-3-hydroxydecanoyl-[acyl-carrier-protein] as substrate, pTO202 [11]) [11]
0.00004 <7> (<7> (R,S)-3-hydroxydodecanoyl-[acyl-carrier-protein] as substrate, pTO9 [11]) [11]

0.000054 <1> (<1> (R,S)-3-hydroxydecanoyl-[acyl-carrier-protein] as substrate, pTO1 [11]) [11]
0.000055 <10> (<10> (R,S)-3-hydroxymyristoyl-[acyl-carrier-protein] as substrate [10]) [10]
0.00006 <1> (<1> (R,S)-3-hydroxytetradecanoyl-[acyl-carrier-protein] as substrate, pTO101 [11]) [11]
0.000067 <1> (<1> (R,S)-3-hydroxymyristoyl-[acyl-carrier-protein] as substrate [10]) [10]
0.000068 <1> (<1> (R,S)-3-hydroxytetradecanoyl-[acyl-carrier-protein] as substrate, pTO104 [11]) [11]
0.00007 <1> (<1> (R,S)-3-hydroxydecanoyl-[acyl-carrier-protein] as substrate [10]) [10]
0.00007 <1> (<1> SM 108 [8]) [8]
0.000079 <1> (<1> (R,S)-3-hydroxydecanoyl-[acyl-carrier-protein] as substrate, pTO101 [11]) [11]
0.000087 <1> (<1> (R,S)-3-hydroxytetradecanoyl-[acyl-carrier-protein] as substrate, pTO103 [11]) [11]
0.0002 <1> (<1> (R,S)-3-hydroxytetradecanoyl-[acyl-carrier-protein] as substrate, pTO102 [11]) [11]
0.000247 <10> (<10> (R,S)-3-hydroxydecanoyl-[acyl-carrier-protein] as substrate [10]) [10]
0.000347 <1> (<1> (R,S)-3-hydroxytetradecanoyl-[acyl-carrier-protein] as substrate, pTO10 [11]) [11]
0.00035 <1> (<1> (R,S)-3-hydroxydecanoyl-[acyl-carrier-protein] as substrate, pTO104 [11]) [11]
0.00035 <13> (<13> (R,S)-3-hydroxydodecanoyl-[acyl-carrier-protein] [13]) [13]
0.00047 <11> (<11> (R,S)-3-hydroxydodecanoyl-[acyl-carrier-protein] [13]) [13]
0.00053 <1> (<1> (R,S)-3-hydroxytetradecanoyl-[acyl-carrier-protein] as substrate, pTO102 [11]) [11]
0.00059 <12> (<12> (R,S)-3-hydroxydecanoyl-[acyl-carrier-protein] [13]) [13]
0.00079 <13> (<13> (R,S)-3-hydroxytetradecanoyl-[acyl-carrier-protein] [13]) [13]
0.00087 <13> (<13> (R,S)-3-hydroxydecanoyl-[acyl-carrier-protein] [13]) [13]
0.001 <1> (<1> (R,S)-3-hydroxydecanoyl-[acyl-carrier-protein] as substrate, pTO102 [11]) [11]
0.001 <11> (<11> (R,S)-3-hydroxytetradecanoyl-[acyl-carrier-protein] [13]) [13]
0.001 <7> (<7> (R,S)-3-hydroxydecanoyl-[acyl-carrier-protein] as substrate, pTO9 [11]) [11]
0.0011 <11> (<11> (R,S)-3-hydroxydecanoyl-[acyl-carrier-protein] [13]) [13]
0.0012 <1> (<1> (R,S)-3-hydroxydodecanoyl-[acyl-carrier-protein] as substrate, pTO1 [11]) [11]

0.0012 <7> (<7> (R,S)-3-hydroxytetradecanoyl-[acyl-carrier-protein] as substrate, pTO201 [11]) [11]
0.00139 <1> (<1> (R,S)-3-hydroxylauroyl-[acyl-carrier-protein] as substrate [10]) [10]
0.0015 <1> (<1> SM 105 [8]) [8]
0.00325 <10> (<10> (R,S)-3-hydroxylauroyl-[acyl-carrier-protein] as substrate [10]) [10]
0.011 <7> (<7> (R,S)-3-hydroxydecanoyl-[acyl-carrier-protein] as substrate [12]) [12]
0.074 <1> (<1> (R,S)-3-hydroxytetradecanoyl-[acyl-carrier-protein] as substrate, pTO1 [11]) [11]
2.92 <1> [2, 3]
12.3 <1> [4]

K_m-Value (mM)
0.0016 <1> ((R)-3-hydroxytetradecanoyl-[acyl-carrier-protein]) [4]
0.099 <1> (UDP-N-acetylglucosamine) [4]
0.82 <1> (UDP-N-acetylglucosamine, <1> wild-type [9]) [9]
7 <1> (UDP-N-acetylglucosamine, <1> K76R [9]) [9]
15 <1> (UDP-N-acetylglucosamine, <1> K76A [9]) [9]

pH-Optimum
6.8-8.2 <1> (<1> broad [4]) [4]

Temperature optimum (°C)
30 <1> (<1> assay at [2,4]) [2, 4]

4 Enzyme Structure

Molecular weight
90000 <1> (<1> gel filtration [4]) [4]

Subunits
? <1> (<1> x * 27000, calculated from nucleotide sequence [2,4]) [2, 4]
dimer or trimer <1> (<1> α_2, 2 * 30000, SDS-PAGE [4]) [4]
trimer <1> (<1> α_3, 3 * 30000, SDS-PAGE [4]) [4, 9]

5 Isolation/Preparation/Mutation/Application

Localization
soluble <1> [2, 4]

Purification
<1> (partial, overproducing strain MC1061/pSR1 [2]) [2, 4]

Crystallization

<15> (crystallized at 24°C using $(NH_4)_2SO_4$ and Na/K tartrate as precipitants in the presence of a detergent, space group: P6322 with unit cell-parameters: a = b = 90.69, and c = 148.20 A [15]) [15]

Cloning

<1> (expression in Corynebacterium glutamicum [11]) [11]
<7> (expression in Escherichia coli [12]) [12]
<10> (expression in Escherichia coli [10]) [10]
<11> (overexpression in Escherichia coli [13]) [13]
<12> (overexpression in Escherichia coli [13]) [13]
<13> (overexpression in Escherichia coli [13]) [13]
<14> (overexpression in Escherichia coli [14]) [14]
<15> (overexpression in Escherichia coli [14]) [15]

Engineering

H122A <1> (<1> lower specific activity than wild-type [9]) [9]
H122N <1> (<1> lower specific activity than wild-type [9]) [9]
H125A <1> (<1> lower specific activity than wild-type [9]) [9]
H125N <1> (<1> lower specific activity than wild-type [9]) [9]
H144A <1> (<1> lower specific activity than wild-type [9]) [9]
H144N <1> (<1> lower specific activity than wild-type [9]) [9]
H160A <1> (<1> lower specific activity than wild-type [9]) [9]
H160F <1> (<1> lower specific activity than wild-type [9]) [9]
K76A <1> (<1> lower specific activity than wild-type [9]) [9]
K76R <1> (<1> lower specific activity than wild-type [9]) [9]

6 Stability

Temperature stability

60 <1> (<1> 20 min stable [2,3]) [2, 3]
100 <1> (<1> inactivation after 10 min [2,3]) [2, 3]

General stability information

<1>, bovine serum albumin prevents denaturation [4]
<1>, freeze-thawing inactivates [2]

Storage stability

<1>, -80°C, flash-frozen in liquid nitrogen, at least 3 months [2]
<1>, 4°C, several days [2]

References

[1] Anderson, M.S.; Bulawa, C.E.; Raetz, C.R.: The biosynthesis of gram-negative endotoxin. Formation of lipid A precursors from UDP-GlcNAc in extracts of Escherichia coli. J. Biol. Chem., **260**, 15536-15541 (1985)

[2] Anderson, M.S.; Raetz, C.R.: UDP-N-acetylglucosamine 3-O-acyltransferase from Escherichia coli. Methods Enzymol., **209**, 449-454 (1992)
[3] Anderson, M.S.; Raetz, C.R.: Biosynthesis of lipid A precursors in Escherichia coli. A cytoplasmic acyltransferase that converts UDP-N-acetylglucosamine to UDP-3-O-(R-3-hydroxymyristoyl)-N-acetylglucosamine. J. Biol. Chem., **262**, 5159-5169 (1987)
[4] Anderson, M.S.; Bull, H.G.; Galloway, S.M.; Kelly, T.M.; Mohan, S.; Radika, K.; Raetz, C.R.: UDP-N-acetylglucosamine acyltransferase of Escherichia coli. The first step of endotoxin biosynthesis is thermodynamically unfavorable. J. Biol. Chem., **268**, 19858-19865 (1993)
[5] Williamson, J.M.; Anderson, M.S.; Raetz, C.R.: Acyl-acyl carrier protein specificity of UDP-GlcNAc acyltransferases from gram-negative bacteria: relationship to lipid A structure. J. Bacteriol., **173**, 3591-3596 (1991)
[6] Vaara, M.: Eight bacterial proteins, including UDP-N-acetylglucosamine acyltransferase (LpxA) and three other transferases of Escherichia coli, consist of a six-residue periodicity theme. FEMS Microbiol. Lett., **97**, 249-254 (1992)
[7] Raetz, C.R.H.; Roderick, S.L.: A left-handed parallel β helix in the structure of UDP-N-acetylglucosamine acyltransferase. Science, **270**, 997-1000 (1995)
[8] Sorensen, P.G.; Lutkenhaus, J.; Young, K.; Eveland, S.S.; Anderson, M.S.; Raetz, C.R.H.: Regulation of UDP-3-O-[R-3-hydroxymyristoyl]-N-acetylglucosamine deacetylase in Escherichia coli. The second enzymic step of lipid A biosynthesis. J. Biol. Chem., **271**, 25898-25905 (1996)
[9] Wyckoff, T.J.O.; Raetz, C.R.H.: The active site of Escherichia coli UDP-N-acetylglucosamine acyltransferase. Chemical modification and site-directed mutagenesis. J. Biol. Chem., **274**, 27047-27055 (1999)
[10] Odegaard, T.J.; Kaltashov, I.A.; Cotter, R.J.; Steeghs, L.; Van Der Ley, P.; Khan, S.; Maskell, D.J.; Raetz, C.R.H.: Shortened hydroxyacyl chains on lipid A of Escherichia coli cells expressing a foreign UDP-N-acetylglucosamine O-acyltransferase. J. Biol. Chem., **272**, 19688-19696 (1997)
[11] Wyckoff, T.J.O.; Lin, S.; Cotter, R.J.; Dotson, G.D.; Raetz, C.R.H.: Hydrocarbon rulers in UDP-N-acetylglucosamine acyltransferases. J. Biol. Chem., **273**, 32369-32372 (1998)
[12] Dotson, G.D.; Kaltashov, I.A.; Cotter, R.J.; Raetz, C.R.H.: Expression cloning of a Pseudomonas gene encoding a hydroxydecanoyl-acyl carrier protein-dependent UDP-GlcNAc acyltransferase. J. Bacteriol., **180**, 330-337 (1998)
[13] Sweet, C.R.; Preston, A.; Toland, E.; Ramirez, S.M.; Cotter, R.J.; Maskell, D.J.; Raetz, C.R.H.: Relaxed acyl chain specificity of Bordetella UDP-N-acetylglucosamine acyltransferases. J. Biol. Chem., **277**, 18281-18290 (2002)
[14] Sweet, C.R.; Lin, S.; Cotter, R.J.; Raetz, C.R.H.: A Chlamydia trachomatis UDP-N-acetylglucosamine acyltransferase selective for myristoyl-acyl carrier protein. Expression in Escherichia coli and formation of hybrid lipid A species. J. Biol. Chem., **276**, 19565-19574 (2001)
[15] Lee, B.I.; Lee, J.Y.; Moon, J.; Han, B.W.; Suh, S.W.: Crystallization and preliminary x-ray crystallographic analysis of UDP-N-acetylglucosamine acyltransferase from Helicobacter pylori. Acta Crystallogr. Sect. D, **D58**, 864-866 (2002)

Galactarate O-hydroxycinnamoyltransferase 2.3.1.130

1 Nomenclature

EC number
2.3.1.130

Systematic name
feruloyl-CoA:galactarate O-(hydroxycinnamoyl)transferase

Recommended name
galactarate O-hydroxycinnamoyltransferase

Synonyms
hydroxycinnamoyltransferase, galacturate

CAS registry number
112956-50-0

2 Source Organism

<1> *Secale cereale* (rye, var. Kustro [1]) [1]

3 Reaction and Specificity

Catalyzed reaction
feruloyl-CoA + galactarate = CoA + O-feruloylgalactarate

Reaction type
acyl group transfer

Substrates and products
- **S** caffeoyl-CoA + galactarate <1> (<1> poor substrate [1]) (Reversibility: ? <1> [1]) [1]
- **P** CoA + O-caffeoylgalactarate
- **S** feruloyl-CoA + galactarate <1> (<1> best substrate, no substrates are 1-hydroxycinnamoyl glucose esters, gluconolactone, galactonate, glucuronate, ascorbate, malate, *m*-tartrate [1]) (Reversibility: ? <1> [1]) [1]
- **P** CoA + O-feruloylgalactarate <1> [1]
- **S** *p*-coumaroyl-CoA + galactarate <1> (<1> condensation at 69% the rate of the reaction with feruloyl-CoA [1]) (Reversibility: ? <1> [1]) [1]
- **P** CoA + O-*p*-coumaroylgalactarate

S sinapoyl-CoA + galactarate <1> (<1> condensation at 78% the rate of the reaction with feruloyl-CoA [1]) (Reversibility: ? <1> [1]) [1]
P CoA + O-sinapoylgalactarate

Activating compounds
DTT <1> (<1> activation [1]) [1]

Specific activity (U/mg)
0.0033 <1> [1]

K_m-Value (mM)
0.045 <1> (feruloyl-CoA) [1]
33 <1> (galactarate) [1]

pH-Optimum
7.3 <1> (<1> 25 mM potassium phosphate buffer [1]) [1]

pH-Range
7.1-7.6 <1> (<1> about 80% of maximal activity at pH 7.1 and 7.6 [1]) [1]

Temperature optimum (°C)
30 <1> (<1> assay at [1]) [1]

5 Isolation/Preparation/Mutation/Application

Source/tissue
leaf <1> (<1> primary [1]) [1]

Purification
<1> (partial [1]) [1]

6 Stability

Temperature stability
80 <1> (<1> denaturation after 5 min [1]) [1]

Storage stability
<1>, -20°C, at least 6 months [1]

References

[1] Strack, D.; Keller, H.; Weissenböck, G.: Enzymatic synthesis of hydroxycinnamic acid esters of sugar acids and hydroaromatic acids by protein preparations from rye (Secale cereale) primary leaves. J. Plant Physiol., **131**, 61-73 (1987)

Glucarate O-hydroxycinnamoyltransferase 2.3.1.131

1 Nomenclature

EC number
2.3.1.131

Systematic name
sinapoyl-CoA:glucarate O-(hydroxycinnamoyl)transferase

Recommended name
glucarate O-hydroxycinnamoyltransferase

Synonyms
hydroxycinnamoyltransferase, glucarate

CAS registry number
112956-51-1

2 Source Organism

<1> *Secale cereale* (rye, var. Kustro [1]) [1]

3 Reaction and Specificity

Catalyzed reaction
sinapoyl-CoA + glucarate = CoA + O-sinapoylglucarate

Reaction type
acyl group transfer

Substrates and products

- **S** caffeoyl-CoA + glucarate <1> (<1> condensation at 8% the rate of the reaction with sinapoyl-CoA [1]) (Reversibility: ? <1> [1]) [1]
- **P** CoA + caffeoylglucarate <1> [1]
- **S** feruloyl-CoA + glucarate <1> (<1> condensation at 54% the rate of the reaction with sinapoyl-CoA [1]) (Reversibility: ? <1> [1]) [1]
- **P** CoA + feruloylglucarate <1> [1]
- **S** *p*-coumaroyl-CoA + glucarate <1> (<1> condensation at 85% the rate of the reaction with sinapoyl-CoA [1]) (Reversibility: ? <1> [1]) [1]
- **P** CoA + *p*-coumaroylglucarate <1> [1]
- **S** sinapoyl-CoA + glucarate <1> (<1> best substrate [1]) (Reversibility: ? <1> [1]) [1]

P CoA + O-sinapoylglucarate <1> [1]

S Additional information <1> (<1> no substrates are 1-hydroxycinnamoyl glucose esters, gluconolactone, galactonate, glucuronate, ascorbate, malate, *m*-tartrate [1]) [1]

P ?

Activating compounds

EDTA <1> (<1> activation [1]) [1]

Specific activity (U/mg)

0.01 <1> (<1> feruloyl-CoA as donor [1]) [1]

K_m-Value (mM)

0.018 <1> (sinapoyl-CoA) [1]

8 <1> (glucarate) [1]

pH-Optimum

7.3 <1> (<1> 100 mM potassium phosphate buffer [1]) [1]

pH-Range

6.2-8 <1> (<1> about 80% of maximal activity at pH 6.2 and 8.0 [1]) [1]

Temperature optimum (°C)

30 <1> (<1> assay at [1]) [1]

5 Isolation/Preparation/Mutation/Application

Source/tissue

leaf <1> (<1> primary [1]) [1]

Purification

<1> (partial [1]) [1]

6 Stability

Temperature stability

80 <1> (<1> denaturation after 5 min [1]) [1]

Storage stability

<1>, -20°C, at least 6 months [1]

References

[1] Strack, D.; Keller, H.; Weissenböck, G.: Enzymatic synthesis of hydroxycinnamic acid esters of sugar acids and hydroaromatic acids by protein preparations from rye (Secale cereale) primary leaves. J. Plant Physiol., **131**, 61-73 (1987)

Glucarolactone O-hydroxycinnamoyltransferase 2.3.1.132

1 Nomenclature

EC number
2.3.1.132

Systematic name
sinapoyl-CoA:glucarolactone O-(hydroxycinnamoyl)transferase

Recommended name
glucarolactone O-hydroxycinnamoyltransferase

Synonyms
hydroxycinnamic acid transferase

CAS registry number
112956-52-2

2 Source Organism

<1> *Secale cereale* (rye, var. Kustro [1]) [1]

3 Reaction and Specificity

Catalyzed reaction
sinapoyl-CoA + glucarolactone = CoA + O-sinapoylglucarolactone

Reaction type
acyl group transfer

Natural substrates and products
- **S** sinapoyl-CoA + glucarolactone <1> (<1>, best substrate [1]) (Reversibility: ? <1> [1]) [1]
- **P** CoA + O-sinapoylglucarolactone <1> [1]

Substrates and products
- **S** caffeoyl-CoA + glucarolactone <1> (<1>, poor substrate [1]) (Reversibility: ? <1> [1]) [1]
- **P** CoA + O-caffeoylglucarolactone <1> [1]
- **S** feruloyl-CoA + glucarolactone <1> (<1>, condensation at 37% the rate of the reaction with sinapoyl-CoA [1]) (Reversibility: ? <1> [1]) [1]
- **P** CoA + O-feruloylglucarolactone <1> [1]

S *p*-coumaroyl-CoA + glucarolactone <1> (<1>, condensation at 54% the rate of the reaction with sinapoyl-CoA [1]) (Reversibility: ? <1> [1]) [1]
P CoA + O-*p*-coumaroylglucarolactone <1> [1]
S sinapoyl-CoA + glucarolactone <1> (<1>, best substrate [1]) (Reversibility: ? <1> [1]) [1]
P CoA + O-sinapoylglucarolactone <1> [1]
S Additional information <1> (<1>, no activity with 1-hydroxycinnamoyl glucose esters, gluconolactone, galactonate, glucuronate, ascorbate, malate, *m*-tartrate [1]) [1]
P ?

Activating compounds
DTT <1> (<1>, activates [1]) [1]

Specific activity (U/mg)
0.022 <1> [1]

K_m-Value (mM)
0.029 <1> (sinapoyl-CoA) [1]
10 <1> (glucarolactone) [1]

pH-Optimum
7.8 <1> (<1>, 50 mM potassium phosphate buffer [1]) [1]

pH-Range
7.4-8.2 <1> (<1>, about 80% of maximal activity at pH 7.4 and pH 8.2 [1]) [1]

5 Isolation/Preparation/Mutation/Application

Source/tissue
primary leaf <1> [1]

Purification
<1> (partial [1]) [1]

6 Stability

Temperature stability
80 <1> (<1>, denaturation after 5 min [1]) [1]

Storage stability
<1>, -20°C, at least 6 months [1]

References

[1] Strack, D.; Keller, H.; Weissenboeck, G.: Enzymatic-synthesis of hydroxycinnamic acid-esters of sugar acids and hydroaromatic acids by protein preparation from rye (Secale cereale) primary leaves. J. Plant Physiol., **131**, 61-73 (1987)

Shikimate O-hydroxycinnamoyltransferase 2.3.1.133

1 Nomenclature

EC number
2.3.1.133

Systematic name
4-coumaroyl-CoA:shikimate O-(hydroxycinnamoyl)transferase

Recommended name
shikimate O-hydroxycinnamoyltransferase

Synonyms
CST
HCT
HST
hydroxycinnamoyl-CoA:shikimate O-hydroxycinnamoyl transferase
hydroxycinnamoyl-CoA:shikimate hydroxycinnamoyl transferase
hydroxycinnamoyl-CoA:shikimate/quinate hydroxycinnamoyltransferase
hydroxycinnamoyltransferase, shikimate
p-coumaroyl-CoA:shikimic acid *p*-coumaroyl transferase
p-hydroxycinnamoyl-CoA:shikimate-*p*-hydroxycinnamoyl transferase
shikimate hydroxycinnamoyltransferase

CAS registry number
73904-44-6

2 Source Organism

<1> *Secale cereale* (rye, var. Kustro [1]) [1]
<2> *Cestrum elegans* (D.F.L. von Schlechtendal [2]) [2]
<3> *Cichorium endivia* [3, 5]
<4> *Phoenix dactylifera* (young green dates [4,6]) [4, 6]
<5> *Raphanus sativus* [5]
<6> *Equisetum arvense* [7]
<7> *Ocimum basilicum* (SW and EMX-1 [8]) [8]
<8> *Nicotiana tabacum* [9]

3 Reaction and Specificity

Catalyzed reaction

4-coumaroyl-CoA + shikimate = CoA + 4-coumaroylshikimate (<4> quasi-independent bi-bi random mechanism [6])

Reaction type

acyl group transfer

Natural substrates and products

S 4-coumaroyl-CoA + shikimate <2, 3, 5, 8> (<2> biosynthetic pathway of hydroxycinnamic acid glucaric acid [2]; <3,5> hydroxycinnamic acid-shikimate ester biosynthesis [5]; <7,8> enzyme appears to play a critical role in the phenylpropanoid pathway, both upstream and downstream of the 3-hydroxylation step [8,9]) [2, 5, 9]

P CoA + 4-coumaroylshikimate <2, 3, 5, 8> [2, 5, 9]

Substrates and products

S 4-coumaroyl-CoA + quinate <2, 4, 5, 8> (Reversibility: r <8> [9]; ? <2, 4, 5> [2, 4-6]) [2, 4-6, 9]

P CoA + 4-coumarylquinate

S 4-coumaroyl-CoA + shikimate <1-8> (<1,4> best substrate [1,4]; <6> shikimate cannot be replaced by quinate [7]) (Reversibility: r <3, 8> [3, 9]; ? <1, 2, 4-7> [1, 2, 4-8]) [1-9]

P CoA + 4-coumaroylshikimate <1-8> [1-9]

S caffeoyl-CoA + shikimate <1, 3-8> (<1> condensation at 52% the rate of the reaction with *p*-coumaroyl-CoA [1]; <3> not [3]; <7> the activity is 100fold lower than the activity with coumaroyl-CoA [8]) (Reversibility: r <3, 8> [3, 9]; ? <1, 4-7> [1, 4-8]) [1, 3-9]

P CoA + caffeoylshikimate <6> [7]

S cinnamoyl-CoA + shikimate <4, 6> (<4> not [4]) (Reversibility: ? <4, 6> [4, 7]) [4, 7]

P CoA + cinnamoylshikimate

S feruloyl-CoA + shikimate <1-8> (<1> condensation at 22% the rate of the reaction with *p*-coumaroyl-CoA [1]; <2,3,7> not [2,3,8]) (Reversibility: r <3, 8> [3, 9]; ? <1, 2, 4-7> [1, 2, 4-8]) [1-9]

P CoA + feruloylshikimate

S sinapoyl-CoA + shikimate <1-6> (<1> condensation at 29% the rate of the reaction with *p*-coumaroyl-CoA [1]; <2-6> not [2,4,5,7]) (Reversibility: ? <1-6> [1, 2, 4, 5, 7]) [1, 2, 4, 5, 7]

P CoA + sinapoylshikimate

S Additional information <1-3, 5> (<2> no substrates are 1-hydroxycinnamoyl glucose [2]; <1> 1-hydroxycinnamoyl glucose esters, gluconolactone, galactonate, glucuronate, ascorbate, malate, *m*-tartrate [1]; <3,5> no substrate: cinnamoyl-CoA, L-malate, D(+)-malate, L(+)-tartrate, *m*-tartrate, citrate, glucose, UDP-glucose, myo-inositol, tyramine, agmatine [5];

<8> no substrate: anthranilate, glucose, malate, tyramine, spermidine, spermine, putrescine, agmatine, benzyl alcohol [9]) [1, 2, 5, 9]

P ?

Inhibitors

4,4(-diisothiocyano-2,2)-stilbene disulfonic acid <4> [6]
Ca^{2+} <3, 5> (<3,5> inhibition at 5 mM [5]) [5]
DTT <3, 4, 5> (<3,5> concentration above 5 mM results in a marked inhibition of activity [5]) [5, 6]
Mg^{2+} <3, 5> (<3,5> inhibition at 5 mM [5]) [5]
diethyldicarbonate <4> (<4> 80% reversion after treatment with hydroxylamine [6]) [6]
p-chloromercuribenzenesulfonic acid <4> [6]
sodium diphosphate <4> [4]
tricine <4> (<4> weak, not Tris/HCl [4]) [4]
Additional information <4> (<4> no effect: 2,3-butanedione, phenylmethylsulfonylfluoride and N-methylmaleimide [6]) [6]

Activating compounds

EDTA <1, 3, 5> (<1> activation [1]; <3,5> no effect [5]) [1, 5]
Tris/HCl <3> (<3> activation [3]) [3]

Metals, ions

Ca^{2+} <3, 5> (<3,5> sligthly stimulated by 3.0 mM, inhibition at 5 mM [5]) [5]
Mg^{2+} <3, 5> (<3,5> slightly stimulated by 0.5 mM, inhibition at 5 mM [5]) [5]

Specific activity (U/mg)

0.0056 <1> [1]
0.732 <4> [4]
1.134 <3> [3]
Additional information <3, 6> (<6> change of enzyme activity during a 1-year growth period [7]) [5, 7]

K_m-Value (mM)

0.006 <4> (feruloyl-CoA) [4, 6]
0.0067 <4> (caffeoyl-CoA) [4, 6]
0.0086 <4> (4-coumaroyl-CoA) [4, 6]
0.024 <1> (4-coumaroyl-CoA) [1]
0.039 <3> (shikimate) [3]
0.045 <1> (caffeoyl-CoA) [1]
0.05 <8> (caffeoyl-CoA) [9]
0.07 <8> (quinate) [9]
0.119 <3> (4-coumaroylshikimate) [3]
0.162 <3> (CoA) [3]
0.35 <8> (feruloyl-CoA) [9]
0.41 <6> (caffeoyl-CoA) [7]
0.5 <6> (4-coumaroyl-CoA) [7]
0.59 <1> (shikimate, <1> with 4-coumaroyl-CoA [1]) [1]
0.6 <8> (4-coumaroyl-CoA) [9]

0.75 <8> (shikimate) [9]
1 <4> (shikimate) [4]
1.1 <6> (4-cinnamoyl-CoA) [7]
1.2 <3> (4-coumaroyl-CoA) [3]
1.9 <6> (feruloyl-CoA) [7]
4.32 <6> (shikimate) [7]
5.5 <6> (shikimate) [7]
12.5 <3> (caffeoyl-CoA) [5]
14.3 <3> (4-coumaroyl-CoA) [5]
14.3 <5> (4-coumaroyl-CoA) [5]
15.4 <3> (feruloyl-CoA) [5]
30.5 <5> (caffeoyl-CoA) [5]
55.5 <5> (feruloyl-CoA) [5]
83 <1> (shikimate, <1> with caffeoyl-CoA [1]) [1]

pH-Optimum

6.5 <3, 4> (<3,4> 50 mM potassium phosphate buffer [4,5]; <3> pI: 4.33 [5]) [3-5]
6.8 <5> (<5> 50 mM potassium phosphate buffer [5]) [5]
7 <7> (<7> 50 mM potassium phosphate buffer [8]) [8]
7.1 <1> (<1> 200 mM potassium phosphate buffer [1]) [1]
7.5 <6> (<6> Tris-HCl [7]) [7]
Additional information <4> (<4> pI: 4.63 [4]) [4]

pH-Range

5-9 <3, 5> [3, 5]
5.5-7.5 <4> (<4> about half-maximal activity at pH 5.5 and 7.5 [4]) [4]
6.2-8.5 <6> (<6> about half-maximal activity at pH 6.2 in potassium phosphate buffer and 8.5 in Tris-HCl puffer [7]) [7]
6.3-7.7 <1> (<1> about 80% of maximal activity at pH 6.3 and 7.7 [1]) [1]

Temperature optimum (°C)

37 <7> (<7> enzyme extremely stable [8]) [8]
40 <3, 4, 6> [3, 4, 7]

Temperature range (°C)

Additional information <7> (<7> above 40°C rapid and irreversible loss of activity [8]) [8]

4 Enzyme Structure

Molecular weight

40000 <3, 4, 5> (<3,4,5> gel filtration [4,5]) [4, 5]
48000 <7> (<7> gel filtration [8]) [8]
50000 <6> (<6> gel filtration [7]) [7]
58000 <3> (<3> gel filtration [3]) [3]

5 Isolation/Preparation/Mutation/Application

Source/tissue

cell suspension culture <3> [3]
fruit <4> [4]
gland <7> [8]
leaf <1, 2, 4, 5, 7> (<1> primary [1]) [1, 2, 5, 6, 8]
stem <8> [9]
Additional information <7> (<7> enzyme activity is much higher in tissues that are actively producing eugenol than in tissues that are not [8]) [8]

Purification

<1> (partial [1]) [1]
<3> (partial [3,6]) [3, 5, 6]
<4> (partial [4,6]) [4, 6]
<6> (partial [7]) [7]
<7> (partial [8]) [8]
<8> (recombinant enzyme [9]) [9]

Cloning

<8> (expression in Escherichia coli [9]) [9]

6 Stability

Temperature stability

80 <1> (<1> denaturation after 5 min [1]) [1]

General stability information

<4>, glycerol stabilizes during storage [4]

Storage stability

<1>, -20°C, at least 6 months [1]
<3>, -20°C, 0.1 M phosphate buffer, pH 6.5, 65% of initial activity retained for at least 85 days [3]
<3>, -20°C, for several months, no apparent loss of activity [5]
<4>, -20°C, with 10% glycerol at least 30 days [4]
<6>, -20°C, for several months with 10-20% loss of apparent activities after thawing [7]
<6>, -80°C, for several months with 10-20% loss of apparent activities after thawing [7]
<7>, incubation for days at room temperature, multiple freeze/thaw cycles, holding at -20°C for several months, with or without the addition of 10% glycerol, no loss of activity [8]

References

[1] Strack, D.; Keller, H.; Weissenböck, G.: Enzymatic synthesis of hydroxycinnamic acid esters of sugar acids and hydroaromatic acids by protein preparations from rye (Secale cereale) primary leaves. J. Plant Physiol., **131**, 61-73 (1987)

[2] Strack, D.; Gross, W.; Heilemann, J.; Keller, H.; Ohm, S.: Enzymic synthesis of hydroxycinnamic acid esters of glucaric acid and hydroaromatic acids from the respective 1-O-hydroxycinnamoylglucoside and hydroxycinnamoyl-coenzyme A thioester as acyldonors with a protein preparation from Cestrum elegans leaves. Z. Naturforsch. C, **43c**, 32-36 (1987)

[3] Ulbrich, B.; Zenk, M.H.: Partial purification and properties of *p*-hydroxycinnamoyl-CoA: shikimate-*p*-hydroxy-cinnamoyl transferase from higher plants. Phytochemistry, **19**, 1625-1629 (1980)

[4] Lotfy, L.; Fleuriet, A.; Macheix, J.J.: Partial purification and characterization of hydroxycinnamoyl CoA:transferase from apple and date fruits. Phytochemistry, **31**, 767-772 (1992)

[5] Lotfy, S.; Fleuriet, A.; Macheix, J.J.: Hydroxycinnamoyl-CoA: transferases in higher plants. II. Characterization in Cichorium endivia and Raphanus sativus and comparison with other plants. Plant Physiol. Biochem., **32**, 355-363 (1994)

[6] Lotfy, S.: Inactivation and kinetic characterization of hydroxycinnamoyl-CoA:hydroaromatic acid O-hydroxycinnamoyltransferases from Cichorium endivia and Phoenix dactylifera. Plant Physiol. Biochem., **33**, 423-431 (1995)

[7] Hohlfeld, M.; Veit, M.; Strack, D.: Hydroxycinnamoyltransferases involved in the accumulation of caffeic acid esters in gametophytes and sporophytes of Equisetum arvense. Plant Physiol., **111**, 1153-1159 (1996)

[8] Gang, D.R.; Beuerle, T.; Ullmann, P.; Werck-Reichhart, D.; Pichersky, E.: Differential production of meta hydroxylated phenylpropanoids in sweet basil peltate glandular trichomes and leaves is controlled by the activities of specific acyltransferases and hydroxylases. Plant Physiol., **130**, 1536-1544 (2002)

[9] Hoffmann, L.; Maury, S.; Martz, F.; Geoffroy, P.; Legrand, M.: Purification, cloning, and properties of an acyltransferase controlling shikimate and quinate ester intermediates in phenylpropanoid metabolism. J. Biol. Chem., **278**, 95-103 (2003)

Galactolipid O-acyltransferase 2.3.1.134

1 Nomenclature

EC number
2.3.1.134

Systematic name
mono-β-D-galactosyldiacylglycerol:mono-β-D-galactosyldiacylglycerol acyltransferase

Recommended name
galactolipid O-acyltransferase

Synonyms
galactolipid:galactolipid acyltransferase

CAS registry number
103537-09-3

2 Source Organism

<1> *Spinacia oleracea* [1]
<2> *Vicia faba* [2]

3 Reaction and Specificity

Catalyzed reaction
2 mono-β-D-galactosyldiacylglycerol = acylmono-β-D-galactosyl-diacylglycerol + mono-β-D-galactosylacylglycerol

Reaction type
acyl group transfer
dismutation <2> [2]

Natural substrates and products
S mono-β-D-galactosyldiacylglycerol <1, 2> (Reversibility: ? <1, 2> [1, 2]) [1, 2]
P acylmono-β-D-galactosyl-diacylglycerol + mono-β-D-galactosylacylglycerol

Substrates and products
S di-D-galactosyldiacylglycerol + mono-β-D-galactosyldiacylglycerol <2> (Reversibility: ? <2> [2]) [2]

P ?

S mono-β-D-galactosyldiacylglycerol <1, 2> (Reversibility: ? <1, 2> [1, 2]) [1, 2]

P acylmono-β-D-galactosyl-diacylglycerol + mono-β-D-galactosylacylglycerol

pH-Optimum

5.4 <2> [2]

Temperature optimum (°C)

25 <2> (<2> assay at [2]) [2]

5 Isolation/Preparation/Mutation/Application

Source/tissue

leaf <2> [2]

Localization

chloroplast outer membrane <1> [1]
soluble <2> [2]

Purification

<2> (partial [2]) [2]

References

[1] Heemskerk, J.W.M.; Wintermans, J.F.G.M.; Joyard, J.; Block, M.A.; Dorne, A.J.; Douce, R.: Localization of galactolipid:galactolipid galactosyltransferase and acyltransferase in outer envelope membrane of spinach chloroplasts. Biochim. Biophys. Acta, **877**, 281-289 (1986)

[2] Heinz, E.: Some properties of the acyl galactosyl diglyceride-forming enzyme from leaves. Z. Pflanzenphysiol., **69**, 359-376 (1973)

Phosphatidylcholine-retinol O-acyltransferase 2.3.1.135

1 Nomenclature

EC number
2.3.1.135

Systematic name
phosphatidylcholine:retinol-[cellular-retinol-binding-protein] O-acyltransferase

Recommended name
phosphatidylcholine-retinol O-acyltransferase

Synonyms
LRAT
acyltransferase, lecithin-retinol
lecithin retinol acyl transferase
lecithin-retinol acyltransferase
lecithin:retinol acyltransferase
retinyl ester synthase

CAS registry number
117444-03-8

2 Source Organism

<1> *Bos taurus* [1, 3, 4, 8, 10, 13, 14, 16, 18]
<2> *Rattus norvegicus* (vitamin A deficient and sufficient diet fed [7,9,12]; vitamin A depleted diet fed [7]) [2, 3, 5, 6, 7, 9, 12, 17, 18]
<3> *Homo sapiens* [6, 10, 11, 13, 14, 15, 18]
<4> *Mus musculus* (Balb-C line, vitamin A sufficient and deficient diet fed [12]) [12]

3 Reaction and Specificity

Catalyzed reaction
phosphatidylcholine + retinol-[cellular-retinol-binding-protein] = 2-acylglycerophosphocholine + retinyl-ester-[cellular-retinol-binding-protein]

Reaction type
acyl group transfer

Natural substrates and products

S phosphatidylcholine + all-trans-retinol <1-4> (<1> enzyme essential for the biosynthesis of 11-cis-retinal and for dietary mobilization of vitamin A [4]; <2> important role in absorption and storage of vitamin A [5,18]) [4, 5, 10, 12, 16, 18]

P all-trans-retinyl acyl esters + 2-acylglycerophosphocholine

Substrates and products

S 1-lauroyl-2-myristoylphosphatidylcholine + retinol-(cellular-retinol-binding-protein)type II <2> (Reversibility: ? <2> [2]) [2]

P retinyl laurate-(cellular-retinol-binding-protein)type II + 2-myristoylphosphatidylcholine <2> [2]

S 1-myristoyl-2-lauroylphosphatidylcholine + retinol-(cellular-retinol-binding-protein)type II <2> (Reversibility: ? <2> [2]) [2]

P retinyl myristate-(cellular-retinol-binding-protein)type II + 2-lauroylphosphatidylcholine <2> [2]

S 1-palmitoyl-2-acetyl-sn-glycerol-3-phosphorylcholine + retinol <1> (Reversibility: ? <1> [1]) [1]

P retinyl palmitate + 2-acetyl-sn-glycerol-3-phosphorylcholine <1> [1]

S 1-palmitoyl-2-decanophosphatidylcholine + retinol-(cellular-retinol-binding-protein)type II <2> (Reversibility: ? <2> [2]) [2]

P retinyl palmitate-(cellular-retinol-binding-protein)type II + 2-decanophosphatidylcholine <2> [2]

S 1-palmitoyl-2-linoleoyl-sn-glycerol-3-phosphorylcholine + retinol <1> (<1> most effective substrate in stimulating palmitate transfer above control levels [1]) (Reversibility: ? <1> [1]) [1]

P retinyl palmitate + 2-linoleoyl-sn-glycerol-3-phosphorylcholine <1> [1]

S 1-palmitoyl-2-oleoyl-sn-glycerol-3-phosphorylcholine + retinol <1> (Reversibility: ? <1> [1]) [1]

P retinyl palmitate + 2-oleoyl-sn-glycerol-3-phosphorylcholine <1> [1]

S 1-palmitoyl-2-stearoyl-sn-glycerol-3-phosphorylcholine + retinol <1> (Reversibility: ? <1> [1]) [1]

P retinyl palmitate + 2-stearoyl-sn-glycerol-3-phosphorylcholine <1> [1]

S 1-palmitoyl-sn-glycerol-3-phosphorylcholine + retinol <1> (Reversibility: ? <1> [1]) [1]

P retinyl palmitate + glycerol-3-phosphorylcholine <1> [1]

S dilauroylphosphatidylcholine + retinol-(cellular retinol binding protein)-type II <2> (Reversibility: ? <2> [2]) [2]

P retinyl laurate-(cellular retinol binding protein)type II + 2-lauroyl-phosphatidylcholine <2> [2]

S dimyristoylphosphatidic acid + retinol-(cellular retinol binding protein) type II <2> (<2> little if any retinyl myristate obtained [2]) (Reversibility: ? <2> [2]) [2]

P retinyl myristate + myristoylphosphatidic acid <2> [2]

S dimyristoylphosphatidylcholine + retinol(cellular retinol binding protein)type II <2> (Reversibility: ? <2> [2]) [2]

P retinyl myristate-(cellular retinol binding protein)type II + 2-myristoyl-phosphatidylcholine <2> [2]

S dimyristoylphosphatidylethanolamine + retinol-(cellular retinol binding protein) type II <2> (<2> little if any retinyl myristate obtained [2]) (Reversibility: ? <2> [2]) [2]

P retinyl myristate-(cellular retinol binding protein)type II + myristoyl-phosphatidylethanolamine <2> [2]

S dipalmitoylphosphatidylcholine + all-trans-retinol <1-3> (<3> mechanistic hypothesis [10,11,14]) (Reversibility: r <1,3> [4,10];? <2> [2,18]) [2, 4, 10, 11, 14, 18]

P all-trans-retinyl palmitate + 2-palmitoylglycerophosphocholine <1-3> [2, 4, 10, 11, 14, 16, 18]

S phosphatidylcholine + 11-cis-retinol <1> (Reversibility: r <1> [8]) [8]

P 11-cis-retinyl acyl ester + 2-acylglycerophosphocholine <1> [8]

S phosphatidylcholine + all-trans-retinol <1-3> (Reversibility: r <1,3> [8,13,15]; ? <2> [5]) [5, 8, 13, 15]

P all-trans-retinyl acyl esters + 2-acylglycerophosphocholine <1-3> [5, 8, 13, 15]

S phosphatidylcholine + all-trans-retinol-(bovine serum albumin) <3> (Reversibility: ? <3> [10]) [10]

P all-trans-retinyl acyl ester-(bovine serum albumin) + 2-acylglycerophosphocholine <3> [10]

S phosphatidylcholine + retinol-(cellular-retinol-binding protein) type II <2> (<2> reaction with free fatty acid, fatty acyl-CoA, phosphatidic acid or ethanolamine as acyl donors [2]) (Reversibility: ? <2> [2,5]) [2, 5]

P 2-acylglycerophosphocholine + retinyl ester-(cellular-retinol-binding-protein) type II <2> [2]

S phosphatidylcholine + retinol-(cellular-retinol-binding-protein) <1-4> (Reversibility: r <1> [3,8]; ? <2-4> [5-7,9,12,17]) [3, 5-7, 9, 12, 17, 18]

P 2-acylglycerophosphocholine + retinyl ester-(cellular-retinol-binding-protein) <1-4> [3, 5-7, 9, 12, 17, 18]

S Additional information <1-3> (<1> all-trans-retinol is a far better substrate than 11-cis-retinol [18]; <2,3> little or no ability to transfer acyl groups from lysophosphatidylcholine, phosphatidylethanolamine or phosphatidic acid to retinol-(cellular-retinol-binding protein) [1-6]; <2,3> only fatty acyl group at the sn-1 is transferred [4,5,6,10,18]; <2,3> phosphatidylcholine selective [5,6]; <1-3> the fatty acid in the 2-position is important in substrate recognition [1,3,18]) [1-6, 10, 18]

P ?

Inhibitors

13-desmethyl-13,14-dihydro-all-trans-retinyl trifluoroacetate <1> (<1> reversible inhibitor [4]; <1> competitive with respect to dipalmitoylphosphocholine, uncompetitive with respect to all-trans-retinol [4]) [4]

N-Boc-L-biocytinyl-11-aminoundecan chloromethylketone <1> (<1> time dependent inhibition [10]; <1> 0.002 mM, 45% inhibition [10]; <1> irreversible inhibitor [10]) [10]

N-ethylmaleimide <2-3> (<2> 0.003 mM, complete inhibition of retinol-(cellular-retinol-binding-protein)type II esterification [2]; <3> 0.001 mM, complete inhibition of retinol-(cellular-retinol-binding-protein)type II esterification [6]) [2, 6]
all-trans-retinyl-α-bromoacetate <1> (<1> irreversible inhibition [4]) [4, 10]
apo-cellular-retinol-binding protein <2> [5]
p-chloromercuribenzoic acid <2> (<2> 0.005 mM, complete inhibition of retinol-(cellular-retinol-binding-protein)type II esterification [2]) [2]
phenylmethylsulfonyl fluoride <2-3> (<2> 1 mM, 90% inhibition of retinol-(cellular-retinol-binding-protein)type II esterification [2]; <3> 2 mM, 90% inhibition of retinol-(cellular-retinol-binding-protein)type II esterification [6]) [2, 6]
Additional information <1-3> (<1> inactivated by trivalent arsenicals [4]; <1> not readily inhibitable by serine-directed chemical reagents [18]; <1> readily inactivated by organomercurials in the micromolar range [18]; <1> 1-acyl-lysophosphatidylcholine does not inhibit LRAT up to 0.020 mM [8]; <1> product inhibition by retinyl ester can be ruled out [8]; <2> 1% dimethylsulfoxide does not inhibit LRAT activity [5]; <3> mutant C161S is not significantly inactivated by N-Boc-L-biocytinyl-11-aminoundecan chloromethylketone [11]) [4, 5, 8, 11, 18]

Activating compounds

N-(4-hydroxyphenyl)-retinamide <2> (<2> restores normal level of LRAT activity in vitamin A depleted rats [9]; <2> necessary dose, 0.5 mg [9]) [9]
retinoic acid <2, 4> (<2> restores normal level of liver LRAT activity in vitamin A deficient and depleted rats [7]; <2> optimal dose, 0.020 mg [7]; <2> minimal dose for significant increase in LRAT activity and mRNA expression: 0.005 mg [17]) [7, 12, 17]
retinol <2> [9]
vitamin A <2> (<2> regulates LRAT activity in the liver but not in the small intestine [12]) [7, 9, 12]
Additional information <2> (<2> activation is blocked completely in vivo by cycloheximide [7]; <2> activation and mRNA expression for LRAT is blocked completely in vivo by actinomycin D [7,17]; <2> retinoic acid must be continuously present to maintain LRAT gene expression in the ON position [17]) [17]

Specific activity (U/mg)

0.0000001 <2> (<2> nonparenchymal cell fraction LRAT activity and hepatocyte cell fraction LRAT activity for vitamin A-deficient rats [9]) [9]
0.0000012 <2> (<2> hepatocyte cell fraction LRAT activity for vitamin A-deficient rats treated with retinoic acid [9]) [9]
0.000003 <2> (<2> hepatocyte fraction LRAT activity of vitamin A-sufficient rats [9]) [9]
0.0000054 <2> (<2> hepatocyte cell fraction LRAT activity for vitamin A-deficient rats treated with N-(4-hydroxyphenyl)-retinamide [9]) [9]
0.0000132 <2> (<2> nonparenchymal cell fraction LRAT activity for vitamin A-deficient rats treated with retinoic acid [9]) [9]

0.00003 <2> (<2> intestinal LRAT [3]) [3]
0.0000309 <2> (<2> nonparenchymal cell fraction LRAT activity for vitamin A-sufficient rats [9]) [9]
0.000036 <2> (<2> vitamin A-deficient rats treated orally with retinyl palmitate [7]) [7]
0.0000401 <2> (<2> vitamin A-deficient rats treated intraperitoneally with retinoic acid [7]) [7]
0.0000415 <2> (<2> vitamin A-deficient rats treated intraperitoneally with retinol [7]) [7]
0.0000434 <2> (<2> vitamin A-deficient rats treated orally with retinoic acid [7]) [7]
0.0000553 <2> (<2> nonparenchymal cell fraction LRAT activity for vitamin A-deficient rats treated with N-(4-hydroxyphenyl)-retinamide [9]) [9]
0.00006 <2> (<2> mammary gland LRAT [3]) [3]
0.000094 <2> (<2> testis LRAT [3]) [3]
0.000145 <2> (<2> liver LRAT [3]) [3]
0.0021 <2> (<2> liver LRAT activity in vitamin A-deficient rats [7]) [7]
0.1 <1> (<1> retinal pigment epithelial LRAT [1]) [1]

K_m-Value (mM)

0.0001 <3> (all-trans-retinol, <3> with dipalmitoylphosphatidylcholine, H46Q mutant [14]) [14]
0.000167 <3> (all-trans-retinol, <3> H126Q mutant [14]) [14]
0.0002 <2> (retinol-(cellular-retinol-binding protein), <2> intestinal LRAT, reaction with phosphatidylcholine [5]) [5]
0.00024 <2> (retinol-(cellular-retinol-binding protein) type II, <2> intestinal LRAT, reaction with phosphatidylcholine [5]) [5]
0.000243 <1> (all-trans-retinol, <1> reaction with dipalmitoylphosphatidylcholine [4]) [4]
0.00032 <2> (retinol-cellular-retinol-binding protein type II, <2> liver LRAT, reaction with phosphatidylcholine [5]) [5]
0.00034 <3> (all-trans-retinol, <3> with dipalmitoylphosphatidylcholine, C182A mutant [11]) [11]
0.00036 <3> (all-trans-retinol, <3> with dipalmitoylphosphatidylcholine, cloned enzyme, expressed in HEK-293T cells [14]) [14]
0.00044 <2> (retinol, <2> liver LRAT, reaction with phosphatidylcholine [5]) [5]
0.00046 <1> (all-trans-retinol, <1> with dipalmitoylphosphatidylcholine, chemically dimerized enzyme [16]) [16]
0.00047 <3> (all-trans-retinol, <3> with dipalmitoylphosphatidylcholine, C161S mutant [11]) [11]
0.0005 <3> (retinol-(cellular-retinol-binding-protein), <3> reaction with phosphatidylcholine [6]) [6]
0.00051 <3> (all-trans-retinol, <3> with dipalmitoylphosphatidylcholine, C208A mutant [11]) [11]
0.00052 <1> (all-trans-retinol, <1> with dipalmitoylphosphatidylcholine [16]) [16]

0.00063 <2> (retinol, <2> intestinal LRAT, reaction with phosphatidylcholine [5]) [5]
0.00078 <2> (retinol-(cellular-retinol-binding protein), <2> liver LRAT, reaction with phosphatidylcholine [5]) [5]
0.00082 <3> (all-trans-retinol, <3> with dipalmitoylphosphatidylcholine, C168S mutant [11]) [11]
0.00094 <3> (all-trans-retinol, <3> with dipalmitoylphosphatidylcholine, cloned enzyme, expressed in HEK-296T cells [11]) [11]
0.001 <2> (retinol-(cellular-retinol-binding protein), <2> liver LRAT, reaction with phosphatidylcholine [6]) [6]
0.00108 <3> (all-trans-retinol, <3> with dipalmitoylphosphatidylcholine, H60Q mutant [14]) [14]
0.00138 <1> (dipalmitoylphosphatidylcholine, <1> reaction with all-trans-retinol [4]) [4]
0.002 <1> (all-trans-retinol) [3]
0.0024 <3> (all-trans-retinol, <3> with dipalmitoylphosphatidylcholine, H72Q mutant [14]) [14]
0.0042 <1> (all-trans-retinol, <1> reaction with phosphatidylcholine [8]) [8]
0.006 <1> (11-cis-retinol, <1> reaction with phosphatidylcholine [8]) [8]

pH-Optimum
7.5-8 <1> [3]

Temperature optimum (°C)
37 <1-3> (<1> assay at [3,5,6]) [3, 5, 6]

4 Enzyme Structure

Subunits
homodimer <1> (SDS-PAGE in absence of 2-mercaptoethanol [16]; 2 * 25300, SDS-PAGE in presence of 2-mercaptoethanol [16]) [16]
monomer <1-4> (<1> 1 * 25300, SDS-PAGE in presence of 2-mercaptoethanol, fully active catalytically [10]; <3> 1 * 25000, fully active catalytically [10]; <4> 1 * 25800, fully active catalytically [12]) [10, 12, 13, 16]
Additional information <1> (<1> LRAT monomer interact in membranes and form functional homodimers, the dimer formation is mediated by disulfide bond formation and protein-protein interactions [16]) [16]

5 Isolation/Preparation/Mutation/Application

Source/tissue
adrenal gland <2> (<2> mRNA expression [12]) [12]
brain <3> (<3> mRNA expression in adult stage [13]) [13]
breast adenocarcinoma <3> (<3> breast carcinoma cells have lower LRAT activity [15]) [15]
carcinoma cell <3> (skin, these cells have lower LRAT activity [15]) [15]

intestinal mucosa <2> [2, 3]
intestine <2> [5, 7, 12]
liver <2-4> (<2> undetectable in vitamin A depleted rats [7,12]; <2> nonparenchymal cells LRAT activity is approximately 10 fold greater than the hepatocyte LRAT activity from both vitamin A-sufficient and retinoid acid treated vitamin A-depleted rats [9]; <3> mRNA expression in adult and fetal stage [10,13]) [3, 5, 6, 7, 9, 10, 12, 13, 17]
lung <2> (<2> mRNA expression [12]) [12]
mammary gland <2> [3]
pancreas <3> (<3> mRNA expression in adult stage [10,13]) [10, 13]
prostate <3> (<3> mRNA expression [13]; <3> practically no activity in prostatic cancer cells [15]) [13, 15]
prostate cancer <3> (<3> adenocarcinoma, LRAT is downregulated as a result of oncogenic transformation [15]) [15]
retinal pigment epithelium <1, 3> [1, 3, 4, 8, 10, 13]
skeletal muscle <2> (<2> mRNA expression [12]) [12]
small intestine <2-3> (<3> mRNA expression [10,13]) [2, 12, 13]
testis <2-3> (<3> mRNA expression [13]) [3, 12, 13]

Localization

microsome <1-3> [1, 3, 6, 18]

Purification

<1> (affinity labeling with N-Boc-L-biocytinyl-11-aminoundecan chloromethylketone, precipitation and gel electrophoresis, prior to affinity labelling, incubation with cholesterol chloroacetate to block nucleophilic reagents, or streptavidin column to remove endogenously biotinylated proteins enhance protein purification [10,18]) [10, 18]
<1> (affinity labeling with all-trans-retinyl-α-bromoacetate, SDS electrophoresis, prior to affinity labelling, incubation with cholesterol chloroacetate to block nucleophilic reagents, or streptavidin column to remove endogenously biotinylated proteins enhance protein purification [10,18]) [10, 18]
<1> (labeling with N-Boc-L-biocytinyl-11-aminoundecan chloromethylketone, affinity chromatography, avidin, prior to affinity chromatography or affinity labelling, incubation with cholesterol chloroacetate to block nucleophilic reagents, or streptavidin column to remove endogenously biotinylated proteins enhance protein purification [10,18]) [10, 18]
<1> (solubilization of membranes and centrifugation, chromatography on DEAE-column and Mono-Q column later on, quantitative information [10]) [10]
<1> (solubilization of membranes and centrifugation, chromatography: mono Q column and Green 5-agarose column [4]) [4]
<1> (solubilization, centrifugation and dialysis [16]) [16]
<3> (enzyme transfected in HEK-293T cells partially purificated by solubilization and centrifugation [14]) [14]

Cloning

<2> (expression in HEK-293T cells [12,17]) [12, 17]
<3> (expression in HEK-293 cells [10]; expression in HEK-293T cells [10,11,13,14]) [10, 11, 13, 14]
<4> (expression in HEK-293T cells [12]) [12]

6 Stability

Temperature stability

37 <1> (<1> half-life in absence of substrate: 17 min [3]) [3]

General stability information

<1>, buffers are bubbled with argon prior to use [3]
<1>, dithiothreitol stabilizes [3]
<1-2>, all operation involving retinoids are performed at darkness, or under red light [3]

Storage stability

<1>, -70°C, 20 mM Tris-HCl pH 9, 2 mM dithiothreitol, 1 mM EDTA, 0.1% Triton X-100, 0.1 mg/ml dipalmitoylphosphatidylcholine, NaCl, stable for months [4]
<1>, -80°C, 10 mM Tris acetate, pH 7, 1 mM dithiothreitol, microsomal fraction, stable for at least 6 months [3]
<3>, -70°C, 100 mM Tris-HCl, pH 8.3, 0.4% Triton X-100, stable for several days [14]

References

[1] Saari, J.C.; Bredberg, D.L.: Lecithin:retinol acyltransferase in retinal pigment epithelial microsomes. J. Biol. Chem., **264**, 8636-8640 (1989)
[2] MacDonald, P.N.; Ong, D.E.: Evidence for a lecithin-retinol acyltransferase activity in the rat small intestine. J. Biol. Chem., **263**, 12478-12482 (1988)
[3] Saari, J.C.; Bredberg, D.L.: Acyl-CoA:retinol acyltransferase and lecithin:retinol acyltransferase activities of bovine retinal pigment epithelial microsomes. Methods Enzymol., **190**, 156-163 (1990)
[4] Shi, Y.-Q.; Hubacek, I.; Rando, R.R.: Kinetic mechanism of lecithin retinol acyl transferase. Biochemistry, **32**, 1257-1263 (1993)
[5] Herr, F.M.; Ong, D.E.: Differential interaction of lecithin-retinol acyltransferase with cellular retinol binding proteins. Biochemistry, **31**, 6748-6755 (1992)
[6] MacDonald, P.N.; Ong, D.E.: A lecithin:retinol acyltransferase activity in human and rat liver. Biochem. Biophys. Res. Commun., **156**, 157-163 (1988)
[7] Matsuura, T.; Ross, A.C.: Regulation of hepatic lecithin:retinol acyltransferase activity by retinoic acid. Arch. Biochem. Biophys., **301**, 221-227 (1993)

[8] Saari, J.C.; Bredberg, D.L.; Farrell, D.F.: Retinol esterification in bovine retinal pigment epithelium: reversibility of lecithin:retinol acyltransferase. Biochem. J., **291**, 697-700 (1993)
[9] Matsuura, T.; Gad, M.Z.; Harrison, E.H.; Ross, A.C.: Lecithin:retinol acyltransferase and retinyl ester hydrolase activities are differentially regulated by retinoids and have distinct distributions between hepatocyte and non-parenchymal cell fractions of rat liver. J. Nutr., **127**, 218-224 (1997)
[10] Ruiz, A.; Winston, A.; Lim, Y.-H.; Gilbert, B.A.; Rando, R.R.; Bok, D.: Molecular and biochemical characterization of lecithin retinol acyltransferase. J. Biol. Chem., **274**, 3834-3841 (1999)
[11] Mondal, M.S.; Ruiz, A.; Bok, D.; Rando, R.R.: Lecithin retinol acyltransferase contains cysteine residues essential for catalysis. Biochemistry, **39**, 5215-5220 (2000)
[12] Zolfaghari, R.; Ross, A.C.: Lecithin:retinol acyltransferase from mouse and rat liver: cDNA cloning and liver-specific regulation by dietary vitamin A and retinoic acid. J. Lipid Res., **41**, 2024-2034 (2000)
[13] Ruiz, A.; Bok, D.: Molecular characterization of lecithin-retinol acyltransferase. Methods Enzymol., **316**, 400-413 (2000)
[14] Mondal, M.S.; Ruiz, A.; Hu, J.; Bok, D.; Rando, R.R.: Two histidine residues are essential for catalysis by lecithin retinol acyl transferase. FEBS Lett., **489**, 14-18 (2001)
[15] Guo, X.; Knudsen, B.S.; Peehl, D.M.; Ruiz, A.; Bok, D.; Rando, R.R.; Rhim, J.S.; Nanus, D.M.; Gudas, L.J.: Retinol metabolism and lecithin:retinol acyltransferase levels are reduced in cultured human prostate cancer cells and tissue specimens. Cancer Res., **62**, 1654-1661 (2002)
[16] Jahng, W.J.; Cheung, E.; Rando, R.R.: Lecithin retinol acyltransferase forms functional homodimers. Biochemistry, **41**, 6311-6319 (2002)
[17] Zolfaghari, R.; Wang, Y.; Chen, Q.; Sancher, A.; Ross, A.C.: Cloning and molecular expression analysis of large and small lecithin:retinol acyltransferase mRNAs in the liver and other tissues of adult rats. Biochem. J., **368**, 621-631 (2002)
[18] Rando, R.R.: Membrane-bound lecithin-retinol acyltransferase. Biochem. Biophys. Res. Commun., **292**, 1243-1250 (2002)

Polysialic-acid O-acetyltransferase 2.3.1.136

1 Nomenclature

EC number
2.3.1.136

Systematic name
acetyl-CoA:polysialic-acid O-acetyltransferase

Recommended name
polysialic-acid O-acetyltransferase

Synonyms
LRAT
lecithin retinol acyl transferase
lecithin-retinol acyltransferase
lecithin:retinol acyltransferase
retinyl ester synthase

CAS registry number
116412-21-6

2 Source Organism

<1> *Escherichia coli* (K1 OAc+ substrains [1,2]) [1, 2]

3 Reaction and Specificity

Catalyzed reaction
acetyl-CoA + an α-2,8-linked polymer of sialic acid = CoA + polysialic acid acetylated at O-7 or O-9

Reaction type
acyl group transfer

Natural substrates and products
S acetyl-CoA + sialic acid polymer of length 12-14 <1> (Reversibility: ? <1> [2]) [2]
P CoA + O-acetyl sialic acid polymer

Substrates and products

S acetyl-CoA + colominic acid <1> (<1> substrate is a commercially available mixture of polysialic acid fragments, about 15 sialic acid residues per polymer, polysialic acids with more than 14 residues are acylated [1]) [1]

P CoA + O-(acetyl)-colominic acid <1> (<1> product is polysialic acid 7- or 9-O-acetylated [1]) [1]

S acetyl-CoA + sialic acid polymer of length 12-14 <1> (Reversibility: ? <1> [2]) [2]

P CoA + O-acetyl sialic acid polymer

Inhibitors

CoA <1> [1]

Metals, ions

Additional information <1> (no requirement of a divalent cation) [1]

K_m-Value (mM)

0.3 <1> (acetyl-CoA) [1]

pH-Optimum

7-7.5 <1> [1]

Temperature optimum (°C)

37 <1> (<1> assay at [1]) [1]

5 Isolation/Preparation/Mutation/Application

Localization

Additional information <1> (<1> eventually membrane-bound or membrane-associated protein [1]) [1]

Purification

<1> (partial [1]) [1]

6 Stability

Temperature stability

60 <1> (<1> stable for 1 h [1]) [1]

General stability information

<1>, Triton X-100, 0.05%, stabilizes [1]

<1>, endogenous polysialic acid stabilizes by a close non-covalent association with the enzyme [1, 2]

Storage stability

<1>, 4°C, at least 6 months [1]

References

[1] Higa, H.H.; Varki, A.: Acetyl-coenzyme A:polysialic acid O-acetyltransferase from K1-positive Escherichia coli. The enzyme responsible for the O-acetyl plus phenotype and for O-acetyl form variation. J. Biol. Chem., **263**, 8872-8878 (1988)
[2] Varki, A.; Higa, H.: Studies of the O-acetylation and (in)stability of polysialic acid. Polysialic Acid, **1993**, 165-170 (1993)

Carnitine O-octanoyltransferase 2.3.1.137

1 Nomenclature

EC number
2.3.1.137

Systematic name
octanoyl-CoA:L-carnitine O-octanoyltransferase

Recommended name
carnitine O-octanoyltransferase

Synonyms
COT <2, 9, 12> [5-7, 9, 11, 14-18]
carnitine medium-chain acyltransferase
easily solubilized mitochondrial carnitine palmitoyltransferase
medium-chain/long-chain carnitine acyltransferase
overt mitochondrial carnitine palmitoyltransferase

CAS registry number
39369-19-2

2 Source Organism

<-1> no activity in *Saccharomyces cerevisiae* [16, 18]
<1> *Bos taurus* (beef, GenBank accession number U65745 [16]) [16]
<2> *Bos taurus* (beef [5,6,14,17]; bovine [8,13,14,18]; calf [1,2,5]; ox [8]) [1, 2, 5, 6, 8, 13-18]
<3> *Canis familiaris* (dog [1,6]) [1, 6]
<4> *Columba sp.* (pigeon [1,3]) [1, 3]
<5> *Oryctolagus cuniculus* (rabbit [1]) [1]
<6> *Homo sapiens* (human, GenBank accession number AF168793 [16]) [16]
<7> *Homo sapiens* (human [1,9]; HeLa cells [15]) [1, 9, 15]
<8> *monkey* [1]
<9> *Mus musculus* (mouse [1,6,7,9]; fed lipid-lowering drug diet containing clofibrate, nafenopin or Wy-14,643 [5]; NMRI/Bom strain [2]) [1, 2, 5-9]
<10> *Pichia guilliermondii* (yeast [12]) [12]
<11> *Rattus norvegicus* (rat, GenBank accession number U26033 [16,18]) [16, 18]
<12> *Rattus norvegicus* (rat [1-12,15,16,18-21]; Sprague-Dawley [8,9]; Wistar [7]; Wistar/Moell [2]) [1-12, 15, 19-21]
<13> *Sus scrofa* (pig [1,3]) [1, 3]

3 Reaction and Specificity

Catalyzed reaction

octanoyl-CoA + L-carnitine = CoA + L-octanoylcarnitine (acts on a range of acyl-CoAs, with optimal activity with C_6 or C_8 acyl groups. cf. EC 2.3.1.7 (carnitine O-acetyltransferase) and EC 2.3.1.21 (carnitine O-palmitoyltransferase))

Reaction type

acyl group transfer

Natural substrates and products

S L-carnitine + octanoyl-CoA <2, 7, 9, 12> (Reversibility: r <2, 9, 12> [1-8, 16]) [1-8, 16]
P octanoyl-L-carnitine + CoA
S acyl-CoA + L-carnitine <1-13> (<12> β-oxidation of fatty acids [20]; <2> central role in fatty acid metabolism [13]; <10> metabolic pathway of peroxisomal β-oxidation in yeasts [12]) (Reversibility: r <1-13> [1-21]) [1-21]
P acyl-L-carnitine + CoA
S acyl-L-carnitine + CoA <2, 3, 7, 9, 10, 12> (Reversibility: r <2, 3, 7, 9, 10, 12> [6, 7, 9-13]) [6, 7, 9-13]
P acyl-CoA + L-carnitine
S octanoyl-L-carnitine + CoA <2, 7, 9, 12> (<9> reverse reaction [6]) (Reversibility: r <2, 7, 9, 12> [1-8, 16]) [1-8, 16]
P L-carnitine + octanoyl-CoA

Substrates and products

S 3-methylglutaryl-CoA + L-carnitine <12> (<12> little capacity to use this substrate [11]) (Reversibility: r <12> [11]) [11]
P 3-methylglutaryl-L-carnitine + CoA
S 3-methylglutaryl-L-carnitine + CoA <12> (Reversibility: r <12> [11]) [11]
P 3-methyglutaryl-CoA + L-carnitine
S 4,8-dimethylnonanoyl-CoA + L-carnitine <6> (Reversibility: r <6> [16]) [16]
P 4,8-dimethylnonanoyl-L-carnitine + CoA
S 4,8-dimethylnonanoyl-L-carnitine + CoA <6> (Reversibility: r <6> [16]) [16]
P 4,8-dimethylnonanoyl-CoA + L-carnitine
S L-carnitine + octanoyl-CoA <2, 7, 9, 12> (Reversibility: r <2, 7, 9, 12> [1-8, 16]) [1-8, 16]
P octanoyl-L-carnitine + CoA
S decanoyl-L-carnitine + CoA <12> (Reversibility: r <12> [3, 8, 11, 16, 18, 19]) [3, 8, 11, 16, 18, 19]
P decanoyl-CoA + L-carnitine
S dodecanoyl-L-carnitine + CoA <2, 7, 9, 12> (Reversibility: r <2, 7, 9, 12> [1-3, 16]) [1-3, 16]
P dodecanoyl-CoA + L-carnitine

S heptanoyl-L-carnitine + CoA <9, 12> (Reversibility: r <9, 12> [2, 16]) [2, 16]
P heptanoyl-CoA + L-carnitine
S hexadecanoyl-L-carnitine + CoA <2, 7, 9, 12> (Reversibility: r <2, 7, 9, 12> [1, 2, 16]) [1, 2, 16]
P hexadecanoyl-CoA + L-carnitine
S hexanoyl-L-carnitine + CoA <2, 4, 7, 9, 12> (Reversibility: r <2, 4, 7, 9, 12> [1-3, 6, 16]) [1-3, 6, 16]
P hexanoyl-CoA + L-carnitine
S palmitoyl-L-carnitine + CoA <9, 12> (Reversibility: r <9, 12> [5, 8-10]) [5, 8-10]
P palmitoyl-CoA + L-carnitine
S pivaloyl-L-carnitine + CoA <12> (Reversibility: r <12> [11]) [11]
P pivaloyl-CoA + L-carnitine
S valproyl-L-carnitine + CoA <12> (Reversibility: r <12> [11]) [11]
P valproyl-CoA + L-carnitine
S acetoacetyl-CoA + L-carnitine <12> (Reversibility: r <12> [3]) [3]
P acetoacetyl-L-carnitine + CoA
S acetoacetyl-L-carnitine + CoA <12> (Reversibility: r <12> [3]) [3]
P acetoacetyl-CoA + L-carnitine
S acetyl-CoA + L-carnitine <9, 12> (Reversibility: r <9, 12> [3, 5]) [3, 5]
P acetyl-L-carnitine + CoA
S acetyl-L-carnitine + CoA <9, 12> (Reversibility: r <9, 12> [3, 5]) [3, 5]
P acetyl-CoA + L-carnitine
S acyl-CoA + L-carnitine <1-13> (<9,12> γ-trimethylamino-β-hydroxybutyrate [8]) (Reversibility: r <1-13> [1-21]) [1-21]
P acyl-L-carnitine + CoA
S acyl-L-carnitine + CoA <2, 3, 7, 9, 10, 12> (Reversibility: r <2, 3, 7, 9, 10, 12> [3, 5-7, 9-13]) [3, 5-7, 9-13]
P acyl-CoA + L-carnitine
S β-hydroxy-β-methylglutaryl-CoA + L-carnitine <12> (Reversibility: r <12> [3]) [3]
P β-hydroxy-β-methylglutaryl-L-carnitine + CoA
S β-hydroxy-β-methylglutaryl-L-carnitine + CoA <12> (Reversibility: r <12> [11]) [11]
P β-hydroxy-β-methyglutaryl-CoA + L-carnitine
S butyryl-CoA + L-carnitine <12> (Reversibility: r <12> [3]) [3]
P butyryl-L-carnitine + CoA
S butyryl-L-carnitine + CoA <12> (Reversibility: r <12> [3]) [3]
P butyryl-CoA + L-carnitine
S choline + L-carnitine <2> (<2> extremely poor substrate [14]) (Reversibility: ? <2> [14]) [14]
P ?
S decanoyl-CoA + L-carnitine <6, 12> (Reversibility: r <6, 12> [3, 8, 11, 16, 18, 19]) [3, 8, 11, 16, 18, 19]
P decanoyl-L-carnitine + CoA

S dodecanoyl-CoA + L-carnitine <2, 7, 9, 12> (Reversibility: r <2, 7, 9, 12> [1-3, 16]) [1-3, 16]
P dodecanoyl-L-carnitine + CoA
S heptanoyl-CoA + L-carnitine <6, 9, 12> (Reversibility: r <6, 9, 12> [2, 16]) [2, 16]
P heptanoyl-L-carnitine + CoA
S hexadecanoyl-CoA + L-carnitine <2, 9, 12> (Reversibility: r <2, 9, 12> [1, 2]) [1, 2]
P hexadecanoyl-L-carnitine + CoA
S hexanoyl-CoA + L-carnitine <2, 4, 7, 9, 12> (Reversibility: r <2, 4, 7, 9, 12> [1-3, 6, 16]) [1-3, 6, 16]
P hexanoyl-L-carnitine + CoA
S malonyl-CoA + L-carnitine <12> (Reversibility: r <12> [3]) [3]
P malonyl-L-carnitine + CoA
S malonyl-L-carnitine + CoA <12> (Reversibility: r <12> [3]) [3]
P malonyl-CoA + L-carnitine
S methylricinoleate + ? <10> (Reversibility: ? <10> [12]) [12]
P γ-decalactone + ?
S myristoyl-CoA + L-carnitine <12> (Reversibility: r <12> [3]) [3]
P myristoyl-L-carnitine + CoA
S myristoyl-L-carnitine + CoA <12> (Reversibility: r <12> [3]) [3]
P myristoyl-CoA + L-carnitine
S octanoyl-L-carnitine + CoA <2, 7, 9, 12> (<9> reverse reaction [6]) (Reversibility: r <2, 7, 9, 12> [1-8, 16]) [1-8, 16]
P carnitine + octanoyl-CoA
S palmitoyl-CoA + L-carnitine <9, 12> (Reversibility: r <9, 12> [5, 8-10]) [5, 8-10]
P palmitoyl-L-carnitine + CoA
S pivaloyl-CoA + L-carnitine <12> (<12> little capacity to use this substrate [11]) (Reversibility: r <12> [11]) [11]
P pivaloyl-L-carnitine + CoA
S propionyl-CoA + L-carnitine <2, 12> (Reversibility: r <2, 12> [1, 3]) [1, 3]
P propionyl-L-carnitine + CoA
S propionyl-L-carnitine + CoA <2, 12> (Reversibility: r <2, 12> [1, 3]) [1, 3]
P propionyl-CoA + L-carnitine
S valproyl-CoA + L-carnitine <12> (<12> little capacity to use this substrate [11]) (Reversibility: r <12> [11]) [11]
P valproyl-L-carnitine + CoA
S Additional information <9> (<9> malonyl-CoA is not a substrate [6]; <12> succinyl-CoA and β-hydroxy-β-methylglutaryl-CoA are no substrates [3]; <12,13> palmitoyl-CoA is no substrate [3]; <6> 2,6-dimethylheptanoyl-CoA is not a substrate [16]) [3, 6, 16]
P ?

Inhibitors

2-bromodecanoic acid <10> [12]
2-bromooctanoic acid <10> [12]
2-bromopalmitic acid <10> (<10> potent inhibitor [12]) [12]
2-bromopalmitoyl-CoA <12> [8]
D-carnitine <9, 12> [6, 11]
Zn^{2+} <9> (<9> inhibits enzyme in reverse direction [6]) [6]
aminocarnitine <12> [11]
chlorpromazine <10> [12]
dodecanoyl-CoA <2, 9, 12> (<2,9,12> substrate inhibition [2]) [2]
etomoxir <12> (<12> hypoglycaemia-inducing drug, double mutant H131A/H340A is insensitive to etomoxir [18,19]) [18, 19]
etomoxiryl-CoA <12> [9, 11]
hemiacylcarnitinium <2> [17]
hemicholinium <2> [17]
hexadecanoyl-CoA <2, 9, 12> (<2,9,12> substrate inhibition [2]) [2]
malonyl-CoA <12> (<12> microsomal COT, no inhibition of the mitochondrial enzyme [11]; <12> inhibits the purified enzyme, inhibition is affected by buffer, pH, substrate and presence or absence of bovine serum albumin [10]; double mutant H131A/H340A is insensitive to malonyl-CoA [19]) [4, 7, 9-11, 18-20]
octanoyl-CoA <2, 9, 12> (<2,9,12> substrate inhibition [2]) [2]
palmitoyl-CoA <12> [9]
tetradecylglycidic acid <10> [12]
trypsin <9> [6]

Activating compounds

5,5'-dithiobis(2-nitrobenzoate) <9> (<9> retains its maximum activity when preincubated at pH 7.0 or 8.5 [6]) [6]

Turnover number (min^{-1})

22 <9> (octanoyl-CoA) [5]

Specific activity (U/mg)

0.072 <9> [5]

K_m-Value (mM)

0.0002 <11> (decanoyl-CoA) [18]
0.000206 <2> (decanoyl-CoA, <2> mutant AAA-COT [13]) [13]
0.00024 <12> (decanoyl-CoA, <12> pH 6.8 [10]) [10]
0.00024 <2> (decanoyl-CoA, <2> mutant [Asn505]COT [14]) [14]
0.00034 <12> (decanoyl-CoA, <12> pH 7.4 [10]) [10]
0.00046 <12> (palmitoyl-CoA, <12> peroxisomal COT [10]) [10]
0.00047 <12> (palmitoyl-CoA, <12> pH 6.8 [10]) [10]
0.0006 <12> (palmitoyl-CoA, <12> purified COT, pH 7.4 [10]) [10]
0.00138 <2> (decanoyl-CoA, <2> purified enzyme, wild-type COT [13,14]) [13, 14]
0.00139 <2> (decanoyl-CoA, <2> mutant ATS-COT [13]) [13]
0.002 <12> (acyl-CoA) [20]

0.002 <12> (decanoyl-CoA, <12> recombinant enzyme expressed in Saccharomyces cerevisiae pYESCOTwt [18,19,21]) [18, 19, 21]
0.0026 <11> (decanoyl-CoA, <11> recombinant enzyme expressed in Saccharomyces cerevisiae pYESCOTH340A/H131 [18]) [18]
0.00588 <2> (decanoyl-CoA, <2> mutant AAS-COT [13]) [13]
0.006 <11> (decanoyl-CoA, <11> recombinant enzyme expressed in Saccharomyces cerevisiae pYESCOTH340 [18]) [18]
0.00762 <2> (decanoyl-CoA, <2> mutant SAS-COT [13]) [13]
0.0107 <12> (decanoyl-CoA) [21]
0.013 <12> (decanoyl-CoA, <12> recombinant enzyme expressed in Saccharomyces cerevisiae pYESCOTH340A/H131 [18,19]) [18, 19]
0.015 <12> (octanoyl-CoA) [20]
0.0167 <12> (acyl-CoA, <12> mutant strain expressed in Saccharomyces cerevisiae COT A238D [20]) [20]
0.036 <2> (hexadecanoyl-L-carnitine) [1]
0.0513 <2> (decanoyl-CoA, <2> mutant STA-COT [13]) [13]
0.069 <9> (palmitoyl-CoA) [5]
0.0819 <2> (decanoyl-CoA, <2> wild-type COT [13]) [13]
0.0948 <2> (carnitine, <2> mutant ATS-COT [13]) [13, 14]
0.1 <9> (octanoylcarnitine) [5, 6]
0.102 <12> (L-carnitine, <12> pH 7.4 [10]) [10]
0.104 <9> (palmitoyl-L-carnitine) [5, 6]
0.106 <11> (L-carnitine, <11> recombinant enzyme expressed in Saccharomyces cerevisiae pYESCOTH340A [18]; <12> mutant A332G [21]) [18, 21]
0.106 <12> (L-carnitine, <12> mutant A332G [21]) [21]
0.11 <9> (CoASH) [5, 6]
0.117 <11> (L-carnitine, <11> recombinant enzyme expressed in Saccharomyces cerevisiae pYESCOTH340A/H131 [18]) [18]
0.13 <9> (L-carnitine) [5]
0.155 <9> (acetyl-CoA) [5]
0.172 <12> (L-carnitine, <12> recombinant enzyme expressed in Saccharomyces cerevisiae pYESCOTwt [18-21]) [18-21]
0.218 <11> (L-carnitine) [18]
0.227 <12> (L-carnitine, <12> pH 6.8 [10]) [10]
0.303 <2> (decanoyl-CoA, <2> mutant SAA-COT [13]) [13]
0.34 <2> (decanoyl-CoA) [17]
0.373 <2> (decanoyl-CoA, <2> mutant ATA-COT [13]) [13]
0.6 <2> (palmitoyl-CoA, <2> mutant SAA-COT [13]) [13]
0.783 <9> (acetyl-L-carnitine) [5, 6]
0.794 <2> (decanoyl-CoA, <2> mutant AAA-COT [13]) [13]
0.992 <2> (L-carnitine, <2> mutant AAS-COT [13]) [13]
7.4 <2> (palmitoyl-L-carnitine) [17]
8.4 <2> (decanoyl-L-carnitine) [17]
16 <2> (acetyl-CoA) [17]
108 <2> (L-carnitine) [17]
143 <2> (choline, <2> mutant [Asn505]COT [14,17]) [14, 17]

146 <12> (L-carnitine, <12> recombinant enzyme expressed in Saccharomyces cerevisiae COT238D [20]) [20]
160 <11> (L-carnitine, <11> recombinant enzyme expressed in Saccharomyces cerevisiae pYESCOTH131A [18]) [18]
160 <2> (L-carnitine, <2> mutant [Asn505]COT [14,17]) [14, 17]

K_i-Value (mM)
0.0035 <12> (malonyl-CoA, <12> pH 6.8 [10]) [10]
0.0064 <12> (malonyl-CoA, <12> peroxisomal COT [10]) [10]
0.106 <12> (malonyl-CoA, <12> pH 7.4 [10]) [10]
0.84 <9> (D-carnitine) [6]
113 <2> (hemiacylcarnitinium) [17]

pH-Optimum
7-8 <12> [3]
8 <9> [6]

pH-Range
5.5-7.5 <12> [3]
5.5-10 <9> [6]

4 Enzyme Structure

Molecular weight
50000-55000 <12> (<12> SDS-PAGE [11]) [11]
53000 <12> (<12> Western blot, polyclonal antibody reaction [11]) [11]
59000 <12> (<12> gel filtration [3]) [3]
60000 <9> (<9> gel filtration [5,6]) [5, 6]
64400 <2> (<2> SDS-PAGE [14]) [14]
66000 <6, 12> (<9> gel filtration [6]; <6> recombinant enzyme, expressed in Saccharomyces cerevisiae [16]) [6, 16]
70000 <6, 12> (<12> gel filtration [8]; <6> calculated from cDNA sequence [16]) [8, 16]
70260 <2> (<2> calculated from cDNA sequence [14]) [14]

Subunits
monomer <2, 6, 9> (<9> 1 * 60000, SDS-PAGE [5]; <2> 1 * 64400, SDS-PAGE [14]; <6> 1 * 66000, SDS-PAGE [16]) [5, 8, 14, 16]

5 Isolation/Preparation/Mutation/Application

Source/tissue
brain <9, 12> [7]
fibroblast <6> (<6> skin fibroblast [16]) [16]
heart <2, 12> [4, 5, 18]
hepatocyte <12> [12]
intestine <9> [6]

kidney <9> [6]
liver <2, 3, 5, 7-9, 12, 13> [1-6, 8-11, 14, 15, 17, 18]
muscle <4, 12> (<4> breast muscle [1,3]) [1, 3, 18]
pancreas <11> [18]

Localization

cytosol <10, 12> [11, 12]
endoplasmic reticulum <2, 12> (<12> rough and smooth [9]) [3, 9, 17]
microsome <2, 9, 12> [3, 5, 6, 9, 11]
mitochondrion <2, 9, 12, 13> (<12> synaptic and non-synaptic mitochondria [7]; <9,12> innermitochondrial membrane [2]) [1, 2, 4-11, 14, 20]
peroxisome <2, 9, 10, 12> (<9> matrix of liver peroxisomes [5]) [3, 5, 6, 8-12, 15-20]

Purification

<2> (recombinant enzyme [13,14]; partially [1]) [1, 5, 13, 14, 17]
<9> [5, 6, 8, 9]
<12> [3, 6, 10, 11]

Cloning

<2> (cloned and expressed in Escherichia coli [13]; cDNA cloning by library screening and recombinant expression in Escherichia coli [14]) [13, 14]
<6> (cDNA PCR amplified, cloned in yeast expression vector pEL26 and expressed in Saccharomyces cerevisiae [16]) [16]
<12> (cDNA clone from a rat liver library, 2 forms of COT in peroxisomes results of the trans-splicing mechanism in pre-mRNAs [15]; cDNA encoding full-length wild-type COT cloned and expressed in Saccharomyces cerevisae [18]; pYES2 plasmid containing the wild-type and the double mutant H131A/H340A of COT expressed in Saccharomyces cerevisiae [19]) [15, 18, 19]

6 Stability

General stability information

<12>, quite unstable, labile in Triton X-100 and octylglucoside, stable in 8 mM CHAPS, 200 mM guanidium chloride and 0.5% Tween 20 [9]
<12>, stable for at least 1 week in 100 mM phosphate buffer with 0.4 M KCl, pH 5.5-7.5 [3]

Storage stability

<2>, -18°C, can be kept for several months with only moderate loss of activity [1]
<9>, -20°C, freezing causes a 90% loss of activity within 2 days [5]
<9>, -80°C, freezing causes a 90% loss of activity within 2 days [5]
<9>, 20°C, 10 mM sodium diphosphate buffer at pH 7.5, stable for months [5]
<9>, 4°C, 10 mM sodium diphosphate buffer at pH 7.5, stable for months [5]
<12>, -70°C, stable to freezing at for at least 5 months [9]

References

[1] Solberg, H.E.: Different carnitine acyltransferases in calf liver. Biochim. Biophys. Acta, **280**, 422-433 (1972)

[2] Solberg, H.E.: Acyl group specificity of mitochondrial pools of carnitine acyltransferases. Biochim. Biophys. Acta, **360**, 101-112 (1974)

[3] Markwell, M.A.K.; Tolbert, N.E.; Bieber, L.L.: Comparison of the carnitine acyltransferase activities from rat liver peroxisomes and microsomes. Arch. Biochem. Biophys., **176**, 479-488 (1976)

[4] Saggerson, E.D.; Carpenter, C.A.: Malonyl CoA inhibition of carnitine acyltransferase activities: effects of thiol-group reagents. FEBS Lett., **137**, 124-128 (1982)

[5] Farrell, S.O.; Bieber, L.L.: Carnitine octanoyltransferase of mouse liver peroxisomes: properties and effect of hypolipidemic drugs. Arch. Biochem. Biophys., **222**, 123-132 (1983)

[6] Farrell, S.O.; Fiol, C.J.; Reddy, J.K.; Bieber, L.L.: Properties of purified carnitine acyltransferases of mouse liver peroxisomes. J. Biol. Chem., **259**, 13089-13095 (1984)

[7] Bird, M.I.; Munday, L.A.; Saggerson, E.D.; Clark, J.B.: Carnitine acyltransferase activities in rat brain mitochondria. Bimodal distribution, kinetic constants, regulation by malonyl-CoA and developmental pattern. Biochem. J., **226**, 323-330 (1985)

[8] Healy, M.J.; Kerner, J.; Bieber, L.L.: Enzymes of carnitine acylation. Is overt carnitine palmitoyltransferase of liver peroxisomal carnitine octanoyltransferase?. Biochem. J., **249**, 231-237 (1988)

[9] Lilly, K.; Bugaisky, G.E.; Umeda, P.K.; Bieber, L.L.: The medium-chain carnitine acyltransferase activity associated with rat liver microsomes is malonyl-CoA sensitive. Arch. Biochem. Biophys., **280**, 167-174 (1990)

[10] Nic A'Bhaird, N.; Ramsay, R.R.: Malonyl-CoA inhibition of peroxisomal carnitine octanoyltransferase. Biochem. J., **286**, 637-640 (1992)

[11] Chung, C.D.; Bieber, L.L.: Properties of the medium chain/long chain carnitine acyltransferase purified from rat liver microsomes. J. Biol. Chem., **268**, 4519-4524 (1993)

[12] Pagot, Y.; Belin, J.M.: Involvement of carnitine acyltransferases in peroxisomal fatty acid metabolism by the yeast Pichia guilliermondii. Appl. Environ. Microbiol., **62**, 3864-3867 (1996)

[13] Cronin, C.N.: The conserved serine-threonine-serine motif of the carnitine acyltransferases is involved in carnitine binding and transition-state stabilization: a site-directed mutagenesis study. Biochem. Biophys. Res. Commun., **238**, 784-789 (1997)

[14] Cronin, C.N.: cDNA cloning, recombinant expression, and site-directed mutagenesis of bovine liver carnitine octanoyltransferase. Arg505 binds the carboxylate group of carnitine. Eur. J. Biochem., **247**, 1029-1037 (1997)

[15] Caudevilla, C.; Serra, D.; Miliar, A.; Codony, C.; Asins, G.; Bach, M.; Hegardt, F.G.: Processing of carnitine octanoyltransferase pre-mRNAs by cis and trans-splicing. Adv. Exp. Med. Biol., **466**, 95-102 (1999)

[16] Ferdinandusse, S.; Mulders, J.; IJlst, L.; Denis, S.; Dacremont, G.; Waterham, H.R.; Wanders, R.J.A.: Molecular cloning and expression of human carnitine octanoyltransferase: Evidence for its role in the peroxisomal β-oxidation of branched-chain fatty acids. Biochem. Biophys. Res. Commun., **263**, 213-218 (1999)

[17] Ramsay, R.R.; Gandour, R.D.: Selective modulation of carnitine long-chain acyltransferase activities: Kinetics, inhibitors, and active sites of COT and CPT-II. Adv. Exp. Med. Biol., **466**, 103-109 (1999)

[18] Morillas, M.; Clotet, J.; Rubi, B.; Serra, D.; Arino, J.; Hegardt, F.G.; Asins, G.: Inhibition by etomoxir of rat liver carnitine octanoyltransferase is produced through the co-ordinate interaction with two histidine residues. Biochem. J., **351**, 495-502 (2000)

[19] Hegardt, F.G.; Bach, M.; Asins, G.; Caudevilla, C.; Morillas, M.; Codony, C.; Serra, D.: Post-transcriptional regulation of rat carnitine octanoyltransferase. Biochem. Soc. Trans., **29**, 316-320 (2001)

[20] Morillas, M.; Gomez-Puertas, P.; Roca, R.; Serra, D.; Asins, G.; Valencia, A.; Hegardt, F.G.: Structural model of the catalytic core of carnitine palmitoyltransferase I and carnitine octanoyltransferase (COT). Mutation of CPT 1 histidine 473 and alanine 381 and COT alanine 238 impairs the catalytic activity. J. Biol. Chem., **276**, 45001-45008 (2001)

[21] Morillas, M.; Gomez-Puertas, P.; Rubi, B.; Clotet, J.; Arino, J.; Valencia, A.; Hegardt, F.G.; Serra, D.; Asins, G.: Structural model of a malonyl-CoA-binding site of carnitine octanoyltransferase and carnitine palmitoyltransferase I. Mutational analysis of a malonyl-CoA affinity domain. J. Biol. Chem., **277**, 11473-11480 (2002)

Putrescine N-hydroxycinnamoyltransferase 2.3.1.138

1 Nomenclature

EC number
2.3.1.138

Systematic name
caffeoyl-CoA:putrescine N-(3,4-dihydroxycinnamoyl)transferase

Recommended name
putrescine N-hydroxycinnamoyltransferase

Synonyms
PHT
caffeoyl-CoA putrescine N-caffeoyl transferase
hydroxycinnamoyl-CoA:putrescine hydroxycinnamoyltransferase
putrescine hydroxycinnamoyl transferase
putrescine hydroxycinnamoyltransferase

CAS registry number
120598-69-8

2 Source Organism

<1> *Nicotiana tabacum* (TX1 [3,4]; TX4 [4]) [2, 3, 4, 5]
<2> *Hordeum vulgare* (barley, after 1-4d of infection with Blumeria graminis f. sp. hordei, increased activity [1]) [1]

3 Reaction and Specificity

Catalyzed reaction
caffeoyl-CoA + putrescine = CoA + N-caffeoylputrescine

Reaction type
acyl group transfer

Natural substrates and products
S caffeoyl-CoA + putrescine <1, 2> (Reversibility: ? <1, 2> [1, 2, 3, 4, 5]) [1, 2, 3, 4, 5]
P CoA + N-caffeoylputrescine <1> [5]

Substrates and products

S 4-fluorocinnamoyl-CoA + putrescine <1> [3]
P CoA + N-4-fluorocinnamoylputrescine
S caffeoyl-CoA + cadaverine <1> (<1> cadaverine i.e. 1,5-pentanediamine, low activity [2]) [2, 3, 4, 5]
P CoA + N-caffeoylcadaverine
S caffeoyl-CoA + diaminoheptane <1> (<1> low activity [2]) [2]
P CoA + N-caffeoyl-1,7-diaminoheptane
S caffeoyl-CoA + diaminohexane <1> (<1> low activity [2]) [2]
P CoA + N-caffeoyl-1,6-diaminohexane
S caffeoyl-CoA + diaminopropane <1> [2, 3, 4]
P CoA + N-caffeoyl-1,3-diaminopropane
S caffeoyl-CoA + putrescine <1, 2> (Reversibility: ? <1, 2> [1, 2, 3, 4, 5]) [1, 2, 3, 4, 5]
P CoA + N-caffeoylputrescine <1> [5]
S cinnamoyl-CoA + putrescine <1> [2, 3, 5]
P CoA + N-cinnamoylputrescine
S coumaroyl-CoA + putrescine <1> [3, 5]
P CoA + N-coumaroylputrescine
S feruloyl-CoA + putrescine <1> [2, 3, 4, 5]
P CoA + N-feruloylputrescine
S sinapoyl-CoA + putrescine <1> (<1> low activity [4]) [2, 3, 5]
P CoA + N-sinapoylputrescine
S Additional information <1> (<1> no activity with monoamines [3]; <1> no activity with spermidine [2]) [2, 3]
P ?

Inhibitors

ammonium sulfate <1> [3]
Additional information <1> (<1> no inhibition by thiol reagents [3]) [3]

K_m-Value (mM)

0.0025 <1> (cinnamoyl-CoA, <1> with putrescine [3]) [3]
0.003 <1> (caffeoyl-CoA, <1> pH 10.0 [4]) [4]
0.01 <1> (putrescine, <1> pH 10.0 [4]; <1> caffeoyl-CoA with 1 mM putrescine [5]) [4, 5]
0.0125 <1> (feruloyl-CoA, <1> with caffeoyl-CoA [3]) [3]
0.033 <1> (caffeoyl-CoA, <1> pH 7.8 [4]) [4]
0.05 <1> (putrescine, <1> with 0.25 mM caffeoyl-CoA [5]) [5]
0.4 <1> (putrescine, <1> pH 7.8 [4]) [4]

pH-Optimum

8.5 <1> [3, 5]
8.8 <1> (<1> 4-coumaroyl-CoA [4]) [4]
10 <1> (<1> caffeoyl-CoA, sinapoyl-CoA [4]) [4]
10.3 <1> (<1> feruloyl-CoA [4]) [4]

pH-Range

7.2-10 <1> (<1> about 50% of activity maximum at pH 7.2 and 10 [5]) [5]
7.5-11 <1> (<1> about 50% of activity maximum at pH 7.5 and 11 [3]) [3]
7.8-10 <1> (<1> at 7.8 about 50% of activity maximum, at 10.0 activity maximum, caffeoyl-CoA + putrescine [4]) [4]

Temperature optimum (°C)

30 <1> (<1> assay at [4]) [4]

4 Enzyme Structure

Molecular weight

45000 <1> (<1> HPLC gel filtration [3]) [3]
48000 <1> (<1> gel filtration [5]) [5]

Subunits

monomer <1> (<1> 1 * 50000, SDS-PAGE [3]) [3]

5 Isolation/Preparation/Mutation/Application

Source/tissue

callus <1> (<1> cultured [2,5]) [2, 5]
cell culture <1> [4]
cell suspension culture <1> [3]

Purification

<1> [3]

6 Stability

Oxidation stability

<1>, rapid denaturation in absence of reducing compounds [5]

General stability information

<1>, thawing and refreezing, 20% loss of activity [4]

Storage stability

<1>, -20°C, stable for 2 months [4]

References

[1] Cowley, T.; Walters, D.R.: Polyamine metabolism in barley reacting hypersensitively to the powdery mildew fungus Blumeria graminis f. sp. hordei. Plant Cell Environ., **25**, 461-468 (2002)

[2] Negrel, J.; Javelle, F.; Paynot, M.: Separation of putrescine and spermidine hydroxycinnamoyl transferases extracted from tobacco callus. Phytochemistry, **30**, 1089-1092 (1991)
[3] Negrel, J.; Paynot, M.; Javelle, F.: Purification and properties of putrescine hydroxycinnamoyl transferase from tobacco (Nicotiana tabaccum) cell suspensions. Plant Physiol., **98**, 1264-1269 (1992)
[4] Meurer-Grimes, B.; Berlin, J.; Strack, D.: Hydroxycinnamoyl-CoA:putrescine hydroxycinnamoyltransferase in tobacco cell cultures with high and low levels of caffeoylputrescine. Plant Physiol., **89**, 488-492 (1989)
[5] Negrel, J.: The biosynthesis of cinnamoylputrescines in callus tissue cultures of Nicotiana tabaccum. Phytochemistry, **28**, 477-481 (1989)

Ecdysone O-acyltransferase 2.3.1.139

1 Nomenclature

EC number
2.3.1.139

Systematic name
palmitoyl-CoA:ecdysone palmitoyltransferase

Recommended name
ecdysone O-acyltransferase

Synonyms
acyl-CoA:ecdysone acyltransferase
fatty acyl-CoA:ecdysone acyltransferase

CAS registry number
120038-26-8

2 Source Organism

<1> *Periplaneta americana* [1]

3 Reaction and Specificity

Catalyzed reaction
palmitoyl-CoA + ecdysone = CoA + ecdysone palmitate

Reaction type
acyl group transfer

Substrates and products
S palmitoyl-CoA + ecdysone <1> (Reversibility: ? <1> [1]) [1]
P CoA + ecdysone palmitate <1> [1]

Inhibitors
20-hydroxyecdysone <1> (<1> at 0.001 mM, 30% inhibition, at 0.01 mM, 90% inhibition [1]) [1]
EDTA <1> (<1> at 1 mM, 15% inhibition [1]) [1]

K_m-Value (mM)
0.0004 <1> (ecdysone) [1]

pH-Optimum
6.5-7.5 <1> [1]

pH-Range
6-8 <1> (<1> 75% of maximal activity at pH 6.0, 70% of maximal activity at pH 8.0 [1]) [1]

Temperature optimum (°C)
35 <1> (<1> assay at [1]) [1]

5 Isolation/Preparation/Mutation/Application

Source/tissue
ovary <1> [1]

Localization
microsome <1> (<1> membrane [1]) [1]

Purification
<1> (partial [1]) [1]

6 Stability

General stability information
<1>, bovine serum albumin and DTT have no effect [1]

Storage stability
<1>, -70°C, 48 h, 60% reduction of activity [1]

References

[1] Slinger, A.J.; Isaac, R.E.: Acyl-CoA:ecdysone acyltransferase activity from the ovary of P. americana. Insect Biochem., **18**, 779-784 (1988)

Rosmarinate synthase 2.3.1.140

1 Nomenclature

EC number
2.3.1.140

Systematic name
caffeoyl-CoA:3-(3,4-dihydroxyphenyl)lactate 2'-O-caffeoyl-transferase

Recommended name
rosmarinate synthase

Synonyms
4-coumaroyl-CoA:4-hydroxyphenyllactic acid 4-coumaroyl transferase
RAS <1> [3]
caffeoyl-coenzyme A:3,4-dihydroxyphenyllactic acid caffeoyltransferase
rosmarinic acid synthase

CAS registry number
117590-80-4

2 Source Organism

<1> *Coleus blumei* (strain Benth [2]; enzyme is induced by 4% sucrose content of the medium [1,2,3]) [1, 2, 3]

3 Reaction and Specificity

Catalyzed reaction
caffeoyl-CoA + 3-(3,4-dihydroxyphenyl)lactate = CoA + rosmarinate

Reaction type
acyl group transfer

Natural substrates and products
S Additional information <1> (<1> probably 4-coumaroyl-CoA and 4-hydroxyphenyllactate are the substrates used in vivo [1]) [1]
P ?

Substrates and products
S 4-coumaroyl-CoA + 4-hydroxyphenyllactate <1> (Reversibility: ? <1> [3]) [3]
P CoA + 4-coumaroyl-4-hydroxyphenyllactate <1> [3]

S caffeoyl-CoA + 3-(3,4-dihydroxyphenyl)lactate <1> (<1> only R(+)-stereoisomer of 3-(3,4-dihydroxyphenyl)lactate is accepted [1,2,3]) (Reversibility: r <1> [3]; ? <1> [1, 2]) [1, 2, 3]
P CoA + rosmarinate <1> [1, 2, 3]
S caffeoyl-CoA + 3-methoxy-4-hydroxyphenyllactate <1> (Reversibility: ? <1> [3]) [3]
P CoA + caffeoyl-3-methoxy-4-hydroxyphenyllactate
S caffeoyl-CoA + 4-hydroxyphenyllactate <1> (Reversibility: ? <1> [1, 2, 3]) [1, 2, 3]
P CoA + caffeoyl-4-hydroxyphenyllactate <1> [3]
S cinnamoyl-CoA + 3,4-dihydroxyphenyllactate <1> (Reversibility: ? <1> [3]) [3]
P CoA + cinnamoyl-3,4-dihydroxyphenyllactate

Inhibitors

3,4-dihydroxyphenylpyruvate <1> [3]
4-hydroxymercuribenzoate <1> (<1> complete inhibition at concentrations higher than 0.5 mM [3]) [3]
4-hydroxyphenylpyruvate <1> [3]
CoA <1> (<1> non-competitive inhibitor [3]) [3]
S(-)-3,4-dihydroxyphenyllactate <1> [3]
rosmarinic acid <1> (<1> strong inhibitor, at concentration higher than 0.2 mM, non-competitive [3]) [3]
Additional information <1> (<1> cinnamic acids are not inhibitory [3]) [3]

Activating compounds

ascorbate <1> (<1> at 0.5 mM, 50% activation [1]) [1]

K_m-Value (mM)

0.015 <1> (rosmarinic acid, <1> reverse reaction [3]) [3]
0.02 <1> (4-coumaroyl-CoA) [3]
0.033 <1> (caffeoyl-CoA) [3]
0.17 <1> (4-hydroxyphenyllactate) [3]
0.31 <1> (CoA, <1> reverse reaction [3]) [3]
0.37 <1> (3,4-dihydroxyphenyllactate) [3]

K_i-Value (mM)

1.7 <1> (CoA, <1> non-competitive inhibitor [3]) [3]

pH-Optimum

7-7.5 <1> [3]

pH-Range

6.5-8.5 <1> (<1> half-maximal activity at pH 6.5 and pH 8.5 [3]) [3]

Temperature optimum (°C)

30 <1> (<1> caffeoyl-CoA as substrate [3]) [3]
40 <1> (<1> 4-coumaroyl-CoA as substrate [3]) [3]

5 Isolation/Preparation/Mutation/Application

Localization
soluble <1> [1, 3]

Purification
<1> [1, 2, 3]

6 Stability

pH-Stability
5.5-10 <1> [3]

Temperature stability
30 <1> (<1> 5 h, no loss of activity [1,3]) [1, 3]

Oxidation stability
<1>, O_2-sensitive, 10 mM DTT enhances enzyme activity 600%, 0.5 mM ascorbate enhances activity 200% [1, 3]

Storage stability
<1>, -18°C, several weeks in desalted state [1, 3]

References

[1] Petersen, M.; Alfermann, A.W.: Two new enzymes of rosmarinic acid biosynthesis from cell cultures of Coleus blumei: Hydroxyphenylpyruvate reductase and rosmarinic acid synthase. Z. Naturforsch. C, **43c**, 501-504 (1988)
[2] Petersen, M.; Häusler, E.; Karwatzki, B.; Meinhard, J.: Proposed biosynthetic pathway for rosmarinic acid in cell cultures of Coleus blumei Benth. Planta, **189**, 10-14 (1993)
[3] Petersen, M.S.: Characterization of rosmarinic acid synthase from cell cultures of Coleus blumei. Phytochemistry, **30**, 2877-2881 (1991)

Galactosylacylglycerol O-acyltransferase 2.3.1.141

1 Nomenclature

EC number
2.3.1.141

Systematic name
acyl-[acyl-carrier protein]:D-galactosylacylglycerol O-acyltransferase

Recommended name
galactosylacylglycerol O-acyltransferase

Synonyms
acyl-ACP:lyso-MGDG acyltransferase
acyl-acyl-carrier protein:lysomonogalactosyldiacylglycerol acyltransferase
acyltransferase, lysomonogalactosyldiacylglycerol

CAS registry number
119129-68-9

2 Source Organism

<1> *Anabaena variabilis* [1]

3 Reaction and Specificity

Catalyzed reaction
acyl-[acyl-carrier protein] + sn-3-D-galactosyl-sn-2-acylglycerol = [acyl-carrier protein] + D-galactosyldiacylglycerol

Reaction type
acyl group transfer

Substrates and products
- **S** acyl-[oleoyl-carrier protein] + sn-3-D-galactosyl-sn-2-acylglycerol <1> (Reversibility: ? <1> [1]) [1]
- **P** oleoyl-carrier protein + D-galactosyldiacylglycerol
- **S** acyl-[palmitoyl-carrier protein] + sn-3-D-galactosyl-sn-2-acylglycerol <1> (Reversibility: ? <1> [1]) [1]
- **P** palmitoyl-carrier protein + D-galactosyldiacylglycerol
- **S** acyl-[stearoyl-carrier protein] + sn-3-D-galactosyl-sn-2-acylglycerol <1> (Reversibility: ? <1> [1]) [1]
- **P** stearoyl-carrier protein + D-galactosyldiacylglycerol

Inhibitors

ATP <1> (<1> weak [1]) [1]
CoA <1> (<1> weak [1]) [1]
acyl-carrier protein <1> [1]

pH-Optimum

8 <1> (<1> assay at [1]) [1]

Temperature optimum (°C)

30 <1> (<1> assay at [1]) [1]

5 Isolation/Preparation/Mutation/Application

Localization

membrane <1> [1]

References

[1] Chen, H.H.; Wickrema, A.; Jaworski, J.G.: Acyl-acyl-carrier protein: lysomonogalactosyldiacylglycerol acyltransferase from the cyanobacterium Anabaena variabilis. Biochim. Biophys. Acta, **963**, 493-500 (1988)

Glycoprotein O-fatty-acyltransferase 2.3.1.142

1 Nomenclature

EC number
2.3.1.142

Systematic name
fatty-acyl-CoA:mucus-glycoprotein fatty-acyltransferase

Recommended name
glycoprotein O-fatty-acyltransferase

Synonyms
acyltransferase, protein

CAS registry number
122191-29-1

2 Source Organism

<1> *Rattus norvegicus* (Sprague-Dawley [1]) [1]
<2> *Homo sapiens* [2]
<3> *Bordetella pertussis* [3]
<4> *Drosophila melanogaster* [4]

3 Reaction and Specificity

Catalyzed reaction
palmitoyl-CoA + mucus glycoprotein = CoA + O-palmitoylglycoprotein

Reaction type
acyl group transfer

Natural substrates and products
- **S** palmitoyl-CoA + adenylate cyclase toxin <3> (<3> palmitoylation activates adenylate cyclase toxin [3]) (Reversibility: ? <3> [3]) [3]
- **P** CoA + palmitoyladenylate cyclase toxin <3> [3]
- **S** palmitoyl-CoA + mucus glycoprotein <1, 2> (<1,2> involved in modification of membrane glycoproteins [1,2]) (Reversibility: ? <1, 2> [1, 2]) [1, 2]
- **P** CoA + O-palmitoylglycoprotein <1, 2> [1, 2]

Substrates and products

S palmitoyl-CoA + adenylate cyclase toxin <3> (<3> acyltransferase CyaC palmitoylates the conserved lysines 983 and 860 of adenylate cyclase toxin [3]) (Reversibility: ? <3> [3]) [3]
P CoA + palmitoyladenylate cyclase toxin <3> [3]
S palmitoyl-CoA + mucus glycoprotein <1> (Reversibility: ? <1> [1]) [1]
P CoA + O-palmitoylglycoprotein <1> [1]

Inhibitors

EDTA <1, 2> (<2> complete inactivation [2]) [1, 2]
$MgCl_2$ <1> [1]
$MnCl_2$ <1> [1]
octyl-β-D-glucoside <2> (<2> at concentrations below its critical micellar concentration [2]) [2]

Activating compounds

2-mercaptoethanol <2> (<2> activation [2]) [2]
Triton X-100 <1> (<1> activation [1]) [1]
dithiothreitol <1> (<1> activation [1]) [1]

Metals, ions

NaF <1> (<1> activation [1]) [1]

Specific activity (U/mg)

0.000934 <1> [1]

Temperature optimum (°C)

37 <1> (<1> assay at [1]) [1]

4 Enzyme Structure

Molecular weight

243000 <1> (<1> native PAGE [1]) [1]

Subunits

? <1> (<1> x * 65000 + x * 67000, SDS-PAGE [1]) [1]

5 Isolation/Preparation/Mutation/Application

Source/tissue

gastric mucosa <1> [1]
heart <1> [1]
kidney <1> [1]
liver <1> [1]
lung <1> [1]
lymphoblast <1> [1]
pancreas <1> [1]

placenta <2> [2]
salivary gland <1> [1]

Localization

microsome <1, 2> (<1> rough microsomes, integral part of rough endoplasmic reticulum, exposed to cytoplasmic side [1]; <2> integral membrane protein [2]) [1, 2]

Purification

<1> (gastric mucosa, solubilized with Triton X-100, affinity chromatography [1]) [1]
<2> (solubilized with Triton X-100 or Nonidet P-40 [2]) [2]

Cloning

<3> (expression of wild-type CyaC acyltransferase and various mutants in Escherichia coli [3]) [3]
<4> (cloning of rasp gene encoding a putative O-acyl transferase [4]) [4]

Engineering

A140G <3> (<3> changed selectivity of acyltransferase for the 2 acetylation sites of adenylate cyclase toxin, a mixture of bi and monoacetylated toxins modified either at both Lys860 and Lys983 or only at Lys860 is generated [3]) [3]
A140V <3> (<3> monoacetylation of adenylate cyclase toxin almost exclusively at Lys 983 [3]) [3]
C67S <3> (<3> no effect on adenylate cyclase toxin activation [3]) [3]
H33D <3> (<3> loss of adenlayte cyclase toxin acetylation activity [3]) [3]
H33S <3> (<3> loss of adenlayte cyclase toxin acetylation activity [3]) [3]
R141K <3> (<3> no effect on adenylate cyclase toxin activation [3]) [3]
R141L <3> (<3> no effect on adenylate cyclase toxin activation [3]) [3]
S30R <3> (<3> loss of adenlayte cyclase toxin acetylation activity [3]) [3]
S30W <3> (<3> loss of adenlayte cyclase toxin acetylation activity [3]) [3]
S68T <3> (<3> no effect on adenylate cyclase toxin activation [3]) [3]

6 Stability

General stability information

<1>, glycerol, 25% v/v, stabilizes [1]
<1>, protease inhibitors stabilize [1]
<2>, 2-mercaptoethanol, 2 mM, stabilizes during purification [2]
<2>, SDS inactivates [2]
<2>, dithiothreitol stabilizes during purification [2]

Storage stability

<1>, -80°C, stable [1]

References

[1] Kasinathan, C.; Grzelinska, E.; Okazaki, K.; Slomiany, B.L.; Slomiany, A.: Purification of protein fatty acyltransferase and determination of its distribution and topology. J. Biol. Chem., **265**, 5139-5144 (1990)
[2] Schmidt, M.F.G.; Burns, G.R.: Solubilization of protein fatty acyltransferase from placental membranes and cell-free acyl transfer on to exogenous and endogenous acceptors. Biochem. Soc. Trans., **17**, 859-861 (1989)
[3] Basar, T.; Havlicek, V.; Bezouskova, S.; Hackett, M.; Sebo, P.: Acylation of lysine 983 is sufficient for toxin activity of Bordetella pertussis adenylate cyclase. Substitutions of alanine 140 modulate acylation site selectivity of the toxin acyltransferase CyaC. J. Biol. Chem., **276**, 348-354 (2001)
[4] Micchelli, C.A.; The, I.; Selva, E.; Mogila, V.; Perrimon, N.: Rasp, a putative transmembrane acyltransferase, is required for Hedgehog signaling. Development, **129**, 843-851 (2002)

β-Glucogallin-tetrakisgalloylglucose O-galloyltransferase 2.3.1.143

1 Nomenclature

EC number
2.3.1.143

Systematic name
1-O-galloyl-β-D-glucose:1,2,3,6-tetrakis-O-galloyl-β-D-glucose 4-O-galloyltransferase

Recommended name
β-glucogallin-tetrakisgalloylglucose O-galloyltransferase

Synonyms
β-glucogallin:1,2,3,6-tetra-O-galloyl-β-D-glucose 4-O-galloyltransferase
β-glucogallin:1,2,3,6-tetra-O-galloylglucose 4-O-galloyltransferase
galloyltransferase, β-glucogallin-tetragalloylglucose 4-
tetragalloylglucose 4-O-galloyltransferase

CAS registry number
122653-70-7

2 Source Organism

<1> *Quercus rubra* (oak [1,2]) [1, 2]

3 Reaction and Specificity

Catalyzed reaction
1-O-galloyl-β-D-glucose + 1,2,3,6-tetrakis-O-galloyl-β-D-glucose = D-glucose + 1,2,3,4,6-pentakis-O-galloyl-β-D-glucose (<1> stoichiometry [1])

Reaction type
acyl group transfer

Natural substrates and products
- **S** Additional information <1> (<1> enzyme catalyzes the last common step in the biosynthesis of hydrolyzable tannins [2]) (Reversibility: ? <1> [1]) [2]
- **P** ?

Substrates and products

S 1-O-galloyl-β-D-glucose + 1,2,3,6-tetra-O-galloyl-β-D-glucose <1> (<1> 1-O-galloyl-β-D-glucose is β-glucogallin, no activity with 1,2,4,6-isomer [1]) (Reversibility: ? <1> [1, 2]) [1, 2]

P D-glucose + 1,2,3,4,6-penta-O-galloyl-β-D-glucose <1> [1, 2]

Specific activity (U/mg)

Additional information <1> [2]

K_m-Value (mM)

1 <1> (1,2,3,6-tetra-O-galloyl-β-D-glucose) [1]

2.3 <1> (1-O-galloyl-β-D-glucose) [1]

pH-Optimum

6.3 <1> [1]

pH-Range

4-8 <1> (<1> 50% of maximal activity at pH 4.0 and pH 8.0 [1]) [1]

Temperature optimum (°C)

40 <1> [1]

Temperature range (°C)

Additional information <1> (<1> some activity at 0°C [1]) [1]

4 Enzyme Structure

Molecular weight

260000 <1> (<1> gel filtration [1,2]) [1, 2]

Subunits

homotetramer <1> (<1> 4 * 65000, SDS-PAGE [2]) [2]

5 Isolation/Preparation/Mutation/Application

Source/tissue

leaf <1> [1, 2]

Purification

<1> (partial [1]) [1, 2]

6 Stability

pH-Stability

5-6.5 <1> [1]

Oxidation stability

<1>, O_2-sensitive, buffers have to be supplemented with 5 mM 2-mercaptoethanol [1]

General stability information

<1>, 0-4°C, within 1-2 weeks no significant loss of activity [1]

References

[1] Cammann, J.; Denzel, K.; Schilling, G.; Gross, G.: Biosynthesis of gallotannins:β-glucogallin-dependent formation of 1,2,3,4,6-pentagalloylglucose by enzymatic galloylation of 1,2,3,6-tetragalloylglucose. Arch. Biochem. Biophys., **273**, 58-63 (1989)

[2] Grundhofer, P.; Gross, G.G.: Purification of tetragalloylglucose 4-O-galloyltransferase and preparation of antibodies against this key enzyme in the biosynthesis of hydrolyzable tannins. Z. Naturforsch. C, **55**, 582-587 (2000)

Anthranilate N-benzoyltransferase 2.3.1.144

1 Nomenclature

EC number
2.3.1.144

Systematic name
benzoyl-CoA:anthranilate N-benzoyltransferase

Recommended name
anthranilate N-benzoyltransferase

Synonyms
anthranilate N-hydroxycinammoyl/benzoyltransferase <1> [2]
benzoyltransferase, anthranilate N-

CAS registry number
125498-59-1

2 Source Organism

<1> *Dianthus caryophyllus* (enzyme is induced by addition of crude elicitor from the cell walls of Phythophthora megasperma f. sp. glycinea or by commercial yeast extract [1]) [1, 2, 3]

3 Reaction and Specificity

Catalyzed reaction
benzoyl-CoA + anthranilate = CoA + N-benzoylanthranilate

Reaction type
acyl group transfer

Natural substrates and products
S benzoyl-CoA + anthranilate <1> (<1> cinnamoyl-CoA, salicyl-CoA and 4-coumaroyl-CoA can act as donors with 73%, 70% and 65% efficiency respectively, enzyme is involved in the biosynthesis of phytoalexins, 4-hydroxyanthranilate is the most probable substrate in vivo [1]) (Reversibility: ? <1> [1]) [1]
P CoA + N-benzoylanthranilate <1> [1]

Substrates and products

S 4-coumaryl-CoA + anthranilate <1> (<1> 127% of activity with benzoyl-CoA [2]) (Reversibility: ? <1> [2]) [2]
P CoA + N-4-(coumaroyl)anthranilate <1> [1]
S benzoyl-CoA + 3-hydroxyanthranilate <1> (<1> 20% of activity with anthranilate [1]) (Reversibility: ? <1> [1]) [1]
P CoA + N-benzoyl-3-hydroxyanthranilate <1> [1]
S benzoyl-CoA + 4-hydroxyanthranilate <1> (<1> 10% of activity with anthranilate [1]) (Reversibility: ? <1> [1]) [1]
P CoA + N-benzoyl-4-hydroxyanthranilate <1> [1]
S benzoyl-CoA + anthranilate <1> (<1> cinnamoyl-CoA, salicyl-CoA and 4-coumaroyl-CoA can act as donors [1,2]) (Reversibility: ? <1> [1]) [1, 2]
P CoA + N-benzoylanthranilate <1> [1, 2]
S cinnamoyl-CoA + anthranilate <1> (<1> 140% of activity with benzoyl-CoA [2]) (Reversibility: ? <1> [2]) [2]
P CoA + N-cinammoylanthranilate <1> [1]
S salicyl-CoA + anthranilate <1> (<1> 60% of activity with benzoyl-CoA [2]) (Reversibility: ? <1> [2]) [2]
P CoA + N-salicylanthranilate <1> [1]

Specific activity (U/mg)

0.00000948 <1> [2]

K_m-Value (mM)

0.009 <1> (benzoyl-CoA) [2]
0.02 <1> (anthranilate) [2]
0.033 <1> (anthranilate) [1]
0.05 <1> (benzoyl-CoA) [1]

pH-Optimum

6.8-7 <1> (<1> crude extract, assay at pH 7.0 [1]; <1> activity is completely lost if the pH drops below pH 6.5 [2]) [1, 2]
Additional information <1> (<1> true pH-optimum may be closer to the amino-pK of anthranilate at 4.95 [1]) [1]

pH-Range

5.5-7.8 <1> [1]

Temperature optimum (°C)

20 <1> [1]

4 Enzyme Structure

Molecular weight

44000 <1> (<1> gel filtration [2]) [2]

Subunits

monomer <1> (<1> 1 * 53000, SDS-PAGE [2]) [2]

5 Isolation/Preparation/Mutation/Application

Source/tissue

callus <1> (<1> cell culture [1]) [1]

Purification

<1> (ammonium sulfate, Q-Sepharose, partially purified [1]; ammonium sulfate, Q-Sepharose, Blue Sepharose, Sephacryl S200, Mono Q [2]) [1, 2]

Cloning

<1> (expression in Escherichia coli [2]; cloning of 2 cDNAs [3]) [2, 3]

6 Stability

pH-Stability

6 <1> (<1> crude extract, 4°C, almost all activity irreversibly lost within less than 2 h [1]) [1]
7 <1> (<1> very labile at pH below [1]) [1]

Temperature stability

40 <1> (<1> crude extract, 5 min, pH 7.5, significant loss of activity [1]) [1]

References

[1] Reinhard, K.; Matern, U.: The biosynthesis of phytoalexins in Dianthus caryophyllus L. cell cultures: induction of benzoyl-CoA:anthranilate N-benzoyltransferase activity. Arch. Biochem. Biophys., **275**, 295-301 (1989)

[2] Yang, Q.; Reinhard, K.; Schiltz, E.; Matern, U.: Characterization and heterologous expression of hydroxycinnamoyl/benzoyl-CoA:anthranilate N-hydroxycinnamoyl/benzoyltransferase from elicited cell cultures of carnation, Dianthus caryophyllus L. Plant Mol. Biol., **35**, 777-789 (1997)

[3] Yang, Q.; Grimmig, B.; Matern, U.: Anthranilate N-hydroxycinnamoyl/benzoyltransferase gene from carnation: rapid elicitation of transcription and promoter analysis. Plant Mol. Biol., **38**, 1201-1214 (1998)

Piperidine N-piperoyltransferase 2.3.1.145

1 Nomenclature

EC number
2.3.1.145

Systematic name
(E,E)-piperoyl-CoA:piperidine N-piperoyltransferase

Recommended name
piperidine N-piperoyltransferase

Synonyms
piperidine piperoyltransferase
piperoyl-CoA:piperidine N-piperoyltransferase

CAS registry number
126806-22-2

2 Source Organism

<1> *Piper nigrum* [1]

3 Reaction and Specificity

Catalyzed reaction
(E,E)-piperoyl-CoA + piperidine = CoA + N-[(E,E)-piperoyl]-piperidine

Reaction type
acyl group transfer

Natural substrates and products
S (E,E)-piperoyl-CoA + piperidine <1> (Reversibility: ? <1> [1]) [1]
P CoA + N-[(E,E)-piperoyl]-piperidine

Substrates and products
S (E,E)-piperoyl-CoA + piperidine <1> (Reversibility: ? <1> [1]) [1]
P CoA + N-[(E,E)-piperoyl]-piperidine
S piperoyl-CoA + 3-pyrroline <1> (<1> 25% of activity with piperidine [1]) (Reversibility: ? <1> [1]) [1]
P CoA + piperoyl-(3-pyrroline) <1> [1]

S piperoyl-CoA + pyrrolidine <1> (<1> 76% of activity with piperidine [1]) (Reversibility: ? <1> [1]) [1]
P CoA + piperoylpyrrolidine <1> [1]

K_m-Value (mM)
3.8 <1> (piperoyl-CoA) [1]
10 <1> (piperidine) [1]

Temperature optimum (°C)
30 <1> [1]

5 Isolation/Preparation/Mutation/Application

Source/tissue
shoot <1> [1]

Purification
<1> (partial [1]) [1]

6 Stability

Temperature stability
40-42 <1> (<1> heat-denaturation [1]) [1]

Oxidation stability
<1>, cell homogenate has to be supplemented with 50 mM ascorbic acid as antioxidant [1]

General stability information
<1>, 10 mM 2-mercaptoethanol stabilizes [1]
<1>, preparations are to be carried out at 0-4°C [1]

Storage stability
<1>, 0-4°C, enzyme stored as ammonium sulfate paste loses 20-25% of activity within one week, completely inactive after two weeks [1]
<1>, diluted enzyme stored after anion-exchange chromatography is unstable [1]

References

[1] Geisler, J.G.; Gross, G.G.: The biosynthesis of piperine in Piper nigrum. Phytochemistry, **29**, 489-492 (1990)

Pinosylvin synthase 2.3.1.146

1 Nomenclature

EC number
2.3.1.146

Systematic name
malonyl-CoA:cinnamoyl-CoA malonyltransferase (cyclizing)

Recommended name
pinosylvin synthase

Synonyms
pine stilbene synthase
pinosylvin synthase
stilbene synthase
Additional information (not identical with EC 2.3.1.74 or EC 2.3.1.95)

CAS registry number
72994-49-1

2 Source Organism

<1> *Pinus sylvestris* (Scots pine [2,3,5,6]; four different clones PST-1, PST-2, PST-3 and PST-5 [3]; treatment with Lophodermium seditosum [5]) [2, 3, 5, 6, 7, 8]
<2> *Pinus densiflora* (Japanese red pine, three isoenzymes PDSTS1, PDSTS2 and PDSTS3 [1]) [1]
<3> *Pinus strobus* (Eastern white pine, two encoding genes STS1 and STS2, STS1 has only 3-5% of activity of STS2 [4]) [4]

3 Reaction and Specificity

Catalyzed reaction
3 malonyl-CoA + cinnamoyl-CoA = 4 CoA + pinosylvin + 3 CO_2

Reaction type
acyl group transfer

Natural substrates and products
S cinnamoyl-CoA + malonyl-CoA <1, 2, 3> (<2> plays a key role in the stilbenoid synthesis [1]) [1, 2, 3, 4, 5, 6, 7, 8]
P 3,5-dihydroxystilbene + CoA + CO_2

Substrates and products

- **S** caffeoyl-CoA + malonyl-CoA <3> (Reversibility: ? <3> [4]) [4]
- **P** piceatannol + CoA + CO_2
- **S** cinnamoyl-CoA + malonyl-CoA <1, 2, 3> (Reversibility: ? <1,2,3> [1,2,3,4,5,6,7,8]) [1, 2, 3, 4, 5, 6, 7, 8]
- **P** 3,5-dihydroxystilbene + CoA + CO_2 <1, 2> (<1> i.e. pinosylvin [7,8]) [1, 7, 8]
- **S** dihydrocinnamoyl-CoA + malonyl-CoA <3> (Reversibility: ? <3> [4]) [4]
- **P** dihydropinosylvin + CoA + CO_2
- **S** feruloyl-CoA + malonyl-CoA <3> (Reversibility: ? <3> [4]) [4]
- **P** rhapontigenin + CoA + CO_2
- **S** *p*-coumaroyl-CoA + malonyl-CoA <2, 3> (Reversibility: ? <2,3> [1,4]) [1, 4]
- **P** resveratrol + CoA + CO_2 <3> [4]
- **S** phenylpropionyl-CoA + malonyl-CoA <1> (Reversibility: ? <1> [6]) [6]
- **P** ?

Inhibitors

pinosylvin <2> (<2> potent inhibitor of isozyme PDSTS2 and PDSTS3, non-competitive [1]) [1]

Activating compounds

Additional information <1> (<1> maximal response to treatment with UV-light of PST-1 after 20 h [3]) [3]

Additional information <1> (<1> maximal response to treatment with conidia Botrytis cinerea after 35 h [3]) [3]

K_m-Value (mM)

0.0004 <2> (cinnamoyl-CoA, <2> isozyme PDSTS3 [1]) [1]
0.0005-0.002 <1> (cinnamoyl-CoA) [6]
0.0009 <2> (*p*-coumaroyl-CoA, <2> isozyme PDSTS3 [1]) [1]
0.003-0.008 <1> (phenylpropionyl-CoA) [6]
0.0038 <2> (cinnamoyl-CoA, <2> isozyme PDSTS2 [1]) [1]
0.0053 <2> (*p*-coumaroyl-CoA, <2> isozyme PDSTS2 [1]) [1]
0.006 <2> (malonyl-CoA, <2> isozyme PDSTS2 [1]) [1]
0.024 <2> (malonyl-CoA, <2> isozyme PDSTS3 [1]) [1]

K_i-Value (mM)

0.013 <2> (pinosylvin, <2> isozyme PDSTS2 [1]) [1]
0.152 <2> (pinosylvin, <2> isozyme PDSTS3 [1]) [1]

pH-Optimum

6 <3> (<3> isozyme STS1 [4]) [4]
7 <1, 2, 3> (<2> reaction with cinnamoyl-CoA, isozymes PDSTS2 and PDSTS3 [1]; <3> isozyme STS2 [4]) [1, 4, 6]
8 <2> (<2> reaction with *p*-coumaroyl-CoA, isozymes PDSTS2 and PDSTS3 [1]) [1]

4 Enzyme Structure

Molecular weight

35000 <2> (<2> isozyme PDSTS3, SDS-PAGE [1]) [1]
42000 <2> (<2> isozymes PDSTS1 and PDSTS2, SDS-PAGE [1]) [1]

5 Isolation/Preparation/Mutation/Application

Source/tissue

hypocotyl <1> [5]
needle <1> [3]
seedling <1, 3> (<1> young seedlings are more susceptible to treatment with a suspension of conidia of Botrytis cinerea [3]) [3, 4, 7, 8]

Purification

<2> (of recombinant enzymes [1]) [1]

Cloning

<1> (expressed in Escherichia coli [3,6]) [3, 6, 7]
<1> (transfected into Nicotiana tabacum cv. Petit Havana SR1 [3]) [3]
<2> (expressed in Escherichia coli [1]) [1]
<3> (expressed in Escherichia coli JM109 [4]) [4]

References

[1] Kodan, A.; Kuroda, H.; Sakai, F.: A stilbene synthase from Japanese red pine (Pinus densiflora): implications for phytoalexin accumulation and down-regulation of flavonoid biosynthesis. Proc. Natl. Acad. Sci. USA, **99**, 3335-3339 (2002)
[2] Chiron, H.; Drouet, A.; Lieutier, F.; Payer, H.D.; Ernst, D.; Sandermann, H., Jr.: Gene induction of stilbene biosynthesis in Scots pine in response to ozone treatment, wounding, and fungal infection. Plant Physiol., **124**, 865-872 (2000)
[3] Preisig-Muller, R.; Schwekendiek, A.; Brehm, I.; Reif, H.J.; Kindl, H.: Characterization of a pine multigene family containing elicitor-responsive stilbene synthase genes. Plant Mol. Biol., **39**, 221-229 (1999)
[4] Raiber, S.; Schroeder, G.; Schroeder, J.: Molecular and enzymic characterization of two stilbene synthases from Eastern white pine (Pinus strobus). A single Arg/His difference determines the activity and the pH dependence of the enzymes. FEBS Lett., **361**, 299-302 (1995)
[5] Lange, B.M.; Trost, M.; Heller, W.; Langebartels, C.; Sandermann, H., Jr.: Elicitor-induced formation of free and cell-wall-bound stilbenes in cell-suspension cultures of Scots pine (Pinus sylvestris). Planta, **194**, 143-148 (1994)
[6] Schanz, S.; Schroeder, G.; Schroeder, J.: Stilbene synthase from Scots pine (Pinus sylvestris). FEBS Lett., **313**, 71-74 (1992)

[7] Schwekendiek, A.; Pfeffer, G.; Kindl, H.: Pine stilbene synthase cDNA, a tool for probing environmental stress. FEBS Lett., **301**, 41-44 (1992)
[8] Schöppner, A.; Kindl, H.: Stilbene synthase (pinosylvine synthase) and its induction by ultraviolet light. FEBS Lett., **108**, 349-352 (1979)

Glycerophospholipid arachidonoyl-transferase (CoA-independent) 2.3.1.147

1 Nomenclature

EC number
2.3.1.147

Systematic name
1-organyl-2-arachidonoyl-sn-glycero-3-phosphocholine:1-organyl-2-lyso-sn-glycero-3-phosphoethanolamine arachidonoyltransferase (CoA-independent)

Recommended name
glycerophospholipid arachidonoyl-transferase (CoA-independent)

Synonyms
CoA-independent transacylase
Coenzyme A-independent transacylase
acyltransferase, 1-alkylglycerophosphocholine

CAS registry number
102347-79-5

2 Source Organism

<1> *Homo sapiens* [1-7, 9]
<2> *Mammalia* [3]
<3> *Mus musculus* [8]
<4> *Oryctolagus cuniculus* [10]

3 Reaction and Specificity

Catalyzed reaction
1-organyl-2-arachidonoyl-sn-glycero-3-phosphocholine + 1-organyl-2-lyso-sn-glycero-3-phosphoethanolamine = 1-organyl-2-arachidonoyl-sn-glycero-3-phosphoethanolamine + 1-organyl-2-lyso-sn-glycero-3-phosphocholine (<1>, formation of an acyl-enzyme intermediate [4])

Reaction type
acyl group transfer

Natural substrates and products
S Additional information <1-4> (<1>, EC 2.3.1.147 and 14000 MW phospholipase A act in a cooperative fashion in the production of platelet-ac-

tivating factor in inflammatory cells [1]; <1>, inhibition of the enzyme can be linked to blockade of proliferation and the induction of apoptosis in HL-60 cells [2]; <2>, the enzyme is considered to play an important role in the reacylation of ether-containing phospholipids to provide polyunsaturated fatty acid containing ether phospholipids [3]; <1>, the enzyme is the key mediator of arachidonate remodeling [4]; <1>, the enzyme is important for the production of inflammatory lipid mediators in stimulated cells [4]; <1>, blockade of the enzyme, which leads to inhibition of arachidonate remodelling between phospholipids, results in the attenuation of platelet-activating factor production, arachidonic acid release and the formation of eicosanoid products [5]; <1>, the enzyme mediates the movement of arachidonate into the large pool of 1-ether-linked phospholipids in human neutrophils [6]; <3>, key enzyme responsible for the prolonged generation of lyso-platelet-activating factor that increases capability of CoA-independent transacylation followed by CoA-dependent acetylation of lyso-platelet-activating factor, can sustain the biosynthesis of platelet activating factor in lipopolysaccharide-stimulated IC-21 macrophages [8]; <4>, the reaction is extremely important in the remodeling of phospholipid molecular species and the mobilization of arachidonate into ether-linked lipids. The transfer of arachidonate to 1-alkyl-2-lyso-sn-glycero-3-phosphocholine is important in the final inactivation step for platelet activating factor (1-alkyl-2-acetyl-sn-glycero-3-phosphocholine), whereby 1-alkyl-2-arachidonoyl-sn-glycerol-3-phosphocholine, a stored precursor of both platelet activating factor and arachidonic acid metabolites is formed [10]) [1-6, 8, 10]

P ?

Substrates and products

S 1-acyl-glycerophosphocholine + diacylcholine glycerolipid <2> [3]

P ?

S 1-acyl-glycerophosphoethanolamine <2> (<2>, transfer of 20:4 from diacylglycerophosphocholine) [3]

P ?

S 1-alkenyl-2-lysoglycerophospholipid + 1-acyl-2-arachidonoyl-glycerophosphocholine <1> [7]

P 1-alkenyl-2-arachidonyl-glycerophospholipid + 1-acyl-2-lyso-glycerophosphocholine

S 1-alkenyl-2-lysoglycerophospholipid + 1-alkyl-2-arachidonoyl-glycerophosphocholine <1> [7]

P 1-alkenyl-2-arachidonoylglycerophospholipid + 1-alkyl-2-lyso-glycerophosphocholine

S 1-alkenyl-glycerophosphoethanolamine + diacylglycerophosphocholine <2> (<2>, transfer of 20:4 from diacylglycerophosphocholine) [3]

P ?

S 1-alkyl-2-lyso-glyceryl-3-phosphorylcholine + 1-acyl-2-arachidonoyl-glyceryl-3-phosphorylcholine <1> [9]

P 1-alkyl-2-arachidonoylglycerophosphocholine + 1-acyl-2-lyso-glycerophosphocholine?
S 1-alkyl-2-lysoglycerophospholipid + 1-acyl-2-arachidonoyl-glycerophosphocholine <1> [7]
P ?
S 1-alkyl-2-lysoglycerophospholipid + 1-alkyl-2-arachidonoyl-glycerophosphocholine <1> [7]
P 1-alkyl-2-arachidonoylglycerophospholipid + 1-alkyl-2-lyso-glycerophosphocholine
S 1-alkyl-glycerophosphocholine + diacylcholine glycerolipid <2> [3]
P ?
S 1-alkyl-glycerophosphoethanolamine + diacylglycerophosphocholine <2> (<2>, transfer of 20:4 from diacylglycerophosphocholine) [3]
P ?
S alkenyl-glycerophosphocholine + diacylcholine glycerolipid <2> [3]
P ?
S Additional information <1, 2, 4> (<1>, selectively transfers arachidonate and other long-chain unsaturated fatty acyl groups. Strong preference for phosphocholine-containing glycerophospholipids and phosphoethanolamine-containing glycerophospholipids [1]; <2>, the acyl donor is a fatty acid esterified at sn-2 position of diacylphospholipids but not a free fatty acid or an acyl-CoA. Only C_{20} and C_{22} polyunsaturated fatty acids are transferred. Both n-6 and n-3 acids can be transferred [3]; <1>, marginal transacylation with 1-acyl-2-lysoglycerophospholipid [7]; <4>, specific for arachidonate [10]) [1, 3, 7, 10]
P ?

Inhibitors

(3S,4R)-[4-(isobutenyloxy)-3-triphenylmethylamino]azetitin-2-one <1> (<1>, i.e. SB 216754 [4]) [4]
1-O-octadecyl-2-O-methyl-sn-glycero-3-phosphocholine <1> [2]
1-[2-[3-(4-chloro-3-trifluoromethylphenyl)ureido]-4-trifluoromethyl phenoxy]-4,5-dichlorobenzene sulfonic acid <1> (<1>, i.e. SK&F 45905 [1, 2, 5, 6]) [1, 2, 5, 6]
4-methoxybenzyl(3S,4R)-6-bromo-6-[(1-methyl-1,2,3-triazol-4-yl)-hydroxymethyl]penicillanate <1> (<1>, i.e. SB 212047 [4]) [4]
N-tosyl-L-phenylalanine chloromethyl ketone <1> [9]
diethyl 7-(1,4,5-triphenyl imidazol-2-yloxy)heptane phosphonate <1> (<1>, SK&F 98628 [2]) [2]
diethyl 7-(3,4,5-triphenyl-2-oxo-2,3-dihydro-imidazol-1-yl)heptane-phosphonate <1> (<1>, i.e. SK&F 98625 [1,2,5,6]) [1, 2, 5, 6]
diethyl dicarbonate <1> [9]
phenylmethyl-sulfonyl fluoride <1> [9]

K_m-Value (mM)

0.0004 <1> (1-alkyl-2-lyso-glyceryl-3-phosphorylcholine, <1>) [9]

pH-Optimum
6.5-9 <1> [9]

5 Isolation/Preparation/Mutation/Application

Source/tissue
Ehrlich ascites carcinoma cell <2> [3]
IC-21 cell <3> [8]
alveolar macrophage <4> [10]
amnion <1> (WISH cell) [7]
brain <2> [3]
heart <2> [3]
macrophage <2> [3]
monocyte <1> (<1>, HL-60 monocytic leukemia cells [2]; <1>, U937 cell line [9]) [1, 2, 4, 9]
neutrophil <2> [3, 5, 6]
platelet <2> [3]
testis <2> [3]

Localization
microsome <1> [9]

Application
pharmacology <1> (the enzyme may be a new therapeutic target to regulate inflammatory mediators) [5]

References

[1] Winkler, J.D.; Sung, C.M.; Marshall, L.A.; Chilton, F.H.: Inhibitors of arachidonate metabolism and effects on PAF production. Adv. Exp. Med. Biol., **416**, 11-45 (1996)
[2] Winkler, J.D.; Eris, T.; Sung, C.M.; Chabot-Fletcher, M.; Mayer, R.J.; Surette, M.E.; Chilton, F.H.: Inhibitors of coenzyme A-independent transacylase induce apoptosis in human HL-60 cells. J. Pharmacol. Exp. Ther., **279**, 956-966 (1996)
[3] Yamashita, A.; Sugiura, T.; Waku, K.: Acyltransferases and transacylases involved in fatty acid remodeling of phospholipids and metabolism of bioactive lipids in mammalian cells. J. Biochem., **122**, 1-16 (1997)
[4] Winkler, J.D.; Sung, C.M.; Chabot-Flecher, M.; Griswold, D.E.; Marshall, L.A.; Chilton, F.H.; Bondinell, W.; Mayer, R.J.: β-Lactams SB 212047 and SB 216754 are irreversible, time-dependent inhibitors of coenzyme A-independent transacylase. Am. Soc. Pharm. Exp. Ther., **53**, 322-329 (1998)
[5] Winkler, J.D.; Fonteh, A.N.; Sung, C.M.; Heravi, J.D.; Nixon, A.B.; Chabot-Fletcher, M.; Griswold, D.; Marshall, L.A.; Chilton, F.H.: Effects of CoA-independent transacylase inhibitors on the production of lipid inflammatory mediators. Am. Soc. Pharm. Exp. Ther., **274**, 1338-1347 (1995)

[6] Chilton, F.H.; Fonteh, A.N.; Sung, C.M.; Hickey, D.M.B.; Torphy, T.J.; Mayer, R.J.; Marshall, L.A.; Heravi, J.D.; Winkler, J.D.: Inhibitors of CoA-independent transacylase block the movement of arachidonate into 1-ether-linked phospholipids of human neutrophils. Biochemistry, **34**, 5403-5410 (1995)
[7] Toyoshima, K.; Narahara, H.; Frenkel, R.A.; Johnston, J.M.: Coenzyme A-independent transacylation in amnion-derived (WISH) cells. Arch. Biochem. Biophys., **314**, 224-228 (1994)
[8] Svetlov, S.I.; Liu, H.; Chao, W.; Olson, M.S.: Regulation of platelet-activating factor (PAF) biosynthesis via coenzyme A-independent transacylase in the macrophage cell line IC-21 stimulated with lipopolysaccharide. Biochim. Biophys. Acta, **1346**, 120-130 (1997)
[9] Winkler, J.D.; Sung, C.M.; Bennett, C.F.; Chilton, F.H.: Characterization of CoA-independent transacylase activity in U937 cells. Biochim. Biophys. Acta, **1081**, 339-346 (1991)
[10] Robinson, M.; Blank, M.L.; Snyder, F.: Acylation of lysophospholipids by rabbit alveolar macrophages. J. Biol. Chem., **260**, 7889-7895 (1985)

Glycerophospholipid acyltransferase (CoA-dependent) 2.3.1.148

1 Nomenclature

EC number
2.3.1.148

Systematic name
1-organyl-2-acyl-sn-glycero-3-phosphocholine:1-organyl-2-lyso-sn-glycero-3-phosphoethanolamine acyltransferase (CoA-dependent)

Recommended name
glycerophospholipid acyltransferase (CoA-dependent)

Synonyms
acylcoenzyme A acyltransferase

CAS registry number
9054-54-0

2 Source Organism

<1> *Rattus norvegicus* [1, 3]
<2> *Oryctolagus cuniculus* [2, 3]
<3> *Mus musculus* [3]
<4> *Homo sapiens* [3]
<5> *Canis familiaris* [3]

3 Reaction and Specificity

Catalyzed reaction
1-organyl-2-acyl-sn-glycero-3-phosphocholine + 1-organyl-2-lyso-sn-glycero-3-phosphoethanolamine = 1-organyl-2-acyl-sn-glycero-3-phosphoethanolamine + 1-organyl-2-lyso-sn-glycero-3-phosphocholine

Reaction type
acyl group transfer

Substrates and products
S 1-acyl-2-arachidonoyl-glycerophosphocholine + lyso-phosphatidylethanolamine <3> (<3>, transfer of arachidonate occurs through a bidirectional movement [3]) [3]

P 1-acyl-2-lyso-glycerophosphocholine + arachidonoylphosphatidylethanolamine
S 1-acyl-2-arachidonoyl-sn-glycero-3-phosphoinositol + lyso-phosphatidylcholine <3> [3]
P 1-acyl-2-lyso-sn-glycero-3-phosphoinositol + ?
S 1-acyl-2-arachidonoyl-sn-glycero-3-phosphophoinositol + lyso-phosphatidylethanolamine <3> (<3>, unidirectional transfer process of arachidonate [3]) [3]
P 1-acyl-2-lyso-sn-glycero-3-phosphoinositol + arachidonoylphosphatidylethanolamine
S 1-alkyl-2-lyso-sn-glycero-3-phosphocholine + phospholipid <2> [2]
P ?
S 1-hexadecyl-2-lyso-glycerophosphocholine + phospholipid <2> [3]
P ?
S lyso-pasmenylethanolamine + phospholipid <4> [3]
P ?
S lyso-phosphatidylethanolamine + phospholipid <4> [3]
P phosphatidylethanolamine + lyso-phospholipid
S lyso-phosphatidylinositol + phospholipid <4> [3]
P phosphatidylinositol + lyso-phospholipid
S lyso-phosphatidylserine + phospholipid <4> [3]
P phosphatidylserine + lyso-phospholipid
S phosphatidylcholine + lyso-phosphatidylethanolamine <3> [3]
P lyso-phosphatidylcholine + phosphatidylethanolamine
S phosphatidylcholine + lyso-phosphatidylinositol <1> (<1>, the linoleoyl transfer takes place at about 25% of the rate of arachidonoyl transfer [1]) [1]
P lyso-phosphatidylcholine + phosphatidylinositol
S phosphatidylinositol + lyso-phosphatidylcholine <1> (<1>, the linoleoyl transfer takes place at about 25% of the rate of arachidonoyl transfer [1]) [1]
P lyso-phosphatidylcholine + phosphatidylinositol
S Additional information <1, 4> (<4>, more efficient transfer of arachidonate from phosphatidylcholine to the ethanolamine-containing phospholipids than from phosphatidylinositol [3]; <1>, arachidonate and linoleate at the sn-2 position of phosphatidylcholine can be transferred to lysophosphatidylethanolamine. The transfer of 16:0, 18:0, and 18:1 acyl moieties at the sn-2-position of phosphatidylcholine is negligible) [3]
P ?

Inhibitors

Zn^{2+} <1> (<1>, strong) [3]

Cofactors/prosthetic groups

CoA <1-5> (<1>, Co-A mediated transfer does not involve a free fatty acid intermediate [1]; <1-4>, required [1-3]; <2,4>, K_m: 0.0014 mM [3]; <1>, K_m: 0.0015 mM, lung enzyme, with lyso-phosphatidylethanolamine as substrate

[3]; <1>, K_m: 0.014 mM, liver enzyme, with 2-lyso-sn-phosphatidylinositol as substrate [3]) [1-3]

K_m-Value (mM)
0.0015 <2> (1-hexadecyl-2-lyso-glycerophosphocholine, <2>) [3]
0.0038 <2> (1-acyl-2-lyso-glycerophosphocholine, <2>) [3]
0.017 <1> (palmitoyl-lyso-glycerophosphocholine, <1>) [3]
0.076 <4> (lyso-phosphatidylserine, <4>) [3]

pH-Optimum
4.5 <5> (<5>, bimodal pH-optimum, at pH 4.5 and at pH 7.5. The activity is 4-5times higher at pH 4.5 than at pH 7.5) [3]
7.5 <1, 5> (<5>, bimodal pH-optimum, at pH 7.5 and at pH 4.5) [3]

5 Isolation/Preparation/Mutation/Application

Source/tissue
alveolar macrophage <2> [2, 3]
heart <5> [3]
liver <1, 2> [1, 3]
lung <1> [3]
lymphocyte <3> [3]
macrophage <3> [3]
platelet <1, 4> [3]

Localization
membrane <2> [2]
microsome <1, 2, 5> [1, 3]

References

[1] Irvine, R.L.; Dawson, R.M.C.: Transfer of arachidonic acid between phospholipids in rat liver microsomes. Biochem. Biophys. Res. Commun., **91**, 1399-1405 (1979)
[2] Robinson, M.; Blank, M.L.; Snyder, F.: Acylation of lysophospholipids by rabbit alveolar macrophages. J. Biol. Chem., **260**, 7889-7895 (1985)
[3] Snyder, F.; Lee, T.C.; Blank, M.L.: The role of transacylases in the metabolism of arachidonate and platelet activating factor. Prog. Lipid Res., **31**, 65-86 (1992)

Platelet-activating factor acetyltransferase 2.3.1.149

1 Nomenclature

EC number
2.3.1.149

Systematic name
1-alkyl-2-acyl-sn-glycero-3-phosphocholine:1-organyl-2-lyso-sn-glycero-3-phospholipid acetyltransferase

Recommended name
platelet-activating factor acetyltransferase

Synonyms
PAF acetyltransferase
PAF: acyllyso-GPC transacetylase

CAS registry number
9012-30-0

2 Source Organism

<1> *Rattus norvegicus* [1]
<2> *Bos taurus* (calf) [2]

3 Reaction and Specificity

Catalyzed reaction
1-alkyl-2-acetyl-sn-glycero-3-phosphocholine + 1-organyl-2-lyso-sn-glycero-3-phospholipid = 1-organyl-2-lyso-sn-glycero-3-phosphocholine + 1-alkyl-2-acetyl-sn-glycero-3-phospholipid

Reaction type
acyl group transfer

Natural substrates and products
S Additional information <1, 2> (<1>, the enzyme may participate in the biosynthesis of ethanolamine plasmalogen and acyl analogs of platelet-activating factor, in vivo, finetuning of platelet-activating factor biological responses and cross-talk between de novo and remodeling pathways of platelet biosynthesis [1]; <2>, ATP regulates the activity of the enzyme

by reversible activation and inactivation via the phosphorylation and dephosphorylation cycle [2]) [1, 2]

P ?

Substrates and products

S 1-alkyl-2-acetyl-sn-glycero-3-phosphocholine + acyllyso-sn-glycero-3-phosphate <1> [1]

P ?

S 1-alkyl-2-acetyl-sn-glycero-3-phosphocholine + acyllyso-sn-glycero-3-phosphoinositol <1> [1]

P ?

S 1-alkyl-2-acetyl-sn-glycero-3-phosphocholine + acyllyso-sn-glycero-3-phosphoserine <1> [1]

P ?

S 1-alkyl-2-acetyl-sn-glycero-3-phosphocholine + alkyllyso-sn-glycero-3-phosphate <1> [1]

P ?

S 1-alkyl-2-acetyl-sn-glycero-3-phosphocholine + cis-9-octadecen-1-ol <1> [1]

P ?

S 1-alkyl-2-acetyl-sn-glycero-3-phosphocholine + lysoplasmalogen <1> (<1>, 1-alkyl-2-acetyl-sn-glycero-3-phosphocholine is platelet-activating factor [1]) [1]

P 1-alk-1-enyl-2-acetyl-sn-glycero-3-phosphoethanolamine + ? <1> [1]

S 1-alkyl-2-acetyl-sn-glycero-3-phosphocholine + radyllyso-sn-glycero-3-phosphocholine <1> [1]

P ?

S 1-alkyl-2-acetyl-sn-glycero-3-phosphocholine + radyllyso-sn-glycero-3-phosphoethanolamine <1> [1]

P ?

S hexadecyl-acetyl-sn-glycero-3-phosphocholine + alk-1-enyllyso-sn-glycero-phosphoethanolamine <1> [1]

P ?

Inhibitors

mepacrine <1> [1]
p-bromophenacyl bromide <1> [1]

K_m-Value (mM)

0.012 <1> (hexadecyl-acetyl-sn-glycero-3-phosphocholine, <1>) [1]
0.1064 <1> (alk-1-enyllyso-sn-glycero-phosphoethanolamine, <1>) [1]

pH-Optimum

7.4 <1> [1]

Temperature range (°C)

15-24 <1> (<1>, active between 15°C and 24°C [1]) [1]

5 Isolation/Preparation/Mutation/Application

Source/tissue

HL-60 cell <1> [1]
pulmonary artery endothelial cell <1, 2> [1, 2]

Localization

membrane <1> [1]

References

[1] Lee, T.C.; Uemura, Y.; Snyder, F.: A novel CoA-independent transacetylase produces the ethanolamine plasmalogen and acetyl analogs of platelet-activating factor (PAF) with PAF as the actetate donor in HL-60 cells. J. Biol. Chem., **267**, 19992-20001 (1992)
[2] Balestrieri, M.L.; Servillo, L.; Lee, T.: The role of platelet-activating factor-dependent transacetylase in the biosynthesis of 1-acyl-2-acetyl-sn-glycero-3-phosphocholine by stimulated endothelial cells. J. Biol. Chem., **272**, 17431-17437 (1997)

Salutaridinol 7-O-acetyltransferase 2.3.1.150

1 Nomenclature

EC number
2.3.1.150

Systematic name
acetyl-CoA:salutaridinol 7-O-acetyltransferase

Recommended name
salutaridinol 7-O-acetyltransferase

Synonyms
acetyl-CoA:salutaridinol-7-O-acetyltransferase
acetyltransferase, salutaridinol 7-O-

CAS registry number
156859-13-1

2 Source Organism

<-1> no activity in *Papaver rhoeas* (cell cultures) [1]
<1> *Papaver somniferum* [1]
<2> *Papaver bracteatum* [1]
<3> *Papaver sp.* [2]

3 Reaction and Specificity

Catalyzed reaction
acetyl-CoA + salutaridinol = CoA + 7-O-acetylsalutaridinol

Reaction type
acyl group transfer

Natural substrates and products
S acetyl-CoA + salutaridinol <1, 3> (<1>, the enzyme catalyzes the formation of thebaine in morphine biosynthesis [1]; <3>, the enzyme is involved in the biosynthetic pathway from L-Tyr to morphine [2]) [1, 2]
P ?

Substrates and products
S acetyl-CoA + nudaurine <1> [1]
P ?

S acetyl-CoA + salutaridinol <1-3> [1, 2]
P CoA + 7-O-acetylsalutaridinol <1-3> (<3>, at higher pH values, the product 7-O-acetylsalutaridinol, spontaneously closes the 4-5 oxide bridge by allylic elimination to form the morphine precursor thebaine [2]) [1, 2]
S n-hexanoyl-CoA + salutaridinol <1> [1]
P ?
S n-propionyl-CoA + salutaridinol <1> [1]
P ?
S Additional information <3> (<3>, 7-epi-salutaridinol is inactive as substrate) [2]
P ?

Inhibitors
1,10-phenanthroline <1> (<1>, 0.1 mM, 56% inhibition) [1]
8-hydroxyquinoline <1> (<1>, 0.1 mM, 85% inhibition) [1]
Cd^{2+} <1> (<1>, 1 mM, strong) [1]
Cu^{2+} <1> (<1>, 1 mM, strong) [1]
NEM <1> (<1>, weak) [1]
Zn^{2+} <1> (<1>, 1 mM, strong) [1]
codeinone <1> (<1>, 0.5 mM, 55% inhibition) [1]
diethyldicarbamate <1> (<1>, 0.05 mM, 87% inhibition) [1]
p-chloromercuribenzoate <1> (<1>, weak) [1]
salutaridine <1> (<1>, 0.5 mM, 74% inhibition) [1]
thebaine <1> (<1>, 0.5 mM, 76% inhibition) [1]

Specific activity (U/mg)
Additional information <1> [1]

K_m-Value (mM)
0.007 <1> (salutaridinol, <1>) [1]
0.046 <1> (acetyl-CoA, <1>) [1]
0.068 <1> (n-propionyl-CoA, <1>) [1]
0.093 <1> (n-hexanoyl-CoA, <1>) [1]

pH-Optimum
6-9 <1> [1]
7-9 <3> [2]

Temperature optimum (°C)
47 <1, 3> [1, 2]

4 Enzyme Structure

Subunits
? <1> (<1>, 1 * 50000, SDS-PAGE) [1]

5 Isolation/Preparation/Mutation/Application

Source/tissue

capsule <1> [1]
cell culture <1> [1]
cell suspension culture <2, 3> [1, 2]
leaf <1> [1]
root <1> [1]
shoot <1> [1]

Purification

<1> [1]
<3> [2]

References

[1] Lenz, R.; Zenk, M.H.: Acetyl coenzyme A:salutaridinol-7-O-acetyltransferase from Papaver somniferum plant cell cultures. J. Biol. Chem., **270**, 31091-31096 (1995)
[2] Lenz, R.; Zenk, M.H.: Closure of the oxide bridge in morphine biosynthesis. Tetrahedron Lett., **35**, 3897-3900 (1994)

Benzophenone synthase 2.3.1.151

1 Nomenclature

EC number
2.3.1.151

Systematic name
malonyl-CoA:3-hydroxybenzoyl-CoA malonyltransferase

Recommended name
benzophenone synthase

Synonyms
2,3',4,6-tetrahydroxybenzophenone synthase
synthase, benzophenone

CAS registry number
175780-21-9

2 Source Organism

<1> *Centaurium erythrea* [1]
<2> *Hypericum androsaemum* [2]

3 Reaction and Specificity

Catalyzed reaction
3 malonyl-CoA + 3-hydroxybenzoyl-CoA = 4 CoA + 2,3',4,6-tetrahydroxybenzophenone + 3 CO_2

Reaction type
acyl group transfer
decarboxylation
intramolecular cyclization

Natural substrates and products
S malonyl-CoA + 3-hydroxybenzoyl-CoA <1> (<1>, central step in xanthone biosynthesis [1]) [1]
P ?

Substrates and products
S malonyl-CoA + 3-hydroxybenzoyl-CoA <1, 2> (<2>, 49% of the activity with benzyol-CoA [2]) [1, 2]

P CoA + 2,3',4,6-tetrahydroxybenzophenone + CO_2 <1, 2> [1, 2]
S malonyl-CoA + 4-hydroxybenzoyl-CoA <2> (<2>, 5% of the activity with benzoyl-CoA [2]) [2]
P ?
S malonyl-CoA + benzoyl-CoA <1, 2> (<1>, 44% of the activity with 3-hydroxybenzoyl-CoA [1]) [1, 2]
P CoA + 2,4,6-trihydroxybenzophenone + CO_2 <1, 2> [1, 2]

pH-Optimum
7 <2> [2]
7.5 <1> [1]

pH-Range
6-8 <1> (<1>, pH 6.0: about 20% of maximal activity, pH 8.0: about 45% of maximal activity) [1]

Temperature optimum (°C)
35 <2> [2]
40-45 <1> [1]

Temperature range (°C)
25-55 <1> (<1>, about 55% of maximal activity, 5°C: about 30% of maximal activity) [1]

5 Isolation/Preparation/Mutation/Application

Source/tissue
cell culture <1> [1]

References

[1] Beerhues, L.: Benzophenone synthase from cultured cells of Centaurium erythraea. FEBS Lett., **383**, 264-266 (1996)
[2] Schmidt, W.; Beerhues, L.: Alternative pathways of xanthone biosynthesis in cell cultures of Hypericum androsaemum. FEBS Lett., **420**, 143-146 (1997)

Alcohol O-cinnamoyltransferase 2.3.1.152

1 Nomenclature

EC number
2.3.1.152

Systematic name
1-O-trans-cinnamoyl-β-D-glucopyranose:alcohol O-cinnamoyltransferase

Recommended name
alcohol O-cinnamoyltransferase

Synonyms
EC 2.4.1.120 (formerly)
HCA-GT
UDP-glucose:sinapic acid glucosyltransferase
glucosyltransferase, uridine diphosphoglucose-sinapate
sinapate glucosyltransferase
sinapic acid glucosyltransferase
uridine 5'-diphosphoglucose-hydroxycinnamic acid acylglucosyltransferase
uridine 5'-diphosphoglucose:hydroxycinnamic acid acyl-glucosyltransferase

CAS registry number
74082-53-4

2 Source Organism

<1> *Raphanus sativus* (var. sativus) [1]

3 Reaction and Specificity

Catalyzed reaction
1-O-trans-cinnamoyl-β-D-glucopyranose + ROH = alkyl cinnamate + glucose (<1>, sequential binding mechanism [1])

Reaction type
glycosyl group transfer

Natural substrates and products
S Additional information <1> (<1>, formation of 1-O-acylglucosides of hydroxycinnamic acid, which are high-energy acyl donors, with a high

group-transfer potential in 1-O-acylglucose-dependent acytransferase reactions leading to various hydroxycinnamic acid O-esters in plants) [1]

P ?

Substrates and products

S UDP-glucose + 4-coumaric acid <1> [1]

P ?

S UDP-glucose + caffeic acid <1> [1]

P ?

S UDP-glucose + ferulic acid <1> [1]

P ?

S UDP-glucose + sinapic acid <1> (<1>, r [1]) [1]

P 1-O-sinapoyl-β-D-glucose + UDP <1> [1]

Inhibitors

1-O-sinapoyl-β-D-glucose <1> (<1>, noncompetitive with both substrates) [1]

UDP <1> (<1>, competitive against UDP-glucose and noncompetitive against sinapic acid) [1]

Specific activity (U/mg)

Additional information <1> [1]

5 Isolation/Preparation/Mutation/Application

Source/tissue

seedling <1> [1]

Purification

<1> (partial) [1]

References

[1] Mock, H.P.; Strack, D.: Energetics of the uridine 5'-diphosphoglucose:hydroxycinnamic acid acyl-glucosyltransferase reaction. Phytochemistry, **32**, 575-579 (1993)

Anthocyanin 5-aromatic acyltransferase 2.3.1.153

1 Nomenclature

EC number
2.3.1.153

Systematic name
hydroxycinnamoyl-CoA:anthocyanidin 3,5-diglucoside 5-O-glucoside-6'''-O-hydroxycinnamoyltransferase

Recommended name
anthocyanin 5-aromatic acyltransferase

Synonyms
acyltransferase, anthocyanin
anthocyanidin 3,5-diglucoside 5-O-glucoside-6''-O-hydroxycinnamoyltransferase
anthocyanin acyltransferase
anthocyanin hydroxycinnamoyltransferase
gentian 5AT
hydroxycinnamoyl-CoA-anthocyanidin 3,5-diglucoside 5-O-glucoside-6'''-O-hydroxycinnamoyltransferase
hydroxycinnamoyltransferase, anthocyanidin 3,5-diglucoside 5-O-glucoside-6''-O-
hydroxycinnamoyltransferase, anthocyanin

CAS registry number
179466-49-0
182238-32-0
198841-53-1

2 Source Organism

<1> *Gentiana triflora* [1]

3 Reaction and Specificity

Catalyzed reaction
hydroxycinnamoyl-CoA + anthocyanidin-3,5-diglucoside = CoA + anthocyanidin 3-glucoside-5-hydroxycinnamoylglucoside

Reaction type

acyl group transfer

Substrates and products

S cyanidin 3,5-diglucoside + caffeoyl-CoA <1> [1]
P CoA + cyanidin 3-glucoside 5-caffeoylglucoside
S cyanidin 3,5-diglucoside + *p*-coumaroyl-CoA <1> [1]
P CoA + cyanidin 3-glucoside 5-coumaroylglucoside
S delphinidin 3,5-diglucoside + caffeoyl-CoA <1> [1]
P CoA + delphinidin 3-glucoside 5-caffeoylglucoside <1> [1]
S delphinidin 3,5-diglucoside + *p*-coumaroyl-CoA <1> [1]
P CoA + delphinidin 3-glucoside 5-coumaroylglucoside

Inhibitors

PCMB <1> (<1>, 90% inhibition at 0.1 mM, complete inhibition at 1 mM [1]) [1]

Metals, ions

Mn^{2+} <1> (<1>, strongly enhances activity [1]) [1]
Zn^{2+} <1> (<1>, moderately enhances activity [1]) [1]

Specific activity (U/mg)

Additional information <1> [1]

K_m-Value (mM)

0.065 <1> (caffeoyl-CoA, with delphinidin 3,5-diglucoside as cosubstrate, <1>) [1]
0.087 <1> (cyanidin 3,5-diglucoside, with caffeoyl-CoA as cosubstrate, <1>) [1]
0.15 <1> (delphinidin 3,5-diglucoside, with caffeoyl-CoA as cosubstrate, <1>) [1]
0.19 <1> (*p*-coumaroyl-CoA, with delphinidin 3,5-diglucoside as cosubstrate, <1>) [1]

pH-Optimum

8-8.5 <1> [1]

pH-Range

5.5-10 <1> (<1>, active in this range) [1]

4 Enzyme Structure

Molecular weight

49000 <1> (<1>, gel filtration) [1]

Subunits

monomer <1> (<1>, 1 * 52000, SDS-PAGE [1]) [1]

5 Isolation/Preparation/Mutation/Application

Source/tissue
petal <1> (blue) [1]

Purification
<1> [1]

6 Stability

Storage stability
<1>, -20°C, 20 mM Tris/HCl, pH 7.0, 0.01 mM (*p*-aminidinophenyl)methanesulfonyl fluoride, 1 mM DTT, stable for at least 2 years [1]

References

[1] Fujiwara, H.; Tanaka, Y.; Fukui, Y.; Nakao, M.; Ashikari, T.; Kusumi, T.: Anthocyanin 5-aromatic acyltransferase from Gentiana triflora. Eur. J. Biochem., **249**, 45-51 (1997)

Propionyl-CoA C^2-trimethyltridecanoyltransferase 2.3.1.154

1 Nomenclature

EC number
2.3.1.154

Systematic name
4,8,12-trimethyltridecanoyl-CoA:propanoyl-CoA C^2-4,8,12-trimethyltridecanoyltransferase

Recommended name
propionyl-CoA C^2-trimethyltridecanoyltransferase

Synonyms
3-oxopristanoyl-CoA hydrolase
3-oxopristanoyl-CoA thiolase
SCP-2/thiolase <2> [2]
SCPx
oxopristanoyl-CoA thiolase
peroxisomal 3-oxoacyl coenzyme A thiolase
peroxisome sterol carrier protein thiolase
sterol carrier protein 2/3-oxoacyl-CoA thiolase <2> [2]
sterol carrier protein x

CAS registry number
195740-63-7

2 Source Organism

<1> *Homo sapiens* [1, 7]
<2> *Rattus norvegicus* [1, 2, 3, 5, 6, 8, 11]
<3> *Mus musculus* [1]
<4> *Caenorhabditis elegans* [4]
<5> *Saccharomyces cerevisiae* [9]
<6> *Candida tropicalis* [10]

3 Reaction and Specificity

Catalyzed reaction

4,8,12-trimethyltridecanoyl-CoA + propanoyl-CoA = 3-oxopristanoyl-CoA + CoA (<2> active on medium and long straight chain 3-oxoacyl-CoAs and 2-methyl-branched 3-oxoacyl-CoAs [3])

Reaction type

transfer of 3-oxoacyl-CoAs

Natural substrates and products

S 3-oxo-(2R)-methylpalmitoyl-CoA + CoA <2> (Reversibility: r <2> [3]) [3]
P tetradecaneoyl-CoA + propanoyl-CoA <2> [3]
S $3\alpha,7\alpha,12\alpha$-trihydroxy-24-keto-5β-cholestanoyl-CoA + CoA <1, 2, 3> (Reversibility: r <1, 2, 3> [1]) [1, 3]
P choloyl-CoA + propanoyl-CoA <1, 2, 3> [1]

Substrates and products

S 3-oxo-(2R)-methylpalmitoyl-CoA + CoA <2> (Reversibility: r <2> [3]) [3]
P tetradecanoyl-CoA + propanoyl-CoA <2> [3]
S 3-oxodecanoyl-CoA + CoA <6> (Reversibility: r <6> [10]) [10]
P octanoyl-CoA + acetyl-CoA <6> [10]
S 3-oxododecanoyl-CoA + CoA <6> (Reversibility: r <6> [10]) [10]
P decanoyl-CoA + acetyl-CoA <6> [10]
S 3-oxohexadecanoyl-CoA + CoA <2> (Reversibility: r <2> [3]) [3]
P tetradecanoyl-CoA + acetyl-CoA <2> [3]
S 3-oxohexanoyl-CoA + CoA <6> (Reversibility: r <6> [10]) [10]
P butanoyl-CoA + acetyl-CoA <6> [10]
S 3-oxooctanoyl-CoA + CoA <2> (Reversibility: r <2> [2]) [2, 8, 10]
P hexanoyl-CoA + acetyl-CoA <2> [2, 8, 10]
S 3-oxopalmitoyl-CoA + CoA <2> (Reversibility: r <2> [3]) [3, 5, 8]
P tetradecanoyl-CoA + acetyl-CoA <2> [3, 5]
S 3-oxopentanoyl-CoA + CoA <6> (Reversibility: r <6> [10]) [10]
P propanoyl-CoA + acetyl-CoA <6> [10]
S 3-oxopristanoyl-CoA + CoA <1, 2> (Reversibility: r <1, 2> [1, 5]) [1, 5]
P 4,8,12-trimethyltridecanoyl-CoA + propanoyl-CoA <1, 2> [1, 5]
S 3-oxotetradecanoyl-CoA + CoA <4> (Reversibility: r <4> [4]) [4]
P didecanoyl-CoA + acetyl-CoA <4> [4]
S $3\alpha,7\alpha,12\alpha$-trihydroxy-24-keto-5β-cholestanoyl-CoA + CoA <1, 2, 3> (Reversibility: r <1, 2, 3> [1]) [1, 3]
P choloyl-CoA + propanoyl-CoA <1, 2, 3> [1]
S acetoacetyl-CoA + CoA <6> (Reversibility: r <6> [10]) [10]
P acetyl-CoA + acetyl-CoA <6> [10]

Inhibitors

acetyl-CoA <4> (<4> competitive to CoA, non-competitive to 3-oxoacyl-CoA [4]) [4, 8]

Cofactors/prosthetic groups

CoA <1-6> [1-11]

Specific activity (U/mg)

0.000214 <1> (<1> skin fibroblasts culture of patients with Zellweger syndrome [1]) [1]
0.00074 <1> (<1> skin fibroblasts culture [1]) [1]
0.04 <2> (<2> whole peroxisomes, 3-oxopristanoyl-CoA as substrate [5]) [5]
2.5 <2> (<2> acetoacetyl-CoA as substrate [8]) [8]
16 <1> (<1> 3-oxoacyl-CoA thiolase activity of mitochondrial long-chain trifunctional enzyme, 3-oxopalmitoyl-CoA as substrate [7]) [7]
26.7 <2> (<2> 3-oxostearoyl-CoA as substrate [8]) [8]
33.3 <2> (<2> 3-oxooleoyl-CoA as substrate [8]) [8]
41 <4> (<4> 3-oxotetradecanoyl-CoA as substrate [4]) [4]
42 <6> [10]
61.5 <2> (<2> 3-oxooctanoyl-CoA as substrate [2]) [2]
75 <2> [11]
120 <2> (<2> 3-oxooctanoyl-CoA as substrate [8]) [8]
240 <5> [9]

K_m-Value (mM)

0.0007 <4> (CoA) [4]
0.0028 <2> ($3\alpha,7\alpha,12\alpha$-trihydroxy-24-keto-5β-cholestanoyl-CoA) [3]
0.0029 <2> (3-oxohexadecanedioyl-CoA) [3]
0.0029 <2> (CoA, <2> 0.0025 mM 3-oxooctanoyl-CoA [8]) [8]
0.003 <2> (3-oxo-2-methylpalmitoyl-CoA) [3]
0.004 <2> (3-oxopalmitoyl-CoA) [3]
0.0046 <2> (3-oxooctanoyl-CoA) [3]
0.0054 <2> (3-oxooctanoyl-CoA) [8]
0.0063 <2> (CoA, <2> 0.01 mM 3-oxooctanoyl-CoA [8]) [8]
0.0079 <2> (3-oxopalmitoyl-CoA) [8]
0.0081 <2> (acetoacetyl-coA) [11]
0.0083 <2> (3-oxooctanoyl-CoA) [11]
0.0088 <6> (3-oxooctanoyl-CoA) [10]
0.0093 <6> (3-oxododecanoyl-CoA) [10]
0.01 <6> (3-oxodecanoyl-CoA) [10]
0.028 <6> (3-oxohexanoyl-CoA) [10]
0.04 <2> (CoA, <2> liver homogenate [1]) [1]
0.059 <6> (3-oxopentanoyl-CoA) [10]
0.08 <4> (3-oxooctanoyl-CoA) [4]
0.093 <6> (acetoacetyl-CoA) [10]

pH-Optimum

7.6 <2> (<2> 3-oxooctanoyl-CoA as substrate [3]) [3]
9.5 <2> (<2> liver homogenate [1]) [1]

4 Enzyme Structure

Molecular weight

86000 <5> (<5> gel filtration [9]) [9]
90000-110000 <2> (<2> three isoforms, homo- and heterodimeric combinations of 58- and 46 kDa subunits, 58 kDa polypeptide is SCP2/thiolase, 46 kDa polypeptide is the thiolase domain of SCP2/thiolase, gel filtration, cross-linking, immunochemical [2,3]) [2, 3]
98000 <6> (<6> gel filtration [10]) [10]

Subunits

dimer <2> (<2> α,α, 2 * 46000, β,β, 2 * 58000, α,β, 1 * 46000 + 1 * 58000, three isoforms, gel filtration, chemical cross-linking, immunochemical [2]) [2]
dimer <5> (<5> α,α, 2 * 45000, SDS-PAGE [9]) [9]
dimer <6> (<6> α,α, 2 * 43000, SDS-PAGE [10]) [10]

5 Isolation/Preparation/Mutation/Application

Source/tissue

fibroblast <1> (<1,3> cell culture [1]) [1]
liver <1, 2> [1, 2, 3, 5, 7, 8, 11]

Localization

mitochondrion <1> [7]
peroxisome <1, 2, 5, 6> [1, 2, 3, 8, 9, 10, 11]

Purification

<2> (GST-SCPx fusion protein expressed in Escherichia coli, glutathione-Sepharose 4B, thrombin cleavage [6]) [6]
<2> (coding sequence expressed in Escherichia coli [8]) [8]
<2> (phosphocellulose, DEAE-Sepharose, hydroxylapatite, Blue-Sepharose, chromatofocusing, yield: 0.7% [2]) [2]
<2> (phosphocellulose, ammonium sulfate, calcium phosphate gel/cellulose, Sephadex G-100, isoelectric focusing [11]) [11]
<4> (recombinant protein with His-tag, expressed in Escherichia coli, affinty chromatography with Ni^{2+} [4]) [4]
<5> (ammonium sulfate precipitation, phosphocellulose, matrix gel Red A, hydroxylapatite, 308fold purification [9]) [9]
<6> (DEAE Sepharose, Cellulofine GCL, 17fold purification [10]) [10]

Cloning

<2> (full-length cDNA expressed in Escherichia coli [5]) [5, 6, 8]
<4> (full-length cDNA expressed in Escherichia coli, His-tagged [4]) [4]

6 Stability

Storage stability

<5>, -20°C, 50% glycerol [9]

References

[1] Ferdinandusse, S.; Denis, S.; van Berkel, E.; Dacremont, G.; Wanders, R.J.A.: Peroxisomal fatty acid oxidation disorders and 58 kDa sterol carrier protein X (SCPx): activity measurements in liver and fibroblasts using a newley developed method. J. Lipid Res., **41**, 336-342 (2000)

[2] Antonenkov, V.D.; van Veldhoven, P.P.; Mannaerts, G.P.: Isolation and subunit composition of native sterol carrier protein 2/3-oxoacyl-coenzyme A thiolase from normal rat liver peroxisomes. Protein Expr. Purif., **18**, 249-256 (2000)

[3] Antonenkov, V.D.; van Veldhoven, P.P.; Waelkens, E.; Mannaerts, G.P.: Substrate specifities of 3-oxoacyl-CoA thiolase A and sterol carrier protein 2/3-oxoacyl-CoA thiolase purified from normal rat liver peroxisomes. J. Biol. Chem., **272**, 26023-26031 (1997)

[4] Bun-Ya, M.; Maebuchi, M.; Hashimoto, T.; Yokota, S.; Kamiryo, T.: A second isoform of 3-ketoacyl-CoA thiolase found in Caenorhabditis elegans, which is similar to sterol carrier protein x but lacks the sequence of sterol carrier protein 2. Eur. J. Biochem., **245**, 252-259 (1997)

[5] Wanders, R.J.A.; Denis, S.; Wouters, F.; Wirtz, K.W.A.; Seedorf, U.: Sterol carrier protein X (SCPx) is a peroxisomal branched-chain β-ketothiolase specifically reacting with 3-oxo-pristanoyl-CoA: a new, uniquie role for SCPx in branched-chain fatty acid metabolism in peroxisomes. Biochem. Biophys. Res. Commun., **236**, 565-569 (1997)

[6] Manfra, D.J.; Baum, C.L.; Reschley, E.; Lundell, D.; Zavodny, P.; Dalie, B.: Expression and purification of two recombinant sterol-carrier proteins: SCPX and SCP2. Protein Expr. Purif., **6**, 196-205 (1995)

[7] Middleton, B.: The mitochondrial long-chain trifunctional enzyme: 2-enoyl-CoA hydratase, 3-hydroxyacyl-CoA dehydrogenase and 3-oxoacyl-CoA thiolase. Biochem. Soc. Trans., **22**, 427-431 (1994)

[8] Seedorf, U.; Brysch, P.; Engel, T.; Schrage, K.; Assmann, G.: Sterol carrier protein x is peroxisomal 3-oxoacyl coenzyme A thiolase with intrinsic sterol carrier and lipid transfer activity. J. Biol. Chem., **269**, 21277-21283 (1994)

[9] Erdmann, R.; Kunau, W.H.: Purification and immunolocalization of the peroxisomal 3-oxoacyl-CoA thiolase from Saccharomyces cerevisiae. Yeast, **10**, 1173-1182 (1994)

[10] Kurihara, T.; Ueda, M.; Tanaka, A.: Peroxisomal acetoacetyl-CoA thiolase and 3-ketoacyl-CoA thiolase from an n-alkane-utilizing yeast, Candida tropicalis: purification and characterization. J. Biochem., **106**, 474-478 (1989)

[11] Miyazawa, S.; Osumi, T.; Hashimoto, T.: The presence of a new 3-oxoacyl-CoA thiolase in rat liver peroxisomes. Eur. J. Biochem., **103**, 589-596 (1980)

Acetyl-CoA C-myristoyltransferase 2.3.1.155

1 Nomenclature

EC number
2.3.1.155

Systematic name
myristoyl-CoA:acetyl-CoA C-myristoyltransferase

Recommended name
acetyl-CoA C-myristoyltransferase

Synonyms
3-oxopalmitoyl-CoA hydrolase
3-oxopalmitoyl-CoA-CoA acetyltransferase
peroxisomal 3-ketoacyl-CoA thiolase <1> [1]

CAS registry number
9027-13-8
9029-97-4

2 Source Organism

<1> *Rattus norvegicus* (Wistar [4]) [1, 3-5]
<2> *Sus scrofa* (pig [2]) [2]
<3> *Homo sapiens* (human [4]) [4]
<4> *Cucumis sativus* (cucumber [4]) [4]
<5> *Escherichia coli* [4]
<6> *Pseudomonas fragi* [4]
<7> *Brassica napus* [5]
<8> *Saccharomyces cerevisiae* (yeast [5]) [5]

3 Reaction and Specificity

Catalyzed reaction
myristoyl-CoA + acetyl-CoA = 3-oxopalmitoyl-CoA + CoA (A peroxisomal enzyme involved in branched chain fatty acid β-oxidation in peroxisomes. It differs from EC 2.3.1.154, propionyl-CoA C^2-trimethyldecanoyltransferase, in not being active towards 3-oxopristanoyl-CoA)

Reaction type
acyl group transfer

Natural substrates and products

S myristoyl-CoA + acetyl-CoA <1> (Reversibility: ? <1> [1]) [1]
P 3-oxopalmitoyl-CoA + CoA

Substrates and products

S 3-ketodecanoyl-CoA + H_2O + CoA <1> (Reversibility: ? <1> [1]) [1]
P acetyl-CoA + octanoyl-CoA
S 3-ketododecanoyl-CoA + H_2O + CoA <1> (Reversibility: ? <1> [1]) [1]
P acetyl-CoA + decanoyl-CoA
S 3-ketohexadecanoylCoA + H_2O <1> (Reversibility: r <1> [1]) [1]
P acetyl-CoA + tetradecanoyl-CoA
S 3-ketohexanoyl-CoA + H_2O + CoA <1> (Reversibility: ? <1> [1]) [1]
P acetyl-CoA + butyryl-CoA
S 3-ketooctanoyl-CoA + H_2O + CoA <1> (Reversibility: ? <1> [1, 4]) [1, 4]
P acetyl-CoA + hexanoyl-CoA
S acetoacetyl-CoA + H_2O <1-7> (Reversibility: ? <1-7> [1, 2, 4]) [1, 2, 4, 5]
P acetyl-CoA
S tetradecanoyl-CoA + acetyl-CoA <1> (<1> myristoyl-CoA [1]) (Reversibility: r <1> [1]) [1]
P 3-ketohexadecanoyl-CoA + CoA <1> (<1> 3-oxopalmitoyl-CoA [1]) [1]

K_m-Value (mM)

0.0019 <1> (3-ketohexadecanoyl-CoA) [1]
0.003 <1> (3-ketododecanoyl-CoA) [1]
0.008 <1> (3-ketodecanoyl-CoA) [1]
0.0086 <1> (3-ketoocatanoyl-CoA) [1]
0.0106 <1> (3-ketohexanoyl-CoA) [1]
0.014 <1> (acetoacetyl-CoA) [1]

4 Enzyme Structure

Molecular weight

38000 <1> (<1> gel filtration conducted at 6°C [1]) [1]
41070 <1> (<1> mature enzyme, amino acid sequence [3]) [3]
48700 <7> (<7> calculated from amino acid sequence [5]) [5]
53000 <7> (<7> SDS-PAGE [5]) [5]
76000 <1> (<1> gel filtration conducted at 28°C [1]) [1]
78000 <1> (<1> peroxisomal bifunctional protein, gel filtration [1,4]) [1, 4]
89000 <1> (<1> peroxisomal 3-ketoacyl-CoA thiolase, equilibrium centrifugation [1]; <1> gel filtration [4]) [1, 3, 4]

Subunits

dimer <1, 8> (<1> 2 * 44500, SDS-PAGE [1]; <1> 2 identical subunits [3]; 2 * 41000, SDS-PAGE [4]) [1, 3-5]
tetramer <1> (<1> 4 * 38500 [1]) [1]

5 Isolation/Preparation/Mutation/Application

Source/tissue
liver <1-3> [1-4]
seed <4> [4]

Localization
glyoxysome <4, 7> [4, 5]
peroxisome <1> [1, 4, 5]

Purification
<1> [1, 3, 4]
<6> [4]
<7> (<7> recombinant protein, heterologous expressed in Escherichia coli [5]) [5]

Cloning
<1> (cDNA clones of peroxisomal 3-ketoacyl-CoA thiolase isolated and sequenced [3]) [3]
<7> (full-length cDNA of glyoxysomal 3-ketoacyl-CoA thiolase cloned and expressed in Escherichia coli [5]) [5]

6 Stability

General stability information
<1>, cold lability [1]

References

[1] Miyazawa, S.; Furuta, S.; Osumi, T.; Hashimoto, T.; Ui, N.: Properties of peroxisomal 3-ketoacyl-CoA thiolase from rat liver. J. Biochem., **90**, 511-519 (1981)
[2] Schulz, H.; Staack, H.: 3-Ketoacyl-CoA-thiolase with broad chain length specificity from pig heart muscle. Methods Enzymol., **71**, 398-403 (1981)
[3] Hijikata, M.; Ishii, N.; Kagamiyama, H.; Osumi, T.; Hashimoto, T.: Structural analysis of cDNA for rat peroxisomal 3-ketoacyl-CoA thiolase. J. Biol. Chem., **262**, 8151-8158 (1987)
[4] Uchida, Y.; Izai, K.; Orii, T.; Hashimoto, T.: Novel fatty acid β-oxidation enzymes in rat liver mitochondria. II. Purification and properties of enoyl-coenzyme A (CoA) hydratase/3-hydroxyacyl-CoA dehydrogenase/3-ketoacyl-CoA thiolase trifunctional protein. J. Biol. Chem., **267**, 1034-1041 (1992)
[5] Olesen, C.; Thomsen, K.K.; Svendsen, I.; Brandt, A.: The glyoxysomal 3-ketoacyl-CoA thiolase precursor from Brassica napus has enzymic activity when synthesized in Escherichia coli. FEBS Lett., **412**, 138-140 (1997)

Phloroisovalerophenone synthase 2.3.1.156

1 Nomenclature

EC number
2.3.1.156

Systematic name
isovaleryl-CoA:malonyl-CoA acyltransferase

Recommended name
phloroisovalerophenone synthase

Synonyms
3-methyl-1-(trihydroxyphenyl)butan-1-one synthase
VPS <1> [1, 2]
valerophenone synthase

CAS registry number
214265-40-4

2 Source Organism

<1> *Humulus lupulus* (hop [1,2]) [1, 2]

3 Reaction and Specificity

Catalyzed reaction
isovaleryl-CoA + 3 malonyl-CoA = 4 CoASH + 3 CO_2 + 3-methyl-1-(2,4,6-trihydroxyphenyl)butan-1-one (<1> Closely related to EC 2.3.1.74, chalcone synthase. The product, 3-methyl-1-(2,4,6-trihydroxyphenyl)butan-1-one, is chloroisovalerophenone. Also acts on isobutyryl-CoA as substrate to give phlorisobutyrophenone. The products are intermediates in the biosynthesis of the bitter α-acids in hops, Humulus lupulus [1,2])

Reaction type
acyl group transfer

Natural substrates and products
- **S** isobutyryl-CoA + malonyl-CoA <1> (Reversibility: ? <1> [1]) [1]
- **P** phloroisobutyrophenone + CO_2 + CoA <1> (<1> phloroglucinol derivative, intermediates in the biosynthesis of the hop bitter α- and β-acids [1,2]) [1, 2]

S isovaleryl-CoA + malonyl-CoA <1> (Reversibility: ? <1> [1]) [1]
P phloroisovalerophenone + CO_2 + CoA <1> (<1> phloroglucinol derivative, intermediates in the biosynthesis of the hop bitter α- and β-acids [1,2]) [1, 2]

Substrates and products

S isobutyryl-CoA + malonyl-CoA <1> (Reversibility: ? <1> [1]) [1]
P phloroisobutyrophenone + CO_2 + CoA
S isovaleryl-CoA + malonyl-CoA <1> (Reversibility: ? <1> [1, 2]) [1, 2]
P phloroisovalerophenone + CO_2 + CoA

K_m-Value (mM)

0.004 <1> (isovaleryl-CoA) [2]
0.01 <1> (isobutyryl-CoA) [2]
0.033 <1> (malonyl-CoA) [2]

pH-Optimum

7 <1> [2]

4 Enzyme Structure

Molecular weight

110000 <1> (<1> native PAGE of purified VPS [2]) [2]

Subunits

dimer <1> (<1>, homodimer, 2 * 45000, SDS-PAGE [2]) [2]

5 Isolation/Preparation/Mutation/Application

Source/tissue

cone <1> (<1> glandular hairs [2]) [2]
inflorescence <1> (<1> glands [2]) [2]

Purification

<1> [2]

Cloning

<1> (characterization of the VPS cDNA clone [2]) [2]

Application

nutrition <1> (<1> cones of the hop plant used in the beer-brewing process [2]) [2]

6 Stability

General stability information

<1>, low stability [2]

References

[1] Zuurbier, K.W.M.; Leser, J.; Berger, T.; Hofte, A.J.P.; Schroder, G.; Verpoorte, R.; Schroder, J.: 4-Hydroxy-2-pyrone formation by chalcone and stilbene synthase with nonphysiological substrates. Phytochemistry, **49**, 1945-1951 (1998)

[2] Paniego, N.B.; Zuurbier, K.W.M.; Fung, S.Y.; Van der Heijden, R.; Scheffer, J.J.C.; Verpoorte, R.: Phlorisovalerophenone synthase, a novel polyketide synthase from hop (Humulus lupulus L.) cones. Eur. J. Biochem., **262**, 612-616 (1999)

Glucosamine-1-phosphate N-acetyltransferase 2.3.1.157

1 Nomenclature

EC number
2.3.1.157

Systematic name
acetyl-CoA:α-D-glucosamine-1-phosphate N-acetyltransferase

Recommended name
glucosamine-1-phosphate N-acetyltransferase

Synonyms
GlcNAc-1-P uridyltransferase <1-3> [2]
GlmU enzyme <1> [1, 2]
GlmU uridyltransferase <1> [1]
N-acetylglucosamine-1-phosphate pyrophosphorylase <1> [3]
N-acetylglucosamine-1-phosphate uridyltransferase <1> [4]
UDP-GlcNAc pyrophosphorylase <1> [1]
bifunctional GlmU protein <1> [2]

CAS registry number
9023-06-7
9031-91-8

2 Source Organism

<1> *Escherichia coli* (JM83 [1,2,4]) [1-4]
<2> *Bacillus subtilis* [1, 2]
<3> *Neisseria gonorrhoeae* [2]

3 Reaction and Specificity

Catalyzed reaction
acetyl-CoA + α-D-glucosamine 1-phosphate = CoA + N-acetyl-α-D-glucosamine 1-phosphate (<1> the enzyme from several bacteria (e.g., Escherichia coli, Bacillus subtilis and Haemophilus influenzae) has been shown to be bifunctional and also to possess the activity of EC 2.7.7.23, UDP-N-acetylglucosamine diphosphorylase [1])

Reaction type
acyl group transfer

Natural substrates and products

S D-glucosamine 1-phosphate + acetyl-CoA <1-3> (Reversibility: ? <1-3> [1]) [1-4]

P N-acetyl-D-glucosamine 1-phosphate + CoA

Substrates and products

S D-glucosamine 1-phosphate + acetyl-CoA <1> (Reversibility: ? <1, 2> [1-4]) [1-4]

P N-acetyl-D-glucosamine 1-phosphate + CoA

S N-acetyl-D-glucosamine 1-phosphate + UTP <1, 2> (<1> glmU gene product, bifunctional enzyme catalyzing 2 subsequent steps in the pathway for UDP-GlcNAc synthesis [1-4]) (Reversibility: r <1> [1-4]; ? <2> [1]) [1-4]

P UDP-N-acetylglucosamine + ?

S UDP-N-acetylglucosamine + H_2O <1> (Reversibility: r <1> [1]) [1]

P N-acetyl-D-glucosamine 1-phosphate + UMP

Inhibitors

2-nitro-5-thiocyanobenzoic acid <1> [1, 2]
DTNB <1> [1, 2]
N-acetylglucosamine-1-phosphate <1> (<1> acetyltransferase activity inhibited by its reaction product [1]) [1]
N-ethylmaleimide <1> [2]
UDP-MurNAc <1> (<1> 1 mM, relative enzyme activity 2% [1]) [1]
iodoacetamide <1> [2]
p-hydroxymercuribenzoate <1> [2]

Metals, ions

Mg^{2+} <1> (<1> absolute requirement for both activities in the sharp range from 1-10 mM with an optimal value of 3 mM [1]) [1]

Turnover number (min^{-1})

742 <1> (N-acetyl-D-glucosamine 1-phosphate, <1> GmlU uridyltransferase [1]) [1]

Specific activity (U/mg)

0.004 <1> (<1> mutant JM83(pFP3-Tr331), acetyltransferase activity [4]) [4]
0.014 <1> (<1> mutant JM83(pTrcHis30), uridyltransferase activity [4]) [4]
0.11 <1> (<1> mutant JM83(pTrcHis30), acetyltransferase activity [4]) [4]
0.12 <1> (<1> mutant JM83(pFP3-Tr331), uridyltransferase activity [4]) [4]
2.2 <1> (<1> GlcN-1-P acetyltransferase activity [1]) [1]
2.85 <1> (<1> mutant JM83(pFP3-Tr331) + IPTG, uridyltransferase activity [4]) [4]
15.1 <1> (<1> GlcNAc-1-P uridyltransferase activity [1]) [1]

K_m-Value (mM)

0.07 <1> (D-glucosamine 1-phosphate, <1> mutant His6-GlmU C296A [2]) [2]
0.07 <1> (N-acetyl-D-glucosamine 1-phosphate, <1> uridyltransferase activity [1]) [1]

0.09 <1> (acetyl-CoA, <1> mutant His6-GlmU C307A [2]) [2]
0.1 <1> (UTP, <1> uridyltransferase activity [1]) [1]
0.12 <1> (D-glucosamine 1-phosphate, <1> mutant His6-GlmU C324A [2]) [2]
0.14 <1> (acetyl-CoA, <1> mutant His6-GlmU C324A [2]) [2]
0.15 <1> (D-glucosamine 1-phosphate) [1, 2]
0.2 <1> (D-glucosamine 1-phosphate, <1> mutant His6GlmU [2]) [2]
0.2 <1> (acetyl-CoA, <1> mutant His6GlmU, mutant His6-GlmU C296A, mutant His6-GlmU C385A [2]) [2]
0.25 <1> (D-glucosamine 1-phosphate, <1> mutant His6-GlmU C307A [2]) [2]
0.25 <1> (acetyl-CoA, <1> mutant His6-GlmU C385A [2]) [2]
0.6 <1> (acetyl-CoA) [1, 2]

pH-Optimum

8.2 <1> (<1> for both acetyltransferase and uridyltransferase [1]) [1]

4 Enzyme Structure

Molecular weight

49000 <1> (<1> gel filtration [1]) [1]
50100 <1> [4]

Subunits

trimer <1> (<1> 2 * 49000 or 3 * 49000, dimer or trimer of identical subunits, gel filtration [1]; 1 * 37100 + 1 * 24700, uridyltransferase and acetyltransferase in separate domains, SDS-PAGE [4]) [1, 3, 4]

5 Isolation/Preparation/Mutation/Application

Purification

<1> (GlmU gene product, bifunctional enzyme with glucosamine-1-phosphate acetyltransferase and uridyltransferase activity [1]) [1, 2, 4]

Crystallization

<1> [3, 4]

Cloning

<1> (GlmU gene cloned from genomic DNA by PCR and inserted into pET3a [3]) [3, 4]

6 Stability

General stability information

<1>, acetyl-CoA protects from inactivation [2]

<1>, acetyltransferase activity rapidly lost when the enzyme is stored in the absence of reducing thiols or acetyl coenzyme A or is treated with thiol-alkylating agents [1]

References

[1] Mengin-Lecreulx, D.; van Heijenoort, J.: Copurification of glucosamine-1-phosphate acetyltransferase and N-acetylglucosamine-1-phosphate uridyltransferase activities of Escherichia coli: characterization of the glmU gene product as a bifunctional enzyme catalyzing two subsequent steps in the pathway for UDP-N-acetylglucosamine synthesis. J. Bacteriol., **176**, 5788-5795 (1994)

[2] Pompeo, F.; Van Heijenoort, J.; Mengin-Lecreulx, D.: Probing the role of cysteine residues in glucosamine-1-phosphate acetyltransferase activity of the bifunctional GlmU protein from Escherichia coli: site-directed mutagenesis and characterization of the mutant enzymes. J. Bacteriol., **180**, 4799-4803 (1998)

[3] Olsen, L.R.; Tian, Y.; Roderick, S.L.: Purification, crystallization and preliminary X-ray data for Escherichia coli GlmU: a bifunctional acetyltransferase/uridyltransferase. Acta Crystallogr. Sect. D, **57**, 296-297 (2001)

[4] Pompeo, F.; Bourne, Y.; Van Heijenoort, J.; Fassy, F.; Mengin-Lecreulx, D.: Dissection of the bifunctional Escherichia coli N-acetylglucosamine-1-phosphate uridyltransferase enzyme into autonomously functional domains and evidence that trimerization is absolutely required for glucosamine-1-phosphate acetyltransferase activity and cell growth. J. Biol. Chem., **276**, 3833-3839 (2001)

Phospholipid:diacylglycerol acyltransferase 2.3.1.158

1 Nomenclature

EC number
2.3.1.158

Systematic name
phospholipid:1,2-diacyl-sn-glycerol O-acyltransferase

Recommended name
phospholipid:diacylglycerol acyltransferase

Synonyms
PDAT

CAS registry number
288587-47-3

2 Source Organism

<1> *Ricinus communis* [1, 2]
<2> *Crepis palaestina* [1, 2]
<3> *Arabidopsis thaliana* [1]
<4> *Euphorbia lagascae* [1]
<5> *Saccharomyces cerevisiae* [2]
<6> *Helianthus annuus* [2]

3 Reaction and Specificity

Catalyzed reaction
phospholipid + 1,2-diacylglycerol = lysophospholipid + triacylglycerol

Reaction type
acyl group transfer

Natural substrates and products
S phospholipid + 1,2-diacylglycerol <1-6> (<1,2>, the enzyme plays a major role in removing ricinoleic acid and vernolic acid from phospholipids in seeds [1]) (Reversibility: ? <1-6> [1, 2]) [1, 2]
P glycerolphosphocholine + triacylglycerol <1-6> [1, 2]

Substrates and products

S dioleoyl-phosphatidylcholine + 1,2-diacylglycerol <5> (Reversibility: ? <5> [2]) [2]

P lyso-phosphatidylcholine + triacylglycerol <5> [2]

S dioleoylphosphatidylcholine + 1,2-dioleoylglycerol <5> (Reversibility: ? <5> [2]) [2]

P 1-oleoyl-2-lyso-phosphatidylcholine + trioleoylglycerol <5> [2]

S dioleoylphosphatidylcholine + 1-ricinoleoylglycerol <5> (Reversibility: ? <5> [2]) [2]

P 1-oleoyl-2-lysophosphatidylcholine + 1-ricinoleoyl-2-oleoyl-glycerol <5> [2]

S dioleoylphosphatidylcholine + 1-vernoloylglycerol <5> (Reversibility: ? <5> [2]) [2]

P 1-oleoyl-2-lysophosphatidylcholine + 1-vernoloyl-2-oleoylglycerol <5> [2]

S oleoyl-phosphatidylcholine + 1,2-diacylglycerol <1-3, 5, 6> (Reversibility: ? <1-3, 5, 6> [1, 2]) [1, 2]

P lyso-phosphatidylcholine + triacylglycerol <1-3, 5, 6> [1, 2]

S oleoyl-phosphatidylethanolamine + 1,2-dioleoylglycerol <5> (Reversibility: ? <5> [2]) [2]

P 1-oleoyl-2-lysophosphatidylethanolamine + trioleoylglycerol <5> [2]

S ricinoleoyl-phosphatidylcholine + 1,2-diacylglycerol <1-6> (Reversibility: ? <1-6> [1, 2]) [1, 2]

P lyso-phosphatidylcholine + triacylglycerol <1-6> [1, 2]

S sn-2-acyl-phosphatidylcholine + 1,2-diacylglycerol <1-4, 6> (Reversibility: ? <1-4, 6> [1, 2]) [1, 2]

P triacylglycerol + glycerophosphocholine <1-4, 6> [1, 2]

S vernoloyl-phosphatidylcholine + 1,2-diacylglycerol <1, 2, 4-6> (Reversibility: ? <1, 2, 4-6> [1, 2]) [1, 2]

P lyso-phosphatidylcholine + triacylglycerol <1, 2, 4-6> [1, 2]

5 Isolation/Preparation/Mutation/Application

Source/tissue

seed <1, 2, 4, 6> [1, 2]
whole plant <3> (vegetative tissue) [1]

Cloning

<5> (the gene encoding the enzyme is YNR008w, overexpression of the enzyme-encoding gene increases triacylglycerol content in yeast cells [2]) [2]

References

[1] Banas, A.; Dahlqvist, A.; Stahl, U.; Lenman, M.; Stymne, S.: The involvement of phospholipid:diacylglycerol acyltransferases in triacylglycerol production. Biochem. Soc. Trans., **28**, 703-705 (2000)

[2] Dahlqvist, A.; Stahl, U.; Lenman, M.; Banas, A.; Lee, M.; Sandager, L.; Ronne, H.; Stymne, S.: Phospholipid:diacylglycerol acyltransferase: an enzyme that catalyzes the acyl-CoA-independent formation of triacylglycerol in yeast and plants. Proc. Natl. Acad. Sci. USA, **97**, 6487-6492 (2000)

Acridone synthase 2.3.1.159

1 Nomenclature

EC number
2.3.1.159

Systematic name
malonyl-CoA:N-methylanthraniloyl-CoA malonyltransferase (cyclizing)

Recommended name
acridone synthase

Synonyms
ACS <1, 2> [3-7]
synthatase, acridone (9Cl)

CAS registry number
99085-53-7

2 Source Organism

<1> *Ruta graveolens* (2 isozymes I and II [3,6,7]) [1-4, 6, 7]
<2> *Ruta graveolens* [5]

3 Reaction and Specificity

Catalyzed reaction
3 malonyl-CoA + N-methylanthraniloyl-CoA = 4 CoA + 1,3-dihydroxy-N-methylacridone + 3 CO_2 (Belongs to a superfamily of plant polyketide synthases. Has many similarities to chalcone and stilbene synthases; <1> two-step mechanism [6,7])

Reaction type
acyl group transfer
condensation <1,2> [3-7]

Natural substrates and products
S malonyl-CoA + N-methylanthraniloyl-CoA <1, 2> (<1,2> activity is increased upon treatment with crude elicitors from the cell wall of Phytophthora megasperma f. sp. glycinea, syn. Phytophthora sojae, and decreased upon light irradiation [4,5]; <1> involved in the biosynthesis of

all acridone alkaloids [3,4]) (Reversibility: ir <1, 2> [5-7]; ? <1> [1-4]) [1-7]

P CoA + 1,3-dihydroxy-N-methylacridone + CO_2 <1, 2> [3-7]

Substrates and products

S malonyl-CoA + 4-coumaroyl-CoA <1> (<1> wild-type shows low activity, while mutants show highly increased activity [7]; <1> isozyme I shows 16% activity and isozyme II shows 12% activity compared to the activity with N-methylanthraniloyl-CoA as substrate, respectively [6]) (Reversibility: ? <1> [6, 7]) [6, 7]

P CoA + naringenin chalcone <1> [6, 7]

S malonyl-CoA + N-methylanthraniloyl-CoA <1, 2> (Reversibility: ir <1, 2> [5-7]; ? <1> [1-4]) [1-7]

P CoA + 1,3-dihydroxy-N-methylacridone + CO_2 <1, 2> [1-7]

Inhibitors

N-ethylmaleimide <1> (<1> 1 mM: 100% inhibition [1]) [1]
iodoacetamide <1> (<1> 0.5 mM: 95% inhibition [1]) [1]
p-chloromercuribenzoate <1> (<1> 0.5 mM: 95% inhibition [1]) [1]
Additional information <1> (<1> light irradiation decreases activity [4]) [4]

Activating compounds

Additional information <1> (<1> induced activity by addition of crude elicitors from the cell wall of Phytophthora megasperma f. sp. glycinea, syn. Phytophthora sojae [4]) [4]

Specific activity (U/mg)

0.18 <1> (<1> crude extract [3]) [3]
6.035 <1> (<1> purified enzyme [1]) [1]
9.54 <1> (<1> purified recombinant isozyme I [3,6]) [3, 6]
14.16 <1> (<1> purified recombinant isozyme II [3,6]) [3, 6]
Additional information <1> [6]

K_m-Value (mM)

0.005 <1> (malonyl-CoA, <1> isozyme II [3]) [3]
0.0106 <1> (N-methylanthraniloyl-CoA, <1> partially purified enzyme [1]) [1]
0.013 <1> (malonyl-CoA, <1> isozyme I [3]) [3]
0.0328 <1> (malonyl-CoA, <1> partially purified enzyme [1]) [1]
0.062 <1> (N-methylanthraniloyl-CoA, <1> isozyme I [3]) [3]
0.077 <1> (N-methylanthraniloyl-CoA, <1> isozyme II [3]) [3]

pH-Optimum

7 <1> (<1> isozyme II [3]) [3]
7.5 <1> (<1> isozyme I [3]) [3]

pH-Range

6-7.5 <1> (<1> isozyme II, assay range [3]) [3]
6-9 <1> (<1> isozyme I, assay range [3]) [3]

Temperature optimum (°C)

32 <1> (<1> isozyme I [3]; <1> assay at [1]) [1, 3]
40 <1> (<1> isozyme II [3]) [3]

4 Enzyme Structure

Molecular weight

44000 <1, 2> (<2> recombinant enzyme from E. coli, gel filtration [5]; <1> isozyme I, gel filtration [3]) [3, 5]
45000 <1> (<1> about, isozyme II, gel filtration [3]) [3]
69000 <1> (<1> gel filtration [1]) [1]
81000 <1> (<1> isozyme I, analytical ultracentrifugation [3]) [3]
82000 <1> (<1> isozyme II, analytical ultracentrifugation [3]) [3]
Additional information <1> (<1> structure model of isozyme II [7]; <1> amino acid sequence determination [1]) [1, 7]

Subunits

dimer <1> (<1> 2 * 44000-45000, isozyme I and II, analytical ultracentrifugation and SDS-PAGE [3]; <1> 2 * 40000, SDS-PAGE [1]) [1, 3]
monomer <2> (<2> 1 * 42000, recombinant enzyme from E. coli, SDS-PAGE and western blot analysis [5]) [5]

5 Isolation/Preparation/Mutation/Application

Source/tissue

cell suspension culture <1, 2> (<1> strain R-20 [3,4]) [1, 3-6]
hypocotyl <1> (<1> endodermis and vascular tissue, mRNA and protein [4]) [4]
root <1> (<1> adjacent to the rhizodermis, mRNA and protein [4]) [4]

Purification

<1> [1]

Cloning

<1> (expression of wild-type and mutants in Escherichia coli [7]; expression in Escherichia coli strain BL21 (DE3)pLys S and in lambda phages via Escherichia coli host, DNA sequence comparison of 4 clones and also with the DNA sequence of Ruta graveolens chalcone synthase [6]; expression in Escherichia coli strain M15 [3]) [3, 6, 7]
<2> (expression in Escherichia coli strain BL21 (DE3)pLys S and in lambda phages via Escherichia coli host, DNA sequence determination [5]; expression level in Escherichia coli is decreased by light irradiation and increased by treatment with elicitors from the cell wall of Phytophthora megasperma f. sp. glycinea [5]) [5]

Engineering

S132T/A133S/V265F <1> (<1> site-directed mutagenesis, exchange of 3 residues, responsible for substrate activity, by the corresponding amino acids of chalcone synthase results in transformation of the enzyme to a chalcone synthase, 25fold increased activity, with only 36% remaining acridone synthase activity compared to wild-type [7]) [7]

V265F <1> (<1> site-directed mutagenesis, 75% reduction of ACS catalytic activity, but 2fold increased chalcone synthase activity [7]) [7]

6 Stability

Storage stability

<1>, -70°C, partially purified, stable for at least 4 weeks [1]

References

[1] Baumert, A.; Maier, W.; Gröger, D.; Deutzmann, R.: Purification and properties of acridone synthase from cell suspension cultures of Ruta graveolens L.. Z. Naturforsch. C, **49**, 26-32 (1994)

[2] Maier, W.; Baumert, A.; Schumann, B.; Furukawa, H.; Gröger, D.: Synthesis of 1,3-dihydroxy-N-methylacridone and its conversion to rutacridone by cell-free extracts of Ruta-graveolens cell cultures. Phytochemistry, **32**, 691-698 (1993)

[3] Lukacin. R.; Springob, K.; Urbanke, C.; Ernwein, C.; Schröder, G.; Schröder, J.; Matern, U.: Native acridone synthases I and II from Ruta graveolens L. form homodimers. FEBS Lett., **448**, 135-140 (1999)

[4] Junghanns, K.T.; Kneusel, R.E.; Groger, D.; Matern, U.: Differential regulation and distribution of acridone synthase in Ruta graveolens. Phytochemistry, **49**, 403-411 (1998)

[5] Junghanns, K.T.; Kneusel, R.E.; Baumert, A.; Maier, W.; Groeger, D.; Matern, U.: Molecular cloning and heterologous expression of acridone synthase from elicited Ruta graveolens L. cell suspension cultures. Plant Mol. Biol., **27**, 681-692 (1995)

[6] Springob, K.; Lukacin, R.; Ernwein, C.; Groning, I.; Matern, U.: Specificities of functionally expressed chalcone and acridone synthases from Ruta graveolens. Eur. J. Biochem., **267**, 6552-6559 (2000)

[7] Lukacin, R.; Schreiner, S.; Matern, U.: Transformation of acridone synthase to chalcone synthase. FEBS Lett., **508**, 413-417 (2001)

Vinorine synthase 2.3.1.160

1 Nomenclature

EC number
2.3.1.160

Systematic name
acyl-CoA:16-epivellosimine O-acetyltransferase (cyclizing)

Recommended name
vinorine synthase

CAS registry number
88844-97-7

2 Source Organism

<1> *Rauwolfia serpentina* [1, 2]

3 Reaction and Specificity

Catalyzed reaction
acetyl-CoA + 16-epivellosimine = CoA + vinorine (The reaction proceeds in two stages. The indole nitrogen of 16-epivellosimine interacts with its aldehyde group giving an hydroxy-substituted new ring. This alcohol is then acetylated. Also acts on gardneral, 11-methoxy-16-epivellosimine. Generates the ajmalan skeleton, which forms part of the route to ajmaline)

Reaction type
cyclization

Natural substrates and products
S acetyl-CoA + 16-epivellosimine <1> (Reversibility: ? <1> [1, 2]) [1, 2]
P CoA + vinorine

Substrates and products
S 11-methoxy-16-epivellosimine + acetyl-CoA <1> (Reversibility: ? <1> [1]) [1]
P CoA + 11-methoxy-vinorine
S acetyl-CoA + 16-epivellosimine <1> (Reversibility: ? <1> [1, 2]) [1, 2]
P CoA + vinorine

Specific activity (U/mg)

2.4 <1> (<1> nkat/mg [1]) [1]

K_m-Value (mM)

0.0194 <1> (16-epivellosimine) [1]
0.039 <1> (11-methoxy-16-epivellosimine) [1]
0.064 <1> (acetyl-CoA) [1]

pH-Optimum

8.5 <1> [1, 2]

Temperature optimum (°C)

35 <1> [1]

Temperature range (°C)

5-10 <1> [1]

4 Enzyme Structure

Molecular weight

31000 <1> (<1> gel filtration [1]) [1]

5 Isolation/Preparation/Mutation/Application

Purification

<1> [1, 2]

References

[1] Pfitzner, A.; Polz, L.; Stöckligt, J.: Properties of vinorine synthase the Rauwolfia enzyme involved in the formation of the ajmaline skeleton. Z. Naturforsch. C, **41**, 103-114 (1986)
[2] Pfitzner, A.; Stoeckigt, J.: Biogenetic link between sarpagine and ajmaline type alkaloids. Tetrahedron Lett., **24**, 5197-5200 (1983)

Lovastatin nonaketide synthase 2.3.1.161

1 Nomenclature

EC number
2.3.1.161

Systematic name
acyl-CoA:malonyl-CoA C-acyltransferase (decarboxylating, oxoacyl- and enoyl-reducing, thioester-hydrolysing)

Recommended name
lovastatin nonaketide synthase

Synonyms
LNKS
synthase, lovastatin nonaketide

CAS registry number
235426-97-8

2 Source Organism

<1> *Aspergillus terreus* (mutant BX102 and other mutants [2]) [1-4]

3 Reaction and Specificity

Catalyzed reaction
acetyl-CoA + 8 malonyl-CoA + 11 NADPH + 10 H^+ + S-adenosyl-L-methionine = dihydrononacolin L + 9 CoA + 8 CO_2 + 11 $NADP^+$ + S-adenosyl-L-homocysteine + 6 H_2O (<1> reaction mechanism [4])

Reaction type
cyclization
decarboxylation
dehydration
reduction

Natural substrates and products
S acetyl-CoA + malonyl-CoA + NADPH + H_2O + S-adenosyl-L-methionine <1> (<1> reaction mechanism [1]; <1> reaction mechanism, enzyme interacts with LovC to catalyze 35 separate reactions in the biosynthesis of

dihydrononacolin [3]; <1> reaction mechanism, interaction with LovC [4]) (Reversibility: ? <1> [1, 3, 4]) [1, 3, 4]

P dihydrononacolin L + CoA + CO_2 + $NADP^+$ + S-adenosyl-L-homocysteine + H_2O

Substrates and products

S acetyl-CoA + malonyl-CoA + NADPH + H_2O + S-adenosyl-L-methionine <1> (<1> enzyme also displays Diels-Alderase activity in vitro [1,3]; <1> enzyme contains six active sites: ketosynthase, acyltransferase, dehydratase, enoyl reductase, ketoreductase and acyl carrier protein [3,4]) (Reversibility: ? <1> [1, 3, 4]) [1, 3, 4]

P dihydrononacolin L + CoA + CO_2 + $NADP^+$ + S-adenosyl-L-homocysteine + H_2O <1> [1, 4]

Specific activity (U/mg)

0.0109 <1> [1]

K_m-Value (mM)

0.5 <1> (malonyl-CoA) [1]

4 Enzyme Structure

Molecular weight

335000 <1> (<1> SDS-PAGE [1]) [1]
335000 <1> (<1> calculated from amino acid sequence [3]) [3]

5 Isolation/Preparation/Mutation/Application

Purification

<1> (homogeneitiy [1]; mutant BX102 [2]) [1, 2]

Cloning

<1> (lovB and lovC protein [1]; wild type and different lov mutants [3]) [1, 3]

Application

medicine <1> (<1> production of cholesterol lowering drugs [1]; <1> antihypercholesterolemic activity [4]) [1, 4]

6 Stability

General stability information

<1>, inactivated by heating [1]

References

[1] Auclair, K.; Sutherland, A.; Kennedy, J.; Witter, D.J.; Van den Heever, J.P.; Hutchinson, C.R.; Vederas, J.C.: Lovastatin nonaketide synthase catalyzes an intramolecular Diels-Alder reaction of a substrate analogue. J. Am. Chem. Soc., **122**, 11519-11520 (2000)
[2] Hendrickson, L.; Davis, C.R.; Roach, C.; Nguyen, D.K.; Aldrich, T.; McAda, P.C.; Reeves, C.D.: Lovastatin biosynthesis in Aspergillus terreus: characterization of blocked mutants, enzyme activities and a multifunctional polyketide synthase gene. Chem. Biol., **6**, 429-439 (1999)
[3] Hutchinson, C.R.; Kennedy, J.; Park, C.; Kendrew, S.; Auclair, K.; Vederas, J.: Aspects of the biosynthesis of non-aromatic fungal polyketides by iterative polyketide synthases. Antonie Leeuwenhoek, **78**, 287-295 (2000)
[4] Kennedy, J.; Auclair, K.; Kendrew, S.G.; Park, C.; Vederas, J.C.; Hutchinson, C.R.: Modulation of polyketide synthase activity by accessory proteins during lovastatin biosynthesis. Science, **284**, 1368-1372 (1999)

Taxadien-5α-ol O-acetyltransferase 2.3.1.162

1 Nomenclature

EC number
2.3.1.162

Systematic name
acetyl-CoA:taxa-4(20),11-dien-5α-ol O-acetyltransferase

Recommended name
taxadien-5α-ol O-acetyltransferase

Synonyms
acetyl coenzyme A:taxa-4(20),11(12)-dien-5α-ol O-acetyl transferase

CAS registry number
229032-29-5

2 Source Organism

<1> *Taxus cuspidate* (methyl jasmonate-induced cells [1]) [1, 2]
<2> *Taxus canadensis* (methyl jasmonate-induced cells [1]) [1, 2]
<3> *Taxus sp.* [2, 3]

3 Reaction and Specificity

Catalyzed reaction
acetyl-CoA + taxa-4(20),11-dien-5α-ol = CoA + taxa-4(20),11-dien-5α-yl acetate (<3> summary of taxol biosynthesis [3])

Reaction type
acyl group transfer

Natural substrates and products
S Additional information <1-3> (<1-3> catalyzes the first acylation step of taxol biosynthesis [1-3]) (Reversibility: ? <1-3> [1-3]) [1-3]
P ?

Substrates and products
S acetyl-CoA + farnesol <1> (Reversibility: ? <1> [1]) [1]
P CoA + ?
S acetyl-CoA + generylgeraniol <1> (Reversibility: ? <1> [1]) [1]
P CoA + ?

S acetyl-CoA + phytol <1> (Reversibility: ? <1> [1]) [1]
P CoA + ?
S acetyl-CoA + taxa-4(20),11(12)-dien-5α-ol <1-3> (Reversibility: ? <1-3> [1-3]) [1-3]
P CoA + taxa-4(20),11-dien-5α-yl acetate <1-3> [1-3]
S acetyl-CoA + taxadienol <1> (Reversibility: ? <1> [1]) [1]
P CoA + ?

Inhibitors

CoA <1, 2> (<1,2> weak [1]) [1]
N-ethylmaleimide <1, 2> (<1,2> weak [1]) [1]
p-hydroxymercuribenzoate <1, 2> (<1,2> weak [1]) [1]

Metals, ions

Additional information <1, 2> (<1,2> monovalent and divalent cations show no effect on the enzyme [1]) [1]

Specific activity (U/mg)

Additional information <2> [2]

K_m-Value (mM)

0.0042 <1, 2> (taxa-4(20),11-dien-5α-ol) [1]
0.0055 <1, 2> (acetyl-CoA) [1]

pH-Optimum

9 <1, 2> (<1,2> pI: 4.7 [1]) [1]

pH-Range

7.5-10 <1, 2> (<1,2> 50% of maximal activity at pH 7.50 and pH 10.0, enzyme inactive at pH 11.0 [1]) [1]

4 Enzyme Structure

Molecular weight

49080 <2> (<2> recombinant enzyme, calculated from amino acid composition [2]) [2]
50000 <1, 2> (<1,2> gel chromatography [1]) [1]

5 Isolation/Preparation/Mutation/Application

Purification

<2> (800fold purification [2]) [2]
<1, 2> (partially [1]) [1]

Cloning

<1, 2> (expression in Escherichia coli [2]) [2]

Application

medicine <1-3> (<1-3> understanding of biosynthetic pathway of taxol, target: possibly biological production of taxol, taxol or paclitaxel is now well established as a potent chemotherapeutic drug [1,2]) [1, 2]

References

[1] Walker, K.; Ketchum, R.E.; Hezari, M.; Gatfield, D.; Goleniowski, M.; Barthol, A.; Croteau, R.: Partial purification and characterization of acetyl coenzyme A:taxa-4(20),11(12)-dien-5α-ol O-acetyl transferase that catalyzes the first acylation step of taxol biosynthesis. Arch. Biochem. Biophys., **364**, 273-279 (1999)

[2] Walker, K.; Schoendorf, A.; Croteau, R.: Molecular cloning of a taxa-4(20),11(12)-dien-5a-ol-O-acetyl transferase cDNA from taxus and functional expression in Escherichia coli. Arch. Biochem. Biophys., **374**, 371-380 (2000)

[3] Walker, K.; Croteau, R.: Taxol biosynthetic genes. Phytochemistry, **58**, 1-7 (2001)

10-Hydroxytaxane O-acetyltransferase 2.3.1.163

1 Nomenclature

EC number
2.3.1.163

Systematic name
acetyl-CoA:taxan-10β-ol O-acetyltransferase

Recommended name
10-hydroxytaxane O-acetyltransferase

Synonyms
acetyl coenzyme A:10-hydroxytaxane O-acetyltransferase
acetyl-coenzyme A:10-hydroxytaxan-O-acetyltransferase

CAS registry number
227465-96-5

2 Source Organism

<1> *Taxus cuspidata* [1]

3 Reaction and Specificity

Catalyzed reaction
acetyl-CoA + 10-desacetyltaxuyunnanin C = CoA + taxuyunnanin C (Acts on a number of related taxane diterpenoids with a free 10β-hydroxy group. May be identical to EC 2.3.1.167, 10-deacetylbaccatin III 10-O-acetyltransferase)

Reaction type
acyl group transfer

Natural substrates and products
- **S** acetyl-CoA + 10-desacetyltaxuyunnanin C <1> (Reversibility: ? <1> [1]) [1]
- **P** CoA + taxuyunnanin C <1> [1]

Substrates and products
- **S** 2,5,10,14-desacetyltaxuyunnanin C <1> (Reversibility: ? <1> [1]) [1]
- **P** CoA + 2,5,10-desacetyltaxuyunnanin C <1> [1]
- **S** acetyl-CoA + 10-desacetyltaxuyunnanin C <1> (Reversibility: ? <1> [1]) [1]
- **P** CoA + taxuyunnanin C <1> [1]

K_m-Value (mM)
0.061 <1> (acetyl-CoA) [1]

4 Enzyme Structure

Molecular weight
71000 <1> (<1> gel filtration [1]) [1]

Subunits
monomer <1> (<1> 1 * 70900, SDS-PAGE [1]) [1]

5 Isolation/Preparation/Mutation/Application

Purification
<1> [1]

Application
pharmacology <1> (<1> use in synthesis of taxol for anticnacer treatment [1]) [1]

References

[1] Menhard, B.; Zenk, M.H.: Purification and characterization of acetyl coenzyme A: 10-hydroxytaxane O-acetyltransferase from cell suspension cultures of Taxus chinensis. Phytochemistry, **50**, 763-774 (1999)

Isopenicillin-N N-acyltransferase 2.3.1.164

1 Nomenclature

EC number
2.3.1.164

Systematic name
acyl-CoA:isopenicillin N N-acyltransferase

Recommended name
isopenicillin-N N-acyltransferase

Synonyms
6-aminopenicillanate acyltransferase
6-aminopenicillanic acid acyltransferase
6-aminopenicillinanic acid phenylacetyltransferase
6-aminopenicillinanic acylase
acyl coenzyme A:6-aminopenicillanic acid acyltransferase
acyl-CoA:6-APA acyltransferase
acyl-CoA:6-aminopenicillanate acyltransferase
acyl-CoA:isopenicillin N acyltransferase
acyl-coenzyme A:isopenicillin N acyltransferase
acyl-coenzyme A:isopenicillin N-acyltransferase
acyltransferase, 6-aminopenicillanate
isopenicillin N acyltransferase
isopenicillin N:acyl-CoA:acyltransferase

CAS registry number
54576-90-8

2 Source Organism

<1> *Penicillium chrysogenum* (penDE genes [1,3,5]; Wis 54-1255 [2]; SC6140, ATCC 2044, source of native enzyme [3]; high yielding strain [4]; C103A mutant enzyme which has lost its ability to be processed to the mature heterodimeric protein and thus represents the proenzyme form [6]) [1, 2, 3, 4, 5, 6]

3 Reaction and Specificity

Catalyzed reaction
phenylacetyl-CoA + isopenicillin N + H_2O = CoA + penicillin G + L-2-aminohexanedioate

Reaction type
acylation
hydrolysis

Natural substrates and products
S phenylacetyl-CoA + isopenicillin N + H_2O <1> (Reversibility: ? <1> [3]) [3]
P CoA + benzylpenicillin + L-2-aminohexanedioate <1> [3]

Substrates and products
S phenylacetyl-CoA + 6-aminopenicillanic acid + H_2O <1> (<1> acyl-CoA:6-aminopenicillanate acyltransferase activity [3]) (Reversibility: ? <1> [2, 3]) [2, 3]
P CoA + benzylpenicillin <1> [2, 3]
S phenylacetyl-CoA + isopenicillin N + H_2O <1> (<1> acyl-coenzyme A:isopenicillin N acyltransferase activity, S309 is involved in substrate acylation [3]) (Reversibility: ? <1> [3]) [3]
P CoA + benzylpenicillin + L-2-aminohexanedioate <1> [3]

Specific activity (U/mg)
Additional information <1> [4]

4 Enzyme Structure

Molecular weight
40000 <1> (<1> unprocessed precursor polypeptide, proenzyme, SDS-PAGE [1,2]; recominant enzyme SDS-PAGE and Western blot analysis [5]) [1, 2, 5]

Subunits
heterodimer <1> (<1> α,1 * 11000 + β,1 * 29000, derived from cleavage of the 40 kDa polypeptide, SDS-PAGE [1]; <1> β,1 * 29000, α subunit not detected [2]; <1> α,1 * 11000 + β,1 * 29000, wild-type and recombinant enzyme, SDS-PAGE [3]) [1, 2, 3]

5 Isolation/Preparation/Mutation/Application

Purification
<1> [6]

Crystallization

<1> (hanging-drop vapour-diffusion method, X-ray diffraction, monoclinic space group C_2, unit-cell parameteres: a = 231.36 A, b = 68.27 A, c = 151.31 A, β = 129.6° at 100 K [6]) [6]

Cloning

<1> (expressed in Escherichia coli JM109 [1,3,5,6]) [1, 3, 5, 6]

Engineering

C103S or A or W <1> (<1> C103 is specifically required for recombinant proenzyme cleavage [5]) [5]
S227A <1> (<1> produces uncleaved proenzyme lacking enzyme activity [3]) [3]
S309A <1> (<1> recombinant proenzyme containing S309A is cleaved, however enzyme activity is not observed [3]) [3]
T105V <1> (<1> results in a significant inhibition of proenzyme cleavage [5]) [5]

References

[1] Tobin, M.B.; Baldwin, J.E.; Cole, S.C.J.; Miller, J.R.; Skatrud, P.L.; Sutherland, J.D.: The requirement for subunit interaction in the production of Penicillium chrysogenum acyl-coenzyme A:isopenicillin N acyltransferase in Escherichia coli. Gene, **132**, 199-206 (1993)

[2] Fernandez, F.J.; Gutierrez, S.; Velasco, J.; Montenegro, E.; Marcos, A.T.; Martin, J.F.: Molecular characterization of three loss-of-function mutations in the isopenicillin N-acyltransferase gene (penDE) of Penicillium chrysogenum. J. Bacteriol., **176**, 4941-4948 (1994)

[3] Tobin, M.B.; Cole, S.C.J.; Kovacevic, S.; Miller, J.R.; Baldwin, J.E.; Sutherland, J.D.: Acyl-coenzyme A:isopenicillin N acyltransferase from Penicillium chrysogenum: effect of amino acid substitutions at Ser227, Ser230 and Ser309 on proenzyme cleavage and activity. FEMS Microbiol. Lett., **121**, 39-46 (1994)

[4] Nielsen, J.; Jorgensen, H.S.: Metabolic control analysis of the penicillin biosynthetic pathway in a high-yielding strain of Penicillium chrysogenum. Biotechnol. Prog., **11**, 299-305 (1995)

[5] Tobin, M.B.; Cole, S.C.J.; Miller, J.R.; Baldwin, J.E.; Sutherland, J.D.: Amino-acid substitutions in the cleavage site of acyl-coenzyme A:isopenicillin N acyltransferase from Penicillium chrysogenum: effect on proenzyme cleavage and activity. Gene, **162**, 29-35 (1995)

[6] Hensgens, C.M.H.; Kroezinga, E.A.; van Montfort, B.A.; van der Laan, J.M.; Sutherland, J.D.; Dijkstra, B.W.: Purification, crystallization and preliminary X-ray diffraction of Cys103Ala acyl coenzyme A:isopenicillin N acyltransferase from Penicillium chrysogenum. Acta Crystallogr. Sect. D, **58**, 716-718 (2002)

6-Methylsalicylic acid synthase 2.3.1.165

1 Nomenclature

EC number
2.3.1.165

Systematic name
acyl-CoA:malonyl-CoA C-acyltransferase (decarboxylating, oxoacyl-reducing, thioester-hydrolysing and cyclizing)

Recommended name
6-methylsalicylic acid synthase

CAS registry number
9045-37-8

2 Source Organism

<1> *Penicillium patulum* (NRRL 2159A [1,2,3]; also known as Penicillium urticae [2,3]) [1-8]

3 Reaction and Specificity

Catalyzed reaction
acetyl-CoA + 3 malonyl-CoA + NADPH + H^+ = 6-methylsalicylate + 4 CoA + 3 CO_2 + $NADP^+$ (<1> a multienzyme complex with a 4'-phosphopantetheine prosthetic group on the acyl carrier protein, it has a similar sequence to vertebrate type I fatty acid synthase, acetoacetyl-CoA can also act as a starter molecule [3]; <1> steric course of the reaction, mechanism [4])

Reaction type
Claisen condensation
reduction

Natural substrates and products
S acetyl-CoA + malonyl-CoA + NADPH <1> (<1> first enzyme of patulin biosynthetic pathway [2]) (Reversibility: ? <1> [1]) [1, 2]
P 6-methylsalicylate + CoA + CO_2 + $NADP^+$ <1> [1, 2]

Substrates and products
S acetoacetyl-CoA + malonyl-CoA <1> (<1> 5% of the rate of 6-methylsalicylic acid synthesis [7]) (Reversibility: ? <1> [7]) [7]

P 6-methyl triacetic acid lactone + CoA <1> [7]

S acetoacetyl-CoA + malonyl-CoA + NADPH <1> (Reversibility: ? <1> [3]) [3]

P ?

S acetyl-CoA + N-acetylcysteamine + NADPH <1> (Reversibility: ? <1> [8]) [8]

P ?

S acetyl-CoA + malonyl-CoA <1> (<1> exclusive product in the absence of NADPH, 5:1 ratio of 6-methylsalicylic acid to triacetic acid lactone at 0.0005 mM NADPH, less then 1% triacetic acid lactone at 0.012 mM NADPH [3]) (Reversibility: ? <1> [3]) [3, 7]

P triacetic lactone <1> [3, 7]

S acetyl-CoA + malonyl-CoA + NADPH <1> (Reversibility: ? <1> [1]) [1-8]

P 6-methylsalicylate + CoA + CO_2 + $NADP^+$ <1> [1-8]

S acetyl-CoA + pantetheine <1> (Reversibility: ? <1> [1]) [1]

P acetyl pantetheine + CoA <1> [1]

S butyryl-CoA + malonyl-CoA <1> (<1> 5% of the rate of 6-methylsalicylic acid synthesis [7]) (Reversibility: ? <1> [7]) [7]

P 6-propyl triacetic acid lactone + CoA <1> [7]

S crotonyl-CoA + malonyl-CoA <1> (<1> 5% of the rate of 6-methylsalicylic acid synthesis [7]) (Reversibility: ? <1> [7]) [7]

P 6-(2-propene) triacetic acid lactone + CoA <1> [7]

S heptanoyl-CoA + malonyl-CoA <1> (<1> 5% of the rate of 6-methylsalicylic acid synthesis [7]) (Reversibility: ? <1> [7]) [7]

P 6-hexyl triacetic acid lactone + CoA <1> [7]

S hexanoyl-CoA + malonyl-CoA <1> (<1> 5% of the rate of 6-methylsalicylic acid synthesis [7]) (Reversibility: ? <1> [7]) [7]

P 6-pentyl triacetic acid lactone + CoA <1> [7]

S propionyl-CoA + malonyl-CoA <1> (<1> 5% of the rate of 6-methylsalicylic acid synthesis [7]) (Reversibility: ? <1> [7]) [7]

P 6-ethyl triacetic acid lactone + CoA <1> [7]

S propionyl-CoA + malonyl-CoA + NADPH <1> (Reversibility: ? <1> [1]) [1]

P 6-ethylsalicylate + CoA + CO_2 + $NADP^+$ <1> [1]

S propionyl-CoA + pantetheine <1> (<1> 13 times slower rate compared to acetyl-CoA [1]) (Reversibility: ? <1> [1]) [1]

P propionyl pantetheine + CoA <1> [1]

S valeryl-CoA + malonyl-CoA <1> (<1> 5% of the rate of 6-methylsalicylic acid synthesis [7]) (Reversibility: ? <1> [7]) [7]

P 6-butyl triacetic acid lactone + CoA <1> [7]

Inhibitors

(2S,3R)-2,3-epoxy-4-oxo-7,10-dodecadienoylamide <1> (<1> mycotoxin produced by Cephalosporium caerulens, trivial name cerulenin, 0.2 mM, complete inactivation after approx. 30 min, second-order rate constant for the reaction with 6-methylsalicylic acid synthase: 13.8 M/s, acetyl-CoA protects,

site of modification: Cys204 [5]; <1> 0.0014 mM, complete inactivation after 40 min, acetyl-CoA protects [6]) [5, 6]
1,3-dibromo-propane-2-one <1> (<1> 0.02 mM, complete inactivation after 40 min, acetyl-CoA protects [3]; <1> very rapid inactivation, inactivation half-life: 7 s [6]) [3, 6]
5,5'-dithiobis(2-nitrobenzoic acid) <1> (<1> 0.01 mM, rapid inactivation, inactivation half-life: approx. 15 s, both acetyl-CoA and malonyl-CoA protect [6]) [6]
NAD^+ <1> (<1> 0.3 mM, 21% inhibition in crude extracts [2]) [2]
NADH <1> (<1> 0.3 mM, 26% inhibition in crude extracts [2]) [2]
$NADP^+$ <1> (<1> 0.3 mM, 21% inhibition in crude extracts [2]) [2]
iodoacetamide <1> (<1> 0.5 mM, complete inactivation of synthetase activity after 15 min at 0°C, enzyme is then able to decarboxlate malonyl-CoA [1]; <1> 100 mM, complete inactivation after 45 min, site of modification: Cys204 [5]; <1> 100fold molar excess, complete inactivation after 30 min [6]) [1, 5, 6]
nicotinamide hypoxanthine dinucleotide phosphate <1> (<1> reduced form, 0.3 mM, 32% inhibition in crude extracts, oxidized form, 0.3 mM, 24% inhibition in crude extracts [2]) [2]

Cofactors/prosthetic groups

NADPH <1> [1, 2]

Specific activity (U/mg)

0.245 <1> [3]

K_m-Value (mM)

0.007 <1> (malonyl-CoA) [3]
0.01 <1> (acetyl-CoA) [3]
0.012 <1> (NADPH) [3]
0.065 <1> (acetoacetyl-coA) [3]

pH-Optimum

7.6 <1> [3]

pH-Range

6.5-9 <1> [3]

4 Enzyme Structure

Molecular weight

750000 <1> (<1> gel filtration [3]) [3]

Subunits

tetramer <1> (<1> α_4, 4 * 180000, SDS-PAGE [3]) [3]

5 Isolation/Preparation/Mutation/Application

Purification

<1> (ammonium sulfate, poly(ethylene glycol) 6000, DEAE-Sepharose, hydroxyapatite, Mono Q [3]) [3]

Cloning

<1> (expression in Escherichia coli and Saccharomyces cerevisiae can be used for 4-hydroxy-6-methyl-2-pyrone production [8]) [8]

Application

biotechnology <1> (<1> production of unnatural polyketides e.g. 4-hydroxy-6-methyl-2-pyrone in E. coli and yeast after heterologous expression of the enzyme [8]) [8]

6 Stability

pH-Stability

7.6 <1> (<1> more than 60% activity remains after 120 min at 30°C in the presence of PMSF, NADPH, acetyl-CoA and dithiothreitol [2]) [2]

Temperature stability

30 <1> (<1> half-life in crude extracts in the presence of 1 mM PMSF: 53 min [2]) [2]
30 <1> (<1> half-life in crude extracts in the presence of 1.5 mM NAD^+: 27 min [2]) [2]
30 <1> (<1> half-life in crude extracts in the presence of 1.5 mM NADH: 27 min [2]) [2]
30 <1> (<1> half-life in crude extracts in the presence of 1.5 mM $NADP^+$: 36 min [2]) [2]
30 <1> (<1> half-life in crude extracts in the presence of 1.5 mM NADPH, 1 mM PMSF, 0.25 mM acetyl-CoA, 0.75 mM malonyl-CoA and 10 mM dithiothreitol: 188 min [2]) [2]
30 <1> (<1> half-life in crude extracts in the presence of 1.5 mM NADPH: 38 min [2]) [2]
30 <1> (<1> half-life in crude extracts: 16 min [2]) [2]

General stability information

<1>, PMSF and benzamidine in disruption buffer extend half-life from 20 min to 6 h, 15% glycerol and proteinase inhibitors further improve stability [3]
<1>, activity is lost upon storage [1]
<1>, very sensitive to proteolysis, significant stabilization by a combination of NADPH, acetyl-CoA, malonyl-CoA, PMSF, and dithitothreitol [2]

Storage stability

<1>, -70°C, ammonium sulfate precipitate, loss of 5% activity per month [3]

References

[1] Dimroth, P.; Ringelmann, E.; Lynen, F.: 6-Methylsalicylic acid synthetase from Penicillium patulum. Some catalytic properties of the enzyme and its relation to fatty acid synthetase. Eur. J. Biochem., **68**, 591-596 (1976)

[2] Lam, K.S.; Neway, J.O.; Gaucher, G.M.: In vitro stabilization of 6-methylsalicylic acid synthetase from Penicillium urticae. Can. J. Microbiol., **34**, 30-37 (1988)

[3] Spencer, J.B.; Jordan, P.M.: Purification and properties of 6-methylsalicylic acid synthase from Penicillium patulum. Biochem. J., **288**, 839-846 (1992)

[4] Spencer, J.B.; Jordan, P.M.: Investigation of the mechanism and steric course of the reaction catalyzed by 6-methylsalicylic acid synthase from Penicillium patulum using (R)-[1-^{13}C;2-^{2}H] and (S)-[1-^{13}C; 2-^{2}H]malonates. Biochemistry, **31**, 9107-9116 (1992)

[5] Child, C.J.; Shoolingin-Jordan, P.M.: Inactivation of the polyketide synthase, 6-methylsalicylic acid synthase, by the specific modification of Cys-204 of the β-ketoacyl synthase by the fungal mycotoxin cerulenin. Biochem. J., **330**, 933-937 (1998)

[6] Child, C.J.; Spencer, J.B.; Bhogal, P.; Shoolingin-Jordan, P.M.: Structural similarities between 6-methylsalicylic acid synthase from Penicillium patulum and vertebrate type I fatty acid synthase: evidence from thiol modification studies. Biochemistry, **35**, 12267-12274 (1996)

[7] Campuzano, I.D.G.; Shoolingin-Jordan, P.M.: Incubation of 6-methylsalicylic acid synthase with alternative starter units in the absence of NADPH and the identification of the resulting triaceticacid lactones. Biochem. Soc. Trans., **26**, S284 (1998)

[8] Richardson, M.T.; Pohl, N.L.; Kealey, J.T.; Khosla, C.: Tolerance and specificity of recombinant 6-methylsalicylic acid synthase. Metab. Eng., **1**, 180-187 (1999)

2α-Hydroxytaxane 2-O-benzoyltransferase 2.3.1.166

1 Nomenclature

EC number
2.3.1.166

Systematic name
benzoyl-CoA:taxan-2α-ol O-benzoyltransferase

Recommended name
2α-hydroxytaxane 2-O-benzoyltransferase

Synonyms
benzoyl-CoA:taxane 2α-O-benzoyltransferase
benzoyltransferase, taxane 2α-O-
taxane 2α-O-benzoyltransferase

CAS registry number
329318-50-5

2 Source Organism

<1> *Taxus cuspidata* [1]

3 Reaction and Specificity

Catalyzed reaction
benzoyl-CoA + 10-deacetyl-2-debenzoylbaccatin III = CoA + 10-deacetylbaccatin III

Reaction type
acylation

Natural substrates and products
S benzoyl-CoA + 10-deacetyl-2-debenzoylbaccatin III <1> (Reversibility: ? <1> [1]) [1]
P CoA + 10-deacetylbaccatin III <1> [1]

Substrates and products
S benzoyl-CoA + 10-deacetyl-2-debenzoylbaccatin III <1> (Reversibility: ? <1> [1]) [1]
P CoA + 10-deacetylbaccatin III <1> [1]

S benzoyl-CoA + 2-debenzoyl-7,13-diacetylbaccatin III <1> (<1> 2-debenzoyl-7,13-diacetylbaccatin III is a semisynthetic substrate [1]) (Reversibility: ? <1> [1]) [1]

P CoA + 7,13-diacetylbaccatin III <1> [1]

S Additional information <1> (<1> enzyme does not benzoylate the 1β-, 7β-, 10β- or 13α-hydroxyl groups of 10-deacetylbaccatin III, nor does it benzoylate the 2α- or 5α-hydroxyl groups of taxa-4(20),11(12)-dien-2α,5α-diol, acetyl-CoA no cosubstrate [1]) [1]

P ?

K_m-Value (mM)

0.3 <1> (benzoyl-CoA) [1]
0.64 <1> (2-debenzoyl-7,13-diacetylbaccatin III) [1]

pH-Optimum

8 <1> [1]

pH-Range

6.5-9.9 <1> [1]

Temperature optimum (°C)

31 <1> (<1> enzyme assay [1]) [1]

4 Enzyme Structure

Molecular weight

50090 <1> (<1> calculated from sequence of cDNA [1]) [1]

Subunits

? <1> (<1> ca. x * 50000, SDS-PAGE [1]) [1]

5 Isolation/Preparation/Mutation/Application

Purification

<1> (partially [1]) [1]

Cloning

<1> (expressed in Escherichia coli JM109 [1]) [1]

References

[1] Walker, K.; Croteau, R.: Taxol biosynthesis: molecular cloning of a benzoyl-CoA:taxane 2α-O-benzoyltransferase cDNA from Taxus and functional expression in Escherichia coli. Proc. Natl. Acad. Sci. USA, **97**, 13591-13596 (2000)

10-Deacetylbaccatin III 10-O-acetyltransferase 2.3.1.167

1 Nomenclature

EC number
2.3.1.167

Systematic name
acetyl-CoA:taxan-10β-ol O-acetyltransferase

Recommended name
10-deacetylbaccatin III 10-O-acetyltransferase

Synonyms
10-deacetylbaccatin III-10β-O-acetyltransferase
10-hydroxytaxane 10-O-acetyltransferase
10-hydroxytaxane O-acetyltransferase
acetyl CoA:10-deacetylbaccatin-III 10-O-acetyltransferase
acetyl coenzyme A:10-hydroxytaxane O-acetyltransferase

CAS registry number
220946-63-4

2 Source Organism

<1> *Taxus cuspidata* [1-3]

3 Reaction and Specificity

Catalyzed reaction
acetyl-CoA + 10-deacetylbaccatin III = CoA + baccatin III (The enzyme will not acylate the hydroxy group at 1β, 7β or 13α of 10-deacetyl baccatin III, or at 5α of taxa-4(20),11-dien-5α-ol. May be identical to EC 2.3.1.163, 10-hydroxytaxane O-acetyltransferas)

Reaction type
acyl group transfer

Natural substrates and products
S acetyl-CoA + 10-deacetylbaccatin III <1> (Reversibility: ? <1> [1-3]) [1-3]
P CoA + baccatin III <1> [1, 3]

Substrates and products

S acetyl-CoA + 10-deacetylbaccatin III <1> (Reversibility: ? <1> [1-3]) [1-3]

P CoA + baccatin III <1> [1, 3]

K_m-Value (mM)

0.008 <1> (acetyl-CoA) [1, 3]
0.01 <1> (10-deacetylbaccatin III) [1, 3]

pH-Optimum

7.4 <1> [1, 2]

pH-Range

6.4-7.8 <1> [1]

4 Enzyme Structure

Molecular weight

49050 <1> (<1> calculation from nucleotide sequence [1]) [1]
50000 <1> (<1> gel filtration [1]) [1]

5 Isolation/Preparation/Mutation/Application

Source/tissue

leaf <1> [1-3]

Purification

<1> [1]

Cloning

<1> (expression in Escherichia coli [1]) [1]

Application

pharmacology <1> (<1> use in synthesis of taxol for anticnacer treatment [1]) [1]

References

[1] Walker. K.; Croteau, R.: Molecular cloning of a 10-deacetylbaccatin III-10-O-acetyl transferase cDNA from Taxus and functional expression in Escherichia coli. Proc. Natl. Acad. Sci. USA, **97**, 583-587 (2000)

[2] Walker, K.; Croteau, R.: Taxol biosynthetic genes. Phytochemistry, **58**, 1-7 (2001)

[3] Pennington, J.J.; Fett-Neto, A.G.; Nicholson, S.A.; Kingston, D.G.I.; Dicosmo, F.: Acetyl CoA: 10-deacetylbaccatin-III-10-O-acetyltransferase activity in leaves and cell suspension cultures of Taxus cuspidata. Phytochemistry, **49**, 2261-2266 (1998)

Dihydrolipoyllysine-residue (2-methylpropanoyl)transferase 2.3.1.168

1 Nomenclature

EC number
2.3.1.168

Systematic name
enzyme-dihydrolipoyllysine:2-methylpropanoyl-CoA S-(2-methylpropanoyl)-transferase

Recommended name
dihydrolipoyllysine-residue (2-methylpropanoyl)transferase

Synonyms
dihydrolipoyl transacetylase
Additional information <1-16> (<6-15> enzyme is the E_2 component of the multienzyme complex pyruvate dehydrogenase [8]; <1-5,16> enzyme is the E2 component of the multienzyme complex branched-chain α-keto acid dehydrogenase, i.e. branched-chain 2-oxo acid dehydrogenase [1-11]) [1-11]

2 Source Organism

<-2> no activity in *Desulfovibrio africanus* [8]
<-1> no activity in *archaebacteria* [8]
<1> *Bos taurus* [1-4, 7, 9, 11]
<2> *Homo sapiens* (gene E2 [5]) [4, 5]
<3> *Homo sapiens* (pseudogene E2 [5]) [5]
<4> *Homo sapiens* (gene E2, complete sequence [6]) [6]
<5> *Homo sapiens* [7, 8]
<6> *Escherichia coli* [8]
<7> *Azotobacter vinelandii* [8]
<8> *Haemophilus influenzae* [8]
<9> *Neisseria meningitidis* [8]
<10> *Alcaligenes eutrophus* [8]
<11> *Thiobacillus ferrooxidans* [8]
<12> *Zymomonas mobilis* [8]
<13> *Streptococcus faecalis* (i.e. Enterococcus faecalis [8]) [8]
<14> *Bacillus stearothermophilus* [8]
<15> *Saccharomyces cerevisiae* [8]
<16> *Onchorhynchus mykiss* (rainbow trout [10]) [10]
<17> *Gallus gallus* [10]

3 Reaction and Specificity

Catalyzed reaction

2-methylpropanoyl-CoA + enzyme N^6-(dihydrolipoyl)lysine = CoA + enzyme N^6-(S-[2-methylpropanoyl]dihydrolipoyl)lysine (<1> random bi bi mechanism [1]; <1> His391 is involved in the catalytic reaction [4]; <1> the active site is located in the inner core domain [9])

Reaction type

alkyl group transfer

Natural substrates and products

S 2-methylpropanoyl-CoA + enzyme N^6-(dihydrolipoyl)lysine <1, 2> (<1> multienzyme complex, scheme of the reaction steps [4]) (Reversibility: r <1> [1]; ? <1-4> [2-4]) [1-4]

P CoA + enzyme N^6-(S-[2-methylpropanoyl]dihydrolipoyl)lysine <1, 2> [1-4]

S Additional information <5> (<5> enzyme is a mitochondrial autoantigen, epitope mapping is performed to define the recognition sites by sera and T cells of patients suffering idopathic dilated cardiomyopathy and dilated cardiomyopathy [7]) [7]

P ?

Substrates and products

S 2-methylpropanoyl-CoA + dihydrolipoamide <1> (Reversibility: r <1> [1]) [1]

P CoA + S-2-methylpropanoyldihydrolipoamide <1> (<1> identification by mass spectrometry [1]) [1]

S 2-methylpropanoyl-CoA + enzyme N^6-(dihydrolipoyl)lysine <1-5, 16> (<1> recombinant apo-E_2 devoid of lipoic acid is fully active [9]; <1,5> enzyme contains 1 lipoyl residue per E_2 chain of branched-chain 2-oxo acid dehydrogenase [3,8]) (Reversibility: r <1> [1]; ? <1-5,16> [2-11]) [1-11]

P CoA + enzyme N^6-(S-[2-methylpropanoyl]dihydrolipoyl)lysine <1-5, 16> [1-11]

S 2-oxoglutaryl-CoA + dihydrolipoamide <16> (<16> low activity [10]) (Reversibility: ? <16> [10]) [10]

P CoA + S-(2-oxoglutaryl)dihydrolipoamide

S 3-methylisovaleryl-CoA + dihydrolipoamide <16> (Reversibility: ? <16> [10]) [10]

P CoA + S-(3-methylisovaleryl)dihydrolipoamide

S 4-methylthiobutyryl-CoA + dihydrolipoamide <16> (<16> low activity [10]) (Reversibility: ? <16> [10]) [10]

P CoA + S-(4-methylthiobutyryl)dihydrolipoamide

S acetyl-CoA + dihydrolipoamide <1, 5-15> (<5-15> part of the pyruvate dehydrogenase reaction [8]; <1> low activity [1]; <5,13> enzyme contains 2 lipoyl residues per E_2 chain of pyruvate dehydrogenase complex [8]; <6-11> enzyme contains 3 lipoyl residues per E_2 chain of pyruvate dehydrogenase complex [8]; <12,15> enzyme contains 1 lipoyl residue per E_2 chain of pyruvate dehydrogenase complex [8]) (Reversibility: r <1> [1]; ? <5-15> [8]) [1, 8]

P CoA + S-acetyldihydrolipoamide <1, 5-15> [1, 8]

S butyryl-CoA + dihydrolipoamide <16> (Reversibility: ? <16> [10]) [10]

P CoA + S-butyryldihydrolipoamide

S isobutyryl-CoA + dihydrolipoamide <1> (Reversibility: r <1> [1]) [1]

P CoA + S-isobutyryldihydrolipoamide <1> [1]

S isocaproyl-CoA + dihydrolipoamide <16> (Reversibility: ? <16> [10]) [10]

P CoA + S-isohexanoyldihydrolipoamide

S isovaleryl-CoA + dihydrolipoamide <1, 16> (<1> recombinant enzyme, i.e. E_2 domain, overexpressed in Escherichia coli [4]) (Reversibility: r <1> [1,4]; ? <16> [10]) [1, 4, 10]

P CoA + S-isovaleryldihydrolipoamide <1> [1]

S Additional information <16> (<16> pyruvate is a poor substrate [10]) [10]

P ?

Inhibitors

CoA <1> (<1> product inhibition, competitive to acyl-CoA, noncompetitive to dihydrolipoamide [1]) [1]

Mg^{2+} <1> [1]

arsenite <1> (<1> strong inhibition [1]) [1]

Additional information <1, 5> (<5> inhibition tests of the multienzyme complex with sera from patients suffering the possibly autoimmune diseases idopathic dilated cardiomyopathy and dilated cardiomyopathy to perform epitope mapping, overview [7]; <1> NEM and thiamine diphosphate have no effect [1]; <1> limited proteolysis by trypsin results in the complete loss of the overall activity of the enzyme complex, but does not affect dihydrolipoyllysine-residue (2-methylpropanoyl)transferase activity [1]) [1, 7]

Specific activity (U/mg)

0.173 <1> (<1> purified recombinant apo-enzyme [9]) [9]

0.185 <1> (<1> purified native enzyme [9]) [9]

2.38 <1> (<1> partially purified E_2 enzyme [1]) [1]

3.1 <16> (<16> purified enzyme, as part of the multienzyme complex, substrate isovaleryl-CoA [10]) [10]

Additional information <1> (<1> recombinant enzyme [4]) [4]

K_m-Value (mM)

0.05 <1> (isovaleryl-CoA, <1> pH 7.4, 37°C [1]) [1]

0.1 <1> (isobutyryl-CoA, <1> pH 7.4, 37°C [1]) [1]

0.11 <1> (acetyl-CoA, <1> pH 7.4, 37°C [1]) [1]

2 <1> (dihydrolipoanide, <1> pH 7.4, 37°C [1]) [1]

Additional information <1, 16> (<1> kinetics [1]) [1, 10]

pH-Optimum

7.4 <1> (<1> assay at [1]) [1]

Temperature optimum (°C)

37 <1> (<1> assay at [1,2]) [1, 2]

4 Enzyme Structure

Molecular weight

84000 <1> (<1> E_2-trimer, sucrose density gradient centrifugation in presence of the chaotropic reagent guanidinium-HCl [11]) [11]
1116000 <1> (<1> amino acid determination and sedimentation equilibrium analysis [4]) [4]

Subunits

24-mer <1> (<1> 24 * 52000, SDS-PAGE [4]) [4]
? <1, 16> (<16> x * 54000, SDS-PAGE [10]; <1> x * 52600, SDS-PAGE [1-3]) [1-3, 10]
Additional information <1, 2, 5-7, 11-14, 16> (<1> the 24-mer can be separated into active trimers of MW 84 kDa by incubation in 1.5 M guanidinium-HCl at 25°C, process is reversible, and removal of guanidinium-HCl leads to spontenaous reassembly to an active 24-mer [11]; <11> C-terminus of E_2 is joined to the N-terminus of E_3 in the pyruvate dehydrogenase complex, subunit organization [8]; <5-7,11-14> lipoyl domains and inner core in the structure of multienzyme complex, overview [8]; <1> enzyme shows the E_2 structure of 3 folded domains: lipoyl-bearing, E_3-binding, and inner core, typical for all E_2 protein of α-keto acid dehydrogenases [4]; <1,2> assembly of 24 E_2 subunits into a cubic structure, forming the core of the mammalian branched-chain α-keto acid dehydrogenase multienzyme complex [4]; <1> lipoate-free inner E_2 core: 26 kDa fragment contains the active site, 22 kDa fragment B is the subunit-binding domain, fragments are gained by tryptic digest [3]; <1> the 26 kDa fragment from tryptic digest is the catalytically active part of the enzyme [2]; <1> subunit structure analysis by tryptic digest [2]; <1> in vitro reconstitution of the 24-meric E_2 inner core requires the chaperonins GroEL and GroES [11]; recombinant apo-E_2 is unable to reconstitute with recombinant E_1 and E_3 to an active branched-chain α-keto dehydrogenase, but recombinant holo-E_2 is able to [9]; <16> reconstitution of the multienzyme complex branched-chain α-keto dehydrogenase [10]) [2-4, 8-11]

5 Isolation/Preparation/Mutation/Application

Source/tissue

fibroblast <2> [5]
kidney <1, 2> [1, 5, 7]
liver <1, 2, 5, 16> [1-4, 7, 10]
lymphoblast <2> [5]

Localization

mitochondrion <1, 2, 5, 15, 16> [2, 5, 7, 8, 10]

Purification

<1> (recombinant E_{2c}, residues 161-421, fused to maltose-binding protein from Escherichia coli [11]; recombinant apo-enzyme E_2 from Escherichia coli strain XL1-Blue [9]; recombinant wild-type, apo-E_2 enzyme, and mutants from Escherichia coli [4]) [1, 4, 7, 9, 11]
<5> (recombinant E_2 fusion protein from Escherichia coli, the tag is cleaved off by factor Xa [7]) [7]
<16> [10]

Renaturation

<1> (E_{2c}, completely unfolded in 4.5 M guanidinium chloride, is diluted 100fold at 25°C and refolded in 5 mM $MgATP^{2-}$ and a 4fold molar excess of chaperonins GroEL and GroES at pH 7.5, full activity is recovered after 45 min, an active GroEL-E_2 24-mer is formed [11]; in vitro reconstitution of the 24-meric E_2 inner core requires the chaperonins GroEL and GroES [11]; recombinant apo-E_2 is unable to reconstitute with recombinant E_1 and E_3 to an active branched-chain α-keto dehydrogenase, but recombinant holo-E_2 is able to [9]) [9, 11]
<16> (reconstitution of the multienzyme complex branched-chain α-keto dehydrogenase [10]) [10]

Cloning

<1> (expression of E_{2c}, residues 161-421, fused to maltose-binding protein, which enhances the yield of the recombinant enzyme, in Escherichia coli [11]; expression of apo-enzyme in Escherichia coli strain XL1-Blue, can be lipoylated in vitro [9]; DNA sequence determination, overexpression of wild-type, lipol-free apo-enzyme, and mutants in Escherichia coli [4]) [4, 9, 11]
<2> (gene E_2, DNA sequence determination and analysis [4,5]; gene contains a mitochondrial targeting presequence [5]; gene E_2, mapping to chromosome 1p31 [4]) [4, 5]
<3> (pseudogene E_2, DNA sequence determination and analysis, retroposon, gene corresponds to the complete mitochondrial presequence and the lipoyl-bearing domain encoded by exon I through IV of the functional gene E_2 [5]; detection of a pseudogene on chromosome 3q24 [4]) [4, 5]
<4> (gene E_2, DNA sequence determination and analysis [6]) [6]
<5> (cloning of full length E_2 and fragments, expression in Escherichia coli JM109 as fusion protein [7]) [7]
<16> (DNA and amino acid sequence determination anad analysis [10]) [10]

Engineering

F214C <2> (<2> natural splicing mutant from WG-34 cells, related to the maple syrup urine disease, phenotype of homo- and heterozygous mutations in humans [4]) [4]
H391N <1> (<1> enzymatically inactive [4]) [4]
H391Q <1> (<1> enzymatically inactive [4]) [4]

References

[1] Chuang, D.T.; Hu, C.C.; Ku, L.S.; Niu, W.L.; Myers, D.E.; Cox, R.P.: Catalytic and structural properties of the dihydrolipoyl transacylase component of bovine branched-chain α-keto acid dehydrogenase. J. Biol. Chem., **259**, 9277-9284 (1984)

[2] Chuang, D.T.; Hu, C.W.; Ku, L.S.; Markovitz, P.J.; Cox, R.P.: Subunit structure of the dihydrolipoyl transacylase component of branched-chain α-keto acid dehydrogenase complex from bovine liver. Characterization of the inner transacylase core. J. Biol. Chem., **260**, 13779-13786 (1985)

[3] Hu, C.W.; Griffin, T.A.; Lau, K.S.; Cox, R.P.; Chuang, D.T.: Subunit structure of the dihydrolipoyl transacylase component of branched-chain α-keto acid dehydrogenase complex from bovine liver. Mapping of the lipoyl-bearing domain by limited proteolysis. J. Biol. Chem., **261**, 343-349 (1986)

[4] Chuang, D.T.; Fisher, C.W.; Lau, K.S.; Griffin, T.A.; Wynn, R.M.; Cox, R.P.: Maple syrup urine disease: domain structure, mutations and exon skipping in the dihydrolipoyl transacylase (E_2) component of the branched-chain α-keto acid dehydrogenase complex. Mol. Biol. Med., **8**, 49-63 (1991)

[5] Lau, K.S.; Herring, W.J.; Chuang, J.L.; McKean, M.; Danner, D.J.; Cox, R.P.; Chuang, D.T.: Structure of the gene encoding dihydrolipoyl transacylase (E_2) component of human branched chain α-keto acid dehydrogenase complex and characterization of an E_2 pseudogene. J. Biol. Chem., **267**, 24090-24096 (1992)

[6] Lau, K.S.; Chuang, J.L.; Herring, W.J.; Danner, D.J.; Cox, R.P.; Chuang, D.T.: The complete cDNA sequence for dihydrolipoyl transacylase (E_2) of human branched-chain α-keto acid dehydrogenase complex. Biochim. Biophys. Acta, **1132**, 319-321 (1992)

[7] Ansari, A.A.; Neckelmann, N.; Villinger, F.; Leung, P.; Danner, D.J.; Brar, S.S.; Zhao, S.; Gravanis, M.B.; Mayne, A.; Gershwin, M.E.: Epitope mapping of the branched chain α-ketoacid dehydrogenase dihydrolipoyl transacylase (BCKD-E_2) protein that reacts with sera from patients with idiopathic dilated cardiomyopathy. J. Immunol., **153**, 4754-4765 (1994)

[8] Perham, R.N.: Swinging arms and swinging domains in multifunctional emzymes: catalytic machines for multistep reactions. Annu. Rev. Biochem., **69**, 961-1004 (2000)

[9] Chuang, J.L.; Davie, J.R.; Wynn, R.M.; Chuang, D.T.: Production of recombinant mammalian holo-E_2 and E_3 and reconstitution of functional branched-chain α-keto acid dehydrogenase complex with recombinant E_1. Methods Enzymol., **324**, 192-200 (2000)

[10] Hakozaki, M.; Ono, K.; Suzuki, T.; Hata, H.; Mori, T.; Kochi, H.: Characterization of rainbow trout branched-chain α-keto acid dehydrogenase complex: inter-domain segments of the E_2 component affect the overall activity. Comp. Biochem. Physiol. B Biochem. Mol. Biol., **132**, 433-442 (2002)

[11] Wynn, R.M.; Davie, J.R.; Zhi, W.; Cox, R.P.; Chuang, D.T.: Invitro reconstitution of the 24-meric E_2 inner core of bovine mitochondrial branched-chain α-keto acid dehydrogenase complex: requirements for chaperonins GroEL and GroES. Biochemistry, **33**, 8962-8968 (1994)

CO-methylating acetyl-CoA synthase 2.3.1.169

1 Nomenclature

EC number
2.3.1.169

Systematic name
acetyl-CoA:corrinoid protein O-acetyltransferase

Recommended name
CO-methylating acetyl-CoA synthase

Synonyms
CO dehydrogenase
CO dehydrogenase enzyme complex
CODH
CODH/ASC
acetyl-CoA synthase
acetyl-coenzyme A synthase
actyl-CoA decarboxylase/synthase
carbon monoxide dehydrogenase
carbon monoxide dehydrogenase-corrinoid enzyme complex
carbon monoxide dehydrogenase/acetyl-CoA synthase
multienzyme CO dehydrogenase/acetyl-CoA synthase complex
multienzyme carbon monoxide dehydrogenase complex

CAS registry number
64972-88-9

2 Source Organism

<1> *Clostridium thermoaceticum* (DSM 521 [2,4]) [1-4, 6-8, 12, 15]
<2> *Methanosarcina thermophila* (TM1 [9]) [5, 9]
<3> *Methanosarcina barkeri* [10, 11, 15]
<4> *Moorella thermoacetica* [13, 14, 15]

3 Reaction and Specificity

Catalyzed reaction
acetyl-CoA + corrinoid protein = CO + methylcorrinoid protein + CoA (<1> Enzyme accepts the methyl group from the methylated corrinoid/iron-sulfur

protein, binds a carbonyl group from CO, CO_2, or the carboxyl of pyruvate, and binds coenzyme A. Then the enzyme catalyses the synthesis of acetyl-CoA from these enzyme bound groups. Additionally the enzyme catalyses two exchange reactions between the methylated corrinoid/iron-sulfur protein and methylated enzyme and between methylated enzyme and the methyl moiety of acetyl-CoA. [1,3]; <1,3,4> pathway [1,3,4,6,11,12,13]; <1> the enzyme-bound complex is an [$NiFe_3$-$4S_4$]-acetyl complex [3]; <1> enzyme contains binding sites for the methyl, carbonyl, and CoA moieties of acetyl-CoA and catalyses the assembly of acetyl-CoA from these enzyme-bound groups, under optimal conditions the rate-limiting step involves methylation of enzyme by the methylated corrinoid/iron-sulfur protein [4]; <2> the multienzyme complex contains at least two protein components: a CO-oxidizing Ni/Fe-S component and a Co/Fe-S component [5]; <1> the FeS cluster is present to relay electrons from enzyme to CO [7,8]; <1> the transfer of methyl group to enzyme occurs by S_N2-type nucleophilic displacement, not a radical, reaction [7,8]; <3> the corronoid protein functions as methyl group carrier during acetyl-CoA synthesis and decomposition [11]; <1> kinetics of methyl group transfer between the cobalt of the corrinoid/iron-sulfur protein and the nickel of Ni-X-Fe_4S_4 cluster, called the A-cluster of enzyme, the reaction is reversible [12,13]; <4> the Ni-Fe_4S_4-5C cluster of enzyme catalyses the reversible reduction of CO_2 to CO and is located in the β-subunit. CO generated at this site migrates through the tunnel to the A-cluster, located in the α-subunit, where it reacts with CoA and a methyl group to generate acetyl-CoA. During catalysis, the two sites are mechanistically coupled. [13]; <1,3,4> The transfer of Co bound methyl group from methylated corrinoid/iron-sulfur protein to acetyl-CoA synthase is an S_N2 attack of a nucleophilic center of enzyme, presumably Ni_4, on the methyl-Co(III) stat of the corrinoid/iron-sulfur protein, generating Co(I) and methylating acetyl-CoA synthase. [15])

Reaction type

demethylation

Natural substrates and products

S Additional information <1, 3, 4> (<1> key enzyme in the autotrophic acetyl-CoA pathway, i.e. Wood pathway, enzyme catalyses the final steps in this pathway [1,2,3,6-8,12]; <1> enzyme and a corrinoid/iron-sulfur protein, methyltransferase and an electron transfer protein such as ferredoxin II play a pivotal role in the conversion of methylhydrofolate, CO, and CoA to acetyl-CoA [4]; <1,3,4> the bifuctional enzyme CO dehydrogenase/acetyl-CoA synthase is central to the Wood-Ljungdahl pathway of autotrophic CO_2 fixation [15]) [1-4, 6-8, 12, 15]

P ?

Substrates and products

S CH_3-(corrinoid/iron-sulfur protein) + CO + HS-CoA <1, 4> (<1> under anaerobic conditions [3,4,7]) (Reversibility: ? <1,4> [3,4,7,14]) [3, 4, 7, 14]

P CH_3-CO-S-CoA + corrinoid/iron-sulfur protein <1, 4> [3, 4, 14]

S CH_3-CO-S-CoA + H^+ + tetrahydromethanopterin <2> (Reversibility: ? <2> [9]) [9]

P CH_3-tetrahydromethanopterin + CO + HS-CoA <2> [9]

S CH_3-CO-S-CoA + tetrahydrosarcinapterin + H_2O <3> (<3> the carbon monoxide dehydrogenase-corrinoid enzyme complex catalyzes the cleavage of acetyl-CoA, tetrahydrosarcinapterin functions as the methyl group acceptor, the major products of reaction are methyltetrahydrosarcinapterin and CO_2, free CoA is identified as an additional product [10,11]) (Reversibility: ? <3> [10]) [10, 11]

P CH_3-tetrahydrosarcinapterin + CO_2 + H^+ + electron <3> [10]

S CH_3-tetrahydrofolate + CO + HS-CoA <1> (<1> this multistep reaction involves four proteins: CO dehydrogenase, methyltransferase, the corrinoid/iron-sulfur protein and ferredoxin [4]; <1> The methyltransferase catalyses the reaction of CH_3-H_4folate with the corrinoid/iron-sulfur protein to form a methylcobalt species. The Ni/Fe-S enzyme CO dehydrogease then catalyses the final steps in the formation of acetyl-CoA. [6]) (Reversibility: ? <1> [1,4,6]) [1, 4, 6-8]

P CH_3-CO-S-CoA + tetrahydrofolatefolate <1> [4, 6-8]

S CH_3-tetrahydrosarcinapterin + CO + HS-CoA <3> (Reversibility: ? <3> [11]) [11]

P CH_3-CO-S-CoA + tetrahydrosarcinapterin <3> [11]

S CH_3I + CO + HS-CoA <1, 2> (<2> the multienzyme complex catalyses the acetyl-CoA synthesis from CH_3I, CO and CoA as well as to cleave acetyl-CoA into its methyl, carbonyl, and CoA components as the first step in the catabolism of acetyl-CoA to methane and CO_2 [5]) (Reversibility: ? <1,2> [1,4,5]) [1, 4, 5]

P CH_3-CO-S-CoA + HI <1> [1, 4]

S CO + H_2O <1-4> (<1,2> the multienzyme complex catalyses the reversible oxidation of CO to CO_2 [9,12]; <4> the $NiFe_4S_4$-5C cluster catalyses the reversible oxidation of CO to CO_2 [13]) (Reversibility: r <1-4> [9,11-13]) [9, 11-13]

P CO_2 + H^+ + electron <1-4> [9, 11-13]

S CO + methyl-X + HS-CoA <1, 3, 4> (<1,3,4> acetyl-CoA synthase catalyses acetyl-CoA synthesis, an intermediate step is the transfer of the cobalt-bound methyl group from methylated corrinoid/iron-sulfur protein to the acetyl-CoA synthase [15]) (Reversibility: ? <1,3,4> [1,4,15]) [1, 3, 4, 6-8, 11, 15]

P CH_3-CO-S-CoA + HX <1, 3, 4> [1, 3, 4, 6-8, 11, 15]

S CO_2 + H^+ + electron <1, 3, 4> (<1,3,4> CO dehydrogenase catalyses the two-electron reduction of CO_2 to CO [15]) (Reversibility: ? <1,3,4> [15]) [15]

P CO + H_2O <1, 3, 4> [15]

S Additional information <1-4> (<1> enzyme catalyses the CoA/acetyl-CoA exchange [3]; <2> the multienzyme complex catalyses the exchange between free CO and carbonyl group of acetyl-CoA, and the exchange between CoA and the CoA moiety of acetyl-CoA [9]; <1,3,4> methylcobinamide, methylcobalamin, and CH_3-(Me3-benzimidazolyl)cobamide are substrates of the acetyl-CoA synthase, methylcobalamin is 2000fold less reactive than methylcobinamide, CO dehydrogenase catalyses the CO-dependent reduction of methylcobinamide 10000fold faster than that of methylcobalamin [15]) [3, 9, 15]

P ?

Inhibitors

CN^- <1> (<1> inhibitor on the CoA/acetyl-CoA exchange, 98% inhibition at 1.2 mM [3]) [3]

CO <1> (<1> non-competitive inhibitor on the CoA/acetyl-CoA exchange, the Ni-Fe-C-center appears to be the inhibitor site for CO [3]; <1> inhibits the methyl group transfer reaction and synthesis of acetyl-CoA [12]) [3, 12]

CO_2 <1> (<1> inhibitor on the CoA/acetyl-CoA exchange [3]) [3]

CoA <1> (<1> at concentration above 10 mM, 50% inhibition of acetyl-CoA synthesis from methyl iodide at 15 mM [1]) [1]

N_2O <1> (<1> inhibitor on the CoA/acetyl-CoA exchange [3]) [3]

Ti^{3+}-citrate <1> (<1> inhibits reverse methyl group transfer, when it is preincubated with methylated enzyme but not when it is preincubated with Co^{2+}-iron-sulfur protein [12]) [12]

dephospho-CoA <1> (<1> inhibitor on the CoA/acetyl-CoA exchange, 75% inhibition at 0.44 mM [3]) [3]

desulfo-CoA <1> (<1> inhibitor on the CoA/acetyl-CoA exchange, 30% mM at 2.1 mM [3]) [3]

dithionite <1> (<1> inhibits reverse methyl group transfer, when it is preincubated with methylated enzyme but not when it is preincubated with Co^{2+}-iron-sulfur protein [12]) [12]

Activating compounds

CO <1> (<1> two effects: stimulation and inhibition on CoA/acetylCoA exchange [3]) [3]

ferredoxin <1> (<1> stimulation of acetyl-CoA synthesis from methyl iodide, maximum at 1 nmol ferredoxin/15 nmol of enzyme [1]; <1> stimulates the rate of synthesis of acetyl-CoA 4fold, K_m is 0.0034 mM [4]) [1, 4]

Metals, ions

Co <1-4> (<1,2> a cobalt-containing Co/Fe-S component of multienzyme complex serves as a methyl carrier in the pathway of methane synthesis from acetate [5,7,8]; <1,3,4> cobalt is the active site for the methyl-transfer reaction [15]) [5, 7, 8, 15]

Fe <1-4> (<1-4> corrinoid/iron-sulfur protein required [1-14]; <1> the enzyme-bound complex can be described as an [$NiFe_3$-$4S_4$]-acetyl complex [3]; <2> the multienzyme complex contains at least two protein components: a CO-oxidizing Ni/Fe-S component and a cobalt-containing Co/Fe-S component [5]; <2> a Ni/Fe-S cluster of multienzyme CO dehydrogenase/acetyl-CoA synthase complex is the active site of acetyl-CoA cleavage and synthesis [9]; <4> the Ni-Fe_4S_4-5C cluster of enzyme catalyses the reversible reduction of CO_2 to CO and is located in the β-subunit. [13]) [1-14]

Ni <1, 2, 4> (<1> the enzyme-bound complex can be described as an [$NiFe_3$-$4S_4$]-acetyl complex [3]; <2> the multienzyme complex contains at least two protein components: a CO-oxidizing Ni/Fe-S component and a cobalt-containing Co/Fe-S component [5]; <2> a Ni/Fe-S cluster of multienzyme CO dehydrogenase/acetyl-CoA synthase complex is the active site of acetyl-CoA cleavage and synthesis [9]; <1,4> enzyme contains nickel in the A-cluster of the enzyme [12-14]; <4> the Ni-Fe_4S_4-5C cluster of enzyme catalyses the reversible reduction of CO_2 to CO and is located in the β-subunit. [13]) [3, 5, 9, 12-14]
Additional information <1, 4> (<1> a nucleophilic metal center on enzyme is the active site which accepts the methyl group from the methylated corrinoid/iron-sulfur protein [1]; <4> Cu is not required for enzyme activity [14]) [1, 14]

Turnover number (min^{-1})
Additional information <1> [3, 13]

Specific activity (U/mg)
0.12 <1> (<1> pH 6.8, acetyl-CoA synthesis, in absence of ferredoxin II [4]) [4]
0.41 <1> (<1> pH 6.8, acetyl-CoA synthesis, in presence of 1 mM ferrous ammonium sulfate [4]) [4]
0.49 <1> (<1> pH 6.8, acetyl-CoA synthesis, in presence of ferredoxin II [4]) [4]
0.74 <1> (<1> pH 6.8, acetyl-CoA synthesis, in presence of 4 mM ATP [4]) [4]
0.8 <1> (<1> pH 6.8, acetyl-CoA synthesis, in absence of ATP and Fe^{2+} [4]) [4]
28 <1> (<1> 40°C, CoA/acetyl-CoA exchange [3]) [3]

K_m-Value (mM)
4.3 <1> (CoA, <1> pH 7.3, 22°C, acetyl-CoA synthesis from methyl iodide [1]) [1]
14.7 <1> (CH_3I, <1> pH 7.3, 22°C [1]) [1]

K_i-Value (mM)
0.4 <1> (CO, <1> pH 7.0, 25°C, under anaerobic conditions, noncompetitive with respect to acetyl-CoA [3]) [3]

pH-Optimum
5.4 <1> (<1> CO/acetyl-CoA exchange reaction [4]) [4]
5.8 <1> (<1> Tris-maleate buffer, the rate of acetyl-CoA synthesis increases with decreasing pH, at pH values below 5.8 the rate of acetyl-CoA synthesis decreases slightly [1]) [1]
6.7-7 <1> (<1> optimum for CoA/acetyl-CoA exchange [3]) [3]

Temperature range (°C)
25-35 <1> (<1> in 10 mM Tris-maleate buffer and pH 5.8, the rate of acetyl-CoA synthesis is increased 2fold at 25°C to 35°C [4]) [4]

4 Enzyme Structure

Molecular weight

150000 <1> [2, 3]
1600000 <3> (<3> gel filtration [10]) [10]

Subunits

dimer <1> (<1> α,β [2,3]) [2, 3]
polymer <3> (<3> $\alpha,\beta,\gamma,\delta,\varepsilon$, 6 * 19700 + 6 * 84500 + 6 * 63200 + 6 * 53000 + 6 * 51400, SDS-PAGE [10]) [10]
tetramer <4> (<4> α,β, containing two unique Ni-Fe-S active sites connected by a molecular tunnel [13,14]) [13, 14]
Additional information <2> (<2> $\alpha,\beta,\gamma,\delta,\varepsilon$, the enzyme complex is part of a five-subunit complex, the α and ε subunits are required for CO oxidation, the γ end δ subunits constitute a corrinoid/iron-sulfur protein, the β subunit harbors the Ni/Fe-S cluster, that is the active site of acetyl-CoA cleavage/synthesis, the interaction between the α,ε dimer and the β subunit is necessary for breaking and forming the C-C bond of acetyl-CoA, SDS-PAGE [9]) [9]

5 Isolation/Preparation/Mutation/Application

Purification

<1> [1, 4]
<3> [10]

Cloning

<1> (the gene has been cloned into Escherichia coli and found to be within an 11 kb gene cluster, recombinant enzyme is inactive [2]) [2]

6 Stability

Temperature stability

67 <1> (<1> the recombinant enzyme: 10 min, 10% loss of activity, the wild-type enzyme: 10 min, no loss of activity [2]) [2]
75 <1> (<1> the wild-type enzyme: 10 min, 50% loss of activity [2]) [2]

General stability information

<1>, reactions in presence of DTT, since enzyme requires strictly anaerobic conditions for stability [3]
<1>, the acetyl-CoA synthesis is dependent on ionic strength, the CO/acetyl-CoA exchange is independent of ionic strength [4]

References

[1] Lu, W.P.; Harder, S.R.; Ragsdale, S.W.: Controlled potential enzymology of methyl transfer reactions involved in acetyl-CoA synthesis by CO dehydrogenase and the corrinoid/iron-sulfur protein from Clostridium thermoaceticum. J. Biol. Chem., **265**, 3124-3133 (1990)
[2] Roberts, D.L.; James-Hagstrom, J.E.; Garvin, D.K.; Gorst, C.M.; Runquist, J.A.; Baur, J.R.; Haase, F.C.; Ragsdale, S.W.: Cloning and expression of the gene cluster encoding key proteins involved in acetyl-CoA synthesis in Clostridium thermoaceticum: carbon monoxide dehydrogenase, the corrinoid/iron-sulfur protein, and methyltransferase. Proc. Natl. Acad. Sci. USA, **86**, 32-36 (1989)
[3] Lu, W.P.; Ragsdale, S.W.: Reductive activation of the coenzyme A/acetyl-CoA isotopic exchange reaction catalyzed by carbon monoxide dehydrogenase from Clostridium thermoaceticum and its inhibition by nitrous oxide and carbon monoxide. J. Biol. Chem., **266**, 3554-3564 (1991)
[4] Roberts, J.R.; Lu, W.P.; Ragsdale, S.W.: Acetyl-coenzyme A synthesis from methyltetrahydrofolate, carbon monoxide, and coenzyme A by enzymes purified from Clostridium thermoaceticum: attainment of in vivo rates and identification of rate-limiting steps. J. Bacteriol., **174**, 4667-4676 (1992)
[5] Jablonski, P.E.; Lu, W.P.; Ragsdale, S.W.; Ferry, J.G.: Characterization of the metal centers of the corrinoid/iron-sulfur component of the carbon monoxide dehydrogenase enzyme complex from Methanosarcina thermophila by EPR spectroscopy and spectroelectrochemistry. J. Biol. Chem., **268**, 325-329 (1993)
[6] Kasmi, A.E.; Rajasekharan, S.; Ragsdale, S.W.: Anaerobic pathway for conversion of the methyl group of aromatic methyl ethers to acetic acid by Clostridium thermoaceticum. Biochemistry, **33**, 11217-11224 (1994)
[7] Menon, S.; Ragsdale, S.W.: Role of the [4Fe-4S] cluster in reductive activation of the cobalt center of the corrinoid iron-sulfur protein from Clostridium thermoaceticum during acetate biosynthesis. Biochemistry, **37**, 5689-5698 (1998)
[8] Menon, S.; Ragsdale, S.W.: The role of an iron-sulfur cluster in an enzymic methylation reaction: Methylation of CO dehydrogenase/acetyl-CoA synthase by the methylated corrinoid iron-sulfur protein. J. Biol. Chem., **274**, 11513-11518 (1999)
[9] Murakami, E.; Ragsdale, S.W.: Evidence for intersubunit communication during acetyl-CoA cleavage by the multienzyme CO dehydrogenase/acetyl-CoA synthase complex from Methanosarcina thermophila. Evidence that the β subunit catalyzes C-C and C-S bond cleavage. J. Biol. Chem., **275**, 4699-4707 (2000)
[10] Grahame, D.A.: Catalysis of acetyl-CoA cleavage and tetrahydrosarcinapterin methylation by a carbon monoxide dehydrogenase-corrinoid enzyme complex. J. Biol. Chem., **266**, 22227-22233 (1991)
[11] Grahame, D.A.: Substrate and cofactor reactivity of a carbon monoxide dehydrogenase-corrinoid enzyme complex: Stepwise reduction of iron-sulfur

and corrinoid centers, the corrinoid cobalt(2+/1+) redox midpoint potential, and overall synthesis of acetyl-CoA. Biochemistry, **32**, 10786-10793 (1993)
[12] Tan, X.S.; Sewell, C.; Lindahl, P.A.: Stopped-flow kinetics of methyl group transfer between the corrinoid-iron-sulfur protein and acetyl-Coenzyme A synthase from Clostridium thermoaceticum. J. Am. Chem. Soc., **124**, 6277-6284 (2002)
[13] Tan, X.; Sewell, C.; Yang, Q.; Lindahl, P.A.: Reduction and methyl transfer kinetics of the a subunit from acetyl Coenzyme A synthase. J. Am. Chem. Soc., **125**, 318-319 (2003)
[14] Bramlett, M.R.; Tan, X.; Lindahl, P.A.: Inactivation of acetyl-CoA synthase/ carbon monoxide dehydrogenase by copper. J. Am. Chem. Soc., **125**, 9316-9317 (2003)
[15] Seravalli, J.; Brown, K.L.; Ragsdale, S.W.: Acetyl Coenzyme A synthesis from unnatural methylated corrinoids: Requirement for "Base-Off" coordination at cobalt. J. Am. Chem. Soc., **123**, 1786-1787 (2001)

D-Glutamyltransferase 2.3.2.1

1 Nomenclature

EC number
2.3.2.1

Systematic name
glutamine:D-glutamyl-peptide 5-glutamyltransferase

Recommended name
D-glutamyltransferase

Synonyms
D-γ-glutamyl transpeptidase
D-glutamyl transpeptidase
γ-GT
glutamyltransferase, D-

CAS registry number
9030-02-8

2 Source Organism

<1> *Bacillus subtilis* (ATCC 9945 [1]) [1]
<2> *Homo sapiens* [2]

3 Reaction and Specificity

Catalyzed reaction
L(or D)-glutamine + D-glutamyl-peptide = NH_3 + 5-glutamyl-D-glutamyl-peptide

Reaction type
aminoacyl group transfer

Substrates and products
S D-glutamine + D-glutamyl-peptide <1> (Reversibility: ? <1> [1]) [1]
P NH_3 + 5-glutamyl-D-glutamyl-peptide <1> [1]
S L-glutamine + D-glutamic acid <1> (Reversibility: ? <1> [1]) [1]
P NH_3 + γ-L-glutamyl-D-glutamate <1> [1]

pH-Optimum
8.8 <1> [1]

5 Isolation/Preparation/Mutation/Application

Source/tissue

kidney <2> (<2>, neonatal [2]) [2]
liver <2> (<2>, neonatal [2]) [2]

Localization

membrane <2> [2]
soluble <1> [1]

References

[1] Williams, W.J.; Litwin, J.; Thorne, C.B.: Further studies on the biosynthesis of γ-glutamyl peptides by transfer reactions. J. Biol. Chem., **212**, 427-438 (1955)
[2] Lackmann, G.M.: Reference values for selected enzyme activities in serum from healthy human neonates. Clin. Biochem., **29**, 599-602 (1996)

γ-Glutamyltransferase 2.3.2.2

1 Nomenclature

EC number
2.3.2.2

Systematic name
(5-L-glutamyl)-peptide:amino-acid 5-glutamyltransferase

Recommended name
γ-glutamyltransferase

Synonyms
GGT <1, 2> [2, 68, 70]
L-γ-glutamyl transpeptidase
L-γ-glutamyltransferase
L-glutamyltransferase
α-glutamyl transpeptidase
γ-GPT
γ-GT
γ-GTP
γ-GTase <24> [71]
γ-glutamyl peptidyltransferase
γ-glutamyl transpeptidase
glutamyl transpeptidase
glutamyltransferase, γ-
Additional information (cf. EC 2.3.2.14)

CAS registry number
9046-27-9

2 Source Organism

<1> *Homo sapiens* (6 isoforms in hepatic cellular carcinoma [60]) [1, 2, 5-8, 10, 13, 36-39, 41-46, 53, 60, 65, 67]
<2> *Rattus norvegicus* (streptozotocin-induced diabetic, female and male [73]; multiparous rats [18]; male, Donryu [17]; Wistar, female [24,68]; Wistar [19,20,23]; male, Sprague-Dawley [21,22,64]) [3, 9, 16-24, 33-36, 40, 42, 61, 62, 64, 68, 70, 73, 75]
<3> *Mus musculus* (Swiss albino and RIII strain, both infected with mammary tumour virus, uninfected: C57BL strain [26]) [26, 34, 35]
<4> *Oryctolagus cuniculus* (male, Fauve de Bourgogne strain [25]) [25, 36]

<5> *Sus scrofa* [11, 31-33]
<6> *Ovis aries* [29, 30, 34]
<7> *Bos taurus* (Holstein breed [27]) [12, 14, 15, 27, 28]
<8> *mammalia* [34, 35]
<9> *Marthasterias glacialis* [4]
<10> *Phaseolus vulgaris* [58, 59]
<11> *Lentinus edodes* (shiitake mushroom [51,52]) [51, 52]
<12> *Penicillium roqueforti* (strain IFO 4622 [54]) [54]
<13> *Morchella esculenta* (3 isoforms I, II and III [55]) [55]
<14> *Saccharomyces cerevisiae* (wild-type and enzyme-deficient strain, nitrogen starvation-induced [74]; strain 1278b [56]) [56, 74]
<15> *Aspergillus oryzae* (strain MA-27-IM [57]) [57]
<16> *Bacillus subtilis* (natto strain NR-1 [47]) [47]
<17> *Escherichia coli* (2 isozymes A and B [48]; K-12, strain MG1655 [48,49]) [48, 49]
<18> *Proteus mirabilis* [50]
<19> *Oncornavirinae of various origins* [26]
<20> *Canis familiaris* [53]
<21> *Lepidochelys kempi* (Kemp's ridley sea-turtle [63]) [63]
<22> *Ascaris suum* [66]
<23> *Bacillus sp.* (strain KUN-17 [69]) [69]
<24> *Lycopersicon esculentum* (2isoforms I and II [71]) [71]
<25> *Treponema denticola* (strain ATCC 35405 [72]) [72]
<26> *Arabidopsis thaliana* [76]
<27> *Homo sapiens* [76]
<28> *Arabidopsis thaliana* [76]
<29> *Arabidopsis thaliana* (SwissProt-accession number Q9M0G0 [76]) [76]
<30> *Arabidopsis thaliana* (SwissProt-accession number O65652 [76]) [76]
<31> *Mus musculus* [76]

3 Reaction and Specificity

Catalyzed reaction

(5-L-glutamyl)-peptide + an amino acid = peptide + 5-L-glutamyl amino acid (<2> almost the entire glutamyl moiety is necessary for recognition in the binding site of the donor substrate [75]; <2> active site [75]; <1,2,4-6,14,22> kinetic study [25,29,32-34,56,66-68,70,73]; <4> kinetic mechanism, pH-dependence [25]; <1,2> active site on small subunit [20,22,36]; <5> ping-pong bi bi mechanism [11]; <2-8,10> mechanism [25,27,29,33-35,58]; <1-3,8> structure [20,35,43])

Reaction type

aminoacyl group transfer

Natural substrates and products

S (5-L-glutamyl)-peptide + acceptor <2, 3, 6, 8, 22, 24, 25> (<2,6,8,24> concurrent reactions: autotranspeptidation with another donor molecule as acceptor, and hydrolase reaction with H_2O as acceptor [34,71]; <2,3,6,8> involved in amino acid transport systems, processing of mercapturic acids, and in pathways of metabolism of prostaglandins, estrogens and leukotrienes [34,35]) (Reversibility: r <2, 3, 8> [35]; ? <2, 6, 8, 22, 24, 25> [34, 64, 66, 71-73]) [34, 35, 64, 66, 71-73]

P peptide + 5-L-glutamyl amino acid <2, 6, 8, 22, 24, 25> [34, 66, 71, 72]

S 5-D/L-glutamyl-phenylhydrazine + acceptor <2> (Reversibility: ir <2> [70]) [70]

P phenylhydrazine + 5-L-glutamyl-acceptor <2> (<2> product is cytotoxic [70]) [70]

S L-Glu-4-nitroanilide + glutathione <26> (<26> L-Glu-4-nitroanilide is a suicide substrate [76]) (Reversibility: ? <26> [76]) [76]

P 4-nitroaniline + 5-L-glutamyl-glutathione

S glutathione + H_2O <14> (Reversibility: ? <14> [74]) [74]

P L-cysteinylglycine + glutamate <14> [74]

S glutathione + an amino acid <1, 4, 14, 16, 22, 24-26> (<2> ectoenzyme protects the epithelial cells against oxidants by replenishing intracellular glutathione [64]; <14,22,25,26> initial step of glutathione degradation [66,72,74,76]; <1,4> involved in 5-L-glutamyl cycle of glutathione metabolism [36]; <16> involved in polyglutamic acid synthesis [47]) (Reversibility: ? <1, 4, 14, 16, 22, 24-26> [36, 47, 64, 66, 71, 72, 74, 76]) [36, 47, 64, 66, 71, 72, 74, 76]

P L-cysteinylglycine + a 5-L-glutamyl-amino acid <25, 26> [72, 76]

S glutathione + an amino acid <2, 3, 8> (<2,3,8> initial step of glutathione degradation [35]; <2,3,8> involved in 5-L-glutamyl cycle of glutathione metabolism [35]) (Reversibility: ir <2, 3, 8> [35]) [35]

P L-cysteinylglycine + L-glutamate + glutamate <2, 3, 8> [32, 35]

S lentinic acid + amino acid <11> (<11> function in aroma evolution from lentinic acid [52]) (Reversibility: ? <11> [52]) [52]

P desglutamyl lentinic acid + 5-L-glutamylamino acid <11> [52]

S Additional information <2, 14, 25> (<14> enzyme activates the yeast cadmium factor 1, responsible for glutathione import into the vacuole [74]; <2> activity in induced diabetic rats is higher than in control rats and higher in male than in female rats [73]; <25> purified recombinant enzyme added to Porphyromonas gingivalis results in H_2S, ammonia and pyruvate release in this bacterium [72]; <25> L-Glu-4-nitroanilide competes with glutathione and inhibits production of H_2S, ammonia and pyruvate [72]) [72-74]

P ?

Substrates and products

S (5-L-glutamyl)-peptide + acceptor <1-26> (<25> L-cysteine is no substrate [72]; <24> α-aminoisobutyric acid is no donor [71]; <17> isozyme A prefers basic and aromatic amino acids as acceptors [48]; <2,3,6,8> stereospecificity [34]; <2> imino acids are no substrates [34]; <1,2,4,17> Glu-donor is 5-L-Glu-4-nitroanilide [17,20,36,42,48,75]; <1,2,4-8,11,13,14,17,24> Glu-donor is GSH [20,28,33,34,36,42,48,51,55,56,71];

<1,2,4> donor: GSH S-substituted derivatives [20,36]; <1,2,5> donor: S-acetophenoneglutathione [33,42]; <1> donor: S-acetylglutathione [42]; <2,3,5,6,8,13,24> donor: glutathione disulfide [33-35,55,71]; <2,5> donors: ophthalmic acid, poly-5-Glu-derivatives, L-5-Glu-7-amino-4-methylcoumarin [33]; <5> donors: 5-ethyl-Glu and β-benzyl-Glu, hydrolyzed by pig enzyme [33]; <2,5> donor: leukotriene C [32,33]; <1,2,4,5> donor: 5-L-Glu-naphthylamide [31,36]; <1,2,4> donor: 5-L-Glu-3-sulfonic-4-nitroanilide [36]; <2, 6, 10, 13, 14, 16-18, 21, 24> donor specificity, overview [30,47,48,50,55,56,58,63,70,71]; <1-3,8> strict L-stereospecificity of 5-Glu-acceptor site [35, 38]; <1-3,8> donor binding site reacts with L- and to a lesser extent D-enantiomer [35, 38]; <1-4, 6-8, 11, 12, 16, 18, 21-24> acceptor specificity, overview [27-30, 34-36, 40, 42, 46, 47, 50, 51, 54, 63, 66, 69-71]; <1, 2, 4, 6, 7, 17, 18> acceptors are amino acids or dipeptides [27, 30, 34, 36, 42, 48, 50]; <16> dipeptides are better acceptors than free amino acids [47]; <1,2,6,8,10-18> concurrent reaction: autotranspeptidation with another donor molecule as acceptor [34,47-58]; <1,2,6,8,10-18,21,24> concurrent reaction: hydrolase reaction with H_2O as acceptor [34,47-58,63,70,71]; <2> poor substrates are β-substituted amino acids [40]; <15,21,23> best acceptors: free amino acids [57,63,69]; <7,23> best donor: glutamate [27,69]; <13,16> donors are Glu, α-L-Glu-L-Ala [47,55]; <24> acceptors: L-methionine, L-tryptophan, L-alanine, 1-aminocyclopropane-1-carboxylic acid, glutathione [71]; <24> donor: glutathione S-conjugates [71]) (Reversibility: ir <1, 2> [17, 38, 70]; ? <1-26> [1-16, 18-37, 39-69, 71-76]) [1-76]

P peptide + 5-L-glutamyl amino acid <1, 2, 6, 8, 10, 13, 14, 16, 17, 21, 25, 26> [7, 34, 47, 48, 55, 56, 58, 60, 63, 70, 72, 74, 76]

S 5-L-glutamyl-4-nitroanilide + L-isoleucine <2, 6, 12, 13> (<2> glutamyl transfer at 17% the rate of the reaction with glycylglycine [24]) (Reversibility: ? <2, 6, 12, 13> [24, 30, 54, 55]) [24, 30, 54, 55]

P 4-nitroaniline + 5-L-glutamyl-L-isoleucine

S 5-D-glutamyl-4-nitroanilide + acceptor <2, 25, 26> (Reversibility: ? <2, 25, 26> [34, 72, 76]) [34, 72, 76]

P 4-nitroaniline + 5-D-glutamyl-acceptor <25, 26> [72, 76]

S 5-D/L-glutamyl-phenylhydrazine + acceptor <2> (Reversibility: ir <2> [70]) [70]

P phenylhydrazine + 5-L-glutamyl-acceptor <2> (<2> product is cytotoxic [70]) [70]

S 5-L-glutamine + 5-L-glutamine <16> (Reversibility: ? <16> [47]) [47]

P 5-poly-L-glutamic acid + glutamate <16> [47]

S 5-L-glutamyglycylglycine + acceptor <16> (Reversibility: ? <16> [47]) [47]

P glycylglycine + 5-L-glutamyl-acceptor

S 5-L-glutamyl-3-carboxy-4-nitroanilide + 5-L-glutamyl-3-carboxy-4-nitroanilide <1> (Reversibility: ? <1> [37, 38]) [37, 38]

P 5-L-glutamyl-5-L-glutamyl-3-carboxy-4-nitroanilide + 3-carboxy-4-nitroaniline <1> [37, 38]

S 5-L-glutamyl-3-carboxy-4-nitroanilide + glycylglycine <1, 2, 4> (Reversibility: ? <1, 2, 4> [1, 25, 36-38]) [1, 25, 36-38]
P 5-L-glutamyl-glycylglycine + 3-carboxy-4-nitroaniline <1> [37, 38]
S 5-L-glutamyl-4-nitroanilide + 5-L-glutamyl-4-nitroanilide <1, 2, 5-7> (<1> autotranspeptidase [7]; <6> autotranspeptidation, L- or, to a lesser extent, D-enantiomer [30]) (Reversibility: ? <1, 2, 5-7> [7, 23, 27, 30, 32]) [7, 23, 27, 30, 32]
P 5-L-glutamyl-5-L-glutamyl-4-nitroanilide + 4-nitroaniline <1, 2, 5-7> [7, 23, 27, 29, 30, 32]
S 5-L-glutamyl-4-nitroanilide + H_2O <1, 6, 24> (<1,6,24> hydrolase reaction [29,67,71]) (Reversibility: ir <6> [29]; ? <1, 24> [67, 71]) [29, 67, 71]
P 4-nitroaniline + L-glutamate <1, 6, 24> [29, 67, 71]
S 5-L-glutamyl-4-nitroanilide + L-alanine <1, 2, 6, 7, 11-13> (<2> glutamyl transfer at 32% the rate of the reaction with glycylglycine [24]; <6> L-Ala or D-Ala [30]) (Reversibility: ? <1, 2, 6, 7, 11-13> [6, 18, 24, 27, 28, 30, 42, 51, 54, 55, 73]) [6, 18, 24, 27, 28, 30, 42, 51, 54, 55, 73]
P 4-nitroaniline + 5-L-glutamyl-L-alanine
S 5-L-glutamyl-4-nitroanilide + L-alanylglycine <2> (Reversibility: ? <2> [73]) [73]
P 4-nitroaniline + 5-L-glutamyl-L-alanylglycine
S 5-L-glutamyl-4-nitroanilide + L-arginine <2, 6, 7, 11, 12> (<2> glutamyl transfer at 30% the rate of the reaction with glycylglycine [24]) (Reversibility: ? <2, 6, 7, 11, 12> [18, 24, 28, 30, 51, 54]) [18, 24, 28, 30, 51, 54]
P 4-nitroaniline + 5-L-glutamyl-L-arginine
S 5-L-glutamyl-4-nitroanilide + L-asparagine <1, 6, 12, 17> (Reversibility: ? <1, 6, 12, 17> [30, 42, 48, 54]) [30, 42, 48, 54]
P 4-nitroaniline + 5-L-glutamyl-L-asparagine
S 5-L-glutamyl-4-nitroanilide + L-aspartate <2, 6, 11-13> (<2> glutamyl transfer at 8.6% the rate of the reaction with glycylglycine [24]) (Reversibility: ? <2, 6, 11-13> [24, 30, 40, 51, 54, 55, 73]) [24, 30, 40, 51, 54, 55, 73]
P 4-nitroaniline + 5-L-glutamyl-L-aspartate
S 5-L-glutamyl-4-nitroanilide + L-cysteine <2, 7, 11-13> (Reversibility: ? <2, 7, 11-13> [28, 40, 51, 54, 55]) [28, 40, 51, 54, 55]
P 4-nitroaniline + 5-L-glutamyl-L-cysteine
S 5-L-glutamyl-4-nitroanilide + L-cysteinylglycine <2, 3, 6, 8> (Reversibility: ? 2, 3, 6, 8 [34]) [34]
P 4-nitroaniline + glutathione
S 5-L-glutamyl-4-nitroanilide + L-cystine <1, 2, 4, 5, 12> (<1> best acceptor, low molecular weight enzyme variant [7]; <2,5> best acceptor [6,33,34]) (Reversibility: ? <1, 2, 4, 5, 12> [6, 7, 18, 33, 36, 40, 54]) [6, 7, 18, 33, 34, 36, 40, 54]
P 4-nitroaniline + 5-L-glutamyl-L-cystine
S 5-L-glutamyl-4-nitroanilide + L-glutamate <1, 2, 6, 7, 11-13> (Reversibility: ? <1, 2, 6, 7, 11-13> [18, 27, 28, 30, 42, 51, 54, 55, 73]) [18, 27, 28, 30, 42, 51, 54, 55, 73]
P 4-nitroaniline + 5-L-glutamyl-L-glutamate

S 5-L-glutamyl-4-nitroanilide + L-glutamine <1, 2, 5-7, 11, 13> (<1> best acceptor, high molecular weight enzyme variant [7]; <2> L-methionine and L-glutamine are best acceptors [3]; <1> best acceptor [7]; <2> glutamyl transfer at 43% the rate of the reaction with glycylglycine [24]) (Reversibility: ? <1, 2, 5-7, 11, 13> [3, 6, 7, 11, 15, 18, 24, 27, 28, 30, 40-42, 46, 51, 55, 73]) [3, 6, 7, 11, 15, 18, 24, 27, 28, 30, 40-42, 46, 51, 55, 73]

P 4-nitroaniline + 5-L-glutamyl-L-glutamine

S 5-L-glutamyl-4-nitroanilide + L-histidine <2, 6, 11, 12> (<2> glutamyl transfer at 8.6% the rate of the reaction with glycylglycine [24]) (Reversibility: ? <2, 6, 11, 12> [24, 30, 51, 54]) [24, 30, 51, 54]

P 4-nitroaniline + 5-L-glutamyl-L-histidine

S 5-L-glutamyl-4-nitroanilide + L-hydroxyproline <2, 11> (<2> glutamyl transfer at 17% the rate of the reaction with glycylglycine [24]) (Reversibility: ? <2, 11> [24, 51]) [24, 51]

P 4-nitroaniline + 5-L-glutamyl-L-hydroxyproline

S 5-L-glutamyl-4-nitroanilide + L-leucine <2, 6, 11, 12> (<2> glutamyl transfer at 20% the rate of the reaction with glycylglycine [24]) (Reversibility: ? <2, 6, 11, 12> [24, 30, 51, 54]) [24, 30, 51, 54]

P 4-nitroaniline + 5-L-glutamyl-L-leucine

S 5-L-glutamyl-4-nitroanilide + L-leucyl-L-alanine <1> (Reversibility: ? <1> [46]) [46]

P 4-nitroaniline + 5-L-glutamyl-L-leucyl-L-alanine

S 5-L-glutamyl-4-nitroanilide + L-lysine <2, 6, 7, 11-13> (<2> glutamyl transfer at 26% the rate of the reaction with glycylglycine [24]) (Reversibility: ? <2, 6, 7, 11-13> [18, 24, 28, 30, 51, 54, 55]) [18, 24, 28, 30, 51, 54, 55]

P 4-nitroaniline + 5-L-glutamyl-L-lysine

S 5-L-glutamyl-4-nitroanilide + L-methionine <1, 2, 5-7, 11, 13, 19> (<2> L-methionine and glutamine are best acceptors [3]; <7> best acceptor [27]; <2> glutamyl transfer at 39% of the rate with glycylglycine [24]; <3> 44% activity, viral enzyme [26]; <6> 44% the rate of the reaction with glycylglycine, L-Met or D-Met [30]) (Reversibility: ? <1, 2, 5-7, 11, 13, 19> [3, 6, 11, 15, 18, 24, 26-28, 30, 40, 42, 51, 55, 73]) [3, 6, 11, 15, 18, 24, 26-28, 30, 40, 42, 51, 55, 73]

P 4-nitroaniline + 5-L-glutamyl-L-methionine

S 5-L-glutamyl-4-nitroanilide + L-phenylalanine <2, 6, 7, 11-13> (<2> glutamyl transfer at 26% the rate of the reaction with glycylglycine [24]) (Reversibility: ? <2, 6, 7, 11-13> [24, 28, 30, 51, 54, 55, 73]) [24, 28, 30, 51, 54, 55, 73]

P 4-nitroaniline + 5-L-glutamyl-L-phenylalanine

S 5-L-glutamyl-4-nitroanilide + L-proline <7, 11, 13> (Reversibility: ? <7, 11, 13> [27, 51, 55]) [27, 51, 55]

P 4-nitroaniline + 5-L-glutamyl-L-proline

S 5-L-glutamyl-4-nitroanilide + L-serine <1, 2, 7, 11-13> (<2> glutamyl transfer at 28% the rate of the reaction with glycylglycine [24]) (Reversibility: ? <1, 2, 7, 11-13> [24, 27, 42, 51, 54, 55]) [24, 27, 42, 51, 54, 55]

P 4-nitroaniline + 5-L-glutamyl-L-serine

S 5-L-glutamyl-4-nitroanilide + L-threonine <2, 11, 13> (<2> glutamyl transfer at 20% the rate of the reaction with glycylglycine [24]) (Reversibility: ? <2, 11, 13> [24, 51, 55]) [24, 51, 55]
P 4-nitroaniline + 5-L-glutamyl-L-threonine
S 5-L-glutamyl-4-nitroanilide + L-tryptophan <11-13> (Reversibility: ? <11-13> [51, 54, 55]) [51, 54, 55]
P 4-nitroaniline + 5-L-glutamyl-L-tryptophan
S 5-L-glutamyl-4-nitroanilide + L-valine <2, 6, 11-13> (<2> glutamyl transfer at 20% the rate of the reaction with glycylglycine [24]) (Reversibility: ? <2, 6, 11-13> [24, 30, 51, 54, 55]) [24, 30, 51, 54, 55]
P 4-nitroaniline + 5-L-glutamyl-L-valine
S 5-L-glutamyl-4-nitroanilide + S-alkylcysteine sulfoxide <11> (<11> best substrates [51]) (Reversibility: ? <11> [51]) [51]
P 4-nitroaniline + 5-L-glutamyl-S-alkylcysteine sulfoxide
S 5-L-glutamyl-4-nitroanilide + acceptor <11> (Reversibility: ? <11> [52]) [52]
P 4-nitroaniline + 5-L-glutamyl-acceptor
S 5-L-glutamyl-4-nitroanilide + aminobutyrate <7, 11> (<7,11> α-enantiomer of aminobutyrate [27,51]; <7> β-enantiomer of aminobutyrate [27]) (Reversibility: ? <7, 11> [27, 51]) [27, 51]
P 4-nitroaniline + 5-L-glutamylaminobutyrate
S 5-L-glutamyl-4-nitroanilide + citrulline <2, 5> (Reversibility: ? <2, 5> [11, 18]) [11, 18]
P 4-nitroaniline + 5-L-glutamyl-citrulline
S 5-L-glutamyl-4-nitroanilide + glycine <2, 6, 7, 11-13, 19> (<2> glutamyl transfer at 17.2% [24]; <19> 19% viral enzyme [26]; <6,7> 19% activity the rate of the reaction with glycylglycine, glycyl-tri- to hexapeptides can act as acceptors [27,30]) (Reversibility: ? <2, 6, 7, 11-13, 19> [24, 26, 27, 30, 51, 54, 55]) [24, 26, 27, 30, 51, 54, 55]
P 4-nitroaniline + 5-L-glutamyl-glycine
S 5-L-glutamyl-4-nitroanilide + glycyl-L-proline <1> (Reversibility: ? <1> [46]) [46]
P 4-nitroaniline + 5-L-glutamyl-glycyl-L-proline
S 5-L-glutamyl-4-nitroanilide + glycylglycine <1-8, 11-13, 15-20, 26> (<1> GGT activity [7]; <1,2,4,7> best acceptor [18,24,27,36]) (Reversibility: ir <1> [38]; ? <1-8, 11-13, 15-20, 26> [6, 7, 11, 12, 15, 18, 20, 22-28, 30, 35, 36, 39, 41, 42, 45-51, 53-55, 57, 60, 65, 67, 70, 73, 76]) [6, 7, 11, 12, 15, 18, 20, 22-28, 30, 35, 36, 38, 39, 41, 42, 45-51, 53-55, 57, 60, 65, 67, 70, 73, 76]
P 4-nitroaniline + 5-L-glutamylglycylglycine <1-4, 7, 8, 19, 20, 26> (<2> 4-nitroaniline is cytotoxic [70]) [26-28, 35, 36, 38, 39, 41, 53, 60, 65, 67, 70, 76]
S 5-L-glutamyl-7-amino-4-methylcoumarin + acceptor <2, 22, 26> (<22> glycylglycine is best acceptor [66]; <2,22> acceptors are amino acids and peptides, overview [66,73]) (Reversibility: ? <2, 22, 26> [66, 73, 76]) [66, 73, 76]
P 7-amino-4-methylcoumarin + 5-L-glutamyl-acceptor

S 5-L-glutamyl-7-amino-4-methylcoumarin + glycylglycine <2, 3, 5, 8> (Reversibility: ? <2, 3, 5, 8> [19, 21, 33, 35, 40]) [19, 21, 33, 35, 40]
P 7-amino-4-methylcoumarin + 5-L-glutamyl-glycylglycine
S 5-L-glutamyl-L-alanine + 5-L-glutamyl-L-alanine <21> (<21> shows also hydrolase activity with this substrate [63]) (Reversibility: ? <21> [63]) [63]
P L-glutamine + 5-L-glutamyl-L-glutamyl-L-alanine <21> [63]
S 5-L-glutamyl-L-lysine + acceptor <21> (<21> shows also hydrolase activity with this substrate [63]) (Reversibility: ? <21> [63]) [63]
P L-lysine + 5-L-glutamyl-acceptor <21> [63]
S 5-L-glutamyl-cis-3-amino-L-proline + acceptor <13> (Reversibility: ? <13> [55]) [55]
P 5-L-glutamyl-acceptor + cis-3-amino-L-proline <13> [55]
S 5-L-glutamylaniline + 5-L-glutamylaniline <10> (Reversibility: ir <10> [58]) [58]
P aniline + 5-L-glutamyl-5-L-glutamylaniline <10> [58]
S 5-L-glutamylaniline + H_2O <10> (Reversibility: ir <10> [58]) [58]
P aniline + L-glutamic acid <10> [58]
S 5-L-glutamylaniline + S-methyl-L-cysteine <10> (Reversibility: ir <10> [58]) [58]
P aniline + 5-L-glutamyl-S-methyl-L-cysteine <10> [58]
S L-α-methyl-5-glutamyl-L-α-aminobutyrate + acceptor <6> (<6> acceptor specificity, overview [29]; <6> hydrolysis in the absence of acceptor, methyl group at α-C prevents autotranspeptidation [29]) (Reversibility: ? <6> [29]) [29]
P L-α-methyl-5-glutamyl acceptor + L-2-aminobutyrate <6> [29]
S S-methylglutathione + acceptor <21> (<21> shows also hydrolase activity with this substrate [63]) (Reversibility: ? <21> [63]) [63]
P S-methyl-L-cysteinylglycine + 5-L-glutamyl-acceptor <21> [63]
S glucana + glucana <1> (Reversibility: ir <1> [38]) [38]
P cana + 5-L-glutamyl-glucana <1> [38]
S glucana + glycylglycine <1> (Reversibility: ir <1> [38]; ? <1> [37]) [37, 38]
P cana + 5-L-glutamylglycylglycine <1> [38]
S glutathione + 1-aminocyclopropane-1-carboxylic acid <24> (<24> isoforms I and II [71]) (Reversibility: ? <24> [71]) [71]
P L-cysteinylglycine + 5-L-glutamyl-1-aminocyclopropane-1-carboxylic acid <24> [71]
S glutathione + H_2O <1-3, 6, 8, 14, 24, 25> (<1-3,6,8,14,24,25> hydrolase reaction, concurrent to (auto-)transpeptidation [29,30,35,41,71,72,74]) (Reversibility: ir <2, 3, 6, 8> [30, 35]; ? <1, 6, 14, 24, 25> [29, 41, 71, 72, 74]) [29, 30, 35, 41, 71, 72, 74]
P cysteinylglycine + glutamate <2, 3, 6, 8, 14, 25> [29, 30, 35, 72, 74]
S glutathione + L-cysteinylglycine <2, 3, 6, 8> (Reversibility: ? 2, 3, 6, 8 [34]) [34]
P L-cysteinylglycine + glutathione

S glutathione + acceptor <1-6, 8, 11, 17, 19, 21> (<2,3,8,17> acceptor: amino acids [35,48]; <2,5> best donor [33]; <11> acceptor: hydroxylamine [51]) (Reversibility: r <2, 3, 6, 8> [30, 35]; ? <1-6, 8, 11, 17, 19, 21> [11, 17, 26, 31, 33, 34, 36, 42, 48, 51, 63]) [11, 17, 26, 30, 31, 33, 34-36, 42, 48, 51, 63]
P cysteinylglycine + 5-L-glutamyl-acceptor <2, 3, 8> [17, 35]
S glutathione + daunomycin <2> (<2> detoxification of daunomycin by transpeptidation [70]) (Reversibility: ir <2> [70]) [70]
P L-cysteinylglycine + 5-L-glutamyl-daunomycin <2> [70]
S glutathione + glutathione <2, 3, 5, 6, 8> (<2,3,5,6,8> autotranspeptidation [29,30,32,34,35]) (Reversibility: r <2, 3, 6, 8> [30, 35]; ? <2, 3, 5, 6, 8> [29, 32, 34]) [29, 30, 32, 34, 35]
P L-cysteinylglycine + 5-glutamyl-glutathione <2, 3, 6, 8> [29, 30, 35]
S glutathione + leukotriene D <5> (Reversibility: ? <5> [33]) [33]
P cysteinylglycine + leukotriene C <5> [33]
S lentinic acid + amino acid <11> (<11> best acceptor is S-alkylcysteine sulfoxide [52]) (Reversibility: ? <11> [52]) [52]
P desglutamyl lentinic acid + 5-L-glutamylamino acid <11> [52]
S leukotriene C_4 + acceptor <2, 5> (<2,5> i.e. glutathione containing derivative of arachidonic acid, acceptors are H_2O, amino acids or dipeptides [33]) (Reversibility: ? <2, 5> [32, 33]) [32, 33]
P leukotriene D_4 + glutamyl-acceptor <2, 5> [32, 33]
S Additional information <2, 6> (<2> ester and amide derivatives of L-Glu-4-nitroanilide are no substrates, overview [75]; <6> the donor substrate determines the acceptor specificity [34]) [34, 75]
P ?

Inhibitors

1-chloro-3-tosylamido-7-amino-2-heptanone <17> (<17> weak [48]) [48]
4-aminobutyramide <2> [75]
4-carboxybutyramide <2> (<2> and derivatives [75]) [75]
4-nitroaniline <2, 26> (<26> strong, competitive against glutathione [76]; <2> cytotoxic, weakly mutagenic [70]) [70, 76]
5,5'-dithiobis(2-nitrobenzoate) <2, 6> (<2> weak [17]; <2,6,8,21> no inhibition [34,63]) [17, 24, 29]
5-glutamylhydrazones <2, 3, 6, 8> [34, 35]
5-glutamylphenylhydrazides <2, 3, 8> (<2> no inhibition and cytotoxity [70]) [35]
5-iodoacetamidofluorescein <1, 2> (<1> inactivation, active site modification [39]) [23, 39]
6-diazo-5-oxo-L-norleucine <1, 2, 18> (<1> complete inactivation, accelerated by maleate, prevented by S-methylglutathione [39]; <2> inactivation of kidney enzyme [20]; <2> irreversible inactivation of pancreatic enzyme, S-methylglutathione prevents at 10 mM, maleate increases inhibition at 0.25-1 M [23]) [20, 23, 39, 50]
6-diazo-5-oxo-L-norleucine <1-3, 6, 8, 17, 18> (<1> i.e. DON, strong, irreversible modification of glutamyl-binding site of light subunit [39,42]; <1> mal-

eate accelerates inactivation [39]; <1> S-methylglutathione [39]; <1> 5-glutamyl-donor, protect transpeptidase and hydrolase activity, not 5-carbon derivative [42]) [20, 23, 34, 35, 39, 42, 48, 50]
Ca^{2+} <2, 7> (<2> weak [24]; <7> at 0.15 M [15]; <2,6,11> no inhibition [3,29,51]) [15, 24]
Co^{2+} <1, 2> (<6> not affected [29]; <1,2> weak [24,41]) [24, 41]
Cu^{2+} <1, 11> (<11> strong [51]) [2, 51]
EDTA <2, 6> (<2,6> weak [24,29]; <2,11> no inhibition [17,19,21,51]) [24, 29]
Hg^{2+} <2, 11, 17, 18> (<11> strong [51]; <17> hydrolase, not transferase reaction [48]) [24, 48, 50, 51]
L- or D-serine/borate <2, 3, 5, 6, 8, 11, 13, 17-19, 22> (<2,6,8> formation of a tetrahedral complex at the active site [34]; <5> transition-state inhibitor [33]; <11> 1:1 mixture, strong [51]; <2,3,17-19> L-serine/borate, not D-serine/borate [24,26,48,50]; <2> reversible [23]; <2,22> competitive [66,70]) [23, 24, 26, 32-35, 48, 50-52, 55, 66, 70]
L-(αS,5S)-α-amino-3-chloro-4,5-dihydro-5-isoxazolacetic acid <5, 22> (<5,22> i.e. acivicin [33,66]; <22> in vivo inhibition and accumulation of glutathione [66]; <22> kinetics [66]; <5,22> strong, irreversible, reacts with glutamyl binding site [33,66]) [33, 66]
L-(αS,5S)-α-amino-3-chloro-4,5-dihydro-5-isoxazole acetic acid <2, 3, 6, 8, 17> (<2,3,8,17> i.e. AT-125 [35,48]; <2,3,8> strong, irreversible [35]; <17> strong, irreversible, 1 mM [48]) [34, 35, 48]
L-1-tosylamido-2-phenylethylchloromethylketone <17> [48]
L-5-glucana <1> (<1> enzyme from human tissues, not serum [36,38]) [36, 38]
L-alanine <17, 18> [48, 50]
L-α-glucana <1> (<1> weak [38]) [38]
L-azaserine <1-3, 6, 8, 17> (<1-3,8> irreversible inactivation [35,42]; <1> transpeptidase and hydrolase reaction, glutamyl-donor protects [42]; <1> not 5-carbon derivative [42]) [23, 34, 35, 39, 42, 48]
L-glutamine <17, 18> (<17,18> not D- [48,50]) [48, 50]
5-L-glutamyl-2-(2-carboxyphenyl)hydrazine <1-3, 6, 8> (<2, 3, 6, 8> i.e. anthglutin, strong, isolated from Penicillium oxalicum [34]; <2,3,8> i.e. anthglutin, strong, in vivo and in vitro, *p*-derivative less effective [35]; <1> methionine protects in vivo [53]; <1> transpeptidation, competitive to the glutamyl-donor, kinetics [53]) [34, 35, 53]
Mg^{2+} <5, 7> (<5> above 0.1 M [11]; <7> at 0.15 M [15]; <2,6,11> no inhibition [19,21,24,29,51]) [11, 15]
Mn^{2+} <2> (<2> weak [24]) [24]
N-acetyl-L-glutamine <2> (<2> poor inhibitor, competitive [75]) [75]
N-ethylmaleimide <1, 2, 6, 8, 17> (<2,6,8,17> weak [17,24,34,48]; <18,21> no inhibition [50,63]) [17, 24, 29, 34, 41, 48]
NH_4^+ <1> (<1> weak [41]) [41]
Na^+ <5, 7> (<6> not affected [29]; <5> above 0.1 M [11]; <7> 0.1 M [15]) [11, 15]
N^{α}-4-tosyl-L-lysine chloromethyl ketone <25> (<25> inactivation [72]) [72]

PMSF <2, 17> (<17> weak [48]; <25> no effect [72]) [35, 48]
Pb^{2+} <12> [54]
Zn^{2+} <1, 2, 11> (<2> 50% inhibition at 0.4 mM [21]; <2> 95% inhibition at 5 mM [19]; <2,11> strong [17,19,51]; <2> reversible by EDTA [19,21]) [2, 17, 19, 21, 24, 51]
acetazolamide maleate <2, 6, 8, 1-3, 6-8> (<2,3,8> transpeptidase, not hydrolase reaction [27,35]) [7, 27, 34, 35]
α-ketoglutarate-5-glutamylhydrazone <3, 19> (<3,19> strong, i.e. GSH-analogue [26]) [26]
aminooxyacetate <18> (<18> hydrolase reaction [50]) [50]
antiserum to rat kidney glutamyltransferase <2> [23]
β-chloro-L-alanine <18> [50]
brefeldin A <1> (<1> inhibition of recombinant mutant enzyme secretion into cell culture medium from Sf21 cells, accumulation in the cells [65]) [65]
bromocresol green <2, 3, 6, 8> [34]
daunomycin <2> (<2> cytotoxic and mutagenic [70]) [70]
free bile acids and their glycine and taurine conjugates <2, 3, 8> [35]
glutathione <1-3, 7, 19> (<7> complete inhibition [28]; <19> 50% inhibition at 0.375 mM [26]; <2> L-Glu-4-nitroanilide as substrate [18]; <18> no inhibition [50]) [18, 26, 28, 41]
glycine <1, 2, 18> (<2> strong [3]; <18> weak [50]) [3, 36, 37, 50]
glycylglycine <1, 4, 5> (<4> high concentration, pH 6.0-7.5 [25]; <5> high concentration, with L-Glu-4-nitroanilide, not glutathione as donor [11]) [11, 25, 36]
hexobarbital <2> [22]
hippurate <2, 3, 8> (<2,3,8> transpeptidase, not hydrolase reaction [35]) [35]
iodoacetamide <1-3, 6, 8, 18, 19> (<1> inhibition of transpeptidase reaction is more efficient than that of autotranspeptidase reaction [7]; <2,6,8> irreversible [34]; <2> no inhibition [17]) [7, 26, 29, 34, 50]
iodoacetate <1, 6, 17, 18> (<1,6,17,18> weak [29,41,48,50]; <2> no inhibition [17]) [29, 41, 48, 50]
maleate <1, 7> [15, 41]
p-chloromercuribenzoate <2, 6, 11> (<11> strong [51]; <2> weak [19,21]; <2,6,8,13,17,18,21> no inhibition [17,34,48,50,55,63]) [19, 21, 29, 51]
p-hydroxymercuribenzoate <1> (<1> weak [41]) [41]
phenobarbital <2, 5> (<2> active site-directed, irreversible inactivation, optimal at pH 9.0, serine/borate or GSH protects, maleate protects slightly, kinetics [22]) [22, 33]
phenylhydrazides <2, 3, 6, 8> (<2> cytotoxic and mutagenic [70]) [34, 70]
phosphate <11, 15> [51, 57]
pronase <25> (<25> inactivation [72]) [72]
proteinase K <25> (<25> inactivation [72]) [72]
sulfobromophthalein derivatives <2, 3, 6, 8> [34]
sulfophthalein derivatives <2, 3, 6, 8> [34]
thiobarbituric acid <2, 5> (<2> active site-directed, irreversible inactivation, optimal at pH 9.5, serine/borate or GSH protects, maleate protects slightly, kinetics [22]) [22, 33]

tosyl fluoride <21> (<21> inhibition by tosyl fluoride only in presence of maleate [63]) [63]
tris(hydroxymethyl)aminomethane <1> [38]
urea <1> (<1> complete inactivation of wild-type and mutant at 6 M, low activity at 4 M [65]) [65]
Additional information <1, 2, 3, 8, 6, 11, 14, 17, 18, 23-25> (<2> inhibition mechanism [75]; <25> not affected by benzamidine [72]; <24> not affected by KCl, KNO_3, KBr, NaCl, $CaCl_2$, $MgCl_2$, $MnCl_2$, $CoCl_2$, $ZnCl_2$, DTT, EDTA [71]; <11> no inhibition by arsenite, citrate, cyanide and borate [51]; <1, 2, 10, 18, 23> inhibition by diverse amino acids, overview [34,41,46,50,58,69]; <6> not affected by Ni^{2+}, K^+, phtalate, pyruvate and fumarate [29]; <14> product inhibiton [56]; <2> no inhibiton by phorbol ester, polyamines, cAMP [3]) [3, 29, 34, 35, 38, 41, 48, 50, 51, 56, 63, 69-72, 75]

Activating compounds

2-mercaptoethanol <25> (<25> activation [72]) [72]
carboxylic acids <10> (<10> activation, e.g. sodium citrate [59]) [58, 59]
dithiothreitol <25> (<25> activation [72]) [72]
free bile acids <2, 3, 5, 8> (<2,3,8> and their glycine and taurine conjugates [35]; <2,3,5,8> activation [33,35]) [33, 35]
hippurate <2, 3, 6, 8> (<2,3,6,8> stimulates hydrolase reaction, inhibits transpeptidation [34,35]) [34, 35]
maleate <1-3, 5-8, 21> (<6> 1.6fold [29]; <2,7> stimulates hydrolase activity [27,34]; <1> stimulates hydrolase reaction 3fold, inhibits transpeptidase reaction, with L-Glu as substrate [42]; <13> no stimulation [55]) [27, 29, 33-35, 42, 63]
Additional information <1, 6, 13, 14, 24> (<14> nitrogen starvation induces enzyme activity [74]; <24> no effects by KCl, KNO_3, KBr, NaCl, $CaCl_2$, $MgCl_2$, $MnCl_2$, $CoCl_2$, $ZnCl_2$, DTT, EDTA [71]; <1> effect of amino acids on activity [41,46]; <6> no effect by phthalate, pyruvate and fumarate [29]; <13> phosphate, arsenate, carbonate [55]) [29, 41, 46, 55, 71, 74]

Metals, ions

Ca^{2+} <2, 6> (<2,6> activation, transpeptidase [17,30]; <2,6,13> not hydrolase [17,30,55]; <2> slight activation [17]; <1,6,24> no effect [29,41,46,71]) [17, 30]
K^+ <6> (<6> activation, transpeptidase [30]; <6,13> not hydrolase [30,55]; <1,6,24> no effect [29,41,46,71]) [30]
Li^+ <6> (<6> activation [30]; <1> no effect [41,46]) [30]
Mg^{2+} <2, 6> (<2,6> activation, transpeptidase [17,30]; <2,6,13> not hydrolase [17, 30, 55]; <2> slight activation [17]; <1, 2, 6, 24> no effect [21, 29, 41, 46, 71]) [17, 30]
Na^+ <6> (<6> activation, transpeptidase [30]; <6,13> not hydrolase [30,55]; <1,6,24> no effect [29,41,46,71]) [30]
Additional information <1, 6, 13, 24> (<24> no effects by $MnCl_2$, $CoCl_2$, $ZnCl_2$ [71]; <6> not affected by Co^{2+} and Ni^{2+} [29]; <1> no activation by Zn^{2+} and Mn^{2+} [41,46]) [29, 41, 46, 55, 71]

Turnover number (min^{-1})

69000 <1> (4-nitroanilide) [42]
75000 <2> (4-nitroanilide) [42]

Specific activity (U/mg)

1.3 <2> (<2> partially purified from mammary gland [18]) [18]
1.4 <17> (<17> purified isozymes A and B [48]) [48]
3.1 <10> (<10> purified enzyme [58]) [58]
6-7.7 <19> (<19> partially purified viral enzyme [26]) [26]
8.175 <24> (<24> purified isoform II, transpeptidation [71]) [71]
8.9 <12> (<12> purified enzyme [54]) [54]
12 <1> (<1> purified recombinant mutant H383A/H505A [67]) [67]
13.41 <24> (<24> purified isoform I, transpeptidation [71]) [71]
16 <1, 18> (<1,18> purified enzyme [39,50]) [39, 50]
17.37 <24> (<24> purified isoform II, hydrolase reaction [71]) [71]
21.2 <3> (<3> partially purified milk fat-globule membrane enzyme [26]) [26]
24.03 <24> (<24> purified isoform I, hydrolase reaction [71]) [71]
25.6 <1> (<1> purified enzyme, normal and cirrhotic liver [60]) [60]
47 <1> (<1> purified recombinant mutant H505A [67]) [67]
62.9 <16> (<16> purified enzyme [47]) [47]
75.7 <1> (<1> purified enzyme [44]) [44]
81 <1> (<1> purified enzyme [46]) [46]
81.5 <16> (<16> purified enzyme [47]) [47]
102.8 <1> (<1> purified enzyme of cellular hepatic carcinoma [60]) [60]
104.5 <2> (<2> purified enzyme [24]) [24]
120 <1, 14> (<1> hepatoma [43]) [43, 56]
150 <1> (<1> purified recombinant mutant H383A [67]) [67]
158.4 <25> (<25> purified enzyme [72]) [72]
165 <2> (<2> purified enzyme [40]) [40]
181 <2> (<2> purified enzyme [17]) [17]
200.4 <1> (<1> purified enzyme [41]) [41]
218 <2> (<2> partially purified enzyme from liver [21]) [21]
250 <1> (<1> purified enzyme from liver [43]) [43]
398 <2> (<2> purified enzyme from biliary tract [21]) [21]
407 <7> (<7> purified enzyme [27]) [27]
423 <1> (<1> purified enzyme [45]) [45]
440 <1> (<1> purified recombinant wild-type enzyme [65,67]) [65, 67]
450 <1> (<1> purified recombinant mutant enzyme [65]) [65]
510 <6> (<6> purified enzyme, without acceptor [30]) [30]
540 <2> (<2> kidney [3]) [3]
630 <2> (<2> purified enzyme [23]) [23]
738 <2> (<2> purified Triton X-100 solubilized kidney enzyme [35]) [35]
792 <2> (<2> purified enzyme [33]) [33]
800 <1> (<1> purified enzyme [42]) [42]
810 <2> (<2> purified papain solubilized kidney enzyme [35]) [35]
956 <7> (<7> purified enzyme [28]) [28]

1009 <22> (<22> purified enzyme [66]) [66]
1320 <2> (<2> purified enzyme from mammary carcinoma [18]) [18]
2213 <21> (<21> purified enzyme [63]) [63]
Additional information <1, 2, 4, 11, 13, 15, 16, 22> (<2> activity in induced diabetic rats is higher than in control rats and higher in male than in female rats [73]; <2> activity during developmental stages in lung epithelial cells type I and II [64]; <1> assay method [37,38]; <2,16,22> activities with diverse substrates [34,47,66]; <2> activity in subcellular fractions [24]) [19, 24, 25, 34, 38, 47, 51, 55, 57, 64, 66, 73]

K_m-Value (mM)

0.005-0.0068 <2, 12> (L-Glu-4-nitroanilide, <2,12> hydrolase reaction [35,54]) [34, 35, 54]
0.0056 <5> (leukotriene C) [32]
0.0057 <5> (glutathione) [32]
0.0068 <12> (L-Glu-4-nitroanilide, <12> hydrolase reaction [54]) [54]
0.008 <1> (L-Glu-4-nitroanilide, <1> hydrolase reaction [67]) [67]
0.029-0.035 <2, 17> (L-cystine, <2> kidney enzyme [34]; <17> glutathione [48]; <2> D-Glu-4-nitroanilide [35]; <2,17> hydrolase reaction [35,48]) [34, 35, 48]
0.03 <22> (5-L-glutamyl-7-amino-4-methylcoumarin) [66]
0.031 <2> (5-D-Glu-4-nitroanilide) [34]
0.033 <18> (glutathione, <18> hydrolase reaction [50]) [50]
0.035 <17> (L-Glu-4-nitroanilide) [48]
0.035 <17> (glutathione) [48]
0.068 <17> (L-Glu-4-nitroanilide, <17> hydrolase reaction [48]) [48]
0.09 <24> (glutathione, <24> isoform II [71]) [71]
0.11 <24> (glutathione, <24> isoform I [71]) [71]
0.13-0.18 <2, 12, 18> (L-Glu-4-nitroanilide, <18> with GSH [50]; <2> with L-Cys-bis-glycine, rat kidney enzyme [34]) [34, 50, 54]
0.18 <18> (glutathione, <18> transpeptidation [50]) [50]
0.21 <2> (L-Glu-4-nitroanilide, <2> with 8 mM L-Gly-Gly, kidney enzyme [3]) [3]
0.21-0.32 <2> (L-Glu-4-nitroanilide, <2> with L-Met-Gly, kidney enzyme [34]) [34]
0.25 <26> (glutathione) [76]
0.31 <18> (L-Glu-4-nitroanilide, <18> hydrolase reaction [50]) [50]
0.32 <2> (L-Glu-4-nitroanilide, <2> with 16 mM L-Gly-Gly , kidney enzyme [3]) [3]
0.39 <2> (5-L-glutamyl-7-amino-4-methylcoumarin, <2> liver enzyme [21]) [21]
0.4 <18> (L-Glu-4-nitroanilide, <18> transpeptidation [50]) [50]
0.41 <2> (5-L-glutamyl-7-amino-4-methylcoumarin) [19]
0.49 <2> (5-L-glutamyl-7-amino-4-methylcoumarin, <2> biliary tract enzyme [21]) [21]
0.65 <1> (L-Glu-3-carboxy-4-nitroanilide) [38]
0.67 <3> (L-Glu-4-nitroanilide) [26]

0.67 <3> (glutathione) [26]
0.76 <2> (L-Gln, <2> rat kidney enzyme [34]) [34]
0.8 <1, 7, 26> (L-Glu-4-nitroanilide, <26> recombinant enzyme [76]; <7> in presence of glycylglycine [28]) [28, 41, 76]
0.81 <1> (L-Glu-4-nitroaninilide) [45]
0.9 <2> (L-Glu-4-nitroanilide, <2> carcinoma enzyme [18]) [18, 23]
1 <1, 2> (L-Glu-4-nitroanilide) [24, 39]
1.09 <1, 2> (L-Glu-3-carboxy-4-nitroanilide, <2> L-Cys-Gly, rat kidney enzyme [34]) [1, 34]
1.09 <2> (L-cysteinylglycine) [34]
1.1 <2> (L-Glu-4-nitroanilide) [17]
1.25 <7> (L-Glu-4-nitroanilide, <7> in absence of glycylglycine [28]) [28]
1.28 <22> (glycylglycine) [66]
1.3 <7> (L-Glu-4-nitroanilide, <7> enzyme from ciliary body [15]) [12, 15]
1.4 <1> (L-Glu-4-nitroanilide, <1> recombinant wild-type, transpeptidation [67]) [67]
1.5 <1> (L-Glu-4-nitroanilide, <1> recombinant wild-type [65]; <1> bile enzyme [46]) [46, 65]
1.7 <24> (L-Glu-4-nitroanilide, <24> isoform I, hydrolase reaction [71]) [71]
2.1 <1, 24> (L-Glu-4-nitroanilide, <24> isoform II, hydrolase reaction [71]; <1> recombinant mutant enzyme [65]) [65, 71]
2.9 <6> (L-α-methyl-5-glutamyl-L-α-aminobutyrate, <6> hydrolase reaction [29]) [29]
2.9-2.96 <1, 2, 18> (glycylglycine, <1> recombinant wild-type enzyme [65]; <2> rat kidney enzyme [34]) [34, 50, 65]
3.4 <1> (glycylglycine, <1> recombinant mutant enzyme [65]) [65]
3.61 <1> (D-Glu-3-carboxy-4-nitroanilide) [38]
4 <11> (L-Glu-4-nitroanilide) [51]
4 <2> (glycylglycine, <2> with 0.8 mM L-Glu-4-nitroanilide, kidney enzyme [3]) [3]
4-4.7 <2> (glycylglycine, <2> L-Met, rat kidney enzyme [34]) [34]
4.35 <2> (glycylglycine, <2> with 1.6 mM L-Glu-4-nitroanilide, kidney enzyme [3]) [3]
7.6 <2> (glycylglycine) [24]
10 <18> (glutamine, <18> hydrolase reaction [50]) [50]
10 <7> (glycylglycine) [12]
12.4 <1> (glycylglycine) [45]
18 <14> (glutathione) [74]
20 <2> (glycylglycine) [23]
32 <18> (L-Phe) [50]
48 <12> (glycylglycine) [54]
210 <17> (L-Arg) [48]
590 <17> (glycylglycine) [48]
Additional information <1-4, 6, 8, 10, 13-15, 22, 24> (<2> K_m-values for several acceptors in diabetic and control rats, female and male, with donors L-Glu-4-nitroanilide and L-5-glutamyl-(7-amido-4-methylcoumarin) [73]; <24> biphasic kinetics in transpeptidase reaction with 1-aminocyclopropane-1-

carboxylic acid as acceptor, K_m-values [71]; <2,14,22> K_m values for several amino acids and dipeptides, overview [34,56,66]; <1> enzyme from seminal plasma, kidney, prostate, and testis [2]; <2> kinetic study of renatured large subunit [20]; <1,2,4-6,14,22> kinetic study [25,29,32-34,56,66-68]; <13> hydrolysis data [55]) [2, 20, 25, 29, 32, 33-35, 55-58, 66-68, 71, 73]

K_i-Value (mM)

0.0057 <20> (5-L-glutamyl-2-(2-carboxyphenyl)hydrazine) [53]
0.0102 <1> (5-L-glutamyl-2-(2-carboxyphenyl)hydrazine, <1> membrane-bound liver enzyme [53]) [53]
0.0136 <1> (5-L-glutamyl-2-(2-carboxyphenyl)hydrazine, <1> soluble liver enzyme [53]) [53]
0.0183 <1> (5-L-glutamyl-2-(2-carboxyphenyl)hydrazine, <1> kidney enzyme [53]) [53]
0.022 <2> (L-serine, <2> in presence of borate [24]) [24]
0.25 <26> (4-nitroaniline) [76]
0.3 <22> (L-serine, <22> in presence of borate [66]) [66]
0.3 <7> (glutathione) [28]
0.4 <19> (α-ketoglutarate-5-glutamylhydrazone) [26]
0.42 <22> (acivicin) [66]
0.61 <22> (D-serine, <22> in presence of borate [66]) [66]
1.17 <2> (glutathione, <2> carcinoma enzyme [18]) [18]
1.2 <2> (6-diazo-5-oxo-L-norleucine) [23]
2.3 <1> (glycine) [37]
12.6 <2> (4-carboxybutyramide) [75]
20 <2> (thiobarbituric acid) [22]
23.2 <2> (4-aminobutyramide) [75]
30 <2> (hexobarbital) [22]
43 <2> (phenobarbital) [22]
Additional information <2, 3, 8> (<2> K_i-values for derivatives of 4-acrboxybutyramide, overview [75]; <1> 5-L-glutamyl-2-(2-carboxyphenyl)hydrazine: K_i values in different tissues, overview [53]; <2,3,8> K_i values for diverse amino acids, overview [34,35]) [34, 35, 53, 75]

pH-Optimum

5 <9> (<9> hydrolase reaction [4]) [4]
6 <18> (<18> hydrolase reaction [50]) [50]
6.5 <11> (<11> hydrolase reaction [52]) [52]
7 <23> (<23> transpeptidation [69]) [69]
7-9 <24> (<24> hydrolysis and transpeptidation, isoforms I and II [71]) [71]
7.5-8 <18> (<18> transpeptidation [50]) [50]
7.6 <1, 11> (<1> assay at [45]) [45, 51]
7.9 <1> [38]
8 <1, 2> (<1,2> assay at [34,41,65,67]) [23, 34, 39, 41, 65, 67]
8-9 <12> [54]
8.1 <1> [46]
8.2 <1, 2, 13> (<13> transpeptidation, isoform II and III [55]) [24, 41, 55]

8.2-8.5 <1> (<1> enzyme from seminal plasma, kidney, prostate, and testis [2]) [1, 2]
8.4 <23> (<23> hydrolase reaction [69]) [69]
8.5 <2, 7, 16, 21, 22> (<2,21,22> assay at [20,63,66]; <7> with acceptor glycylglycine [27]) [20, 27, 47, 63, 66]
8.5-9 <15> [57]
8.6 <1, 13> (<1> L-Glu-4-nitroanilide + glycylglycine [7]; <13> transpeptidation, isoform I [55]) [7, 55]
8.6-9 <13> (<13> hydrolase reaction, isoform III [55]) [55]
8.8 <6, 17> (<6,17> transpeptidation [30,48]) [30, 48]
8.8-9 <2> [17]
9 <4, 7, 11, 13> (<4> transferase activity [25]; <7> broad [27]; <11> above, transpeptidation [52]; <7> without acceptor [27]; <13> hydrolase reaction, isoform II [55]) [25, 27, 52, 55]
9-9.2 <13> (<13> hydrolase reaction, isoform I [55]) [55]
9.2 <9> (<9> transfer reaction [4]) [4]
9.4 <1> (<1> L-Glu-4-nitroanilide + L-Glu-4-nitroanilide [7]) [7]
9.5 <6, 10, 17> (<6,17> hydrolase reaction [30,48]) [30, 48, 58]
10 <4> (<4> hydrolase reaction [25]) [25]
Additional information <1, 2, 7, 10, 25> (<25> pI: 5.9 [72]; <1> pI: 4.2, liver enzyme [60]; <1> pIs of 6 isoforms in hepatic cellular carcinoma [60]; <4,10> pH-dependence of reaction kinetics [25,58]; <1> several isozymic forms with different content of sialic acid [42]; <2> 12 isozymes with pI-values ranging from 5 to 8 [35]; <2> pI: 3.0 [17]; <1> pI: 3.4-3.45 [5]; <7> pI: 3.63 [12]; <7> pI: 3.85 [27]; <1> pI: 4.5 [41]; <1> pI: 6.0 [1]) [1, 5, 12, 17, 25, 27, 35, 41, 42, 58, 60, 72]

pH-Range

6-12 <4> [25]
6.5-9.5 <11> (<11> about half-maximal activity at pH 6.5 and 9.5 [51]) [51]
7-9 <1> [46]
7.3-8.5 <7> (<7> about half-maximal activity at pH 7.3, maximal activity at pH 8.5 [27]) [27]
7.5-9 <16> (<16> about half-maximal activity at pH 7.5, about 90% of maximal activity at pH 9.0 [47]) [47]
8.1-9.6 <2> (<2> about 70% of maximal activity at pH 8.1, about 75% of maximal activity at pH 9.6 [17]) [17]

Temperature optimum (°C)

25 <7> (<7> assay at [28]) [28]
30 <1, 4, 15> (<1,4,15> assay at [25,38,57]) [25, 38, 57]
37 <1-3, 5-7, 10, 11, 16-19, 21, 22> (<1-3,5-7,10,11,16-19,21,22> assay at [17,20-24,26,27,29,30,32-34,39-42,44-48,50,52,53,58,63,65-67]) [17, 20-24, 26, 27, 29, 30, 32-34, 39-42, 44-48, 50, 52, 53, 58, 63, 65-67]
40 <23> (<23> hydrolase reaction [69]) [69]
45 <12> [54]
50 <23> (<23> transpeptidation [69]) [69]
60 <16> [47]

Temperature range (°C)

45-80 <16> (<16> about half-maximal activity at 45°C and about 70% of maximal activity at 80°C [47]) [47]

4 Enzyme Structure

Molecular weight

43000 <24> (<24> gel filtration [71]) [71]
54000 <1> (<1> gel filtration [1]) [1]
57000 <17> (<17> isoform B, gel filtration [48]) [48]
58000 <16, 17> (<17> isoform A, gel filtration [48]; <16> gel filtration [47]) [47, 48]
68000 <2> (<2> gel filtration [20,23]) [20, 23]
70000 <2, 22, 23> (<2,22,23> gel filtration [24,66,69]) [24, 66, 69]
71000 <1> (<1> papain-solubilized, gel filtration [39]) [39]
78000 <1> (<1> seminal plasma, prostate, testis [2]) [2]
80000 <1, 5, 7, 18> (<1> bile enzyme, gel filtration [10]; <7,18> gel filtration [27,50]; <5> sucrose gradient centrifugation [31]) [10, 27, 31, 50]
82000 <1> (<1> papain-solubilized enzyme, gel filtration [39]) [39]
86000 <1> (<1> gel filtration [42]) [36, 42]
90000 <1, 6, 14> (<1> PAGE [13]; <1,14> gel filtration [8,41,56]; <6> PAGE in 8 M urea, after cross-linkage of subunits with dimethylsuberimidate [30]) [8, 13, 30, 41, 45, 56]
98000 <1> (<1> Triton X-100-solubilized, PAGE [13]) [13]
100000 <7> (<7> gel filtration [12]) [12]
102000 <13> (<13> isoform III, gel filtration [55]) [55]
110000 <1, 7> (<1> bromelain solubilized [5]; <7> PAGE [28]; <1> gel filtration [45]) [5, 28, 45]
113000 <2> (<2> analytical ultracentrifugation [17]) [17]
120000 <12> (<12> gel filtration [54]) [54]
155000 <13> (<13> isoform I, gel filtration [55]) [55]
160000 <1> (<1> liver enzyme, gel filtration [10]; <1> Triton X-100 solubilized, gel filtration [39]) [10, 39]
180000 <10> (<10> gel filtration [58]) [58]
200000 <1, 21> (<21> gel filtration [63]; <1> Triton X-100 solubilized [6]) [6, 63]
210000 <1> (<1> Triton X-100 solubilized [5]) [5]
219000 <13> (<13> isoform II, gel filtration [55]) [55]
229000 <9> [4]
250000 <1, 2> (<1,2> gel filtration in the presence of Triton X-100 [21,44]) [21, 44]
300000 <1> (<1> serum enzyme, gel filtration [10]) [10]
500000 <1> (<1> gel filtration without Triton X-100 [44]) [44]
Additional information <1, 2, 7, 16, 19, 25-31> (<26-31> amino acid sequence alignment, enzyme gene family [76]; <25> amino acid sequence [72]; <16> N-terminal sequence [47]; <1> enzyme from kidney, native MW:

79000 and 105000 [2]; <1> multiple forms of different molecular weights [10]; <1,2> in the presence of Triton X-100: molecular aggregation [21,39,41]; <1> protease solubilized enzymes have a reduced MW [39]; <3> viral enzyme of at least MW 400000 [26]; <1,2,7,25> amino acid composition [17,18,23,28,39,41,42,44,72]; <1> immunological characterization [44]) [2, 10, 17, 18, 21, 23, 26, 28, 39, 41, 42, 44, 47, 72, 76]

Subunits

? <1, 21, 25> (<25> x * 31000, SDS-PAGE [72]; <1> x * 43000-44000 + x * 24000, wild-type and mutant enzyme, SDS-PAGE [65]; <21> x * 54000 + x * 21000, SDS-PAGE [63]) [63, 65, 72]

dimer <1, 2, 6, 7, 12, 14, 16-18, 22, 23> (<1> 1 * 38000 + 1 * 14000, SDS-PAGE [1]; <17> 1 * 38600 + 1 * 22000, isoform B, SDS-PAGE [48]; <17> 1 * 39200 + 1 * 22000, isoform A, SDS-PAGE [48]; <23> 1 * 42000 + 1 * 22000, SDS-PAGE [69]; <2> 1 * 43000 + 1 * 25000, SDS-PAGE [23]; <22> 1 * 43000 + 1 * 30000, SDS-PAGE [66]; <16> 1 * 45000 + 1 * 22000, SDS-PAGE [47]; <2> 1 * 45000 + 1 * 23000, SDS-PAGE [24]; <2> 1 * 46000 + 1 * 22000, papain solubilized, SDS-PAGE [35]; <1> 1 * 47000 + 1 * 22000, SDS-PAGE [45]; <18> 1 * 47000 + 1 * 28000, SDS-PAGE [50]; <2> 1 * 48000 + 1 * 20000, about, gel filtration of denatured and native enzyme [20]; <2> 1 * 50000 + 1 * 22000, kidney, SDS-PAGE [3]; <2> 1 * 51000 + 1 * 22000, Triton X-100 solubilized, SDS-PAGE [35]; <1> 1 * 53000 + 1 * 20000, papain-solubilized, SDS-PAGE [39]; <7> 1 * 55000 + 1 * 25000, SDS-PAGE [27]; <1> 1 * 57000 + 1 * 21000, SDS-PAGE, presence of urea, 2-mercaptoethanol [41]; <7> 1 * 57000 + 1 * 25500, SDS-PAGE [14]; <1> 1 * 61000 + 1 * 27000, SDS-PAGE [8]; <1> 1 * 62000 + 1 * 20000, SDS-PAGE [39]; <1> 1 * 62000 + 1 * 22000, SDS-PAGE [42]; <14> 1 * 64000 + 1 * 23000, SDS/2-mercaptoethanol-PAGE [56]; <1> 1 * 64000 + 1 * 29000, SDS-PAGE [43]; <6> 1 * 65000 + 1 * 27000, SDS-PAGE, 8 M urea-PAGE [30]; <12> 1 * 66000 + 1 * 55000, SDS-PAGE [54]; <7> 1 * 68000 + 1 * 27000, SDS-PAGE [28]; <7> 1 * 71000 + 1 * 28000, SDS-PAGE [12]; <1> 1 * 150000 + 1 * 95000, SDS-PAGE [44]) [1, 3, 8, 12, 14, 20, 23, 24, 27, 28, 30, 35, 39, 41-45, 47, 48, 50, 54, 56, 66, 69]

monomer <24> (<24> 1 * 43000, isoforms I and II, SDS-PAGE [71]) [71]

Additional information <1, 2> (<2> small subunit MW: 20000 [22]; <2> the enzyme is composed of two non-identical catalytically active subunits [20]; <1,2> the active site resides on the small subunit [20,22,36]; <1> protease solubilization leads to a reduced MW of the larger subunit [39]) [20, 22, 36, 39]

Posttranslational modification

glycoprotein <1, 2, 7, 8, 13, 14, 24> (<24> isoforms I and II [71]; <2> heterogenous, tissue-dependent glycosylation pattern [68]; <1> recombinant wild-type and mutant [65]; <14> 31.4% carbohydrates, binds lentil lectin [56]; <1> comparison of sugar content of enzymes from tumour and normal tissue source, overview [43]; <2> post-translational multiple forms differing in sugar portion [16]; <7> both subunits are glycoproteins [14,27]; <1> binding of lentil lectin [60]; <1,2,24> binding of concanavalin A [5,13,21,24,45,60,68,71]; <1, 2, 7> sialoglycoprotein [8, 15, 16, 27, 28, 60]; <1,2,4,7> carbohydrate-rich

[28,36]; <2> 36% carbohydrate [17]; <7> 20% carbohydrate [27]; <7> contains neutral and amino sugars and sialic acid [28]; <2> both subunits are glycoproteins, structure [35,43]) [1, 5, 8, 12-17, 21, 24, 27, 28, 35, 36, 39, 42, 43, 45, 55, 56, 60, 64, 65, 68, 71, 74]

5 Isolation/Preparation/Mutation/Application

Source/tissue

Koji culture <12, 15> (<12,15> solid culture on wheat bran [54,57]) [54, 57]
WI-38 cell <1> (<1> whole body, fetal [39]) [39]
bile <1> [10, 46, 53]
bile duct <2> (<2> 80-90% of activity [21]) [21]
blood <1> (<1> leakage of enzyme by proteolysis from liver to blood serum [13]; <1> 6 isoforms, serum of patients with cellular hepatic carcinoma [60]; <1> serum [10,13,36-38,53]) [10, 13, 36-38, 53, 60]
brain <1, 2> (<2> microvessel of endothelial cells, immortalized cell line [62]; <1> cortex microvessels [6]) [6, 62, 68]
cell culture <1, 2, 12, 15> [5, 39, 54, 57, 62, 64]
cerebellum <3> [34]
ciliary body <7> (<7> highest activity [15]) [15]
colostrum <7> [27]
deciduoma cell <2> [24]
duodenum <1, 2, 4> [36]
epididymis <2, 3, 8> [35]
epithelium <2> (<2> alveolar and bronchial, type I and type II cells [64]; <2> proximal tubular [36]) [36, 64]
fruit <10> [58, 59]
fruitbody <11> [51, 52]
gland <21> (<21> secretions of Rathke's gland [63]) [63]
hepatoma cell <2> (<2> azo dye-induced [17]; <2> primary and Yoshida ascites (AH 13) [16]) [16, 17]
ileum <1> [53]
intestine <1-5, 8> [11, 35, 36, 53]
iris <7> [15]
jejunum <2, 3, 8> (<2,3,8> outer space of epithelial membranes [35]) [35]
kidney <1-3, 5-8, 20> (<6> cortex [30]; <6> brush border of convoluted tubules [30]; <1,2> luminal membrane of proximal tubules [36]) [2, 3, 9, 15, 16, 20, 22, 29-33, 34-36, 40-42, 53, 68, 75]
lens <7> (<7> very low activity [15]) [15]
liver <1-4, 7, 8> (<2> normal and induced enzyme activity, not expression level, by streptozotocin diabetes induced in rats [73]; <1> leakage of enzyme by proteolysis from liver to blood serum [13]; <1> tissue, serum and bile contain multiple enzyme forms of different molecular weights [10]; <2> 80-90% of activity in biliary tract, distribution in liver tissues [21]; <2> hyperplastic hepatic nodules, primary and Yoshida ascites (AH 13) hepatomas [16];

<1> normal and cirrhotic liver [60]; <1> hepatocellular carcinoma [43,60]) [10, 13, 16, 17, 19, 21, 25, 28, 35, 36, 43, 45, 53, 60, 61, 73]
lung <1, 2> (<2> alveolar and ciliated bronchial epithelium [64]) [39, 64]
lung fibroblast <1, 2> (<1> WI-38 fetal [39]) [39, 64]
lymphoid cell <1> [36]
mammary carcinoma cell <2> (<2> transplantable, MT 13762 [18]) [18]
mammary gland <2, 3, 7, 19> (<7> lactating, milk-membranes [14]; <3,19> mammary cell products formed by exocytosis: mammary tumour virus [26]) [14, 18, 26]
milk <3, 7> (<3> in milk of virus infected and uninfected mice, milk fat-globule membrane [26]) [14, 26]
mycelium <12, 13> [54, 55]
pancreas <1-4, 8> (<1> acinar cells, apical part [36]) [5, 8, 23, 35, 36, 68]
pancreatic cancer cell <1> [5]
pancreatic cancer cell <1> (<1> HPC-Y1 [5]) [5]
parotid gland <7> [12]
pericarp <24> (<24> 2 isoforms I and II [71]) [71]
prostate gland <1> (<1> secretion into seminal plasma [44]) [2, 36, 44]
pyloric region <9> (<9> pyloric cecum [4]) [4]
rectum <1> [53]
reproductive system <1> [2, 3]
retinal pigment epithelium <7> [15]
seminal plasma <1-3, 8> [2, 35, 44]
small intestine <5> [11]
spleen <2, 3, 8> [35, 68]
testis <1> [1, 2]
urine <1> (<1> high- and low-molecular weight variants [7]) [7, 53]
uterus <2> [24]
Additional information <2, 3, 8> (<2> blood-brain-barrier-associated isozyme activity induces the cell to form three-dimensional structures in cell culture, inducible by angiogenic and astroglial factors, e.g. bFGF, formation is reduced by cAMP [62]; <2,3,8> distribution [35]) [35]

Localization

brush border <2, 3, 6, 8> (<2,6> kidney [3,30]; <2,3,8> jejunum, microvilli [35]) [3, 30, 35]
cytoplasm <14> [56]
extracellular <1-3, 5, 7, 8, 16, 21, 23, 26> (<26> recombinant enzyme is extracellular in transgenic tabacco plants [76]; <23> secretion into culture medium [69]; <1> mutant lacking the membrane anchor domain, secretion into the medium of recombinant Sf21 cell culture [65]; <21> secretion of Rathke's gland [63]; <2,5> extrinsic to plasma membrane [32,64]; <3,7> in milk [14,26]) [14, 26, 32, 35-38, 44, 47, 60, 63-65, 69, 76]
membrane <1-8, 14, 19, 22, 24, 26> (<14> bound to vascular membrane [74]; <2,26> membrane-bound [68,76]; <1> wild-type, membrane-bound [65]; <2> polarized, outer apical surface alveolar epithelium [64]; <7> associated with milk-membranes [14]; <1,2,4,5> extrinsic to plasma membrane [32,36];

<2,3,8> intrinsic membrane protein [35]) [5, 6, 11, 14, 15, 17-19, 24-28, 30, 32, 33, 35, 36, 39, 40, 64-66, 68, 71, 74, 76]
microsome <1, 2, 7, 11> (<1> recombinant wild-type in Sf21 cells [65]) [15, 17, 51, 65]
peripheral membrane protein <24> (<24> isoforms I and II [71]) [71]
periplasm <17> [48, 49]
vacuole <14, 26> (<14> regulation of glutathione import into the vacuoles, kinetics [74]; <14,26> membrane-bound [74,76]) [74, 76]
Additional information <2> (<2> blood-brain-barrier-associated isozyme [62]) [62]

Purification

<1> (recombinant wild-type and mutants from Sf21 insect cells [65,67]; from normal and cirrhotic liver and 6 isoforms from serum of patients with cellular hepatic carcinoma [60]; partial from bile [46]; partial from pancreas and pancreatic carcinoma and pancreatic carcinoma cell line HPC-Y1 [5]; from seminal plasma, kidney, prostate, and testis [2]; Triton X-100-solubilized liver enzyme is hydrophobic, papain-solubilized liver enzyme is hydrophilic [13]; partial from liver [10,13]; partial [6,45]) [1, 2, 5, 6, 8, 10, 13, 39, 41-46, 60, 65, 67]
<2> (from normal and diabetic rat liver [73]; partial [68]; from carcinoma and partially from normal mammary gland [18]; partial from liver, highly from biliary tract [21]; from hyperplastic hepatic nodules, Yoshida ascites hepatoma, primary hepatoma, kidney [16]; solubilized with either detergents or proteinases [35]; solubilized by deoxycholate [17]; Triton X-100 [18,24,33,68]; Lubrol WX/deoxycholate [19]; papain [24,33]) [3, 16-19, 21, 23, 24, 33, 35, 40, 68, 73, 75]
<3> (partial from mammary-tumour virus and from milk fat-globule membranes, in presence of Triton X-100 [26]) [26]
<4> (solubilized with Lubrol/deoxycholate [25]) [25]
<5> (solubilized with ficin [31]) [31, 33]
<6> [30]
<7> (partial from ciliary body, solubilized with Emulphogene BC720 [15]) [12, 14, 15, 27, 28]
<9> [4]
<10> [58]
<11> (partial [51,52]) [51, 52]
<12> [54]
<13> (partial, 3 isoforms [55]) [55]
<16> [47]
<17> (isozymes A and B [47]) [48]
<18> [50]
<21> (from secretion [63]) [63]
<22> (solubilized by Triton X-100 [66]) [66]
<23> (partial [69]) [69]
<24> (2 isoforms I and II [71]) [71]
<25> (recombinant from E. coli [72]) [72]
<26> (recombinant His-tagged enzyme from E. coli [76]) [76]

Renaturation

<2> (renaturation as catalytically active enzyme after inactivation with 6-diazo-5-oxo-L-norleucine, requires glutathione or S-methyl derivative as a substrate ligand, circular dichroic spectra [20]) [20]
<23> (refolding after denaturation with SDS and urea by dialysis [69]) [69]
<2, 3, 8> [35]

Crystallization

<17> (ammonium sulfate precipitation from 20 mM Tris-HCl, pH 8.0, 14.2 mg/ml protein, refrigerator, 1 week [48]) [48]
<18> (ammonium sulfate precipitation, refrigerator, 2 weeks [50]) [50]

Cloning

<1> (expression of recombinant wild-type and mutants in Spodoptera frugiperda Sf21 cells via baculovirus infection [67]; expression of wild-type and mutant lacking the putative signal sequence/anchor domain, amino acid 1-27, in Spodoptera frugiperda Sf21 cells via baculovirus infection [65]) [65, 67]
<2> (cloning of liver specific isoform III from fetal mRNA, screening of rat genetic library, single copy gene, DNA, including enhancer and promotor, sequence determination, plasmid construction and expression in Escherichia coli, transient enzyme expression in hepatoma cell lines, dependent on differentiation state of the cells [61]) [61]
<25> (expression in Escherichia coli [72]) [72]
<26> (overexpression as His-tagged protein in Escherichia coli JM109, DNA sequence analysis [76]) [76]

Engineering

H383A <1> (<1> site-directed mutagenesis of conserved His383 residue, 3fold reduced activity and 62% reduced V_{max}, altered binding of acceptor [67]) [67]
H383A/H505A <1> (<1> site-directed mutagenesis, 37fold reduced activity [67]) [67]
H505A <1> (<1> site-directed mutagenesis, 10fold reduced activity [67]) [67]
Additional information <26> (<26> construction of transgenic Nicotiana tabacum plants via Agrobacterium tumefaciens transformation, functional extracellular overexpression of the Arabidopsis thaliana enzyme in leaves, transgenic and control plants show the same amount of glutathione degradation [76]) [76]

Application

medicine <2> (<2> bioconversion of cytotoxic and mutagenic agents to inactive 5-L-glutamyl-derivatives [70]) [70]

6 Stability

pH-Stability

4 <18> (<18> 50% loss of activity after 60 min, 37°C [50]) [50]
5-9 <18> (<18> at least 60 min stable, 37°C [50]) [50]
6-9 <12> (<12> stable [54]) [54]
7-8 <16> (<16> stable [47]) [47]
10 <18> (<18> 50% loss of activity after 60 min, 37°C [50]) [50]

Temperature stability

4-30 <12> (<12> at least 10 min stable [54]) [54]
40 <11, 16, 18> (<16> at least 15 min stable [47]; <18> 10% loss of activity within 10 min [50]; <11> stable below [51]) [47, 50, 51]
45 <17, 18> (<17> 12% loss of activity within 15 min [48]; <18> 37% loss of activity within 10 min [50]) [48, 50]
48 <17> (<17> $t_{1/2}$: 15 min [48]) [48]
50 <17, 18> (<18> 95% loss of activity within 10 min [50]; <17> 92% loss of activity within 15 min [48]) [48, 50]
55 <16> (<16> inactivation within 15 min at pH 8.0 [47]) [47]
56 <25> (<25> inactivation [72]) [72]
58 <2> (<2> $t_{1/2}$: 5 min [24]; <2> 12 min [17]; <2> 20% of initial activity retained after 60 min [17]; <2> in the presence of glutathione, 0.02 M, at least 60 min stable [17]) [17, 24]
65 <11> (<11> inactivation, GSH and serine/borate protect [51]) [51]

General stability information

<1>, PMSF stabilizes [7]
<10>, freezing or lyophilization inactivates purified enzyme [58]
<13>, dilution inactivates [55]
<13>, glycerol, 5% v/v, stabilizes [55]

Storage stability

<1>, -30°C, freeze-dried, at least 3 months [44]
<1>, -80°C, 9 months [38]
<1>, 2-8°C, 7 days [37]
<1>, 4°C, at least 5 days [38]
<1>, 4°C, $t_{1/2}$: 2 months [41]
<2>, -22°C, at least 3 months [17]
<2>, -70°C, at least 30 days without loss of activity [24]
<10>, 0°C, at least 3 weeks [58]
<13>, -20°C, crude, at least 1 year [55]
<14>, -19°C, at least 2 months [56]
<18>, 4°C, crystalline, at least 3 months [50]
<25>, -20°C, stable [72]

References

[1] Yoshida, K.I.; Arai, K.; Kobayashi, N.; Saitoh, H.: Purification and properties of γ-glutamyl transpeptidase from human testis. Andrologia, **22**, 239-246 (1990)
[2] Arai, K.; Yoshida, K.; Komoda, T.; Sakagishi, Y.: Comparative studies on the properties of purified γ-glutamyl transferase from human reproductive system and the kidney. Clin. Biochem., **23**, 105-112 (1990)
[3] Kimm, S.W.; Kim, E.G.; Park, S.C.: Partial purification and characterization of γ-glutamyl transpeptidase from rat kidney. Korean J. Biochem., **18**, 135-143 (1986)
[4] Glynn, B.P.; Johnson, D.B.: γ-Glutamyltransferase from Marthasterias glacialis: purifiication procedures and enzyme characterization. Comp. Biochem. Physiol. B Comp. Biochem., **80**, 941-948 (1985)
[5] Sugimoto, M.; Yamaguchi, N.; Keiichi, K.: Characterization of γ-GTP in a human pancreatic cancer cell line. Gastroenterol. Jpn., **19**, 227-231 (1984)
[6] Vesely, J.; Cernoch, M.: Solubilization and some properties of γ-glutamyltransferase from human brain microvessels. Neurochem. Res., **9**, 927-934 (1984)
[7] Rambabu, K.; Pattabiraman, T.N.: Studies on the properties of the variants of γ-glutamyl transpeptidase in human urine. J. Biosci., **4**, 287-294 (1982)
[8] Masuike, M.; Ogawa, M.; Kosaki, G.; Minamiura, N.; Yamamoto, T.: Purification and characterization of γ-glutamyl transpeptidase from human pancreas. Enzyme, **27**, 163-170 (1982)
[9] Frielle, T.; Curthoys, N.P.: Characterization of the amphipathic structure of γ-glutamyltranspeptidase F13. Biophys. J., **37**, 193-195 (1982)
[10] Echetebu, Z.O.; Moss, D.W.: Multiple forms of human γ-glutamyltransferase: preparation and characterization of different molecular weight fractions. Enzyme, **27**, 1-8 (1982)
[11] Nakamura, Y.; Kato, H.; Suzuki, F.; Nagata, Y.: Some properties of γ-glutamyltransferase from hog small intestine. Biomed. Res., **2**, 509-516 (1981)
[12] Hata, K.; Hayakawa, M.; Abiko, Y.; Takiguchi, H.: Purification and properties of γ-glutamyl transpeptidase from bovine parotid gland. Int. J. Biochem., **13**, 681-692 (1981)
[13] Tsuji, A.; Matsuda, Y.; Katunuma, N.: Characterization of human serum γ-glutamyltranspeptidase. Clin. Chim. Acta, **104**, 361-364 (1980)
[14] Baumrucker, C.R.: Purification and identification of γ-glutamyl transpeptidase of milk membranes. J. Dairy Sci., **63**, 49-54 (1980)
[15] Das, N.D.; Shichi, H.: γ-Glutamyl transpeptidase of bovine ciliary body: purification and properties. Exp. Eye Res., **29**, 109-121 (1979)
[16] Tsuchida, S.; Hoshino, K.; Sato, T.; Ito, N.: Purification of γ-glutamyltransferases from rat hepatomas and hyperplastic hepatic nodules, and comparison with the enzyme from rat kidney. Cancer Res., **39**, 4200-4205 (1979)
[17] Taniguchi, N.: Purification and some properties of γ-glutamyl transpeptidase from azo dye-induced hepatoma. J. Biochem., **75**, 473-480 (1974)

[18] Jaken, S.; Mason, M.: Purification and comparison of several catalytic parameters of the γ-glutamyltranspeptidase of rat mammary adenocarcinoma (13762) and of normal rat mammary gland. Biochim. Biophys. Acta, **568**, 331-338 (1979)
[19] Ding, J.L.; Smith, G.D.; Peters, T.J.: The purification of γ-glutamyltransferase from normal rat liver. Biochem. Soc. Trans., **8**, 77 (1980)
[20] Horiuchi, S.; Inoue, M.; Morino, Y.: Latent active site in rat-kidney γ-glutamyl transpeptidase. The refolding process of the large subunit and characterization of the renatured enzyme. Eur. J. Biochem., **105**, 93-102 (1980)
[21] Ding, J.L.; Smith, G.D.; Peters, T.J.: Purification and properties of γ-glutamyl transferase from normal rat liver. Biochim. Biophys. Acta, **657**, 334-343 (1981)
[22] Sachdev, G.P.; Leahy, D.S.; Chace, K.V.: Phenobarbital and related compounds as novel inhibitors of γ-glutamyltranspeptidase. Biochim. Biophys. Acta, **749**, 125-129 (1983)
[23] Takahashi, S.; Steinman, H.M.; Ball, D.: Purification and characterization of γ-glutamyltransferase from rat pancreas. Biochim. Biophys. Acta, **707**, 66-73 (1982)
[24] Tarachand, U.: Purification and properties of γ-glutamyl transpeptidase from rat deciduoma. J. Appl. Biochem., **6**, 278-288 (1984)
[25] Bagrel, D.; Petitclerc, C.; Schiele, F.; Siest, G.: Some kinetic properties of γ-glutamyltransferase from rabbit liver. Biochim. Biophys. Acta, **658**, 220-231 (1981)
[26] Franois, C.; Calberg-Bacq, C.M.; Gosselin, L.; Kozma, S.; Osterrieth, P.M.: Identification, partial purification and biochemical characterization of γ-glutamyltranspeptidase present as a membrane component in skimmed milk and milk fat-globule membranes, and in mammary-tumour virus from the milk of infected mice. Biochim. Biophys. Acta, **567**, 106-115 (1979)
[27] Yasumoto, K.; Iwami, K.; Fushiki, T.; Mitsuda, H.: Purification and enzymatic properties of γ-glutamyltransferase from bovine colostrum. J. Biochem., **84**, 1227-1236 (1978)
[28] Furukawa, M.; Higashi, T.; Tateishi, N.; Ochi, K.; Sakamoto, Y.: Purification and properties of bovine liver γ-glutamyltransferase. J. Biochem., **93**, 839-846 (1983)
[29] Karkowsky, A.M.; Bergamini, M.V.W.; Orlowski, M.: Kinetic studies of sheep kidney γ-glutamyl transpeptidase. J. Biol. Chem., **251**, 4736-4743 (1976)
[30] Zelazo, P.; Orlowski, M.: γ-Glutamyl transpeptidase of sheep-kidney cortex. Isolation, catalytic properties and dissociation into two polypeptide chains. Eur. J. Biochem., **61**, 147-155 (1976)
[31] Leibach, F.H.; Binkley, F.: γ-Glutamyl transferase of swine kidney. Arch. Biochem. Biophys., **127**, 292-301 (1968)
[32] Írning, L.; Hammarström, S.: Kinetics of the conversion of leukotriene C by γ-glutamyl transpeptidase. Biochem. Biophys. Res. Commun., **106**, 1304-1309 (1982)
[33] Bernström, K.; Orning, L.; Hammarström, S.: γ-Glutamyl transpeptidase, a leukotriene metabolizing enzyme. Methods Enzymol., **86**, 38-45 (1982)

[34] Allison, D.: γ-Glutamyl transpeptidase: kinetics and mechanism. Methods Enzymol., **113**, 419-437 (1985)
[35] Tate, S.S.; Meister, A.: γ-Glutamyl transpeptidase from kidney. Methods Enzymol., **113**, 400-419 (1985)
[36] Shaw, L.M.: γ-Glutamyltransferase. (γ-Glutamyl)-peptide:amino acid γ-glutamyltransferase, EC 2.3.2.2. 1. General. Methods Enzym. Anal., 3rd Ed. (Bergmeyer, H.U., ed.), **3**, 349-352 (1983)
[37] Wahlefeld, A.W.; Bergmeyer, H.U.: γ-Glutamyltransferase. (γ-Glutamyl)-peptide:amino acid γ-glutamyltransferase, EC 2.3.2.2. 2. Routine method. Methods Enzym. Anal., 3rd Ed. (Bergmeyer, H.U., ed.), **3**, 352-356 (1983)
[38] Shaw, L.M.; Stromme, J.H.: γ-Glutamyltransferase. (γ-Glutamyl)-peptide:amino acid γ-glutamyltransferase, EC 2.3.2.2. 3. IFCC reference method (provisional). Methods Enzym. Anal., 3rd Ed. (Bergmeyer, H.U., ed.), **3**, 357-364 (1983)
[39] Takahashi, S.; Zukin, R.S.; Steinman, H.M.: γ-Glutamyl transpeptidase from WI-38 fibroblasts: purification and active site modification studies. Arch. Biochem. Biophys., **207**, 87-95 (1981)
[40] Cook, N.D.; Peters, T.J.: Purification of γ-glutamyltransferase by phenyl boronate affinity chromatography. Studies on the acceptor specificity of transpeptidation by rat kidney γ-glutamyltransferase. Biochim. Biophys. Acta, **828**, 205-212 (1985)
[41] Miller, S.P.; Awasthi, Y.C.; Srivastava, S.K.: Studies of human kidney γ-glutamyl transpeptidase. Purification and structural, kinetic and immunological properties. J. Biol. Chem., **251**, 2271-2278 (1976)
[42] Tate, S.S.; Ross, M.E.: Human kidney γ-glutamyl transpeptidase. Catalytic properties, subunit structure, and localization of the γ-glutamyl binding site on the light subunit. J. Biol. Chem., **252**, 6042-6045 (1977)
[43] Yamashita, K.; Totani, K.; Iwaki, Y.; Takamisawa, I.; Tateishi, N.; Higashi, T.; Sakamoto, Y.; Kobata, A.: Comparative study of the sugar chains of γ-glutamyltranspeptidases purified from human hepatocellular carcinoma and from human liver. J. Biochem., **105**, 728-735 (1989)
[44] Abe, S.; Kochi, H.; Hiraiwa, K.: Purification and immunological characterization of a new form of γ-glutamyltransferase of human semen. Biochim. Biophys. Acta, **1077**, 259-264 (1991)
[45] Huseby, N.: Purification and some properties of γ-glutamyltransferase from human liver. Biochim. Biophys. Acta, **483**, 46-56 (1977)
[46] Indirani, N.; Hill, P.G.: Partial purification and some properties of γ-glutamyl transpeptidase from human bile. Biochim. Biophys. Acta, **483**, 57-62 (1977)
[47] Ogawa, Y.; Hosoyama, H.; Hamano, M.; Motai, H.: Purification and properties of γ-glutamyltranspeptidase from Bacillus subtilis (natto). Agric. Biol. Chem., **55**, 2971-2977 (1991)
[48] Suzuki, H.; Kumagai, H.; Tochikura, T.: γ-Glutamyltranspeptidase from Escherichia coli K-12: purification and properties. J. Bacteriol., **168**, 1325-1331 (1986)

[49] Suzuki, H.; Kumagai, H.; Tochikura, T.: γ-Glutamyltranspeptidase from Escherichia coli K-12: formation and localization. J. Bacteriol., **168**, 1332-1335 (1986)
[50] Nakayama, R.; Kumagai, H.; Tochikura, T.: Purification and properties of γ-glutamyltranspeptidase from Proteus mirabilis. J. Bacteriol., **160**, 341-346 (1984)
[51] Iwami, K.; Yasumoto, K.; Nakamura, K.; Mitsuda, H.: Properties of γ-glutamyltransferase from Lentinus edodes. Agric. Biol. Chem., **39**, 1933-1940 (1975)
[52] Iwami, K.; Yasumoto, K.; Nakamura, K.; Mitsuda, H.: Reactivity of Lentinus γ-glutamyltransferase with lentinic acid as the principal endogenous substrate. Agric. Biol. Chem., **39**, 1941-1946 (1975)
[53] Minato, S.: Isolation of anthglutin, an inhibitor of γ-glutamyl transpeptidase from Penicillum oxalicum. Arch. Biochem. Biophys., **192**, 235-240 (1979)
[54] Tomita, K.; Yano, T.; Tsuchida, T.; Kumagai, H.; Tochikura, T.: Purification and properties of γ-glutamyltranspeptidase from Penicillium roqueforti IFO 4622. J. Ferment. Bioeng., **70**, 128-130 (1990)
[55] Moriguchi, M.; Yamada, M.; Suenaga, S.; Tanaka, H.; Wakasugi, A.; Hatanaka, S.I.: Partial purification and properties of γ-glutamyltranspeptidase from mycelia of Morchella esculenta. Arch. Microbiol., **144**, 15-19 (1986)
[56] Penninckx, M.J.; Jaspers, C.J.: Molecular and kinetic properties of purified γ-glutamyl transpeptidase from yeast (Saccharomyces cerevisiae). Phytochemistry, **24**, 1913-1918 (1985)
[57] Tomita, K.; Ito, M.; Yano, T.; Kumagai, H.; Tochikura, T.: γ-Glutamyltranspeptidase activity and the properties of the extracellular glutaminase from Aspergillus oryzae. Agric. Biol. Chem., **52**, 1159-1163 (1988)
[58] Goore, M.Y.; Thompson, J.F.: γ-Glutamyl transpeptidase from kidney bean fruit. I. Purification and mechanism of action. Biochim. Biophys. Acta, **132**, 15-26 (1967)
[59] Goore, M.Y.; Thompson, J.F.: γ-Glutamyl transpeptidase of kidney bean fruit. II. Studies on the activating effect of sodium citrate. Biochim. Biophys. Acta, **132**, 27-32 (1967)
[60] Ohta, H.; Sawabu, N.; Kawakami, H.; Watanabe, H.; Ozaki, K.; Toya, D.; Hattori, N.: Characterization of γ-glutamyltranspeptidase from human hepatocellular carcinoma, compared with enzymes from normal liver and cirrhotic liver. Clin. Chim. Acta, **214**, 83-92 (1993)
[61] Brouillet, A.; Darbouy, M.; Okamoto, T.; Chobert, M.N.; Lahuna, O.; Garlatti, M.; Goodspeed, D.; Laperche, Y.: Functional characterization of the rat γ-glutamyl transpeptidase promoter that is expressed and regulated in the liver and hepatoma cells. J. Biol. Chem., **269**, 14878-14884 (1994)
[62] Roux, F.; Durieu-Trautmann, O.; Chaverot, N.; Claire, M.; Mailly, P.; Bourre, J.M.; Strosberg, A.D.; Couraud, P.O.: Regulation of γ-glutamyl transpeptidase and alkaline phosphatase activities in immortalized rat brain microvessel endothelial cells. J. Cell. Physiol., **159**, 101-113 (1994)
[63] Krishna, R.G.; Chin, C.C.Q.; Weldon, P.J.; Wold, F.: Characterization of γ-glutamyl transpeptidase from the Rathke's gland secretions of Kemp's ridley

sea turtles (Lepidochelys kempi). Comp. Biochem. Physiol. B, **111**, 257-264 (1995)

[64] Ingbar, D.H.; Hepler, K.; Dowin, R.; Jacobsen, E.; Dunitz, J.M.; Nici, L.; Jamieson, J.D.: γ-Glutamyl transpeptidase is a polarized alveolar epithelial membrane protein. Am. J. Physiol., **269**, L261-L271 (1995)

[65] Ikeda, Y.; Fujii, J.; Taniguchi, N.; Meister, A.: Expression of an active glycosylated human γ-glutamyl transpeptidase mutant that lacks a membrane anchor domain. Proc. Natl. Acad. Sci. USA, **92**, 126-130 (1995)

[66] Hussein, A.S.; Walter, R.D.: Purification and characterization of γ-glutamyl transpeptidase from Ascaris suum. Mol. Biochem. Parasitol., **77**, 41-47 (1996)

[67] Ikeda, Y.; Fujii, J.; Taniguchi, N.: Effects of substitutions of the conserved histidine residues in human γ-glutamyl transpeptidase. J. Biochem., **119**, 1166-1170 (1996)

[68] Dvorakova, L.; Krusek, J.; Stastny, F.; Lisy, V.: Relationship between kinetic properties of γ-glutamyl transpeptidase and the structure of its saccharide moiety. Biochim. Biophys. Acta, **1292**, 163-167 (1996)

[69] Hwang, S.Y.; Ryang, J.H.; Lim, W.J.; Yoo, I.D.; Oishi, K.: Purification and properties of γ-glutamyl transpeptidase from Bacillus sp. KUN-17. J. Microbiol. Biotechnol., **6**, 238-244 (1996)

[70] Keren, R.; Stark, A.A.: γ-Glutamyl transpeptidase-dependent mutagenicity and cytotoxicity of γ-glutamyl derivatives: a model for biochemical targeting of chemotherapeutic agents. Environ. Mol. Mut., **32**, 377-386 (1998)

[71] Martin, M.N.; Slovin, J.P.: Purified γ-glutamyl transpeptidases from tomato exhibit high affinity for glutathione and glutathione S-conjugates. Plant Physiol., **122**, 1417-1426 (2000)

[72] Chu, L.; Xu, X.; Dong, Z.; Cappelli, D.; Ebersole, J.L.: Role for recombinant γ-glutamyltransferase from Treponema denticola in glutathione metabolism. Infect. Immun., **71**, 335-342 (2003)

[73] Cornwell, P.D.; Watkins, J.B.: Changes in the kinetic parameters of hepatic γ-glutamyltransferase from streptozotocin-induced diabetic rats. Biochim. Biophys. Acta, **1545**, 184-191 (2001)

[74] Mehdi, K.; Thierie, J.; Penninckx, M.J.: γ-Glutamyl transpeptidase in the yeast Saccharomyces cerevisiae and its role in the vacuolar transport and metabolism of glutathione. Biochem. J., **359**, 631-637 (2001)

[75] Castonguay, R.; Lherbet, C.; Keillor, J.W.: Mapping of the active site of rat kidney γ-glutamyl transpeptidase using activated esters and their amide derivatives. Bioorg. Med. Chem., **10**, 4185-4191 (2002)

[76] Storozhenko, S.; Belles-Boix, E.; Babiychuk, E.; Herouart, D.; Davey, M.W.; Slooten, L.; Van Montagu, M.; Inze, D.; Kushnir, S.: γ-Glutamyl transpeptidase in transgenic tobacco plants. Cellular localization, processing, and biochemical properties. Plant Physiol., **128**, 1109-1119 (2002)

Lysyltransferase 2.3.2.3

1 Nomenclature

EC number
2.3.2.3

Systematic name
L-lysyl-tRNA:phosphatidylglycerol 3-O-lysyltransferase

Recommended name
lysyltransferase

CAS registry number
37257-20-8

2 Source Organism

<1> *Staphylococcus aureus* [1]

3 Reaction and Specificity

Catalyzed reaction
L-lysyl-tRNA + phosphatidylglycerol = tRNA + 3-phosphatidyl-1'-(3'-O-L-lysyl)glycerol

Reaction type
aminoacyl group transfer

Substrates and products

S L-lysyl-tRNA + phosphatidylglycerol <1> (<1> 2'- (not 3'-) deoxy-analogue of phosphatidylglycerol can also act as L-lysyl-acceptor [1]) (Reversibility: ? <1> [1]) [1]

P tRNA + 3-phosphatidyl-1'-(3'-O-L-lysyl)glycerol <1> [1]

Activating compounds
Additional information <1> (<1> anionic surfactant, e.g. sodium-salt of a fatty acid, and high ionic strength lead to activation [1]) [1]

5 Isolation/Preparation/Mutation/Application

Localization

membrane <1> [1]

Purification

<1> (partial [1]) [1]

6 Stability

Storage stability

<1>, -20°C, membrane-bound enzyme extracted with organic solvents, $t_{1/2}$: 3-5 days [1]

<1>, -20°C, native membrane-bound enzyme, several weeks [1]

References

[1] Lennarz, W.J.; Bonsen, P.P.M.; Van Deenen, L.L.M.: Substrate specificity of O-L-lysylphosphatidylglycerol synthetase. Enzymatic studies on the structure of O-L-lysylphosphatidylglycerol. Biochemistry, **6**, 2307-2312 (1967)

γ-Glutamylcyclotransferase 2.3.2.4

1 Nomenclature

EC number
2.3.2.4

Systematic name
(5-L-glutamyl)-L-amino-acid 5-glutamyltransferase (cyclizing)

Recommended name
γ-glutamylcyclotransferase

Synonyms
L-glutamic cyclase
cyclotransferase, γ-glutamyl
γ-L-glutamylcyclotransferase
γ-glutamyl-amino acid cyclotransferase
γ-glutamyltranspeptidase <3> [14]

CAS registry number
9045-44-7

2 Source Organism

<1> *Nicotiana tabacum* (tobacco, cv. Samsun [1]) [1]
<2> *Musca domestica* (house fly [13]) [13]
<3> *Homo sapiens* [2-4, 9, 10, 12, 14]
<4> *Ovis aries* [2, 8, 9, 11]
<5> *Sus scrofa* [10]
<6> *Rattus norvegicus* (Sprague-Dawley [5]; Holtzman [7]; female Wistar [10]) [5-7, 9, 10, 15]
<7> *Mus musculus* [6, 8, 9]
<8> *Cavia porcellus* (guinea pig [9]) [9]
<9> *Bos taurus* [15]

3 Reaction and Specificity

Catalyzed reaction
(5-L-glutamyl)-L-amino acid = 5-oxoproline + L-amino acid (<3> mechanism [12])

Reaction type

aminoacyl group transfer

Natural substrates and products

S γ-L-glutamyl-L-amino acid <2, 3> (<3> enzyme of L-glutamyl-cycle [4]; <2> involved in sclerotization process of white pupae [13]) (Reversibility: ? <2, 3> [4, 13]) [4, 13]
P 5-oxoproline + L-amino acid <2, 3> [4, 13]

Substrates and products

S 2-N-(5-L-glutamyl)-L-lysine <6> (Reversibility: ? <6> [5]) [5]
P 5-oxoproline + L-lysine <6> [5]
S 5-(4-methyl)glutamyl-L-2-aminobutyrate <5> (<5> substrate analog, 90% as effective as 5-glutamyl-L-2-aminobutyrate [10]) (Reversibility: ? <5> [10]) [10]
P 3-methyl-5-oxoproline + L-α-aminobutyrate <5> [10]
S 5-L-(threo-4-methyl)glutamyl-L-2-aminobutyrate <4, 6, 7> (<6,7> model substrate, in vivo and in vitro [6,8]) (Reversibility: ? <4, 6, 7> [6, 8]) [6, 8]
P 3-methyl-5-oxoproline + L-α-aminobutyrate <4, 6, 7> [6, 8]
S 5-L-glutamyl-1-naphthylamide <3, 5> (<3,5> best substrate [10]) (Reversibility: ? <3, 5> [10]) [10]
P 5-oxoproline + naphthylamine <3, 5> [10]
S 5-L-glutamyl-5-L-glutamyl-L-alanine <6> (<6> high activity [6]) (Reversibility: ? <6> [6]) [6, 7]
P 5-oxoproline + 5-L-glutamyl-L-alanine <6> [6, 7]
S 5-L-glutamyl-5-L-glutamyl-L-amino acid <6> (Reversibility: ? <6> [6]) [6, 7]
P 5-oxoproline + 5-L-glutamyl-L-amino acid <6> [6, 7]
S 5-L-glutamyl-5-L-glutamyl-L-glutamate <6> (Reversibility: ? <6> [6]) [6, 7]
P 5-oxoproline + 5-L-glutamyl-L-glutamate <6> [6, 7]
S 5-L-glutamyl-5-L-glutamyl-L-glutamine <6> (<6> high activity [6]) (Reversibility: ? <6> [6]) [6, 7]
P 5-oxoproline + 5-L-glutamyl-L-glutamine <6> [6, 7]
S 5-L-glutamyl-5-L-glutamyl-L-glycine <6> (<6> high activity [6]) (Reversibility: ? <6> [6]) [6, 7]
P 5-oxoproline + 5-L-glutamyl-L-glycine <6> [6, 7]
S 5-L-glutamyl-5-L-glutamyl-L-leucine <6> (Reversibility: ? <6> [6]) [6, 7]
P 5-oxoproline + 5-L-glutamyl-L-leucine <6> [6, 7]
S 5-L-glutamyl-5-L-glutamyl-L-lysine <6> (Reversibility: ? <6> [6]) [6, 7]
P 5-oxoproline + 5-L-glutamyl-L-lysine <6> [6, 7]
S 5-L-glutamyl-5-L-glutamyl-L-phenylalanine <6> (Reversibility: ? <6> [6]) [6, 7]
P 5-oxoproline + 5-L-glutamyl-L-phenylalanine <6> [6, 7]
S 5-L-glutamyl-5-L-glutamyl-L-proline <6> (Reversibility: ? <6> [6]) [6, 7]
P 5-oxoproline + 5-L-glutamyl-L-proline <6> [6, 7]
S 5-L-glutamyl-5-L-glutamyl-L-threonine <6> (<6> high activity [6]) (Reversibility: ? <6> [6]) [6, 7]

P 5-oxoproline + 5-L-glutamyl-L-threonine <6> [6, 7]
S 5-L-glutamyl-5-L-glutamyl-L-tyrosine <6> (Reversibility: ? <6> [6]) [6, 7]
P 5-oxoproline + 5-L-glutamyl-L-tyrosine <6> [6, 7]
S 5-L-glutamyl-5-L-glutamyl-L-valine <6> (Reversibility: ? <6> [6]) [6, 7]
P 5-oxoproline + 5-L-glutamyl-L-valine <6> [6, 7]
S 5-L-glutamyl-L-5-L-glutamyl-L-4-nitroanilide <2-4, 6> (<3,4> best substrate [2,9]; <2> 25% as effective as 5-L-glutamyl-L-phenylalanine, [13]; <2> no activity with D-glutamyl-isomer [13]) (Reversibility: ? <2-4, 6> [2, 5, 6, 9, 13]) [2, 5, 6, 9, 13]
P 5-L-oxoproline + 5-L-glutamyl-L-*p*-nitroanilide <2-4, 6> [2, 5, 6, 9, 13]
S 5-L-glutamyl-L-6-N-benzyloxycarbonyl-L-lysine <6> (Reversibility: ? <6> [5]) [5]
P 5-oxoproline + L-6-N-benzyloxycarbonyl-L-lysine <6> [5]
S 5-L-glutamyl-L-alanine <3, 4, 6> (Reversibility: ? <3, 4, 6> [2-4, 6, 7, 9, 12]) [2-4, 6, 7, 9, 12]
P 5-L-oxoproline + L-alanine <3, 4, 6> [2-4, 6, 7, 9, 12]
S 5-L-glutamyl-L-α-aminobutyrate <3-7> (Reversibility: ? <3-7> [2, 5-11]) [2, 5-11]
P 5-L-oxoproline + L-α-aminobutyrate <3-7> [2, 5-11]
S 5-L-glutamyl-L-amino acid <1, 3-7> (<3,5,6> stereospecific [10]; <3,5> highly specific for 5-L-glutamyl peptides [10]; <3,6> preferred substrates: 5-L-glutamyl-5-L-glutamyl-amino acids [7,10]; <2> no activity with: α-L-glutamyl-glycine, α-L-glutamyl-L-alanine, α-L-glutamyl-L-tyrosine, α-L-glutamyl-2-naphthylamide, α-L-glutamyl-4-methoxy-2-β-naphthylamide [13]; <2-4> no activity with: glutathione, i.e. γ-L-glutamyl-L-cysteinylglycine [2,13]; <3,4> no activity with: 5-glutamyl-4-nitroanilide [2]; <3,4> no activity with LD-, DD- and DL-isomers of 5-L-glutamyl-5-L-glutamyl-4-nitroanilide [2]; <2-5> no activity with L-glutamine [9,10,13]; <3,4> no activity with L-glutamyl-β-alanine, L-glutamyl-3-aminoisobutyrate [9]; <4> no activity with D-glutamyl-2-aminobutyrate, L-(3-methyl)glutamyl-2-aminobutyrate, 3-aminoglutaryl-2-aminobutyrate, L-5-(N-methyl)-glutamyl-2-aminobutyrate, 5-(5-methyl)glutamyl-2-aminobutyrate [11]; <2> no activity with L-glutamyl-L-glutamyl-L-phenylalanine and L-glutamyl-L-glutamyl-1-naphthylamide [13]) (Reversibility: ? <1, 3-7> [1, 2, 5-13]) [1, 2, 5-13]
P 5-oxoproline + L-amino acid <1, 3-7> (<1,3,4,6,7> i.e. 2-pyrrolidone-5-carboxylic acid [1,2,5-8,12]) [1, 2, 5-13]
S 5-L-glutamyl-L-aspartate <6> (Reversibility: ? <6> [6, 7]) [6, 7]
P 5-oxoproline + L-aspartate <6> [6, 7]
S 5-L-glutamyl-L-cysteine <1, 6> (Reversibility: ? <1, 6> [1, 6]) [1, 6]
P 5-oxoproline + L-cysteine <1> [1]
S 5-L-glutamyl-L-glutamate <5, 6> (<6> poor substrate [5]) (Reversibility: ? <5, 6> [5-7, 10]) [5-7, 10]
P 5-oxoproline + L-glutamate <5, 6> [5-7, 10]
S 5-L-glutamyl-L-glutamine <3, 4, 6> (<6> best substrate [5-7]) (Reversibility: ? <3, 4, 6> [2, 5-7, 9]) [2, 5-7, 9]
P 5-oxoproline + L-glutamine <3, 4, 6> [2, 5-7, 9]

S 5-L-glutamyl-L-isoleucine <6> (<6> very low activity [6]) (Reversibility: ? <6> [6]) [6]
P 5-oxoproline + L-isoleucine <6> [6]
S 5-L-glutamyl-L-leucine <3, 4, 6> (<6> poor substrate [5]) (Reversibility: ? <3, 4, 6> [5, 9]) [5, 9]
P 5-oxoproline + L-leucine <3, 4, 6> [5, 9]
S 5-L-glutamyl-L-leucine <6> (<6> very low activity [6]) (Reversibility: ? <6> [6]) [6]
P 5-oxoproline + L-leucine <6> [6]
S 5-L-glutamyl-L-methionine <1, 2, 6> (<1> highly specific for, no activity with D-glutamyl-L-methionine [1]; <6> best substrate [5]; <2> 70% as effective as L-glutamyl-L-phenylalanine [13]) (Reversibility: ? <1, 2, 6> [1, 5-7, 13]) [1, 5-7, 13]
P 5-oxoproline + L-methionine <1> [1]
S 5-L-glutamyl-L-phenylalanine <2-6> (<2> best substrate [13]; <3-6> poor substrate [6,7,9,10]) (Reversibility: ? <2-6> [6, 7, 9, 10, 13]) [6, 7, 9, 10, 13]
P 5-oxoproline + L-phenylalanine <2-6> [6, 7, 9, 10, 13]
S 5-L-glutamyl-L-proline <6> (<6> very low activity [6]) (Reversibility: ? <6> [6]) [6]
P 5-oxoproline + L-proline <6> [6]
S 5-L-glutamyl-L-tyrosine <6> (<6> very low activity [6]) (Reversibility: ? <6> [6]) [6]
P 5-oxoproline + L-tyrosine <6> [6]
S 5-L-glutamyl-L-valine <2, 6> (<2> 10% as effective as 5-L-glutamyl-L-phenylalanine [13]; <6> poor substrate [6]) (Reversibility: ? <2, 6> [6, 13]) [6, 13]
P 5-oxoproline + L-valine <2, 6> [6, 13]
S 5-L-glutamyl-L-valine <6> (<6> very low activity [6]) (Reversibility: ? <6> [6]) [6]
P 5-oxoproline + L-valine <6> [6]
S 5-L-glutamyl-S-methyl-L-cysteine <3, 4> (Reversibility: ? <3, 4> [2, 9]) [2, 9]
P 5-oxoproline + S-methyl-L-cysteine <3, 4> [2, 9]
S 5-L-glutamyl-cysteine <6> (Reversibility: ? <6> [6]) [6]
P 5-oxoproline + L-cysteine <6> [6]
S 5-L-glutamyl-glycine <3, 4, 6> (<6> poor substrate [5-7]) (Reversibility: ? <3, 4, 6> [2, 5-7, 9]) [2, 5-7, 9]
P 5-oxoproline + glycine <3, 4, 6> [2, 5-7, 9]
S 5-glutamyl-glutathione <3, 5> (Reversibility: ? <3, 5> [10]) [10]
P 5-oxoproline + glutathione <3, 5> [10]
S N^{ε}-(5-L-glutamyl)-L-lysine <6, 9> (Reversibility: ? <6, 9> [15]) [15]
P 5-oxoproline + L-lysine <6, 9> [15]

Inhibitors

2-aminoglutaryl-L-1-aminobutyrate <4, 6, 8> (<4> 8 mM, 96% inhibition [11]) [6, 8, 9, 11]
4-hydroxymercuribenzoate <6> (<6> 0.5 mM, 43% inhibition of isoenzyme with pI 5.1 [7]) [7]

5-D-glutamyl-L-1-aminobutyrate <4> (<4> 8 mM, 69% inhibition [11]) [11]
β-aminoglutaryl-L-alanine <3> [12]
cystamine <6> (<6> 10 mM, 57% inhibition of isoenzyme with pI 5.1 [7]) [7]
γ-glutamyl cysteine disulfide <6> (<6> 10 mM, 64% inhibition of isoenzyme with pI 5.1 [7]) [7]

Specific activity (U/mg)

0.06 <3> (<3> activity in normal lung [14]) [14]
0.24 <3> (<3> activity in bronchoalveo carcinoma [14]) [14]
0.36 <3> (<3> activity in squamous cell carcinoma [14]) [14]
0.77 <3> (<3> activity in adenocarcinoma [14]) [14]
8.9 <6> (<6> isoenzyme B, substrate 5-L-glutamyl-5-L-glutamyl-*p*-nitroanilide [5]) [5]
8.9 <6> (<6> isoenzyme B, substrate 5-L-glutamyl-5-L-glutamyl-*p*-nitroanilide [5]) [5]
9.7 <6> (<6> isoenzyme B, substrate 5-L-glutamyl-L-α-aminobutyrate [5]) [5]
33.8 <3> [3]
38 <6> (<6> isoenzyme A, substrate 5-L-glutamyl-L-α-aminobutyrate [5]) [5]
40 <6> (<6> isoenzyme A, substrate 5-L-glutamyl-5-L-glutamyl-*p*-nitroanilide [5]) [5]
70.7 <2> [13]
100 <3> [2]
158 <6> (<6> pI 4.6 isoenzyme [6]) [6]
181 <6> (<6> pI 5.1 isoenzyme [6]) [6]
Additional information <6, 9> (<6,9> spectrometric assay [15]) [15]

K_m-Value (mM)

0.4 <3> (5-L-glutamyl-5-L-glutamyl-*p*-nitroanilide) [2]
0.6 <6> (5-L-glutamyl-5-L-glutamyl-*p*-nitroanilide) [5]
2 <3> (5-L-glutamyl-L-alanine) [4]
2.2 <3> (5-L-glutamyl-L-alanine) [3]
2.94 <2> (5-L-glutamyl-L-phenylalanine) [13]
4.48 <3> (5-L-glutamyl-L-alanine) [12]
5 <7> (5-L-(threo-2-methyl)glutamyl-L-α-aminobutyrate) [8]
6 <6> (5-L-glutamyl-L-α-aminobutyrate, <6> isoenzyme with pI 5.1, in extract, K_m increases 4-10fold during purification [6]) [6]
6.6 <6> (5-L-glutamyl-L-α-aminobutyrate, <6> isoenzyme with pI 4.6 [6]) [6]
8 <6> (5-L-glutamyl-L-α-aminobutyrate, <6> isoenzyme B [5]) [5]
10 <6> (5-L-glutamyl-L-glutamine, <6> isoenzyme B [5]) [5]
12 <6> (5-L-glutamyl-L-α-aminobutyrate, <6> isoenzyme A [5]) [5]
18 <6> (5-L-glutamyl-L-glutamine, <6> isoenzyme A [5]) [5]
100 <6> (α-N-(γ-L-glutamyl)-L-lysine, <6> isoenzyme A [5]) [5]
Additional information <3> (<3> kinetics [12]) [12]

K_i-Value (mM)

0.3 <3> (D-β-aminoglutaryl-L-alanine) [12]
0.46 <6> (2-aminoglutaryl-L-α-aminobutyrate) [6]
0.6 <4> (2-aminoglutaryl-L-α-aminobutyrate) [8]

pH-Optimum

7.1-7.3 <2> [13]
7.5-8 <6> [5-7]
7.8-8.2 <3, 4> (<3,4> borate buffer [9]) [2, 9]
9 <3> [3]
Additional information <6> (<6> two isozymic forms of different pI-values: pI 4.6 and 5.1 [6,7]) [6, 7]

pH-Range

6-9.5 <6> (<6> approx. half-maximal activity at pH 6 and 9.5 [7]) [7]
6.5-8.1 <2> (<2> approx. half-maximal activity at pH 6.5 and 8.1 [13]) [13]
6.5-11.5 <3> (<3> approx. half-maximal activity at pH 6.5 and 11.5 [3]) [3]

Temperature optimum (°C)

35 <1> [1]

4 Enzyme Structure

Molecular weight

25250 <3> (<3> gel filtration [3]) [3]
27500 <6> (<6> gel filtration [5]) [5]
30000 <2> (<2> gel filtration [13]) [13]
Additional information <6> (<6> amino acid analysis [7]) [7]

Subunits

monomer <6> (<6> 1 * 27000, SDS-PAGE [6]) [6]

5 Isolation/Preparation/Mutation/Application

Source/tissue

adenocarcinoma cell <3> [14]
brain <3, 4, 6> (<3> high activity [10]; <6> low activity [10]) [2, 5, 8, 10, 11]
cell suspension culture <1> [1]
erythrocyte <3, 6> [3, 4, 10, 12]
intestine <3, 6, 7, 8> [9]
kidney <3, 6, 7> (<6> very high activity [10]) [5-8, 10]
liver <5-8> [5, 8, 9, 10]
lung <3, 6, 7, 8> [9, 14]
lung cancer cell <3> [14]
pupa <2> [13]
skin <3, 7, 8> [9]

squamous cell carcinoma <3> [14]
Additional information <3-7> (<3,6> tissue distribution [10]) [5, 8-10]

Localization

cytoplasm <1-6> [1-7, 9, 10, 13]

Purification

<2> (ammonium sulfate, Sephadex G-75, DEAE-cellulose, hydroxylapatite [13]) [13]
<3> (pH 4.2, ammonium sulfate, heat, Sephadex G-75, carboxymethyl cellulose, DEAE-cellulose [2]; ammonium sulfate, Sephadex G-75, DEAE-cellulose, hydroxyapatite [3]) [2, 3, 9]
<4> [2, 9]
<5> [10]
<6> (immunologically identical isoenzymes A and B, ammonium sulfate, gel filtration, carboxymethylcellulose, DEAE-cellulose [5]; isoenzyme with pI 5.1: ammonium sulfate, heat, Sephadex γ-75, DEAE-cellulose, thiol-Sepharose 4B, isoenzyme with pI 4.6: ammonium sulfate, heat, CM-cellulose, Sephadex γ-75, DEAE-cellulose, ioselectric focusing, gel filtration [6]) [5-7]

6 Stability

Temperature stability

57 <3> (<3> purified enzyme, complete loss of activity after 5 min [2]) [2]

Storage stability

<3, 4>, -20°C, stable [2]
<3, 4>, 0°C, at least 1 month [2, 9]
<6>, 0°C, pI 4.6-isoform, at least 2 months [6]
<6>, 0°C, pI 5.1-isoform, 3 weeks, 30% loss of activity [6]

References

[1] Steinkamp, R.; Schweihofen, B.; Rennenberg, H.: γ-Glutamylcyclotransferase in tobacco suspension cultures: catalytic properties and subcellular localization. Physiol. Plant., **69**, 499-503 (1987)

[2] Orlowski, M.; Richman, P.G.; Meister, A.: Isolation and properties of γ-L-glutamylcyclotransferase from human brain. Biochemistry, **8**, 1048-1055 (1969)

[3] Board, P.G.; Moore, K.A.; Smith, J.E.: Purification and properties of γ-glutamylcyclotransferase from human erythrocytes. Biochem. J., **173**, 427-431 (1978)

[4] York, M.J.; Kuchel, P.W.; Chapman, B.E.: A proton nuclear magnetic resonance study of γ-glutamyl-amino acid cyclotransferase in human erythrocytes. J. Biol. Chem., **259**, 15085-15088 (1984)

[5] Orlowski, M.; Meister, A.: γ-Glutamyl cyclotransferase. Distribution, isozymic forms, and specificity. J. Biol. Chem., **248**, 2836-2844 (1973)

[6] Meister, A.: γ-Glutamylcyclotransferase from rat kidney. Methods Enzymol., **113**, 428-445 (1981)

[7] Taniguchi, N.; Meister, A.: γ-Glutamyl cyclotransferase from rat kidney. Sulfhydryl groups and isolation of a stable form of the enzyme. J. Biol. Chem., **253**, 1799-1806 (1978)

[8] Bridges, R.J.; Griffith, O.W.; Meister, A.: L-γ-(threo-β-methyl)glutamyl-L-α-aminobutyrate, a selective substrate of α-glutamyl cyclotransferase. J. Biol. Chem., **255**, 10787-10792 (1980)

[9] Orlowski, M.; Meister, A.: Enzymology of pyrrolidone carboxylic acid. The Enzymes, 3rd Ed. (Boyer, P.D., ed.), **4**, 123-151 (1971)

[10] Szweczuk, A.; Connell, G.E.: Specificity of γ-glutamyl cyclotransferase. Can. J. Biochem., **53**, 706-712 (1975)

[11] Griffith, O.W.; Meister, A.: Selective inhibition of γ-glutamyl-cycle enzymes by substrate analogs. Proc. Natl. Acad. Sci. USA, **74**, 3330-3334 (1977)

[12] York, M.J.; Crossley, M.J.; Hyslop, S.J.; Fisher, M. L.; Kuchel, P.W.: γ-Glutamylcyclotransferase: inhibition by D-β-aminoglutaryl-L-alanine and analysis of the solvent kinetic isotope effect. Eur. J. Biochem., **184**, 97-101 (1989)

[13] Bodnaryk, R.P.; McGirr, L.: Purification, properties and function of a unique γ-glutamyl cyclotransferase from the housefly, Musca domestica. Biochim. Biophys. Acta, **315**, 352-362 (1973)

[14] Korotkina, R.N.; Matskevich, G.N.; Devlikanova, A.S.; Vishnevskii, A.A.; Kunitsyn, A.G.; Karelin, A.A.: Activity of glutathione-metabolizing and antioxidant enzymes in malignant and benign tumors of human lungs. Bull. Exp. Biol. Med., **133**, 606-608 (2002)

[15] Danson, J.W.; Trawick, M.L.; Cooper, A.J.: Spectrophotometric assays for L-lysine α-oxidase and γ-glutamylamine cyclotransferase. Anal. Biochem., **303**, 120-130 (2002)

Glutaminyl-peptide cyclotransferase 2.3.2.5

1 Nomenclature

EC number
2.3.2.5

Systematic name
L-glutaminyl-peptide γ-glutamyltransferase (cyclizing)

Recommended name
glutaminyl-peptide cyclotransferase

Synonyms
QC
cyclotransferase, glutaminyl-transfer ribonucleate
glutaminyl cyclase
glutaminyl-tRNA cyclotransferase

CAS registry number
37257-21-9

2 Source Organism

<1> *Rattus norvegicus* (neonatal [1]) [1, 2]
<2> *Homo sapiens* [1, 8, 9, 10]
<3> *Sus scrofa* [1]
<4> *Bos taurus* [1, 2, 4]
<5> *Carica papaya* [3, 5, 6, 7, 11, 12]

3 Reaction and Specificity

Catalyzed reaction
L-glutaminyl-peptide = 5-oxoprolyl-peptide + NH_3

Reaction type
aminoacyl group transfer

Natural substrates and products
S L-glutaminyl-peptide <1-4> (<1-4>, involved in posttranslational modification of the N-terminal glutamine of peptide hormones or neurotransmitters, such as thyrotropin releasing hormone, luteinizing hormone releasing hormone, gastrin, heavy chain of γ-globulin [1]; <4>, the enzyme

may participate in the posttranslational processing of hormonal precursors to pyroglutamyl peptides [2]; <2>, the N-terminal formation of 5-oxoproline is a common post-translational event during biosynthesis of a number of peptides [10]) (Reversibility: ? <1-4> [1, 2, 10]) [1, 2, 10]

P 5-oxoprolyl-peptide + NH_3

Substrates and products

S Gln <5> (Reversibility: ? <5> [7]) [7]
P L-5-oxoproline + NH_3
S Gln-Ala <2, 4, 5> (Reversibility: ? <2, 4, 5> [4, 7, 8]) [4, 7, 8]
P Glu-Ala + NH_3
S Gln-Gln <2, 4, 5> (Reversibility: ? <2, 4, 5> [4, 7, 8, 10]) [4, 7, 8, 10]
P Glu-Gln + NH_3
S Gln-Gln-Gln <2, 4, 5> (Reversibility: ? <2, 4, 5> [4, 7, 8]) [4, 7, 8]
P Glu-Gln-Gln + NH_3
S Gln-Glu <4, 5> (Reversibility: ? <4, 5> [4, 7]) [4, 7]
P Glu-Glu + NH_3
S Gln-Gly <2, 4, 5> (Reversibility: ? <2, 4, 5> [4, 7, 8]) [4, 7, 8]
P Glu-Gly + NH_3
S Gln-Gly-Pro <4> (Reversibility: ? <4> [4]) [4]
P Glu-Gly-Pro + NH_3
S Gln-His-Pro-NH_2 <2> (Reversibility: ? <2> [10]) [10]
P L-5-oxoprolyl-His-Pro-NH_2 + NH_3
S Gln-NH_2 <2, 4, 5> (Reversibility: ? <2, 4, 5> [4, 7, 8, 10]) [4, 7, 8, 10]
P L-Glu-amide + NH_3
S Gln-Tyr-Ala-OH <2> (Reversibility: ? <2> [10]) [10]
P L-5-oxoprolyl-Tyr-Ala + NH_3
S Gln-Val <5> (Reversibility: ? <5> [7]) [7]
P Glu-Val + NH_3
S H-Gln-Gln-OH <2> (Reversibility: ? <2> [10]) [10]
P L-5-oxoprolyl-Gln + NH_3
S L-glutamine tert-butyl ester <5> (Reversibility: ? <5> [5, 7, 11]) [5, 7, 11]
P L-glutarate-tert-butyl ester + NH_3
S L-glutaminyl-2-naphthylamide <2> [10]
P L-5-oxopropyl-2-naphthylamide + NH_3
S L-glutaminyl-4-methylcoumarinylamide <2> [10]
P L-5-oxoprolyl-4-methylcoumarinylamide + NH_3
S L-glutaminyl-peptide <1-5> (<1-4>, thyrotropin-releasing hormone [1,2]; <1,4>, Gln-His-Pro-Glythyrotropin-releasing hormone, gonadotropin-releasing hormone and L-Gln-Tyr-Ala [2]; <5>, simple intramolecular cyclization. The mechanism consists of the following main steps: 1. intramolecular nucleophilic attack on the γ-C=O carbon by the nitrogen of the α-amino group, 2. transfer of a proton from the α-amino group to the nitrogen of the amide group, facilitated by an acidic group of the enzyme, 3. expulsion of the ammonia-leaving group promoted by this or another acidic enzyme group [5]) [1-3]
P 5-oxoprolyl-peptide + NH_3 <1-5> [1-3]

S Additional information <1, 4, 5> (<1,4>, no substrates are D-Gln-Tyr-Ala and Lys-Arg-Gln-His-Pro-Gly-Lys-Arg, i.e. thyrotropin releasing hormone precursor [2]; <5>, the second amino acid residue in N-terminal glutaminyl peptides significantly accelerates activity while the third residue provides no further rate enhancement. Substrate binding is the main specificity-determining step [7]; <5>, Cys residues do not participate to the catalytic events [11]) [2, 7, 11]

P ?

Inhibitors

1,10-phenanthroline <3> (<3>, 2 mM, complete inhibition [1]) [1]
6-methylpterin <3> [1]
$CuCl_2$ <3> (<3>, 0.1 mM [1]) [1]
FADH <3> (<3>, 2 mM [1]) [1]
NEM <3> [1]
NH_4Cl <4> (<4>, 1.0 M [2]) [2]
$NiCl_2$ <3> (<3>, 0.1 mM [1]) [1]
$ZnCl_2$ <3> (<3>, 0.1 mM [1]) [1]
ascorbate <3> [1]
diethyl dicarbonate <2> (<2>, rapid inactivation by modification of three essential His residues, at neutral pH, partial reactivation with hydroxylamine [9]) [9]

Activating compounds

EDTA <3> (<3>, stimulates [1]) [1]

Turnover number (min^{-1})

21.6 <2> (Gln-NH_2, <2>, mutant enzyme H319Q [9]) [9]
22.1 <2> (Gln-Gly) [8]
23.3 <2> (Gln-NH_2, <2>, mutant enzyme H307Q [9]) [9]
26.1 <2> (Gln-NH_2, <2>, unmutated enzyme [9]) [9]
33 <2> (Gln-NH_2) [8]
45.3 <2> (Gln-Gln-Gln) [8]
50.9 <2> (Gln-Gln) [8]
76.2 <2> (Gln-Ala) [8]
324 <2> (L-glutaminyl-4-methylcoumarinylamide, <2>, 30°C, recombinant enzyme, expressed in Pichia pastoris [10]) [10]
768 <2> (Gln-Gln-OH, <2>, 30°C, recombinant enzyme, expressed in Pichia pastoris [10]) [10]
1128 <2> (L-glutaminyl-2-naphthylamide, <2>, 30°C, recombinant enzyme, expressed in Pichia pastoris [10]) [10]
1242 <2> (Gln-NH_2, <2>, 30°C, recombinant enzyme, expressed in Pichia pastoris [10]) [10]
1980 <5> (L-glutamine tert-butyl ester) [11]
4980 <2> (Gln-His-Pro-NH_2, <2>, 30°C, recombinant enzyme, expressed in Pichia pastoris [10]) [10]
7500 <2> (Gln-Tyr-Ala-OH, <2>, 30°C, recombinant enzyme, expressed in Pichia pastoris [10]) [10]

13200 <2> (Gln-Tyr-Ala-OH, <2>, 37°C, recombinant enzyme, expressed in Pichia pastoris [10]) [10]

Specific activity (U/mg)

18.7 <2> [10]
Additional information <2> [8]

K_m-Value (mM)

0.051 <2> (L-glutaminyl-4-methylcoumarinylamide, <2>, 30°C, recombinant enzyme, expressed in Pichia pastoris [10]) [10]
0.06 <2> (L-glutaminyl-2-naphthylamide, <2>, 30°C, recombinant enzyme, expressed in Pichia pastoris [10]) [10]
0.063 <4> (gonadotropin releasing hormone) [2]
0.088 <4> ([Gln1,Gly4]-thyrotropin releasing hormone) [2]
0.09 <5> (Gln-Gln) [7]
0.09 <2> (Gln-His-Pro-NH_2, <2>, 30°C, recombinant enzyme, expressed in Pichia pastoris [10]) [10]
0.101 <2> (Gln-Tyr-Ala-OH, <2>, 30°C, recombinant enzyme, expressed in Pichia pastoris [10]) [10]
0.12 <5> (Gln-Val) [7]
0.132 <4> (Gln-Thr-Ala) [2]
0.148 <2> (Gln-Gln-OH, <2>, 30°C, recombinant enzyme, expressed in Pichia pastoris [10]) [10]
0.153 <2> (Gln-Tyr-Ala-OH, <2>, 37°C, recombinant enzyme, expressed in Pichia pastoris [10]) [10]
0.17 <5> (Gln-Glu) [7]
0.21 <5> (Gln-Ala) [7]
0.31 <5> (Gln-Gln-Gln) [7]
0.32 <5> (glutamine tert-butyl ester) [7]
0.379 <5> (L-glutamine tert-butyl ester, <5>, pH 6.5 [5]) [5]
0.4 <5> (L-glutamine tert-butyl ester, <5>, pH 6.75 [5]) [5]
0.409 <2> (Gln-NH_2, <2>, 30°C, recombinant enzyme, expressed in Pichia pastoris [10]) [10]
0.43 <5> (L-glutamine tert-butyl ester, <5>, pH 7.45 [5]) [5]
0.44 <5> (L-glutamine tert-butyl ester, <5>, pH 8.05 [5]) [5]
0.46 <5> (L-glutamine tert-butyl ester, <5>, pH 7.0 or pH 8.6 [5]) [5]
0.53 <2> (Gln-Gln-Gln) [8]
0.55 <5> (L-glutamine tert-butyl ester, <5>, pH 6.2 [5]) [5]
0.6 <5> (L-glutamine tert-butyl ester, <5>, pH 9.0 [5]) [5]
0.62 <5> (L-glutamine tert-butyl ester, <5>, pH 6.3 [5]) [5]
0.64 <2> (Gln-Gln) [8]
0.9 <5> (L-glutamine tert-butyl ester, <5>, pH 9.55 [5]) [5]
1.1 <5> (Gln-Gly) [7]
1.1 <2> (Gln-NH_2, <2>, wild-type enzyme [9]) [9]
1.3 <2> (Gln-Ala) [8]
1.3 <5> (Gln-NH_2) [7]
1.3 <5> (L-glutamine tert-butyl ester, <5>, pH 9.95 [5]) [5]
1.88 <2> (Gln-Gly) [8]

4.3 <2> (Gln-NH_2, <2>, mutant enzyme H307Q [9]) [9]
4.9 <2> (Gln-NH_2, <2>, mutant enzyme H319Q [9]) [9]
7.15 <2> (Gln-NH_2) [8]
16 <5> (Gln) [7]
Additional information <3> [1]

pH-Optimum

7.2-7.5 <3> [1]
7.3 <4> (<4>, pituitary enzyme [4]) [4]
8 <4> [2]
8.5 <4> (<4>, enzyme from spleen [4]) [4]

pH-Range

6.5-8.3 <4> (<4>, pH 6.5: about 50% of maximal activity, pH 8.3: about 40% of maximal activity, pituitary enzyme [4]) [4]
7.5-9.3 <4> (<4>, pH 7.5: about 55% of maximal activity, pH 9.3: about 45% of maximal activity, enzyme from spleen [4]) [4]
Additional information <5> (<5>, activity changes very slightly in the pH range between 4.5 and 10 [5]) [5]

4 Enzyme Structure

Molecular weight

48500 <4> (<4>, gel filtration [4]) [4]
55000 <3> (<3>, gel filtration [1]) [1]

Subunits

? <2, 5> (<5>, x * 32980, native, glycosylated enzyme, SDS-PAGE [5]; <5>, x * 33000, native, glycosylated enzyme, SDS-PAGE [5]; <2>, x * 38795, deglycosylated enzyme, MALDI-TOF mass analysis [10]; <2>, x * 40000, SDS-PAGE [8]; <2>, x * 40876, calculation from nucleotide sequence [8]) [5, 8, 10]
Additional information <5> (<5>, enzyme contains extensive β-sheet structure and is likely to have only short immobile loops connecting ist β-strands [3]) [3]

Posttranslational modification

glycoprotein <1-4, 5> (<5>, presence of terminal mannose residues, mannose and N-acetylglucosamine in a 3:1 molar ratio [5]; <2>, deglycosylation has no effect on enzymatic activity [10]) [1, 5, 10, 12]

5 Isolation/Preparation/Mutation/Application

Source/tissue

B-lymphocyte <2> (<2>, secretory and non-secretory cell line P3X63Ag8 [1]) [1]
P3X63Ag8 cell <2> (<2>, secretory and non-secretory cell line P3X63Ag8 [1]) [1]

adrenal medulla <4> [1]
brain <1, 4> [1, 2, 4]
hypothalamus <1> [2]
latex <5> [3, 7, 11, 12]
laticifer <5> [3]
pituitary gland <2, 3, 4> [1, 2, 4, 8, 9]
spleen <4> [4]

Localization

chromaffin granule <4> [1]
secretory granule <4> [2]
soluble <4, 5> (<4>, a 40000 Da enzyme form is exclusively detected in soluble pituitary extract, a 32000 Da enzyme form is detected in soluble fraction of all tissues which show immunoreactivity. The difference in the two immunoreacive proteins lies in the proteolytic cleavage within a hydrophilic region approximately 80 amino acids from the carboxy-terminus [4]) [4, 5]
vesicular fraction <4> [1]

Purification

<2> (of 6His-tagged enzyme expressed in Pichia pastoris and enzyme expressed in E. coli [10]) [10]
<4> (partial [4]) [4]
<5> [3, 5]

Cloning

<2> (cloned into the Escherichia coli expression vectors pMALc2 and pET19b. Expression of this cDNA in either vector results in the production of a glutaminyl cyclase fusion protein which is enzymatically active and reacts with anti-bovine glutaminyl cyclase antisera [8]) [8]
<2> (expression in Pichia pastoris. In Escherichia coli only 50% of the protein does not contain a disulfide bond that is present in the enzyme expressed in Pichia pastoris [10]) [10]
<5> (expression in Escherichia coli as either His-tagged enzyme with three different signal peptides and in fusions with three different signal peptides and in fusion with thioredoxin, glutathione S-transferase, and (pre-)maltose-binding protein. In all cases, the expressed protein is either undetectable or insoluble. Expression in Pichia pastoris of the enzyme fused to the α-factor leader results in low levels of activity. Extracellular expression of the enzyme in the insect cell/baculovirus system is sucessfull. Enzyme N-terminally fused to a combined secretion signal/His-tag peptide is correctly processed by the host signal peptidase and the His-tag can subsequently be removed with dipeptidyl peptidase I [6]; large-scale expression in Pichia pastoris [10]) [6, 10]

Engineering

H140Q <2> (<2>, inactive enzyme [9]) [9]
H307Q <2> (<2>, mutant enzyme with increased K_M-value [9]) [9]
H319Q <2> (<2>, mutant enzyme with increased K_M-value [9]) [9]
H330Q <2> (<2>, inactive enzyme [9]) [9]

Additional information <5> (<5>, N53A and N123A: no expression of either of the mutants is detected [5]) [5]

6 Stability

Temperature stability

65 <5> (<5>, pH 8.8, several hours, no inactivation [5]) [5]

General stability information

<5>, complete unfolding of the enzyme requires a combination of an acidic medium and chemical denaturant such as urea or guanidine hydrochloride [12]

<5>, resistance to denaturation induced by guanidine-HCl [11]

<5>, treatment with trypsin and chymotrypsin at pH 8.2 and 8.0 respectively, for 16 h at 37°C: enzyme is recovered in ist native form [3]

Storage stability

<4>, stable for several weeks [4]

References

[1] Busby, W.H.; Quackenbush, G.E.; Humm, J.; Youngblood, W.W.; Kizer, J.S.: An enzyme(s) that converts glutaminyl-peptides into pyroglutamyl-peptides. Presence in pituitary, brain, adrenal medulla, and lymphocytes. J. Biol. Chem., **262**, 8532-8536 (1987)

[2] Fischer, W.H.; Spiess, J.: Identification of a mammalian glutaminyl cyclase converting glutaminyl into pyroglutamyl peptides. Proc. Natl. Acad. Sci. USA, **84**, 3628-3632 (1987)

[3] Oberg, K.A.; Ruysschaert, J.M.; Azarkan, M.; Smolders, N.; Zerhouni, S.; Wintjens, R.; Amrani, A.; Looze, Y.: Papaya glutamine cyclase, a plant enzyme highly resistant to proteolysis, adopts an all-β conformation. Eur. J. Biochem., **258**, 214-222 (1998)

[4] Sykes, P.A.; Watson, S.J.; Temple, J.S.; Bateman, R.C., Jr.: Evidence for tissue-specific forms of glutaminyl cyclase. FEBS Lett., **455**, 159-161 (1999)

[5] Gololobov, M.Y.; Song, I.; Wang, W.; Bateman, R.C.M.: Steady-state kinetics of glutamine cyclotransferase. Arch. Biochem. Biophys., **309**, 300-307 (1994)

[6] Dahl, S.W.; Slaughter, C.; Lauritzen, C.; Bateman, R.C., Jr.; Connerton, I.; Pedersen, J.: Carica papaya glutamine cyclotransferase belongs to a novel plant enzyme subfamily: cloning and characterization of the recombinant Enzyme. Protein Expr. Purif., **20**, 27-36 (2000)

[7] Gololobov, M.Y.; Wang, W.; Bateman, R.C., Jr.: Substrate and inhibitor specificity of glutamine cyclotransferase (QC). Biol. Chem. Hoppe-Seyler, **377**, 395-398 (1996)

[8] Song, I.; Chuang, C.Z.; Bateman, J.R.C.: Molecular cloning, sequence analysis and expression of human pituitary glutaminyl cyclase. J. Mol. Endocrinol., **13**, 77-86 (1994)
[9] Bateman, R.C., Jr.; Temple, J.S.; Misquitta, S.A.; Booth, R.E.: Evidence for essential histidines in human pituitary glutaminyl cyclase. Biochemistry, **40**, 11246-11250 (2001)
[10] Schilling, S.; Hoffmann, T.; Rosche, F.; Manhart, S.; Wasternack, C.; Demuth, H.U.: Heterologous expression and characterization of human glutaminyl cyclase: evidence for a disulfide bond with importance for catalytic activity. Biochemistry, **41**, 10849-10857 (2002)
[11] Zerhouni, S.; Amrani, A.; Nijs, M.; Vandermeers, A.; Looze, Y.: Purification and characterization of the plant glutaminyl-peptide cyclotransferase isolated from papaya latex. Int. J. Bio-Chromatogr., **3**, 189-206 (1997)
[12] Azarkan, M.; Amrani, A.; Zerhouni, S.; Oberg, K.A.; Ruysschaert, J.M.; Wintjens, R.; Looze, Y.: Evidence that thermodynamic stability of papaya glutamine cyclase is only marginal. Biopolymers, **65**, 325-335 (2002)

Leucyltransferase 2.3.2.6

1 Nomenclature

EC number
2.3.2.6

Systematic name
L-leucyl-tRNA:protein leucyltransferase

Recommended name
leucyltransferase

Synonyms
L/F-transferase
leucyl, phenylalanine-tRNA-protein transferase
leucyl, phenylalanyl transfer ribonucleic acid-protein transferase
leucyl-phenylalanine-transfer ribonucleate-protein aminoacyltransferase
leucyl-phenylalanine-transfer ribonucleate-protein transferase
leucyl/phenylalanyl-tRNA-protein transferase

CAS registry number
37257-22-0

2 Source Organism

<1> *Escherichia coli* (B [1-3, 5]; K12 strain W4977 and revertant strain R18 of mutant strain MS845 [5]) [1-8]

3 Reaction and Specificity

Catalyzed reaction
L-leucyl-tRNA + protein = tRNA + L-leucyl-protein

Reaction type
aminoacyl group transfer

Substrates and products
S L-Trp-tRNA + acceptor protein <1> (Reversibility: ? <1> [6]) [6]
P L-Trp-acceptor protein + tRNA
S L-leucyl-tRNA + acceptor protein <1> (<1>, acceptor protein: bovine serum albumin [1,2]; <1>, incorporation of Leu into the peptide linkage with the amino-terminal aspartic acid of albumin [2]; <1>, acceptor protein:

α_{S1}-casein [3]; <1>, all peptides containing a NH_2-terminal L-Arg or Lys residue function as acceptor, however D-Arg-D-Val is inactive [3]; <1>, dipeptide specificity [5]; <1>, acceptors with arginine or lysine as initial NH_2-terminal residue [4]; <1>, basic NH_2-terminal is absolute determinant of specificity [3, 5]; <1>, the association of the Leu-tRNA-enzyme complex is diffusion controlled [7]) (Reversibility: ? <1> [1-8]) [1-8]

P tRNA + L-leucyl-protein <1> [1-8]

S L-methionyl-tRNA + acceptor protein <1> (<1>, methionyl-$tRNA_m^{Met}$ preferred to methionyl-$tRNA_f^{Met}$. Peptides containing a basic amino acid at the NH_2-terminus function as acceptors [5]; <1>, wild-type Met-$tRNA_m^{Met}$ (CAU anticodon) and mischarged Met-$tRNA^{Val-1}$ (CAU anticodon) are substrates for LF-transferase during the NH_2-terminal aminoacylation of α-casein [7]) (Reversibility: ? <1> [5, 6, 8]) [5, 6, 7, 8]

P tRNA + L-methionyl-protein

S L-phenylalanyl-tRNA + Lys-Ala-Ala <1> (Reversibility: ? <1> [3]) [3]

P Phe-Lys-Ala-Ala + tRNA <1> [3]

S L-phenylalanyl-tRNA + acceptor protein <1> (<1>, incorporation of Phe into the peptide linkage with the amino-terminal aspartic acid of albumin [2]) (Reversibility: ? <1> [1-8]) [1-8]

P tRNA + L-phenylalanyl-protein

S Additional information <1> (<1>, no activity with Val-$tRNA^{Val-1}$ (UAC anticodon), Val-$tRNA_m^{Met}$ (UAC anticodon), and Arg-$tRNA_m^{Met}$ [7]; <1>, substrate recognition. Vatiants of the enzyme, lacking either 33 or 78 N-terminal residues, retain measurable peptidyltransferase activity and wild type substrate specificity [8]) [7, 8]

P ?

Inhibitors

Arg <1> (<1>, inhibition of α_{S1}-casein-dependent reaction [3]) [3]
Arg-Ala <1> (<1>, inhibition of α_{S1}-casein-dependent reaction [3]) [3]
Arg-Arg <1> (<1>, inhibition of α_{S1}-casein-dependent reaction [3]) [3]
Arg-Asp <1> (<1>, inhibition of α_{S1}-casein-dependent reaction [3]) [3]
Arg-Glu <1> (<1>, inhibition of α_{S1}-casein-dependent reaction [3]) [3]
Arg-Gly <1> (<1>, inhibition of α_{S1}-casein-dependent reaction [3]) [3]
Arg-Gly-Gly <1> (<1>, inhibition of α_{S1}-casein-dependent reaction [3]) [3]
Arg-Leu <1> (<1>, inhibition of α_{S1}-casein-dependent reaction [3]) [3]
Arg-Lys <1> (<1>, inhibition of α_{S1}-casein-dependent reaction [3]) [3]
Arg-Lys <1> (<1>, inhibition of α_{S1}-casein-dependent reaction [3]) [3]
Arg-Phe <1> (<1>, inhibition of α_{S1}-casein-dependent reaction [3]) [3]
Arg-Tyr <1> (<1>, inhibition of α_{S1}-casein-dependent reaction [3]) [3]
Arg-Val <1> (<1>, inhibition of α_{S1}-casein-dependent reaction [3]) [3]
$CaCl_2$ <1> (<1>, 20 mM, 35% inhibition [2]) [2]
Lys <1> (<1>, inhibition of α_{S1}-casein-dependent reaction [3]) [3]
Lys-Ala <1> (<1>, inhibition of α_{S1}-casein-dependent reaction [3]) [3]
Lys-Ala-Ala <1> (<1>, inhibition of α_{S1}-casein-dependent reaction [3]) [3]
Lys-Glu <1> (<1>, inhibition of α_{S1}-casein-dependent reaction [3]) [3]
Lys-Leu <1> (<1>, inhibition of α_{S1}-casein-dependent reaction [3]) [3]

Lys-Phe <1> (<1>, inhibition of α_{S1}-casein-dependent reaction [3]) [3]
Lys-Sr <1> (<1>, inhibition of α_{S1}-casein-dependent reaction [3]) [3]
Lys-Tyr-Thr <1> (<1>, inhibition of α_{S1}-casein-dependent reaction [3]) [3]
Lys-Val <1> (<1>, inhibition of α_{S1}-casein-dependent reaction [3]) [3]
Mg^{2+} <1> (<1>, 50 mM, 80% inhibition [2]) [2]
$MnCl_2$ <1> (<1>, 91% inhibition [2]) [2]
arginine methyl ester <1> (<1>, inhibition of α_{S1}-casein-dependent reaction [3]) [3]
puromycin <1> (<1>, aminonucleoside derivative, which lacks the amino-acyl analogus structure, fails to inhibit the reaction [2]) [2, 7]

Metals, ions

K^+ <1> (<1>, monovalent cation required, stimulation of Leu transfer is somewhat greater than that of Phe [1]; <1>, monovalent cation required, maximal activity is 0.15 M KCl for the transfer of Phe and 0.2 M for the transfer of Leu [2]) [1, 2]
NH_4^+ <1> (<1>, monovalent cation required, stimulation of Leu transfer is somewhat greater than that of Phe [1]) [1]
Na^+ <1> (<1>, requirement [1]) [1]
Additional information <1> (<1>, no Mg^{2+}-requirement [2]) [2]

Specific activity (U/mg)

Additional information <1> [1, 2, 4, 5]

K_i-Value (mM)

0.2 <1> (Arg-Lys, <1>, inhibition of reaction with α_{S1}-casein as acceptor [3]) [3]
0.4 <1> (Arg-Arg, <1>, inhibition of reaction with α_{S1}-casein as acceptor [3]) [3]
0.4 <1> (Arg-Gly-Gly, <1>, inhibition of reaction with α_{S1}-casein as acceptor [3]) [3]
1 <1> (Arg-Val, <1>, inhibition of reaction with α_{S1}-casein as acceptor [3]) [3]
1.3 <1> (Arg-Gly, <1>, inhibition of reaction with α_{S1}-casein as acceptor [3]) [3]
1.4 <1> (Arg-Phe, <1>, inhibition of reaction with α_{S1}-casein as acceptor [3]) [3]
1.5 <1> (Arg-Ala, <1>, inhibition of reaction with α_{S1}-casein as acceptor [3]) [3]
1.5 <1> (arginine methyl ester, <1>, inhibition of reaction with α_{S1}-casein as acceptor [3]) [3]
1.6 <1> (Arg-Tyr, <1>, inhibition of reaction with α_{S1}-casein as acceptor [3]) [3]
3 <1> (Lys-Ala-Ala, <1>, inhibition of reaction with α_{S1}-casein as acceptor [3]) [3]
3.5 <1> (Arg-Leu, <1>, inhibition of reaction with α_{S1}-casein as acceptor [3]) [3]

3.6 <1> (Lys-Ala, <1>, inhibition of reaction with α_{S1}-casein as acceptor [3]) [3]
4 <1> (Lys-Phe, <1>, inhibition of reaction with α_{S1}-casein as acceptor [3]) [3]
4 <1> (Lys-Val, <1>, inhibition of reaction with α_{S1}-casein as acceptor [3]) [3]
6 <1> (Arg-Asp, <1>, inhibition of reaction with α_{S1}-casein as acceptor [3]) [3]
6 <1> (Lys-Gly, <1>, inhibition of reaction with α_{S1}-casein as acceptor [3]) [3]
6.3 <1> (Lys-Tyr-Thr, <1>, inhibition of reaction with α_{S1}-casein as acceptor [3]) [3]
7.5 <1> (Arg-Glu, <1>, inhibition of reaction with α_{S1}-casein as acceptor [3]) [3]
7.8 <1> (Lys-Leu, <1>, inhibition of reaction with α_{S1}-casein as acceptor [3]) [3]
10 <1> (Lys-Ser, <1>, inhibition of reaction with α_{S1}-casein as acceptor [3]) [3]
10 <1> (arginine, <1>, inhibition of reaction with α_{S1}-casein as acceptor [3]) [3]
15 <1> (Lys-Glu, <1>, inhibition of reaction with α_{S1}-casein as acceptor [3]) [3]

pH-Optimum
7.6-8.2 <1> (<1>, transfer of Leu [2]) [2]
8.2-8.6 <1> (<1>, transfer of Phe [2]) [2]

pH-Range
7-9 <1> (<1>, about 80% of maximal activity at pH 7.0 and 9.0 [2]) [2]

5 Isolation/Preparation/Mutation/Application

Localization
soluble <1> [1-5]

Purification
<1> (partial [1]) [1-6]

Cloning
<1> [6]

6 Stability

Temperature stability
60 <1> (<1>, $t_{1/2}$: 1.5 min, complete inactivation after 8.5 min [2]) [2]

General stability information

<1>, stable to repeated freeze-thawing [3]

Storage stability

<1>, -20°C, stable for at least 2 months [2]
<1>, -20°C, stable for at least 6 months [3]
<1>, -20°C, stable for several months [5]

References

[1] Leibowitz, M.J.; Soffer, R.L.: A soluble enzyme from Escherichia coli which catalyzes the transfer of leucine and phenylalanine from tRNA to acceptor proteins. Biochem. Biophys. Res. Commun., **36**, 47-53 (1969)
[2] Leibowitz, M.J.; Soffer, R.L.: Enzymatic modification of proteins. 3. Purification and properties of a leucyl, phenylalanyl transfer ribonucleic acid protein transferase from Escherichia coli. J. Biol. Chem., **245**, 2066-2073 (1970)
[3] Soffer, R.L.: Peptide acceptors in the leucine, phenylalanine transfer reaction. J. Biol. Chem., **248**, 8424-8428 (1973)
[4] Deutch, C.E.: Aminoacyl-tRNA: protein transferases. Methods Enzymol., **106**, 198-205 (1984)
[5] Scarpulla, R.C.; Deutch, C.E.; Soffer, R.L.: Transfer of methionyl residues by leucyl, phenylalanyl-tRNA-protein transferase. Biochem. Biophys. Res. Commun., **71**, 584-589 (1976)
[6] Abrahamochkin, G.; Shrader, T.E.: The leucyl/phenylalanyl-tRNA-protein transferase. Overexpression and characterization of substrate recognition, domain structure, and secondary structure. J. Biol. Chem., **270**, 20621-20628 (1995)
[7] Abramochkin, G.; Shrader, T.E.: Aminoacyl-tRNA recognition by the leucyl/phenylalanyl-tRNA-protein transferase. J. Biol. Chem., **271**, 22901-22907 (1996)
[8] Ichetovkin, I.E.; Abramochkin, G.; Shrader, T.E.: Substrate recognition by the leucyl/phenylalanyl-tRNA-protein transferase. Conservation within the enzyme family and localization to the trypsin-resistant domain. J. Biol. Chem., **272**, 33009-33014 (1997)

Aspartyltransferase 2.3.2.7

1 Nomenclature

EC number
2.3.2.7

Systematic name
L-asparagine:hydroxylamine γ-aspartyltransferase

Recommended name
aspartyltransferase

Synonyms
aspartotransferase
β-aspartyl transferase

CAS registry number
37257-23-1

2 Source Organism

<1> *Mycobacterium tuberculosis* (H37Ra, strain number 7417) [1]

3 Reaction and Specificity

Catalyzed reaction
L-asparagine + hydroxylamine = NH_3 + L-aspartylhydroxamate

Reaction type
aminoacyl group transfer

Natural substrates and products

S L-asparagine + hydroxylamine <1> (<1>, involved in initial step of asparagine metabolism [1]) (Reversibility: ? <1> [1]) [1]

P NH_3 + β-L-aspartohydroxamic acid

Substrates and products

S asparagine + hydroxylamine <1> (<1>, strict specificity, L- and D-enantiomer equally effective, no substrate: L-glutamine [1]) (Reversibility: ? <1> [1]) [1]

P NH_3 + β-aspartohydroxamic acid <1> [1]

Inhibitors

Co^{2+} <1> [1]
Cu^{2+} <1> [1]
L-aspartate <1> (<1>, 20 mM, 30% inhibition [1]) [1]
L-cysteine <1> (<1>, 5 mM, 50% inhibition [1]) [1]
PCMB <1> (<1>, reversible by GSH, 2-mercaptoethanol or 2,3-dimercapto-propanol, not L-cysteine [1]) [1]
streptomycin <1> (<1>, 5 mM, 50% inhibition [1]) [1]
Additional information <1> (<1>, no inhibition by NaN_3, isonicotinic acid hydrazide, L-homoserine, L-glutamic acid, terminal amino acids of aspartic acid pathway, i.e. lysine, threonine, methionine and isoleucine, up to 20 mM [1]) [1]

Activating compounds

L-cystine <1> (<1>, 20 mM, slight activation [1]) [1]
isonicotinic acid hydrazide <1> (<1>, activation [1]) [1]

Metals, ions

Additional information <1> (<1>, no metal ion requirement [1]) [1]

Specific activity (U/mg)

0.0245 <1> [1]

K_m-Value (mM)

7.14 <1> (L-asparagine) [1]
19.3 <1> (hydroxylamine) [1]

pH-Optimum

9 <1> [1]

pH-Range

8.5-9.4 <1> (<1>, about half-maximal activity at pH 8.5 and 9.4 [1]) [1]

Temperature optimum (°C)

50 <1> [1]

Temperature range (°C)

40-60 <1> (<1>, 40°C: about 50% of maximal activity, 60°C: about 80% of maximal activity [1]) [1]

5 Isolation/Preparation/Mutation/Application

Purification

<1> [1]

6 Stability

Temperature stability

0 <1> (<1>, 6 h, 23% loss of activity [1]) [1]
28 <1> (<1>, 6 h, stable [1]) [1]
37 <1> (<1>, 6 h, about 8% loss of activity [1]) [1]
60 <1> (<1>, 10 min, about 45% loss of activity [1]) [1]
65 <1> (<1>, 10 min, about 66% loss of activity [1]) [1]
70 <1> (<1>, $t_{1/2}$: 1 min, about 55% loss of activity after 2 min [1]) [1]

General stability information

<1>, dialysis against 0.05 M Tris buffer, pH 7.0, 75% loss of activity [1]

References

[1] Jayaram, H.N.; Ramakrishnan, T.; Vaidyanathan, C.S.: Aspartotransferase from Mycobacterium tuberculosis H37Ra. Indian J. Biochem., **6**, 106-110 (1969)

Arginyltransferase 2.3.2.8

1 Nomenclature

EC number
2.3.2.8

Systematic name
L-arginyl-tRNA:protein arginyltransferase

Recommended name
arginyltransferase

Synonyms
L-arginyltransferase
R-transferase
arginyl tRNA transferase
arginyl-tRNA protein transferase
arginyl-transfer ribonucleate-protein aminoacyltransferase
arginyl-transfer ribonucleate-protein transferase

CAS registry number
37257-24-2

2 Source Organism

<1> *Oryctolagus cuniculus* (rabbit [1]) [1-3, 6, 7]
<2> *Ovis aries* [8]
<3> *Sus scrofa* (hog [4,5,15]) [4, 5, 15]
<4> *Rattus norvegicus* (male Sprague-Dawley [14]) [4, 10, 14]
<5> *Saccharomyces cerevisiae* [4, 13]
<6> *Mus musculus* (Balb/c [9]; deletion mutant [11]) [9, 11, 12]
<7> *Homo sapiens* [12]
<8> *Arabidopsis thaliana* (delayed-leaf-senescence mutant [16]) [16]

3 Reaction and Specificity

Catalyzed reaction
L-arginyl-tRNA + protein = tRNA + L-arginyl-protein

Reaction type
acyl group transfer

Natural substrates and products

- **S** L-arginyl-tRNA + acceptor protein <1, 3-6, 7> (<1> catalyzes post-translational ribosome-independent modification of certain acceptor proteins [1,7]; <1> possibly involved in degradation of proteins with acidic NH_2-termini [7]; <6,5> involved in ubiquitin mediated protein degradation [11,13]) (Reversibility: ? <1, 3-6, 7> [1-7, 9, 11, 12, 13, 14]) [1-7, 9, 11, 12, 13, 14]
- **P** tRNA + L-arginyl-acceptor protein

Substrates and products

- **S** L-arginyl-tRNA + Asp-β-galactosidase <6> (Reversibility: ? <6> [11]) [11]
- **P** tRNA + L-arginyl-Asp-β-galactosidase
- **S** L-arginyl-tRNA + Glu-β-galactosidase <6> (Reversibility: ? <6> [11]) [11]
- **P** tRNA + L-arginyl-Glu-β-galactosidase
- **S** L-arginyl-tRNA + L-Asp-L-Ala <1> (<1> other dipeptides: overview [3]) (Reversibility: ? <1> [3]) [3]
- **P** tRNA + L-Arg-L-Asp-L-Ala
- **S** L-arginyl-tRNA + L-Glu-L-Ala <1> (<1> other dipeptides: overview, not dipeptides with D-Glu [3]) (Reversibility: ? <1> [3]) [3]
- **P** tRNA + L-Arg-L-Glu-L-Ala <1> [3]
- **S** L-arginyl-tRNA + L-Glu-L-Ala-L-Ala <1> (<1> other tripeptides: overview [3]) (Reversibility: ? <1> [3]) [3]
- **P** tRNA + L-Arg-L-Glu-L-Ala-L-Ala
- **S** L-arginyl-tRNA + L-aspartic acid <1> (<1> poor substrate [3]) (Reversibility: ? <1> [3]) [3]
- **P** tRNA + L-Arg-L-Asp <1> [3]
- **S** L-arginyl-tRNA + L-cystinyl-bis-L-Ala <1> (<1> poor substrate [3]) (Reversibility: ? <1> [3]) [3]
- **P** tRNA + L-Arg-L-cystinyl-bis-L-Ala
- **S** L-arginyl-tRNA + L-glutamic acid <1> (<1> poor substrate [3]) (Reversibility: ? <1> [3]) [3]
- **P** tRNA + L-Arg-L-Glu <1> [3]
- **S** L-arginyl-tRNA + N-aspartyl-N'-dansylamido-1,4-butanediamine <3, 4> (<3, 4> i.e. Asp(4)DNS [14, 15]) (Reversibility: ? <3, 4> [14, 15]) [14, 15]
- **P** tRNA + N-arginylaspartyl-N'-dansylamido-1,4-butanediamine <3> (<3> i.e. ArgAsp(4)DNS [15]) [15]
- **S** L-arginyl-tRNA + acceptor protein <1, 3-6, 7> (<1,6> addition to amino-terminus [1,9]; <1> acceptor proteins require an acidic amino terminus, an Asp- or Glu-residue at the acceptor site [6]; <6> ATE1-2p is less active than ATE1-1p [12]) (Reversibility: ? <1, 3-6, 7> [1-7, 9, 11, 12, 13, 14]) [1-7, 9, 11, 12, 13, 14]
- **P** tRNA + L-arginyl-acceptor protein <1, 3-5> [1-7, 13]

S L-arginyl-tRNA + albumin <1> (<1> from bovine serum, accepts 1 mol arginine/mol [2]; <1> from bovine serum [1,3,7]) (Reversibility: ? <1> [1, 2, 3, 7]) [1, 2, 3, 7]
P tRNA + L-arginyl-albumin
S L-arginyl-tRNA + α lactalbumin <1, 3, 6> (<1> bovine [4,7]) (Reversibility: ? <1, 3, 6> [4, 7, 11]) [4, 7, 11]
P tRNA + L-arginyl-lactalbumin
S L-arginyl-tRNA + fructose diphosphatase <1> (<1> from rabbit liver [1,2]) (Reversibility: ? <1> [1, 2]) [1, 2]
P tRNA + L-Arg-fructose diphosphatase
S L-arginyl-tRNA + immunoglobulin <1> (<1> κ-light chain of immunoglobulin [7]) (Reversibility: ? <1> [7]) [7]
P tRNA + L-arginyl-immunoglobulin
S L-arginyl-tRNA + insulin <1> (<1> less effective [2]) (Reversibility: ? <1> [1, 2]) [1, 2]
P tRNA + L-arginyl-insulin
S L-arginyl-tRNA + isoasparagine <1> (<1> non-peptide-derivatives of dicarboxylic amino acids with blocked α-carboxyl group and unsubstituted β- or γ-carboxyl group, such as isoasparagine and isoglutamine [3]) (Reversibility: ? <1> [3]) [3]
P tRNA + L-arginyl-isoasparagine
S L-arginyl-tRNA + isoglutamate <1> (<1> non-peptide-derivatives of dicarboxylic amino acids with blocked α-carboxyl group and unsubstituted β- or γ-carboxyl group, such as isoasparagine and isoglutamine [3]) (Reversibility: ? <1> [3]) [3]
P tRNA + L-arginyl-isoglutamate
S L-arginyl-tRNA + thyroglobulin <1> (<1> accepts 2 mol arginine/mol [2]) (Reversibility: ? <1> [1, 2]) [1, 2]
P tRNA + L-arginyl-thyroglobulin
S L-arginyl-tRNA + transaldolase <1> (<1> I and III from Candida [2]) (Reversibility: ? <1> [2]) [2]
P tRNA + L-arginyl-transaldolase
S L-arginyl-tRNA + trypsin inhibitor <1> (<1> from soybean [7]) (Reversibility: ? <1> [7]) [7]
P tRNA + L-Arg-trypsin inhibitor

Inhibitors

DNAse <6> (<6> [9]) [9]
L-Glu-L-Val-L-Phe <1> [7]
N-ethylmaleimide <5> (<5> rapid inactivation at 0.2 mM [13]) [13]
RNAse <6> (<6> complete inhibition [9]) [9]
aminophenylarsenoxide <5> (<5> 50% inhibition at 0.0025 mM [13]) [13]

di- and tripeptides <1> (<1> with glutamyl-, asparagyl- and to a lesser extent cystinyl-NH_2-terminal residues, competitive to bovine serum albumin acceptor [3]; <1> most potent inhibitor is Glu-Val-Phe, inhibits transfer of arginine to acceptor proteins [7]) [3, 7]
heparin <3-5> (<3-5> 50% inhibition at 0.05 mg/ml, kinetics, competitive to L-arginyl-tRNA [4]) [4]
isoasparagine <1> [3]
isoglutamine <1> [3]
p-[(bromoacetyl)amino]phenylarsenoxide <5> (<5> irreversible, potent inhibitor [13]) [13]
phenylarsenoxide <5> (<5> inhibition below 0.005 mM is fully reversed by excess dithiothreitol [13]) [13]
putrescine <3> (<3> above 30 mM [4]) [4]
spermidine <3-5> (<3-5> competitive to L-arginyl-tRNA [4]) [4]
spermine <3-5> (<3-5> competitive to L-arginyl-tRNA [4]) [4]
Additional information <1, 3-5> (<1> no inhibition by puromycin [1]; <3-5> no inhibition by chondroitinsulfate A, B or C, hyaluronic acid, D-glucosamine N-sulfate, D-glucose 6-sulfate, D-glucosamine, D-galactosamine, N-acetyl-D-glucosamine, N-acetyl-D-galactosamine, D-xylose, D-glucuronic acid [4]; <3> antibodies to hog enzyme prepared in rabbit [5]) [1, 4, 5]

Activating compounds

2-mercaptoethanol <1, 6> (<1> requirement, 0.1 M, stabilizes the enzyme in a reduced state [2]; <6> essential for arginylation of chromatin [9]) [1, 2, 9]
dithiothreitol <1> (<1> requirement, 0.01 M stabilizes the enzyme in a reduced state [2]) [2]

Metals, ions

K^+ <1, 6> (<1> requirement, optimal concentration is 0.2 M [2]; <1> requirement for monovalent cations [1,2]; <6> required for arginylation of chromatin [9]) [1, 2, 9]
NH_4^+ <1> (<1> requirement for monovalent cations [1]) [1]
Na^+ <1> (<1> requirement for monovalent cations [1]) [1]
Additional information <1> (<1> inactive with Mg^{2+} [1]) [1]

Turnover number (min^{-1})

4 <5> (L-arginyl-tRNA) [13]

Specific activity (U/mg)

0.00084 <1> [1]
0.0047 <1> [7]
0.114 <1> [2, 6]
1.3 <1> [7]
2.3 <3> (<3> after addition of inhibitors during purification [4]) [4]
7.6 <3> [5]
Additional information <4> (<4> specific activity depends on type and age of tissue, picomolar range [14]) [14]

K_m-Value (mM)
0.0005 <3> (arginyl-tRNA) [4]
0.03 <3> (N-aspartyl-N'-dansylamido-1,4-butanediamine) [15]

K_i-Value (mM)
0.0008 <3> (spermine) [4]
0.0015 <3> (heparin) [4]
0.02 <1> (Glu-Val-Phe) [7]
Additional information <1> (<1> overview: di- and tripeptide inhibitors [3]) [3]

pH-Optimum
8.6-9.4 <1> [2]

pH-Range
7.2-7.8 <6> [9]
8-9.8 <1> (<1> about half-maximal activity at pH 8.0, about 80% of maximal activity at pH 9.8) [2]

Temperature optimum (°C)
37 <1> (<1> assay at [1,2]) [1, 2]

4 Enzyme Structure

Molecular weight
58000 <5> (<5> SDS-PAGE [13]) [13]
360000 <1> (<1> multienzyme complex of several molecules of arginyl-tRNA synthetase and arginyl-tRNA-protein transferase, gel filtration [7]) [7]

Subunits
? <3> (<3> x * 35000, SDS-PAGE [5]) [5]
Additional information <1> (<1> multienzyme complex of several molecules of arginyl-tRNA synthetase and arginyl-tRNA-protein transferase [7]) [7]

5 Isolation/Preparation/Mutation/Application

Source/tissue
HEK-293 cell <7> [12]
blood <4> [14]
brain <4> [14]
eye <4> [14]
fibroblast <6> (<6> embryonic [12]) [12]
kidney <3, 4> [4, 5, 14, 15]
liver <1, 4> [1-3, 6, 9, 14]
lung <4> [14]
prostate gland <4> [14]
reticulocyte <1> [7]

spleen <4> [4, 14]
stomach <4> [14]
testis <4> [14]
thymus <4> [14]
thyroid gland <2> [8]

Localization

cytoplasm <1-5> (<1-5> soluble [1,2,4,5,6,8]) [1, 2, 4-6, 8]
cytosol <6> (<6> ATE1-1p [12]) [12]
nucleus <6> (<6> ATE1-1p and ATE1-2p [12]) [12]
soluble <5> [13]

Purification

<1> (partial [1,7]; 7000fold [2]) [1, 2, 6, 7]
<3> (affinity chromatography on heparin-Sepharose, homogeneity [4]; affinity chromatography on heparin-Sepharose and angiotensin II-Sepharose combined with affinity elution with lactalbumin [5]) [4, 5, 15]
<5> (homogeneity [13]) [13]

Cloning

<5> [13]
<6> (2 species: Ate1-1 Ate1-2, fusion protein with green fluorescent protein [12]) [12]
<8> [16]

Engineering

C315A <5> (<5> full enzyme activity [13]) [13]

6 Stability

General stability information

<1>, 2-mercaptoethanol or dithiothreitol stabilize during storage [2]
<1>, freezing inactivates [1]
<1>, glycerol, 20%, stabilizes concentrated enzyme solutions during storage [2]
<1>, polyethylene glycol, 30%, stabilizes concentrated solutions during storage [2]
<3>, phosphate, 0.04 M, stabilizes [5]

Storage stability

<1>, -20°C, crude enzyme, at least 6 months [1]
<1>, 0°C, concentrated enzyme solution, 20% glycerol, $t_{1/2}$: about 1 week [2, 6]
<1>, 0°C, partially purified, 75% loss of activity within 1 week [1]
<3>, 4°C, concentrated enzyme solution, 0.5 M phosphate-KOH-buffer, pH 7.8, 0.1 M 2-mercaptoethanol, up to 6 months [5]
<3>, 4°C, purified, 0.05 M Tris-HCl buffer, pH 8.0, 0.1 M 2-mercaptoethanol, 0.09 M KCl, 0.02 M spermine, 1 week [5]

References

[1] Soffer, R.L.; Horinishi, H.: Enzymic modification of proteins. I. General characteristics of the arginine-transfer reaction in rabbit liver cytoplasm. J. Mol. Biol., **43**, 163-175 (1969)

[2] Soffer, R.L.: Enzymatic modification of proteins. II. Purification and properties of the arginyl transfer ribonucleic acid-protein transferase from rabbit liver cytoplasm. J. Biol. Chem., **245**, 731-737 (1970)

[3] Soffer, R.L.: Peptide acceptors in the arginine transfer reaction. J. Biol. Chem., **248**, 2918-2921 (1973)

[4] Kato, M.: Heparin as an inhibitor of L-arginyl-tRNA: protein arginyltransferase. J. Biochem., **94**, 2015-2022 (1983)

[5] Kato, M.; Nozawa, Y.: Complete purification of arginyl-tRNA:protein arginyltransferase from hog kidney and production of its antibody. Anal. Biochem., **143**, 361-367 (1984)

[6] Deutch, C.E.: Aminoacyl-tRNA: protein transferases. Methods Enzymol., **106**, 198-205 (1984)

[7] Ciechanover, A.; Ferber, S.; Ganoth, D.; Elias, S.; Avram, H.; Arfin, S.: Purification and characterization of arginyl-tRNA-protein transferase from rabbit reticulocytes. Its involvement in post-translational modification and degradation of acidic NH_2 termini substrates of the ubiquitin pathway. J. Biol. Chem., **263**, 11155-11167 (1988)

[8] Soffer, R.L.: The arginine transfer reaction. Biochim. Biophys. Acta, **155**, 228-240 (1968)

[9] Kaji, H.: Amino-terminal arginylation of chromosomal proteins by arginyl-tRNA. Biochemistry, **15**, 5121-5125 (1976)

[10] Kaji, H.; Novelli, G. D.; Kaji, A.: A soluble amino acid incorporation system from rat liver. Biochim. Biophys. Acta, **76**, 474-477 (1963)

[11] Kwon, Y.T.; Kashina, A.S.; Davydov, I.V.; Hu, R.G.; An, J.Y.; Seo, J.W.; Du, F.; Varshavsky, A.: An essential role of N-terminal arginylation in cardiovascular development. Science, **297**, 96-99 (2002)

[12] Kwon, Y.T.; Kashina, A.S.; Varshavsky, A.: Alternative splicing results in differential expression, activity, and localization of the two forms of arginyl-tRNA-protein transferase, a component of the N-end rule pathway. Mol. Cell. Biol., **19**, 182-193 (1999)

[13] Li, J.; Pickart, C.M.: Inactivation of arginyl-tRNA protein transferase by a bifunctional arsenoxide: Identification of residues proximal to the arsenoxide site. Biochemistry, **34**, 139-147 (1995)

[14] Takao, K.; Samejima, K.: Arginyl-tRNA-protein transferase activities in crude supernatants of rat tissues. Biol. Pharm. Bull., **22**, 1007-1009 (1999)

[15] Takao, K.; Xu, Y.J.; Samejima, K.; Shirahata, A.; Nitsu, M.: Preparation and usefulness of some fluorogenic substrates for assay of arginyl-tRNA-Protein transferase by HPLC. Anal. Biochem., **267**, 373-381 (1999)

[16] Yoshida, S.; Ito, M.; Callis, J.; Nishida, I.; Watanabe, A.: A delayed leaf senescence mutant is defective in arginyl-tRNA: protein arginyltransferase, a component of the N-end rule pathway in Arabidopsis. Plant J., **32**, 129-137 (2002)

Agaritine γ-glutamyltransferase 2.3.2.9

1 Nomenclature

EC number
2.3.2.9

Systematic name
(γ-L-glutamyl)-N^1-(4-hydroxymethylphenyl)hydrazine:(acceptor) γ-glutamyltransferase

Recommended name
agaritine γ-glutamyltransferase

Synonyms
glutamyl transferase, agaritine γ-

CAS registry number
37257-25-3

2 Source Organism

<-18> no activity in *Helvella sp.* [1]
<-17> no activity in *Pleurotus sp.* [1]
<-16> no activity in *Xeromphalina sp.* [1]
<-15> no activity in *Cortinarius sp.* [1]
<-14> no activity in *Discina sp.* [1]
<-13> no activity in *Colosyphia sp.* [1]
<-12> no activity in *Boletus sp.* [1]
<-11> no activity in *Caprinus sp.* (in sporophores [1]) [1]
<-10> no activity in *Neurospora crassa* [1]
<-9> no activity in *Eremothecium ashbyii* [1]
<-8> no activity in *Escherichia coli* [1]
<-7> no activity in *pigeon* [1]
<-6> no activity in *Rattus norvegicus* [1]
<-5> no activity in *Persea americana* [1]
<-4> no activity in *Daucus carote* [1]
<-3> no activity in *Spinacia oleracea* [1]
<-2> no activity in *Agaricus subrutilescens* [1]
<-1> no activity in *Agaricus sterlingii* [1]
<1> *Agaricus bisporus* (basidiomycete [1]) [1, 2]
<2> *Agaricus edulis* [1]
<3> *Agaricus pattersonii* [1]

<4> *Agaricus perrarus* [1]
<5> *Agaricus xanthoderma* [1]

3 Reaction and Specificity

Catalyzed reaction

agaritine + acceptor = 4-hydroxymethylphenylhydrazine + γ-L-glutamyl-acceptor

Reaction type

aminoacyl group transfer

Natural substrates and products

S β-N-(γ-L(+)-glutamyl)-4-hydroxymethylphenylhydrazine + acceptor <1-5> (<1-5> i.e. agaritine, unique enzyme for γ-L-glutamyl transfer in the basidiomycete family Agaricaceae [1]) (Reversibility: ? <1-5> [1]) [1]
P 4-hydroxymethylphenylhydrazine + γ-L-glutamyl-acceptor <1-5> [1]

Substrates and products

S (γ-L-glutamyl)hydrazine + H_2O <1-5> (Reversibility: ? <1-5> [1]) [1]
P L-glutamat + hydrazine <1-5> [1]
S 1-methyl-2-(γ-L-glutamyl)hydrazine + H_2O <1-5> (Reversibility: ? <1-5> [1]) [1]
P L-glutamate + methylhydrazine <1-5> [1]
S L-glutamine + H_2O <1-5> (Reversibility: ? <1-5> [1]) [1]
P L-glutamate + NH_3 <1-5> [1]
S N-(γ-L-glutamyl)ethylamine + H_2O <1-5> (Reversibility: ? <1-5> [1]) [1]
P L-glutamat + ethylamine <1-5> [1]
S N-γ-L-glutamylcyclohexylamine + NH_4^+ <1-5> (<1-5> at high NH_4^+ concentrations [1]) (Reversibility: ? <1-5> [1]) [1]
P cyclohexylamine + glutamine <1-5> [1]
S N-γ-L-glutamylcyclohexylamine + phenylhydrazine <1-5> (<1> most effective donor [1]) (Reversibility: ? <1-5> [1]) [1]
P cyclohexylamine + γ-L-glutamylphenylhydrazine <1-5> [1]
S β-N-(γ-L(+)-glutamyl)-4-hydroxymethylphenylhydrazine + acceptor <1-5> (<1-5> trivial name agaritine, can be replaced by γ-L-glutamyl-1-naphthylhydrazine, donor-specificity: γ-L-glutamyl-residue is essential and not replaceable by acetyl-, β-aspartyl-, δ-homoglutaryl-, glutaryl-, γ-D-glutamyl-, γ-L-(α-N-acetyl)glutamyl and γ-L-(α-glutamyl)glycine-residues, acceptors may be hydroxylamine, phenylhydrazine, *p*-hydroxyaniline, NH_4^+, not glycine, phenylalanine and aspartic acid, in the absence of acceptor the γ-glutamyl-moiety is irreversibly transferred to H_2O, hydrolyzes γ-glutamyl-amides and esters as well as hydrazides, no donors are γ-glutamyl-derivatives of ammonia or hydrazine and their corresponding N-alkylated analogues, glutathione, γ-glutamylphenylalanine, γ-glutamyl-β-aminoisobutyric acid [1]) (Reversibility: ? <1-5> [1]) [1]
P 4-hydroxymethylphenylhydrazine + γ-L-glutamyl-acceptor <1-5> [1]

S β-N-(γ-L-glutamyl)-1-naphthylhydrazine + H_2O <1-5> (Reversibility: ? <1-5> [1]) [1]
P L-glutamate + 1-naphthylhydrazine <1-5> [1]
S β-N-(γ-L-glutamyl)-4-hydroxymethylphenylhydrazine + H_2O <1-5> (Reversibility: ? <1-5> [1]) [1]
P L-glutamate + 4-hydroxymethylphenylhydrazine <1-5> [1]
S β-N-(γ-L-glutamyl)cyclohexylamine + H_2O <1-5> (Reversibility: ? <1-5> [1]) [1]
P L-glutamate + cyclohexylamine <1-5> [1]
S β-N-(γ-L-glutamyl)p-hydroxyaniline + H_2O <1-5> (Reversibility: ? <1-5> [1]) [1]
P L-glutamate + 4-hydroxyaniline <1-5> [1]
S β-N-(γ-L-glutamyl)phenylhydrazine + H_2O <1-5> (<1-5> hydrolysis is probably irreversible [1]) (Reversibility: ? <1-5> [1]) [1]
P L-glutamate + phenylhydrazine <1-5> [1]
S γ-(O-benzyl)-L-glutamic acid + H_2O <1-5> (Reversibility: ? <1-5> [1]) [1]
P L-glutamate + benzoic acid <1-5> [1]
S γ-L-glutamyl-*p*-hydroxyaniline + phenylhydrazine <1-5> (Reversibility: ? <1-5> [1]) [1]
P *p*-hydroxyaniline + γ-L-glutamylphenylhydrazine <1-5> [1]
S γ-benzyl-L-glutamate + phenylhydrazine <1-5> (Reversibility: ? <1-5> [1]) [1]
P benzoic acid + γ-L-glutamylphenylhydrazine <1-5> (<1-5> L-glutamate in the absence of phenylhydrazine [1]) [1]
S γ-glutamyl-*p*-nitroanilide + H_2O <1> (Reversibility: ? <1> [2]) [2]
P L-glutamate + *p*-nitroaniline <1> [2]
S glutamine + phenylhydrazine <1-5> (Reversibility: ? <1-5> [1]) [1]
P β-N-(γ-L-glutamyl)phenylhydrazine <1-5> [1]

Inhibitors

4-hydroxyaniline <1> (<1> agaritine hydrolysis [1]) [1]
Cu^{2+} <1> (<1> 1 mM, complete inhibition [1]) [1]
Hg^{2+} <1> (<1> 5 mM, complete inhibition [1]) [1]
L-glutamine <1> (<1> competitive inhibition of agaritine hydrolysis [1]) [1]
N-(γ-L-glutamyl)-4-hydroxyaniline <1> (<1> competitive inhibition of agaritine hydrolysis [1]) [1]
N-(γ-L-glutamyl)-cyclohexylamine <1> (<1> competitive inhibition of agaritine hydrolysis [1]) [1]
NaCN <1> (<1> 5 mM, 40% inhibition [1]) [1]
Zn^{2+} <1> (<1> 1 mM, complete inhibition [1]) [1]
β-N-(γ-L-glutamyl)-1-naphthylhydrazine <1> (<1> competitive inhibition of agaritine hydrolysis [1]) [1]
diisopropylfluorophosphate <1> (<1> 5 mM, 40% inhibition [1]) [1]
γ-(O-benzyl)-L-glutamic acid <1> (<1> competitive inhibition of agaritine hydrolysis [1]) [1]
γ-L-glutamylhydrazine <1> (<1> competitive inhibition of agaritine hydrolysis [1]) [1]

glycerol <1> (<1> 40% inhibition [1]) [1]
glycine <1> (<1> hydroxylamine as substrate [1]) [1]
iodoacetate <1> (<1> 5 mM, complete inhibition [1]) [1]
p-hydroxymercuribenzoate <1> (<1> 2 mM, complete inhibition [1]) [1]
phenylalanine <1> (<1> hydroxylamine as substrate [1]) [1]
Additional information <1> (<1> not inhibited by NH_4^+ at high concentrations, leucine, aspartic acid [1]) [1]

Activating compounds
Additional information <1> (<1> not activated by ADP or ATP [1]) [1]

Metals, ions
Additional information <1> (<1> not activated by Mg^{2+} or Mn^{2+} [1]) [1]

Specific activity (U/mg)
1.4 <1> [1]

pH-Optimum
7 <1> [1]

pH-Range
6.6-7.8 <1> (<1> approx. half-maximal activity at pH 6.6 and 7.8 [1]) [1]

Temperature optimum (°C)
30 <1> (<1> assay at [1]) [1]

5 Isolation/Preparation/Mutation/Application

Source/tissue
fruitbody <1> (<1> present in stip base, gills and skin [2]) [2]
sporophore <1-5> [1]

Localization
cytoplasm <1-5> [1]
soluble <1> [2]

Purification
<1> (partial [1]) [1]

6 Stability

Organic solvent stability
glycerol <1> (<1> 40%, inactivates [1]) [1]

General stability information
<1>, 2-mercaptoethanol stabilizes during purification [1]
<1>, freeze-thawing inactivates [1]

References

[1] Gigliotti, H.J.; Levenberg, B.: Studies on the γ-glutamyltransferase of Agaricus bisporus. J. Biol. Chem., **239**, 2274-2284 (1964)
[2] Jolivet, S.; Mooibroek, H.; Wichers, H.J.: Space-time distribution of γ-glutamyltransferase activity in Agaricus bisporus. FEMS Microbiol. Lett., **163**, 263-267 (1998)

UDP-N-acetylmuramoylpentapeptide-lysine N^6-alanyltransferase 2.3.2.10

1 Nomenclature

EC number
2.3.2.10

Systematic name
L-alanyl-tRNA:UDP-N-acetylmuramoyl-L-alanyl-D-glutamyl-L-lysyl-D-alanyl-D-alanine N^6-alanyltransferase

Recommended name
UDP-N-acetylmuramoylpentapeptide-lysine N^6-alanyltransferase

Synonyms
UDP-N-acetylmuramoylpentapeptide lysine N^6-alanyltransferase
alanyl-transfer ribonucleate-uridine diphosphoacetylmuramoylpentapeptide transferase
alanyltransferase, uridine diphosphoacetylmuramoylpentapeptide lysine N^6-
uridine diphosphoacetylmuramoylpentapeptide lysine N^6-alanyltransferase

CAS registry number
37257-26-4

2 Source Organism

<1> *Lactobacillus viridescens* (femX gene [2]) [1, 2]

3 Reaction and Specificity

Catalyzed reaction
L-alanyl-tRNA + UDP-N-acetylmuramoyl-L-alanyl-D-glutamyl-L-lysyl-D-alanyl-D-alanine = tRNA + UDP-N-acetylmuramoyl-L-alanyl-D-glutamyl-N^6-(L-alanyl)-L-lysyl-D-alanyl-D-alanine (also acts on L-seryl-tRNA; <1> transfer of L-Ala to ε-amino group of L-lysine of UDP-UDP-acetylmuramoylpentapeptide in an ordered sequential mechanism [2]; <1> catalysis via a ternary complex [2])

Reaction type
aminoacyl group transfer

Natural substrates and products

S L-alanyl-tRNA + UDP-N-acetylmuramoyl-L-alanyl-D-glutamyl-L-lysyl-D-alanyl-D-alanine <1> (<1> involved in peptidoglucan metabolism, responsible for interpeptide bridges [1]) (Reversibility: ir <1> [1]) [1]

P tRNA + UDP-N-acetylmuramoyl-L-alanyl-D-glutamyl-N^{6}-(L-alanyl)-L-lysyl-D-alanyl-D-alanine

Substrates and products

S L-alanyl-tRNA^{Ala} + UDP-N-acetylmuramoyl-L-alanyl-D-glutamyl-L-lysyl-D-alanyl-D-alanine <1> (<1> no strict specificity for tRNA-carrier: tRNA^{Cys} can replace tRNA^{Ala}, tRNA from other species can substitute to some extent for the homologous tRNA from Lactobacillus viridescens, higher specificity for transferred amino acid: amino acids with larger substituents are no substrates, UDP-acetylmuramyl-L-Ala-D-Glu-meso-diaminopimelic acid-D-Ala-D-Ala can replace UDP-acetylmuramyl-L-Ala-D-Glu-L-Lys-D-Ala-D-Ala at 15% the transfer rate [1]; <1> poor or no acceptors are acetylmuramoyl- and phosphoacetylmuramoyl-L-Ala-L-Glu-L-Lys-D-Ala-D-Ala or UDP-acetylmuramoyl-L-Ala-L-Glu-L-Lys [1]) (Reversibility: ir <1> [1]; ? <1> [2]) [1, 2]

P tRNA^{Ala} + UDP-N-acetylmuramoyl-L-alanyl-D-glutamyl-N^{6}-(L-alanyl)-L-lysyl-D-alanyl-D-alanine <1> [1, 2]

S L-alanyl-tRNA^{Ser} + UDP-N-acetylmuramoyl-L-alanyl-D-glutamyl-L-lysyl-D-alanyl-D-alanine <1> (Reversibility: ? <1> [2]) [2]

P tRNA^{Ser} + UDP-N-acetylmuramoyl-L-alanyl-D-glutamyl-N^{6}-(L-alanyl)-L-lysyl-D-alanyl-D-alanine <1> [2]

S L-cysteinyl-tRNA + UDP-N-acetylmuramoyl-L-alanyl-D-glutamyl-L-lysyl-D-alanyl-D-alanine <1> (Reversibility: ir <1> [1]) [1]

P tRNA + UDP-N-acetylmuramoyl-L-alanyl-D-glutamyl-N^{6}-(L-cysteinyl)-L-lysyl-D-alanyl-D-alanine

S L-glycyl-tRNA + UDP-N-acetylmuramoyl-L-alanyl-D-glutamyl-L-lysyl-D-alanyl-D-alanine <1> (<1> poor substrate [1]) (Reversibility: ir <1> [1]) [1]

P tRNA + UDP-N-acetylmuramoyl-L-alanyl-D-glutamyl-N^{6}-(L-glycyl)-L-lysyl-D-alanyl-D-alanine

S L-seryl-tRNA + UDP-N-acetylmuramoyl-L-alanyl-D-glutamyl-L-lysyl-D-alanyl-D-alanine <1> (<1> about half as effective as L-alanyl-tRNA [1]) (Reversibility: ir <1> [1]) [1]

P tRNA + UDP-N-acetylmuramoyl-L-alanyl-D-glutamyl-N^{6}-(L-seryl)-L-lysyl-D-alanyl-D-alanine <1> [1]

S L-seryl-tRNA^{Ser} + UDP-N-acetylmuramoyl-L-alanyl-D-glutamyl-L-lysyl-D-alanyl-D-alanine <1> (Reversibility: ? <1> [2]) [2]

P tRNA + UDP-N-acetylmuramoyl-L-alanyl-D-glutamyl-N^{6}-(L-seryl)-L-lysyl-D-alanyl-D-alanine <1> [2]

S Additional information <1> (<1> also utilizes UDP-N-acetylmuramoylpentapeptide containing L-ornithine as substrate [2]) [2]

P ?

Inhibitors

Ser-tRNA <1> (<1> versus UDP-acetylmuramoylpentapeptide [2]) [2]
UDP-N-acetylmuramoyl-L-alanyl-D-glutamyl-L-lysyl-D-alanyl-D-alanine <1> (<1> product inhibition [2]) [2]
Additional information <1> (<1> no inhibition by EDTA or metal ions [1]) [1]

Metals, ions

Additional information <1> (<1> no activation by EDTA or metal ions [1]) [1]

Turnover number (min^{-1})

11 <1> (L-alanyl-$tRNA^{Ala}$, <1> recombinant mutant D109N [2]) [2]
55 <1> (L-alanyl-$tRNA^{Ala}$, <1> recombinant mutant E320Q [2]) [2]
660 <1> (L-alanyl-$tRNA^{Ala}$, <1> recombinant wild-type [2]) [2]

Specific activity (U/mg)

276 <1> (<1> purified enzyme [1]) [1]

K_m-Value (mM)

0.0002 <1> (UDP-acetylmuramoylpentapeptide) [1]
0.015 <1> (Ala-$tRNA^{Ala}$, <1> Ala-$tRNA^{Ala}$ from E. coli, recombinant wild-type [2]) [2]
0.042 <1> (UDP-acetylmuramoylpentapeptide, <1> recombinant wild-type [2]) [2]
0.054 <1> (Ala-$tRNA^{Ala}$, <1> Ala-$tRNA^{Ala}$ from E. coli, recombinant mutant E320Q [2]) [2]
0.061 <1> (Ala-$tRNA^{Ala}$, <1> Ala-$tRNA^{Ala}$ from E. coli, recombinant mutant D109N [2]) [2]
0.067 <1> (UDP-acetylmuramoylpentapeptide, <1> recombinant mutant E320Q [2]) [2]
0.165 <1> (UDP-acetylmuramoylpentapeptide, <1> recombinant mutant D109N [2]) [2]
Additional information <1> [2]

K_i-Value (mM)

0.048 <1> (UDP-N-acetylmuramoyl-L-alanyl-D-glutamyl-L-lysyl-D-alanyl-D-alanine, <1> product inhibition [2]) [2]
0.248 <1> (Ser-tRNA, <1> versus UDP-acetylmuramoylpentapeptide [2]) [2]
Additional information <1> [2]

pH-Optimum

6.8-7.2 <1> (<1> L-alanyl-tRNA [1]) [1]
7.2-7.6 <1> (<1> L-seryl-tRNA [1]) [1]
Additional information <1> (<1> broad maximum [2]) [2]

pH-Range

5.2-8.6 <1> (<1> L-alanyl-tRNA, about 80% of maximal activity at pH 5.2 and pH 8.6 [1]) [1]
5.5-9.1 <1> [2]

6.1-8.6 <1> (<1> L-seryl-tRNA, about half-maximal activity at pH 6.1 and about 90% of maximal activity at pH 8.6 [1]) [1]
Additional information <1> (<1> pK values of 5.5 and 9.3 [2]) [2]

Temperature optimum (°C)
25 <1> (<1> assay at [2]) [2]
30 <1> (<1> assay at [1]) [1]

4 Enzyme Structure

Molecular weight
40000 <1> (<1> PAGE [1]) [1]

5 Isolation/Preparation/Mutation/Application

Localization
cytosol <1> [1]

Purification
<1> (recombinant His-tagged wild-type and mutants from E. coli [2]) [1, 2]

Cloning
<1> (expression of wild-type and mutants as His-tagged proteins in Escherichia coli, in vitro transcription of tRNAs [2]) [2]

Engineering
D109N <1> (<1> site-directed mutagenesis, loss of 99% activity compared to the wild-type [2]) [2]
E205Q <1> (<1> site-directed mutagenesis, unaltered activity compared to the wild-type [2]) [2]
E215Q <1> (<1> site-directed mutagenesis, loss of 15% activity compared to the wild-type [2]) [2]
E316Q <1> (<1> site-directed mutagenesis, loss of 60% activity compared to the wild-type [2]) [2]
E320Q <1> (<1> site-directed mutagenesis, loss of 96% activity compared to the wild-type [2]) [2]

References

[1] Plapp, R.; Strominger, J.L.: Biosynthesis of the peptidoglycan of bacterial cell walls. J. Biol. Chem., **245**, 3673-3682 (1970)
[2] Hegde, S.S.; Blanchard, J.S.: Kinetic and mechanistic characterization of recombinant Lactobacillus viridescens FemX (UDP-N-acetylmuramoyl pentapeptide-lysine N^6-alanyltransferase). J. Biol. Chem., **278**, 22861-22867 (2003)

Alanylphosphatidylglycerol synthase 2.3.2.11

1 Nomenclature

EC number
2.3.2.11

Systematic name
L-alanyl-tRNA:phosphatidylglycerol alanyltransferase

Recommended name
alanylphosphatidylglycerol synthase

Synonyms
O-alanylphosphatidylglycerol synthase
alanyl phosphatidylglycerol synthetase
synthase, O-alanylphosphatidylglycerol

CAS registry number
37257-27-5

2 Source Organism

<1> *Clostridium welchii* [1]

3 Reaction and Specificity

Catalyzed reaction
L-alanyl-tRNA + phosphatidylglycerol = tRNA + 3-O-L-alanyl-1-O-phosphatidylglycerol

Reaction type
aminoacyl group transfer

Substrates and products
S L-alanyl-$tRNA^{Ala}$ + phosphatidylglycerol <1> (<1> other L-alanyl-tRNAs than L-alanyl-$tRNA^{Ala}$ are no substrates [1]) (Reversibility: ? <1> [1]) [1]
P $tRNA^{Ala}$ + 3-O-L-alanyl-1-O-phosphatidylglycerol <1> [1]
S Additional information <1> (<1> no substrates are N-acetylalanyl-tRNA, lactyl-tRNA, alanyl-$tRNA^{Cys}$ and phenylalanyl-$tRNA^{Ala}$ [1]) [1]
P ?

pH-Optimum
5.7 <1> [1]

Temperature optimum (°C)

30 <1> (<1> assay at [1]) [1]

5 Isolation/Preparation/Mutation/Application

Localization

membrane <1> [1]

6 Stability

General stability information

<1>, glycerol stabilizes during storage [1]

Storage stability

<1>, -20°C, 3 weeks, 10 mM Tris/HCl, pH 8.0, 10 mM 2-mercaptoethanol, 30% glycerol [1]

<1>, -60°C, several months [1]

References

[1] Gould, R.M.; Thornton, M.P.; Liepkalns, V.; Lennarz, W.J.: Participation of aminoacyl transfer ribonucleic acid in aminoacyl phosphatidylglycerol synthesis. II. Specificity of alanyl phosphatidylglycerol synthetase. J. Biol. Chem., **243**, 3096-3104 (1968)

Peptidyltransferase 2.3.2.12

1 Nomenclature

EC number
2.3.2.12

Systematic name
peptidyl-tRNA:aminoacyl-tRNA N-peptidyltransferase

Recommended name
peptidyltransferase

Synonyms
ribosomal peptidyltransferase
transpeptidase

CAS registry number
9059-29-4

2 Source Organism

<1> *Escherichia coli* (B cells [1,3,4,7,14,15]; B cells, frozen [13,16,18]; probably, organism not mentioned [6]; MRE-600 [8,10,12]) [1, 2, 3, 4, 6-16, 18, 21]
<2> *Rattus norvegicus* [5]
<3> *Oryctolagus cuniculus* [17, 19, 20]
<4> *Geobacillus stearothermophilus* [22]
<5> *Saccharomyces cerevisiae* (wild type strain 2D-J809, strain 2A-J809 lacking the two genes encoding L24, strain L1726 lacking the single gene encoding L39, strain L1725 lacking the genes for both proteins L24 and L39 [23]; wild type strain YKS99-2A, strain YKS99-2C missing the protein L41, the two genes RPL41A and RPL41B [24]) [23, 24]

3 Reaction and Specificity

Catalyzed reaction
peptidyl-$tRNA_1$ + aminoacyl-$tRNA_2$ = $tRNA_1$ + peptidyl-aminoacyl-$tRNA_2$

Reaction type
aminoacyl group transfer
peptidyl group transfer

Natural substrates and products

S peptidyl-$tRNA_1$ + α-aminoacyl-$tRNA_2$ <1> (Reversibility: ? <1> [1, 2, 3, 4, 5, 6, 8, 9]) [1, 2, 3, 4, 5, 6, 8, 9]

P $tRNA_1$ + peptidyl-amino-$tRNA_2$ <1> [1, 2, 3, 4, 5, 6, 8, 9]

Substrates and products

S 2'(3')-O-(N-formylmethionyl)-adenosine-5'-phosphate + CACCA-Phe <1> (Reversibility: ? <1> [11]) [11]

P ?

S AcPhe-tRNA + puromycin <1, 2, 3, 5> (<1> reaction only in the presence of 70S ribosomes and the appropriate mRNA [4]; <1> AcPhe-tRNA-polyU-70S ribosome complex [12, 14, 15]; <1> AcPhe-tRNA-polyU-ribosome complex [16, 18]; <3> AcPhe-tRNA-polyU-80S ribosome complex [17, 19, 20, 23, 24]) (Reversibility: ? <1, 2, 3, 5> [1, 4, 5, 11, 12, 13, 14, 15, 16, 17, 18, 19, 20, 21, 23, 24]) [1, 4, 5, 11-21, 23, 24]

P tRNA + AcPhe-puromycin <1, 2, 3, 5> [1, 4, 5, 11-21, 23, 24]

S CACCA-AcLeu + puromycin <1> (<1> fragment reaction [4,7,11]) (Reversibility: ? <1> [4, 7, 10, 11]) [4, 7, 10, 11]

P CACCA + AcLeu-puromycin <1> [4, 7, 10, 11]

S N-AcMet-tRNA + puromycin <4> (Reversibility: ? <4> [22]) [22]

P tRNA + N-AcMet-puromycin <4> [22]

S N-AcPhe-tRNA + puromycin <1> (Reversibility: ? <1> [9]) [9]

P tRNA + N-AcPhe-puromycin <1> [9]

S formylmethionyl-tRNA + α-hydroxy-puromycin <1> (<1> ester linkage [6]) (Reversibility: ? <1> [6]) [6]

P tRNA + formylmethionyl-α-hydroxy-puromycin <1> [6]

S formylmethionyl-tRNA + puromycin <1> (<1> formylmethionyl-tRNA-AUG-70S ribosome complex [15]) (Reversibility: ? <1> [6, 15]) [6, 15]

P tRNA + formylmethionyl-puromycin <1> [6, 15]

S peptidyl-tRNA1 + α-aminoacyl-$tRNA_2$ <1> (Reversibility: ? <1> [1, 2, 3, 4]) [1, 2, 3, 4]

P tRNA1 + peptidyl-amino-$tRNA_2$ <1> [1, 2, 3, 4]

S phenylalanyl-tRNA + puromycin <1> (<1> poly-U-directed translation system [8]) (Reversibility: ? <1> [8]) [8]

P tRNA + phenylalanyl-puromycin <1> [8]

S polylysyl-tRNA + puromycin <1> (<1> reaction only in the presence of 70S ribosomes and the appropriate mRNA [4]) (Reversibility: ? <1> [3, 4]) [3, 4]

P tRNA + polylysyl-puromycin <1> [3, 4]

S polyphenylalanyl-tRNA + puromycin <1> (Reversibility: ? <1> [2]) [2]

P tRNA + polyphenylalanyl-puromycin <1> [2]

S Additional information <1> (<1> in the presence of elongation factor EF-G with GTP the poly-U-directed translation is much more resistant to inhibitors of the peptidyl-transferase [8]) [8]

P ?

Inhibitors

40S subunits of ribosomes <2> (<2> inhibition proportional to the 40S-subunit-concentration [5]) [5]
EDTA <1> [2]
Gly-Phe-chloramphenicol <1> (<1> competitive inhibition with acetylphenylalanyl-tRNA-polyU-ribosome complex, new formed complex is inactive toward puromycin [16]) [16]
Gly-chloramphenicol <1> (<1> competitive inhibition with acetylphenylalanyl-tRNA-polyU-ribosome complex, new formed complex is inactive toward puromycin [16]) [16]
L-Phe-Gly-chloramphenicol <1> (<1> competitive inhibition with acetylphenylalanyl-tRNA-polyU-ribosome complex, new formed complex is inactive toward puromycin [16]) [16]
L-Phe-chloramphenicol <1> (<1> competitive inhibition with acetylphenylalanyl-tRNA-polyU-ribosome complex, new formed complex is inactive toward puromycin [16]) [16]
N^1,N^{12}-diacetylspermine <1> (<1> both stimulatory and inhibitory effects at the kinetic phase in the presence of the factors washable from ribosomes, depending on ligand concentration [21]) [21]
N^1,N^{12}-dipivaloylspermine <1> (<1> both stimulatory and inhibitory effects at the kinetic phase in the presence of the factors washable from ribosomes, depending on ligand concentration [21]) [21]
N^1-acetylspermine <1> (<1> both stimulatory and inhibitory effects at the kinetic phase in the presence of the factors washable from ribosomes, depending on ligand concentration [21]) [21]
amicetin <1> (<1> complete inhibition with AcPhe-tRNA as donor, with polylysyl-tRNA as donor less active [4]; <1> less potent than oxamicetin [7]) [4, 7, 8]
anisomycin <3> (<3> inhibition of formation of AcPhe-puromycin catalyzed by rabbit reticulocyte ribosomes [17]) [17, 20]
bamicetin <1> (<1> complete inhibition with AcPhe-tRNA as donor, with polylysyl-tRNA as donor less active [4]) [4]
blasticidin S <1, 3> (<3> inhibition of formation of Ac-Phe-puromycin catalyzed by rabbit reticulocyte ribosomes [17]) [13, 17]
chloramphenicol <1, 4> (<1> 50% inhibition [2]; <1> competitive inhibition with acetylphenylalanyl-tRNA-polyU-ribosome complex, newly formed complex is inactive toward puromycin [16]; <4> weak peptide bond formation inhibition only by ribosomes with A2451C, A2451U or A2451G, peptide bond formation inhibition by ribosomes with G2447A is essentially unimpaired [22]) [2, 3, 8, 14, 16, 22]
chlorotetracycline <1> (<1> 59% inhibition [2]) [2]
cytidylyl(3'-5')2'(3')-O-(α-aminoisobutyryl)adenosine <1> [12]
cytidylyl(3'-5')2'(3')-O-cycloleucyladenosine <1> [12]
cytidylyl-3'-5'-/2'(3')-O-L-phenylalanyl/L-adenosine <1> (<1> 50% inhibition of peptidyltransferase, inhibition can be reversed by increasing concentration of puromycin [9]) [9]
erythromycin <1> [3]

gougerotin <1> [8]
griseoviridin <1> [8]
lincomycin <1> [8]
m-nitrophenylboric acid <1> (<1> more potent inhibitor than phenylboric acid [11]) [11]
oxamicetin <1> (<1> more potent than amicetin [7]) [7]
phenylboric acid <1> [11]
sparsomycin <3> (<3> inhibition of formation of AcPhe-puromycin catalyzed by rabbit reticulocyte ribosomes [17]) [17, 20]
spermine <1> (<1> competitive inhibitor of peptide bond formation at the kinetic phase of the puromycin reaction [1]; <1> in the absence of factors washable from ribosomes [18]) [1, 18, 21]
tevenel <1> [14]
thiamphenicol <1> [14]

Activating compounds

K^+ <1> (<1> reaction is completely dependent upon potassium ions [2]; <1> stimulation not so marked as that caused by NH_4^+ ions [18]) [2, 18, 19]
N^1,N^{12}-diacetylspermine <1> (<1> both stimulatory and inhibitory effects at the kinetic phase in the presence of the factors washable from ribosomes, depending on ligand concentration [21]) [21]
N^1,N^{12}-dipivaloylspermine <1> (<1> both stimulatory and inhibitory effects at the kinetic phase in the presence of the factors washable from ribosomes, depending on ligand concentration [21]) [21]
N^1-acetylspermine <1> (<1> both stimulatory and inhibitory effects at the kinetic phase in the presence of the factors washable from ribosomes, depending on ligand concentration [21]) [21]
NH_4^+ <1> (<1> reaction is completely dependent upon ammonium ions [2]; <1> combination of NH_4^+ ions with spermine produces an additive activity increase [18]) [2, 3, 18]
methanol <1, 2> (<2> about 30-40% concentration [5]; <1> optimal concentration: 45-50% [10]) [5, 10]
poly-A <1> [3]
poly-U <2> (<2> with acetylphenylalanyl-tRNA as donor only [5]) [5]

Metals, ions

Mg^{2+} <1, 2, 5> [2, 5, 6, 8, 9, 10, 21, 23]

Turnover number (min^{-1})

0.61 <5> (AcPhe-tRNA, <5> strain YKS99-2C missing the protein L41, the two genes RPL41A and RPL41B [24]) [24]
0.69 <5> (AcPhe-tRNA, <5> strain 2A-J809 lacking the two genes encoding L24 [23]) [23]
0.8 <1> (AcPhe-tRNA, <1> when NH_4Cl ribosomal wash is absent [15]) [15]
1.07 <5> (AcPhe-tRNA, <5> triple mutant strain L1725 lacking the proteins L24 and L39 [23]) [23]
1.7 <3> (AcPhe-tRNA, <3> reaction in solution [17]) [17]
1.8 <5> (AcPhe-tRNA, <5> wild type strain YKS99-2A [24]) [24]

2 <1> (AcPhe-tRNA, <1> when NH_4Cl ribosomal wash is present [15]) [15]
2.18 <5> (AcPhe-tRNA, <5> wild type strain 2D-J809 [23]) [23]
2.9 <3> (AcPhe-tRNA, <3> disk reaction [17]) [17]
3.05 <5> (AcPhe-tRNA, <5> strain L1729 lacking the single gene encoding L39 [23]) [23]
8.3 <1> (formylmethionyl-tRNA, <1> when NH_4Cl ribosomal wash is present [15]) [15]
20 <3> (AcPhe-tRNA) [19]

K_m-Value (mM)

0.000058 <2> (acetylphenylalanyl-tRNA) [5]

K_i-Value (mM)

0.000045 <1> (thiamphenicol) [14]
0.00007 <1> (chloramphenicol, <1> initial phase of competitive inhibition [14]) [14]
0.00013 <3> (sparsomycin, <3> from the competitive phase of the inhibition [20]) [20]
0.00017 <1> (tevenel) [14]
0.0002 <1> (blasticidin S) [13]
0.00022 <3> (sparsomycin, <3> [20]) [20]
0.00067 <3> (anisomycin) [20]
0.002 <1> (L-Phe-Gly-chloramphenicol) [16]
0.0058 <1> (Gly-chloramphenicol) [16]
0.007 <1> (chloramphenicol) [16]
0.02 <1> (L-Phe-chloramphenicol) [16]
0.09 <1> (Gly-Phe-chloramphenicol) [16]
0.46 <1> (*m*-nitrophenylboric acid) [11]
0.52 <1> (phenylboric acid) [11]
3.2 <1> (spermine) [21]
5 <1> (N^1-acetylspermine) [21]
12.9 <1> (N^1,N^{12}-dipivaloylspermine) [21]
33 <1> (N^1,N^{12}-diacetylspermine) [21]

pH-Optimum

7.1-7.4 <1> [8]
7.5-8.25 <2> [5]

Temperature optimum (°C)

0 <1> (<1> enzyme assay [10]) [10]
20 <2> (<2> enzyme assay [5]) [5]
25 <1, 3> (<1> enzyme assay [8,13,15,16,18]; <3> enzyme assay, reaction in solution [17]; <3> enzyme assay, disk reaction [19,20]) [8, 13, 15, 16, 17, 18, 19, 20, 21]
30 <1, 5> (<1> enzyme assay [2,6]; <5> enzyme assay, disk reaction and reaction in solution [2,23]) [2, 6, 23]
35 <1> (<1> enzyme assay [3,4]) [3, 4]
37 <1, 3> (<1> enzyme assay [9]; <3> enzyme assay, disk reaction [17]) [9, 17]
45 <4> (<4> enzyme assay [22]) [22]

Temperature range (°C)

15-30 <2> [5]

5 Isolation/Preparation/Mutation/Application

Source/tissue

blood <3> (<3> salt-washed ribosomes [17]) [17]
liver <2> [5]
reticulocyte <3> (<3> unwashed ribosomes [17,20]) [17, 19, 20]

Localization

ribosome <1, 2, 3, 4> (<1> additionally factors washable from ribosomes [1,16,18]; <1> 50S subunit [2]; <4> 50S subunit, wild type ribosomes, ribosomes with site-directed mutated residues: A2451, G2447 [22]; <2> 60S subunit [5]; <1> 70S ribosomes [7,9,12,15]; <1> 50S and 70S ribosomes respectively [10]; <3> 80S ribosomes [17,19,20,23,24]) [1-24]

6 Stability

Storage stability

<5>, -70°C, both ribosomes and factors washable from ribosomes [23]

References

[1] Kalpaxis, D.L.; Drainas D.: Inhibitory effect of spermine on ribosomal peptidyltransferase. Arch. Biochem. Biophys., **300**, 629-634 (1993)
[2] Traut, R.R.; Monro, R.E.: The puromycin reaction and its relation to protein synthesis. J. Mol. Biol., **10**, 63-72 (1964)
[3] Rychlik, I.: Release of lysine peptides by puromycin from polylysyl-transfer ribonucleic acid in the presence of ribosomes. Biochim. Biophys. Acta, **114**, 425-427 (1966)
[4] Cerna, J.; Rychlik, I.; Lichtenthaler, F.W.: The effect of the aminoacyl-4-aminohexosyl-cytosine group of antibiotics on ribosomal peptidyl transferase. FEBS Lett., **30**, 147-150 (1973)
[5] Thompson, H.A.; Moldave, K.: Characterization of the peptidyltransferase reaction catalyzed by rat liver 60S ribosomal subunits. Biochemistry, **13**, 1348-1353 (1974)
[6] Fahnestock, S.; Neumann, H.; Rich, A.: Assay of ester and polyester formation by the ribosomal peptidyltransferase. Methods Enzymol., **30 F**, 489-497 (1974)
[7] Lichtenthaler, F.W.; Cerna, J.; Rychlik, I.: The effect of oxamicetin and some amicetin analogs on ribosomal peptidyl transferase. FEBS Lett., **53**, 184-187 (1975)

[8] Spirin, A.S.; Asatryan, L.S.: The effect of ribosomal peptidyl-transferase inhibitors is antagonized by elongation factor G with GTP. FEBS Lett., **70**, 101-104 (1976)

[9] Bhuta, P.; Zemlicka, J.: Inhibition of ribosomal peptidyltransferase with cytidyl-3 leads to 5-[2(3)-O-L-phenylalanyl]-L-adenosine. Biochem. Biophys. Res. Commun., **83**, 414-420 (1978)

[10] Streltsov, S.; Kosenjuk, A.; Kukhanova, M.; Krayevsky, A.; Gottikh, B.: Kinetic constants for model substrates of peptidyltransferase donor site of Escherichia coli ribosomes. FEBS Lett., **104**, 279-283 (1979)

[11] Cerna, J.; Rychlik, I.: Phenylboric acids - A new group of peptidyl transferase inhibitors. FEBS Lett., **119**, 342-348 (1980)

[12] Chladek, S.; Bhuta, P.: Inhibition of the peptidyltransferase acceptor site by 2(3)-O-cycloleucyl- and α-aminoisobutyryl derivatives of cytidylyl-(3-5)adenosine. Biochim. Biophys. Acta, **696**, 212-217 (1982)

[13] Kalpaxis, D.L.; Theocharis, D.A.; Coutsogeorgopoulos, C.: Kinetic studies on ribosomal peptidyltransferase. The behaviour of the inhibitor blasticidin S. Eur. J. Biochem., **154**, 267-271 (1986)

[14] Drainas D.; Kalpaxis, D.L.; Coutsogeorgopoulos, C.: Inhibition of ribosomal peptidyltransferase by chloramphenicol. Kinetic studies. Eur. J. Biochem., **164**, 53-58 (1987)

[15] Synetos, D.; Coutsogeorgopoulos, C.: Studies on the catalytic rate constant of ribosomal peptidyltransferase. Biochim. Biophys. Acta, **923**, 275-285 (1987)

[16] Michelinaki, M.; Mamos, P.; Coutsogeorgopoulos, C.; Kalpaxis, D.L.: Aminoacyl and peptidyl analogs of chloramphenicol as slow-binding inhibitors of ribosomal peptidyltransferase: a new approach for evaluating their potency. Mol. Pharmacol., **51**, 139-146 (1997)

[17] Ioannou, M.; Coutsogeorgopoulos, C.; Drainas, D.: Determination of eukaryotic peptidyltransferase activity by pseudo-first-order kinetic analysis. Anal. Biochem., **247**, 115-122 (1997)

[18] Michelinaki, M.; Spanos, A.; Coutsogeorgopoulos, C.; Kalpaxis, D.L.: New aspects on the kinetics of activation of ribosomal peptidyltransferase-catalyzed peptide bond formation by monovalent ions and spermine. Biochim. Biophys. Acta, **1342**, 182-190 (1997)

[19] Ioannou, M.; Coutsogeorgopoulos, C.: Kinetic studies on the activation of eukaryotic peptidyltransferase by potassium. Arch. Biochem. Biophys., **345**, 325-331 (1997)

[20] Ioannou, M.; Coutsogeorgopoulos, C.; Synetos, D.: Kinetics of inhibition of rabbit reticulocyte peptidyltransferase by anisomycin and sparsomycin. Mol. Pharmacol., **53**, 1089-1096 (1998)

[21] Karahalios, P.; Mamos, P.; Karigiannis, G.; Kalpaxis, D.L.: Structure/function correlation of spermine-analog-induced modulation of peptidyltransferase activity. Eur. J. Biochem., **258**, 437-444 (1998)

[22] Thompson, J.; Kim, D.F.; O'Connor, M.; Lieberman, K.R.; Bayfield, M.A.; Gregory, S.T.; Green, R.; Noller, H.F.; Dahlberg, A.E.: Analysis of mutations at residues A2451 and G2447 of 23S rRNA in the peptidyltransferase active

site of the 50S ribosomal subunit. Proc. Natl. Acad. Sci. USA, **98**, 9002-9007 (2001)
[23] Dresios, J.; Panopoulos, P.; Frantziou, C.P.; Synetos, D.: Yeast ribosomal protein deletion mutants possess altered peptidyltransferase activity and different sensitivity to cycloheximide. Biochemistry, **40**, 8101-8108 (2001)
[24] Dresios, J.; Panopoulos, P.; Suzuki, K.; Synetos, D.: A dispensable yeast ribosomal protein optimizes peptidyltransferase activity and affects translocation. J. Biol. Chem., **278**, 3314-3322 (2003)

Protein-glutamine γ-glutamyltransferase 2.3.2.13

1 Nomenclature

EC number
2.3.2.13

Systematic name
protein-glutamine:amine γ-glutamyltransferase

Recommended name
protein-glutamine γ-glutamyltransferase

Synonyms
Laki-Lorand factor <2> [12]
R-glutaminyl-peptide:amine γ-glutamyl transferase
factor XIIIa
fibrin stabilizing factor
fibrinoligase
glutaminylpeptide γ-glutamyltransferase
glutamyltransferase, glutaminylpeptide γ-
polyamine transglutaminase
tissue transglutaminase
transglutaminase
transglutaminase C
TGC

CAS registry number
80146-85-6

2 Source Organism

<1> *Rattus norvegicus* (tissue-type transglutaminase, abbrevation TGC [18]; distribution [26]; mature male Wistar [30]; male Sasco/King (SD)BR strain [28]) [12, 18, 19, 22, 26-30, 41, 48, 60]
<2> *Homo sapiens* (coagulation factor XIIIa [15,40]; zymogen factor XIII i.e. proenzyme of XIIIa [40]) [2, 5-8, 9, 10-15, 17, 18, 21, 27, 29, 40, 41, 44, 49, 51, 52, 57, 58, 60, 62]
<3> *Mus musculus* (BALB/c strain [20, 33]; newborn, CF57 strain [34]) [12, 20, 33, 34]
<4> *Bos taurus* (calf [25]; higher activity in 100% confluent endothelial cells than in 50% confluent cells [22]) [22, 25]
<5> *Cavia porcellus* (Hartley strain [1]; cloned in E. coli [37]) [1, 3-5, 10, 12, 15-17, 23, 24, 27, 31, 37, 38, 40, 46, 49, 51, 53, 54, 59]

<6> *Oryctolagus cuniculus* [9, 42, 49]
<7> *Mesocricetus auratus* [26]
<8> *Columba livia* (pigeon [32]) [32]
<9> *Gallus gallus* [8]
<10> *Tachypleus tridentatus* (japanese horseshoe crab [35]) [35]
<11> *Pisum sativum* (pea, var. Kelvedon wonder [36]) [36]
<12> *Streptoverticillium sp.* (tentatively classified, strain S-8112 [38]) [38, 63, 64]
<13> *Physarum polycephalum* (slime mould, strain M3cV [39]) [39, 47, 61]
<14> *Glycine max* [43]
<15> *Ovis aries* (merino sheep [45]) [45]
<16> *Streptoverticillium mobaraense* [46, 53]
<17> *Caenorhabditis elegans* [50]
<18> *squid* [55]
<19> *Patinopecten yessoensis* (scallop [56]) [56]
<20> *mammalia* (8 distinct transglutaminase isoenzymes have been identified on the genomic level, 6 have been isolated and characterized: factor XIII, type 1 or keratinocyte transglutaminase, type 2 or tissue transglutaminase, type 3 or epidermal transglutaminase, type 4 or prostate transglutaminase and type 5 transglutaminase [60]) [60]

3 Reaction and Specificity

Catalyzed reaction

protein glutamine + alkylamine = protein N^5-alkylglutamine + NH_3 (<2,5> mechanism and structure [2,16]; <5> putative incorporation site of putrescine in β-casein [23]; <5> liver transglutaminase, Q155 and Q159 are putative sites in bovine β-lactoglobulin that incorporate putrescine [24]; <1-3,5> structure, function, evolution [12,17,40]; <5> modified double displacement mechanism i.e. modified ping pong reaction [15,16])

Reaction type

aminoacyl group transfer

Natural substrates and products

S protein-bound γ-glutamine + alkylamine <2, 3, 5, 8, 9, 13, 17> (<2> involved in a wide variety of cellular processes, including growth, differentiation, stabilization of cytoskeleton [13]; <2,5,9> last enzyme in blood coagulation forming intermolecular γ-glutamyl-ε-lysine crosslinks between fibrin molecules [2,8,13,15]; <5> production of vaginal plug by postejaculatory clotting of rodent seminal plasma, formation of chemically resistant envelope of the stratum [15,17]; <8> mediates membrane-structural changes [32]; <3> epidermal enzyme involved in formation of cornified envelope [33]; <2> dimerization of fibrin γ-chains, cross-linking of α_2-plasmin inhibitior to fibrin α-chain and cross-linking of fibronectin to fibrin α-chains by factor XIIIa [17]; <13> actin is probably the major endogenous substrate [39]; <13> plasmodia-specific 40000 Da protein

LAV1-2 is the preferred in situ substrate [47]; <17> transglutaminase is probably involved in cell death program [50]) (Reversibility: ? <2, 3, 5, 8, 9, 13, 17> [2, 8, 13, 15, 17, 32, 33, 39, 50]) [2, 8, 13, 15, 17, 32, 33, 39, 47, 50]

P protein N^5-alkylglutamine + NH_3 <2, 3, 5, 8, 9, 13, 17> [2, 8, 13, 15, 17, 32, 33, 39, 47, 50]

Substrates and products

S 1-N-(carbobenzoxy-L-glutaminylglycyl)-5-N-(5'-N',N'-dimethylamino-1'-naphthalenesulfonyl)diamidopentane + α-carbobenzoxy-lysine <5, 16> (<16> fluorescent substrate for detection and characterization of glutamine acceptor compounds [46]) (Reversibility: ? <5, 16> [46]) [46]

P ?

S 1-N-(carbobenzoxy-L-glutaminylglycyl)-5-N-(5'-N',N'-dimethylamino-1'-naphthalenesulfonyl)diamidopentane + butylamine <5, 16> (<16> fluorescent substrate for detection and characterization of glutamine acceptor compounds [46]) (Reversibility: ? <5, 16> [46]) [46]

P ?

S 1-N-(carbobenzoxy-L-glutaminylglycyl)-5-N-(5'-N',N'-dimethylamino-1'-naphthalenesulfonyl)diamidopentane + ethylamine <5, 16> (<16> fluorescent substrate for detection and characterization of glutamine acceptor compounds [46]) (Reversibility: ? <5, 16> [46]) [46]

P ?

S 1-N-(carbobenzoxy-L-glutaminylglycyl)-5-N-(5'-N',N'-dimethylamino-1'-naphthalenesulfonyl)diamidopentane + propylamine <5, 16> (<16> fluorescent substrate for detection and characterization of glutamine acceptor compounds [46]) (Reversibility: ? <5, 16> [46]) [46]

P ?

S 1-N-(carbobenzoxy-L-glutaminylglycyl)-5-N-(5'-N',N'-dimethylamino-1'-naphthalenesulfonyl)diamidopentane + α_{S1}-casein <5, 16> (<16> fluorescent substrate for detection and characterization of glutamine acceptor compounds [46]) (Reversibility: ? <5, 16> [46]) [46]

P ?

S EAQQIVM + monodansylcadaverine <5> (<5> liver transglutaminase, peptide derived from the N-terminal sequence of fibronection, first modified residue is mainly Q3 [54]) (Reversibility: ? <5> [54]) [54]

P ?

S GTP + H_2O <5, 6> (<5> intrinsic GTPase activity [5]; <6> Mg^{2+}-dependent GTP hydrolytic activity [42]) (Reversibility: ? <5, 6> [5, 42]) [5, 42]

P GDP + phosphate <5, 6> [5, 42]

S N-(5-aminopentyl)-5-dimethylamino-1-naphtalenesulfonamide + F-actin <2, 9> (<2,9> trivial name dansylcadaverine [8]) (Reversibility: ? <2, 9> [8]) [8]

P ?

S N-(5-aminopentyl)-5-dimethylamino-1-naphthalenesulfonamide + casein <2, 6, 9> (<6> trivial name dansylcadaverine, casein can be replaced by various synthetic peptide acceptors [9]) (Reversibility: ? <2, 6, 9> [6, 8, 9]) [6, 8, 9]

P ?

S N-(5-aminopentyl)biotinamide + N,N'-dimethylcasein <17> (Reversibility: ? <17> [50]) [50]

P ?

S α-difluoroornithine + casein <5> (<5> suicide substrate [3]) (Reversibility: ? <5> [3]) [3]

P ?

S benzyloxycarbonyl-L-glutaminglycine + glycine ethyl ester <5> (Reversibility: r <5> [40]) [40]

P benzyloxycarbonyl-α-L-glutamyl(γ-glycine ethyl ester)glycine + NH_3 <5> [40]

S biotinylated TVQQEL + calcium binding protein S100A7 <2> (<2> transglutaminase 2 [58]) (Reversibility: ? <2> [58]) [58]

P ?

S histamine + acetyl-α_{S1}-casein <5> (Reversibility: ? <5> [37]) [37]

P ?

S histamine + maleyl-bovine serum albumin <5> (Reversibility: ? <5> [37]) [37]

P ?

S hydroxylamine + carbobenzoxy-Gln-Gly <12> (<12> other substrates are carbobenzoxy-Gln-Gln-Gly, carbobenzoxy-Gly-Gln-Gln-Gly, carbobenzoxy-Gly-Gly-Gln-Gly with 38%, 13% and 28% efficiency, respectively [38]) (Reversibility: ? <12> [38]) [38]

P carbobenzoxy-Gln-Gly-hydroxamate + ? <12> [38]

S methylamine + succinyl-β-casein <1> (<1> transglutaminase B [27]) (Reversibility: ? <1> [27]) [27]

P ?

S mono-6-amino-6-deoxy-α-cyclodextrin + bovine pancreatic trypsin <12> (Reversibility: ? <12> [63]) [63]

P ?

S mono-6-amino-6-deoxy-β-cyclodextrin + bovine pancreatic trypsin <12> (Reversibility: ? <12> [63]) [63]

P ?

S mono-6-amino-6-deoxy-γ-cyclodextrin + bovine pancreatic trypsin <12> (Reversibility: ? <12> [63]) [63]

P ?

S monodansylcadaverine + N,N-dimethylcasein <6, 13> (Reversibility: ? <6, 13> [47, 49]) [47, 49]

P ?

S ornithine + casein <5> (Reversibility: ? <5> [3]) [3]

P ?

S *p*-nitrophenyl acetate + alanine ethylester <5> (Reversibility: ? <5> [40]) [40]

P N-acetylalanine ethylester + *p*-nitrophenol <5> [40]

S *p*-nitrophenyl trimethylacetate + H_2O <5> (<5> liver transglutaminase, ester hydrolysis in the presence of Ca^{2+} [16]) (Reversibility: ? <5> [16]) [16]

P *p*-nitrophenol + trimethylacetate <5> [16]

S protein-bound γ-glutamine + alkylamine <1-6, 9-12, 14, 15, 16, 17, 20> (<2,5> acyl-transfer reaction [5,40]; <5> hydrolysis and aminolysis of certain aliphatic amides and active esters e.g. *p*-nitrophenyl esters and thiolesters [15,16]; <1-3,5> catalyzes post-translational protein modifications by transamidation of glutamine residues [12]; <1-3,5,9> forms intramolecular isopeptide bonds between fibrin molecules [2,8,12]; <2,5,9> donors: γ-carboxamide groups of protein-bound glutamine, acceptors: ε-amino groups of protein-bound lysine [8,40]; <5> liver transglutaminase, amine donors: putrescine, phenylethylamine, glycinamide, histamine, methylamine, ethanolamine, amonia [40]; <1> transglutaminase B: simultaneously γ-polymer and α-polymer formation [27]; <2> amine donors: primary amines [15]; <11> amine donors: diamines and polyamines [36]; <2,5> no amine donors are tyrosinamide, glycine, Gly-Leu, γ-aminobutyric acid [40]; <2,5> broad specificity towards amine acceptor [15,40]; <2,5> H_2O acts as substrate in the absence of amine acceptors [15]; <2> synthetic peptide acceptors for factor XIIIa in descending order of affinity: pyroglutamic acidEAQQIV, tert-butyloxycarbonylAQQIV [9]; <6> synthetic peptide acceptors for transglutaminase in descending order: tert-butyloxycarbonylQQIV, tert-butyloxycarbonylAQQIV, pyroglutamic acidQQIV [9]; <10> identification of natural protein substrates [35]; <2> substrates are membrane-associated erythrocyte proteins [7]; <2> substrates are coagulation factor V, α_2-macroglobulin, platelet myosin, actin [2]; <1-3,5> substrate fibronectin [2,12]; <2> recombinant factor XIIIa: substrate plasminogen-activator inhibitor type-2 [10]; <5> liver transglutaminase: substrate plasminogen-activator inhibitor type-2 [10]; <5> substrates are acetylated B-chains of oxidized insulin [15]; <11> substrates are pepsin, thrombin, cellulase, creatine kinase [36]; <2,5,11> substrates are fibrinogen, β-lactoglobulin, casein, insulin [36,40]; <5> substrate carbobenzoxy-L-Gln-Gly [15]; <11> no activity with catalase [36]; <11,12> no activity with native bovine serum albumin [36,38]; <12> no activity with bovine myosin, histone mixture, human serum fibronectin, spinach ribulose 1,5-diphosphate carboxylase-oxygenase, carbobenzoxyglutamine, carbobenzoxy-Asn-Gly [38]; <5> no activity with benzyloxycarbonyl-L-glutaminylglycine, benzyloxycarbonyl-α-L-glutamyl(γ-*p*-nitrophenyl ester) glycine, guinea pig hair follicle enzyme [40]; <5> liver transaminase catalyzes also the hydrolysis and aminolysis of certain aliphatic amides and of active and some inactive esters [15]; <4> lens transglutaminase, endogenous substrate β-crystallin [25]; <1> chondrosarcoma transglutaminase B, no activity with type I collagen and fibronectin [27]; <15> crosslinking of Hammersten casein, crosslinking between Ac-IB and Bz-Gly-Lys [45]; <6> lens transglutaminase, crosslinking of β-crystallin [49]; <5,16> aliphatic amine donors incorporated into benzyloxycarbonyl-L-Gln-Gly: hydroxylamine, methylamine, ethylamine, n-propylamine, n-butylamine, n-pentylamine, n-hexylamine, amino acids incorporated: L-lysine and D-lysine, amino acid esters incorporated: Gly, Ala, Val, and Met ethyl esters, Lys-analogs incorporated: L-ornithine, aliphatic amines with ω-carboxyl groups incorporated: 5-aminovaleric acid, ε-amino-n-caproic acid, 7-aminoheptanoic acid, ω-aminocaprylic acid, amines with functional groups incorporated: carbonyl, phosphate, sulfo groups and saccharides [53]; <2> substrates of recombinant full-length transglutaminase 5: loricrin, small proline rich proteins 1, 2 and 3 and involucrin [57]; <2> substrates of transglutaminase I and II: EF-hand-containing calcium binding proteins S100A11, S100A10 and S100A07 [58]; <20> endogenous substrates: cellular proteins e.g. aldolase, glyceraldehyde-3-phosphate dehydrogenase, phosphorylase kinase, crystallins, gluthathione S-transferase, actin, myosin, troponin, β-tubulin, tau, Rho A, histone, α-oxoglutarate dehydrogenase, cytochromes, erythrocyte band III, CD38, acetylcholine esterase, collagen, fibronectin, fibrinogen, vitronectin, osteopontin, nidogen, laminin, LTBP-1, osteonectin, osteocalcin, substance P, phospholipase A2, midkine, exogenous substrates: wheat gliadin, whey proteins, soy proteins, pea legumin, Candida albicans surface proteins, HIV envelope glycoproteins gp120 and gp41, HIV aspartyl proteinase, hepatitis C virus core protein [60]) (Reversibility: ? <1-6, 9-12, 14, 15, 16, 17, 20> [2, 4, 5, 7-10, 12, 15, 16, 25, 27, 35, 36, 38, 40, 43, 45, 50, 53, 60]) [2, 4, 5, 7-10, 12, 15, 16, 25, 27, 35, 36, 38, 40, 43, 45, 49, 50, 53, 54, 57, 58, 60]

P protein N^5-alkylglutamine + NH_3 <1-6, 9-12, 14, 15, 16, 17, 20> (<1-3,5> resulting bonds are covalent and stable to proteolysis [12]; <2,5> peptide bound glutamic acid with H_2O as acceptor [15]) [2, 4, 5, 7-10, 12, 15, 16, 25, 27, 35, 36, 38, 40, 43, 45, 49, 50, 53, 54, 57, 58, 60]

S putrescine + N,N'-dimethylcasein <1-4, 8, 11, 13, 14> (<8,11,13> spermidine can replace putrescine [32,36,39]; <11,13> spermine can replace putrescine [36,39]; <8,11,13> diaminopropane and cadaverine can replace putrescine [32,36,39]; <14> spermine and spermidine can replace putrescine [43]; <2> transglutaminase 5 [57]) (Reversibility: ? <1-4, 8, 11, 13, 14> [7, 13, 14, 22, 28-30, 32, 34, 36, 39, 43]) [7, 13, 14, 22, 28-30, 32, 34, 36, 39, 43, 54]

P ?

S putrescine + bovine muscle actin <13> (<13> preferred substrate [39]) (Reversibility: ? <13> [39]) [39]

P ?

S putrescine + casein <2, 3, 5, 13> (<5> in vitro acceptor [3]; <2,5> α- or β-casein [15]) (Reversibility: ? <2, 3, 5, 13> [3-5, 15, 31, 33, 39]) [3-5, 15, 31, 33, 39]

P ?

S putrescine + fibronectin <5, 6> (<5> in vivo acceptor [3]) (Reversibility: ? <5, 6> [3, 9]) [3, 9]

P ?

S vimentin + 5-(biotinamido)pentylamine <2> (<2> transglutaminase 5 [57]) (Reversibility: ? <2> [57]) [57]

P ?

Inhibitors

1,5-iodoacetyl-5'-(sulfonyl-1-naphthyl)-ethylenediamine <1> (<1> 0.01 mM, 91% and 92% inhibition of chondrosarcoma transglutaminases B and C respectively [27]) [27]
5,5'-dithiobis(2-nitrobenzoic acid) <5> (<5> irreversible, carbobenzoxy-Phe protects [4]; <5> 1 mol per mol enzyme, 85% inactivation, not reversed by glutathione [1]; <5> reversed by dithiothreitol [40]) [1, 4, 40]
ADP <1, 2> (<1,2> complete inhibition of rat liver and human brain transglutaminase, reversible, non-competitive to putrescine [18,29]) [18, 29]
AMP <1> (<1> weak inhibition of liver transglutaminase [18,29]) [18, 29]
ATP <1, 2, 13> (<1,2> 3 mM, complete inhibition of rat liver and human brain tissue-type transglutaminase, reversible [29]; <1,2> 3 mM, complete inhibition of rat liver and human brain transglutaminase, reversible, non-competitive to putrescine [18]; <13> 0.5 mM, approx. 60% inhibition of recombinant tranglutaminase [61]) [18, 29, 61]
CTP <1> (<1,2> 3 mM, complete inhibition of rat liver and human brain transglutaminase, reversible, non-competitive to putrescine [18]) [18, 29]
Ce^{3+} <2> (<2> not reversible by Ca^{2+} [11]) [11]
Cu^{2+} <2, 5, 10-12> (<10> 1 mM, complete inhibition [35]; <11> trace amounts, 0.45 mM diethyldithiocarbamate stimulates crude preparation [36]; <12> weak [38]; <5> KCN or dithiothreitol restore activity [40]; <5> mechanism [40]) [35, 36, 38, 40]
Cu^{2+} <19> (<19> strong inactivation [56]) [56]
EDTA <1, 2, 5, 10, 13> (<10> 5 mM, complete inhibition [35]; <1> complete inactivation above 10 mM [30]; <10,13> 5 mM [35,39]; <13> reversible [39]) [11, 30, 35, 39, 40]
EGTA <11, 13, 17> (<13> 2 mM, irreversible [39]; <11> weak, reversible by Ca^{2+} [36]; <17> 2.5 mM, 94% inhibition [50]) [36, 39, 50]
Fe^{2+} <5> [40]
GDP <2, 5> (<2> at low levels of Ca^{2+} [14]; <2,5> inhibits hydrolysis of GTP [5]) [5, 14, 43]
GMP <2> (<2> at low levels of Ca^{2+} [14]) [14]
GTP <1-3, 5, 6, 13, 14, 17> (<1,2> 50% as effective as ATP [29]; <1-3,5> inhibition of tissue transglutaminase [12,13]; <2> 0.1 mM, complete inhibition at suboptimal Ca^{2+}-levels [14]; <6> 0.005 mM, complete inhibition [42]; <6> 0.05 mM, 50% inhibition of lens transglutaminase, 0.5 mM, complete inhibition in the presence of 0.5 mM Ca^{2+}, increasing the Ca^{2+} concentration to 3 mM reverses inhibition [49]; <17> weak inhibition in the millimolar range [50]; <2> 0.02-0.1 mM, inhibition of transglutaminases 2 and 3 [52]; <13> 0.5 mM, almost complete inhibition of recombinant transglutaminase in the presence of 0.5 mM Ca^{2+}, no inhibition in the presence of 2 mM Ca^{2+} [61]) [12-14, 18, 29, 42, 43, 49, 50, 52, 61]
GTP-γ-S <2, 5> (<2,5> inhibition of GTP-hydrolysis [5]) [5]
Gd^{3+} <2> (<2> not reversible by Ca^{2+} [11]) [11]
Hg^{2+} <5> [40]
$HgCl_2$ <17> (<17> 5 mM, 96% inhibition in the presence of 10 mM Ca^{2+} [50]) [50]

KCl <10, 13> (<10> above 500 mM [35]; <13> 45% activity at 400 mM [47]) [35, 47]
La^{3+} <2> (<2> not reversible by Ca^{2+} [11]) [11]
$MgCl_2$ <17> (<17> 5 mM, 88% inhibition, 30% inhibition in the presence of 10 mM Ca^{2+} [50]) [50]
N-ethylmaleimide <1, 2, 5, 10, 12, 13, 14, 15, 19> (<12> 1 mM, 90% inhibition [38]; <13> 0.1 mM [39]; <5> 3.6 mol per mol enzyme, 91% inhibition in the presence of Ca^{2+}, not inhibited in the absence of Ca^{2+}, substrate protects [1]; <2> strong inactivation in the presence of Ca^{2+}, not inhibited in the absence of Ca^{2+} [7]; <1> 79% inactivation [19]; <10> 1 mM, 98% inhibition [35]; <5> 1 mM, 75% inhibition [38]; <14> 0.1 mM, 60% inhibition [43]; <15> 1 mM, complete inhibition of DEAE-unabsorbed transglutaminase, 93% inhibition of DEAE-absorbed transglutaminase [45]; <13> irreversible inhibition, inhibition increases with increasing Ca^{2+} concentrations, 50% inhibition at 1 mM Ca^{2+} [47]; <19> strong inactivation [56]) [1, 7, 19, 35, 38, 39, 43, 45, 47, 56]
NH_4^+ <1, 17> (<1> complete inhibition of coagulating gland transglutaminase with more than 1 mM NH_4Cl or 20 mM $(NH_4)_2SO_4$ [30]; <17> 5 mM, 36% inhibition [50]) [30, 50]
NaCl <10, 19> (<10> above 500 mM [35]; <19> 500 mM, 2 h at 20°C in the absence of substrate, 86% inhibition, reversible to some extent by dilution [56]) [35, 56]
$Pb(CH_3COO)_2$ <12> (<12> 1 mM, 56% inhibition [38]) [38]
S-nitroso-N-acetylpenicillamine <2> (<2> NO-donor, 8-16 mM, almost complete inhibition of transglutaminases 1 and 3, weak inhibition of transglutaminase 3 [52]) [52]
Tb^{3+} <2> (<2> noncompetitive inhibition of factor XIIIa, at high Ca^{2+}-levels, not reversed by Ca^{2+} [11]) [11]
UTP <1> (<1> 50% as effective as ATP [18,29]) [18, 29]
Zn^{2+} <5, 12> (<12> 1 mM, 89% inhibition [38]) [38, 40]
$ZnCl_2$ <17, 19> (<17> 5 mM, 97.5% inhibition, 43% inhibition in the presence of 10 mM Ca^{2+} [50]) [50, 56]
α-difluoromethylornithine <5> (<5> 2.6 mM, 50% inhibition of putrescine transfer to casein, suicide substrate, irreversible, competitive to putrescine or fibrinonectin [3]) [3]
cadaverine <11> (<11> strong, putrescine as substrate [36]) [36]
chlorpromazine <2, 9> (<2,9> reverses calmodulin enzyme stimulation [8]) [8]
cysteine <11> (<11> 85% inhibition [36]) [36]
dansylcadavarin <1, 4> (<1,4> 0.0019 mM, 50% inhibition [22]) [22]
diethyl dicarbonate <2> (<2> not without Ca^{2+} [7]) [7]
dithiothreitol <11> (<11> 1.5 mM, 42% inhibition, 16.5 mM, 40% inhibition [36]) [36]
hydroxylamine <10> (<10> 100 mM, complete inhibition [35]) [35]
hydroxymercuribenzoate <1> (<1> 99% inactivation [19]) [19]
iodoacetamide <1, 2, 5, 9, 15, 17> (<2,9> factor XIIIa [8]; <5> 0.1 mM, pH 6.8, in the presence of Ca^{2+}, complete inhibition, irreversible, incorporation

of 1 mol carbamidomethyl/mol enzyme, substrate protects [1,40]; <5> mechanism [40]; <2,9> not inhibited in the absence of Ca^{2+}, calmodulin regulated transglutaminases is not inhibited [8]; <1> 86% and 91% inhibition of chondrosarcoma transglutaminase B and C respectively [27]; <15> 1 mM, 89% inhibition of DEAE-absorbed transglutaminase [45]; <17> 10 mM, 62% inhibition in the presence of 10 mM Ca^{2+} [50]) [1, 8, 27, 40, 45, 50]
lysine <15> (<15> 1 mM, 43% inhibition of DEAE-unabsorbed transglutaminase, 90% inhibition of DEAE-absorbed transglutaminase [45]) [45]
methylamine <1, 4> (<1,4> 1.8 mM, 50% inhibition [22]) [22]
monodansylcadaverine <8, 13> (<13> partial inhibition above 0.5 mM [47]) [32, 47]
monoiodoacetate <5, 10, 12, 15> (<12> 1 mM, 24% inhibition [38]; <10> 1 mM, 94% inhibition [35]; <5> 1 mM, 97% inhibition [38]; <15> 1 mM, 93% inhibition of DEAE-unabsorbed transglutaminase, 85% inhibition of DEAE-absorbed transglutaminase [45]) [35, 38, 45]
o-phenanthroline <11> (<11> not reversible by Ca^{2+} [36]) [36]
ornithine <5> (<5> weak, suicide substrate in the presence of casein [3]) [3]
p-chloromercuribenzoate <1, 5, 12, 13, 19> (<13> 0.1 mM [39]; <5> reversible by glutathione [1]; <1> 21% inhibition of chondrosarcoma transglutaminase B [27]; <12> 1 mM, 44% inhibition [38]; <5> 1 mM, 98% inhibition [38]; <19> strong inactivation [56]) [1, 27, 38, 39, 56]
putrescine <13> (<13> 0.1-0.2 mM, substrate inhibition [39]) [39]
sodium citrate <1> (<1> above 10 mM, complete inactivation [30]) [30]
spermidine <13> (<13> 0.1-0.2 mM, substrate inhibition [39]) [39]
spermine <1, 4, 13> (<13> 0.1-0.2 mM, substrate inhibition [39]; <1,4> 0.17 mM, 50% inhibition [22]) [22, 39]
tetrathionate <5> (<5> inactivation, not reversible by dithiothreitol [40]) [40]
Additional information <1, 2, 4, 5, 6, 10, 12> (<1> not inhibited by adenosine or adenine [18,29]; <1> not inhibited by PMSF [27]; <1,10> not inhibited by diisopropylfluorophosphate [27,35]; <12> not inhibited by Ba^{2+}, Co^{2+}, Fe^{3+}, K^{+}, Mg^{2+}, Mn^{2+}, Na^{+}, Ni^{2+}, Sr^{2+} [38]; <1> not inhibited by sulfhydryl-reagents in the absence of Ca^{2+} [27]; <2> rapid inactivatin of factor XIIIa by trypsin and thrombin in the absence of metal ions [11]; <1,4> not inhibited by mono- and dimethylated dansylcadaverine [22]; <1> transglutaminase of coagulating gland, not inhibited by GTP [30]; <6> not inhibited by GDP [42]; <5> rapid degradation of liver tissue transglutaminase in the presence of micro-calpain, GTP-γ-S inhibits degradation [51]; <2> transglutaminase 1, not inhibited by GTP [52]) [11, 18, 22, 27, 29, 30, 35, 38, 42, 51, 52]

Activating compounds

NaCl <18> (<18> 700 mM, 15fold activation [55]) [55]
calmodulin/Ca^{2+} <2, 9> (<2,9> 0.01-0.2 mM, 3fold activation, inhibition above 0.3 mM, kinetics [8]) [8]
chaotropic salts <2> (<2> strong activation of epidermal transglutaminase in the presence of Ca^{2+} [15]) [15]

dithiothreitol <1, 3, 5, 6, 7, 10, 14> (<3> requirement [20]; <10> weak activation [35]; <1> 6-8fold activation of lung transglutaminase B, no activation of transglutaminase C [26]; <14> 10 mM, approx. 1.8fold activation [43]; <6> required for activity of lens transglutaminase [49]) [20, 24, 26, 28, 35, 43, 49]
papain <2> (<2> limited proteolysis of plasma factor XIII [40]) [40]
reptilase <2> (<2> limited proteolysis of plasma factor XIII [40]) [40]
thrombin <1-3, 5> (<5> activation, of proenzyme [40]; <2> slight activation [8]; <2,5> Ca^{2+}-dependent proteolytical activation, in the presence of fibrin [15]; <2> removes blocked NH_2-terminal peptide of factor XIII and unmasks reactive thiol-group at Cys-314 to yield catalytically active factor XIIIa, i.e. fibrinoligase, from inactive zymogen XIII [2,15]; <1> particulate enzyme [27]; <1> guinea pig epidermal enzyme [27]) [2, 8, 12, 15, 27, 40, 45]
trypsin <2, 3> (<2> strong activation of epidermal tranglutaminase in the presence of Ca^{2+} [15]; <3> 3fold activation of epidermal transglutaminase, activation is blocked by trypsin-inhibitors [20]) [15, 20, 40]
Additional information <2> (<2> 25fold increase in activity by heating at 56°C in the presence of Ca^{2+} [15]; <2> 3-16fold increase in activity of 72000 Da epidermal transglutaminase isoform by heating at 56°C in the presence of calcium [21]) [15, 21]

Metals, ions

Ba^{2+} <5> (<5> activation, can replace Ca^{2+} to a lesser extent [40]) [40]
Ca^{2+} <1-11, 13, 15, 17, 18, 19> (<1-11,13,15> requirement [1-3,5,7-9,11-15,19-22,26-37,39,40,45,47,49]; <1> 0.5 mM, maximum activity of transglutaminase from coagulating gland [30]; <2> crucial during thrombin cleavage of factor XIII for the formation of factor XIIIa and factor XIIIa activation, Ln3+ can replace Ca^{2+} during trypsin activation of factor XIII but not in activation of factor XIIIa [11]; <2,5> 1 mM [40]; <10> 8 mM, approx. 10fold activation [35]; <2> Cys-thiol active binding-site identified [2]; <2,5> catalytically active monomeric metal-enzyme complex [15]; <5> mechanism [40]; <4> no activity below 1.25 mM, maximal activity at 2.5 mM and above, half-maximal activity at approx. 1.5 mM [22]; <2> Ca^{2+} leads to plasma factor XIIIa dissociation into α'- and β-dimers [15]; <5> activation of GTPase activity [5]; <11> only with N,N'-dimethylcasein as substrate, no activation with plant proteins as substrate [36]; <13> half-maximal activity at 0.7 mM, maximal activity at about 2 mM [47]; <1> brain transglutaminase NI, maximal activity at 0.1 mM, transglutaminase NII, maximal activity at 0.01 mM, inhibition above [48]; <17> maximal activity at 10 mM [50]; <18> maximal activity at 10 mM Ca^{2+} in the presence of 700 mM NaCl [55]; <2> transglutaminase 3 binds 3 Ca^{2+} ions, Er^{3+}, Sm^{3+}, Tb^{3+} and Lu^{3+} can substitute for Ca^{2+} to retain activity [62]) [1-3, 5, 7-9, 11-15, 19-22, 26-37, 39, 40, 41, 45, 47, 48, 49, 50, 55, 56, 61, 62]
Ga^{3+} <2> (<2> requirement [7]; <2> no activation [10]) [7, 10]
La^{3+} <2> (<2> requirement, 0.01-0.1 mM [11]) [11]
Mg^{2+} <1, 2, 5, 6> (<6> activation [9]; <1,2,5> slight activation [19,40]; <5> stimulation of GTPase activity, 5 mM [5]) [5, 9, 19, 40]

Mn^{2+} <1, 2, 5> (<2> requirement [7]; <1> activation, 3.5% as effective as Ca^{2+} [19]; <1> slight activation [27]) [7, 19, 27, 40]
Sr^{2+} <1, 5> (<1> activation, 27% as effective as Ca^{2+} [19]; <1> slight activation [27]) [19, 27, 40]
Tb^{3+} <2> (<2> requirement [7]; <2> no activation [10]) [7, 10]
Zn^{2+} <1> (<1> activation, 24.6% as effective as Ca^{2+} [19]) [19]
Additional information <1, 2, 12, 14> (<2> not activated by trivalenic lanthanide ions, 0.01-0.1 mM [11]; <1> not activated by Cu^{2+} [27]; <12> not activated by Ca^{2+} [38,64]; <14> Ca^{2+} is not required for activity [43]) [11, 27, 38, 43]

Turnover number (min^{-1})

0.021 <16> (α-carbobenzoxy-lysine, <16> 1-N-(carbobenzoxy-L-glutaminylglycyl)-5-N-(5'-N'-N'-dimethylamino-1'naphthalenesulfonyl)diamidopentane as glutamine donor [46]) [46]
0.024 <16> (propylamine, <16> 1-N-(carbobenzoxy-L-glutaminylglycyl)-5-N-(5'-N',N'-dimethylamino-1'-naphthalenesulfonyl)diamidopentane as glutamine donor [46]) [46]
0.033 <16> (butylamine, <16> 1-N-(carbobenzoxy-L-glutaminylglycyl)-5-N-(5'-N',N'-dimethylamino-1'-naphthalenesulfonyl)diamidopentane as glutamine donor [46]) [46]
0.033 <16> (ethylamine, <16> 1-N-(carbobenzoxy-L-glutaminylglycyl)-5-N-(5'-N',N'-dimethylamino-1'-naphthalenesulfonyl)diamidopentane as glutamine donor [46]) [46]
0.14 <2> (involucrin, <2> recombinant transglutaminase 5 [57]) [57]
0.21 <2> (loricrin, <2> recombinant transglutaminase 5 [57]) [57]
0.26 <2> (small proline-rich protein 3, <2> recombinant transglutaminase 5 [57]) [57]
19.2 <1> (methylamine, <1> chondrosarcoma transglutaminase B [27]) [27]
28 <5> (methylamine, <5> transglutaminase C [27]) [27]
31.9 <1> (N-dimethylated casein) [19]
111.4 <2> (methylamine, <2> factor XIIIa [27]) [27]

Specific activity (U/mg)

0.00017 <5> (<5> follicle transglutaminase [15]) [15]
0.0013 <2> (<2> epidermal transglutaminase [15]) [15]
0.0037 <2> (<2> plasma factor XIII [15]) [15]
0.005 <13> [39]
0.0052 <15> (<15> DEAE-unabsorbed transglutaminase [45]) [45]
0.0072 <2> (<2> platelet factor XIII [15]) [15]
0.07 <2> (<2> erythrocyte transglutaminase [6]) [6]
0.092 <15> (<15> DEAE-absorbed transglutaminase [45]) [45]
0.156 <2> (<2> tissue transglutaminase [13]) [13]
0.16 <14> [43]
0.526 <5> [31]
0.546 <5> [4]
1 <2> [14]
1.2 <5> (<5> recombinant transglutaminase [37]) [37]

2.5-3 <2> [7]
8.8 <13> (<13> incorporation of monodansylcadaverine into N,N-dimethyl-casein [47]) [47]
12.5 <5> [1]
14 <5> [15]
22.6 <12> [38]
25 <16> (<16> formation of L-glutamic acid γ-monohydroxamate [53]) [53]
Additional information <5, 10> (<10> 938 amine incorporation units/min [35]; <5> catalytic properties of recombinant liver transglutaminase differ from tat of the native enzyme [59]) [35, 59]

K_m-Value (mM)

0.0029 <5> (acetyl-α_{S1}-casein, <5> recombinant enzyme [37]) [37]
0.0032 <5> (acetyl-α_{S1}-casein, <5> native enzyme [37]) [37]
0.0037 <2> (involucrin, <2> recombinant transglutaminase 5 [57]) [57]
0.0044 <5> (GTP, <5> GTPase activity [5]) [5]
0.0044 <2> (loricrin, <2> recombinant transglutaminase 5 [57]) [57]
0.006 <9> (casein, <9> plasma factor XII, actin [8]) [8]
0.007 <2> (actin, <2> platelet transglutaminase [8]) [8]
0.0077 <2> (small proline-rich protein 3, <2> recombinant transglutaminase 5 [57]) [57]
0.011 <2> (casein, <2> platelet transglutaminase [8]) [8]
0.012 <9> (casein, <9> gizzard transglutaminase [8]) [8]
0.0214 <13> (spermidine) [39]
0.024 <5> (methylamine, <5> transglutaminase C [27]) [27]
0.0317 <13> (spermine) [39]
0.034 <13> (monodansylcadaverine) [47]
0.04 <1> (putrescine, <1> soluble liver transglutaminase [28]) [28]
0.04-0.05 <1, 5> (putrescine) [28, 37]
0.042 <14> (spermidine) [43]
0.044 <1> (putrescine, <1> membrane-associated liver transglutaminase [28]) [28]
0.049 <13> (putrescine) [39]
0.05 <1> (N-dimethyled casein) [19]
0.051 <1> (methylamine, <1> chondrosarcoma transglutaminase B [27]) [27]
0.051 <3> (putrescine, <3> retinoic-acid-induced enzyme [33]) [33]
0.061 <2> (methylamine, <2> factor XIIIa [27]) [27]
0.065 <3> (putrescine, <3> retinoic acid induced transglutaminase [33]) [33]
0.069 <14> (spermine) [43]
0.098-0.106 <3> (putrescine, <3> enzyme induced by 12-O-tetradecanoyl-phorbol-13-acetate or Ca^{2+} [33]) [33]
0.11 <14> (putrescine) [43]
0.17 <5> (putrescine) [3]
0.17 <1> (putrescine, <1> brain transglutaminase NII [48]) [48]
0.203 <4> (putrescine, <4> tissue transglutaminase from endothelial cells [22]) [22]
0.28 <1> (putrescine, <1> brain transglutaminase NI [48]) [48]

0.38 <5> (histamine, <5> recombinant enzyme [37]) [37]
0.52 <5> (histamine, <5> native enzyme [37]) [37]
0.6 <2> (putrescine) [7]
1.35 <5> (ornithine) [3]
2.1 <5> (α-difluoromethylornithine) [3]
2.1 <5> (α-difluoroornithine, <5> suicide substrate [3]) [3]
7 <5> (benzyloxycarbonyl-L-glutaminylglycine) [40]
9.63 <11> (putrescine) [36]
66 <5> (carbobenzoxy-L-glutaminylglycine) [1]
Additional information <1, 2, 5> (<5> hair follicle and liver enzyme with identical kinetic features [15]; <5> kinetic mechanism [16,40]; <1> kinetic studies of membrane-associated and soluble enzyme [28]; <1> kinetic study with various transglutaminases: human factor XIIIa, guinea pig liver TGC, rat chondrosarcoma TGB [27]; <1> K_m of N,N'-dimethylcasein: 0.44-0.52 mg/ml [28]; <2> K_m of dimethyl casein: 1.3 mg/ml [7]; <1> membrane-associated liver transglutaminase, K_m for N,N'-dimethylcasein: 0.44 mg/ml, soluble transglutaminase, 0.52 mg/ml [28]) [7, 15, 16, 27, 28, 40]

K_i-Value (mM)

0.071 <2> (Tb^{3+}) [11]
0.9 <14> (GTP) [43]
2.28 <5> (α-difluoromethylornithine) [3]

pH-Optimum

6 <2, 5> (<2,5> approx. value, hydrolysis reaction [40]) [40]
6-7 <12> [38]
7.5 <13> (<13> less than 50% of maximal activity at pH 5,5 and pH 9.5 [47]) [47]
7.6 <14> [43]
8 <17, 18> [50, 55]
8.5-9 <15> [45]
9 <1> [19]
9.5 <3> [33]
10 <3> [20]
Additional information <1, 2, 5, 12, 13> (<5> optima depend on amine donor substrate [40]; <2> pI: 5.6 [7]; <13> pI: 6.1 [39]; <1> pI: 7.6, crude enzyme [30]; <1> pI: 8.5-8.7, purified enzyme [30]; <12> pI: 8.9 [38]) [7, 30, 38-40]

Temperature optimum (°C)

20 <18> [55]
25 <5> (<5> assay at [24]) [24]
30 <2> (<2> assay at [6,7]) [6, 7]
35 <11, 13> (<11,13> assay at [36,39]) [36, 39]
37 <1-5, 8, 10, 12, 14> (<1-5,8,10,12> assay at [3-5,22,29,32,34,35,37,38]) [3-5, 22, 29, 32, 34, 35, 37, 38, 43]
50 <12> (<12> at pH 6.0 [38]) [38]
60 <15> [45]

4 Enzyme Structure

Molecular weight

40000 <12> (<12> gel filtration [38]) [38]
50000 <3> (<3> epidermal and hair follicle cationic isozyme, gel filtration [34]) [34]
54000 <5> (<5> hair follicle transglutaminase, gel filtration [15,40]) [15, 40]
55000 <2, 3, 5> (<2> epidermal transglutaminase, gel filtration [15]; <3> epidermal transglutaminase, gel filtration [20]; <5> hair follicle transglutaminase, native PAGE [40]) [15, 20, 40]
65000 <1, 2> (<2> gel filtration and sucrose gradient centrifugation [7]; <1> transglutaminase C, gel filtration [30]) [7, 30]
76900 <5> (<5> sedimentation and diffusion [1,40]) [1, 40]
77000 <13> (<13> gel filtration [39]) [39]
80000 <1, 2, 14, 15> (<2> gel filtration [6]; <1> liver and chondrosarcoma transglutaminase C, gel filtration [27]; <1> liver transglutaminase, gel filtration [22]; <14> gel filtration [43]; <15> DEAE-absorbed transglutaminase, gel filtration [45]) [6, 22, 27, 43, 45]
82000 <15> (<15> DEAE-unabsorbed transglutaminase, gel filtration [45]) [45]
83010 <2> (<2> primary structure [2]) [2]
86000-94000 <5> (<5> sedimentation equilibrium [1,40]) [1, 40]
88000 <4> (<4> tissue transglutaminase, gel filtration [22]) [22]
90000 <2, 3, 5> (<5> sedimentation equilibrium, meniscus depletion method, iodoacetamide-incorporation studies [1]; <3> epidermis anionic isozyme, gel filtration [34]; <2> recombinant transglutaminase 1, gel filtration [52]) [1, 34, 52]
100000 <1> (<1> lung transglutaminase B, gel filtration [26]; <1> liver and chondrosarcoma transglutaminase B [27]) [26, 27]
101000 <13> (<13> gel filtration [47]) [47]
280000-330000 <2> (<2> plasma factor XIII, gel filtration and sedimentation equilibrium [40]) [40]

Subunits

? <1, 2, 6, 17, 18> (<2> x * 83005, α subunit of placental factor XIIIa, deduced from amino acid sequence [2]; <2> x * 73183, factor XIII β-subunit, deduced from nucleotide sequence [17]; <6> x * 80000, SDS-PAGE [42]; <1> x * 29000, brain transglutaminse NII, SDS-PAGE [48]; <1> x * 45000, brain transglutaminase NI, SDS-PAGE [48]; <6> x * 78000, lens transglutaminase, SDS-PAGE [49]; <17> x * 61000 Da, SDS-PAGE, immunoblot [50]; <2> x * 84000, recombinant transglutaminase 5, immunoblot [57]; <18> x * 94000, SDS-PAGE [55]; <19> x * 95000, SDS-PAGE [56]) [1, 2, 17, 42, 48, 49, 50, 55, 56, 57]
dimer <5, 13, 15> (<5> 2 * 27000, hair follicle transglutaminase, SDS-PAGE [15,40]; <13> 2 * 39600, SDS-PAGE [39]; <15> 2 * 40000, DEAE-absorbed transglutaminase, SDS-PAGE [45]) [15, 39, 40, 45]
monomer <2, 3, 5, 10, 12, 13, 14, 15> (<12> 1 * 40000, SDS-PAGE [38]; <3> 1 * 50000, hair follicle transglutaminase, SDS-PAGE [34]; <2> 1 * 55000, epi-

dermal transglutaminase, SDS-PAGE [15]; <5> 1 * 75000, SDS-PAGE [15]; <5> 1 * 80000, SDS-PAGE [5]; <2> 1 * 80000, SDS-PAGE [6]; <5> 1 * 80000-90000, gel filtration in 6 M guanidine [40]; <5> 1 * 85000, SDS-PAGE [40]; <10> 1 * 86000, SDS-PAGE [35]; <2> 1 * 92000, SDS-PAGE with or without 2-mercaptoethanol [7]; <5> 1 * 76620, liver transglutaminase, deduced from nucleotide sequence [17]; <14> 1 * 80000, SDS-PAGE [43]; <15> 1 * 77000, DEAE-unabsorbed transglutaminase, SDS-PAGE [45]; <13> 1 * 96000, gel filtration in the presence of 1 mM Ca^{2+} [47]) [5-7, 15, 17, 34, 35, 38, 40, 43, 45, 47]

tetramer <2> (<2> $\alpha'_2\beta_2$, 4 * 71000, plasma factor XIIIa, α' is the proteolytically modified catalytic subunit of factor XIII, SDS-PAGE [15]; <2> $\alpha_2\beta_2$, 2 * 75000 + 2 * 80000, plasma factor XIII, 2 catalytic α subunits and 2 noncatalytic β subunits [15]; <2> $\alpha_2\beta_2$, 2 * 83005 + 2 * 80000, plasma factor XIII, deduced from amino acid sequence [12]) [12, 15]

Additional information <2> (<2> zymogens of plasma and platelet-coagulation factor XIIIa with different subunit structure: $\alpha_2\beta_2$ and α_2, respectively [12,15,17]; <2> 75000 Da catalytically active α-subunits of plasma enzyme are stabilized by 80000 Da β-subunits [17]; <2> tissue-type and epidermal TG are monomers, hair follicle TG and coagulation factor XIIIa, i.e. fibrinoligase, are dimers [17]; <2> autocatalytic crosslinking of transglutaminase 5, molecular mass approx. 200000 Da [57]) [12, 15, 17, 57]

Posttranslational modification

glycoprotein <1> (<1> chondrosarcoma transglutaminase [27]; <1> coagulating gland transglutaminase, glycosylated with mannosyl-residues, no terminal position, substituted with saturated acyl residues and phosphoinositol [30]) [27, 30]

lipoprotein <2> (<2> membrane-bound enzyme of keratocytes is anchored via palmitate and myristate [12]) [12]

proteolytic modification <2> (<2> activation of platelet factor XIII through thrombin cleavage at Arg37-Gly38 peptide bond leading to factor XIIIa [11]) [11]

Additional information <1-3, 5> (<2,5> no glycoprotein [15,40]; <1-3,5> no glycoprotein [2,12]; <2> α-subunit of factor XIIIa is most probably not glycosylated [17]) [2, 12, 15, 17, 40]

5 Isolation/Preparation/Mutation/Application

Source/tissue

A-431 cell <2> (<2> epidermal carcinoma cell line [13]) [13]

CACO-2 cell <2> (<2> cell line derived from human colon carcinoma [41]) [41]

IEC-6 cell <1> (<1> cell line derived from small intestine crypt cells [41]) [41]

SH-SY5Y cell <2> (<2> neuroblastoma SH-SY5Y cells [51]) [51]

blood plasma <2> (<2> identical with enzyme from placenta or platelets [2]; <2> coagulation factor XIIIa [15]) [2, 15, 40]
blood platelet <2> (<2> identical with enzyme from plasma [2]; <2> placenta [2,17]; <2> uterus, macrophages [17]; <2> coagulation factor XIIIa [15]) [2, 8, 15, 17, 40]
brain <1, 2> (<1> brain specific transglutaminases NI and NII [48]) [18, 48]
cell suspension culture filtrate <12> [38]
chondrocyte <1> (<1> malignant, swarm chondrosarcoma [27]) [27]
coagulating gland <1> (<1> secretion, not immunologically related to tissue-type enzyme or blood factor XIIIa [30]) [30]
cytotrophoblast <2> (<2> 3rd trimester of pregnancy [10]) [10]
endothelium <2, 4> (<4> aortic cell suspension culture [22]) [12, 22]
epidermal cell <3> (<3> primary and MCA3A1-cell line, retinoic acid induced enzyme differs from normal epidermal enzyme [33]) [33]
epidermis <2, 3, 5> (<2> stratum corneum [13,21]; <2> 2 epidermal transglutaminases [21]; <2> callus, keratinocytes [12]; <3> subcellular distribution [33]; <3> 2 isoforms: anionic and cationic [34]) [12, 20, 21, 33, 34]
erythrocyte <2, 5, 8> (<8> ghosts [32]) [5-7, 10, 14, 32]
gill <18> [55]
gizzard <9> (<9> smooth muscle [8]) [8]
hair follicle <3, 5, 15> (<2> outer root shear cells [12]; <3> 2 cationic isoforms, [34]; <5> not related to liver enzyme [15,40]) [12, 15, 34, 40, 45]
hemocyte <10> [35]
hepatocyte <1> [12]
keratinocyte <2> (<2> transglutaminase 5 [57]) [57]
leaf <14> [43]
lens <4, 6> (<4> cortex [25]) [9, 25, 49]
liver <1-5, 6, 12> [1, 3-5, 12, 15, 16, 18, 22-24, 27-29, 31, 37, 38, 40, 42, 54]
lung <1> [26]
macrophage <3> [12]
muscle <19> (<19> striated adductor muscle [56]) [56]
ovary <7> [26]
placenta <2> (<2> identical with enzyme from plasma [2]; <2> platelets [2,17]; <2> macrophages [17]) [2, 12, 17, 40]
seedling <11> (<11> apical meristematic tissue [36]) [36]
skin <2> (<2> transglutaminase 1, required for the formation of a cornified envelope in stratified squamous epithelia [52]) [52]
syncytiotrophoblast <2> [10]
uterus <2> [40]
Additional information <1-3, 5, 7> (<1-3,5> tissue distribution [12]) [12, 26]

Localization

cytoplasm <1-3, 5> (<1-3,5> tissue transglutaminase, factor XIIIa [12]) [12, 13, 17]
extracellular <12> [38]

membrane <1, 2> (<2> keratocytes [12]; <1> liver transglutaminase, plasma-membrane associated, lateral domain [28]; <2> transglutaminase 1, membrane-anchored [52]) [12, 28, 52]
nucleus <6> [42]
particle-bound <1-3, 5> (<1> tissue transglutaminase of rat hepatocytes [12]; <3> on the surface of monocytes and tissue macrophages [12]) [12, 26, 27]
soluble <1-5, 7, 8, 10-12> (<4> depending on state of cell-proliferation, intracellular distribution [22]; <3> retinoic acid induced enzyme [33]) [1, 6, 13, 22, 26, 32, 33, 35, 36, 38, 40]
spheroplast <13> [39]
Additional information <1> (<1> soluble in cells and organs devoid of significant association with extensive filamentous structure or extracellular matrix, particulate in organs with extensive filamentous structure or extracellular matrix [26]) [26]

Purification

<1> (partial [26-28]; lung, matrix-bound enzyme solubilized, 3 isoforms: A, B and C [26]; selectively solubilized with glycerol [28]; Triton X-100, deoxycholate or n-octylglucoside only 20-30% effective [28]; tissue-type transglutaminase [29]; brain transglutaminases NI and NII, DEAE-Sephacel, CM-cellulose, Q-Sepharose, FPLC gel filtration [48]) [18, 19, 26-30, 48]
<2> (cytosolic tissue transglutaminase, DE52, heparin-agarose, casein-agarose [13]; epidermal transglutaminase, DEAE-cellulose, Sephadex G-75, CM-cellulose, gel filtration [15]; transglutaminase C [15]; plasma factor XIII, $BaCl_2$, glycine, heat treatment, gel filtration, platelet factor XIII, DEAE-cellulose, hydroxyapatite [15]; partial [18,29]; 2 forms of epidermal enzyme [21]; erythrocyte transglutaminase [14]; erythrocyte transglutaminase, DE-52 gel, Sephacryl S-300, Blue Sepharose, DEAE bio-gel [6]; erythrocyte transglutaminase, DEAE cellulose, AcA 44, heparin-Sepharose, gel filtration [7]; recombinant factor XIII-tissue transglutaminase chimeras [44]; recombinant transglutaminase 1, DEAE-Sephacel, heparin-Sepharose, gel filtration [52]) [6, 7, 13-15, 18, 21, 29, 40, 44, 52]
<3> (CM-cellulose, Sephadex G-150 [20]; DEAE-cellulose, CM-cellulose, PO4-cellulose, Sephacryl S-200, Mono-S, cationic dermal isozymes A and B [34]) [20, 34]
<4> (partial [22]) [22]
<5> (one-step purification by monoclonal antibody immunoadsorbent [31]; affinity chromatography on phenylalanine-Sepharose [4]; affinity chromatography on GTP-agarose [5]; liver transglutaminase, DEAE-cellulose, protamine, gel filtration [15]; recombinant enzyme, immunoaffinity chromatography, Mono-Q [37]; follicle transglutaminase, 6% agarose, DEAE-cellulose, CM-cellulose, 10% agarose [15]) [1, 4, 5, 15, 31, 37, 40]
<6> (DEAE-Sephacel, Mono-Q, Ultrogel AcA34 [42]; lens transglutaminase, DE-52, affinity chromatography on 42000 Da thermolytic fragment of human plasma fibronectin [49]) [42, 49]
<9> (co-purified with α-actinin [8]) [8]

<10> (CM-Sepharose CL-6B, ammonium sulfate, DEAE-cellulose, Sephacryl S-300, DEAE-cosmogel, Zn-chelating Sepharose [35]) [35]
<12> (amberlite CG-50, Blue Sepharose [38]) [38]
<13> (15% polyethylene glycol, DEAE-cellulos, isoelectric focusing [39]; streptomycin sulfate, DEAE-cellulose, phenyl-Sepharose [47]) [39, 47, 61]
<14> (ammonium sulfate, DEAE-Sepharose, Blue-Sepharose CL-6B, ω-aminohexyl agarose, α-casein agarose [43]) [43]
<15> (DEAE-unabsorbed transglutaminase, CM-Sepharose, heparin-Sepharose, Superdex 200, DEAE-absorbed transglutaminase, DEAE-Sephacel, heparin-Sepharose, hydroxyapatite, Superdex 200 [45]) [45]
<16> (strong-acid ion-exchange column [46]) [46, 53]
<18> (Sephacryl S-300, hydroxyapatite [55]) [55]
<19> (tissue-type transglutaminase, DE-52 cellulose, Sephacryl S-300, Mono Q [56]) [56]

Crystallization

<2> (hanging drop vapour diffusion, 15 mg/ml transglutaminase 3, 0.1 mM β-octylglucoside in 0.1 ml of enzyme in 20 mM Tris-HCl, pH 8.0, 1 mM EDTA, 125 mM NaCl and 3 mM $CaCl_2$, well solution containing 4-12% polyethylene glycol 6K, 100 mM bicine, pH 9.0, and 1% dioxane at 21°C, X-ray structure of zymogen and activated transglutaminase 3 at 2.2 and 2.1 A resolution [62]) [62]

Cloning

<2> (cloning of placenta FXIIIa-gene cDNA [2,12]; factor XIII, transglutaminase C and transglutaminase K [12]; cloning of factor XIII β-subunit cDNA [17]; cloning of factor XIII-tissue transglutaminase chimeras in Escherichia coli [44]; expression of transglutaminase 1 in Sf9 insect cells [52]; cloning and expression of full-length and exon 3,5 and 11 deleted versions of transglutaminase 5 in Sf9 cells [54]; expression of transglutaminase 3 in Sf9 cells [62]) [2, 12, 17, 44, 52, 54, 62]
<3> (transglutaminase C from macrophage [12]) [12]
<5> (liver transglutaminase C [12,17]; expression in Escherichia coli [37]; overexpression of liver transglutaminase in Escherichia coli, coexpression of Dank, DnaJ, GrpE or trigger factor increases solubility [59]) [12, 17, 37, 59]
<13> (expression in Escherichia coli [61]) [61]

Application

synthesis <12> (<12> selective introduction of aminated compounds into proteins [63]; <12> soy flour as a source of transglutaminase substrates to prepare hydrocolloid films together with pectin [64]) [63, 64]

6 Stability

pH-Stability

5-9 <12> (<12> 10 min stable, 37°C [38]) [38]
6-9 <1> (<1> stable at 25°C [19]) [19]

7.5-9 <18> (<18> in the absence of Ca^{2+} [55]) [55]
9 <2, 3> (<2> $t_{1/2}$: 1-2 min stable at 37°C, at least 20 min stable at 4°C [13]; <3> thermolabile: $t_{1/2}$: 3-4 min, 37°C, retinoic acid induced enzyme, 20% loss of activity within 20 min, epidermal enzyme [33]) [13, 33]

Temperature stability

4 <2> (<2> at least 20 min stable, pH 9.0 [13]) [13]
25 <1, 3> (<3> at least 30 min stable, pH 9.0, retinoic acid induced enzyme [33]; <1> stable at pH 6.0-9.0 [19]) [19, 33]
37 <2, 3, 12> (<2> $t_{1/2}$: 1-2 min, pH 9.0 [13]; <3> retinoic acid induced transgutaminase, loss of approx. 75% activity after 10 min at pH 9, epidermal transglutaminase, 25% loss of activity after 20 min, pH 9.0 [33]; <12> 10 min stable, pH 5-9 [38]) [13, 33, 38]
40 <12> (<12> 10 min stable, pH 7 [38]) [38]
44 <1> (<1> inactivation within 20 min [19]) [19]
50 <12, 18> (<12> 26% loss of activity within 10 min, pH 7.0 [38]) [38, 55]
52 <1> (<1> inactivation within 4 min [19]) [19]
56 <2, 3> (<2> heating in the presence of Ca^{2+} increases activity 25-fold, human epidermal enzyme [15]; <3> epidermal transglutaminase, stable for 45 min in the presence of Ca^{2+} [20]) [15, 20]
60 <1> (<1> inactivation within 1 min [19]) [19]

General stability information

<1>, Ca^{2+} stabilizes in combination with dithiothreitol [28]
<1>, EDTA stabilizes in combination with dithiothreitol [28]
<1>, brain transglutaminases NI and NI, very unstable [48]
<1>, dialysis against Ca^{2+}-free buffers inactivates [28]
<1>, glycerol, 50% v/v, solubilizes and stabilizes [28]
<1>, ion-exchange chromatography on DEAE-cellulose, with 0.4 M NaCl containing buffer with or without Ca^{2+}, stable to [28]
<2>, gel electrophoresis inactivates [6]
<3>, chromatography on DEAE-cellulose, anionic isozyme, rapid decrease of activity [34]
<3>, lyophilization, cationic isozyme, stable to [34]
<3>, repeated freeze-thawing cycles, cationic isozyme, stable to [34]
<5>, EDTA, 1-2 mM, stabilizes [1]
<5>, alkaline conditions destabilize [31]
<10>, loss of activity after freezing [35]
<1, 5>, DTT stabilizes [28, 37]
<2, 5>, repeated freeze-thawing results in some loss of activity, dithothreitol restores activity [15]

Storage stability

<1>, 4°C, membrane preparation, rapid inactivation, DTT protects in combination with Ca^{2+} or EDTA [28]
<1>, brain transglutaminase NI, 4°C, 2 weeks, complete loss of activity [48]
<2>, -30°C, human zymogen factor XIII, over 1 year [15]
<2>, -30°C, lyophilized human zymogen factor XIII, several years [15]

<2>, 4°C, 50 mM Tris, pH 7.5, 1 mM EDTA/0.5 mM DTT, up to 2 weeks, no loss of activity [13]
<2>, 4°C, at least 4 months [6]
<2>, 4°C, epidermal transglutaminase, at least 1 month [15]
<2>, dilute enzyme solutions are unstable to storage at 4°C [13]
<2>, human factor XIIIa is more stable to storage, when Ca^{2+} is omitted [15]
<3>, 20°C, pH 6-8, cationic isozyme, long periods [34]
<5>, -20°C, concentrated guinea pig enzyme solutions, several months [15]
<5>, -70°C, concentrated enzyme solution, at least 3 months, no loss of activity [31]
<5>, 4°C, 5 mM Tris-HCl/2 mM EDTA buffer, pH 7.5, up to 3 months [1]
<5>, follicle transglutaminase, -20°C, 3 months, no loss of activity [15]
<10>, 4°C, at least 3 months [35]
<13>, -20°C, partially purified, 3 years [39]
<13>, -70°C, 6 months, little loss in activity [47]

References

[1] Folk, J.E.; Cole, P.W.: Mechanism of action of guinea pig liver transglutaminase. I. Purification and properties of the enzyme: identification of a functional cysteine essential for activity. J. Biol. Chem., **241**, 5518-5525 (1966)

[2] Takahashi, N.; Takahashi, Y.; Putnam, F.W.: Primary structure of blood coagulation factor XIIIa (fibrinoligase, transglutaminase) from human placenta. Proc. Natl. Acad. Sci. USA, **83**, 8019-8023 (1986)

[3] Delcros, J.G.; Roch, A.M.; Quash, G.: The competitive inhibition of tissue transglutaminase by α-difluoromethylornithine. FEBS Lett., **171**, 221-226 (1984)

[4] Brookhart, P.P.; McMahon, P.L.; Takahashi, M.: Purification of guinea pig liver transglutaminase using a phenylalanine-sepharose 4B affinity column. Anal. Biochem., **128**, 202-205 (1983)

[5] Lee, K.N.; Birckbichler, P.J.; Patterson, M.K.: GTP hydrolysis by guinea pig liver transglutaminase. Biochem. Biophys. Res. Commun., **162**, 1370-1375 (1989)

[6] Ando, Y.; Imamura, S.; Yamagata, Y.; Kikuchi, T.; Murachi, T.; Kannagi, R.: High-performance liquid chromatographic assay of transglutaminase and its application to the purification of human erythrocyte transglutaminase and platelet factor XIII. J. Biochem., **101**, 1331-1337 (1987)

[7] Signorini, M.; Bortolotti, F.; Poltronieri, L.; Bergamini, C.M.: Human erythrocyte transglutaminase: purification and preliminary characterisation. Biol. Chem. Hoppe-Seyler, **369**, 275-281 (1988)

[8] Puzkin, E.G.; Raghuraman, V.: Catalytic properties of a calmodulin-regulated transglutaminase from human platelet and chicken gizzard. J. Biol. Chem., **260**, 16012-16020 (1985)

[9] Parameswaran, K.N.; Velasco, P.T.; Wilson, J.; Lorand, L.: Labeling of ε-lysine crosslinking sites in proteins with peptide substrates of factor XIIIa and transglutaminase. Proc. Natl. Acad. Sci. USA, **87**, 8472-8475 (1990)

[10] Jensen, P.H.; Lorand, L.; Ebbesen, P.; Gliemann, J.: Type-2 plasminogen-activator inhibitor is a substrate for trophoblast transglutaminase and factor XIIIa. Transglutaminase-catalyzed cross-linking to cellular and extracellular structures. Eur. J. Biochem., **214**, 141-146 (1993)
[11] Achyuthan, K.E.; Mary, A.; Greenberg, C.S.: Tb(III)-ion-binding-induced conformational changes in platelet factor XIII. Biochem. J., **257**, 331-338 (1989)
[12] Greenberg, C.S.; Birckbichler, P.J.; Rice, R.H.: Transglutaminases: multifunctional cross-linking enzymes that stabilize tissues. FASEB J., **5**, 3071-3077 (1991)
[13] Dadabay, C.Y.; Pike, L.J.: Purification and characterization of a cytosolic transglutaminase from a cultured human tumour-cell line. Biochem. J., **264**, 679-685 (1989)
[14] Bergamini, C.M.; Signorini, M.; Poltronieri, L.: Inhibition of erythrocyte transglutaminase by GTP. Biochim. Biophys. Acta, **916**, 149-151 (1987)
[15] Folk, J.E.; Chung, S.I.: Transglutaminases. Methods Enzymol., **113**, 358-375 (1985)
[16] Folk, J.E.: The trimethylacetyl-transglutaminase complex. Methods Enzymol., **87**, 36-42 (1982)
[17] Ichinose, A.; Bottenus, R.E.; Davie, E.W.: Structure of transglutaminases. J. Biol. Chem., **265**, 13411-13414 (1990)
[18] Kawashima, S.: Inhibition of rat liver transglutaminase by nucleotides. Experientia, **47**, 709-712 (1991)
[19] Wong, W.S.D.; Batt, C.; Kinsella, J.E.: Purification and characterization of rat liver transglutaminase. Int. J. Biochem., **22**, 53-59 (1990)
[20] Nakayama, J.; Osaki, M.; Nagae, S.; Asahi, M.; Urabe, H.: Properties of partially purified mouse epidermal transglutaminase. J. Dermatol., **13**, 448-455 (1986)
[21] Negi, M.; Colbert, M.C.; Goldsmith, L.A.: High-molecular-weight human epidermal transglutaminase. J. Invest. Dermatol., **85**, 75-78 (1985)
[22] Korner, G.; Schneider, D.E.; Purdon, M.A.; Bjornsson, T.D.: Bovine aortic endothelial cell transglutaminase. Enzyme characterization and regulation of activity. Biochem. J., **262**, 633-641 (1989)
[23] Coussons, P.J.; Price, N.C.; Kelly, S.M.; Fothergill-Gilmore, L.A.: The modification of bovine β-caesin using transglutaminae purified from guinea pig liver. Biochem. Soc. Trans., **20**, 48S (1991)
[24] Coussons, P.J.; Price, N.C.; Kelly, S.M.; Smith, B.; Sawyer, L.: Transglutaminase catalyses the modification of glutamine side chains in the C-terminal region of bovine β-lactoglobulin. Biochem. J., **283**, 803-806 (1992)
[25] Berbers, G.A.M.; Bentlage, H.C.M.; Brans, A.M.M.; Bloemendal, H.; de Jong, W.W.: β-Crystallin: endogenous substrate of lens transglutaminase. Characterization of the acyl-donor site in the β Bp chain. Eur. J. Biochem., **135**, 315-320 (1983)
[26] Cocuzzi, E.T.; Chung, S.I.: Cellular transglutaminase. Lung matrix-associated transglutaminase: characterization and activation with sulfhydryls. J. Biol. Chem., **261**, 8122-8127 (1986)

[27] Chang, S.K.; Chung, S.I.: Cellular transglutaminase. The particulate-associated transglutaminase from chondrosarcoma and liver: partial purification and characterization. J. Biol. Chem., **261**, 8112-8121 (1986)
[28] Slife, C.W.; Morris, G.S.; Snedeker, S.W.: Solubilization and properties of the liver plasma membrane transglutaminase. Arch. Biochem. Biophys., **257**, 39-47 (1987)
[29] Kawashima, S.: Inhibition of rat liver transglutaminase by nucleotides. Experientia, **47**, 709-712 (1991)
[30] Seitz, J.; Keppler, C.; Huntemann, S.; Rausch, U.; Aumuller, G.: Purification and molecular characterization of a secretory transglutaminase from coagulating gland of the rat. Biochim. Biophys. Acta, **1078**, 139-146 (1991)
[31] Ikura, K.; Sakurai, H.; Okumura, K.; Sasaki, R.; Chiba, H.: One-step purification of guinea pig liver transglutaminase using a monoclonal-antibody immunoadsorbent. Agric. Biol. Chem., **49**, 3527-3531 (1985)
[32] Porta, R.; De Santis, A.; Esposito, C.; Draetta, G.F.; Di Donato, A.; Illiano, G.: Inhibition of adenylate cyclase by transglutaminase-catalyzed reactions in pigeon erythrocyte ghosts. Biochem. Biophys. Res. Commun., **138**, 596-603 (1986)
[33] Lichti, U.; Ben, T.; Yuspan, S.H.: Retinoic acid-induced transglutaminase in mouse epidermal cells is distinct from epidermal transglutaminase. J. Biol. Chem., **260**, 1422-1426 (1985)
[34] Martinet, N.; Kim, H.C.; Girard, J.E.; Nigra, D.H.; Strong, D.H.; Chung, S.I.; Folk, J.E.: Epidermal and hair follicle transglutaminases. Partial characterization of soluble enzymes in newborn mouse skin. J. Biol. Chem., **263**, 4236-4241 (1988)
[35] Tokunaga, F.; Yamada, M.; Miyata, T.; Ding, Y.L.; Hiranaga-Kawabata, M.; Muta, T.; Iwanaga, S.: Limulus hemocyte transglutaminase. Its purification and characterization, and identification of the intracellular substrates. J. Biol. Chem., **268**, 252-261 (1993)
[36] Icekson, I.; Apelbaum, A.: Evidence for transglutaminase activity in plant tissue. Plant Physiol., **84**, 972-974 (1987)
[37] Ikura, K.; Tsuchiya, Y.; Sasaki, R.; Chiba, H.: Expression of guinea-pig liver transglutaminase cDNA in Escherichia coli. Amino-terminal N α-acetyl group is not essential for catalytic function of transglutaminase. Eur. J. Biochem., **187**, 705-711 (1990)
[38] Ando, H.; Adachi, M.; Umeda, K.; Matsuura, A.; Nonaka, M.; Uchio, K.; Tanaka, H.; Motoki, M.: Purification and characteristics of a novel transglutaminase derived from microorganisms. Agric. Biol. Chem., **53**, 2613-2617 (1989)
[39] Klein, J.D.; Guzman, E.; Kuehn, G.D.: Purification and partial characterization of transglutaminase from Physarum polycephalum. J. Bacteriol., **174**, 2599-2605 (1992)
[40] Folk, J.E.; Chung, S.I.: Molecular and catalytic properties of transglutaminases. Adv. Enzymol. Relat. Areas Mol. Biol., **38**, 109-191 (1973)
[41] McCormack, S.A.; Wang, J.Y.; Viar, M.J.; Tague, L.; Davies, P.J.A.; Johnson, L.R.: Polyamines influence transglutaminase activity and cell migration in two cell lines. Am. J. Physiol., **267**, C706-C714 (1994)

[42] Singh, U.S.; Erickson, J.W.; Cerione, R.A.: Identification and biochemical characterization of an 80 kilodalton GTP-binding/transglutaminase from rabbit liver nuclei. Biochemistry, **34**, 15863-15871 (1995)
[43] Kang, H.; Cho, Y.D.: Purification and properties of transglutaminase from soybean (Glycine max) leaves. Biochem. Biophys. Res. Commun., **223**, 288-292 (1996)
[44] Hettasch, J.M.; Peoples, K.A.; Greenberg, C.S.: Analysis of factor XIII substrate specificity using recombinant human factor XIII and tissue transglutaminase chimeras. J. Biol. Chem., **272**, 25149-25156 (1997)
[45] Kumazawa, Y.; Ohtsuka, T.; Ninomiya, D.; Seguro, K.: Purification and calcium dependence of transglutaminases from sheep hair follicles. Biosci. Biotechnol. Biochem., **61**, 1086-1090 (1997)
[46] Pasternack, R.; Laurent, H.P.; Ruth, T.; Kaiser, A.; Schon, N.; Fuchsbauer, H.L.: A fluorescent substrate of transglutaminase for detection and characterization of glutamine acceptor compounds. Anal. Biochem., **249**, 54-60 (1997)
[47] Mottahedeh, J.; Marsh, R.: Characterization of 101-kDa transglutaminase from Physarum polycephalum and identification of LAV1-2 as substrate. J. Biol. Chem., **273**, 29888-29895 (1998)
[48] Kwak, S.J.; Kim, S.Y.; Kim, Y.S.; Song, K.Y.; Kim, I.G.; Park, S.C.: Isolation and characterization of brain-specific transglutaminases from rat. Exp. Mol. Med., **30**, 177-185 (1998)
[49] Murthy, S.N.P.; Velasco, P.T.; Lorand, L.: Properties of purified lens transglutaminase and regulation of its transamidase/crosslinking activity by GTP. Exp. Eye Res., **67**, 273-281 (1998)
[50] Madi, A.; Punyiczki, M.; di Rao, M.; Piacentini, M.; Fesus, L.: Biochemical characterization and localization of transglutaminase in wild-type and cell-death mutants of the nematode Caenorhabditis elegans. Eur. J. Biochem., **253**, 583-590 (1998)
[51] Zhang, J.; Guttmann, R.P.; Johnson, G.V.W.: Tissue transglutaminase is an in situ substrate of calpain: regulation of activity. J. Neurochem., **71**, 240-247 (1998)
[52] Hitomi, K.; Yamagiwa, Y.; Ikura, K.; Yamanishi, K.; Maki, M.: Characterization of human recombinant transglutaminase 1 purified from baculovirus-infected insect cells. Biosci. Biotechnol. Biochem., **64**, 2128-2137 (2000)
[53] Ohtsuka, T.; Sawa, A.; Kawabata, R.; Nio, N.; Motoki, M.: Substrate specificities of microbial transglutaminase for primary amines. J. Agric. Food Chem., **48**, 6230-6233 (2000)
[54] Sato, H.; Yamada, N.; Shimba, N.; Takahara, Y.: Unique substrate specificities of two adjacent glutamine residues in EAQQIVM for transglutaminase: identification and characterization of the reaction products by electrospray ionization tandem mass spectrometry. Anal. Biochem., **281**, 68-76 (2000)
[55] Nozawa, H.; Cho, S.Y.; Seki, N.: Purification and characterization of transglutaminase from squid gill. Fish. Sci., **67**, 912-919 (2001)
[56] Nozawa, H.; Seki, N.: Purification of transglutaminase from scallop striated adductor muscle and NaCl-induced inactivation. Fish. Sci., **67**, 493-499 (2001)

[57] Candi, E.; Oddi, S.; Terrinoni, A.; Paradisi, A.; Ranalli, M.; Finazzi-Agro, A.; Melino, G.: Transglutaminase 5 cross-links loricrin, involucrin, and small proline-rich proteins in vitro. J. Biol. Chem., **276**, 35014-35023 (2001)
[58] Ruse, M.; Lambert, A.; Robinson, N.; Ryan, D.; Shon, K.J.; Eckert, R.L.: S100A7, S100A10, and S100A11 are transglutaminase substrates. Biochemistry, **40**, 3167-3173 (2001)
[59] Ikura, K.; Kokubu, T.; Natsuka, S.; Ichikawa, A.; Adachi, M.; Nishihara, K.; Yanagi, H.; Utsumi, S.: Co-overexpression of folding modulators improves the solubility of the recombinant guinea pig liver transglutaminase expressed in Escherichia coli. Prep. Biochem. Biotechnol., **32**, 189-205 (2002)
[60] Griffin, M.; Casadio, R.; Bergamini, C.M.: Transglutaminases: Nature's biological glues. Biochem. J., **368**, 377-396 (2002)
[61] Wada, F.; Nakamura, A.; Masutani, T.; Ikura, K.; Maki, M.; Hitomi, K.: Identification of mammalian-type transglutaminase in Physarum polycephalum. Evidence from the cDNA sequence and involvement of GTP in the regulation of transamidating activity. Eur. J. Biochem., **269**, 3451-3460 (2002)
[62] Ahvazi, B.; Kim, H.C.; Kee, S.H.; Nemes, Z.; Steinert, P.M.: Three-dimensional structure of the human transglutaminase 3 enzyme: binding of calcium ions changes structure for activation. EMBO J., **21**, 2055-2067 (2002)
[63] Villalonga, R.; Fernandez, M.; Fragoso, A.; Cao, R.; Di Pierro, P.; Mariniello, L.; Porta, R.: Transglutaminase-catalyzed synthesis of trypsin-cyclodextrin conjugates: Kinetics and stability properties. Biotechnol. Bioeng., **81**, 732-737 (2003)
[64] Mariniello, L.; DiPierro, P.; Esposito, C.; Sorrentino, a.; Masi, P.; Porta, R.: Preparation and mechanical properties of edible pectin-soy flour films obtained in the absence or presence of transglutaminase. J. Biotechnol., **102**, 191-198 (2003)

D-Alanine γ-glutamyltransferase 2.3.2.14

1 Nomenclature

EC number
2.3.2.14

Systematic name
L-glutamine:D-alanine γ-glutamyltransferase

Recommended name
D-alanine γ-glutamyltransferase

Synonyms
Additional information (cf. EC 2.3.2.2)

CAS registry number
9046-27-9

2 Source Organism

<1> *Pisum sativum* (cv. Alaska [1]) [1]

3 Reaction and Specificity

Catalyzed reaction
L-glutamine + D-alanine = NH_3 + γ-L-glutamyl-D-alanine

Reaction type
aminoacyl group transfer

Substrates and products

S L-glutamine + D-2-amino-n-butanoic acid <1> (Reversibility: ? <1> [1]) [1]
P ?
S L-glutamine + D-alanine <1> (<1> best substrate [1]) (Reversibility: ? <1> [1]) [1]
P NH_3 + γ-L-glutamyl-D-alanine <1> [1]
S L-glutamine + D-phenylalanine <1> (Reversibility: ? <1> [1]) [1]
P NH_3 + γ-L-glutamyl-L-phenylalanine
S γ-L-glutamylethylester + D-alanine <1> (<1> equally effective as L-glutamine [1]) (Reversibility: ? <1> [1]) [1]
P ?

S reduced glutathione + D-alanine <1> (<1> 50% as effective as L-glutamine [1]) (Reversibility: ? <1> [1]) [1]

P ?

S Additional information <1> (<1> no substrates are D-valine, D-leucine, D-aspartic acid, D-glutamic acid, D-serine, D-proline [1]) [1]

P ?

Inhibitors

6-diazo-5-oxo-L-norleucine <1> [1]
L-serine borate <1> [1]
citrate <1> (<1> to some extent [1]) [1]
Additional information <1> (<1> no inhibition by Ca^{2+}, Mg^{2+}, NH_4^+, and maleate [1]) [1]

Specific activity (U/mg)

0.245 <1> [1]

K_m-Value (mM)

2 <1> (L-glutamine) [1]
2.9 <1> (D-alanine) [1]

pH-Optimum

9.5 <1> [1]

Temperature optimum (°C)

37 <1> (<1> assay at [1]) [1]

5 Isolation/Preparation/Mutation/Application

Source/tissue

seedling <1> (<1> decotyledonized [1]) [1]

Purification

<1> (partial [1]) [1]

References

[1] Kawasaki, Y.; Ogawa, T.; Sasaoka, K.: Occurence and some properties of a novel γ-glutamyltransferase responsible for the synthesis of γ-L-glutamyl-D-alanine in pea seedlings. Biochim. Biophys. Acta, **716**, 194-200 (1982)

Glutathione γ-glutamylcysteinyltransferase 2.3.2.15

1 Nomenclature

EC number
2.3.2.15

Systematic name
glutathione:poly(4-glutamyl-cysteinyl)glycine 4-glutamylcysteinyltransferase

Recommended name
glutathione γ-glutamylcysteinyltransferase

Synonyms
PCS <1> [1]
γ-glutamylcysteine dipeptidyl transpeptidase
phytochelatin synthase

CAS registry number
125390-02-5

2 Source Organism

<1> *Arabidopsis thaliana* (strain Heyn. [1]; ecotype Columbia [3,4,6,7,10,11]; has a second gene (AtPCS2) encoding another functional phytochelatin synthase [7]) [1, 3, 4, 6, 7, 10, 11]
<2> *Silene cucubalus* (strain Wib. [1]) [1, 13]
<3> *Dunaliella tertiolecta* (ATCC 30929 [2]) [2]
<4> *Nicotiana tabacum* (cv. Bright Yellow-2 [5]) [5]
<5> *Caenorhabditis elegans* (strain N2 [9]) [8, 9]
<6> *Lycopersicon esculentum* (tomato, strain Mill. cv. VFNT-Cherry [12]) [12]
<7> *Podophyllum peltatum* [13]
<8> *Eschscholtzia californica* [13]
<9> *Beta vulgaris* [13]
<10> *Equisetum giganteum* [13]

3 Reaction and Specificity

Catalyzed reaction
glutathione + $[Glu(-Cys)]_n$-Gly = Gly + $[Glu(-Cys)]_{n+1}$-Gly

Reaction type
γ-glutamylcysteinyl transfer

Natural substrates and products

S glutathione + [Glu(-Cys)]$_n$-Gly <1, 2, 5> (<1> cellular functions are formation of heavy-metal binding peptides and degradation of glutathione-S-conjugates [1]; <1> n = 2-11 [1]; <7> n = 2-5 [13]; <2,5> critical for heavy metal tolerance [9,13]) (Reversibility: ? <1, 2, 5> [1, 9]) [1, 9, 13]

P Gly + [Glu(-Cys)]$_{n+1}$-Gly <1> [1]

Substrates and products

S S-butylglutathione + [Glu(-Cys)]$_n$-Gly <1> (Reversibility: ? <1> [10]) [10]

P Gly + [Glu(-Cys)]$_n$-Glu-S-butyl-Cys-Gly

S S-ethylglutathione + [Glu(-Cys)]$_n$-Gly <1> (Reversibility: ? <1> [10]) [10]

P Gly + [Glu(-Cys)]$_n$-Glu-S-ethyl-Cys-Gly

S S-hexylglutathione + [Glu(-Cys)]$_n$-Gly <1> (Reversibility: ? <1> [10]) [10]

P Gly + [Glu(-Cys)]$_n$-Glu-S-hexyl-Cys-Gly

S S-methylglutathione + [Glu(-Cys)]$_n$-Gly <1> (<1> n = 2,3 [10]) (Reversibility: ? <1> [10]) [10]

P Gly + [Glu(-Cys)]$_n$-Glu-S-methyl-Cys-Gly

S S-propylglutathione + [Glu(-Cys)]$_n$-Gly <1> (Reversibility: ? <1> [10]) [10]

P Gly + [Glu(-Cys)]$_n$-Glu-S-propyl-Cys-Gly

S acetamido-fluorescein-glutathione + [Glu(-Cys)]$_n$-Gly <1> (Reversibility: ? <1> [1]) [1]

P Gly + [Glu(-Cys)]$_n$-Glu-S-acetamido-fluorescein-Cys-Gly

S benzyl-glutathione + [Glu(-Cys)]$_n$-Gly <1> (Reversibility: ? <1> [1]) [1]

P Gly + [Glu(-Cys)]$_n$-Glu-S-benzyl-Cys-Gly

S bimane-glutathione + [Glu(-Cys)]$_n$-Gly <1, 2> (Reversibility: ? <1, 2> [1]) [1]

P Gly + [Glu(-Cys)]$_n$-Glu-S-bimane-Cys-Gly

S glutathione + [Glu(-Cys)]$_n$-Gly <1, 2, 3, 4, 5, 6, 7, 8, 9, 10> (Reversibility: ? <1, 2, 3, 4, 5, 6, 7, 8, 9, 10> [1, 2, 3, 4, 5, 7, 8, 10, 11, 12, 13]) [1, 2, 3, 4, 5, 7, 8, 10, 11, 12, 13]

P Gly + [Glu(-Cys)]$_{n+1}$-Gly <1, 2, 7> (<1> n = 2-11 [1]; <7> n = 2-5 [13]) [1, 13]

S monobromobimane-glutathione + [Glu(-Cys)]$_n$-Gly <2> (Reversibility: ? <2> [13]) [13]

P Gly + [Glu(-Cys)]$_n$-Glu-S-monobromobimane-Cys-Gly

S nitrobenzyl-glutathione + [Glu(-Cys)]$_n$-Gly <1> (Reversibility: ? <1> [1]) [1]

P Gly + [Glu(-Cys)]$_n$-Glu-S-nitrobenzyl-Cys-Gly

S uracil-glutathione + [Glu(-Cys)]$_n$-Gly <1> (Reversibility: ? <1> [1]) [1]

P Gly + [Glu(-Cys)]$_n$-Glu-S-uracil-Cys-Gly

Metals, ions

Ag^+ <1, 2, 4, 6> (<1,2,4,6> activation [4,5,12,13]) [4, 5, 12, 13]

As^{3+} <1> (<1> activation [10]) [10]

AsO_4^{3-} <1> (<1> activation, wild-type and AtPCS2-enzyme [7]) [7, 10]

Au^{+} <6> (<6> activation [12]) [12]
Bi^{3+} <2> (<2> activation [13]) [13]
Cd^{2+} <1, 2, 3, 4, 6> (<1, 2, 3, 4, 6> activation [1, 2, 3, 4, 5, 7, 10, 12, 13];<1> strongly dependent, optimal concentration 0.03 mM [1]; <3> best at 0.4 mM [2]; <1> under 0.085 mM $CdCl_2$ stress for 3 days, 1.3- to 2.1-fold increase when compared with wild-type [3]; <4> best at 0.05 mM [5]; <1> activation of AtPCS2-enzyme [7]; <6> best at 0.5 mM [12]) [1, 2, 3, 4, 5, 7, 10, 12, 13]
Co^{2+} <1> (<1> activation [10]) [10]
Cu^{2+} <1, 2, 4, 6> (<1,2,4,6> activation [1,4,5,10,12,13]; <1> optimal at 0.04 mM, can replace Cd^{2+} [1]) [1, 4, 5, 12, 13]
Fe^{2+} <6> (<6> activation [12]) [12]
Hg^{2+} <1, 2, 4, 6> (<1, 2, 4, 6> activation [4, 5, 12, 13]) [4, 5, 10, 12, 13]
Mg^{2+} <1> (<1> activation [10]) [10]
Ni^{2+} <1> (<1> activation [10]) [10]
Pb^{2+} <1, 2, 4, 6> (<1, 2, 4, 6> activation [4, 5, 12, 13]) [4, 5, 12, 13]
Zn^{2+} <1, 2, 3, 4, 6> (<1, 2, 4, 6> activation [1, 2, 4, 5, 10, 12, 13]; <1> optimal at 0.1 mM, can replace Cd^{2+} [1]; <3> best at 0.2 mM [2]) [1, 2, 4, 5, 10, 12, 13]
Additional information <1> (<1> free metal ions are not essential for catalysis [10]) [10]

Specific activity (U/mg)
0.0055 <6> [12]

K_m-Value (mM)
1.5 <2> (bimane-glutathione) [1]
1.5 <2> (monobromobimane-glutathione) [13]
6.7 <2> (glutathione) [13]
7.5 <2> (glutathione) [1]
7.7 <6> (glutathione) [12]
8.6 <4> (glutathione) [5]

pH-Optimum
7.9 <2> [13]
7.9-8.5 <1> [1]
8 <4, 6> (<6> half maximal activity at pH 6.2 and 9.2 [12]) [5, 12]

pH-Range
6-9 <1> [1]

Temperature optimum (°C)
35 <2, 4, 6> [5, 12, 13]

Temperature range (°C)
20-47 <2> (<2> half-maximal activity at 20°C and 47°C [13]) [13]

4 Enzyme Structure

Molecular weight

56300 <1> (<1> recombinant enzyme, SDS-PAGE [4]) [4]
95000 <2> (<2> gel filtration [13]) [13]

Subunits

homotetramer <2> (<2> 4 * 25000, SDS-PAGE, dimer also catalytically active [13]) [13]

5 Isolation/Preparation/Mutation/Application

Source/tissue

leaf <1> [4]
root <6> [12]
stem <6> [12]

Purification

<2> [1, 13]

Cloning

<1> (AtPCS2-gene expressed in Saccharomyces cerevisiae strain INVScl [7]) [7]
<1> (AtPCS2-gene expressed in Schizosaccharomyces pombe strain FY254, a phytochelatin synthase knockout strain [7]) [7]
<1> (heterologously expressed in Escherichia coli [1,4]; fused to a C-terminal Flag epitope [4,10]; transformed into Saccharomyces cerevisiae DTY167 [10,11]) [1, 4, 10, 11]
<1> (overexpressed in transgenic Arabidopsis thaliana [3,6]) [3, 6]
<5> (expressed in Schizosaccharomyces pombe strain SP27, a phytochelatin synthase knockout strain [8]) [8]
<5> (transformed into Saccharomyces cerevisiae strain DTY67, hypersensitive to Cd^{2+}-stress [9]) [9]

6 Stability

pH-Stability

7.5 <4> (<4> most stable [5]) [5]

Temperature stability

-20 <2> (<2> 30% glycerol, 500 h, small activity decrease [13]) [13]
4 <2, 4> (<4> 50% activity after 6 h [5]; <2> 50% activity after 140 h [13]) [5, 13]
22 <2> (<2> 50% activity after 34 h [13]) [5, 13]
35 <2, 4> (<4> 50% activity after 0.5 h [5]; <2> 50% activity after 0.5 h [13]) [5, 13]

General stability information

<4>, at 30% glycerol,m/v, stability enhancement [5]

Storage stability

<1>, -20°C, 10 mM Tris-HCl, pH 8.0, 1 mM mercaptoethanol, 20% glycerol and 1% bovine serum albumin [1]

<2>, -20°C, 10 mM Tris-HCl, pH 8.0, 10 mM 2-mercaptoethanol, 0.1% NaN_3 [13]

References

[1] Beck, A.; Lendzian, K.; Oven, M.; Christmann, A.; Grill, E.: Phytochelatin synthase catalyzes key step in turnover of glutathione conjugates. Phytochemistry, **62**, 423-431 (2003)
[2] Tsuji, N.; Hirayanagi, N.; Iwabe, O.; Namba, T.; Tagawa, M.; Miyamoto, S.; Miyasaka, H.; Takagi, M.; Hirata, K.; Miyamoto, K.: Regulation of phytochelatin synthesis by zinc and cadmium in marine green alga, Dunaliella tertiolecta. Phytochemistry, **62**, 453-459 (2003)
[3] Lee, S.; Moon, J.S.; Ko, T.S.; Petros, D.; Goldsbrough, P.B.; Korban, S.S.: Overexpression of Arabidopsis phytochelatin synthase paradoxically leads to hypersensitivity to cadmium stress. Plant Physiol., **131**, 656-663 (2003)
[4] Sauge-Merle, S.; Cuine, S.; Carrier, P.; Lecomte-Pradines, C.; Luu, D.T.; Peltier, G.: Enhanced toxic metal accumulation in engineered bacterial cells expressing Arabidopsis thaliana phytochelatin synthase. Appl. Environ. Microbiol., **69**, 490-494 (2003)
[5] Nakazawa, R.; Kato, H.; Kameda, Y.; Takenaga, H.: Optimum assay conditions of the activity of phytochelatin synthase from tobacco cells. Biol. Plant., **45**, 311-313 (2002)
[6] Lee, S.; Moon, J.S.; Domier, L.L.; Korban, S.S.: Molecular characterization of phytochelatin synthase expression in transgenic Arabidopsis. Plant Physiol. Biochem., **40**, 727-733 (2002)
[7] Cazale, A.C.; Clemens, S.: Arabidopsis thaliana expresses a second functional phytochelatin synthase. FEBS Lett., **507**, 215-219 (2001)
[8] Clemens, S.; Schroeder, J.I.; Degenkolb, T.: Caenorhabditis elegans expresses a functional phytochelatin synthase. Eur. J. Biochem., **268**, 3640-3643 (2001)
[9] Vatamaniuk, O.K.; Bucher, E.A.; Ward, J.T.; Rea, P.A.: A new pathway for heavy metal detoxification in animals. Phytochelatin synthase is required for cadmium tolerance in Caenorhabditis elegans. J. Biol. Chem., **276**, 20817-20820 (2001)
[10] Vatamaniuk, O.K.; Mari, S.; Lu, Y.P.; Rea, P.A.: Mechanism of heavy metal ion activation of phytochelatin (PC) synthase: blocked thiols are sufficient for PC synthase-catalyzed transpeptidation of glutathione and related thiol peptides. J. Biol. Chem., **275**, 31451-31459 (2000)

[11] Vatamaniuk, O.K.; Mari, S.; Lu, Y.P.; Rea, P.A.: AtPCS1, a phytochelatin synthase from Arabidopsis: isolation and in vitro reconstitution. Proc. Natl. Acad. Sci. USA, **96**, 7110-7115 (1999)
[12] Chen, J.; Zhou, J.; Goldsbrough, P.B.: Characterization of phytochelatin synthase from tomato. Physiol. Plant., **101**, 165-172 (1997)
[13] Grill, E.; Löffler, S.; Winnacker, E.L.; Zenk, M. H.: Phytochelatins, the heavy-metal-binding peptides of plants, are synthesized from glutathione by a specific γ-glutamylcysteine dipeptidyl transpeptidase (phytochelatin synthase). Proc. Natl. Acad. Sci. USA, **86**, 6838-6842 (1989)

Citrate (Si)-synthase 2.3.3.1

1 Nomenclature

EC number
2.3.3.1

Systematic name
acetyl-CoA:oxaloacetate C-acetyltransferase [thioester-hydrolysing, (pro-S)-carboxymethyl forming]

Recommended name
citrate (Si)-synthase

Synonyms
(R)-citric synthase
CS <2, 9, 19, 43, 46, 53-57, 61> [1, 38, 43, 48, 53-56, 63]
EC 4.1.3.7 (formerly)
citrate condensing enzyme
citrate oxaloacetate-lyase ((pro-3S)-CH_2COO → acetyl-CoA)
citrate oxaloacetate-lyase (CoA-acetylating)
citrate oxaloacetate-lyase, CoA-acetylating
citrate synthase
citrate synthetase
citric synthase
citric-condensing enzyme
citrogenase
condensing enzyme
oxalacetic transacetase
oxaloacetate transacetase
synthase, citrate
Additional information <40> (<40> enzyme participates in multienzyme complexes of enzymes belonging to tricarboxylic acid cycle, probably substrate channeling, in vivo NMR measurements [45]) [45]

CAS registry number
9027-96-7

2 Source Organism

<1> *Penicillium spiculisporum* [2]
<2> *Azotobacter vinelandii* (strain OP [3]) [1, 3]
<3> *Acinetobacter anitratus* [3, 42]
<4> *Bacillus subtilis* (strain S2A1 [17]; strain HS2A2 [3]) [3, 17]

<5> *Halobacterium cutirubrum* [6]
<6> *Aspergillus niger* [4]
<7> *Xanthochymus quttiferae* [4]
<8> *Mycobacterium tuberculosis* [4]
<9> *Sus scrofa* (crystalline [55]) [4, 5, 29, 31, 55]
<10> *Dictyostelium discoideum* (strain NC-4, ATCC 24697 [7]) [7]
<11> *Pseudomonas sp.* (marine, 2 enzyme forms with different molecular weight, kinetic and regulatory properties, dissociated CS I /native CS II [8]) [8]
<12> *Thermus aquaticus* (recombinant enzyme [61]) [9, 61]
<13> *Anthocidaris crassispina* (sea urchin [10]) [10]
<14> *Agave americana* [11]
<15> *Pisum sativum* (cv. Progress 9 [26]) [12, 26]
<16> *Escherichia coli* (isozyme CS II [58]; wild-type strains and citrate synthase deficient strain [29]) [13, 29, 33, 42, 58]
<17> *Bacillus megaterium* (strain D101 [14]) [14]
<18> *Halobacterium halobium* [15]
<19> *Thermoplasma acidophilum* (strain DSM 1728 [22]; strain 122-1B3, ATCC 27658 [16]) [15, 16, 22, 40, 48, 49, 55]
<20> *Sulfolobus acidocaldarius* (strain 98-3, DSM 639 [16,22]) [15, 16, 22]
<21> *Methylophilus methylotrophus* [17]
<22> *Hyphomicrobium sp.* (strains X and Hanham [17]) [17]
<23> *Arthrobacter sp.* (strain DS2-3R [57]; strain 2B2 [17]) [17, 57]
<24> *Lycopersicon esculentum* [19]
<25> *Citrullus vulgaris* (water melon, mitochondrial and glyoxysomal isoenzyme [20]) [20]
<26> *Sulfolobus solfataricus* [21]
<27> *Halobacterium vallismortis* [15]
<28> *Halobacterium volcanii* [15]
<29> *Natronobacterium pharaonis* [15]
<30> *Natronobacterium gregoryi* [15]
<31> *Natronococcus occultus* [15]
<32> *Methanosarcina barkeri* [15]
<33> *Pyrococcus furiosus* [18, 30, 49, 57]
<34> *Ralstonia eutropha* (basonym Alcaligenes eutrophus [23]) [23]
<35> *Acetobacter europaeus* [24]
<36> *Nitrobacter agilis* (strain ATCC 14123 [25]) [25]
<37> *Solanum tuberosum* (mitochondrial isozyme [27]; cv. Desiree [27]) [27]
<38> *Yarrowia lipolytica* [28]
<39> *Drosophila melanogaster* [32]
<40> *Saccharomyces cerevisiae* (peroxisomal isozyme CS II and mitochondrial isozyme CS I [41,59]; 3 isozymes [45]; mitochondrial isozyme [34]; wild-type strains and citrate synthase deficient strain [34,59]) [34, 41, 45, 59]
<41> *Pseudomonas aeruginosa* (2 enzyme isoforms with different molecular weight, kinetic and regulatory properties, CS I and CS II [35]) [35]
<42> *Tetrahymena pyriformis* (strain W [36]; identical with 14-nm filament protein, i.e. 49K protein [36,37]) [36, 37]

<43> *Bos taurus* [38, 52]
<44> *Haloferax volcanii* [39]
<45> *Nitrosomonas sp.* (strain TK794 [44]) [44]
<46> *Methylobacterium extorquens* [43]
<47> *Cavia porcellus* [4]
<48> *Rattus norvegicus* (Wistar-Furth, citrate synthase deficient strain, 28% decreased activity compared to wild-type [46]) [46]
<49> *Aspergillus niger* (strain N400, CBS 120.49, gene citA [47]) [47]
<50> *yeast* (recombinant fusion protein of yeast citrate synthase and mitochondrial malate dehydrogenase [50]) [50]
<51> *Mesorhizobium ciceri* (strain CC 1192, bacteroids formed in symbiosis with Cicer arietinum plants [51]) [51]
<52> *Cicer arietinum* [51]
<53> *Idotea baltica* (isopod, crustacea [53]) [53]
<54> *Idotea emarginata* (isopod, crustacea [53]) [53]
<55> *Daucus carota* [54]
<56> *Arabidopsis thaliana* (strain WS [63]) [54, 63]
<57> *Homo sapiens* [56]
<58> *Podospora anserina* (recessive gene cit1, mitochondrial isozyme [59]; filamentous fungus [59]) [59]
<59> *Rhodothermus marinus* (strain DSM 4252 [60]; 2 enzyme forms [60]) [60]
<60> *Aspergillus nidulans* (methylcitrate synthase, also exhibiting citrate synthase activity [62]) [62]
<61> *Daucus carota* [63]

3 Reaction and Specificity

Catalyzed reaction

acetyl-CoA + H_2O + oxaloacetate = citrate + CoA (The stereospecificity of this enzyme is opposite to that of EC 2.3.3.4, citrate (Re)-synthase, which is found in some anaerobes; <50> free diffusion mechanism, no channeling of oxaloacetate from malate dehydrogenase [50]; <9, 16, 19, 33> active site [18, 55, 58]; <9> active site His320 [5]; <2-4> random sequential reaction order mechanism [3]; <10> ordered sequential mechanism [7]; <6-9, 16, 39> mechanism [4, 5, 32, 58])

Reaction type

Claisen condensation <6-9, 19, 20, 39, 47> [4, 22, 32, 55]
addition
elimination <9> (<9> of H_2O, C-O bond cleavage [4]) [4]
hydrolysis <9, 19> (<9,19> of intermediate citryl-CoA [55]) [55]

Natural substrates and products

S acetyl-CoA + oxaloacetate + H_2O <2-4, 6-10, 15, 16, 21-23, 25, 34, 37, 40, 42-44, 48, 49, 53, 54, 56, 61> (<37> regulation [27]; <10, 43> key enzyme of tricarboxylic acid cycle [7, 52]; <2-4, 6-8, 15, 16, 37, 40, 43-44> entry step to tricarboxylic acid cycle [3, 4, 12, 27, 29, 36, 37, 39, 45, 52]; <2>

high activity necessary for N_2 fixation [1]; <9> stoichiometry [4]; <21-23> only anabolic pathway [17]; <25> key enzyme of glyoxylate cycle in fat-storing seedlings [20]; <34,46> involved in poly 3-hydroxybutyrate synthesis [23,43]; <49> contributes little to flux control in the pathway [47]; <40> regulation of mitochondrial uptake and efflux by citrate synthase activity [34]; <42> also acting as structural protein in oral morphogenesis and conjugation [36,37]) (Reversibility: r <6-10> [4, 7]; ? <2-4, 15, 16, 21-23, 25, 34, 37, 40, 42-44, 48, 49, 53, 54, 56, 61> [1, 3, 12, 17, 20, 23, 27, 29, 34, 36, 37, 39, 43, 45-47, 52-54, 63]) [1, 3, 4, 7, 12, 17, 20, 23, 27, 29, 34, 36, 37, 39, 43, 45-47, 52-54, 63]

P citrate + CoA <37, 40, 43, 48, 49, 53, 54, 56, 61> [27, 45-47, 52-54, 63]

Substrates and products

S acetyl-CoA + oxaloacetate + H_2O <1-61> (<19> condensation step nearly irreversible, reversibility is increased at 70°C [55]; <9> condensation step is reversible [55]; <50> no channeling of oxaloacetate between malate dehydrogenase and citrate synthase in a recombinant fusion protein using a coupled assay [50]; <9> intermediate citryl-CoA [5]; <6-9> stereospecific reaction [4]; <9> reverse reaction: equilibrium constants, stereospecificity [4]) (Reversibility: r <6-10, 14, 19> [4, 5, 7, 11, 55]; ir <50> [50]; ? <1-5, 11-13, 15-49, 51-61> [1-4, 6, 8-10, 12-49, 51-54, 56-63]) [1-63]

P citrate + CoA <1-61> [1-63]

S fluoroacetyl-CoA + oxaloacetate + H_2O <9, 47> (Reversibility: r <9, 47> [4]) [4]

P fluorocitrate + CoA <9, 47> [4]

S propionyl-CoA + oxaloacetate + H_2O <23, 59, 60> (<60> methylcitrase synthase, also exhibiting citrate synthase activity [62]; <59> only the dimeric enzyme form, no activity with the hexameric enzyme form [60]) (Reversibility: ? <23, 59, 60> [57, 60, 62]) [57, 60, 62]

P 2-methylcitrate + CoA <59, 60> [60, 62]

S Additional information <9> (<9> acetylpantheteine and propionyl-CoA are no substrates [4]) [4]

P ?

Inhibitors

2-oxoglutarate <16, 19, 20, 27, 28, 30-32, 39, 45, 52> (<52> 30% inhibition at 1 mM [51]; <16> wild-type and mutants [33]; <39,45> competitive against oxaloacetate [32,44]; <10,18,29,36,45,46,51> no inhibition [7,15,25,43,44,51]) [13, 15, 32, 33, 44, 51]

5,5'-dithiobis(2-nitrobenzoate) <11, 25> (<25> inactivation of the glyoxysomal isozyme, half-life: approx. 1 min, not mitochondrial isozyme [20]; <15> no inhibition [12]; <11> inactivation, half-lifes at 0.4 mM: 2.8 min for CS 1, 50 min for CS II [8]) [8, 20]

ADP <11, 13, 15, 17, 40> (<46,51> no inhibition [43,51]; <11> CS II insensitive, CS I inhibited competitively with acetyl-CoA, noncompetitively with oxaloacetate [8]; <15> 25% inhibition at 5 mM [12]; <17> 64% inhibition at 10 mM [14]; <41>, 30% inhibition at 1 mM, only CS II [35]) [2, 3, 8-12, 14, 16, 20, 21, 35, 41]

AMP <11, 13, 15, 17, 40> (<51> no inhibition [51]; <40> only CS II [41]; <11> CS II insensitive, CS I inhibited competitively with acetyl-CoA, noncompetitively with oxaloacetate [8]; <15> 8% inhibition at 5 mM [12]; <17> 36% inhibition at 10 mM [14]) [8, 10, 12, 14, 41]

ATP <1-4, 11-15, 17-20, 24-32, 35, 38-42, 45, 46, 52> (<52> 30% inhibition at 1 mM [51]; <40> strong inhibition [41]; <38> noncompetitive against acetyl-CoA [28]; <19, 20, 26, 39, 45> competitive against acetyl-CoA [16, 21, 32, 44]; <15> feed back inhibition [12]; <10, 36, 45, 51> no inhibition [7,25,44,51]; <1> competitive against acetyl-CoA, inhibition reduced by Mg^{2+} [2]; <2-4> possibly involved in enzyme regulation [3]; <11> CS II insensitive, CS I inhibited [8]; <14> 50% inhibition at 7.5 mM, acetyl-CoA protects [11]; <15> 50% inhibition at 5 mM, competitive against acetyl-CoA [12]; <12> 20%, 40%, and 60% inhibition at 1 mM, 2 mM, and 5 mM ATP, respectively [9]; <13> 44% inhibition at 5 mM [10]; <17> 79% inhibition at 10 mM [14]; <35> 25% inhibition at 10 mM [24]) [2, 3, 8-12, 14-17, 19-21, 24, 28, 32, 35, 36, 41, 43, 44, 51]

CTP <13> (<13> poor [10]) [10]

$CaCl_2$ <13, 51> (<51> 50% inhibition at 10 mM, not reversible by KCl [51]; <13> 29% inhibition at 10 mM [10]) [10, 51]

CoA <2-4, 10, 51> (<51> 50% inhibition at 0.2 mM [51]; <2-4,10> competitive with acetyl-CoA, noncompetitive with oxaloacetate [3,7]) [3, 7, 51]

$CuSO_4$ <36> [25]

GTP <13> (<13> poor [10]) [10]

$HgCl_2$ <15, 36, 45> (<15> weak inhibition [12]) [12, 25, 44]

KCl <11, 13> (<11>, CS I inhibited competitively with acetyl-CoA, noncompetitively with oxaloacetate [8]; <13>, 9% inhibition at 50 mM [10]) [8, 10]

L-malate <9> (<51> no inhibition [51]; <9> slight inhibition [4]) [4]

Mg^{2+} <6> [4]

$MgCl_2$ <13, 51> (<51> 50% inhibition at 7 mM, not reversible by KCl [51]; <13> 32% inhibition at 10 mM [10]) [10, 51]

$MnCl_2$ <13> (<13> 56% inhibition at 10 mM [10]) [10]

N-ethylmaleimide <45> (<45> strong [44]) [44]

NADH <2, 3, 11, 12, 16-20, 22, 26, 28-30, 32, 40, 41, 51> (<16> CS II, strong and specific allosteric inhibition [58]; <51> complete inhibition at 1 mM, completely reversible by NAD^+ [51]; <19,20,26> competitive against acetyl-CoA [16,21]; <3,16> allosteric [13,42,58]; <2> 70% inhibition at 2 mM, completely reversed by 0.17 mM 5'-AMP [1]; <3> 12% inhibition at 1 mM [3]; <11> weak inhibition, reversible by AMP, hinders the activation by AMP, CS II [8]; <21-23> not, only gram-negative facultative methylotrophs [17]; <11> CS II inhibited, CS II insensitive [8]; <12> inhibition only with gram-negative bacterial enzymes [9]; <10,12,15,24,36,46,52> no effect [7,9,12,19,25,43,51]; <17> non-specific, 31% inhibition at 10 mM [14]; <41> 78% inhibition by 0.1 mM of CS I, 15% inhibition of CS II at 1 mM [35]; <40> 51% inhibition of peroxisomal isozyme, 12% inhibition of mitochondrial isozyme at 5 mM [41]) [1, 3, 8, 9, 13-17, 21, 35, 41, 42, 51, 58]

$NADP^+$ <40> (<51> no inhibition [51]; <40> only CS II slightly [41]) [41]

NADPH <3, 17, 19, 20, 40> (<46,51> no inhibition [43,51]; <19,20> competitive against acetyl-CoA [16]; <3> 29% inhibition at 0.25 mM [3]; <17> 57% inhibition at 10 mM [14]; <40> 40% inhibition by 5 mM, both isoenzymes [41]) [3, 14, 16, 41]
NaCl <13> (<51> no effect [51]; <13> 15% inhibition at 50 mM [10]) [10]
SDS <36, 45> (<36,45> strong inhibition [25,44]) [25, 44]
TTP <13> (<13> poor [10]) [10]
UTP <13> (<13> poor [10]) [10]
Zn^{2+} <45> [44]
acetyl-CoA <4> (<4> substrate inhibition at high concentration [3]; <2,3,10,51> no inhibition [3,7,51]) [3]
cAMP <51> (<51> no inhibition [51]) [51]
cations <13> (<13> monovalent and divalent [10]) [10]
citrate <2-4, 38, 51> (<51> 50% inhibition at 2.5 mM [51]; <2-4> competitive with acetyl-CoA, noncompetitive with oxaloacetate [3]) [3, 28, 51]
dithionitrobenzoate <1> [2]
elongation factor 1α <42> (<42> causes polymerization of 49K protein, reduced activity [37]) [37]
fluoroacetyl-CoA <9, 47> (<9,47> competitive against acetyl-CoA [4]) [4]
guanidinium chloride <19> (<19> irreversible inactivation of recombinant wild-type at 1.6 M and of recombinant mutant G196V at 0.2 M, at 0.5 M activation of the wild-type [48]) [48]
iodoacetamide <1, 13> (<13> 10% inhibition at 0.002 mM [10]) [2, 10]
lauroyl-CoA <1> [2]
maleate <9> (<9> slight inhibition [4]) [4]
oxaloacetate <4> (<4> substrate inhibition [3]; <2,3,10,51> no inhibition [3,7,51]) [3]
p-chloromercuribenzoate <9, 10, 13, 45> (<45> strong [44]; <10> can be restored by dithiothreitol treatment [7]; <13> 6% loss at 0.002 mM [10]) [4, 7, 10, 44]
p-hydroxymercuribenzoate <15> (<6-9> no inhibition [4]; <15> 60% inhibition at 0.1 mM, protection by oxaloacetate [12]) [12]
propionyl-CoA <39> (<39> competitive against acetyl-CoA [32]) [32]
succinyl-CoA <39> (<39> mixed-type inhibition [32]) [32]
tartrate <9> (<9> slight inhibition [4]) [4]
urea <19> (<19> irreversible inactivation of recombinant wild-type at 9.3 M and of recombinant mutant G196V at 5 M, at up to 8 M activation of the wild-type [48]) [48]
Additional information <5, 16, 39, 40, 42, 45, 51> (<16> allosteric inhibition mechanism, structure relationship [58]; <51> isocitrate, (R)-3-hydroxybutyrate, malonate, pyruvate, acetoacetyl, NaCl, NH_4Cl, CsCl, and RbCl have no effect [51]; <45> not affected by EDTA [44]; <40> no inhibition by NAD^+ [41]; <42> inhibited by a specific antibody against 49K protein [36]; <5> low salt concentrations inactivate reversibly [6]; <39> tricarboxylic acid cycle intermediates [32]) [6, 32, 36, 41, 44, 51, 58]

Cofactors/prosthetic groups

AMP <2, 11> (<2> positive effector [1]; <11> only CS II, counteracted by NADP [8]) [1, 8]
acetyl-CoA <6-8> [4]

Activating compounds

AMP <3, 11> (<11> CS II, counteracted by NADH [8]; <3> slight activation [3]) [3, 8]
acetate <35> [24]
aldosterone <48> [46]
α-ketoglutarate <3> [3]
ammonium sulfate <10> (<10> activates at 0.5 mM [7]) [7]
guanidinium chloride <19> (<19> activation of recombinant wild-type 1.6fold at 0.5 M, inactivation at 1.6 M [48]) [48]
light <15> (<15> light activates up to 1.4fold, regulation in correlation with photosynthesis [26]) [26]
urea <19, 51> (<51> isocitrate, (R)-3-hydroxybutyrate, malonate, pyruvate, acetoacetyl have no effect [51]; <19> activation of recombinant wild-type by 35% at up to 8 M, inactivation at 9.3 M [48]) [48, 51]

Metals, ions

$(NH_4)_2SO_4$ <10> (<10> activates at at least 0.5 mM [7]) [7]
Ca^{2+} <5> (<5> as $CaCl_2$, inhibition at concentrations higher than 50 mM [6]) [6]
K^+ <5> (<5> activation, Cl^- most efficient anion [6]) [6]
KCl <2, 3, 5, 11, 16, 33, 42, 45, 46, 51> (<51> 40% activation at 50 mM, 30% activation at 100 mM [51]; <3,16> mutants, overview [33,42]; <16> 39fold activation at saturating concentration, wild-type [33]; <5> maximal activation at 3 M [6]; <33> 5fold increase at 100-200 mM [30]; <42> maximum activity at 150 mM [36]; <46> up to 80 mM [43]; <45> at 0.1-100 mM [44]) [1, 3, 6, 8, 30, 33, 36, 42-44, 51]
Mg^{2+} <1, 5> (<1> reduces ATP inhibition [2]; <6-9> not required [4]; <5> as $MgCl_2$, inhibition at concentrations higher than 50 mM [6]) [2, 6]
NaCl <5, 33> (<33> 5fold increase at 100-200 mM [30]) [6, 30]
phosphate <11> [8]
Additional information <36, 51> (<51> NaCl, NH_4Cl, CsCl, and RbCl have no effect [51]; <36> no activation by metal ions [25]) [25, 51]

Turnover number (min^{-1})

0.8 <19> (acetyl-CoA, <19> recombinant mutant G196V [48]) [48]
9 <19> (acetyl-CoA, <19> recombinant wild-type [48]) [48]
16.5 <9> (oxaloacetate, <9> mutant H320G [5]) [5]
240 <23> (propionyl-CoA, <23> mutant loop/K313L/A361R [57]) [57]
360 <23> (propionyl-CoA, <23> mutant A361R [57]) [57]
420 <23> (acetyl-CoA, <23> mutant A10E [57]) [57]
480 <23> (propionyl-CoA, <23> wild-type [57]) [57]
550 <19> (acetyl-CoA, <19> recombinant wild-type, pH 8.0 [55]) [55]
550 <19> (citrate, <19> recombinant enzyme, pH 8.0 [55]) [55]

720 <59> (propionyl-CoA, <59> dimeric enzyme form [60]) [60]
780 <23> (acetyl-CoA, <23> mutant A361R/A10E [57]) [57]
900 <23> (acetyl-CoA, <23> mutant A361R [57]) [57]
1080 <23> (acetyl-CoA, <23> wild-type [57]) [57]
1260 <59> (acetyl-CoA, <59> hexameric enzyme form [60]) [60]
1320 <59> (acetyl-CoA, <59> dimeric enzyme form [60]) [60]
4680 <3> (acetyl-CoA, <3> wild-type, + 0.1 M KCl [42]) [42]
4860 <16> (acetyl-CoA, <16> wild-type, + 0.1 M KCl [42]) [42]
6642 <2> (oxaloacetate) [1]
7000 <9> (citrate) [55]
10000 <9> (acetyl-CoA, <9> pH 7.5 [55]) [55]
10000 <9> (oxaloacetate, <9> wild-type [5]) [5]
10220 <2> (oxaloacetate, <2> with 0.1 mM KCl [1]) [1]
10440 <2> (acetyl-CoA, <2> with 0.1 M KCl [1]) [1]
56500 <3> (citrate) [3]
126000 <2> (citrate) [3]
Additional information <3, 9, 16, 19, 23> (<23> values for acetyl-CoA and oxaloacetate of mutants with a loop introduced into the active site [57]; <9,19> kinetics [55]; <19> k_{cat} is stable over pH-range 6.0-8.0 [55]; <9,19> k_{cat} with citrate and CoA [55]; <3,16> chimeric mutants, overview [42]) [42, 55, 57]

Specific activity (U/mg)

0.00023-0.00028 <56> (<56> wild-type [54]) [54]
0.00078 <55> (<55> recombinant in transgenic Arabidopsis thaliana plants [54]) [54]
0.002 <32> [15]
0.008 <19> (<19> recombinant purified mutant G196V [48]) [48]
0.014 <30> [15]
0.02 <37> (<37> recombinant enzyme in crude E. coli enzyme extract [27]) [27]
0.026 <29> [15]
0.051 <18> [15]
0.061 <28> [15]
0.076 <48> (<48> citrate synthase deficient strain [46]) [46]
0.083 <19> (<19> recombinant purified wild-type [48]) [48]
0.172 <19> [15]
0.23-0.28 <56> [63]
0.28 <20> [15]
0.764 <25> (<25> purified glyoxysomal isozyme [20]) [20]
0.78 <61> (<61> transgenic Arabidopsis thaliana plants overexpressing the Daucus carota citrate synthase [63]) [63]
0.87 <52> (<52> partially purified enzyme [51]) [51]
1.25 <51> (<51> partially purified enzyme [51]) [51]
2.6 <4> (<4> partially purified enzyme [3]) [3]
4.7 <41> (<41> isozymes CS I and CS II [35]) [35]
9 <19> (<19> S43C mutant [40]) [40]

14 <20> (<20> partially purified enzyme [16]) [16]
18 <26> (<26> purified enzyme [21]) [21]
18.8 <19> (<19> partially purified enzyme [16]) [16]
21.6 <17> (<17> purified enzyme [14]) [14]
24 <36, 46> (<36,46> purified enzyme [25,43]) [25, 43]
25.8 <44> (<44> purified enzyme [39]) [39]
27.45 <5> [6]
28 <59> (<59> purified dimeric and hexameric enzyme form [60]) [60]
30 <33> (<33> purified recombinant enzyme from E. coli [30]) [30]
35.1 <42> (<42> purified enzyme [36]) [36]
41 <19> [22]
41.5 <60> (<60> citrate synthase activity of a purified methylcitrate synthase [62]) [62]
45 <16> (<16> purified wild-type enzyme [42]) [42]
50-60 <1> (<1> partially purified enzyme [2]) [2]
51.4 <19> (<19> wild-type [40]) [40]
52.3 <2> (<2> purified enzyme [3]) [3]
53 <20> [22]
57 <3> (<3> purified recombinant chimeric Acinetobacter-type protein with small E. coli domain [42]) [42]
60 <59> (<59> dimeric and hexameric enzyme [60]) [60]
62.6 <16> (<16> purified enzyme [13]) [13]
64 <34> (<34> purified enzyme [23]) [23]
70 <40> (<40> purified recombinant isozyme CS II [41]) [41]
79 <49> (<49> purified enzyme [47]) [47]
88 <3> (<3> purified wild-type enzyme [42]) [42]
102.7 <11> (<11> purified enzyme [8]) [8]
108 <24> (<24> purified enzyme [19]) [19]
112.5 <33, 39> (<33> recombinant enzyme [30]; <39> purified enzyme [32]) [30, 32]
120 <16> (<16> purified recombinant chimeric E. coli-type protein with small Acinetobacter domain [42]) [42]
161.2 <38> [28]
213 <50> (<50> purified fusion protein of citrate synthase and mitochondrial malate dehydrogenase, coupled assay [50]) [50]
230 <35> (<35> purified enzyme [24]) [24]
235 <3> (<3> purified enzyme [3]) [3]
288 <45> (<45> purified enzyme [44]) [44]
471.7 <10> (<10> partially purified enzyme [7]) [7]
523 <2> [3]
664.2 <15> [26]
1190 <13> (<13> partially purified enzyme [10]) [10]
Additional information <3, 5, 9, 14, 16, 37, 40, 42, 57> (<57> activity in trained and untrained skeletal muscle, highest in acutely trained muscle due to mitochondrial membrane breakdown, mechanism [56]; <3,16> influence of KCl on wild-type and mutants [42]; <9,16> allosteric and nonallosteric activity [29]; <37> northern blot analysis, activity in tissues [27]; <42> ac-

tivity increased by depolymerization, decreased by polymerization, possible mode of regulation [37]) [4-6, 11, 27, 29, 37, 42, 45, 56]

K_m-Value (mM)

0.0017 <33> (acetyl-CoA, <33> recombinant enzyme [30]) [30]
0.0017 <60> (propionyl-CoA, <60> methylcitrate synthase exhibiting also citrate synthase activity [62]) [62]
0.002 <19, 23, 41, 59> (oxaloacetate, <59> dimeric enzyme form with propionyl-CoA [60]; <23> mutant A361R [57]; <19> wild-type [49]; <41> CS II [35]) [35, 49, 57, 60]
0.0025 <60> (acetyl-CoA, <60> methylcitrate synthase exhibiting also citrate synthase activity [62]) [62]
0.003 <40, 59> (oxaloacetate, <59> dimeric enzyme form with acetyl-CoA [60]; <40> mitochondrial isozyme CS I [41]) [41, 60]
0.003 <59> (propionyl-CoA, <59> dimeric enzyme form [60]) [60]
0.0031 <39> (oxaloacetate) [32]
0.0032 <13> (oxaloacetate) [10]
0.004 <33, 40> (acetyl-CoA, <33> wild-type [49]; <40> mitochondrial isozyme CS I [41]) [41, 49]
0.004 <23> (propionyl-CoA, <23> mutant A361R [57]) [57]
0.0041 <19> (oxaloacetate) [16]
0.0045 <19> (acetyl-CoA, <19> recombinant mutant G196V [48]) [48]
0.005 <9, 19> (acetyl-CoA, <19> recombinant wild-type [48]; <9,19> wild-type [5,49]) [5, 48, 49]
0.005 <19, 33, 38> (oxaloacetate, <33> recombinant enzyme [30]) [15, 28, 30]
0.0057 <19> (oxaloacetate, <19> recombinant mutant G196V [48]) [48]
0.0059 <9> (oxaloacetate, <9> wild-type [5]) [5]
0.006 <19> (acetyl-CoA) [15, 16]
0.006 <23, 59> (oxaloacetate, <59> hexameric enzyme form [60]; <23> mutant A361R/A10E [57]) [57, 60]
0.006-0.007 <40> (oxaloacetate, <40> native and recombinant peroxisomal isozyme CS II [41]) [41]
0.0066 <19> (oxaloacetate, <19> recombinant wild-type [48]) [48]
0.0067 <39> (acetyl-CoA) [32]
0.007 <12, 26, 32, 59> (acetyl-CoA, <59> dimeric enzyme form [60]) [9, 15, 21, 60]
0.007 <10, 23, 49> (oxaloacetate, <23> wild-type [57]; <49> pH 7.5 [47]) [7, 47, 57]
0.008 <23> (oxaloacetate, <23> mutant A10E [57]) [57]
0.008 <23> (propionyl-CoA, <23> mutant loop/K313L/A361R [57]) [57]
0.01 <10, 20, 38> (acetyl-CoA) [7, 15, 16, 28]
0.01 <19> (oxaloacetate, <19> recombinant wild-type and mutant S34C [40]) [40]
0.011 <19, 40> (acetyl-CoA, <19> recombinant wild-type and mutant S43C [40]; <40> native peroxisomal isozyme CS II [41]) [40, 41]
0.011 <1, 3, 45> (oxaloacetate, <3> wild-type [42]; <1> in presence of KCL [2]) [2, 42, 44]

0.012 <59> (acetyl-CoA, <59> dimeric enzyme form [60]) [60]
0.012 <17, 46> (oxaloacetate) [14, 43]
0.013-0.014 <40> (acetyl-CoA, <40> recombinant peroxisomal isozyme CS II [41]) [41]
0.014 <2, 49> (acetyl-CoA, <49> pH 7.5 [47]; <2> with 0.1 M KCl [1]) [1, 47]
0.015 <23, 33> (oxaloacetate, <23> mutant A361R [57]; <33> wild-type [49]) [49, 57]
0.016 <15> (oxaloacetate) [12]
0.016 <23> (propionyl-CoA, <23> wild-type [57]) [57]
0.018 <24> (acetyl-CoA) [19]
0.018 <5, 32> (oxaloacetate, <5> minimum at 1 M KCl [6]) [6, 15]
0.019 <24> (oxaloacetate) [19]
0.02 <2, 11, 16, 20, 23, 26, 35> (oxaloacetate, <23> mutant mutant loop/K313L/A361R [57]; <16> wild-type [42]; <11> CS I [8]) [3, 8, 15, 21, 24, 42, 57]
0.026 <16, 20> (oxaloacetate, <16> wild-type [33]) [16, 33]
0.027 <36> (oxaloacetate) [25]
0.0283 <42> (oxaloacetate) [36]
0.0286 <42> (acetyl-CoA) [36]
0.03 <1> (acetyl-CoA) [2]
0.031 <15> (acetyl-CoA) [12]
0.033 <13> (acetyl-CoA) [10]
0.035 <11> (acetyl-CoA, <11> CS I [8]) [8]
0.037 <23> (acetyl-CoA, <23> mutant A361R/A10E [57]) [57]
0.038 <23> (acetyl-CoA, <23> mutant A361R [57]) [57]
0.038 <28> (oxaloacetate) [15]
0.043 <9> (oxaloacetate, <9> mutant H320G [5]) [5]
0.05 <1, 8, 41> (acetyl-CoA, <1> in presence of 2.5 mM ATP [2]; <41> CS II [35]) [2, 4, 35]
0.05 <11> (oxaloacetate, <11> CS II [8]) [8]
0.051 <35> (acetyl-CoA) [24]
0.058 <18> (oxaloacetate) [15]
0.063 <30> (oxaloacetate) [15]
0.07 <1> (oxaloacetate) [2]
0.075 <28> (acetyl-CoA) [15]
0.076 <17> (acetyl-CoA) [14]
0.08 <1, 2> (acetyl-CoA, <1> in presence of 5 mM ATP [2]) [2, 3]
0.084 <46> (acetyl-CoA) [43]
0.088 <51> (acetyl-CoA, <51> in presence of 100 mM KCl [51]) [51]
0.089 <9> (acetyl-CoA, <9> mutant H320G [5]) [5]
0.094 <34> (acetyl-CoA) [23]
0.1 <3> (acetyl-CoA) [3]
0.11 <3, 30> (acetyl-CoA, <3> wild-type [42]) [15, 42]
0.11 <14> (oxaloacetate) [11]
0.12 <16> (acetyl-CoA, <16> wild-type [33]) [33]
0.12 <34> (oxaloacetate) [23]
0.143-0.161 <22> (acetyl-CoA, <22> strains X and Hanham [17]) [17]

0.174 <51, 51> (acetyl-CoA, <51> in absence of KCl [51]) [51]
0.2 <14, 23> (acetyl-CoA, <23> wild-type [57]) [11, 57]
0.21 <29> (oxaloacetate) [15]
0.211 <18> (acetyl-CoA) [15]
0.22 <9> (acetyl-CoA) [4]
0.247 <45> (acetyl-CoA) [44]
0.26 <5> (acetyl-CoA, <5> minimum at 1 M KCl [6]) [6]
0.41 <36> (acetyl-CoA) [25]
0.45 <11, 29> (acetyl-CoA, <11> CS II [8]) [8, 15]
0.6 <4> (acetyl-CoA) [3]
0.63 <60> (oxaloacetate, <60> with propionyl-CoA, methylcitrate synthase exhibiting also citrate synthase activity [62]) [62]
0.7 <16> (acetyl-CoA, <16> wild-type [42]) [42]
0.86 <41> (acetyl-CoA, <41> CS I [35]) [35]
0.96 <41> (oxaloacetate, <41> CS I [35]) [35]
1.24 <23> (acetyl-CoA, <23> mutant A10E [57]) [57]
5 <38> (oxaloacetate) [28]
10 <38> (acetyl-CoA) [28]
Additional information <2, 5, 12, 16, 22, 23, 54> (<23> K_m values for acetyl-CoA and oxaloacetate of mutants with a loop introduced into the active site [57]; <54> affinity for oxaloacetate is increased at adaption of growth temperature from 5 to 15°C [53]; <16> mutants, with and without KCl, overview [33]; <22> influence of NADH and AMP on K_m for acetyl-CoA [17]; <2> allosteric, Hill coefficient: 1.5 [1]; <5> K_m increases 5fold with salt concentration from 1 to 3 M, overview [6]) [1, 4, 6, 9, 17, 33, 53, 57]

K_i-Value (mM)

0.0024 <39> (propionyl-CoA, <39> versus acetyl-CoA, competitive [32]) [32]
0.012 <2> (NADH, <2> in 20 mM Tris-HCl [1]) [1]
0.0225 <39> (propionyl-CoA, <39> versus oxaloacetate, noncompetitive [32]) [32]
0.023 <51> (NADH) [51]
0.1 <32> (NADH, <32> below [15]) [15]
0.11 <10> (CoA) [7]
0.125 <39> (ATP, <39> versus acetyl-CoA [32]) [32]
0.3 <32> (2-oxoglutarate, <32> below [15]) [15]
0.3 <32> (ATP, <32> below [15]) [15]
0.33 <25> (ATP, <25> mitochondrial isozyme [20]) [20]
0.36 <26> (ATP) [21]
0.38 <39> (2-oxoglutarate, <39> versus oxaloacetate, competitive [32]) [32]
0.4 <20> (ATP) [16]
0.4-12 <22> (NADH, <22> strains X and Hanham [17]) [17]
0.5 <19> (ATP) [16]
0.66 <2> (NADH, <2> in 0.1 M Tris-HCl [1]) [1]
0.75 <2> (ATP, <2> versus acetyl-CoA [3]) [3]
0.76 <16> (2-oxoglutarate, <16> wild-type [33]) [33]
0.8 <20> (NADPH) [16]

0.9 <20> (ATP) [15]
1 <4> (ATP) [3]
1.9 <28> (ATP) [15]
2.2 <20> (2-oxoglutarate) [15]
2.2 <19> (ATP) [15]
2.6 <25> (ATP, <25> glyoxysomal isozyme [20]) [20]
3-4 <26> (NADH) [21]
3.2 <30> (NADH) [15]
3.5 <1> (ATP) [2]
3.5 <26> (NADH) [21]
4.2 <30> (ATP) [15]
4.4 <20> (NADH) [16]
4.6 <20> (NADH) [15]
4.9 <19> (NADPH) [16]
5 <15> (ATP) [12]
5.2 <29> (NADH) [15]
5.4 <19> (2-oxoglutarate) [15]
5.4 <19> (NADH) [15]
5.6 <18> (ATP) [15]
7.5 <14> (ATP) [11]
10 <18, 28> (NADH, <18,28> above [15]) [15]
10.4 <19> (NADH) [16]
20 <28, 30> (2-oxoglutarate, <28,30> above [15]) [15]
30 <29> (ATP, <29> above [15]) [15]
Additional information <9, 16, 47> (<16> mutants, overview [33]; <9,47> fluoroacetyl-CoA [4]) [4, 33]

pH-Optimum

7.3-7.8 <51> [51]
7.4 <37, 57> (<37,57> enzyme assay at [27,56]) [27, 56]
7.5 <2> (<2> enzyme assay at [1]) [1]
7.5-8 <14, 36, 45> [11, 25, 44]
7.6 <5, 11> (<5> assay at [6]; <11> enzyme assay at [8]) [6, 8]
7.8 <13, 16, 42> (<16> recombinant chimeric protein [42]; <42> assay at [37]) [10, 36, 37, 42]
7.9 <24> (<24> enzyme assay at [19]) [19]
8 <8, 12, 17, 19, 23, 25, 33, 34, 40, 48> (<19> recombinant wild-type and mutant [40]; <12, 17, 19, 23, 25, 33, 34, 48> enzyme assay at [9, 14, 20, 23, 30, 46, 48, 49, 57]; <40> both mitochondrial and peroxisomal enzymes [41]) [4, 9, 14, 20, 23, 30, 40, 41, 46, 48, 49, 57]
8-9 <39> [32]
8.1 <1, 35> (<1> enzyme assay at [2]) [2, 24]
8.5 <38, 45, 49> [28, 44, 47]
8.6 <15> [12]
8.6-9 <46> [43]
Additional information <19, 36, 45> (<19> pI: 8.9 [55]; <45> pI: 5.0 [44]; <36> pI: 5.4 [25]) [25, 44, 55]

pH-Range

7-9 <14> [11]
7.2-9 <10> [7]

Temperature optimum (°C)

23 <10> (<10> enzyme assay at [7]) [7]
25 <1, 2, 6-8, 12, 13, 15, 17, 25, 41> (<1, 2, 9, 12, 13, 15, 17, 25, 41> enzyme assay at [1, 2, 4, 9, 10, 12, 14, 20, 35]) [1, 2, 4, 9, 10, 12, 14, 20, 35]
30 <2-5, 11, 18, 28-30, 32, 37, 46, 57> (<2-5, 11, 18, 28, 30, 32, 37, 46, 57> enzyme assay at [3, 6, 8, 15, 27, 43, 56]) [3, 6, 8, 15, 27, 43, 56]
30-35 <36> [25]
34 <34> (<34> enzyme assay at [23]) [23]
35 <14> [11]
37 <9, 42> (<42> assay at [37]) [37, 55]
40 <45> [44]
45 <39> [32]
55 <19, 20, 33, 59> (<19,20,33,59> enzyme assay at [15,22,30,48,60]) [15, 22, 30, 48, 60]
65 <33> (<33> enzyme mutant lacking 13 amino acid residues of the C-terminal [49]) [49]
65-70 <19, 33> (<19,33> chimeric mutant containing the large subunit of Pyrcoccus furiosus and the small subunit of Thermoplasma acidophilum [49]) [49]
70 <19> [55]
70-75 <19> (<19> wild-type [49]) [49]
75 <33> (<12> recombinant enzyme [61]; <33> enzyme mutant lacking 2 amino acid residues of the C-terminal [49]) [49]
80 <12, 33> (<33> mutants D113S and D113A [49]) [49, 61]
85 <59> (<59> dimeric and hexameric enzyme forms [60]) [60]
90 <19, 33> (<19,33> chimeric mutant containing the large subunit of Thermoplasma acidophilum and the small subunit of Pyrcoccus furiosus [49]; <33> above, wild-type [49]) [49]

Temperature range (°C)

20-50 <14> [11]
40-90 <19, 33, 59> [49, 60]

4 Enzyme Structure

Molecular weight

60000-70000 <1> (<1> gel filtration [2]) [2]
60000-95000 <4> (<4> gel filtration, sedimentation equilibrium centrifugation [3]) [3]
68000 <19> (<19> gel filtration [16]) [16]
70000 <38> (<38> gel filtration [28]) [28]
73000 <20> (<20> gel filtration [16]) [16]
78000 <52> (<52> gel filtration [51]) [51]

80000 <26, 49> (<26,49> gel filtration [21,35,47]) [21, 35, 47]
80300 <41> (<41> CS II, gel filtration [35]) [35]
81000 <39> (<39> gel filtration [32]) [32]
83000 <20> (<20> gel filtration [15,22]) [15, 22]
84000 <17> (<17> Stokes radius, sedimentation coefficient [14]) [14]
85000 <19> (<19> gel filtration [15,22]) [15, 22]
89700 <33> (<33> gel filtration [30]) [30]
98000 <12> (<12> gel filtration [9]) [9]
100000 <11, 13, 40, 59> (<59> dimeric enzyme form [60]; <40> CS II [41]; <11> CS I [8]; <11,13,40,59> gel filtration [8,10,41,60]) [8, 10, 41, 60]
104000 <24> (<24> gel filtration [19]) [19]
105000 <28, 31> (<28,31> gel filtration [15]) [15]
106000 <29> (<29> gel filtration [15]) [15]
110000 <10> (<10> gel filtration [7]) [7]
111000 <18> (<18> gel filtration [15]) [15]
112000 <27, 30> (<27,30> gel filtration [15]) [15]
150000 <21-23> (<21-23> above: gram-negative methylotrophs, below: gram-positive methylotrophs [17]) [17]
227000 <34> (<34> gel filtration [23]) [23]
228000 <51> (<51> gel filtration [51]) [51]
240000 <41> (<41> gel filtration, CS I [35]) [35]
242000 <3> (<3> sedimentation equilibrium centrifugation [3]) [3]
249000-275000 <2> (<2> equilibrium sedimentation [1]) [1]
250000 <2, 36> (<2> sedimentation equilibrium centrifugation [3]; <36> gel filtration [25]) [3, 25]
260000 <46> (<46> gel filtration [43]) [43]
269000 <16> (<16> gel filtration [13]) [13]
280000 <35, 59> (<59> hexameric enzyme form [60]; <35,59> gel filtration [24,60]) [24, 60]
287000 <45> (<45> gel filtration [44]) [44]
300000 <11> (<11> CS II, gel filtration [8]) [8]
330000 <2> (<2> gel filtration [1]) [1]
Additional information <2-4, 12, 16, 19, 33, 35, 37, 39-41, 43, 44, 49, 59, 60> (<19> free enzyme forms various binary and ternary complexes with substrates [55]; <40> enzyme participates in multienzyme complexes of enzymes belonging to tricarbonic acid cycle, probably substrate channeling, in vivo NMR measurements [45]; <2,3,16> alignment of partial amino acid sequence [1,42]; <16,33> amino acid sequence alignment [30,58]; <37> amino acid sequence comparison [27]; <35,59> N-terminal protein sequence [24,60]; <49> C-terminal amino acid sequence alignment [47]; <19,49> N-terminal protein sequence alignment [22,47]; <4,21-23> small and large enzyme [17]; <12> gram-negative bacteria: approximately 250000, gram-positive-bacteria and eucaryotes: approximately 100000 [9]; <33> secondary and tertiary structure [18,30]; <39,41> amino acid composition [32,35]; <43,44,60> amino acid sequence [39,52,62]) [1, 9, 17, 18, 22, 24, 27, 30, 32, 35, 39, 42, 45, 47, 52, 55, 58, 60, 62]

Subunits

? <9, 25, 42, 44> (<44> x * 45000, SDS-PAGE [39]; <42> x * 49000, SDS-PAGE [36]; <9> x * 47000, SDS-PAGE [31]; <25> x * 48000, glyoxysomal isozyme, SDS-PAGE [20]) [20, 31, 36, 39]
dimer <11, 17, 19, 20, 24, 26, 33, 38-41, 49, 59> (<59> 2 * 42000, dimeric enzyme form, SDS-PAGE [60]; <41> 2 * 36500, CS II, SDS-PAGE [35]; <49> 2 * 48000, mitochondrial enzyme, SDS-PAGE [47]; <26,38> 2 * 40000, SDS-PAGE [21,28]; <17> 2 * 40300, SDS-PAGE, gel filtration in guanidine-HCl, amino acid composition [14]; <20> 2 * 41000, SDS-PAGE [22]; <33> 2 * 42600, SDS-PAGE [30]; <19> 2 * 43000, SDS-PAGE [22]; <39> 2 * 48700, SDS-PAGE [32]; <40> 2 * 50000, CS II, SDS-PAGE [41]; <24> 2 * 50000, SDS-PAGE [19]; <11> 2 * 53000, SDS-PAGE, CS I, reassociation to CS II possible [8]) [8, 14, 19, 21, 22, 28, 30, 32, 35, 41, 47-49, 60]
hexamer <2, 11, 16, 34, 35, 41, 45, 59> (<59> 6 * 45000, hexameric enzyme form [60]; <41> 6 * 42000, CS i, SDS-PAGE [35]; <2> 6 * 48000, SDS-PAGE [1]; <16> 6 * 44000, SDS-PAGE, 6 * 49000, guanidine-HCl gel filtration [13]; <45> 6 * 44700, SDS-PAGE [44]; <35> 6 * 46000, SDS-PAGE, amino-terminal amino acid sequence [24]; <34> 6 * 47000, SDS-PAGE, or homopentamer [23]; <11> 6 * 53000, SDS-PAGE, CS II [8]) [1, 8, 13, 23, 24, 35, 44, 60]
tetramer <2, 3, 36, 46> (<36> 2 * 45000 + 2 * 80000, α, β, SDS-PAGE [25]; <2> 4 * 58500, sedimentation equilibrium in guanidine-HCl and dithiothreitol [3]; <3> 4 * 69000, sedimentation equilibrium in guanidine-HCl and dithiothreitol [3]; <46> 4 * 66000, SDS-PAGE [43]) [3, 25, 43]
Additional information <3, 9, 12, 16, 19, 20, 33, 41> (<12> inter-subunit ionic network, molecular modeling [61]; <16> subunit organization and crystal structure, amino acid residues involved in subunit interaction [58]; <19,33> chimeric mutants with exchanged large and small subunits between Thermoplasma acidophilum and Pyrococcus furiosus, the large subunit is responsible for subunit interaction, the small subunit is responsable for catalytic activity [49]; <33> subunit organisation from crystal structure, dimer [18]; <9> molecular basis of subunit interactions [5]; <41> CS I expressed from different structural gene than CS II [35]; <3,16> recombinant chimeric enzymes, domain functions, subunit organisation [42]) [5, 18, 22, 35, 42, 49, 58, 61]

5 Isolation/Preparation/Mutation/Application

Source/tissue

cotyledon <25> [20]
egg <13> [10]
embryo <43> (<43> highest at 4-cell stage [52]) [52]
epicotyl <15> [12]
flower bud <37> [27]
fruit <24> [19]
heart <9, 43> (<43> ventricular tissue with greater activity than right atrium [38]) [4, 5, 29, 31, 38, 55]

heart ventricle <43> (<43> left free wall shows higher activity than that at the apex [38]) [38]
kidney <48> [46]
leaf <14, 15, 37> (<37> 3fold higher in fully mature leaves than immature, declines in senescent leaves, expression and activity under developmental control [27]) [11, 26, 27]
mycelium <49> [47]
nodule <51, 52> (<51> bacteroids formed in symbiosis with Cicer arietinum plants [51]) [51]
root <56> (<56> wild-type and transgenis plants, the latter expressing the Daucus carota enzyme [54]) [54]
skeletal muscle <57> (<57> trained and untrained [56]) [56]
stolon <37> [27]
tuber <37> [27]

Localization

glyoxysome <25> [20]
mitochondrial inner membrane <40> (<40> 45% of CS II is bound to mitochondrial inner membrane and 2% of CS I [41]) [41]
mitochondrion <9, 15, 25, 37, 40, 42, 43, 48, 49, 57, 58, 61> (<49> N-terminal mitochondrial import signal sequence [47]; <40,58> mitochondrial isozyme [45,59]; <40> citrate uptake into mitochondria [34]; <9> encoded by nuclear DNA, precursor is synthesized in the cytosol, import into mitochondrion [31]; <43> mitochondrial adaptation [38]) [12, 20, 26, 27, 31, 34, 36, 38, 41, 45-47, 56, 59, 63]
peroxisome <40> (<40> peroxisomal isozyme [59]) [41, 59]
Additional information <14, 49> (<49> N-terminal mitochondrial signal and C-terminal peroxisomal target sequence, unclear if both are functional [47]; <14> distribution [11]) [11, 47]

Purification

<1> (partial [2]) [2]
<2> (partial [1]) [1, 3]
<3> (recombinant chimeric protein from E. coli [42]) [3, 42]
<4> (partial [3]) [3]
<5> [6]
<6> [4]
<7> [4]
<8> [4]
<10> (partial [7]) [7]
<11> [8]
<13> (partial [10]) [10]
<15> (mitochondrial enzyme, antibody studies [26]) [26]
<16> (recombinant protein expressed from plasmid [13]; recombinant chimeric protein from E. coli [42]) [13, 42, 58]
<17> [14]
<19> (recombinant from E. coli [55]; recombinant chimeric mutants from E. coli [49]; recombinant wild-type and mutant G196V from E. coli [48]; partial

[16]; native and recombinant wild-type, the latter from E. coli and recombinant S43C mutant enzyme from E. coli [40]) [16, 22, 40, 48, 49, 55]
<20> (partial [16]) [16, 22]
<24> [19]
<25> (mitochondrial isozyme partially, glyoxysomal isozyme to homogeneity [20]) [20]
<26> [21]
<33> (recombinant wild-type, chimeric mutants, C-terminal deletion mutants, and D113 mutants from E. coli [49]; recombinant enzyme from E. coli [18,30]) [18, 30, 49]
<34> [23]
<35> [24]
<36> [25]
<38> [28]
<39> (antibody studies [32]) [32]
<40> (mitochondrial isozyme [45]; recombinant peroxisomal isozyme CS II from E. coli [41]) [41, 45]
<41> (both isoenzymes [35]) [35]
<42> [36, 37]
<44> [39]
<45> [44]
<46> [43]
<49> [47]
<50> (recombinant His-tagged fusion protein of citrate synthase and mitochondrial malate dehydrogenase from E. coli [50]) [50]
<51> (partial [51]) [51]
<52> (partial [51]) [51]
<59> (2 enzyme forms [60]) [60]

Renaturation

<9> (only after removal of 6 M denaturing guanidinium chloride by dialysis, not by dilution, dithiothreitol-dependent [31]) [31]

Crystallization

<16> (hanging drop vapour diffusion method, from 2-2.3 M ammonium sulfate, 2% v/v polyethylene glycol 400, 0.1 M HEPS, pH 6.0, X-ray diffraction analysis, structure determination and modeling: 3 identical dimer units arranged about a central 3-fold axis [58]) [58]
<33> (hanging drop vapour diffusion method, room temperature, 2 mg/ml protein concentration, precipitation by 0.1 M sodium citrate and 0.1 M ammonium phosphate, 20 mM Tris-HCl, pH 8.0, 25 mM KCl, X-ray structure analysis, structural basis of high thermostability [18]) [18]
<19,33> (crystal structure analysis of chimeric mutants with exchange of large and small subunits between Thermoplasma acidophilum and Pyrococcus furiosus [49]) [49]

Cloning

<3> (expression of a chimeric protein with one Escherichia coli domain in Escherichia coli, domain interactions, subunit interactions [42]) [42]
<9> (expression in citrate synthase deficient Escherichia coli strain [29]) [29]
<12> [61]
<16> (expression in citrate synthase deficient Escherichia coli strain [29]; expression of a chimeric protein with one Acinetobacter domain in Escherichia coli, domain interactions, subunit interactions [42]) [13, 29, 42]
<19> (expression in Escherichia coli [55]; expression of chimeric mutant in Escherichia coli citrate synthase deficient strain MOB154 [49]; expression of wild-type and mutant in Escherichia coli [40,48]) [40, 48, 49, 55]
<23> (cloning and expression of wild-type and mutants in Escherichia coli [57]) [57]
<25> (in vitro translation of glyoxysomal isozyme, heterologous expression in Xenopus laevis oocytes [20]) [20]
<33> (expression of wild-type, chimeric mutant and site-directed mutants in Escherichia coli citrate synthase deficient strain MOB154 [49]; overexpression in Escherichia coli JM105, DNA and amino acid sequence analysis [30]) [30, 49]
<37> (functional expression in citrate synthase deficient Escherichia coli mutant strain K214 under control of lacZ promotor, DNA and amino acid sequence analysis [27]) [27]
<40> (expression of labeled mitochondrial isozyme in Saccharomyces from plasmid, NMR measurements in vivo [45]; functional expression of peroxisomal isozyme CS II in citrate synthase deficient Escherichia coli mutant and wild-type Escherichia coli strain [41]) [41, 45]
<43> (cDNA cloning and sequencing, mRNA expression pattern during embryonic development, DNA and amino acid sequence analysis [52]) [52]
<49> (cloning and expression of citA in Aspergillus niger, 11fold overexpression does not increase the citrate activity level in vivo [47]) [47]
<50> (cloning of His-tagged fusion protein of citrate synthase and mitochondrial malate dehydrogenase, expression in Escherichia coli [50]) [50]
<58> (cloning and sequencing [59]) [59]
<60> (cloning and expression in Aspergillus nidulans, DNA sequence [62]) [62]
<61> (ectopic overexpression in Arabidopsis thaliana strain WS via infection with Agrobacterium tumefaciens, DNA sequence determination [63]) [63]

Engineering

A10E <23> (<23> site-directed mutagenesis, reduced k_{cat}, increased K_m for acetyl-CoA [57]) [57]
A361R <23> (<23> site-directed mutagenesis, slightly reduced k_{cat} , reduced K_m for acetyl-CoA and increased K_m for oxaloacetate, enhanced activity with propionyl-CoA [57]) [57]
A361R/A10E <23> (<23> site-directed mutagenesis, reduced k_{cat}, reduced K_m for acetyl-CoA [57]) [57]

D113A <33> (<33> site-directed mutagenesis, deletion of C-terminal amino acid residues which arrange the subunit contact, slightly increased K_m for acetyl-CoA and oxaloacetate, slightly increased reaction velocity, reduced thermostability [49]) [49]
D113S <33> (<33> site-directed mutagenesis, deletion of C-terminal amino acid residues which arrange the subunit contact, slightly decreased K_m for acetyl-CoA and oxaloacetate, slightly increased reaction velocity, reduced thermostability [49]) [49]
D362A <16> (<16> acetyl-CoA binding site mutant, reduced turnover, increased K_i for oxaloacetate and 2-oxoglutarate [33]) [33]
F383A <16> (<16> acetyl-CoA binding site mutant, reduced turnover [33]) [33]
G196V <19> (<19> site-directed mutagenesis, mutation interferes with dimerization, improper dimerization or dissociation of the dimer, reduced enzyme activity and conformational stability [48]) [48]
H229Q <16> (<16> active site mutant, reduced turnover, increased K_i for 2-oxoglutarate [33]) [33]
H264A <16> (<16> acetyl-CoA binding site mutant, reduced turnover, increased K_i for oxaloacetate and 2-oxoglutarate [33]) [33]
H305A <16> (<16> active site mutant, reduced turnover [33]) [33]
H309G <58> (<58> site-directed mutagenesis, mutant allelic strain, altered developmental phenotype [59]) [59]
H320G <9> (<9> effect of H320G substitution, solvent accessibility, and conformational changes on catalysis and activity [5]) [5]
K313L <23> (<23> site-directed mutagenesis [57]) [57]
K313L/A361R <23> (<23> site-directed mutagenesis [57]) [57]
K313L/A361R/A10E <23> (<23> site-directed mutagenesis [57]) [57]
R314L <16> (<16> active site mutant, reduced turnover [33]) [33]
R387L <16> (<16> active site mutant, reduced turnover [33]) [33]
R407L <16> (<16> active site mutant, reduced turnover, increased K_i for oxaloacetate and 2-oxoglutarate [33]) [33]
S43C <19> (<19> site-directed mutagenesis, 5.7fold reduced activity, unaltered K_m values for the substrates and unaltered thermostability [40]) [40]
Additional information <3, 9, 16, 19, 23, 33, 50, 55, 58, 61> (<61> construction of transgenic Arabidopsis thaliana plants ectopically overexpressing the Daucus carota citrate synthase gene via transformation with Agrobacterium tumefaciens, transgenic plants process the enzyme to its mature form, targeting into mitochondria, increased growth and phosphate accumulation [63]; <58> deletion mutants, altered developmental phenotypes, complete deletion reveals that the enzyme is more important for completion of meiosis than for catalytic activity [59]; <23> introduction of a loop into the active site of wild-type, mutant K313L, K313L/A361R/A10E , and mutant K313L/A361R, the latter showing increased thermostability and decreased temperature optimum for catalytic activity, K_m and activities [57]; <55> construction of transgenic Arabidopsis thaliana plants via Agrobacterium infection, functional expression and targeting to the mitochondria [54]; <50> no channeling of oxaloacetate between malate dehydrogenase and citrate synthase in a recombinant

fusion protein using a coupled assay [50]; <33> construction of mutant with disrupted inter-subunit ionic network by partly eleminating the C-terminal end amino acid residues, increased K_m for substrates, reduced thermostability [49]; <19,33> construction of chimeric enzyme mutants with mix of the large and small subunits of Thermoplasma acidophilum and Pyrococcus furiosus, functional analysis of the subunits [49]; <16> effect of active-site mutantions on substrate binding [33]; <9,16> engineering of non-allosteric pig heart enzyme in citrate synthase mutant of E. coli, studies on regulation [29]; <3,16> chimeric proteins [42]) [29, 33, 42, 49, 50, 54, 57, 59, 63]

Application

agriculture <55, 56, 61> (<56,61> overexpression of Daucus carota enzyme in Arabidopsis thaliana plant mitochondria improves the ability of the plants to grow in phosphorus limited soil due to enhanced citrate excretion from the roots [54,63]) [54, 63]
analysis <43> (<43> cellular stress marker [38]) [38]
biotechnology <9, 16> (<9,16> in vivo metabolic engineering [29]) [29]
synthesis <34> (<34> involved in poly 3-hydroxybutyrate formation [23]) [23]

6 Stability

pH-Stability

6-8.5 <45> [44]
6-9 <36> [25]
7 <13> (<13> unstable below [10]) [10]

Temperature stability

40 <45> (<45> stable [44]) [44]
47 <33> (<33> half-life 4.7 min [18]) [18]
50 <13, 14> (<13> in presence of oxaloacetate 70% of activity retained after 10 min, in absence of oxaloacetate complete loss of activity, acetyl-CoA does not protect [10]; <14> rapid inactivation [11]) [10, 11]
55 <36> [25]
70 <33> (<33> half-life 56.8 min [18]) [18]
80 <12, 19, 20> (<19> $t_{1/2}$ of recombinant wild-type is 10 min, $t_{1/2}$ of recombinant mutant G196V is 8 min, inactivation of the mutant occurs in 2 stages: a first slow and a second rapid one [48]; <12> stable up to 5 min [9]; <19> 65% of activity retained after 10 min [16]; <20> 90% of activity after 10 min [16]) [9, 16, 48]
90 <19, 26, 59> (<59> hexameric enzyme form, half-life: 18 min, dimeric enzyme form, half-life: 12 min [60]; <19> mutant G196V: $t_{1/2}$: below 1 min, $t_{1/2}$ wild-type: 5 min [48]; <26> 100% stable, 10 min [21]) [21, 48, 60]
95 <26> (<26> 90% loss of activity after 10 min [21]) [21]
100 <33> (<33> half-life: 17 min, structural basis [18]; <33> half-life: 20 min [30]) [18, 30]
101 <12> (<12> recombinant enzyme, stable [61]) [61]

104 <33> (<33> half-life: 1 min [30]) [30]
Additional information <5, 13, 19, 23, 24, 33> (<23> thermal stability of mutants [57]; <19,33> thermostability of wild-type and chimeric mutants [49]; <5> cold-lability in presence of 3 M KCl, but not with NaCl up to 5 M [6]; <13> oxaloacetate protects against heat denaturation [10, 19]; <19> effect of C43 mutation [40]) [6, 10, 19, 40, 49, 57]

General stability information

<1>, very labile, 97% loss of activity after 2 d at 0°C due to aggregation [2]
<4>, very labile, glycerol, 20% v/v or more stabilizes [3]
<5>, dilute solutions instable, best stabilized by high concentrations of NaCl or KCl [6]
<10>, destabilization by all salts, exception is ammonium sulfate, which stabilizes at low concentration [7]
<10>, dithiothreitol stabilizes for about 5 days [7]
<11>, CS II much more stable than CS I [8]
<17>, very instable without glycerol [14]
<40>, peroxisomal enzyme protected by oxaloacetate [41]
<51>, KCl stabilizes [51]
<51>, freeze-drying results in 70% loss of activity [51]

Storage stability

<5>, 4°C, Tris-HCl buffer, pH 7.6, high protein concentration, below 20% loss of activity in 2 months [6]
<9>, -16°C, saturated ammonium sulfate, crystalline enzyme, several years with minor loss of activity [4]
<9>, 0°C, below, phosphate buffer, pH 7.4, several weeks [4]
<10>, 0°C, below, Tris-HCl buffer, pH 7.0, 1 mM dithiothreitol and 5 mg/ml bovine serum albumin, several months [7]
<10>, 4°C, Tris-HCl buffer, pH 7.0, 1 mM dithiothreitol, 1 day [7]
<11>, 4°C, Tris-HCl buffer, pH 7.6, 0.2 M KCl, 20% loss of activity after 1 month, only CS II [8]
<13>, 4°C, phosphate buffer, pH 7.4, 20% loss of activity after 2 weeks [10]
<17>, 4°C, Tris-HCl buffer, pH 8.0, 20% v/v glycerol and 0.1 M KCl, at least 6 months [14]
<26>, -20°C, Tris-HCl buffer, pH 8.0, 1 year [21]
<35>, -20°C, potassium phosphate buffer, pH 7.2, 15% glycerol, several weeks [24]
<40>, -20°C, Tris-HCl buffer, pH 7.4, 50% glycerol, 1 month [41]
<46>, 4°C, Tris-HCl buffer, pH8.0, 50 mM glycerol, at least 1 week [43]
<51>, 4°C, 25 mM Tris-HCl, pH 7.7, loss of 80% activity within 3 days, 50 mM KCl protects [51]

References

[1] Rault-Leonardon, M.; Atkinson, M.A.L.; Slaughter, C.A.; Moomaw, C.R.; Srere, P.A.: Azotobacter vinelandii citrate synthase. Biochemistry, **34**, 257-263 (1995)

[2] Mahlen, A.: Purification and some properties of citrate synthase from Penicillium spiculisporum. Eur. J. Biochem., **29**, 60-66 (1972)

[3] Johnson, D.E.; Hanson, R.S.: Bacterial citrate synthases: purification, molecular weight and kinetic mechanism. Biochim. Biophys. Acta, **350**, 336-353 (1974)

[4] Stern, J.R.: Oxalacetate transacetase, (condensing enzyme, citrogenase). The Enzymes, 2nd Ed. (Boyer, P.D., Lardy, H., Myrbäck, K., eds.), **5**, 367-380 (1961)

[5] Kurz, L.C.; Shah, S.; Frieden, C.; Nakra, T.; Stein, R.E.; Drysdale, G.R.; Evans, C.T.; Srere, P.A.: Catalytic strategy of citrate synthase: subunit interactions revealed as a consequence of a single amino acid change in the oxaloacetate binding site. Biochemistry, **34**, 13278-13288 (1995)

[6] Higa, A.; Cazzulo, J.J.: Some properties of the citrate synthase from the extreme halophile, Halobacterium cutirubrum. Biochem. J., **147**, 267-274 (1975)

[7] Porter, J.S.; Wright, B.E.: Partial purification and characterization of citrate synthase from Dictyostelium discoideum. Arch. Biochem. Biophys., **181**, 155-163 (1977)

[8] Higa, A.I.; Massarini, E.; Cazzulo, J.J.: Purification and some properties of the citrate synthase from a marine Pseudomonas. Can. J. Microbiol., **24**, 215-221 (1978)

[9] Weitzman, P.D.J.: Anomalous citrate synthase from Thermus aquaticus. J. Gen. Microbiol., **106**, 383-386 (1978)

[10] Okabayashi, K.; Nakano, E.: Partial purification and properties of citrate synthase from sea urchin eggs. J. Biochem., **85**, 1061-1066 (1979)

[11] Alejandre, M.J.; Segovia, J.I.; Zaera, M.F.; Garcia-Peregrin, E.: Characteristics of citrate synthase from Agave americana L. leaves. Z. Pflanzenphysiol., **94**, 85-93 (1979)

[12] Iredale, S.E.: Properties of citrate synthase from Pisum sativum mitochondria. Phytochemistry, **18**, 1057 (1979)

[13] Robinson, M.S.; Easom, R.A.; Danson, M.J.; Weitzman, P.D.J.: Citrate synthase of Escherichia coli. Characterization of the enzyme from a plasmid-cloned gene and amplification of the intracellular levels. FEBS Lett., **154**, 51-54 (1983)

[14] Robinson, M.S.; Danson, M.J.; Weitzman, P.D.: Citrate synthase from a gram-positive bacterium. Biochem. J., **213**, 53-59 (1983)

[15] Danson, M.J.; Black, S.C.; Woodland, D.L.; Wood, P.A.: Citric acid cycle enzymes of the archaebacteria: citrate synthase and succinate thiokinase. FEBS Lett., **179**, 120-124 (1985)

[16] Grossebuter, W.; Görisch, H.: Partial purification and properties of citrate synthase from the thermoacidophilic archaebacteria Thermoplasma acido-

philum and Sulfolobus acidocaldarius. Syst. Appl. Microbiol., **6**, 119-124 (1985)
[17] Otto, R.: Regulatory and molecular properties of citrate synthases from methylotrophs. FEMS Microbiol. Lett., **34**, 191-194 (1986)
[18] Russell, R.J.M.; Ferguson, J.M.C.; Hough, D.W.; Danson, M.J.; Taylor, G.L.: The crystal structure of citrate synthase from the hyperthermophilic archaeon Pyrococcus furiosus at 1.9 Angstroem resolution. Biochemistry, **36**, 9983-9994 (1997)
[19] Jeffery, D.; Goodenough, P.W.; Weitzman, P.D.J.: Citrate synthase and malate dehydrogenase from tomato fruit. Phytochemistry, **27**, 41-44 (1988)
[20] Sautter, C.; Keller, G.; Hock, B.: Glyoxysomal citrate synthase from watermelon cotyledons: immunochemical localization and heterologous translation in Xenopus oocytes. Planta, **173**, 289-297 (1988)
[21] Loehlein-Wehrhahn, G.; Goepfert, P.; Eggerer, H.: Purification and properties of an archaebacterial enzyme: citrate synthase from Sulfolobus solfataricus. Biol. Chem. Hoppe-Seyler, **369**, 109-113 (1988)
[22] Smith, L.D.; Stevenson, K.J.; Hough, D.W.; Danson, M.J.: Citrate synthase from the thermophilic archaebacteria Thermoplasma acidophilum and Sulfolobus acidocaldarius. FEBS Lett., **225**, 277-281 (1987)
[23] Henderson, R.A.; Jones, C.W.: Poly-3-hydroxybutyrate production by washed cells of Alcaligenes eutrophus, purification, characterization and potential regulatory role of citrate synthase. Arch. Microbiol., **168**, 486-492 (1997)
[24] Sievers, M.; Stoeckli, M.; Teuber, M.: Purification and properties of citrate synthase from Acetobacter europaeus. FEMS Microbiol. Lett., **146**, 53-58 (1997)
[25] Takahashi, R.; Usui, K.; Sakuraba, T.; Tokuyama, T.: Purification and some properties of citrate synthase from a nitrite-oxidising chemoautotroph, Nitrobacter agilis ATCC 14123. J. Ferment. Bioeng., **77**, 97-99 (1994)
[26] Unger, E.A.; Vasconcelos, A.C.: Purification and characterization of mitochondrial citrate synthase. Plant Physiol., **89**, 719-723 (1989)
[27] Landschuetze, V.; Mueller-Roeber, B.; Willmitzer, L.: Mitochondrial citrate synthase from potato: predominant expression in mature leaves and young flower buds. Planta, **196**, 756-764 (1995)
[28] Morgunov, I.G.; Sharishev, A.A.; Mikulinskaya, O.V.; Sokolov, D.M.; Finogenova, T.V.: Isolation, purification and some properties of citrate synthase from citric acid-accumulating yeasts Yarrowia (Candida) lipolytica. Biokhimiya, **59**, 1320-1329 (1994)
[29] Evans, C.T.: Metabolic engineering of a non-allosteric citrate synthase in an Escherichia coli citrate synthase mutant. J. Mol. Recognit., **8**, 327-333 (1995)
[30] Muir, J.M.; Russell, R.J.M.; Hough, D.W.; Danson, M.J.: Citrate synthase from the hyperthermophilic archaeon, Pyrococcus furiosus. Protein Eng., **8**, 583-592 (1995)
[31] Kelly, S.M.; Price, N.C.: Reactivation of denatured citrate synthase. Int. J. Biochem., **24**, 627-630 (1992)

[32] Lee, S.; Park, C.; Yim, J.: Characterization of citrate synthase purified from Drosophila melanogaster. Mol. Cells, **7**, 599-604 (1997)

[33] Pereira, D.S.; Donald, L.J.; Hesfield, D.J.; Duckworth, H.W.: Active site mutants of Escherichia coli citrate synthase. J. Biol. Chem., **269**, 412-417 (1994)

[34] Sandor, A.; Johnson, J.H.; Srere, P.A.: Cooperation between enzyme and transporter in the inner mitochondrial membrane of yeast. J. Biol. Chem., **269**, 29609-29612 (1994)

[35] Mitchell, C.G.; Anderson, S.C.K.; El-Mansi, E.M.T.: Purification and characterization of citrate synthase isoenzymes from Pseudomonas aeruginosa. Biochem. J., **309**, 507-511 (1995)

[36] Kojima, H.; Chiba, J.; Watanabe, Y.; Numata, O.: Citrate synthase purified from Tetrahymena mitochondria is identical with Tetrahymena 14-nm filament protein. J. Biochem., **118**, 189-195 (1995)

[37] Takeda, T.; Kurasawa, Y.; Watanabe, Y.; Numata, O.: Polymerization of highly purified Tetrahymena 14-nm filament protein/citrate synthase into filaments and its possible roles in regulation of enzymatic activity. J. Biochem., **117**, 869-874 (1995)

[38] Sylven, C.; Kallner, A.; Jansson, F.: Regional distribution of citrate synthase and lactate dehydrogenase isoenzymes in the bovine heart. Acta Physiol. Scand., **136**, 331-337 (1989)

[39] James, K.D.; Bonete, M.J.; Byrom, D.; Danson, M.J.; Hough, D.W.: Citrate synthase from Haloverax volcanii: enzyme purification and gene cloning. Biochem. Soc. Trans., **20**, 12S (1991)

[40] Kocabryik, S.; Erduran, I.; Russell, R.J.M.; Danson, M.J.; Hough, D.W.: The effect of C43 mutation on thermostability and kinetic properties of citrate synthase from Thermoplasma acidophilum. Biochem. Biophys. Res. Commun., **224**, 224-228 (1996)

[41] Kispal, G.; Srere, P.A.: Studies on yeast peroxisomal citrate synthase. Arch. Biochem. Biophys., **286**, 132-137 (1991)

[42] Molgat, G.F.; Donald, L.J.; Duckworth, H.W.: Chimeric allosteric citrate synthases: construction and properties of citrate synthases containing domains from two different enzymes. Arch. Biochem. Biophys., **298**, 238-246 (1992)

[43] Belova, L.L.; Sokolov, A.P.; Morgunov, I.G.; Trotsenko, YuA.: Purification and characterization of citrate synthase from Methylobacterium extorquens - a methylotrophic producer of polyhydroxybutyrate. Biochemistry, **62**, 71-76 (1997)

[44] Asai, T.; Takahashi, R.; Fujiola, T.; Tokuyama, T.: Purification and properties of a citrate synthase from Nitrosomonas sp. TK794 and a comparison with the enzymes of another nitrifying bacteria. J. Ferment. Bioeng., **79**, 67-69 (1995)

[45] Haggie, P.M.; Brindle, K.M.: Mitochondrial citrate synthase is immobilized in vivo. J. Biol. Chem., **274**, 3941-3945 (1999)

[46] Ullian, M.E.; Robinson, C.J.; Evans, C.T.; Melnick, J.Z.; Fitzgibbon, W.R.: Role of citrate synthase in aldosterone-mediated sodium reabsorption. Hypertension, **35**, 875-879 (2000)

[47] Ruijter, G.J.G.; Panneman, H.; Xu, D.B.; Visser, J.: Properties of Aspergillus niger citrate synthase and effects of citA overexpression on citric acid production. FEMS Microbiol. Lett., **184**, 35-40 (2000)
[48] Kocabiyik, S.; Erduran, I.: The effect of valine substitution for glycine in the dimer interface of citrate synthase from Thermoplasma acidophilum on stability and activity. Biochem. Biophys. Res. Commun., **275**, 460-465 (2000)
[49] Arnott, M.A.; Michael, R.A.; Thompson, C.R.; Hough, D.W.; Danson, M.J.: Thermostability and thermoactivity of citrate synthases from the thermophilic and hyperthermophilic archaea, Thermoplasma acidophilum and Pyrococcus furiosus. J. Mol. Biol., **304**, 657-668 (2000)
[50] Pettersson, H.; Olsson, P.; Bulow, L.; Pettersson, G.: Kinetics of the coupled reaction catalyzed by a fusion protein of yeast mitochondrial malate dehydrogenase and citrate synthase. Eur. J. Biochem., **267**, 5041-5046 (2000)
[51] Tabrett, C.A.; Copeland, L.: Biochemical controls of citrate synthase in chickpea bacteroids. Arch. Microbiol., **173**, 42-48 (2000)
[52] Winger, Q.A.; Hill, J.R.; Watson, A.J.; Westhusin, M.E.: Characterization of a bovine cDNA encoding citrate synthase, and presence of citrate synthase mRNA during bovine pre-attachment development. Mol. Reprod. Dev., **55**, 14-19 (2000)
[53] Salomon, M.; Buchholz, F.: Effects of temperature on the respiration rates and the kinetics of citrate synthase in two species of Idotea (Isopoda, Crustacea). Comp. Biochem. Physiol. B, **125B**, 71-81 (2000)
[54] Koyama, H.; Kawamura, A.; Kihara, T.; Hara, T.; Takita, E.; Shibata, D.: Overexpression of mitochondrial citrate synthase in Arabidopsis thaliana improved growth on a phosphorus-limited soil. Plant Cell Physiol., **41**, 1030-1037 (2000)
[55] Kurz, L.C.; Drysdale, G.; Riley, M.; Tomar, M.A.; Chen, J.; Russell, R.J.M.; Danson, M.J.: Kinetics and mechanism of the citrate synthase from the thermophilic archaeon Thermoplasma acidophilum. Biochemistry, **39**, 2283-2296 (2000)
[56] Leek, B.T.; Mudaliar, S.R.; Henry, R.; Mathieu-Costello, O.; Richardson, R.S.: Effect of acute exercise on citrate synthase activity in untrained and trained human skeletal muscle. Am. J. Physiol., **280**, R441-447 (2001)
[57] Gerike, U.; Danson, M.J.; Hough, D.W.: Cold-active citrate synthase: mutagenesis of active-site residues. Protein Eng., **14**, 655-661 (2001)
[58] Nguyen, N.T.; Maurus, R.; Stokell, D.J.; Ayed, A.; Duckworth, H.W.; Brayer, G.D.: Comparative analysis of folding and substrate binding sites between regulated hexameric type II citrate synthases and unregulated dimeric type I enzymes. Biochemistry, **40**, 13177-13187 (2001)
[59] Ruprich-Robert, G.; Zickler, D.; Berteaux-Lecellier, V.; Velot, C.; Picard, M.: Lack of mitochondrial citrate synthase discloses a new meiotic checkpoint in a strict aerobe. EMBO J., **21**, 6440-6451 (2002)
[60] Karlsson, E.N.; Abou-Hachem, M.; Holst, O.; Danson, M.J.; Hough, D.W.: Rhodothermus marinus: a thermophilic bacterium producing dimeric and hexameric citrate synthase isoenzymes. Extremophiles, **6**, 51-56 (2002)

[61] Nordberg Karlsson, E.; Crennell, S.J.; Higgins, C.; Nawaz, S.; Yeoh, L.; Hough, D.W.; Danson, M.J.: Citrate synthase from Thermus aquaticus: a thermostable bacterial enzyme with a five-membered inter-subunit ionic network. Extremophiles, **7**, 9-16 (2003)
[62] Brock, M.; Fischer, R.; Linder, D.; Buckel, W.: Methylcitrate synthase from Aspergillus nidulans: implications for propionate as an antifungal agent. Mol. Microbiol., **35**, 961-973 (2000)
[63] Koyama, H.; Kawamura, A.; Kihara, T.; Hara, T.; Takita, E.; Shibata, D.: Overexpression of mitochondrial citrate synthase in Arabidopsis thaliana improved growth on a phosphorus-limited soil. Plant Cell Physiol., **41**, 1030-1037 (2000)

Decylcitrate synthase 2.3.3.2

1 Nomenclature

EC number
2.3.3.2

Systematic name
dodecanoyl-CoA:oxaloacetate C-dodecanoyltransferase (thioester-hydrolysing, 1-carboxyundecyl-forming)

Recommended name
decylcitrate synthase

Synonyms
(2S,3S)-2-hydroxytridecane-1,2,3-tricarboxylate oxaloacetate-lyase (CoA-acylating)
2-decylcitrate synthase
EC 4.1.3.23 (formerly)

CAS registry number
9068-72-8

2 Source Organism

<1> *Penicillium spiculisporum* (strain Lehman [1,2]) [1, 2]

3 Reaction and Specificity

Catalyzed reaction
lauroyl-CoA + H_2O + oxaloacetate = (2S,3S)-2-hydroxytridecane-1,2,3-tricarboxylate + CoA

Reaction type
condensation

Natural substrates and products
S lauroyl-CoA + H_2O + oxaloacetate <1> (Reversibility: ? <1> [1]) [1]
P (2S,3S)-2-hydroxytridecane-1,2,3-tricarboxylate + CoA

Substrates and products
S 11-formamidoundecanoyl-CoA + H_2O + oxaloacetate <1> (Reversibility: ? <1> [1]) [1]
P (2S,3S)-2-hydroxy-11-formamidotridecanoyl + CoA

S acetyl-CoA + H_2O + oxaloacetate <1> (Reversibility: ? <1> [2]) [2]
P citrate + CoA <1> [2]
S butanoyl-CoA + H_2O + oxaloacetate <1> (Reversibility: ? <1> [2]) [2]
P (2S,3S)-2-hydroxyhexane-1,2,3-tricarboxylate + CoA
S decanoyl-CoA + H_2O + oxaloacetate <1> (Reversibility: ? <1> [2]) [2]
P (2S,3S)-2-hydroxydodecane-1,2,3-tricarboxylate + CoA
S lauroyl-CoA + H_2O + 2-oxoglutarate <1> (Reversibility: ? <1> [2]) [2]
P ?
S lauroyl-CoA + H_2O + oxaloacetate <1> (Reversibility: ? <1> [1, 2]) [1, 2]
P (2S,3S)-2-hydroxytridecane-1,2,3-tricarboxylate + CoA
S lauroyl-CoA + H_2O + pyruvate <1> (Reversibility: ? <1> [2]) [2]
P ?
S myristoyl-CoA + H_2O + oxaloacetate <1> (Reversibility: ? <1> [2]) [2]
P (2S,3S)-2-hydroxypentadecane-1,2,3-tricarboxylate + CoA
S octanoyl-CoA + H_2O + oxaloacetate <1> (Reversibility: ? <1> [2]) [2]
P (2S,3S)-2-hydroxydecane-1,2,3-tricarboxylate + CoA
S palmitoyl-CoA + H_2O + oxaloacetate <1> (Reversibility: ? <1> [2]) [2]
P (2S,3S)-2-hydroxyheptadecane-1,2,3-tricarboxylate + CoA
S phenylacetyl-CoA + H_2O + oxaloacetate <1> (Reversibility: ? <1> [2]) [2]
P phenylcitrate + CoA <1> [2]

Inhibitors

5,5'-dithio-bis-(2-nitrobenzoic acid) <1> [1]
CoASSCoA <1> [1]
N-ethylmaleimide <1> [1]
NADH <1> (<1> at 5 mM, slight inhibition [1]) [1]
acetyl-CoA <1> [1]
iodoacetamide <1> [1]
lauroyl-CoA <1> (<1> strong substrate inhibition, does not follow Michaelis-Menten kinetics [1,2]) [1, 2]
p-chloromercuribenzoate <1> [1]
palmitoyl-CoA <1> [1]

K_m-Value (mM)

0.0001 <1> (lauroyl-CoA) [1]
0.03 <1> (11-formamidoundecanoyl-CoA) [1]
0.035 <1> (oxaloacetate, <1> at 0.021 mM lauroyl-CoA [1]) [1]
1.2 <1> (oxaloacetate) [2]

pH-Optimum

9 <1> (<1> above this pH [2]) [2]

Temperature optimum (°C)

30-35 <1> [2]

4 Enzyme Structure

Molecular weight

90000 <1> (<1> gel filtration [1]) [1]

Subunits

dimer <1> (<1> 2 * 45000, SDS-PAGE [1]) [1]

5 Isolation/Preparation/Mutation/Application

Purification

<1> [1, 2]

6 Stability

General stability information

<1>, thiols such as mercaptoethanol or cysteine stabilize [2]

Storage stability

<1>, -18°C, 0.5 M $(NH_4)_2SO_4$, 20% activity loss, 2 months [1]
<1>, 0°C, enzyme concentration 2 mg/ml, 80 h, 50% activity [2]

References

[1] Mahlen, A.: Properties of 2-decylcitrate synthase from Penicillium spiculisporum Lehman. Eur. J. Biochem., **22**, 104-114 (1971)
[2] Mahlen, A.; Gatenbeck, S.: A metabolic variation in Penicillium spiculisporum Lehman. Acta Chem. Scand., **22**, 2617-2623 (1968)

Citrate (Re)-synthase 2.3.3.3

1 Nomenclature

EC number
2.3.3.3

Systematic name
acetyl-CoA:oxaloacetate C-acetyltransferase [thioester-hydrolysing, (pro-R)-carboxymethyl-forming]

Recommended name
citrate (Re)-synthase

Synonyms
(R)-citrate synthase
Re-citrate-synthase
citrate oxaloacetate-lyase [(pro-3-R)-CH_2COO-acetyl-CoA]

CAS registry number
9077-70-7

2 Source Organism

<1> *Clostridium acidi-urici* [1, 2, 3, 6, 7]
<2> *Clostridium kluyveri* [3, 4, 5]
<3> *Clostridium cylindrosporum* [4]

3 Reaction and Specificity

Catalyzed reaction
acetyl-CoA + H_2O + oxaloacetate = citrate + CoA

Reaction type
condensation
enolization
hydrolysis

Substrates and products
S (R)-acetyl-CoA + H_2O + oxaloacetate <1> (Reversibility: ? <1> [6]) [6]
P (R)-citrate + CoA <1> [6]
S (R)-citryl-CoA + H_2O <1> (Reversibility: r <1> [7]) [7]
P (R)-citrate + CoA <1> [7]

S (R)-malyl-CoA + H_2O <1> (Reversibility: r <1> [7]) [7]
P (R)-malate + CoA <1> [7]
S acetyl-CoA + H_2O + oxaloacetate <1, 2, 3> (<1> stereospecific for (R)-citrate [1]; <1> no activity with propionyl-CoA, glyoxylate, pyruvate, 2-oxobutyrate, 2-oxoglutarate, 2-oxoisovalerate and 2-oxoisocaproate [2]; <1,2> stereospecific for (R)-citrate, no change from (R)-type to (S)-type in the presence of 4-chloromercuribenzoate [3]; <2> stereospecificity of citrate synthase in extracts can be changed reversibly from the (R)-type to the (S)-type by treatment with 4-chloromercuribenzoate [5]; <1> enzyme has acetyl-CoA enolase activity that is induced by 2-oxoglutarate and Co^{2+} [7]) (Reversibility: ? <1, 2, 3> [1, 3, 4]; r <1> [7]) [1-7]
P (R)-citrate + CoA <1, 2, 3> [1-7]

Inhibitors

2-oxoglutarate <1> (<1> 5 mM, 64% inhibition, competitive vs. oxaloacetate [2]) [2]
4-chloromercuribenzoate <1, 2> (<1> strong inhibition [2]; <1> 0.04 mM, 73% inhibition [3]; <2> 0.04 mM, 27% inhibition [3]; <2> 1 mM and 5 mM, 35 and 42% inhibition in extracts of Barker strain, respectively [5]) [2, 3, 5]
Cd^{2+} <1> (<1> complete inactivation [2]) [2]
H_2O_2 <1> [2]
Hg^{2+} <1> (<1> complete inactivation [2]) [2]
Zn^{2+} <1> (<1> above 0.02 mM, activation below [2]) [2]
iodoacetamide <1> (<1> strong inhibition [2]) [2]
Additional information <1, 2, 3> (<1> not inhibited by 5,5'-dithiobis(2-nitrobenzoate) [2]; <2,3> citrate synthase in extracts is not inhibited by EDTA [4]) [2, 4]

Activating compounds

2-mercaptoethanol <1> (<1> 10-20 mM, reducing reagent is required for maximal activity [1]; <1> purified enzyme does not require 2-mecaptoethanol or anaerobic conditiones for maximal activity [2]) [1, 2]
L-cysteine <1> (<1> reducing reagent is required for maximal activity [1]) [1]
glutathione <1> (<1> reducing reagent is required for maximal activity [1]) [1]

Metals, ions

Co^{2+} <1> (<1> can replace Mn^{2+} [2]) [2]
Mn^{2+} <1> (<1> 0.1 mM, required for maximal activity [1]) [1]
Zn^{2+} <1> (<1> activation below 0.02 mM, inhibition above [2]) [2]
Additional information <2, 3> (<2,3> citrate synthase in extracts is not activated by Mn^{2+} [4]) [4]

Specific activity (U/mg)

5.5 <1> [2]

K_m-Value (mM)
0.0015 <1> ((R)-citryl-CoA) [7]
0.02 <1> ((R)-malyl-CoA) [7]
0.04 <1> (oxaloacetate) [2]
0.05 <1> (acetyl-CoA) [2]

pH-Optimum
8 <1> [7]
8-8.5 <1> (<1> sharp decrease in activity above [1]) [1]

pH-Range
6.5-8.5 <1> (<1> approx. 50% activity at pH 7.0 [1]) [1]

5 Isolation/Preparation/Mutation/Application

Purification
<1> (isoelectric precipitation, DEAE-cellulose, DEAE-Sephadex [1]; isoelectric precipitate, DEAE-cellulose, ammonium sulfate, DEAE-Sephadex, ammonium sulfate, gel filtration [2]) [1, 2]
<2> [3]

6 Stability

Oxidation stability
<1>, reducing reagents and anaerobic conditions are required for maximal activity [1]

Storage stability
<1>, refrigerator, 24 h, 10% loss of activity [1]

References

[1] Gottschalk, G.: Partial purification and some properties of the (R)-citrate synthase from Clostridium acidi-urici. Eur. J. Biochem., **7**, 301-106 (1969)
[2] Goschalk, G.; Dittbrenner, S.: Properties of (R)-citrate synthase from Clostridium acici-urici. Hoppe-Seyler's Z. Physiol. chem., **351**, 1183-1190 (1970)
[3] Dittbrenner, S.; Chowdhury, A.A.; Gottschalk, G.: The stereospecificity of the (R)-citrate synthase in the presence of *p*-chloromercuribenzoate. Biochem. Biophys. Res. Commun., **36**, 802-808 (1969)
[4] Stern, J.R.; O'brien, R.W.: Failure of ethylenediaminetetraacetate or manganese to affect (R)-citrate synthesis in Clostridium kluyveri and Clostridium cylndrosporum. FEBS Lett., **4**, 72-74 (1969)
[5] O'Brien, R.W.; Stern, J.R.: Reversal of the stereospecificity of the citrate synthase of Clostridium kluyveri by *p*-chloromercuribenzoate. Biochem. Biophys. Res. Commun., **34**, 271-276 (1969)

[6] Wunderwald, P.; Buckel, W.; Lenz, H.; Buschmeier, V.; Eggerer, H.; Gottschalk, G.; Cornforth, J.W.; Redmond, J.W.; Mallaby, R.: Stereochemistry of the re-citrate-synthase reaction. Eur. J. Biochem., **24**, 216-221 (1971)
[7] Gottschalk, G.; Dittbrenner, S.; Lenz, H.; Eggerer, H.: re-Citrate synthase reaction. Eur. J. Biochem., **26**, 455-461 (1972)

Decylhomocitrate synthase 2.3.3.4

1 Nomenclature

EC number
2.3.3.4

Systematic name
dodecanoyl-CoA:2-oxoglutarate C-dodecanoyltransferase (thioester-hydrolysing, 1-carboxyundecyl-forming)

Recommended name
decylhomocitrate synthase

Synonyms
2-decylhomocitrate synthase
3-hydroxytetradecane-1,3,4-tricarboxylate 2-oxoglutarate-lyase (CoA-acylating)
EC 4.1.3.29 (formerly)

CAS registry number
51845-40-0

2 Source Organism

<1> *Penicillium spiculisporum* [1]

3 Reaction and Specificity

Catalyzed reaction
dodecanoyl-CoA + H_2O + 2-oxoglutarate = (3S,4S)-3-hydroxytetradecane-1,3,4-tricarboxylate + CoA

Reaction type
condensation

Natural substrates and products
S dodecanoyl-CoA + H_2O + 2-oxoglutarate <1> (Reversibility: ? <1> [1]) [1]
P (3S,4S)-3-hydroxytetradecane-1,3,4-tricarboxylate + CoA

Substrates and products
S decanoyl-CoA + H_2O + 2-oxoglutarate <1> (Reversibility: ? <1> [1]) [1]
P (3S,4S)-3-hydroxydodecane-1,3,4-tricarboxylate + CoA

S dodecanoyl-CoA + H_2O + 2-oxoglutarate <1> (Reversibility: ? <1> [1]) [1]
P (3S,4S)-3-hydroxytetradecane-1,3,4-tricarboxylate + CoA <1> [1]

Inhibitors
5,5'-dithiobis(2-nitrobenzoic acid) <1> (<1> enzyme inactivation, lauroyl-CoA or palmitoyl-CoA protects [1]) [1]
oxaloacetate <1> (<1> competitive to 2-oxoglutarate [1]) [1]
palmitoyl-CoA <1> (<1> partially competitive with lauroyl-CoA [1]) [1]

Specific activity (U/mg)
1.25 <1> [1]

K_m-Value (mM)
0.001 <1> (lauroyl-CoA) [1]
0.75 <1> (2-oxoglutarate) [1]

K_i-Value (mM)
0.4-0.8 <1> (oxaloacetate, <1> competitive inhibitor [1]) [1]

pH-Optimum
8.3 <1> [1]

pH-Range
7.3-9 <1> (<1> 75% of maximal activity at pH 7.5 and 9 [1]) [1]

4 Enzyme Structure

Subunits
? <1> (<1> ? * 43000, SDS-PAGE [1]) [1]

5 Isolation/Preparation/Mutation/Application

Purification
<1> [1]

References

[1] Mahlen, A.: Purification and properties of 2-decylhomocitrate synthase from Penicillium spiculisporum. Eur. J. Biochem., **38**, 32-39 (1973)

2-Methylcitrate synthase — 2.3.3.5

1 Nomenclature

EC number
2.3.3.5

Systematic name
propanoyl-CoA:oxaloacetate C-propanoyltransferase (thioester-hydrolysing, 1-carboxyethyl-forming)

Recommended name
2-methylcitrate synthase

Synonyms
2-methylcitrate oxaloacetate-lyase
EC 4.1.3.31 (formerly)
MCA condensing enzyme
MCS
PrpC <16> [12]
citrate synthase II <22> [7]
methylcitrate synthase
methylcitrate synthetase

CAS registry number
57827-78-8

2 Source Organism

<-2> no activity in *Sus scrofa* (pig [8]) [8]
<-1> no activity in *Candida rugosa* (IFO 0750, NPA-1, IFO 1364 [5]) [5]
<1> *Arthrobacter sp.* (psychrotolerant antarctic eubacterium, strain DS2-3R [8]) [8]
<2> *Aspergillus nidulans* [9, 10, 12]
<3> *Aspergillus niger* (IFO 6662 [5]) [5]
<4> *Burkholderia cepacia* (contig261 [10]) [10]
<5> *Burkholderia sacchari* (strain IPT101 [10]) [10, 12]
<6> *Candida catenulata* (CBS 565, IFO 0720, CBS 1904, CBS 2015, CBS 6174, OM-102 [5]) [5]
<7> *Candida curvata* (IFO 0732 [5]) [5]
<8> *Candida gropengiesseri* (IFO 0569 [5]) [5]
<9> *Candida humicola* (IFO 0753 [5]) [5]
<10> *Candida kefyr* (IFO 0586, IFO 0882 [5]) [5]
<11> *Candida magnoliae* (IFO 0661 [5]) [5]

<12> *Candida pararugosa* (IFO 0966 [5]) [5]
<13> *Citeromyces matritensis* (IFO 0651 [5]) [5]
<14> *Corynebacterium glutamicum* (ATCC 13032, prp gene cluster, GenBank accession no. AF434798 [12]) [12]
<15> *Corynebacterium glutamicum* (ATCC 13032, prp gene cluster, GenBank accession no. AF434799 [12]) [12]
<16> *Corynebacterium glutamicum* (ATCC 13032 [12]) [12]
<17> *Cryptococcus neoformans* (IFO 0545 [5]) [5]
<18> *Debaryomyces hansenii* (IFO 0032 [5]) [5]
<19> *Debaryomyces polymorphus* (IFO 1363 [5]) [5]
<20> *Escherichia coli* (O157:H17 [10]) [10]
<21> *Escherichia coli* (prp gene cluster, GenBank accession no. U73857 [7,12]) [7, 12]
<22> *Escherichia coli* (JM109 [8]; K12 [7,8,10]; O157:H17 [10]) [2, 7-13]
<23> *Escherichia coli* (SWISSPROT acc. No. P31660 [11]) [11]
<24> *Hansenula anomala* (IFO 0121 [5]) [5]
<25> *Hansenula polymorpha* (IFO 0799 [5]) [5]
<26> *Hansenula saturnus* (IAM 4094 [5]) [5]
<27> *Legionella pneumophila* [10]
<28> *Lipomyces starkeyi* (IFO 0678 [5]) [5]
<29> *Metschnikowia pulcherima* (IFO 0561 [5]) [5]
<30> *Mucor rouxianus* (IFO 5773 [5]) [5]
<31> *Neisseria meningitidis* [9, 10]
<32> *Neisseria meningitidis* (serogroup B [10]) [10]
<33> *Neisseria meningitidis* (serogroup A [10]) [10]
<34> *Neurospora crassa* (IFO 6067 [5]) [5]
<35> *Pichia haplophila* (IFO 0947 [5]) [5]
<36> *Pseudomonas aeruginosa* (PAO1 [10]) [10]
<37> *Pseudomonas aeruginosa* (PAO [4]; PAO1 [10]) [4, 5, 9, 10, 12, 13]
<38> *Pseudomonas fluorescens* (contig267 [10]) [10]
<39> *Pseudomonas putida* (KT2440 [9,10]; contig10768 [10]) [4, 9, 10]
<40> *Pseudomonas syringae* (contig5376 [10]) [10]
<41> *Pseudomonas citronellolis* [4]
<42> *Pyrococcus furiosus* [8, 10]
<43> *Ralstonia metallidurans* (formerly Ralstonia eutropha, HF39 prp gene cluster, GenBank accession no. AF32554 [10,12]) [10, 12]
<44> *Ralstonia metallidurans* (formerly Ralstonia eutropha, prp gene cluster, GenBank accession no. AF331923 [12]) [12]
<45> *Ralstonia metallidurans* (formerly Ralstonia eutropha, HF39 [9,10]; CH34 [9,10]) [9-12]
<46> *Rhizopus chinensis* (IAM 6003 [5]) [5]
<47> *Rhodothermus marinus* (R-10 (DSM 4252) [13]) [13]
<48> *Saccharomyces cerevisiae* (SANK 50182 [5]; CBS 8066 [6]) [5-8]
<49> *Saccharomycopsis fibuligera* (IFO 1711, IFO 1744 [5]) [5]
<50> *Salmonella enterica* (serovar Typhimurium [10]) [10]
<51> *Salmonella enterica* (serovar Typhimurium, prp gene cluster, GenBank accession no. U51879 [7,12]) [7, 12]

<52> *Salmonella enterica* (serovar Typhimurium LT2 [9-11]; synonym Salmonella typhimurium [7,8]) [7-11]
<53> *Shewanella putrefaciens* (contig93 [10]) [10]
<54> *Thermoplasma acidophilum* [8]
<55> *Torulaspora delbrueckii* (IFO 0381 [5]) [5]
<56> *Ustilago crus-galli* (F-B-6 [5]) [5]
<57> *Ustilago utriculosa* (F-B-5 [5]) [5]
<58> *Vibrio cholerae* [9, 10]
<59> *Vibrio cholerae* [10]
<60> *Yarrowia lipolytica* (formerly Candida lipolytica [1,2,4,7,11,12]; Candida lipolytica R-2 [1]; synonym Saccharomycopsis lipolytica IFO 1457, IFO 1542 [5]; IFO 1659 [1-3,5]) [1-8, 11, 12]

3 Reaction and Specificity

Catalyzed reaction

propanoyl-CoA + H_2O + oxaloacetate = (2R,3S)-2-hydroxybutane-1,2,3-tricarboxylate + CoA (the enzyme acts on acetyl-CoA, propanoyl-CoA, butanoyl-CoA and pentanoyl-CoA. The relative rate of condensation of acetyl-CoA and oxaloacetate is 140% of that of propanoyl-CoA and oxaloacetate, but the enzyme has been separated from EC 2.3.3.1 citrate (Si) synthase. Oxaloacetate cannot be replaced by glyoxylate, pyruvate or 2-oxoglutarate)

Reaction type

condensation

Natural substrates and products

S propanoyl-CoA + H_2O + oxaloacetate <1-60> (<5, 16, 22, 36, 37, 45, 48, 52, 60> first key enzyme of the methylcitric acid cycle [2-5, 9, 10, 12]; <60> physiological role of the methylcitric acid cycle is a catabolic reaction sequence concerning the oxidation of propionate into pyruvate, regulatory enzyme of this cycle [2]; <48> 2-methylcitrate pathway is the major pathway of propionate metabolism [6]; <5, 22> implicated in the metabolism of propionate [10, 13]) (Reversibility: ? <1-60> [1-13]) [1-13]
P 2-methylcitrate + CoA

Substrates and products

S acetyl-CoA + H_2O + oxaloacetate <34, 47, 60> (<34, 60> uses acetyl-CoA in preference to propionyl-CoA [4, 5]) (Reversibility: ? <34, 47, 60> [2, 4, 5, 13]) [2, 4, 5, 13]
P citrate + CoA
S n-butyryl-CoA + H_2O + oxaloacetate <60> (Reversibility: r <60> [2]) [2]
P 2-ethylcitrate + CoA
S n-valeryl-CoA + H_2O + oxaloacetate <60> (Reversibility: r <60> [2]) [2]
P 2-propylcitrate + CoA
S propanoyl-CoA + H_2O + oxaloacetate <1-60> (Reversibility: ? <1-60> [1-13]) [1-13]
P 2-methylcitrate + CoA

S Additional information <37> (<22,37> enzyme is induced by growth on propionate, propanol and heptanoate [4,8,13]; <37> enzyme is relatively specific for its acetyl-CoA substrate, utilizing propionyl-CoA and not acetyl-CoA [4]; <47,60> constitutive [3,4,13]; <60> enzyme is specific for oxaloacetate, does not catalyze liberation of CoA from acetyl- or propionyl-CoA in presence of glyoxylate, pyruvate or α-ketoglutarate, isobutyryl-CoA and isovaleryl-CoA are no substrates [2]) [2-4, 13]

P ?

Inhibitors

ADP <60> (<60> slight inhibition [2]) [2]
ATP <60> [2]
H_2O_2 <60> [2]
Hg^{2+} <60> [2]
NADH <60> [2]
NADPH <60> [2]
Zn^{2+} <60> [2]
monoiodoacetamide <60> [2]
p-chloromercuribenzoate <60> [2]

Activating compounds

isocitrate <60> (<60> 1 mM, relative activity 101% [2]) [2]

Metals, ions

Ca^{2+} <60> (<60> 1 mM, relative activity 104% [2]) [2]
Fe^{2+} <60> (<60> 0.02 mM, relative activity 104% [2]) [2]
Mg^{2+} <60> (<60> 1 mM, relative activity 102% [2]) [2]

Specific activity (U/mg)

0.002 <6> (<6> CBS 1904, CBS 2015 [5]) [5]
0.005 <6> (<6> OM-102 [5]) [5]
0.0051 <60> (<60> nuclei-free homogenate, glucose grown cells [3]) [3]
0.006 <6> (<6> CBS 6174 [5]) [5]
0.007 <6, 48> (<6> IFO 0720 [5]) [5]
0.008 <17> [5]
0.009 <13> [5]
0.0092 <60> (<60> nuclei-free homogenate, n-alkane grown cells [3]) [3]
0.01 <46> [5]
0.011 <30> [5]
0.012 <3, 6, 60> (<6> CBS 565 [5]; <60> IFO 1659, propionyl-CoA [1]) [1, 5]
0.014 <7, 57> [5]
0.015 <49> (<49> IFO 1711 [5]) [5]
0.016 <18, 49, 55> (<49> IFO 1744 [5]) [5]
0.017 <26> [5]
0.018 <24, 34> [5]
0.019 <60> (<60> heavy particulate fraction, glucose grown cells [3]) [3]
0.022 <12> [5]
0.023 <8> [5]
0.025 <25> [5]

0.027 <9, 29> [5]
0.028 <60> (<60> IFO 1457 [5]) [5]
0.03 <60> (<60> IFO 1659 [5]; <60> heavy particulate fraction, n-alkane grown cells [3]) [3, 5]
0.034 <60> (<60> IFO 1542 [5]) [5]
0.038 <19> [5]
0.042 <10> (<10> CBS 0882 [5]) [5]
0.052 <60> (<60> R-2, propionyl-CoA [1]) [1]
0.053 <56> [5]
0.056 <10> (<10> CBS 0586 [5]) [5]
0.065 <35> [5]
0.086 <11> [5]
0.116 <28> [5]
0.12 <60> (<60> partially purified preparation [1]) [1]
0.21 <60> (<60> IFO 1659, acetyl-CoA [1]) [1]
0.22 <60> (<60> R-2, acetyl-CoA [1]) [1]
0.33 <22> [7]
0.436 <60> (<60> substrate propionyl-CoA [2]) [2]
0.623 <60> (<60> substrate acetyl-CoA [2]) [2]
28 <47> [13]

K_m-Value (mM)

0.002 <47> (oxaloacetate, <47> second substrate propionyl-CoA [13]) [13]
0.0029 <60> (acetyl-CoA) [2]
0.003 <1, 47> (oxaloacetate) [8, 13]
0.003 <47> (propionyl-CoA) [13]
0.005 <22> (oxaloacetate, <22> completed mutant strain W620 [8]) [8]
0.0053 <60> (propionyl-CoA) [2]
0.007 <47> (acetyl-CoA) [13]
0.016 <1> (propionyl-CoA) [8]
0.017 <22> (propionyl-CoA) [7]
0.022 <60> (butyryl-CoA) [2]
0.025 <60> (oxaloacetate) [2]
0.025 <60> (valeryl-CoA) [2]
0.037 <22> (propionyl-CoA, <22> completed mutant strain W620 [8]) [8]
0.101 <22> (acetyl-CoA, <22> completed mutant strain W620 [8]) [8]

pH-Optimum

8-8.5 <60> [2]
9 <22> [7]

pH-Range

6-9.5 <60> [2]

Temperature optimum (°C)

45 <60> [2]
45-50 <22> [7]
85 <47> [13]

Temperature range (°C)
10-50 <60> [2]
40-90 <47> [13]

4 Enzyme Structure

Molecular weight
42570 <16> (<16> prpC1, amino acid sequence calculated from cDNA [12]) [12]
42600 <16> (<16> prpC2, amino acid sequence calculated from cDNA [12]) [12]
42690 <5> (<5> amino acid sequence calculated from cDNA [10]) [10]
42720 <45> (<45> amino acid sequence [9]) [9]
43000 <22> (<22> SDS-PAGE [7]) [7]
43100 <22> (<22> amino acid sequence calculated from cDNA [11]) [11]

Subunits
dimer <22, 47> [13]

5 Isolation/Preparation/Mutation/Application

Localization
mitochondrion <60> [3]

Purification
<1> [8]
<22> [7, 11]
<37> (partially [4]) [4]
<42> [8]
<47> [13]
<54> [8]
<60> (partial [1]) [1, 2]

Cloning
<5> (prpC cloned, sequenced and expressed in E.coli [10]) [10]
<16> (genome sequencing, prpDBC gene cluster, prpC2 gene for 2-methylcitrate synthase, cloned into vector pTrc99a and expressed in Escherichia coli [12]) [12]
<22> (gene prpC sequenced [8,11,13]; pRES1, prpC gene cloned into mutant Escherichia coli W620 [8]) [8, 11, 13]
<37> (gene sequenced [13]) [13]
<45> (prpC encoded enzyme complements a prpC mutant of Salmonella enterica serovar Typhimurium, transduction of genomic DNA ligated into the cosmid pHC79 to Escherichia coli S17-1 [9]) [9]
<52> (gene prpC sequenced [11]) [11]

Application

agriculture <2> (<2> implications for propionate as an antifungal agent [9]) [9]

6 Stability

Temperature stability

50 <60> (<60> activity completely lost after 10 min [2]) [2]
95 <47> (<47> half-life at 90°C is 12 min, active even at mesophilic temperatures, at 40°C enzyme retains 20-30% of the maximum activity [13]) [13]

References

[1] Tabuchi, T.; Uchiyama, H.: Methylcitrate condensing and methylisocitrate cleaving enzymes; evidence for the pathway of oxidation of propionyl-CoA to pyruvate via C7-tricarboxylic acids. Agric. Biol. Chem., **39**, 2035-2042 (1975)

[2] Uchiyama, H.; Tabuchi, T.: Properties of methylcitrate synthase from Candida lipolytica. Agric. Biol. Chem., **40**, 1411-1418 (1976)

[3] Uchiyama, H.; Ando, M.; Toyonaka, Y.; Tabuchi, T.: Subcellular localization of the methylcitric-acid-cycle enzymes in propionate metabolism of Yarrowia lipolytica. Eur. J. Biochem., **125**, 523-527 (1982)

[4] Watson, D.; Lindel, D.L.; Fall, R.: Pseudomonas aeruginosa contains an inducible methylcitrate synthase. Curr. Microbiol., **8**, 17-21 (1983)

[5] Miyakoshi, S.; Uchiyama, H.; Someya, T.; Satoh, T.; Tabuchi, T.: Distribution of the methylcitric acid cycle and β-oxidation pathway for propionate catabolism in fungi. Agric. Biol. Chem., **51**, 2381-2387 (1987)

[6] Pronk, J.T.; van der Linden-Beuman, A.; Verduyn, C.; Scheffers, W.A.; van Dijken, J.P.: Propionate metabolism in Saccharomyces cerevisiae: implications for the metabolon hypothesis. Microbiology, **140**, 717-722 (1994)

[7] Textor, S.; Wendisch, V.F.; De Graaf, A.A.; Mueller, U.; Linder, M.I.; Linder, D.; Buckel, W.: Propionate oxidation in Escherichia coli: evidence for operation of a methylcitrate cycle in bacteria. Arch. Microbiol., **168**, 428-436 (1997)

[8] Gerike, U.; Hough, D.W.; Russell, N.J.; Dyall-Smith, M.L.; Danson, M.J.: Citrate synthase and 2-methylcitrate synthase: structural, functional and evolutionary relationships. Microbiology, **144**, 929-935 (1998)

[9] Braemer, C.O.; Steinbuechel, A.: The methylcitric acid pathway in Ralstonia eutropha: new genes identified involved in propionate metabolism. Microbiology, **147**, 2203-2214 (2001)

[10] Braemer, C.O.; Silva, L.F.; Gomez, J.G.C.; Priefert, H.; Steinbuechel, A.: Identification of the 2-methylcitrate pathway involved in the catabolism of propionate in the polyhydroxyalkanoate-producing strain Burkholderia sacchari IPT101T and analysis of a mutant accumulating a copolyester with

higher 3-hydroxyvalerate content. Appl. Environ. Microbiol., **68**, 271-279 (2002)
[11] Brock, M.; Maerker, C.; Schuetz, A.; Voelker, U.; Buckel, W.: Oxidation of propionate to pyruvate in Escherichia coli: Involvement of methylcitrate dehydratase and aconitase. Eur. J. Biochem., **269**, 6184-6194 (2002)
[12] Claes, W.A.; Puehler, A.; Kalinowski, J.: Identification of two prpDBC gene clusters in Corynebacterium glutamicum and their involvement in propionate degradation via the 2-methylcitrate cycle. J. Bacteriol., **184**, 2728-2739 (2002)
[13] Karlsson, E.N.; Abou-Hachem, M.; Holst, O.; Danson, M.J.; Hough, D.W.: Rhodothermus marinus: a thermophilic bacterium producing dimeric and hexameric citrate synthase isoenzymes. Extremophiles, **6**, 51-56 (2002)

2-Ethylmalate synthase 2.3.3.6

1 Nomenclature

EC number
2.3.3.6

Systematic name
acetyl-CoA:2-oxobutanoate C-acetyltransferase (thioester-hydrolysing, carboxymethyl-forming)

Recommended name
2-ethylmalate synthase

Synonyms
(R)-2-ethylmalate 2-oxobutanoyl-lyase (CoA-acetylating)
2-ethylmalate-3-hydroxybutanedioate synthase
formely 4.1.3.3
propylmalate synthase
propylmalic synthase
Additional information (formerly wrongly included with EC 2.3.3.7, 3-ethylmalate synthase)

CAS registry number
9024-01-5

2 Source Organism

<1> *Saccharomyces cerevisiae* [1]
<2> *Pseudomonas aeruginosa* [2]

3 Reaction and Specificity

Catalyzed reaction
acetyl-CoA + H_2O + 2-oxobutanoate = (R)-2-ethylmalate + CoA (<1> also acts on (R)-2-(n-propyl)-malate [1])

Reaction type
acyl group transfer

Substrates and products
S acetyl-CoA + H_2O + 2-ketoisovalerate <2> (Reversibility: ? <2> [2]) [2]
P ?

S acetyl-CoA + H_2O + 2-ketovalerate <1, 2> (Reversibility: ? <1, 2> [1, 2]) [1, 2]
P (R)-2-(n-propyl)-malate + CoA <1, 2> [1, 2]
S acetyl-CoA + H_2O + 2-oxobutanoate <1, 2> (Reversibility: ir <2> [2]; ? <1> [1]) [1, 2]
P (R)-2-ethylmalate + CoA <1, 2> [1, 2]
S acetyl-CoA + H_2O + pyruvate <2> (Reversibility: ? <2> [2]) [2]
P ?
S Additional information <2> (<2> not: 2-ketocaproate and 2-ketoisocaproate [2]) [2]
P ?

Inhibitors
D-leucine <2> (<2> weak [2]) [2]
L-isoleucine <2> (<2> weak [2]) [2]
L-leucine <2> (<2> 50% inhibition [2]) [2]
Mg^{2+} <1> [1]
cysteine <1> [1]

Activating compounds
$(NH_4)_2SO_4$ <1> (<1> maximum of activity: about 45% $(NH_4)_2SO_4$ and decrease in activity at higher concentrations [1]) [1]

Metals, ions
Mg^{2+} <2> (<2> in the absence of Mg^{2+} product is formed to an extent of 64% in the complete reaction [2]) [2]

Specific activity (U/mg)
Additional information <1> [1]

pH-Optimum
6.5 <2> [2]

pH-Range
Additional information <2> (<2> product formation doubled between pH 6.0 and 6.5, but there is a broud shoulder of activity between pH 6.5 and 7.5 [2]) [2]

6 Stability

Temperature stability
Additional information <1> (<1> heating in boiling water for 1 min results in a total loss of activity [1]) [1]

References

[1] Strassman, M. and Ceci, L.N.: A study of acetyl-CoA condensation with α-keto acids. Arch. Biochem. Biophys., **119**, 420-428 (1967)
[2] Rabin, R.; Salomon, I.I.; Bleiweis, A.S.; Carlin, J.; Ajl, S.J.: Metabolism of ethylmalic acids by Pseudomonas aeruginosa. Biochem.J., **7**, 377-389 (1968)

3-Ethylmalate synthase 2.3.3.7

1 Nomenclature

EC number
2.3.3.7

Systematic name
butanoyl-CoA:glyoxylate C-butanoyltransferase (thioester-hydrolysing, 1-carboxypropyl-forming)

Recommended name
3-ethylmalate synthase

Synonyms
2-ethyl-3-hydroxybutanedioate synthase
3-ethylmalate glyoxylate-lyase (CoA-butanoylating)
formely 4.1.3.10

CAS registry number
9024-01-5

2 Source Organism

<1> *Pseudomonas aeruginosa* [1, 2]
<2> *Escherichia coli* [3]

3 Reaction and Specificity

Catalyzed reaction
butanoyl-CoA + H_2O + glyoxylate = 3-ethylmalate + CoA

Natural substrates and products
S butanoyl-CoA + H_2O + glyoxylate <1, 2> (Reversibility: ? <1, 2> [1-3]) [1-3]
P 3-ethylmalate + CoA <1> [1]

Substrates and products
S butanoyl-CoA + H_2O + glyoxylate <1, 2> (Reversibility: ? <1, 2> [1-3]) [1-3]
P 3-ethylmalate + CoA <1> [1]

Metals, ions
Mg^{2+} <1> (<1> reduced activity in absence of Mg^{2+} [1]) [1]

Specific activity (U/mg)

0.009 <1> (<1> activity in fraction of 40-60% $(NH_4)_2SO_4$ [1]) [1]
Additional information <1> [1]

6 Stability

pH-Stability

7 <1> (<1> complete inactivation [1]) [1]

Storage stability

<1>, -20°C, inactivation in 24 h [1]

References

[1] Rabin, R; Reeves, H.C.; Ajl, S.J.: β-Ethylmalate synthase. J. Bacteriol., **86**, 937-944 (1963)
[2] Wegener, W.S.; Reeves, H.C.; Rabin, R.; Ajl, S.J.: A radioactive assay for malate synthase and other glyoxylate condensing enzymes. Methods Enzymol., **13**, 362-365 (1969)
[3] Wegener, W.S.; Furmanski, P.; Ajl, S.J.: Selection of mutants constitutive for several glyoxylate condensing enzymes during growth on valeric acid. Biochim. Biophys. Acta, **144**, 34-50 (1967)

ATP Citrate synthase 2.3.3.8

1 Nomenclature

EC number
2.3.3.8

Systematic name
acetyl-CoA:oxaloacetate C-acetyltransferase [(pro-S)-carboxymethyl-forming, ADP-phosphorylating]

Recommended name
ATP citrate synthase

Synonyms
ACL
ATP citrate (pro-S)-lyase
ATP-citric lyase
ATP:citrate lyase
ATP:citrate oxaloacetate lyase ((pro-3S)-CH_2COO → acetyl-CoA) (ATP-dephosphorylating)
ATP:citrate oxaloacetate-lyase (pro-3S-CH_2COO → acetyl-CoA, ATP dephosphorylating)
ATP:citrate oxaloacetate-lyase CoA-acetylating and ATP-dephosphorylating
EC 4.1.3.8 (formerly)
adenosine triphosphate citrate lyase
citrate cleavage enzyme
citrate-ATP lyase
citric cleavage enzyme

CAS registry number
9027-95-6

2 Source Organism

<1> *Rhodosporidium toruloides* (CBS 13 [1]) [1]
<2> *Rattus norvegicus* [2-4, 6, 8-11, 13-15, 17-20, 23-26, 28, 30-33, 38, 39, 41]
<3> *Lipomyces starkeyi* (CBS 1809 [5]) [5]
<4> *Chlorobium limicola* (forma thiosulfatophilum, strain IC and strain L [12]) [7, 12, 43]
<5> *Homo sapiens* [10, 37, 38, 39]
<6> *Gallus gallus* [10, 11, 20, 22]
<7> *Columba livia* [11, 20]

<8> *Ricinus communis* [16]
<9> *Bos taurus* [20]
<10> *Sus scrofa* [20]
<11> *Oryctolagus cuniculus* [20]
<12> *Rhodotorula gracilis* [35]
<13> *Penicillium spiculisporum* [21]
<14> *Chlorobium tepidum* [27]
<15> *Hydrogenobacter thermophilus* (strain TK-6 [29]) [29]
<16> *Aspergillus nidulans* (strain R21 [34]) [34, 36]
<17> *Arabidopsis sp.* [40]
<18> *Pisum sativum* [40, 42]
<19> *Nicotiana tabacum* [42]
<20> *Brassica napus* [42]
<21> *Spinacia oleracea* [42]

3 Reaction and Specificity

Catalyzed reaction

ADP + phosphate + acetyl-CoA + oxaloacetate = ATP + citrate + CoA (<2>, double-displacement mechanism with a phosphoenzyme intermediate [4]; <4>, citrate cleavage is of the si-type [7];<6>, cleaves citrate with inversion of configuration [22]; <2>, mechanism [20,31]; <2>, in the absence of the CoA acceptor, the enzyme contains a freely diffusible highly reactive intermediate, with similar properties to other high-free-energy phosphotransfer intermediates such as aminoacyl adenylates [31]; <4>, ordered mechanism where the phosphate group of ATP is first covalently bound to the His273 catalytic residue, and subsequently transferred to the citrate molecule to form citryl phosphate [43])

Reaction type

elimination (of an oxo-acid, C-C bond cleavage)

Natural substrates and products

S ATP + citrate + CoA <2, 4, 6, 7, 8, 14, 16, 17, 18, 19, 20, 21> (<2>, the enzyme catalyzes the first cytoplasmic step in the synthesis of long-chain fatty acids in mammalian tissues [4]; <2>, the enzyme is responsible for production of cytoplasmic acetyl-CoA for lipogenesis [19]; <2,6,7>, one of the key enzymes of lipogenesis [11]; <7>, enzyme is induced by starvation-refeeding regimen [11]; <8>, possibly plays a role in providing acetyl-CoA for lipid biosynthesis [16]; <2>, the enzyme is engaged in the transport of acetyl groups from mitochondria to cytosol. On the metabolic pathway from carbohydrate to lipid it is the first enzyme which is exclusively biosynthetic [24]; <14>, key enzyme of CO_2 fixation by reductive tricarboxylic acid [27]; <16>, the enzyme is regulated by the carbon source present in the medium [36]; <17>, enzyme generates cytosolic acetyl-CoA [40]; <2>, key enzyme for lipid accumulation [41]; <20,21>, the plastidic enzyme is proposed to function for the supply of acetyl-coenzyme A for lipid biosynthesis de novo [42]; <18,19,20,21>, the cytosolic enzyme may provide acetyl-coenzyme A for the mevalonate pathway or fatty acid elongation [42]; <4>, enzyme regulates both the direction and carbon flux of the carbon dioxide-fixing reductive tricarboxylic acid cycle [43]) (Reversibility: ? <2, 6, 7, 8, 14, 16, 17, 18, 19, 20, 21> [4, 11, 16, 19, 24, 27, 36, 40, 41, 42]; ir <4> [43]) [4, 11, 16, 19, 24, 27, 36, 40, 41, 42, 43]

P ADP + phosphate + acetyl-CoA + oxaloacetate

Substrates and products

S ADP + phosphate + acetyl-CoA + oxaloacetate <2, 15> (Reversibility: r <2, 15> [20, 29]) [20, 29]

P ATP + citrate + CoA <2, 15> [20, 29]

S ATP + citrate + CoA <1-21> (<15>, the synthase activity is less than 0.5% of the ATP:citrate lyase activity [29]; <2,6>, specific for citrate [20]; <1,3,8>, specific for ATP [1,5,16]; <8>, specific for CoA [16]; <6>, cleaves citrate with inversion of configuration [22]; <4>, the first step of the reaction is reversible [43]) (Reversibility: r <2, 15> [20, 29]; ?, <1, 2-14, 16, 17, 18, 19, 20, 21> [1-28, 30-42]; ir <4> [43]) [1-43]

P ADP + phosphate + acetyl-CoA + oxaloacetate <1-21> [1-43]

S GTP + citrate + CoA <15> (Reversibility: ? <15> [29]) [29]

P GDP + phosphate + acetyl-CoA + oxaloacetate

S N^6-etheno-adenosine triphosphate + citrate + CoA <13> (Reversibility: ? <13> [21]) [21]

P N^6-etheno-adenosine diphosphate + phosphate + acetyl-CoA + oxaloacetate

S dATP + citrate + CoA <4, 15> (<4>, 40% of the activity with ATP [43]) (Reversibility: ? <4, 15> [29, 43]) [29, 43]

P dADP + phosphate + acetyl-CoA + oxaloacetate

Inhibitors

3,3,14,14-tetramethylhexadecanedioic acid <2> (<2>, i.e. Medica-16 [30]) [30]

ADP <1, 4, 12, 15, 16> (<12>, 1 mM [35]; <16>, 2 mM, 82% inhibition [36]; <4>, competitive inhibitor of ATP [43]) [1, 7, 29, 35, 36, 43]

AMP <12> (<12>, 1 mM [35]) [35]

CTP <4> [7]

Ca^{2+} <4> [12]

D-fructose 2,6-diphosphate <16> (<16>, 18% inhibition [36]) [36]

D-glucose 6-phosphate <1, 12> [1, 35]

DL-isocitrate <4> [7]

GSSG <2> (<2>, inactivation involves formation of a protein-protein disulfide rather than a protein-glutathione complex [30]) [30]

GTP <4> [7]

Glu <2, 12> (<2>, 10 mM, inhibition is unlikely to be due to any direct interaction of L-Glu and ATP citrate lyase [24]) [24, 35]

L-Leu <16> (<16>, 2 mM, 20% inhibition [36]) [36]

UTP <4> [7]

[S(2R,S)]-2-hydroxy-2-[(S-methylsulfonimidoyl)methyl]butanedioic acid <2> (<2>, weak, reversible [26]) [26]
citrate <4> [7]
dATP <4, 15> (<4>, weak [7]) [7, 29]
dTTP <4> [7]
fluorocitrate <4> [7]
lauroyl-CoA <3, 12> (<3>, 10 mM, 52% inhibition [5]) [5, 35]
malonyl-CoA <16> (<16>, 0.4 mM, 70% inhibition [36]) [36]
myristoyl-CoA <3, 12> (<3>, 10 mM, 5% inhibition [5]) [5, 35]
oleoyl-CoA <1, 3, 12> (<3>, 10 mM, 21% inhibition [5]) [1, 5, 35]
oxaloacetate <2, 4> [7, 23]
palmitoyl-CoA <1, 3, 12> (<3>, 10 mM, 5% inhibition [5]) [1, 5, 35]
radicicol <2> (<2>, noncompetitive [38]) [38]
radicicol biotinylated at the C-17 position <2> (<2>, no inhibition with the derivative biotinylated at the C-18 position [38]) [38]
stearoyl-CoA <3, 12> (<3>, 10 mM, 45% inhibition [5]) [5, 35]
tricarballylate <4> [7]

Activating compounds

Cl^- <2> (<2>, activates [15]) [15]
D-fructose 1,6-diphosphate <5> (<5>, activates [39]) [39]
D-fructose 2,6-diphosphate <5> (<5>, activates [39]) [39]
D-fructose 6-phosphate <5> (<5>, potent activator of the unphosphorylated recombinant enzyme, half-maximal activation at 0.16 mM [39]) [39]
D-glucose 1-phosphate <5> (<5>, activates [39]) [39]
D-glucose 6-phosphate <5> (<5>, activates [39]) [39]
D-ribulose 5-phosphate <5> (<5>, activates [39]) [39]
D-xylulose 5-phosphate <5> (<5>, activates [39]) [39]
HCO_3^- <2> (<2>, activates [15]) [15]
NH_4^+ <1, 12> (<1>, 95% stimulation [1]; <12>, stimulates [35]) [1, 35]
acetate <2> (<2>, activates [15]) [15]
phosphoenolpyruvate <5> (<5>, activates [39]) [39]
Additional information <2> (<2>, 6-8 very reactive sulfhydryl groups appear to be essential for activity [20]) [20]

Metals, ions

Co^{2+} <3, 4, 13> (<3>, 37.1% of the activation with Mg^{2+} at 10 mM [5]; <4,13>, can partially replace Mg^{2+} [7,21]) [5, 7, 21]
Mg^{2+} <3, 4, 13-15> (<3,13>, required [5,12,21,27]; <4>, activates [12];<15>, Mn^{2+} or Mg^{2+} required [29]; <4>, optimal concentration 4 mM, divalent cation required [7]; <14>, K_m: 4 mM [27]; <15>, K_m: 8 mM $MgCl_2$ [29]) [5, 7, 21, 27, 29]
Mn^{2+} <3, 4, 13-15> (<3>, 22.6% of the activation with Mg^{2+} at 10 mM [5]; <15>, Mn^{2+} or Mg^{2+} required [29]; <4,13>, can partially replace Mg^{2+} [7,21]; <14>, 85% of the activation with Mg^{2+} [27]) [5, 7, 21, 27, 29]

Specific activity (U/mg)

0.922 <3> [5]
2 <5> (<5>, recombinant enzyme [39]) [39]
3 <2> [32]
5.6 <14> [27]
13.6 <2> [4]
19.6 <16> [34, 36]
26.7 <15> [29]
Additional information <2, 4, 6, 13> [3, 6, 7-9, 15, 17, 21-23]

K_m-Value (mM)

0.001-0.003 <13> (CoA) [21]
0.003 <2> (CoA) [15]
0.0042 <2> (CoA) [30]
0.0098 <2> (acetyl-CoA) [20]
0.011 <2, 5> (CoA, <2>, phosphorylated enzyme [9]) [9, 10]
0.024 <14> (CoA) [27]
0.0408 <15> (CoA) [29]
0.054 <2> (ATP, <2>, phosphoenzyme in Hepes buffer, pH 7.6 [13]) [13]
0.065 <4> (ATP) [7]
0.07 <3> (citrate) [5]
0.077 <4> (CoA) [7]
0.09 <13> (ATP) [21]
0.11 <2> (citrate) [30]
0.12 <5> (citrate) [10]
0.135 <3> (ATP) [5]
0.15 <1> (ATP) [1]
0.175 <12> (ATP) [35]
0.177 <2> (oxaloacetate) [20]
0.178 <2> (ATP) [20]
0.18 <13> (N^6-etheno-adenosine triphosphate) [21]
0.18 <2, 13> (citrate, <2>, phosphorylated enzyme [9]) [9, 21]
0.19 <1> (citrate) [1]
0.2 <2> (ATP) [30]
0.208 <2> (ATP, <2>, dephosphoenzyme in Tris-HCl buffer, pH 8.7 [13]) [13]
0.21 <4> (ATP) [43]
0.21 <14> (citrate) [27]
0.222 <2> (ATP, <2>, phosphoenzyme in Tris-HCl buffer, pH 8.7 [13]) [13]
0.225 <4> (citrate) [7]
0.292 <2> (ATP, <2>, dephosphoenzyme in Hepes buffer, pH 7.6 [13]) [13]
0.3 <2> (ATP) [15]
0.31 <14> (ATP) [27]
0.57 <2> (ATP, <2>, phosphorylated enzyme [9]) [9]
0.588 <2> (citrate) [20]
0.65 <15> (ATP) [29]
1.25 <12> (citrate, <12>, 0.2 M Tris-HCl buffer containing 0.25 M KCl [35]) [35]

1.3 <5> (citrate) [10]
1.49 <2> (phosphate) [20]
2.59 <5> (CoA) [39]
2.63 <12> (citrate, <12>, in 0.1 M Tris-HCl buffer [35]) [35]
3.44 <12> (citrate, <12>, in Tris-acetate buffer [35]) [35]
6.25 <15> (citrate) [29]
41 <5> (ATP) [39]
Additional information <2, 4, 5> (<5>, K_m-value for phosphorylated enzyme forms [39]; <4>, a strong negative cooperativity is observed with respect to citrate binding [43]) [2, 13, 39, 43]

K_i-Value (mM)

0.00005 <2> ((4S)-hydroxycitrate, <2>, phospho-form of enzyme [13]) [13]
0.00007 <2> ((4S)-hydroxycitrate, <2>, dephospho-form of enzyme [13]) [13]
0.018 <4> (ADP) [7]
0.037 <4> (ADP) [43]
0.18 <4> (oxaloacetate) [7]
Additional information <2> [38]

pH-Optimum

6.7-6.9 <15> [29]
7.5 <8> [16]
8 <13> [21]
8-8.5 <1> [1]
8.2-8.6 <3> [5]
8.4 <12> [35]
8.5 <14> [27]
8.5 <4> [7, 12]

pH-Range

6.9-8.3 <8> (<8>, pH 6.9: about 45% of maximal activity, pH 8.3: about 50% of maximal activity [16]) [16]
7-8.8 <13> (<13>, about 50% of maximal activity at pH 7.0 and at pH 8.8 [21]) [21]
7.3-9.5 <3> (<3>, about 50% of maximal activity at pH 7.3 and at pH 9.5 [5]) [5]

Temperature optimum (°C)

37 <12> [35]
48 <14> [27]
80 <15> [29]

Temperature range (°C)

20-57 <14> (<14>, 20°C: about 45% of maximal activity, pH 57°C: about 30% of maximal activity [27]) [27]

4 Enzyme Structure

Molecular weight

200000 <13> (<13>, above, gel filtration [21]) [21]
240000 <4> (<4>, sucrose density gradient centrifugation [7]) [7]
260000 <15> (<15>, gel filtration [29]) [29]
371000 <16> (<16>, equilibrium sedimentation [36]) [36]
380000-385000 <16> (<16>, equilibrium sedimentation [34]) [34]
400000 <8> (<8>, gel filtration [16]) [16]
440000 <2> (<2>, equilibrium sedimentation [18]) [18]
470000 <18> (<18>, gel filtration [40]) [40]
473000 <2> (<2>, equilibrium sedimentation [9]) [9]
480000 <1, 5, 16> (<5>, velocity sedimentation [39]; <1>, gel filtration [1]; <16>, more than 480000 Da, gel filtration [34]) [1, 34, 39]
500000 <2, 17> (<2>, non-denaturing PAGE [4]; <17>, gel filtration [40]) [4, 40]
510000 <3> (<3>, gel filtration [5]) [5]
520000 <12> (<12>, gel filtration [35]) [35]
550000 <14, 16> (<14,16>, gel filtration [27,36]) [27, 36]

Subunits

? <2, 5, 16, 20, 21> (<16>, x * 55000 + x * 70000, may have hexameric structure, SDS-PAGE [34]; <2>, x * 105000, SDS-PAGE [8]; <2>, x * 110000, SDS-PAGE [6]; <5,20,21>, x * 120000, SDS-PAGE [39,42]; <2>, x * 123000, SDS-PAGE [4]; <2>, x * 125000, SDS-PAGE [28]) [4, 6, 8, 28, 34, 39, 42]
hexamer <15> (<15>, 6 * 43000, SDS-PAGE [29]) [29]
octamer <17> (<17>, 4 * 45000 + 4 * 65000, SDS-PAGE [40]) [40]
tetramer <2, 12, 14> (<2>, 4 * 110000, SDS-PAGE [17]; <2>, 4 * 116000, SDS-PAGE [9]; <12>, 4 * 120000, the two minor proteins 51000 Da and 49000 Da can be the result of proteolytic cleavage of ATP:citrate lyase by an endogenous trypsin-like proteinase [35]; <2>, 4 * 12000, gel filtration in presence of 6 M guanidinium chloride [18]; <14>, 4 * 135000, SDS-PAGE [27]) [9, 17, 18, 27, 35]

Posttranslational modification

phosphoprotein <5> (<5>, phosphorylation of recombinant human ATP:citrate lyase by cAMP-dependent protein kinase abolishes homotropic allosteric regulation of the enzyme by citrate and increases the enzyme activity. Cyclic AMP-dependent protein kinase catalyzes the incorporation of 1 mol of phosphate per mol of enzyme homotetramer, and glycogen synthase kinase-3 incorporated an additional 2 mol of phosphate into the phosphorylated protein [39]) [39]
side-chain modification <2, 14> (<2, 14>, phosphoprotein [2, 9, 13, 17, 23, 27, 28]; <2>, maximum level of phosphorylation is 2 mol phosphate per mol of tetramer, phosphorylation can affect the activity of the enzyme [2]; <2>, contains 2 mol phosphate per mol of tetramer [9, 17]; <2>, regulation of the enzyme by reversible phosphorylation. The enzyme which has been phos-

phorylated by cyclic-AMP-dependent protein kinase, can be completely dephosphorylated by incubation with either protein phosphatase 1 or protein phosphatase 2 [9]; <2>, structural phosphate behaves as a serine phosphate [17]; <2>, citrate lyase phosphorylation by cAMP-dependent protein kinase or this kinase plus glycogen synthase kinase-3 decreases the maximal velocity whereas the apparent K_m for citrate is unchanged [23]; <14>, phosphorylation occurs on a His residue at the active site [27]; <2>, histidine phosphorylation of ATP-citrate lyase is inhibited by vanadate [28]; <2>, the phosphorylation occurs by a direct transfer of a phosphoryl group from the catalytic His of nucleoside diphosphate kinase to a His on the ATP-citrate lyase [32]; <2>, phosphorylation on Ser residues by a cytosolic kinase, insulin enhances the Ser phosphorylation by 2-3-fold [33]) [2, 9, 13, 17, 23, 27, 28, 32, 33]

5 Isolation/Preparation/Mutation/Application

Source/tissue

HeLa cell <5> [38]
adipose tissue <2, 5, 6> (<6>, low activity [10,11]) [10, 11, 15]
adrenal gland <2, 5> [10]
brain <2, 6, 9> (<2>, the enzyme from brain is similar or identical to that present in liver [6]) [6, 11, 20]
colon <2, 5, 6> [10, 11, 16]
culture condition:L-glucose-grown cell <4> (as well as in autotrophically grown cells [12]) [12]
culture condition:acetate-grown cell <4> (as well as in autotrophically grown cells [12]) [12]
heart <10> [20]
ileum <2, 5, 6> [10, 16]
intestine <6> [11]
jejunum <2, 5, 6> [10, 16]
leaf <18, 19, 20, 21> (<18,19,20,21>, young [42]) [42]
leukocyte <5> [37]
liver <2, 5, 6, 7> (<2>, the enzyme from brain is similar or identical to that present in liver [6]) [2-4, 6, 8, 10, 13, 15, 17-20, 22-24, 26, 28, 30, 32, 33, 38, 41]
mammary gland <2> (<2>, lactating [9]) [9]
seedling <8, 18> (<8>, maximal activity in 4-day-old to 5-day-old seedlings [16]) [16, 40]
Additional information <2, 6> (<6>, influence of different feeding regimens on the specific activities of the enzyme in different tissues [11]; <2>, no measurable activity in heart, skeletal muscle, and kidney [15]) [11, 15]

Localization

cytosol <18, 19, 20, 21> (<18, 19, 20, 21>, enzyme is mainly cytosolic [42]) [42]
mitochondrion <2> (<2>, the enzyme tends to associate with the mitochondria [13,14]) [13, 14]

plastid <8, 20, 21> (<8>, endosperm tissue [16]; <20,21>, predominantly localized in plastids [42]) [16, 42]
Additional information <2> (<2>, less than 3% of the enzyme is cytosolic [14]) [14]

Purification

<1> (partial) [1]
<2> (liver enzyme [3, 4, 8, 15, 17, 25, 32]; brain enzyme [6]; mammary gland [9]; partial [32]) [3, 4, 6, 8, 9, 15, 17, 18, 19, 25, 31, 32]
<3> [5]
<4> (partial [7]; recombinant enzyme [43]) [7, 43]
<5> (recombinant enzyme [39]) [10, 39]
<6> [22]
<7> (partial [11]) [11]
<12> [35]
<13> [21]
<14> [27]
<15> [29]
<16> [34, 36]

Cloning

<2> (expression in Escherichia coli [39]; a gene encoding a fusion protein of rat liver ATP:citrate lyase with the transit peptide for the small subunit of ribulose bisphosphate carboxylase is constructed and introduced into the genome of tobacco [41]) [25, 39, 41]
<4> (expression in Escherichia coli [43]) [43]
<5> (expression in Escherichia coli [39]) [39]
<17> (expression in Saccharomyces cerevisiae [40]) [40]

Engineering

H274A <4> (<4>, mutation abolishes both the catalytic activity and phosphorylation of the enzyme by ATP [43]) [43]

Application

analysis <2> (coupled fluorometric assay for EC 2.3.1.1 based on coupling coenzyme A production to the oxidation of NADH via ATP-citrate lyase and malate dehydrogenase [3]) [3]
medicine <5> (<5>, renal stone patients have low urinary citrate excretion with high leukocyte ATP citrate (pro-3S)-lyase activity. In northeast Thailand, low potassium status and a high carbohydrate and low fat diet may cause the increased enzyme activity. Hypokaliuria and high leukocyte ATP citrate (pro-3S)-lyase activity can be corrected by potassium-sodium citrate salt therapy [37]) [37]

6 Stability

pH-Stability

5-11 <15> (<15>, stable [29]) [29]
7-7.2 <14> (<14>, maximal stability [27]) [27]

Temperature stability

30 <5> (<5>, 4 h, stable [39]) [39]
40 <2> (<2>, 30 min, protein concentration of 0.25 mg/ml, phosphorylated and dephosphorylated form of enzyme, stable [13]) [13]
45 <2> (<2>, 10 min, protein concentration of 0.16 mg/ml, phosphorylated and dephosphorylated form of enzyme, 50% loss of activity [13]) [13]
65 <14> (<14>, loss of activity after 5 min without addition of citrate, loss of activity after 20 min in presence of citrate [27]) [27]
80 <15> (<15>, stable up to [29]) [29]
95 <15> (<15>, 30 min, more than 90% loss of activity in absence of citrate, less than 10% loss of activity in presence of citrate [29]) [29]

Oxidation stability

<2>, rapid loss of activity in air at 3°C in Tris buffer [20]
<2>, the enzyme is in equilibrium between an oxidized inactive form and a reduced active form [30]

General stability information

<2>, a peptide factor stabilizes [19]
<2>, chymotrypsin and pronase treatment inactivates, trypsin does not affect the enzyme [18]
<8>, citrate and glycerol stabilize [16]
<2, 3>, ATP and citrate stabilize [5, 31]

Storage stability

<3>, 4°C, 50 mM Tris-HCl, pH 7.5, 1 mM $MgCl_2$, 10 mM 2-mercaptoethanol, 83% loss of activity after 48 h. 52% loss of activity in presence of 1 mM ATP. 30% loss of activity in presence of 20 mM citrate, 9% loss of activity in presence of 1 mM ATP and 20 mM citrate [5]
<5>, -20°C, 1 year, stable [39]
<5>, -20°C, pH 8 stable for 4 months [10]
<12>, -20°C, 20% glycerol, stable up to 2 months [35]
<2, 8>, -20°C, stable [16, 31]

References

[1] Evans, C.T.; Ratledge, C.: Possible regulatory roles of ATP:citrate lyase, malic enzyme, and AMP deaminase in lipid accumulation by Rhodsporidium toruloides CBS 14. Can. J. Microbiol., **31**, 1000-1005 (1985)

[2] Houston, B.; Nimmo, H.G.: Effects of phosphorylation on the kinetic properties of rat liver ATP-citrate lyase. Biochim. Biophys. Acta, **844**, 233-239 (1985)
[3] Wraight, C.; Day, A.; Hoogenraad, N.; Scopes, R.: A two-step purification of ATP-citrate lyase from rat liver and its use in a fluorometric assay for N-acetylglutamate synthetase. Anal. Biochem., **144**, 604-609 (1985)
[4] Houston, B.; Nimmo, H.G.: Purification and some kinetic properties of rat liver ATP citrate lyase. Biochem. J., **224**, 437-443 (1984)
[5] Boulton, C.A.; Ratledge, C.: Partial purification and some properties of ATP:citrate lyase from the oleaginous yeast Lipomyces starkeyi. J. Gen. Microbiol., **129**, 2863-2869 (1983)
[6] Szutowicz, A.; Srere, P.A.: Purification and some properties of ATP-citrate lyase from rat brain. Arch. Biochem. Biophys., **221**, 168-174 (1983)
[7] Antranikian, G.; Herzberg, C.; Gottschalk, G.: Characterization of ATP citrate lyase from Chlorobium limicola. J. Bacteriol., **152**, 1284-1287 (1982)
[8] Redshaw, J.C.; Loten, E.G.: A rapid purification of hepatic ATP citrate lyase using blue Sepharose. FEBS Lett., **123**, 261-264 (1981)
[9] Guy, P.S.; Cohen, P.; Hardie, D.G.: Purification and physicochemical properties of ATP citrate (pro-3S) lyase from lactating rat mammary gland and studies of its reversible phosphorylation. Eur. J. Biochem., **114**, 399-405 (1981)
[10] Hoffmann, G.E.; Andres, H.; Weiss, L.; Kreisel, C. Sander, R.: Properties and organ distribution of ATP citrate (pro-3S)-lyase. Biochim. Biophys. Acta, **620**, 151-158 (1980)
[11] Hoffmann, G.E.; Kraupe, P.; Weiss, L.; Wittmann, J.: Avian ATP citrate (pro-3S)-lyase. Hoppe-Seyler's Z. Physiol. Chem., **361**, 1117-1119 (1980)
[12] Ivanovsky, R.N.; Sintsov, N.V.; Kondratieva, E.N.: ATP-linked citrate lyase activity in the green sulfur bacterium Chlorobium limicola forma thiosulfatophilum. Arch. Microbiol., **128**, 239-241 (1980)
[13] Ranganathan, N.S.; Srere, P.A.; Linn, T.C.: Comparison of phospho- and dephospho-ATP citrate lyase. Arch. Biochem. Biophys., **204**, 52-58 (1980)
[14] Janski, A.M.; Cornell, N.W.: Association of ATP citrate lyase with mitochondria. Biochem. Biophys. Res. Commun., **92**, 305-312 (1980)
[15] Hoffmann, G.E.; Schiessl, J.; Weiss, L.: ATP citrate (pro-3S)-lyase in the rat; two-step purification procedure, properties, organ distribution. Hoppe-Seyler's Z. Physiol. Chem., **360**, 1445-1451 (1979)
[16] Fritsch, H.; Beevers, H.: ATP citrate lyase from germinating castor bean endosperm. Plant Physiol., **63**, 687-691 (1979)
[17] Linn, T.; Srere, P.A.: Identification of ATP citrate lyase as a phosphoprotein. J. Biol. Chem., **254**, 1691-1698 (1979)
[18] Singh, M.; Richards, E.G.; Mukherjee, A.; Srere, P.A.: Structure of ATP citrate lyase from rat liver. Physicochemical studies and proteolytic modification. J. Biol. Chem., **251**, 5242-5250 (1976)
[19] Osterlund, B.; Bridger, W.A.: Effect of a peptide stabilizing factor on liver ATP citrate lyase. Biochem. Biophys. Res. Commun., **76**, 1-8 (1977)
[20] Spector, L.B.: Citrate cleavage and related enzymes. The Enzymes, 3rd Ed. (Boyer, P.D., ed.), **7**, 357-389 (1972)

[21] Mahlen, A.: Purification and some properties of ATP citrate lyase from Penicillium spiculisporum. Eur. J. Biochem., **36**, 342-346 (1973)
[22] Lenz, H.; Buckel, W.; Wunderwald, P.; Biedermann, G.; Buschmeier, V.; Eggerer, H.; Cornforth, J.W.; Redmond, J. W.; Mallaby, R.: Stereochemistry of si-citrate synthase and ATP-citrate-lyase reactions. Eur. J. Biochem., **24**, 207-215 (1971)
[23] Pentyala, S.N.; Benjamin, W.B.: Effect of oxaloacetate and phosphorylation on ATP-citrate lyase activity. Biochemistry, **34**, 10961-10969 (1995)
[24] Rider, C.C.; Wilson, A.P.: Inhibition of liver ATP citrate lyase in the presence of L-glutamate. Biochem. Soc. Trans., **13**, 152-153 (1985)
[25] Choi, J.W.; Kim, K.S.; Park, S.W.; Whang, S.K.; Kim, Y.S.: Purification of ATP-citrate lyase in rat liver and molecular cloning of its cDNA. Korean J. Biochem., **23**, 223-230 (1991)
[26] Dolle, R.E.; McNair, D.; Hughes, M.J.; Kruse, J.I.; Eggelston, D.; Saxty, B.A.; Wells, T.N.C.; Groot, P.H.E.: ATP-citrate lyase as a target for hypolipidemic intervention. Sulfoximine and 3-hydroxy-β-lactam containing analogues of citric acid as potential tight-binding inhibitors. J. Med. Chem., **35**, 4875-4884 (1992)
[27] Wahlund, T.M.; Tabita, F.R.: The reductive tricarboxylic acid cycle of carbon dioxide assimilation: initial studies and purification of ATP-citrate lyase from the green sulfur bacterium Chlorobium tepidum. J. Bacteriol., **179**, 4859-4867 (1997)
[28] Krivanek, J.; Novakova, L.: ATP-citrate lyase is another enzyme the histidine phosphorylation of which is inhibited by vanadate. FEBS Lett., **282**, 32-34 (1991)
[29] Ishii, M.; Igarashi, Y.; Kodama, T.: Purification and characterization of ATP:citrate lyase from Hydrogenobacter thermophilus TK-6. J. Bacteriol., **171**, 1788-1792 (1989)
[30] Wells, T.N.C.; Saxty, B.A.: Redox control of catalysis in ATP-citrate lysate from rat liver. Eur. J. Biochem., **204**, 249-255 (1992)
[31] Wells, T.N.C.: ATP-citrate lyase from rat liver. Characterisation of the citryl-enzyme complexes. Eur. J. Biochem., **199**, 163-168 (1991)
[32] Wagner, P.D.; Vu, N.D.: Phosphorylation of ATP-citrate lyase by nucleoside diphosphate kinase. J. Biol. Chem., **270**, 21758-21764 (1995)
[33] Yu, K.T.; Benjamin, W.B.; Ramakrishna, S.; Khalaf, N.; Czech, M.P.: An insulin-sensitive cytosolic protein kinase accounts for the regulation of ATP citrate-lyase phosphorylation. Biochem. J., **268**, 539-545 (1990)
[34] Adams, I.P.; Dack, S.; Dickinson, F.M.; Midgley, M.; Ratledge, C.: ATP: citrate lyase from Aspergillus nidulans. Biochem. Soc. Trans., **25**, S670 (1997)
[35] Shashi, K.; Bachhawat, A.K.; Joseph, R.: ATP:citrate lyase of Rhodotorula gracilis: purification and properties. Biochim. Biophys. Acta, **1033**, 23-30 (1990)
[36] Adams, I.P.; Dack, S.; Dickinson, F.M.; Ratledge, C.: The distinctiveness of ATP:citrate lyase from Aspergillus nidulans. Biochim. Biophys. Acta, **1597**, 36-41 (2002)
[37] Tosukhowong, P.; Borvonpadungkitti, S.; Prasongwatana, V.; Tungsanga, K.; Jutuporn, S.; Dissayabutr, T.; Reungjui, S.; Sriboonlue, P.: Urinary citrate

excretion in patients with renal stone: roles of leucocyte ATP citrate lyase activity and potassium salts therapy. Clin. Chim. Acta, **325**, 71-78 (2002)
[38] Ki, S.W.; Ishigami, K.; Kitahara, T.; Kasahara, K.; Yoshida, M.; Horinouchi, S.: Radicicol binds and inhibits mammalian ATP citrate lyase. J. Biol. Chem., **275**, 39231-39236 (2000)
[39] Potapova, I.A.; El-Maghrabi, M.R.; Doronin, S.V.; Benjamin, W.B.: Phosphorylation of recombinant human ATP:citrate lyase by cAMP-dependent protein kinase abolishes homotropic allosteric regulation of the enzyme by citrate and increases the enzyme activity. Allosteric activation of ATP:citrate lyase by phosphorylated sugars. Biochemistry, **39**, 1169-1179 (2000)
[40] Fatland, B.L.; Ke, J.; Anderson, M.D.; Mentzen, W.I.; Cui, L.W.; Allred, C.C.; Johnston, J.L.; Nikolau, B.J.; Wurtele, E.S.: Molecular characterization of a heteromeric ATP-citrate lyase that generates cytosolic acetyl-coenzyme A in Arabidopsis. Plant Physiol., **130**, 740-756 (2002)
[41] Rangasamy, D.; Ratledge, C.: Genetic enhancement of fatty acid synthesis by targeting rat liver ATP:citrate lyase into plastids of tobacco. Plant Physiol., **122**, 1231-1238 (2000)
[42] Rangasamy, D.; Ratledge, C.: Compartmentation of ATP:citrate lyase in plants. Plant Physiol., **122**, 1225-1230 (2000)
[43] Kanao, T.; Fukui, T.; Atomi, H.; Imanaka, T.: Kinetic and biochemical analyses on the reaction mechanism of a bacterial ATP-citrate lyase. Eur. J. Biochem., **269**, 3409-3416 (2002)

Malate synthase 2.3.3.9

1 Nomenclature

EC number
2.3.3.9

Systematic name
acetyl-CoA:glyoxylate C-acetyltransferase (thioester-hydrolysing, carboxymethyl-forming)

Recommended name
malate synthase

Synonyms
EC 4.1.3.2 (formerly)
L-malate glyoxylate-lyase (CoA-acetylating)
MS
MSG
glyoxylate transacetase
glyoxylic transacetase
malate synthase 1
malate synthase G
malate synthetase
malic synthetase
malic-condensing enzyme

CAS registry number
9013-48-3

2 Source Organism

<1> *Ricinus communis* (castor bean [1,14]) [1, 14, 25]
<2> *Gossypium hirsutum* (cotton [2]) [2, 9, 36]
<3> *Pinus densiflora* [3]
<4> *Cucumis sativus* (cucumber [4]) [4, 7, 8]
<5> *Candida tropicalis* (pK 233 [5]) [5]
<6> *Saccharomyces cerevisiae* (strains KM10-15 [34]) [6, 10, 16, 18, 21, 34]
<7> *Bacillus sp.* (thermophilic [11,12]) [11, 12]
<8> *Neurospora crassa* [11]
<9> *Escherichia coli* [11, 12, 19, 33]
<10> *Euglena gracilis* [13, 20]
<11> *Colwellia maris* (formerly Vibrio sp. strain ABE-1 [35]) [35]

<12> *Caenorhabditis elegans* [15]
<13> *Bacillus stearothermophilus* [17]
<14> *Astasia longa* [20]
<15> *Pseudomonas ovalis* [21]
<16> *Streptomyces arenae* [22]
<17> *Pinus taeda* [23]
<18> *Cucurbita sp.* [24]
<19> *Glycine max* [25]
<20> *Corynebacterium glutamicum* [26]
<21> *Hansenula polymorpha* [27]
<22> *Zea mays* [28, 29, 31]
<23> *Brassica napus* [30]
<24> *Petricola pholadiformis* [32]
<25> *Bacillus licheniformis* [12]
<26> *Pseudomonas indigofera* [12]
<27> *Helianthus annuus* (sunflower [36]) [36]
<28> *Streptomyces coelicolor A3(2)* (M130 [37]) [37]
<29> *Streptomyces clavuligerus* (NRRL3585 [37]) [37]
<30> *Mycobacterium tuberculosis* [38]

3 Reaction and Specificity

Catalyzed reaction

acetyl-CoA + H_2O + glyoxylate = (S)-malate + CoA (<2> sequential bireactant mechanism [9]; <6> sequencial random mechanism [10]; <22> compulsory-order mechanism, glyoxylate being the first-binding substrate, glyoxylate triggers a conformational change in the enzyme and as a consequence, the correctly shaped binding site for acetyl-CoA is created [31]; <9, 30> mechanism [33, 38])

Reaction type

Claisen condensation <9, 30> [33, 38]
aldol condensation

Natural substrates and products

S glyoxylate + acetyl-CoA + H_2O <1, 2, 4, 6, 7, 9-11, 16, 17, 21, 22, 24, 27> (<1, 2, 4, 6, 7, 9-11, 16, 17, 21, 22, 24, 27> enzyme specifically involved in glyoxalate cycle metabolism [1, 2, 4, 9, 11, 13, 22, 23, 27, 29, 32-36]; <11> enzyme essential for acetate use in the bacterial cells [35]) [1, 2, 4, 7, 9, 11, 13, 22, 23, 27, 29, 32-36]
P (S)-malate + CoA

Substrates and products

S glyoxylate + acetyl-CoA + H_2O <1-30> (<6, 30> structure of enzyme-substrate complex [16, 38]; <28, 29> the optimum concentration of acetyl-CoA 0.0001 mM, the optimum concentration of glyoxylate 0.001 mM

[37]) (Reversibility: ir <2, 6, 15> [9, 21]; ? <1, 3-5, 7-14, 16-30> [1-8, 10-20, 22-38]) [1-38]

P (S)-malate + CoA <2, 6, 9, 10, 15, 17, 30> [2, 9, 10, 13, 18, 21, 23, 33, 34, 38]

S glyoxylate + butyryl-CoA + H_2O <2> (Reversibility: ir <2> [9]) [9]

P 2-oxohexanedioic acid + CoA

S glyoxylate + fluoroacetyl-CoA + H_2O <15> (Reversibility: ir <15> [21]) [21]

P 2-oxo-3-fluoro-butanedioic acid + CoA

S glyoxylate + propionyl-CoA + H_2O <2, 21> (<21> not [27]) (Reversibility: ir <2> [9]; ? <21> [27]) [9, 27]

P 2-oxopentanedioic acid + CoA

S Additional information <1> (<1> enzyme undergoes in vitro phosphorylation [1]) [1]

P ?

Inhibitors

ADP <1, 2> (<1> 5 mM, 46% inhibition [14]) [9, 14]
AMP <1> (<1> 5 mM, 46% inhibition [14]) [14]
ATP <1, 2, 3, 20> (<1> 5 mM, 55% inhibition [14]) [3, 9, 14, 26]
EDTA <3> (<3> 1.0 mM, 30% inhibition [3]) [3]
KCl <11> [35]
NaCl <11> [35]
S-acetonyl-CoA <2> [9]
S-ethyl-CoA <2> [9]
acetaldehyde <2> (<2> weak [9]) [9]
acetyl-CoA <28, 29> (<28,29> slight reduction of enzyme activity at high concentrations of acety-CoA [37]) [37]
bromopyruvate <30> [38]
butanedione <2> (<2> slight inhibition at high concentrations [9]) [9]
chloroacetyl-CoA <2> [9]
deamino-acetyl-CoA <2> [9]
deoxycholate <2> [9]
dephosphoacetyl-CoA <2> [9]
diethyl dicarbonate <2> (<2> inhibition prevented by preincubation with acetyl-CoA [9]) [9]
fluoroacetate <6, 15, 22> [21, 31]
glycolate <2, 6, 7, 15, 20, 22, 30> (<30> only at fairly high concentration [38]) [9-11, 21, 26, 31, 38]
glyoxylate <28, 29> (<28,29> slight reduction of enzyme activity at high concentrations of glyoxylate [37]) [37]
malate <30> (<30> 1.0 mM, 50% inhibition [38]) [38]
methylglyoxal <2> [9]
oxalate <2, 3, 6, 7, 15, 20, 30> (<3> competiteive with glyoxylate [3]) [3, 9-11, 21, 26, 38]
oxaloacetate <6> [10]
oxamate <2> [9]

phosphoenolpyruvate <3, 20, 30> (<20> not [26]) [3, 26, 38]
pyridoxal-5'-phosphate <2> (<2> preincubation with glyoxylate but not acetyl-CoA prevents inhibition [9]) [9]
pyruvate <2, 6, 7, 20, 22> (<2> weak [9]; <20> not [26]) [9, 10, 12, 26, 31]
Additional information <20> (<20,30> not: fructose-1,6-biphosphate [26,38]; <20> acetyl-phosphate [26]; <30> 3-phosphoglycerate, 6-phosphogluconate, malonic acid [38]) [26, 38]

Activating compounds

EDTA <3> (<3> 0.01 mM, 10% activation [3]) [3]
Triton X-100 <2> (<2> activation [9]) [9]

Metals, ions

Ba^{2+} <2> (<2> about 30% of the activation with Mg^{2+} [9]) [9]
Co^{2+} <2, 11, 20> (<2> about 30% of the activation with Mg^{2+} [9]; <20> 25% of the activation with Mg^{2+} [26]; <11> about 35% of the activation with Mg^{2+} [35]) [9, 26, 35]
Mg^{2+} <1-3, 5, 6, 9, 11, 15, 20-22, 28, 29> (<1, 3, 5, 6, 9, 11, 15, 21, 22, 28, 29> required [3, 5, 14, 21, 27, 28, 31, 33, 35, 37]; <20> absolute requirement for divalent cation, maximal activity with 40 mM Mg^{2+} [26]; <3> optimal concentration 10 mM $MgCl_2$ [3]; <2> absolute requirement for divalent metal ion, best fulfilled by Mg^{2+} [9]; <6> binding of acetyl-CoA to the synthase is independent of Mg^{2+} but that of glyoxylate is strictly dependent on the presence of Mg^{2+} [10]; <6> causes no structurel effects, suggesting the metal ion to be involved in enzymatic catalysis rather than structural alternations [18]; <5> K_m: 4.7 mM [5]; <2> K_m: 0.59 mM [9]; <6,15> K_m: 0.5 mM [21]; <21> K_m: 0.3 mM [27]; <9> enzyme-substrate complex with glyoxylate and Mg^{2+}, Glu427 and Asp455 bind the magnesium ion [33]; <28> optimal concentration 0.005 mM $MgCl_2$ [37]; <29> optimal concentration 0.01 mM $MgCl_2$ [37]; <30> absolute requirement for divalent cation, maximal activity with 5 mM Mg^{2+} [38]) [3, 5, 8-10, 14, 18, 21, 26-28, 31, 33, 35, 37, 38]
Mn^{2+} <2, 6, 11, 15, 20, 30> (<2> about 30% of the activation with Mg^{2+} [9]; <6,15> can partially replace Mg^{2+} in activation [21]; <20> 15% of the activation with Mg^{2+} [26]; <11> 19% of the activation with Mg^{2+} [35]; <30> 40% of the activation with Mg^{2+} [38]) [9, 21, 26, 35, 38]
Ni^{2+} <11> (<11> about 98% of the activation with Mg^{2+} [35]) [35]
Additional information <30> (<30> Co^{2+}, Fe^{2+}, Ca^{2+}, Ba^{2+}, Ni^{2+}, Cd^{2+}, Zn^{2+}, Cu^{2+}, Hg^{2+} are not able to support the activity of enzyme [38]) [38]

Turnover number (min^{-1})

1600 <13> (acetyl-CoA) [17]
1670 <7> (acetyl-CoA) [11]
3642 <11> (acetyl-CoA, <11> at 20°C [35]) [35]
9672 <11> (acetyl-CoA, <11> at 45°C [35]) [35]
Additional information <6, 15> [21]

Specific activity (U/mg)

0.0012 <29> [37]
0.0262 <28> [37]

0.29 <10> [13]
2.5 <1> [14]
6 <30> (<30> recombinant enzyme [38]) [38]
18.9 <5> [5]
24.5 <22> [28]
26.9 <7> (<7> wild-type enzyme [12]) [12]
27 <7> (<7> mutant enzyme PC2 NG35 [12]) [12]
28.6 <20> [26]
34.3 <9> [12]
36.9 <3> [3]
69.6 <11> [35]
308 <17> [23]
Additional information <2, 6, 13, 15, 16, 18, 19, 21-23> [2, 9, 10, 17, 21, 22, 24, 25, 27, 28, 30]

K_m-Value (mM)
0.00059 <29> (glyoxylate) [37]
0.00349 <28> (glyoxylate) [37]
0.008 <7, 13> (acetyl-CoA) [11, 17]
0.01 <1, 2> (acetyl-CoA) [9, 14]
0.011 <21> (acetyl-CoA) [27]
0.012 <20> (acetyl-CoA) [26]
0.0198 <11> (glyoxylate) [35]
0.02 <22> (acetyl-CoA) [31]
0.0228 <22> (acetyl-CoA) [28]
0.03 <30> (acetyl-CoA, <30> recombinant enzyme [38]) [38]
0.03 <20> (glyoxylate) [26]
0.05 <10> (glyoxylate) [13]
0.052 <3> (acetyl-CoA) [3]
0.052 <2> (glyoxylate) [9]
0.057 <30> (glyoxylate, <30> recombinant enzyme [38]) [38]
0.058 <7> (glyoxylate) [11]
0.06 <21> (glyoxylate) [27]
0.063 <15> (glyoxylate) [21]
0.076 <3> (glyoxylate) [3]
0.08 <5, 10> (acetyl-CoA) [5, 13]
0.083 <6> (acetyl-CoA) [10]
0.088 <13> (acetyl-CoA) [17]
0.093 <6> (glyoxylate) [21]
0.098 <22> (glyoxylate) [28]
0.1 <6> (glyoxylate) [10]
0.104 <22> (glyoxylate) [31]
0.11 <3> (acetyl-CoA) [3]
0.14 <3> (glyoxylate) [3]
1 <5> (glyoxylate) [5]
1.25 <16> (glyoxylate) [22]
2 <1> (glyoxylate) [14]

Additional information <2, 27, 28, 29> (<2,27> thermal dependencies of K_m, Gossypium hirsutum has a minimum value of 0.0083 mM at 27.5°C and higher values at temperatures above or below, the K_m of Helianthus annuus enzyme increases with temperature [36]) [36, 37]

K_i-Value (mM)

0.01 <3> (ATP, <3> enzyme form MSH [3]) [3]
0.019 <6> (oxalate) [10]
0.019 <6, 15> (oxalate) [21]
0.06 <30> (bromopyruvate) [38]
0.07 <7> (glycolate) [11]
0.09 <7> (oxalate) [11]
0.11 <1> (oxalate) [14]
0.15 <1> (glycolate) [14]
0.2 <3> (ATP, <3> enzyme form MSL [3]) [3]
0.2 <30> (phosphoenolpyruvate) [38]
0.246 <6, 15> (fluoroacetate) [21]
0.27 <20> (oxalate) [26]
0.308 <6, 15> (glycolate) [21]
0.31 <6> (glycolate) [10]
0.4 <30> (oxalate) [38]
0.44 <20> (glycolate) [26]
0.54 <6> (pyruvate) [10]
0.9 <30> (glycolate) [38]
1.2 <3> (oxalate, <3> enzyme form MSH [3]) [3]
1.5 <3> (oxalate, <3> enzyme form MSL [3]) [3]
1.5 <6> (oxaloacetate) [10]
4.3 <20> (ATP) [26]

pH-Optimum

7.5 <21, 30> [27, 38]
7.6 <3, 22> [3, 28]
8 <1, 2, 5, 10, 11> (<2> Tris/HCl buffer [9]) [5, 9, 13, 14, 35]
8.2 <2> (<2> phosphate-citrate-borate buffer or MOPS-KOH buffer [9]) [9]
8.5 <6, 15> [21]
8.6 <7, 13> [11, 17]
Additional information <6, 7, 22> (<6> pI 7.5 [10]; <7> 4.6 [11]; <22> 5.0 [28]) [10, 11, 28]

pH-Range

5-9.5 <10> (<10> pH 5.0: about 35% of maximal activity, pH 9.5: about 70% of maximal activity [13]) [13]
6.8-9.7 <2> (<2> pH 6.8 and 9.7: about 50% of maximal activity, phosphate-citrate-borate buffer [9]) [9]
7-8.5 <3> (<3> pH 7.0: about 80% of maximal activity, pH 8.7: about 60% of maximal activity, enzyme form MSL and MSH [3]) [3]

7.5-10.5 <16> (<16> pH 7.5: about 55% of maximal activity, pH 10.5: about 25% of maximal activity, enzyme form MSL and MSH [22]) [22]
8-10 <28, 29> (<28,29> significant reduction of specific activity at pH 7.0 and below [37]) [37]

Temperature optimum (°C)
40 <2> [9]
43 <20> [26]
45 <11> [35]

Temperature range (°C)
20-45 <28, 29> [37]
Additional information <11> (<11> K_m for glyoxylate increases with decreasing temperature [35]) [35]

4 Enzyme Structure

Molecular weight
52000-54000 <9> (<9> gel filtration, equilibrum sedimentation centrifugation, light scattering, two distinct forms of enzyme [19]) [19]
55000 <9> (<9> gel filtration [12]) [12]
56000 <25, 26> (<25,26> gel filtration [12]) [12]
62000 <7, 8, 13> (<7,8,13> gel filtration [11,12,17]) [11, 12, 17]
65000 <3> (<3> gel filtration, enzyme form MSL [3]) [3]
80000 <20> (<20> gel filtration [26]) [26]
81000 <9> [33]
110000 <16> (<16> gel filtration [22]) [22]
120000 <18> (<18> sucrose density gradient centrifugation [24]) [24]
160000 <8, 11> (<8,11> gel filtration [11,35]) [11, 35]
170000 <6> (<6> gel filtration, ultracentrifugation, sucrose density gradient centrifugation [18]) [18]
180000 <6> (<6> high speed equilibrium sedimentation [10]) [10]
186000 <6> (<6> small-angle X-ray scattering technique [16]) [16]
250000 <5, 21> (<5,21> gel filtration [5,27]) [5, 27]
350000 <10> (<10> gel filtation [13]) [13]
480000 <18> (<18> sucrose density gradient centrifugation [24]) [24]
510000 <22> (<22> gel filtration [28]) [28]
520000 <17> (<17> gel filtration [23]) [23]
575000 <1> (<1> sucrose density gradient centrifugation [14]) [14]
630000 <3, 19> (<3> enzyme form MSH [3]; <3,19> gel filtration [3,25]) [3, 25]
730000 <2> (<2> gel filtration [9]) [9]
750000 <2> (<2> sucrose density gradient centrifugation [2,9]) [2, 9]
Additional information <3> (<3> by incubation with 5 mM ATP the high molecular weight enzyme form MSH is converted to the low molecular weight enzyme form MSL [3]) [3]

Subunits

? <1, 3, 17, 23> (<17> x * 62000, SDS-PAGE [23]; <3> x * 62000, SDS-PAGE, enzyme forms MSL and MSH [3]; <23> x * 63000, SDS-PAGE [30]; <1> x * 64000, SDS-PAGE [1,14]) [1, 3, 11, 14, 23, 30]
decamer <19> (<19> 10 * 62000, SDS-PAGE [25]) [25]
dimer <10, 11, 16, 18> (<10> 2 * 175000, SDS-PAGE [13]; <18> 2 * 60000, enzyme exists as dimer and as octamer, SDS-PAGE [24]; <16> 2 * 61360, calculation from nucleotide sequence [22]; <11> 2 * 763000, SDS-PAGE [35]) [13, 22, 24, 35]
dodecamer <2> (<2> 12 * 63000, SDS-PAGE [2]) [2]
monomer <7, 8, 9, 13, 20> (<7,13> 1 * 58000, SDS-PAGE [11,12,17]; <9> 1 * 56000, SDS-PAGE [12]; <20> 1* 82362, calculation from nucleotide sequence [26]; <20> 1 * 90000, SDS-PAGE [26]) [11, 12, 17, 23, 26]
octamer <4, 18, 22> (<18> 8 * 60000, enzyme exists as dimer and as octamer, SDS-PAGE [24]; <4> 8 * 63000, predominant form, SDS-PAGE [7]; <22> 8 * 22000, SDS-PAGE [28]) [7, 24, 28]
tetramer <5, 6, 21> (<5> 4 * 61000, SDS-PAGE [5]; <6> 4 * 70000, SDS-PAGE [18]; <21> 4 * 62000, SDS-PAGE [27]) [5, 18, 27]
trimer <6> (<6> 3 * 66000, SDS-PAGE [10]) [10]
Additional information <8, 9> (<8,9> enzyme monomeric in procaryotes but multimeric in eucaryotes [11]) [11]

Posttranslational modification

no modification <17, 22> (<17> no glycoprotein [23]; <22> contains no covalent linked carbohydrate residues [28]) [23, 28]
side-chain modification <1, 4> (<1> posttranslational phosphorylation at a Ser residue [1]; <4> enzyme contains phospholipid [8]; <4> enzyme synthesized as a monomeric precursor in the cytoplasm, processing which is a prerequisite for oligomerization takes place rapidly in the glyoxysomes [7]) [1, 7, 8]

5 Isolation/Preparation/Mutation/Application

Source/tissue

cotyledon <2, 4, 7, 18, 19, 23> (<2> of 72-hours dark-grown seedlings [2]; <2> of dark-germinated seeds [9]) [2, 4, 7-9, 11, 12, 24, 25, 30]
culture condition:acetate-grown cell <11, 13, 19> [17, 25, 35]
culture condition:alkane-grown cell <3> [3]
culture condition:dark-grown cell <10> [13]
culture condition:ethanol-grown cell <6> (<6> enzyme cytosolic [34]) [34]
culture condition:oleic acid medium-grown cell <6> (<6> enzyme peroxisomal [34]) [34]
endosperm <1> [1, 14]
pollen <3> [3]
scutellum <22> [28, 29]
seed <1, 2, 4, 17> (<4> grown both in light and in dark [4]) [1, 4, 8, 9, 14, 23]
seedling <2, 27> [36]

Localization

cytosol <6> (<6> abundant in ehtanol-grown cells [34]) [34]
glyoxysome <1, 4, 17, 18, 19, 23> (<4> membrane [8]; <18> exclusively localized in [24]) [1, 7, 8, 14, 20, 23-25, 30]
microsome <1> [14]
peroxisome <5, 6, 21> (<5,6,21> matrix [5,27,34]; <6> abundant cells grown on oleic acid [34]) [5, 27, 34]

Purification

<1> [14]
<2> [2, 9]
<3> [3]
<4> [8]
<5> [5]
<6> [10, 21]
<7> [11, 12]
<9> [12, 19, 33]
<10> [13]
<11> [35]
<13> [17]
<15> [21]
<16> [22]
<17> [23]
<18> [24]
<19> [25]
<20> [26]
<21> [27]
<22> [28]
<23> [30]
<30> (recombinant enzyme [38]) [38]
<28, 29> (recombinant enzymes [37]) [37]

Crystallization

<9> (structure of enzyme based on a β_8/α_8 barrel fold [33]) [33]
<30> (mixed α/β structure [38]) [38]

Cloning

<4> [4]
<6> [34]
<20> (expression in Escherichia coli and Corynebacterium glutamicum [26]) [26]
<9, 16, 28, 29, 30> (expression in Escherichia coli [22, 33, 37]) [22, 33, 37, 38]

6 Stability

pH-Stability

5.5 <3> (<3> 4°C, 12 h, about 60% loss of activity, enzyme form MSH [3]) [3]
6 <3> (<3> 4°C, 12 h, about 90% loss of activity, enzyme form MSH [3]) [3]
7 <16> (<16> enzyme inactive at pH 7.0 and below [22]) [22]
7-8 <3> (<3> 4°C, 12 h, stable, enzyme forms MSL and MSH [3]) [3]
9 <3> (<3> 4°C, 12 h, about 50% loss of activity, enzyme form MSL, about 40% loss of activity of enzyme form MSH [3]) [3]

Temperature stability

30 <3> (<3> pH 7.6, 5 min, enzyme form MSL is stable up to [3]) [3]
40 <3> (<3> pH 7.6, 5 min, about 30% loss of activity of enzyme form MSL [3]) [3]
41 <16> (<16> 1 h, little effect [22]) [22]
44 <16> (<16> 1 h, complete inactivation [22]) [22]
45 <3, 7, 9> (<3> pH 7.6, 5 min, about 60% loss of activity of enzyme form MSL, enzyme form MSH is stable up to [3]; <9> 50 mM Tris-HCl, pH 8.0, without KCl or with 0.2 M KCl, half-life: 25 min [12]; <7> pH 7.5-9.5, completely stable for at least 2 h [12]) [3, 12]
50 <3, 9> (<3> pH 7.6, 5 min, complete loss of activity of enzyme form MSL [3]; <9> 50 mM Tris-HCL, pH 8.0, without KCl or with 0.2 M KCl, half-life: 3 min [12]) [3, 12]
55 <3> (<3> pH 7.6, 5 min, complete loss of activity of enzyme form MSH [3]) [3]
60 <7, 13> (<7> 25 mM Tris-HCl, pH 7.0, without KCl, half-life: 168 min, 25 mM Tris-HCL, pH 7.0, with KCl, half-life: 30 min, 25 mM glycine-NaOH buffer, pH 8.5, without KCl, half-life: 750 min, 25 mM glycine-NaOH buffer, pH 8.5, with 0.2 KCl, half-life: 50 min, 25 mM glycine-NaOH buffer, pH 9.0, without KCl, half-life: 522 min, 25 mM glycine-NaOH buffer, pH 9.0, with 0.2 KCl, half-life: 12 min, 25 mM Tris-HCL, pH 6.86, with 0.2 M KCl, half-life: 2 min [12]; <13> less than 10% loss of activity after 2 h at pH 8 [17]) [12, 17]
65 <2> (<2> 2 min, 84% loss of activity [9]) [9]
Additional information <3> (<3> heating for 3 min, complete inactivation of enzyme form MSL and 65% loss of activity of enzyme form MSH, slight increase of heat stability in presence of glyoxylate and/or Mg^{2+} [3]) [3]

General stability information

<3>, trypsin, treatment for 10 min, about 50% loss of activity of enzyme form MSH, 92% loss of activity of enzyme form MSL [3]
<6>, inactivation in air-saturated aqueous solution by X-irradiation, inactivation is mainly due to the action of OH radicals, to a minor extent to O_2 radicals and H_2O_2 [6]
<11>, DTT and glycerol stabilize [35]
<11>, DTT stabilizes [35]
<11>, salt does not stabilize [35]

<22>, limited proteolysis with trypsin results in cleavage of malate synthase into two framents of respectively 45000 Da and 19000 Da [31]
<5, 15>, freezing and thawing inactivates [5, 21]
<6, 15>, Mg^{2+} stabilizes [21]

Storage stability
<5>, 0-4°C, stable for at least 2 weeks [5]
<11>, -20°C, stable for more than one month [35]
<20>, 4°C, stable for more than 3 weeks [26]
<21>, -70°C, stable for at least 3 months [27]
<22>, -70°C, 200 mM Hepes buffer, containing 6 mM $MgCl_2$, 2 mM 2-mercaptoethanol, pH 7.6, stable for several months [28]
<6, 15>, 2°C, stable for at least a month [21]

References

[1] Yang, Y.-P., Randall, D.D., Trelease, R.N.: Phosphorylation of glyoxysomal malate synthase from castor oil seeds Ricinus communis L.. FEBS Lett., **234**, 275-279 (1988)
[2] Trelease, R.N., Hermerath, C.A., Turley, R.B., Kunce, C.M.: Cottonseed malate synthase. Plant Physiol., **84**, 1343-1349 (1987)
[3] Fukawa, H., Ejiri, S.-i., Katsumata, T.: Purification and some properties of malate synthase from the pollen of Pinus densiflora sieb. et zucc.. Agric. Biol. Chem., **51**, 1553-1560 (1987)
[4] Smith, S.M., Leaver, C.J.: Glyoxysomal malate synthase of cucumber: molecular cloning of a cDNA and regulation of enzyme synthesis during germination. Plant Physiol., **81**, 762-767 (1986)
[5] Okada, H., Ueda, M., Tanaka, A.: Purification of peroxisomal malate synthase from alkane-grown Candida troppicalis and some properties of the purified enzyme. Arch. Microbiol., **144**, 137-141 (1986)
[6] Durchschlag, H., Zipper, P.: Post-irradation inactivation of the sulfhydryl enzyme malate synthase. Biochem. Biophys. Res. Commun., **118**, 364-370 (1984)
[7] Kruse, C., Kindl. H.: Oligomerization of malate synthase during glyoxysome biosynthesis. Arch. Biochem. Biophys., **223**, 629-638 (1983)
[8] Kruse, C., Kindl, H.: Malate synthase: aggregation, deaggregation, and binding of phopholipids. Arch. Biochem. Biophys., **223**, 618-628 (1983)
[9] Miernyk, J.A., Trelease, R.N.: Malate synthase from Gossypium hirsutum. Phytochemistry, **20**, 2657-2663 (1981)
[10] Durchschlag, H., Biedermann, G., Eggerer, H.: Large-scale purification and some properties of malate synthase from bakers yeast. Eur. J. Biochem., **114**, 255-262 (1981)
[11] Sundaram, T.K., Chell, R.M., Wilkinson, A.E.: Monomeric malate synthase from a thermophilic Bacillus. Arch. Biochem. Biophys., **199**, 515-525 (1980)

[12] Chell, R.M., Sundaram, T.K.: Structural basis of the thermostability of monomeric malate synthase from a thermophilic Bacillus. J. Bacteriol., **135**, 334-341 (1978)
[13] Woodcock, E., Merrett, M.J.: Purification and immunchemical characterization of malate synthase from Euglena gracilis. Biochem. J., **173**, 95-101 (1978)
[14] Bowden, L., Lord, J.M.: Purification and comparative properties of microsomal and glyoxysomal malate synthase from castor bean endosperm. Plant Physiol., **61**, 259-265 (1978)
[15] Patel, T.R., McFadden, B.A.: Particulate isocitrate lyase and malate synthase in Caenorhabditis elegans. Arch. Biochem. Biophys., **183**, 24-30 (1977)
[16] Zipper, P., Durchschlag, H.: Small-angle X-ray studies on malate synthase from bakers yeast. Biochem. Biophys. Res. Commun., **75**, 394-400 (1977)
[17] Chell, R.M., Sundaram, T.K.: Isolation and characterization of isocitrate lyase and malate synthase from Bacillus stearathermophilus. Biochem. Soc. Trans., **3**, 303-306 (1975)
[18] Schmid, G., Durchschlag, H., Biedermann, G., Eggerer, H., Jaenicke, R.: Molecular structure of malate synthase and structural change upon ligand binding to the enzyme. Biochem. Biophys. Res. Commun., **58**, 419-426 (1974)
[19] Falmagne, P., Wiame, J.-M.: Purification et caracterisation partielle des deux malate synthases d' Escherichia coli. Eur. J. Biochem., **37**, 415-424 (1973)
[20] Begin-Heick, N.: The localization of enzymes of intermediary metabolism in Astia and Euglena. Biochem. J., **134**, 607-616 (1973)
[21] Dixon, G.H., Kornberg, H.L., Lund, P.: Purification and properties of malate synthetase. Biochim. Biophys. Acta, **41**, 217-233 (1960)
[22] Hüttener, S.; Mecke, D.; Fröhlich, K.-U.: Gene cloning and sequencing, and enzyme purification of the malate synthase of Streptomyces arenae. Gene, **188**, 239-246 (1997)
[23] Mullen, R.T.; Gifford, D.J.: Purification and characterization of the glyoxysomal enzyme malate synthase following seed germination in Pinus taeda. Plant Physiol. Biochem., **33**, 639-648 (1995)
[24] Mori, H.; Yokota, S.; Akazawa, T.; Nishimura, M.: : Purification and characterization of glyoxysomal enzymes from germination pumpkin cotyledons. Plant Cell Physiol., **29**, 449-460 (1988)
[25] Henry, H.; Escher, C.-L.; Widmer, F.: : Salt-mediated interconversions and purification of malate synthase from germiating soybean cotyledons (Glycine max.). Plant Sci., **82**, 21-27 (1992)
[26] Reinscheid, D.J.; Eikmanns, B.J.; Sahm, H.: : Malate synthase from Corynebacterium glutamicum: sequence analysis of the gene and biochemical characterization of the enzyme. Microbiology, **140**, 3099-3108 (1994)
[27] Bruinenberg, P.G.; Blaauw, M,; Veenhuis, M.; Ab,G.: Purification and some properties of malate synthase from the methylotropic yeast Hansenula polymorpha. FEMS Microbiol.Lett., **61**, 11-16 (1989)
[28] Khan, A. S., Van Driessche, E., Kanarek, L., Beeckmans, S.: Purification of the glyoxylate cycle enzyme malate synthase from maize (Zea mays L.) and

characterization of a proteolytic fragment. Protein Expression Purif., **4**, 519-528 (1993)

[29] Beeckmans, A., Khan, A.S., Van Driessche, E., Kanarek, L.: A specific association between the glyoxylic-acid-cycle enzymes isocitrate lyase and malate synthase. Eur.J.Biochem., **224**, 197-201 (1994)

[30] Hoppe, A., Theimer, R.R.: Rapid purification of malate synthase from cotyledon of Brassica napus L.. FEBS Lett., **374**, 225-227 (1995)

[31] Beeckmans, S., Khan, A.S., Kanarek, L., Van Driessche, E.: Ligand binding on to maize (Zea mays) malate synthase: a structural study. Biochem. J., **303**, 413-421 (1994)

[32] Benvides, J.M., Tremblay, G. C., Hammen, C. S.: : Determination of isocitrate lyase and malate synthase activities in a marine bivalve mollusk by a new methoid of assay. Comp. Biochem. Physiol., **94B**, 779-782 (1989)

[33] Howard, B.R.; Endrizzi, J.A.; Remington, S.J.: Crystal structure of Escherichia coli malate synthase G complexed with magnesium and glyoxylate at 2.0 Å resolution: Mechanistic implications. Biochemistry, **39**, 3156-3168 (2000)

[34] Kunze, M.; Kragler, F.; Binder, M.; Hartig, A.; Gurvitz, A.: Targeting of malate synthase 1 to the peroxisomes of Saccharomyces cerevisiae cells depends on growth on oleic acid medium. Eur. J. Biochem., **269**, 915-922 (2002)

[35] Watanabe, S.; Takada, Y.; Fukunaga, N.: Purification and characterization of a cold-adapted isocitrate lyase and a malate synthase from Colwellia maris, a psychrophilic bacterium. Biosci. Biotechnol. Biochem., **65**, 1095-1103 (2001)

[36] Mahan, J.R.: Thermal dependence of malate synthase activity and its relationship to the thermal dependence of seedling emergence. J. Agric. Food Chem., **48**, 4544-4549 (2000)

[37] Loke, P.; Goh, L.L.; Soh, B.S.; Yeow, P.; Sim, T.S.: Purification and characterization of recombinant malate synthase enzymes from Streptomyces coelicolor A3(2) and S. clavuligerus NRRL3585. J. Ind. Microbiol. Biotechnol., **28**, 239-243 (2002)

[38] Smith, C.V.; Huang, C.C.; Miczak, A.; Russell, D.G.; Sacchettini, J.C.; Honer zu Bentrup, K.: Biochemical and structural studies of malate synthase from Mycobacterium tuberculosis. J. Biol. Chem., **278**, 1735-1743 (2003)

Hydroxymethylglutaryl-CoA synthase 2.3.3.10

1 Nomenclature

EC number
2.3.3.10

Systematic name
acetyl-CoA:acetoacetyl-CoA C-acetyltransferase (thioester-hydrolysing, carboxymethyl-forming)

Recommended name
hydroxymethylglutaryl-CoA synthase

Synonyms
(S)-3-hydroxy-3-methylglutaryl-CoA acetoacetyl-CoA-lyase (CoA-acetylating)
3-hydroxy-3-methylglutaryl CoA synthetase
3-hydroxy-3-methylglutaryl coenzyme A synthase
3-hydroxy-3-methylglutaryl coenzyme A synthetase
3-hydroxy-3-methylglutaryl-CoA synthase
3-hydroxy-3-methylglutaryl-coenzyme A synthase
EC 4.1.3.5 (formerly)
HMG-CoA synthase
acetoacetyl coenzyme A transacetase
β-hydroxy-β-methylglutaryl-CoA synthase
hydroxymethylglutaryl CoA synthetase
hydroxymethylglutaryl coenzyme A synthase
hydroxymethylglutaryl coenzyme A synthase
hydroxymethylglutaryl coenzyme A-condensing enzyme
hydroxymethylglutaryl-coenzyme A synthase

CAS registry number
9027-44-5

2 Source Organism

<1> *Arabidopsis thaliana* [23]
<2> *Arabidopsis thaliana* [27, 28]
<3> *Blattella germanica* (cockroach, cytosolic isoform 2 [22, 25, 27, 28]) [22, 25, 27, 28]
<4> *Blattella germanica* (cockroach, cytosolic isoform 1 [22, 25, 27, 28]) [22, 25, 27, 28]

<5> *Blattella germanica* (cockroach [26,31,32]) [26, 31, 32]
<6> *Borrelia burgdorferi* [32]
<7> *Borrelia burgdorferi* [27, 28]
<8> *Bos taurus* (ox [11,13,14,24-26,32]) [11, 13, 14, 24-26, 32]
<9> *Caenorhabditis elegans* [27]
<10> *Catharanthus roseus* [23]
<11> *Cavia porcellus* (guinea pig [3]) [3]
<12> *Columba livia* (pigeon [6]) [6]
<13> *Cricetulus griseus* (chinese hamster, cytosolic isoform [22,27]) [22, 27]
<14> *Cricetulus griseus* (chinese hamster [16,21,23,24]) [16, 21, 23, 24]
<15> *Diploptera punctata* [26]
<16> *Enterococcus faecalis* [32]
<17> *Gallus gallus* (chicken, cytosolic isoform [27,28]) [27, 28]
<18> *Gallus gallus* (chicken [3,5,6,9-13,15-17,21-24,26]; white Leghorn cockerels [6]; avian [12,20,22,25,27-29,32,33]) [3, 5, 6, 8-13, 15-17, 20-29, 32, 33]
<19> *Homo sapiens* (human, mitochondrial isoform [27,28]) [27, 28]
<20> *Homo sapiens* (human, cytosolic isoform [27]) [27]
<21> *Homo sapiens* (human [21-23,25,27,28,32]; HeLa cells [16]; HepG2 cells [24]) [16, 21-25, 27, 28, 32]
<22> *Locusta migratoria* [26]
<23> *Meleagris gallopavo* (turkey [6]) [6]
<24> *Mesoricetus auratus* (standard hamster [19,24,25,27,32]) [19, 24, 25, 27, 32]
<25> *Methanobacterium thermoautotrophicum* [27, 28]
<26> *Mus musculus* (mouse, mitochondrial isoform [27,28]) [27, 28]
<27> *Mus musculus* (mouse [19,22,25,32]) [19, 22, 25, 32]
<28> *Pinus sylvestris* (pine [27,28]) [27, 28]
<29> *Raphanus sativus* (radish [23]) [23]
<30> *Rattus norvegicus* (rat, cytosolic isoform [27]) [27]
<31> *Rattus norvegicus* (rat, mitochondrial isoform [27,28]) [27, 28]
<32> *Rattus norvegicus* (rat [3,4,6,9,12,16,18-25,32]; Sprague-Dawley [12,30]) [3, 4, 6, 9, 12, 16, 18-25, 30, 32]
<33> *Saccharomyces cerevisiae* (baker's yeast [27]) [27]
<34> *Saccharomyces cerevisiae* (baker's yeast [1,2,4,6,7,26]) [1, 2, 4-8, 11, 13, 25-27]
<35> *Schizosaccharomyces pombe* [27, 28]
<36> *Staphylococcus aureus* [32]
<37> *Staphylococcus epidermidis* [32]
<38> *Staphylococcus haemolyticus* [32]
<39> *Streptococcus pneumoniae* [32]
<40> *Streptococcus pyogenes* [32]
<41> *Streptomyces sp.* [32]
<42> *Sus scrofa* (pig, mitochondrial isoform [27,28]) [27, 28]
<43> *Sus scrofa* (pig [32]; piglet [30]) [30, 32]

3 Reaction and Specificity

Catalyzed reaction

acetyl-CoA + H_2O + acetoacetyl-CoA = (S)-3-hydroxy-3-methylglutaryl-CoA + CoA

Reaction type

acyl group transfer
condensation
hydrolysis
transacetylation

Natural substrates and products

S acetyl-CoA + acetoacetyl-CoA + H_2O <1-43> (<34> synthesis of isoprenes and steroids [1,2]; <18> acetoacetate biosynthesis, 3-hydroxy-3-methylglutaryl-CoA pathway [3]; <18> potentially favorable position for the regulation of hepatic ketogenesis [10]; <18> cytosolic synthase catalyzes the second step of cholesterogenesis from acetyl-CoA [5]; <18> commited step in the pathways for isoprenoid, cholesterol and ketone body production, cytosolic isoform involved in isoprenoid/cholesterol biosynthesis [33]; <32> major cytoplasmic synthase, synthase II is implicated in cholesterol synthesis [6]; <32> first committed step in isoprenoid biosynthesis [16]; <12,18,23> mitochondrial enzyme of liver functions in ketogenesis, its cytoplasmic counterpart participates in cholesterogenesis [5,6,9]; <18,21,34> key intermediate in cholesterogenic and ketogenic pathways, cholesterogenic isoform in cytosol, ketogenic isoform in mitochondria, committed step for both pathways [27]; <5> regulation of isoprenoid synthesis and metabolism [26]; <18> mitochondrial enzyme is involved in the synthesis of acetoacetate, cytoplasmic enzyme in that of mevalonate [11]; <18> enzyme involved in ketone body production, control of hepatic fatty acid oxidation, potential control point for ketogenesis [12]; <18> produces a key intermediate in steroidogenic and ketogenic metabolic pathways [22,29]; <5> mevalonate pathway [31]; <8> pathway of ketogenesis in the mitochondria, mevalonate synthesis in the cytoplasm [13]; <1> key role in synthesis of various sterols and isoprenoids [23]; <16> isopentenyl diphosphate biosynthesis [32]; <43> pathway converting acetyl-CoA to ketone bodies [30]) (Reversibility: ir <16, 18, 21, 34, 43> [1, 5, 17, 27, 30, 32]) [1-33]

P (S)-3-hydroxy-3-methylglutaryl-CoA + CoA <5, 11, 12, 18, 23, 32, 34> [1-10, 26]

Substrates and products

S 3-oxobutyl-CoA + acetoacetyl-CoA + H_2O <18> (Reversibility: ? <18> [17, 25]) [17, 25]

P 3-hydroxy-3-methyl-5-oxoheptanedioyl-CoA + CoA

S acetyl-3'-dephospho-CoA + acetoacetyl-CoA + H_2O <18, 34> (Reversibility: ? <18, 34> [7, 17]) [7, 17]

P 3-hydroxy-3-methylglutaryl-CoA + 3'-dephospho-CoA

S acetyl-CoA + 3'-dephospho-CoA + H_2O <18> (Reversibility: ? <18> [5, 8]) [5, 8]
P acetyldephospho-CoA + CoA <18> [5, 8]
S acetyl-CoA + H_2O <18> (<18> in absence of co-substrate enzyme hydrolyzes acetyl-CoA [20]) (Reversibility: ? <18> [8, 20, 27, 33]) [8, 20, 27, 33]
P acetate + CoA <18> [8]
S acetyl-CoA + acetoacetyl-[acyl-carrier-protein] + H_2O <34> (Reversibility: ? <34> [7]) [7]
P 3-hydroxy-3-methylglutaryl-[acyl-carrier-protein]
S acetyl-CoA + acetoacetyl-CoA + H_2O <1-43> (Reversibility: ir <16, 18, 21, 34> [1, 5, 17, 27, 32]) [1-29, 31-33]
P (S)-3-hydroxy-3-methylglutaryl-CoA + CoA <11, 12, 18, 23, 32, 34> [1-10, 26]
S acetyl-CoA+ N-acetyl-S-acetoacetylcysteamine + H_2O <18> (Reversibility: ? <18> [17]) [17]
P ? + CoA
S acetyl-CoA+ cysteamine + H_2O <18> (Reversibility: ? <18> [8]) [8]
P ? + CoA
S acetylglutathione + acetoacetyl-CoA + H_2O <34> (Reversibility: ? <34> [7]) [7]
P ?
S acetylpantetheine + acetoacetyl-CoA + H_2O <34> (Reversibility: ? <34> [7]) [7]
P 3-hydroxy-3-methylglutaryl-CoA + pantetheine
S Additional information <5, 18, 32, 34> (<34> acetyl-CoA analogues can act as substrates, acetoacyl-analogues are no substrates [2]; <34> acetyl-CoA homologues do not act as substrates [1]; <34> absolutely specific for acetyl-CoA and acetoacetyl-CoA in the reaction, neither propionyl- nor butyryl-CoA can replace acetyl-CoA, no reaction observed when 3-oxo-hexanoyl-CoA is substituted for acetoacetyl-CoA [1]; <34> absolutely specific for the acyl-moiety of both its substrates, neither propionyl- nor butyryl CoA can replace acetyl-CoA [7]; <18> cholesterol feeding causes a significant reduction in the HMG-CoA synthase activity of the cytosolic fraction of liver [6]; <32> feeding cholestyramine to rats causes a 2.6fold increase in the cytosolic activities of the synthase in liver [6]; <32> reaction is stereospecific and associated with an intramolecular hydrogen isotope effect, condensation proceeds with inversion of configuration at the methyl group [4]; <5> first organism known to have 2 functional cytosolic HMG-CoA synthases [26,31]) [1, 2, 4, 6, 7, 26, 31]
P ?

Inhibitors

(E,E)-11-[3-(hydroxymethyl)-4-oxo-2-oxytanyl]-3,5,7-trimethyl-2,4-undecadienenoic acid <21, 32> [16, 19]
1-ethyl-3-(3-dimethylaminopropyl)carbodiimide <18, 21, 34> [27]
3,3'-dimethylglutaryl-CoA <34> [1]
3-carboxy-2,2,5,5,-tetramethyl-1-pyrrolidinyloxyl-CoA <18> [17, 25, 28]

3-carboxypropionyl-CoA <8> [14]
3-chloropropionyl-CoA <18> [15, 28]
3-oxohexanoyl-CoA <34> [1]
5,5'-dithiobis(2-nitrobenzoate) <18, 34> (<34> residual activity 51% [2]; <18> Ellman's reagent [5]) [2, 5]
Cd^{2+} <34> [7]
$CdSO_4$ <34> (<34> residual activity 9% [2]) [2]
CoA <8, 34> (<8> mixed product inhibition [13]) [1, 7, 11, 13]
Cu^{2+} <18> [3]
DL-3-hydroxy-3-methylglutaryl-CoA <8, 18, 34> (<34> uncompetitive inhibition [1]; <18> competitive inhibition [17]; <18> product inhibition [15]; <8> mixed product inhibition [13]) [1, 13, 15, 17]
DL-3-methylglutaryl-CoA <34> [1]
L-659,699 <21, 32> (<32> antibiotic 1233A [18]; <21> potent inhibitor of the recombinant enzyme [21]) [18, 21]
Mg^{2+} <18> (<18> mitochondrial enzyme [6,9,20]) [6, 9, 20]
$MgCl_2$ <18, 32> (<18> synthase I [6]; <32> mitochondrial enzyme [6]) [6]
N-ethylmaleimide <34> (<34> residual activity 46% [2]) [2, 7]
$NaAsO_2$ <34> (<34> residual activity 90% [2]) [2]
Zn^{2+} <18> [3]
acetoacetyl-CoA <8, 18, 21, 32, 34> (<18,32,34> substrate inhibition [1,12]; <8> competitive inhibition [13]) [1, 5-7, 9, 10, 12, 13, 17, 21]
arsenite + CoA <34> (<34> residual activity 25% [2]) [2]
bromoacetyl-CoA <34> [2]
butyryl-CoA <34> [1]
decanoyl-CoA <34> [1]
desulfo-CoA <34> [1]
glutaryl-CoA <34> [1]
heptanoyl-CoA <34> [1]
hexanoyl-CoA <34> [1]
hexanoyl-CoA <34> [1]
iodoacetamide <34> (<34> residual activity 84% [2]) [2, 7]
iodoacetate <34> [7]
lifibrol <18, 21> [24]
nonanoyl-CoA <34> [1]
octanoyl-CoA <34> [1]
p-chloromercuribenzoate <18, 34> (<34> residual activity 16% [2]) [2, 5, 7]
palmitoyl-CoA <18> (<18> mitochondrial synthase [9]) [9]
phenyl arsenious oxide <34> (<34> residual activity 75% [2]) [2]
phenyl arsenious oxide + CoA <34> (<34> residual activity 41% [2]) [2]
propionyl-CoA <34> [1]
succinyl-CoA <8, 18> (<18> mitochondrial synthase, 50% inhibition [9]) [9, 14]
Additional information <34> (<34> desulfo-CoA protects against bromoacetyl-CoA inhibition, but not against *p*-chloromercuribenzoate and N-ethylmaleimide [2]) [2]

Activating compounds

$MgCl_2$ <16, 18> (<18> small activating effect on synthases II and III [6]; <16> optimal activity in 2.0 mM [32]) [6, 32]
chloride <34> [8]
potassium phosphate <34> [8]
sodium phosphate <34> [8]
sulfate <34> [8]

Metals, ions

Mg^{2+} <18> (<18> cytoplasmic enzyme [6,9]; <18> activity of the recombinant cytosolic isozyme ist stimulated by [20]) [6, 9, 20]
Mn^{2+} <18> (<18> increases acetoacetate synthesis activity [3]) [3]

Turnover number (min^{-1})

2.5 <8> (acetyl-CoA) [13]
4 <18> (acetyl-CoA) [29]

Specific activity (U/mg)

0.00083 <18> (<18> mitochondrial pellet [3]) [3]
0.00133 <18> (<18> mitochondrial pellet, 0.25 M sucrose [3]) [3]
0.00245 <18> (<18> disrupted mitochondria, 0.25 M sucrose [3]) [3]
0.00288 <18> (<18> mitochondrial pellet, 0.50 M sucrose [3]) [3]
0.00325 <18> (<18> mitochondrial pellet, 0.88 M sucrose [3]) [3]
0.00485 <18> (<18> homogenate [3]) [3]
0.00567 <18> (<18> disrupted mitochondria, 0.50 M sucrose [3]) [3]
0.00792 <18> (<18> disrupted mitochondria, 0.88 M sucrose [3]) [3]
0.00922 <18> (<18> mitochondrial supernatant [3]) [3]
0.00958 <18> (<18> mitochondrial supernatant, 0.88 M sucrose [3]) [3]
0.0101 <18> [24]
0.01053 <18> (<18> mitochondrial supernatant, 0.50 M sucrose [3]) [3]
0.0111 <18> (<18> mitochondrial supernatant, 0.25 M sucrose [3]) [3]
0.3 <18> [5]
0.5-1 <18> (<18> liver enzyme [20]) [20]
0.65 <18> (<18> cytoplasmic enzyme, isozyme III and IV [6]) [6]
0.76 <21> [21]
0.8-1 <18> (<18> recombinant enzyme [20]) [20]
0.85 <18> [17]
0.88 <8> [13]
1 <18> [9, 10, 20]
2 <34> [1]
2.1 <34> [2, 7]
10 <16> [32]
40 <5> [26]

K_m-Value (mM)

0.0001 <34> (acetoacetyl-CoA) [26]
0.00035 <18> (acetoacetyl-CoA) [12, 17]
0.0004 <34> (acetoacetyl-CoA, <34> pH 8.0 [1]) [1]
0.00055 <18> (acetoacetyl-CoA, <18> mutant H197N [25]) [25]

0.00062 <18> (acetoacetyl-CoA, <18> mutant H436N [25]) [25]
0.00085 <18> (acetoacetyl-CoA) [25]
0.00119 <18> (acetoacetyl-CoA) [27]
0.0012 <18> (acetoacetyl-CoA) [28, 32]
0.0015 <5> (acetyl-CoA) [32]
0.002 <18> (acetoacetyl-CoA, <18> cytoplasmic isozymes I-IV [6]) [6]
0.003 <34> (acetoacetyl-CoA, <34> pH 8.9 [7]) [7]
0.0032 <34> (acetoacetyl-CoA, <34> pH 8.9 [1]) [1]
0.005 <18> (acetoacetyl-CoA) [1, 9, 33]
0.0078 <18> (acetyl-CoA, <18> mutant D159A [27]) [27]
0.0081 <18> (acetyl-CoA, <18> mutant D159A, acetyl-S-enzyme formation [27]) [27]
0.009 <32> (acetyl-CoA, <32> absence of acetoacetyl-CoA, hydrolysis reaction [12]) [12]
0.01 <16> (acetoacetyl-CoA) [32]
0.011 <18> (acetyl-CoA, <18> acetyl-CoA hydrolysis, wild-type [28]) [28]
0.012 <18> (acetoacetyl-CoA, <18> recombinant enzyme, hydrolase reaction [20]) [20, 28]
0.014 <18> (acetyl-CoA, <18> wild-type, acetyl-CoA hydrolysis [20,27]; <18> acetyl-CoA hydrolysis, mutant E95A [28]) [20, 27, 28]
0.014 <34> (acetyl-CoA, <34> pH 8.0 [1]) [1, 7, 26]
0.015 <18> (acetyl-CoA, <18> acetyl-CoA hydrolysis [25]) [25]
0.0152 <5> (acetyl-CoA) [26]
0.018 <34> (acetyl-CoA, <34> pH 8.9 [1]) [1, 7]
0.019 <18> (oxobutyl-CoA) [17]
0.021 <18> (oxobutyl-CoA) [25]
0.0227 <18> (acetyl-CoA, <18> wild-type, acetyl-S-enzyme formation [27]) [27]
0.028 <18> (acetyl-CoA, <18> mutant Y130F, hydrolysis partial reaction [33]) [33]
0.029 <21> (acetyl-CoA) [21]
0.033 <18> (acetyl-CoA, <18> mutant D159A, acetyl-CoA hydrolysis [27]) [27]
0.036 <18> (acetyl-CoA, <18> mutant H264A, acetyl-CoA hydrolysis [25]) [25]
0.037 <18> (acetyl-CoA, <18> mutant D203A [27]) [27]
0.038 <18> (acetyl-CoA, <18> mutant H264N, acetyl-CoA hydrolysis [25]) [25]
0.049 <18> (acetoacetyl-CoA, <18> mutant D159A [27]) [27]
0.05 <18> (acetyl-CoA, <18> 0.005 mM acetoacetyl-CoA [10]) [10]
0.051 <8> (acetyl-CoA) [26]
0.058 <18> (acetoacetyl-CoA, <18> mutant H264N [25]) [25]
0.066 <18> (acetoacetyl-CoA, <18> mutant D99A [27]) [27]
0.1 <18> (acetyl-CoA) [11, 17]
0.115 <18> (acetoacetyl-CoA, <18> mutant H264A [25]) [25]
0.116 <18> (3-oxobutyl-CoA, <18> mutant H264A [25]) [25]
0.118 <18> (acetyl-CoA, <18> mutant D64N [27]) [27]

0.122 <18> (acetyl-CoA, <18> mutant H264A [25]) [25]
0.172 <18> (acetyl-CoA, <18> mutant H264N [25]) [25]
0.189 <18> (acetyl-CoA, <18> mutant E63Q [27]) [27]
0.2 <15> (acetyl-CoA) [26]
0.2 <32> (acetyl-CoA, <32> 0.01 M acetoacetyl-CoA [12]) [12]
0.207 <18> (acetyl-CoA, <18> mutant D99A [27]) [27]
0.221 <18> (acetyl-CoA, <18> mutant H436N [25]) [25]
0.27 <18> (acetyl-CoA, <18> reombinant enzyme [20]) [20, 32]
0.274 <18> (acetyl-CoA, <18> mutant C59A [22]) [22]
0.28 <18> (acetyl-CoA, <18> mutant C69A [22]) [22]
0.29 <18> (acetyl-CoA, <18> synthase I [6]; <18> wild-type enzyme [27,28]) [6, 27, 28]
0.294 <18> (acetyl-CoA, <18> wild-type [22]) [22, 25]
0.3 <18> (acetyl-CoA, <18> 0.1 mM acetoacetyl-CoA [5]) [5, 20, 26]
0.307 <18> (acetyl-CoA, <18> mutant Y376F [33]) [33]
0.31 <18> (acetyl-CoA, <18> synthases II and III [6]) [6]
0.319 <18> (acetyl-CoA, <18> mutant D217A [27]) [27]
0.35 <16> (acetyl-CoA) [32]
0.354 <18> (acetyl-CoA, <18> mutant E37Q [27]) [27]
0.355 <18> (acetyl-CoA, <18> mutant C232A [22]) [22]
0.358 <18> (acetyl-CoA, <18> mutant D282A [27]) [27]
0.372 <18> (acetyl-CoA, <18> mutant E121A [27]) [27]
0.4 <22> (acetyl-CoA) [26]
0.435 <18> (acetyl-CoA, <18> mutant C268A [22]) [22]
0.44 <18> (acetyl-CoA, <18> mutant H197N [25]) [25]
0.448 <18> (acetyl-CoA, <18> mutant D203A, acetyl-S-enzyme formation [27]) [27]
0.5 <18> (acetyl-CoA, <18> 0.05 mM acetoacetyl-CoA [10]) [10]
0.565 <18> (acetyl-CoA, <18> mutant Y130F [33]) [33]
0.623 <18> (acetyl-CoA, <18> mutant D99A, acetyl-S-enzyme formation [27]) [27]
0.677 <18> (acetyl-CoA, <18> mutant D124A [27]) [27]
0.695 <43> (acetyl-CoA) [30]
0.811 <32> (acetyl-CoA) [30]
1 <18> (acetyl-CoA, <18> mitochondrial synthase [6]) [6]
1.561 <18> (acetyl-CoA, <18> mutant C224A [22]) [22]

K_i-Value (mM)

0.0000537 <21> (L-659,699) [21]
0.0005 <18> (palmityl-CoA) [9]
0.00126 <5> (acetoacetyl-CoA) [26]
0.0027 <34> (decanoyl-CoA, <34> pH 8.0 [1]) [1]
0.0035 <8> (acetoacetyl-CoA) [13, 26]
0.004 <34> (nonanoyl-CoA, <34> pH 8.0 [1]) [1]
0.005 <34> (propionyl-CoA, <34> pH 8.0 [1]) [1]
0.006-0.01 <18> (acetoacetyl-CoA) [12]
0.0071 <34> (3-oxohexanoyl-CoA, <34> pH 8.0 [1]) [1]

0.008 <34> (acetoacetyl-CoA, <34> pH 8.0 [1]) [1, 8, 10]
0.0081 <34> (octanoyl-CoA, <34> pH 8.0 [1]) [1]
0.01 <18> (acetoacetyl-CoA) [17]
0.012 <18> (DL-3-hydroxy-3-methylglutaryl-CoA) [15, 17]
0.012 <18> (acetoacetyl-CoA) [11]
0.012 <21> (acetoacetyl-CoA) [21]
0.0123 <34> (heptanoyl-CoA, <34> pH 8.0 [1]) [1]
0.013 <34> (DL-3-hydroxy-3-methylglutaryl-CoA, <34> pH 8.0, competitive inhibition, substrate acetyl-CoA [1]) [1]
0.014 <34> (3-oxohexanoyl-CoA, <34> pH 8.0, mixed inhibition, substrate acetoacetyl-CoA [1]) [1]
0.015 <18> (3-chloropropionyl-CoA) [15]
0.02 <34> (acetoacetyl-CoA, <34> pH 8.9 [1,7]) [1, 7]
0.022 <34> (hexanoyl-CoA, <34> pH 8.0 [1]) [1]
0.03 <34> (desulfo-CoA, <34> pH 8.0 [1]) [1]
0.033 <34> (butyryl-CoA, <34> pH 8.0 [1]) [1]
0.038 <34> (CoA, <34> pH 8.0, mixed inhibition, substrate acetyl-CoA [1]) [1]
0.06 <34> (CoA, <34> pH 8.0, mixed inhibition, substrate acetoacetyl-CoA [1]) [1]
0.07 <34> (desulfo-CoA, <34> pH 8.0, mixed inhibition, substrate acetoacetyl-CoA [1]) [1]
0.1 <34> (DL-3-methylglutaryl-CoA, <34> pH 8.0 [1]) [1]
0.11 <34> (glutaryl-CoA, <34> pH 8.0 [1]) [1]
0.12 <34> (3,3'-dimethylglutaryl-CoA, <34> pH 8.0 [1]) [1]

pH-Optimum

9.2 <18> (<18> synthase II [6]) [6]
9.3 <18> (<18> synthase I [6]) [6]
9.4 <18> (<18> synthase III [6]) [5, 6, 12]
9.5 <18> [22]
9.8 <16> [32]

pH-Range

7.75-9.5 <18> [22]
8-9.6 <18> [5]

Temperature optimum (°C)

37 <16> [32]

4 Enzyme Structure

Molecular weight

50000 <5, 43> (<5> SDS-PAGE, immunoblot by antibodies, prediction from cDNA nucleotide sequence [26]; <43> recombinant enzyme, epressed in Mev-1 cells [30]) [26, 30]
51040 <1> (<1> predicted from cDNA sequence [23]) [23]

53000 <18> (<18> SDS-PAGE [24]) [24]
55500 <1> (<1> SDS-PAGE [23]) [23]
56000 <43> (<43> recombinant enzyme, epressed in Escherichia coli [30]) [30]
57250 <14> (<14> deduced amino acid sequence [23]) [23]
57290 <21> (<21> deduced amino acid sequence [23]) [23]
57370 <32> (<32> deduced amino acid sequence [23]) [23]
57490 <18> (<18> deduced amino acid sequence [23]) [23]
57600 <18> (<18> SDS-PAGE [20]) [20]
83900 <16> (<16> analytical ultracentrifugation [32]) [32]
84300 <16> (<16> calculated from amino acid sequence [32]) [32]
88000 <18> (<18> gel filtration [11]) [11]
90000 <18> (<18> gel filtration, synthase I [6]) [6]
94000-100000 <18> (<18> gel filtration, synthase II [6]) [6]
94600 <8> (<8> sedimentation equilibrium [13]) [13]
96000 <18> (<18> gel filtration [9]) [9]
96000-105000 <18> (<18> gel filtration, mitochondrial synthase [6]) [6]
100000 <8, 18> (<18> gel filtration, synthases I, II, III, IV, multiple forms of cytoplasmic enzyme [6]; <18> cytosolic enzyme, sedimentation equilibrium [5]) [5, 6, 13]
105000 <18> (<18> sedimentation equlibrium [8-10,17]) [8-10, 17]
111000 <8> (<8> gel filtration [13]) [13]
116000 <18> (<18> gel filtration [28]) [28]
119300 <8> (<8> sedimentation velocity and light-scattering [13]) [13]
120000 <21> (<21> gel filtration [21]) [21]
130000 <34> (<34> gel filtration [2,7]) [2, 7, 13]

Subunits

dimer <8, 16, 18, 21> (<18> 2 * 52000, mitochondrial synthase, synthase I, SDS-PAGE [6]; <18> 2 * 53000, SDS-PAGE [8-10,17]; <18> 2 * 57000, gel filtration [9]; <18> 2 * 55000, synthase II, SDS-PAGE [5,6]; <18> 2 * 58000, synthases III and IV, SDS-PAGE [6]; <18> 2 * 57600 [28]; <8> 2 * 47900, SDS-PAGE [13]; <16> 2 * 42000, calculated from amino acid sequence [32]; <16> 2 * 45000, SDS-PAGE [32]) [5, 6, 8-10, 13, 17, 21, 28, 32]

5 Isolation/Preparation/Mutation/Application

Source/tissue

adrenal gland <21, 32> [21, 24]
brain <18> (<18> synthases I, II and III [6]) [6]
corpus allatum <15, 22> [26]
embryo <5> [31]
fat body <5> [26]
fetus <21> [21]
fibroblast <21> [21, 23, 24]
hepatocyte <8> [14]

kidney <12, 23> [6]
liver <8, 11, 12, 18, 23, 27, 32, 43> [3-6, 8-21, 23-26, 30, 32]
myocardium <32> [12]
ovary <5, 14> [16, 21, 23, 24, 26]
seedling <29> [23]

Localization

cytoplasm <2, 7-9, 11, 12, 18, 21, 23, 25, 26, 28, 31-35, 42> (<18> distinct mitochondrial and cytoplasmic forms [9, 11]; <2, 7, 9, 18, 21, 25, 26, 28, 31, 33-35, 42> cholesterogenic isoform [27]) [3, 5, 6, 9, 11-13, 16, 19, 21, 25, 27, 32]
cytosol <2-5, 7, 9, 13, 18, 20, 21, 24-26, 28, 31-35, 42> (<2, 7, 9, 18, 21, 25, 26, 28, 31, 33-35, 42> cholesterogenic isoform [27]) [5, 6, 9, 16, 19-22, 24, 26, 27, 32]
membrane <1> [23]
mitochondrion <8, 11, 12, 18, 21, 23, 27, 32, 43> (<18> distinct mitochondrial and cytoplasmic forms [9, 11]; <2, 7, 9, 18, 21, 25, 26, 28, 31, 33-35, 42> ketogenic isoform [27]) [3, 5, 6, 9-17, 21, 22, 24, 25, 27, 30, 32]

Purification

<1> [23]
<5> (recombinant enzyme [26]) [26]
<8> [11, 13, 14, 24]
<16> (recombinant enzyme [32]) [32]
<18> (mitochondrial form [9,12]; synthase species I-IV [6]; cytosolic synthase [5]; recombinant enzyme, expressed in Escherichia coli [20]; partially [24]; several mutants partially purified [27]) [5, 6, 9-12, 17, 20, 22, 24, 27-29, 33]
<21> [21, 32]
<32> (partially [16]) [16, 19]
<34> (partially [8]) [1, 2, 4, 7, 8]

Cloning

<1> (cDNA cloned by complementation of a bap1 mutation and expressed in Saccharomyces cerevisiae [23]) [23]
<5> (2 cytosolic HMG-CoA synthase genes, HMGS-1 and HMGs-2 described, HMGS-1 cloned in expression vector pSBLA1 and overexpressed in Escherichia coli BL21DE3 pLys S [26]; intronless gene, encodes for an active enzyme able to complement eukaryotic Mev-1 cells [26,31]) [26, 31]
<16> (mvaS gene isolated, expressed in Escherichia coli from a pET28 vector [32]) [32]
<18> (expression vector pET-3d, cloned and expressed in Escherichia coli BL21 (DE3) [20,22,25,27,28,33]; cDNA cloned and sequenced [24]) [20, 22, 24, 25, 27-29, 32, 33]
<21> (expression of recombinant wild-type cytoplasmic and cys129 mutant enzymes [32]; cDNA cloned and sequenced [24]; cDNA subcloned and expressed from a T7-based vector in Escherichia coli [21]) [21, 24, 28, 32]
<24> (cDNA cloned and sequenced [24]) [24]

<32> (cDNA cloned and sequenced [24]; rat mitochondrial HMG-CoA synthase encodes an active enzyme in the eukaryotic cell line Mev-1 in Escherichia coli BL21 (DE3) [30]) [24, 30]
<41> (cloning of a gene cluster encoding enzymes responsible for the mevalonate pathway [32]) [32]
<43> (pig mitochondrial HMG-CoA synthase encodes an active enzyme in the eukaryotic cell line Mev-1 and in Escherichia coli BL21 (DE3) [30]) [30]

Application

medicine <16, 21> (<21> enzyme is of growing medical interest as it is highly regulated [24]; <21> deficiency of the mitochondrial isoform correlates with human metabolic disease, cytosolic isoform is potentially useful target for drugs aimed at lowering cholesterol levels [21,27]; <21> target of a potent antisteroidogenic drug [25,28]; <16> potential of the mevalonate pathway enzymes of enterococci as targets for antibiotics [32]) [21, 24, 25, 27, 28, 32]

6 Stability

General stability information

<8>, can be rapidly frozen in solid CO_2/acetone and freeze-dried for long-term storage, not stable to freezing and thawing in absence of glycerol [13]
<18>, enzyme is vulnerable to proteolysis, expression in Escherichia coli leads to an quite stable enzyme [20]
<18>, relatively stable dialyzed against 10 mM potassium phosphate in 0.1 mM dithiothreitol and EDTA for 4-10 days [10]
<34>, enzyme is labile to freezing and thawing, stabilized by glycerol [2, 7]

Storage stability

<8>, -20°C, stored in presence of 30% glycerol and 5 mM dithiothreitol, remains active for several months [13]
<18>, -20°C, may be stored in 30% glycerin solution for 2-3 months with no loss of activity [10]
<18>, -20°C, purified synthase may be stored for 2-3 months with little loss of activity by adding 30% glyerol [9]
<18>, -80°C, recombinant enzyme, negligible loss of activity after 1 year [20]
<18>, -85°C, in presence of 30% glycerol stable for over 1 year [17]
<18>, -90°C, purified synthases II-IV in 20 mM potassium phosphate, pH 7.0, 5 mM dithiothreitol, can be stored as pellets following precipitation by 60% saturated ammonium sulfate up to 3 months without loss of activity [6]
<18>, 0°C, purified synthase I stored frozen in 40% glycerol containing 20 mM potassium phosphate, pH 7.0, 5 mM dithiothreitol, stable for 3-4 months [6]
<18>, 4°C, in presence of 30% glycerol stable for several days [17]
<18>, 4°C, purified synthase I stored in 40% glycerol containing 20 mM potassium phosphate, pH 7.0, 5 mM dithiothreitol is unstable [6]

<18>, 4°C, purified synthases II-IV in 20 mM potassium phosphate, pH 7.0, 5 mM dithiothreitol, stable for several weeks [6]
<18>, 4°C, recombinant enzyme, loses 50% activity after 1 year [20]
<34>, -20°, stored in presence of dithiothreitol and 30-50% glycerol purified enzyme retains complete activity for up to 3 months [7]
<34>, 4°C, enzyme loses activity in absence of glycerol [7]

References

[1] Middleton, B.: The kinetic mechanism of 3-hydroxy-3-methylglutaryl-coenzyme a synthase from baker's yeast. Biochem. J., **126**, 35-47 (1972)
[2] Middleton, B.; Tubbs, P.K.: The purification and some properties of 3-hydroxy-3-methylglutaryl-coenzyme A synthase from baker's yeast. Biochem. J., **126**, 27-34 (1972)
[3] Allred, J.B.: Properties and subcellular distribution of enzymes required for acetoacetate biosynthesis in chicken liver. Biochim. Biophys. Acta, **297**, 22-30 (1973)
[4] Cornforth, J.W.; Phillips, G.T.: Substrate stereochemistry of 3-hydroxy-3-methylglutaryl-coenzyme A synthase. Eur. J. Biochem., **42**, 591-604 (1974)
[5] Clinkenbeard, K.D.; Sugiyama, T.; Lane, M.D.: Cytosolic 3-hydroxy-3-methylglutaryl-CoA synthase from chicken liver. Methods Enzymol., **35 B**, 160-167 (1975)
[6] Clinkenbeard, K.D.; Sugiyama, T.; Reed, W.D.; Lane, M.D.: Cyoplasmic 3-hydroxy-3-methylglutaryl coenzyme A synthase from liver.. J. Biol. Chem., **250**, 3124-3135 (1975)
[7] Middleton, B.; Tubbs, P.K.: 3-Hydroxy-3-methylglutaryl-CoA synthase from baker's yeast. Methods Enzymol., **35 B**, 173-177 (1975)
[8] Miziorko, H.M.; Clinkenbeard, K.D.; Reed, W.D.; Lane, M.D.: 3-Hydroxy-3-methylglutaryl coenzyme A synthase. J. Biol. Chem., **250**, 5768-5773 (1975)
[9] Reed, W.D.; Clinkenbeard, K.D.; Lane, M.D.: Molecular and catalytic properties of mitochondrial (ketogenic) 3-hydroxy-3-methylglutaryl coenzyme A synthase of liver. J. Biol. Chem., **250**, 3117-3123 (1975)
[10] Reed, W.D.; Lane, M.D.: Mitochondrial 3-hydroxy-3-methylglutaryl-CoA synthase from chicken liver. Methods Enzymol., **35 B**, 155-160 (1975)
[11] Page, M.A.; Tubbs, P.K.: Some properties of 3-hydroxy-3-methylglutaryl-coenzyme A synthase from ox liver. Biochem. J., **173**, 925-928 (1978)
[12] Menahan, L.A.; Hron, W.T.; Hinkelman, D.G.; Miziorko, H.M.: Interrelationships between 3-hydroxy-3-methylglutaryl-CoA synthase, aetoacetyl-CoA and ketogenesis. Eur. J. Biochem., **119**, 287-296 (1981)
[13] Lowe, D.M.; Tubbs, P.K.: 3-Hydroxy-3-methylglutaryl-coenzyme A synthase from ox liver. Biochem. J., **227**, 591-599 (1985)
[14] Lowe, D.M.; Tubbs, P.K.: Succinylation and inactivation of 3-hydroxy-3-methylglutaryl-CoA synthase by succinyl-CoA and its possible relevance to the control of ketogenesis. Biochem. J., **232**, 37-42 (1985)

[15] Miziorko, H.M.; Behnke, C.E.: Active-site directed inhibition of 3-hydroxy-3-methylglutaryl coenzyme A synthase by 3-chloropropionyl coenzyme A. Biochemistry, **34**, 3174-3179 (1985)
[16] Greenspan, M.D.; Yudkovitz, J.B.; Lo, C.Y.L.; Chen, J.S.; Alberts, A.W.; Hunt, V.M.; Chang, M.N.; Yang, S.S.; Thompson, K.L.; Chiang, Y.C.P.; Chabala, J.C.; Monaghan, R.L.; Schwartz, R.L.: Inhibition of hydroxymethylglutaryl-coenzyme A synthase by L659,699. Proc. Natl. Acad. Sci. USA, **84**, 7488-7492 (1987)
[17] Miziorko, H.M.: 3-Hydroxy-3-methylglutaryl-CoA synthase from chicken liver. Methods Enzymol., **110**, 19-26 (1988)
[18] Mayer, R.J.; Louis-Flamberg, P.; Elliott, J.D.; Fisher, M.; Leber, J.: Inhibition of 3-hydroxy-3-methylglutaryl coenzyme A synthase by antibiotic 1233A and other β-lactones. Biochem. Biophys. Res. Commun., **169**, 610-616 (1990)
[19] Greenspan, M.D.; Bull, H.B.; Yudkovitz, J.B.; Hanf, D.P.; Alberts, A.W.: Inhibition of 3-hydroxy-3-methylglutaryl-CoA synthase and cholesterol biosynthesis by β-lactone inhibitors and binding of these inhibitors to the enzyme. Biochem. J., **289**, 889-895 (1993)
[20] Misra, I.; Narasimhan, C.; Miziorko, H.M.: Avian 3-hydroxy-3-methylglutaryl-CoA synthase. J. Biol. Chem., **268**, 12129-12135 (1993)
[21] Rokosz, L.L.; Boulton, D.A.; Butkiewicz, E.A.; Sanyal, G.; Cueto, M.A.; Lachance, P.A.; Hermes, J.D.: Human cytoplasmic 3-hydroxy-3-methylglutaryl coenzyme A synthase: expression, purification, andcharacterization of recombinant wild-type and Cys-129 mutant enzymes. Arch. Biochem. Biophys., **312**, 1-13 (1994)
[22] Misra, I.; Charlier, H.A.; Miziorko, H.M.: Avian cytosolic 3-hydroxy-3-methylglutaryl-CoA synthase. Evaluation of the role of cysteines in reaction chemistry. Biochim. Biophys. Acta, **1247**, 253-259 (1995)
[23] Montamat, F.; Guilloton, M.; Karst, F.; Delrot, S.: Isolation and characterization of a cDNA encoding Arabidopsis thaliana 3-hydroxy-3-methylglutaryl-coenzyme A synthase. Gene, **167**, 197-201 (1995)
[24] Scharnagl, H.; März, W.; Schliack, M.; Löser, R.; Gross, W.: A novel assay of cytosolic 3-hydroxy-3-methylglutaryl-coenzyme A synthase activity using reversed-phase ion-pair chromatography: demonstration that Lifibrol (K12.148) modulates the enzyme activity. J. Lipid Res., **36**, 622-627 (1995)
[25] Misra, I.; Miziorko, H.M.: Evidence for the interaction of avian 3-hydroxy-3-methylglutaryl-CoA synthase H264 with acetoacetyl-CoA. Biochemistry, **35**, 9610-9616 (1996)
[26] Cabano, J.; Buesa, C.; Hegardt, F.G.; Marrero, P.F.: Catalytic properties of recombinant 3-hydroxy-3-methylglutaryl coenzyme A synthase-1 from Blattella germanica. Insect Biochem. Mol. Biol., **27**, 499-505 (1997)
[27] Chun, K.Y.; Vinarov, D.A.; Miziorko, H.M.: 3-Hydroxy-3-methylglutaryl-CoA synthase: Participation of invariant acidic residues in formation of the acetyl-S-enzyme reaction intermediate. Biochemistry, **39**, 14670-14681 (2000)

[28] Chun, K.Y.; Vinarov, D.A.; Zajicek, J.; Miziorko, H.M.: 3-hydroxy-3-methylglutaryl-CoA synthase. A role for glutamate 95 in general acid/base catalysis of C-C bond formation. J. Biol. Chem., **275**, 17946-17953 (2000)
[29] Vinarov, D.A.; Miziorko, H.M.: 3-Hydroxy-3-methylglutaryl-coenzyme A synthase reaction intermediates: Detection of a covalent tetrahedral adduct by differential isotope shift ^{13}C nuclear magnetic resonance spectroscopy. Biochemistry, **39**, 3360-3368 (2000)
[30] Barrero, M.J.; Alho, C.S.; Ortiz, J.A.; Hegardt, F.G.; Haro, D.; Marrero, P.F.: Low activity of mitochondrial HMG-CoA synthase in liver of starved pigletsis due to low levels of protein despite high mRNA levels. Arch. Biochem. Biophys., **385**, 364-371 (2001)
[31] Casals, N.; Buesa, C.; Marrero, P.F.; Belles, X.; Hegardt, F.G.: 3-Hydroxy-3-methylglutaryl coenzyme A synthase-1 of Blattella germanica has structural and functional features of an active retrogene. Insect Biochem. Mol. Biol., **31**, 425-433 (2001)
[32] Sutherlin, A.; Hedl, M.; Sanchez-Neri, B.; Burgner, J.W.; Stauffacher, C.V.; Rodwell, V.W.: Enterococcus faecalis 3-hydroxy-3-methylglutaryl coenzyme A synthase, an enzyme of isopentenyl diphosphate biosynthesis. J. Bacteriol., **184**, 4065-4070 (2002)
[33] Misra, I.; Wang, C.Z.; Miziorko, H.M.: The influence of conserved aromatic residues in 3-hydroxy-3-methylglutaryl-CoA synthase. J. Biol. Chem., **12**, 12 (2003)

2-Hydroxyglutarate synthase 2.3.3.11

1 Nomenclature

EC number
2.3.3.11

Systematic name
propanoyl-CoA:glyoxylate C-propanoyltransferase (thioester-hydrolysing, 2-carboxyethyl-forming)

Recommended name
2-hydroxyglutarate synthase

Synonyms
2-hydroxyglutarate glyoxylate-lyase (CoA-propanoylating)
2-hydroxyglutaratic synthetase
2-hydroxyglutaric synthetase
EC 4.1.3.9 (formerly)
α-hydroxyglutarate synthetase
hydroxyglutarate synthase
synthase, 2-hydroxyglutarate

CAS registry number
9024-02-6

2 Source Organism

<1> *Escherichia coli* (E-26 [1-3]; mutant E-26V [1]; strain K 12 derivate K2PRP3 [4]) [1-4]

3 Reaction and Specificity

Catalyzed reaction
propanoyl-CoA + H_2O + glyoxylate = 2-hydroxyglutarate + CoA

Reaction type
condensation

Natural substrates and products
S propanoyl-CoA + glyoxylate <1> (<1> the 2-hydroxyglutarate pathway is metabolically significant during growth of Escherichia coli E-26V on propionate by providing an auxiliary mechanism for the formation of C_4

acids [1]; <1> the enzyme may play a role in the biosynthesis of cell constituents [2]; <1> the enzyme is involved in the metabolism of propionate [4]) (Reversibility: ? <1> [1, 2, 4]) [1, 2, 4]

P 2-hydroxyglutarate + CoA <1> [1, 2, 4]

Substrates and products

S propanoyl-CoA + glyoxylate + H_2O <1> (Reversibility: ? <1> [1-4]) [1-4]

P 2-hydroxyglutarate + CoA <1> [1-4]

Specific activity (U/mg)

Additional information <1> (<1> radioactive assay [3]) [3, 4]

5 Isolation/Preparation/Mutation/Application

Source/tissue

culture condition:acetate-grown cell <1> (<1> no activity in E. coli strain K2 [4]) [4]

culture condition:propionate-grown cell <1> [1, 2, 4]

References

[1] Wegener, W.S.; Reeves, H.C.; Ajl, S.J.: Propionate metabolism. III. Studies on the significance of the α-hydroxyglutarate pathway. Arch. Biochem. Biophys., **123**, 62-65 (1968)

[2] Reeves, H.C.; Ajl, S.J.: α-Hydroxyglutaric acid synthetase. J. Bacteriol., **84**, 186-187 (1962)

[3] Wegener, W.S.; Reeves, H.C.; Rabin, R.; Ajl, S.J.: A radioactive assay for malate synthase and other glyoxylate condensing enzymes. Methods Enzymol., **13**, 362-365 (1969)

[4] Kay, W.W.: Genetic control of the metabolism of propionate by Escherichia coli K12. Biochim. Biophys. Acta, **264**, 508-521 (1972)

3-Propylmalate synthase 2.3.3.12

1 Nomenclature

EC number

2.3.3.12

Systematic name

pentanoyl-CoA:glyoxylate C-pentanoyltransferase (thioester-hydrolysing, 1-carboxybutyl-forming)

Recommended name

3-propylmalate synthase

Synonyms

3-(n-propyl)-malate synthase
3-propylmalate glyoxylate-lyase (CoA-pentanoylating)
EC 4.1.3.11 (formerly)
β-n-propylmalate synthase
n-propylmalate synthase
n-propylmalate synthetase
synthase, 3-propylmalate

CAS registry number

37290-62-3

2 Source Organism

<1> *Escherichia coli* (strain E 26 [1]) [1-3]

3 Reaction and Specificity

Catalyzed reaction

pentanoyl-CoA + H_2O + glyoxylate = 3-propylmalate + CoA

Reaction type

condensation

Natural substrates and products

S Additional information <1> (<1> evidence is present that pentanoate does not act to induce or depress enzyme formation, but rather acts to select a mutant population which is constitutive for the formation of the enzyme [2]) [2]

P ?

Substrates and products

S pentanoyl-CoA + glyoxylate + H_2O <1> (<1> i.e. valeroyl-CoA [1-3]) (Reversibility: ? <1> [1-3]) [1-3]

P 3-propylmalate + CoA <1> [1-3]

Inhibitors

glycolaldehyde <1> [1]
glycolate <1> [1]
glyoxal <1> [1]

Metals, ions

Mg^{2+} <1> (<1> accelerates, optimal concentration: 0.0032 mM [1]) [1]

Specific activity (U/mg)

4.8 <1> [1]
Additional information <1> [2]

pH-Optimum

8.5 <1> [1]

pH-Range

7-9 <1> (<1> pH 7: about 60% of maximal activity, inactive at pH 9.5 [1]) [1]

5 Isolation/Preparation/Mutation/Application

Source/tissue

culture condition:acetate-grown cell <1> (<1> low activity [2]) [2]
culture condition:butyrate-grown cell <1> (<1> low activity [2]) [2]
culture condition:pentanoate-grown cell <1> [1, 2]
culture condition:propionate-grown cell <1> (<1> low activity [2]) [2]

Purification

<1> (partial [1]) [1]

References

[1] Imai, K.; Reeves, H.C.; Ajl, S.J.: n-Propylmalate synthetase. J. Biol. Chem., **238**, 3193-3198 (1963)

[2] Wegener, W.S.; Furmanski, P.; Ajl, S.J.: Selection of mutants constitutive for several glyoxylate condensing enzymes during growth on valeric acid. Biochim. Biophys. Acta, **144**, 34-50 (1967)

[3] Wegener, W.S.; Reeves, H.C.; Rabin, R.; Ajl, S.J.: A radioactive assay for malate synthase and other glyoxylate condensing enzymes. Methods Enzymol., **13**, 362-365 (1969)

2-Isopropylmalate synthase 2.3.3.13

1 Nomenclature

EC number
2.3.3.13

Systematic name
acetyl-CoA:3-methyl-2-oxobutanoate C-acetyltransferase (thioester-hydrolysing, carboxymethyl-forming)

Recommended name
2-isopropylmalate synthase

Synonyms
3-hydroxy-4-methyl-3-carboxy-pentanoate 2-oxo-3-methyl-butanoate lyase (CoA-acetylating)
3-hydroxy-4-methyl-3-carboxyvalerate 2-oxo-3-methyl-butyrate lyase (CoA-acetylating)
EC 4.1.3.12 (formerly)
IPM synthase
IPM-synthase
IPMS
α-IPM synthase
α-IPM synthetase
α-isopropylmalate synthase
α-isopropylmalate synthase I (<1> isoenzyme encoded by LEU, subform Ia, MW 68000, which is imported into the mitochondrial matrix, and cytoplasmic synthase Ib [26])
α-isopropylmalate synthase II (<1> minor enzyme form [26])
α-isopropylmalate synthetase
α-isopropylmalic synthetase
isopropylmalate synthase
isopropylmalate synthetase
synthase, 2-isopropylmalate
synthase, α-isopropylmalate

CAS registry number
9030-98-2

2 Source Organism

<1> *Saccharomyces cerevisiae* (strain D273-10B and S288c [5]; strain SK101 [7]; strain 60615 [14]; the Leu4 gene encodes 2 forms: a short cytoplasmic form and a long form that is targeted to the mitochondria [1]; wild type, mutants with resistance to inhibition by Leu, mutants with resistance to inhibition by CoA [23]; 3 enzyme forms, 1. α-isopropylmalate synthase I, isoenzyme encoded by LEU, subform Ia, MW 68000, which is imported into the mitochondrial matrix, and cytoplasmic synthase Ib, 2. α-isopropylmalate synthase II, minor enzyme form [26]) [1, 5, 7, 8, 13, 14, 23, 26, 28]
<2> *Alcaligenes eutrophus* (strain H 16 [2,9-11]; Val-Ile double auxotrophic mutant of strain H 16 [10]) [2, 9-12]
<3> *Bacteroides ruminicola* (strain 23 [3]) [3]
<4> *Bacteroides fragilis* (ATCC 23745 [3]) [3, 6]
<5> *Bacillus sp.* (No. 221 [4]) [4]
<6> *Clostridium thermoaceticum* [6]
<7> *Clostridium formicoaceticum* [6]
<8> *Clostridium pasteurianum* [6]
<9> *Clostridium kluyveri* [6]
<10> *Salmonella typhimurium* (strain CV-123, and mutant strain CV-241 with a feed-back insensitive enzyme [16]; strain CV-19 [18]; strain CV-19, and mutant strain CV-241 with a feed-back insensitive enzyme [20]) [15, 16, 18, 20, 27]
<11> *Hydrogenomonas sp.* (strain H16 [17]) [17, 27]
<12> *Neurospora crassa* [19]
<13> *Zygosaccharomyces rouxii* (mutant M21-10 [21]) [21]
<14> *Corynebacterium glutamicum* (strain H-1204, strain L-76 [22]) [22]
<15> *Candida maltosa* [24]
<16> *Spinacia oleracea* [25]
<17> *Chromatium sp.* (strain D) [27]
<18> *Rhodopseudomonas sphaeroides* [27]
<19> *Pseudomonas aeruginosa* [27]
<20> *Pseudomonas fluorescens* [27]
<21> *Vibrio extorquens* [27]
<22> *Rhizobium japonicum* [27]
<23> *Alcaligenes viscolactis* [27]
<24> *Escherichia coli* (strain B) [27]
<25> *Proteus vulgaris* [27]
<26> *Aerobacter aerogenes* [27]
<27> *Micrococcus sp.* [27]
<28> *Micrococcus lysodeikticus* [27]
<29> *Bacillus polymyxa* [27]
<30> *Bacillus subtilis* [27]
<31> *Nocardia opaca* [27]
<32> *Mucobacterium tuberculosis* [29]

<33> *Arabidopsis thaliana* (<33> IMS 1 [30]) [30]
<34> *Arabidopsis thaliana* (<33> IMS 2 [30]) [30]
<35> *Arabidopsis thaliana* (<33> IMS 3 [30]) [30]

3 Reaction and Specificity

Catalyzed reaction

acetyl-CoA + 3-methyl-2-oxobutanoate + H_2O = 2-hydroxy-2-isopropylsuccinate + CoA

Reaction type

aldol condensation

Natural substrates and products

S 2-oxo-3-methylbutanoate + acetyl-CoA + H_2O <1-4, 6, 10-12, 14, 15, 17-35> (<2-4, 6, 14> involved in Leu biosynthesis [2, 3, 6, 22]; <1, 2, 10-12, 17-35> first enzyme in biosynthesis of L-Leu [7, 10, 14, 17, 18, 19, 24, 26-30]; <14> key enzyme in biosynthesis of Leu [22]; <1> the cytoplasmic form is involved in Leu biosynthesis [1]; <2> the enzyme is regulated by the level of Leu [2]; <6,11> repression by Leu [6, 17]) [1-3, 6, 10, 14, 17-19, 22, 24, 26-30]

P ?

Substrates and products

S 2-oxo-3-methylbutanoate + acetyl-CoA + H_2O <1-35> (Reversibility: ? <1-35> [1-30]) [1-30, 1-30]

P 3-hydroxy-4-methyl-3-carboxypentanoate + CoA <1-35> (<1> i.e. α-isopropylmalate [1-30]) [1-30]

S 2-oxo-3-methylbutanoate + butanoyl-CoA + H_2O <2> (Reversibility: ? <2> [11]) [11]

P 3-ethyl-2-hydroxy-2-isopropylsuccinic acid + CoA

S 2-oxo-3-methylbutanoate + malonyl-CoA + H_2O <2> (Reversibility: ? <2> [11]) [11]

P 2-hydroxy-3-metylbutane-11,2-tricarboxyl acid + CoA

S 2-oxo-3-methylbutanoate + propanoyl-CoA + H_2O <2> (Reversibility: ? <2> [11]) [11]

P 2-hydroxy-2-isopropyl-3-methylsuccinic acid + CoA

S 2-oxo-3-methylbutanoate + valeryl-CoA + H_2O <2> (Reversibility: ? <2> [11]) [11]

P 2-hydroxy-2-isopropyl-3-propylsuccinic acid + CoA

S 2-oxo-3-methylbutanoyl-methylester + H_2O <2> (Reversibility: ? <2> [11]) [11]

P 3-hydroxy-3-(methoxycarbonyl)-4-methylpentanoic acid + CoA

S 2-oxo-butanoate + acetyl-CoA + H_2O <1, 2, 10, 12, 15> (Reversibility: ? <1, 2, 10, 12, 15> [11, 14, 18, 19, 24, 26]) [11, 14, 18, 19, 24, 26]

P 2-ethyl-2-hydroxysuccinic acid + CoA

S 2-oxo-n-pentanoate + acetyl-CoA + H_2O <2> (Reversibility: ? <2> [11]) [11]

P 2-hydroxy-2-propylsuccinic acid + CoA

S pyruvate + acetyl-CoA + H_2O <1, 2, 10, 12, 15> (Reversibility: ? <1, 2, 10, 12, 15> [11, 14, 18, 19, 24, 26]) [11, 14, 18, 19, 24, 26]

P 2-hydroxy-2-methylsuccinic acid + CoA

Inhibitors

2-oxo-isohexanoate <1, 2, 4, 6-10> (<4,6-9> 1 mM [6]; <1> competitive [26]) [6, 11, 12, 18, 26]

2-oxopentanoate <1> (<1> competitive [26]) [26]

5',5',5'-trifluoroleucine <1, 11, 12> (<12> reversed by acetyl-CoA [19]) [14, 17, 19]

Br^- <12> (<12> above 0.02 M [19]) [19]

Cd^{2+} <15> [24]

CoA <1, 2> (<1> Zn^{2+}-dependent reversible inactivation [13,26]; <1> in presence of Zn^{2+}, protection by high concentrations of ATP, and to a much lesser extent, ADP, by a high adenylate charge, by chelators, and by 3'-dephospho-CoA [7]; <2> competitive against both 2-oxo-3-methylbutanoate and acetyl-CoA [11]; <1> two distinct CoA sites on each enzyme subunit: the first site interacts with CoA an desulfo-CoA, the second site is absolutely specific for CoA. Binding of CoA to this site occurs only when Zn^{2+} is present, is independent of the specific activity of the enzyme and does not eliminate CoA binding at the product site [13]) [7, 11, 13, 14, 26]

Cu^{2+} <15> [24]

D-leucine amide <1> (<1> weak [14]) [14]

DL-azaleucine <12> (<12> 10 mM, at pH 6.5 [19]) [19]

DL-fluoroleucine <12> (<12> at pH 7.5 [19]) [19]

EDTA <1, 2, 8, 10> (<8> time-dependent inactivation is not reversible by intensive dialysis [6]; <1> dialysis at an initial concentration of 50 mM reduces the zinc content by more than 80%, complete loss of activity, restored by addition of Zn^{2+}, Mn^{2+}, Fe^{2+}, Co^{2+}, or Cd^{2+} [8]; <2> Mn^{2+} plus dithiothreitol restores activity, 2-oxo-3-methylbutanoate prevents inactivation [9]) [6, 8, 9, 18]

F^- <12> (<12> above 0.02 M [19]) [19]

Hg^{2+} <10, 15> [18, 24]

K^+ <15> (<15> at high concentrations, stimulation at lower concentrations [24]) [24]

K_2SO_4 <2> (<2> 250 mM K_2SO_4, pH 7.5, 20% inhibition [10]) [10]

KCl <2> (<2> 400 mM KCl, pH 8.2, 20% inhibition [10]) [10]

L-2-hydroxyisopentanoate <1> [14]

L-Phe <12> (<12> 10 mM, at pH 6.5 [19]) [19]

L-Val <12> (<12> 10 mM, at pH 6.5 [19]) [19]

Zn^{2+} <12, 15> (<12> 0.1 mM [19]) [19, 24]

desulfo-CoA <1> [13]

isopropylmalate <2> (<2> competitive against both 2-oxo-3-methylbutanoate and acetyl-CoA [11]) [11]

isovalerate <3> (<3> moderate [3]) [3]
leucine <1-4, 6-8, 10-16> (<3> strong [3]; <2> L-Leu [11,12]; <2> at low concentrations of Leu the inhibition mechanism is of the competitive type with respect to substrate acetyl-CoA and of the non-competitive type with respect to 2-oxo-3-methylbutanoate [12]; <1> mixed-type inhibition, strongly pH-dependent, Leu concentration necessary for half-maximal inhibition increases about 10fold as the pH increases from 7.5 to 8.5 [14]; <1> both Fe^{2+} and Co^{2+} lower the inhibition by Leu [14]; <1> D-Leu, weak [14]; <10> the enzyme from the mutant strain CV241 is insensitive to feedback inhibition by Leu due to a mutation at the extreme operator-distal end of leuA [16]; <11> non-competitive, sensitivity is maximal at pH 7.2 and negligible at pH 8.4 [17]; <10> end-product inhibition, non-competitive with respect to 2-oxo-3-methylbutanoate and competitive with respect to acetyl-CoA, more sensitive at pH 6.5 than pH 8.5 [18]; <12> at pH 7.5 [19]; <10> mechanism of feedback inhibition by Leu, binding of Leu to wild-type and feedback-resistant enzyme and its structural consequences [20]; <13> enzyme from mutant M21-10 is not inhibited [21]; <14> enzyme from strain L-76 is not inhibited [22]; <15> competitive with respect to 2-oxo-3-methylbutanoate, non-competitive with respect to acetyl-CoA [24]; <1> pH: 7.2, complete inhibition, pH: 8.0, 70% inhibition, pH: 8.8, 15% inhibition [26]) [1, 3, 6, 11, 12, 14, 16-27]

Activating compounds

Additional information <2> (<2> the addition of 5 mM valine plus isoleucine with and without 5 mM threonine caused a 4-6.6fold increase in the formation of active enzyme [2]) [2]

Metals, ions

K^+ <1, 12, 15> (<1> required [14,19]; <15> strongly dependent on the presence of monovalent cations, K^+ is most effective; K_m: 80 mM, inhibitory at higher concentrations [24]; <1> enzyme requires the presence of monovalent cations, K^+ is most effective [26]) [14, 19, 24, 26]
Li^+ <12, 15> (<12> stimulates at low concentrations [19]; <15> can partially replace K^+ in activation [24]) [19, 24]
Mg^{2+} <16> (<16> stimulates, maximal activity at 2 mM [25]) [25]
Mn^{2+} <2> (<2> 1 mM, 25% activation [10]) [10]
Na^+ <16> (<16> stimulates, maximal activity at 1 mM [25]) [25]
Rb^+ <15> (<15> can partially replace Rb^+ in activation [24]) [24]
zinc <1> (<1> contains approximately 4 gatoms of zinc per dimer of 130000 Da [8]; <1> enzyme of strain S288C contains 2 gatoms of zinc per subunit [26]) [8, 26]
Additional information <2, 8> (<2,8> enzyme contains a metal ion [6,9]) [6, 9]

Specific activity (U/mg)

0.0019 <8> [6]
0.0024 <7> [6]
0.0086 <4> [6]

0.0089 <6> [6]
1.413 <8> [6]
1.58 <1> [14]
1.67 <10> [16]
7.1 <1> [13, 26]
12.05 <2> [10]
14.5 <10> [18]
Additional information <1, 12> [8, 19, 23, 28]

K_m-Value (mM)
0.005 <16> (acetyl-CoA) [25]
0.009 <1> (acetyl-CoA) [14]
0.01 <12> (2-oxo-3-methylbutanoate) [19]
0.016 <1> (2-oxo-3-methylbutanoate) [14]
0.0245 <12> (acetyl-CoA, <12> pH 7.5 [19]) [19]
0.0246 <32> (2-oxo-3-methylbutanoate) [29]
0.028 <12> (acetyl-CoA, <12> pH 7.0 [19]) [19]
0.035 <12> (acetyl-CoA, <12> pH 6.6 [19]) [19]
0.06 <2, 10> (2-oxo-3-methylbutanoate, <2,10> with acetyl-CoA as cosubstrate [11,18]) [11, 18]
0.064 <15> (acetyl-CoA) [24]
0.075 <16> (2-oxo-3-methylbutanoate) [25]
0.075 <2> (acetyl-CoA) [11]
0.1 <2> (propanoyl-CoA) [11]
0.2 <10> (acetyl-CoA) [18]
0.2 <1> (pyruvate) [11, 26]
0.2435 <32> (acetyl-CoA) [29]
0.4 <2> (2-oxo-n-pentanoate, <2> with acetyl-CoA as cosubstrate [11]) [11]
0.4 <2> (butanoyl-CoA) [11]
0.5 <2> (malonyl-CoA) [11]
0.57 <15> (2-oxo-3-methylbutanoate) [24]
0.57 <1> (2-oxobutanoate) [14, 26]
0.9 <2> (pentanoyl-CoA) [11]
1.1 <10> (2-oxobutanoate) [18]
1.8 <2> (2-oxobutanoate, <2> with acetyl-CoA as cosubstrate [11]) [11]
2.1 <12> (2-oxobutanoate) [19]
5 <2> (2-oxo-3-methylbutanoyl-methylester, <2> with acetyl-CoA as cosubstrate [11]) [11]
5.8 <12> (pyruvate) [19]
10 <2, 10> (pyruvate, <2> with acetyl-CoA as cosubstrate [11]) [11, 18]
Additional information <1, 15> (<1> K_m of enzyme with and without Zn^{2+} or Mn^{2+} [8]) [8, 24, 26]

K_i-Value (mM)
0.001 <2> (leucine, <2> pH 7.35, acetyl-CoA varied [12]) [12]
0.002 <2> (leucine, <2> pH 7.6, acetyl-CoA varied [12]) [12]
0.0025 <2> (leucine, <2> pH 8.1, acetyl-CoA varied [12]) [12]

0.008 <10> (leucine, <10> pH 6.8, 0.4 mM acetyl-CoA, 0.1 mM 2-oxo-3-methylbutanoate, wild-type enzyme [20]) [20]
0.022 <10> (leucine, <10> pH 6.8, 0.8 mM acetyl-CoA, 4 mM 2-oxo-3-methylbutanoate, wild-type enzyme [20]) [20]
0.07 <1> (CoA, <1> with respect to acetyl-CoA [13]; <1> interacting with the product site [26]) [13, 26]
0.09 <1> (desulfo-CoA, <1> with respect to acetyl-CoA [13]) [13]
0.1 <2> (2-oxo-isohexanoate, <2> with respect to 2-oxo-3-methylbutanoate [12]) [12]
0.1 <1> (leucine, <1> only isoenzyme I activity [26]) [26]
0.1 <2> (leucine, <2> pH 7.6, 2-oxo-3-methylbutanoate varied [12]) [12]
0.2 <2> (2-oxo-isohexanoate, <2> with respect to acetyl-CoA [12]) [12]
0.2 <1> (leucine, <1> pH: 7.2, half-maximal inhibition [26]) [26]
0.2 <2> (leucine, <2> pH 8.1, 2-oxo-3-methylbutanoate varied [12]) [12]
0.4 <1> (leucine) [1]
0.5 <2> (isopropylmalate, <2> with respect to 2-oxo-3-methylbutanoate at a substrate concentration of 0.07 mM acetyl-CoA [11]) [11]
0.81 <15> (leucine, <15> competitive, with respect to 2-oxo-3-methylbutanoate [24]) [24]
0.9 <2> (isopropylmalate, <2> with respect to acetyl-CoA at a substrate concentration of 0.375 mM 2-oxo-3-methylbutanoate [11]) [11]
1.2 <1> (leucine, <1> only isoenzyme II activity [26]) [26]
1.3 <15> (leucine, <15> non-competitive, with respect to acetyl-CoA [24]) [24]
3 <10> (leucine, <10> pH 6.8, 0.4 mM acetyl-CoA, 0.1 mM 2-oxo-3-methylbutanoate, feedback-resistent enzyme [20]) [20]
4 <2> (isopropylmalate, <2> with respect to acetyl-CoA at a substrate concentration of 0.75 mM 2-oxo-3-methylbutanoate [11]) [11]
20 <10> (leucine, <10> pH 6.8, 0.8 mM acetyl-CoA, 4 mM 2-oxo-3-methylbutanoate, feedback-resistent enzyme [20]) [20]
230 <15> (K^+, <15> at high concentration [24]) [24]
Additional information <1> (leucine, <1> K_i with and without Zn^{2+} or Mn^{2+} [8]) [8]

pH-Optimum

7-8 <12> [19]
7-9 <16> [25]
7.2-8.5 <1> [26]
7.5-8.8 <15> [24]
8 <1, 4> [6, 14]
8-8.5 <2> (<2> Tris-HCl buffer [10]) [10]
8.2 <2> (<2> phosphate buffer [10]) [10]
8.4 <11> [17]
8.5 <2, 8-10, 32> [6, 12, 18, 29]
8.8 <6, 7> [6]

pH-Range

6-9.5 <16> (<16> pH 6.0: about 50% of maximal activity, pH 9.5: about 80% of maximal activity [25]) [25]
7.1-10 <32> (<32> pH 7.1: about 50% of maximal activity, pH 10.0: about 80% of maximal activity [29]) [29]

Temperature optimum (°C)

37-40 <8> [6]
37-50 <32> (<32> almost full activity [29]) [29]
44 <2> (<2> pH: 7.5, temperature optimum increased to 45°C in the presence of 2 mM valine and to 48°C in the presence of leucine [12]) [10, 12]
46 <6, 7> [6]

4 Enzyme Structure

Molecular weight

100000 <2> (<2> in presence of Leu, gel filtration [10]) [10]
114500 <2> (<2> in absence of Leu, gel filtration [10]) [10]
121000 <1> (<1> in absence of Leu, gel filtration [14]) [14]
127000 <1> (<1> sedimentation equilibrium centrifugation [13]) [13]
130000 <1> (<1> cytoplasmic form, gel filtration [1]) [1]
137000 <1> (<1> in presence of substrate, gel filtration [14]) [14]

Subunits

? <1, 10> (<10> x * 50000, SDS-PAGE, ultracentrifugation of the enzyme denatured in 6 M guanidine hydrochloride and 0.1 M 2-mercaptoethanol [15]; <1> x * 65000, equilibrium sedimentation under denaturing conditions [13]; <1> x * 68416, isoenzyme I, calculation from nucleotide sequence [26]) [13, 15, 26]
dimer <1> (<1> 2 * 65000, cytoplasmic enzyme [1]; <1> 2 * 67000, SDS-PAGE [13]; <1> 2 * 65000-68000 [26]) [1, 26]

5 Isolation/Preparation/Mutation/Application

Source/tissue

culture condition:acetate-grown cell <2> (<2> unusually high specific activity of enzyme [2]) [2]
leaf <16, 33-35> [25, 30]
root <34, 35> [30]
stem <34, 35> [30]
Additional information <2> (<2> highest specific activity in cells harvested in the late exponential phase [10]) [10]

Localization

chloroplast <16> [25]
cytoplasm <1> (<1> the Leu4 gene encodes 2 forms: a short cytoplasmic form and a long form that is targeted to the mitochondria [1]; <1> α-isopropylmalate synthase Ib [26]) [1, 26]
mitochondrion <1> (<1> the Leu4 gene encodes 2 forms: a short cytoplasmic form and a long form that is targeted to the mitochondria [1]; <1> large enzyme form, matrix [5]; <1> α-isopropylmalate synthase Ia, MW 68000, is imported into the mitochondrial matrix [26]) [1, 4, 5, 26]
soluble <2> [2]

Purification

<1> [13, 14, 26]
<2> [10]
<8> (partial [6]) [6]
<10> [16, 18]
<12> [19]
<32> (recombinant enzyme [29]) [29]

Renaturation

<2> (irreversible denaturation after 5 min at 60°C [10]) [10]

Cloning

<1> (construction of a LEU4'-'lacZ fusion. The Leu4 gene encodes 2 forms: a short cytoplasmic form and a long form that is targeted to the mitochondria [1]; <1> construction of three knock-out mutants, two carrying a deletion in YOR108w or in Leu4 ORF, and one carrying both the deletions [28]) [1, 23, 26, 28]
<5> (expression in Escherichia coli and Bacillus subtilis [4]) [4]
<32> (expression in Escherichia coli [29]) [29]

6 Stability

pH-Stability

4 <32> (<32> enzyme inactive [29]) [29]
7 <1, 2> (<2> 3 h, about 15% of maximal activity [10]; <1> activity of enzyme drops off rapidly below pH 7.0 [26]) [10, 26]
7.2-8.2 <2> (<2> 3 h, stable [10]; <1> activity of enzyme drops off rapidly above pH 9.0 [26]) [10, 26]
9 <2> (<2> 3 h, about 50% loss of activity [10]) [10]
9.5 <2> (<2> 3 h, about 30% of maximal activity [10]) [10]

Temperature stability

49 <12> (<12> 7 min, 90% loss of activity in absence of Leu, about 50% loss of activity after 15 min in presence of 0.2 mM Leu [19]) [19]
50 <1> (<1> pH 6.9, 5 min, 80% loss of activity, 1 mM Leu provides significant but incomplete protection [14]) [14]

60 <2> (<2> 5 min, in absence of substrates irreversible denaturation [10]) [10]
65 <2> (<2> 5 min, in presence of 2-oxo-3-methylbutanoate and acetyl-CoA, stable [10]) [10]
Additional information <1, 2, 12> (<2> cold lability can be overcome by stabilizing the enzyme by buffer of increased ionic strength [10]; <2> stabilization at low temperature with KCl [10]; <1> Leu, and to a lesser extent, 2-oxo-3-methylbutanoate stabilize against heat inactivation, acetyl-CoA accelerates inactivation [14]; <12> thermal stability at pH 6.5 increases as a function of Leu concentration. The protection afforded by Leu is greatest at pH 6.5, less at pH 7.5, and is entirely lost at pH 8.3 [19]) [10, 14, 19]

Storage stability

<2>, 30°C, pH 7.5, stable for more than 10 days [10]
<2>, freezing at -20°C and thawing, about 10% loss of activity. Storage at -20°C, 100 mM potassium phosphate, 0.2 mM 2-oxo-3-methylbutanoate, pH 7.5 or pH 8.0, no further loss of activity after 5 months [10]
<8>, -18°C, 50 mM potassium phosphate, pH 7.2, loss of activity after 3 months [6]
<8>, 0°C, more than 80% loss of activity after 2 d [6]
<8>, 22°C, in presence of 0.05% sodium azide, 8% loss of activity after 3 months [6]
<12>, -20°C, phosphate leucine buffer, little loss of activity [19]
<32>, in an icebox, pH 7.5-8.5, 30% loss of activity after 48 h [29]

References

[1] Beltzer, J.P.; Morris, S.R.; Kohlhaw, G.B.: Yeast LEU4 encodes mitochondrial and nonmitochondrial forms of α-isopropylmalate synthase. J. Biol. Chem., **263**, 368-374 (1988)
[2] Wiegel, J.: Leucine biosynthesis in Alcaligenes eutrophus H16: influence of amino acid additions on the formation of active α-isopropylmalate synthase and α-acetohydroxy acid synthase. Arch. Microbiol., **142**, 194-199 (1985)
[3] Allison, M.J.; Baetz, A.L.; Wiegel, J.: Alternative pathways for biosynthesis of leucine and other amino acids in Bacteroides ruminicola and Bacteroides fragilis. Appl. Environ. Microbiol., **48**, 1111-1117 (1984)
[4] Honda, H.; Kato, C.; Kudo, T.; Horikoshi, K.: Cloning of leucine genes of alkalophilic Bacillus No. 221 in E. coli and B. subtilis. J. Biochem., **95**, 1485-1490 (1984)
[5] Hampsey, D.M.; Lewin, A.S.; Kohlhaw, G.B.: Submitochondrial localization, cell-free synthesis, and mitochondrial import of 2-isopropylmalate synthase of yeast. Proc. Natl. Acad. Sci. USA, **80**, 1270-1274 (1983)
[6] Wiegel, J.: α-Isopropylmalate synthase as a marker for the leucine biosynthetic pathway in several clostridia and in Bacteroides fragilis. Arch. Microbiol., **130**, 385-390 (1981)

[7] Hampsey, D.M.; Kohlhaw, G.B.: Inactivation of yeast α-isopropylmalate synthase by CoA. Antagonism between CoA and adenylates and the mechanism of CoA inactivation. J. Biol. Chem., **256**, 3791-3796 (1981)
[8] Roeder, P.R.; Kohlhaw, G.B.: α-Isopropylmalate synthase from yeast. A zinc metalloenzyme. Biochim. Biophys. Acta, **613**, 482-487 (1980)
[9] Wiegel, J.: Mn^{2+}-specific reactivation of EDTA inactivated α-isopropylmalate synthase from Alcaligenes eutrophus H 16. Biochem. Biophys. Res. Commun., **82**, 907-912 (1978)
[10] Wiegel, J.; Schlegel, H.G.: α-Isopropylmalate synthase from Alcaligenes eutrophus H 16 I. Purification and general properties. Arch. Microbiol., **112**, 239-246 (1977)
[11] Wiegel, J.; Schlegel, H.G.: α-Isopropylmalate synthase from Alcaligenes eutrophus H 16. II. Substrate specificity and kinetics. Arch. Microbiol., **112**, 247-254 (1977)
[12] Wiegel, J.; Schlegel, H.G.: α-Isopropylmalate synthase from Alcaligenes eutrophus H 16. III. Endproduct inhibition and its relief by valine and isoleucine. Arch. Microbiol., **114**, 203-210 (1977)
[13] Tracy, J.W.; Kohlhaw, G.B.: Evidence for two distinct CoA binding sites on yeast α-isopropylmalate synthase. J. Biol. Chem., **252**, 4085-4091 (1977)
[14] Ulm, E.H.; Boehme, R.; Kohlhaw, G.: α-Isopropylmalate synthase from yeast: purification, kinetic studies and effect of ligands on stability. J.Bacteriol., **110**, 1118-1126 (1972)
[15] Leary, T.R.; Kohlhaw, G.B.: α-Isopropylmalate synthase from Salmonella typhimurium. J. Biol. Chem., **247**, 1089-1095 (1972)
[16] Bartholomew, J.C.; Calvo, J.M.: α-Isopropylmalate synthase from Salmonella typhimurium. Carboxypeptidase digestion studies of parent and feedback-insensitive enzymes. Biochim. Biophys. Acta, **250**, 568-576 (1971)
[17] Hill, F.; Schlegel, H.G.: The α-isopropylmalate synthetase of Hydrogenomonas H 16. Arch. Mikrobiol., **68**, 1-17 (1969)
[18] Kohlhaw, G.; Leary, T.R.; Umbarger, H.E.: α-Isopropylmalate synthase from Salmonella typhimurium. Purification and properties. J. Biol. Chem., **244**, 2218-2225 (1969)
[19] Webster, R.E.; Gross, S.R.: The α-isopropylmalate synthetase of Neurospora crassa. I. The kinetics and end product control of α-isopropylmalate synthetase function. Biochemistry, **4**, 2309-2318 (1965)
[20] Teng-Leary, E.; Kohlhaw, G.B.: Mechanism of feedback inhibition by leucine. Binding of leucine to wild-type and feedback-resistant α-isopropylmalate synthases and its structural consequences. Biochemistry, **12**, 2980-2986 (1973)
[21] Yoshikawa, S.; Oguri, I.; Kondo, K.; Fukuzawa, M.; Shimosaka, M.; Okazaki, M.: Enhanced formation of isoamyl alcohol in Zygosaccharomyces rouxii due to elimination of feedback inhibition of α-isopropylmalate synthase. FEMS Microbiol. Lett., **127**, 139-143 (1995)
[22] Azuma, T.; Nakanishi, T.: Enzymatic background for the reversion or stabilization of an L-leucine producing strain of Corynebacterium glutamicum. Agric. Biol. Chem., **52**, 1525-1528 (1988)

[23] Cavalieri, D.; Casalone, E.; Bendoni, B.; Fia, G.; Polsinelli, M.; Barberio, C.: Trifluoroleucine resistance and regulation of α-isopropyl malate synthase in Saccharomyces cerevisiae. Mol. Gen. Genet., **261**, 152-160 (1999)
[24] Bode, R.; Birnbaum, D.: Some properties of the leucine-biosynthesizing enzymes from Candida maltosa. J. Basic Microbiol., **31**, 21-26 (1991)
[25] Hagelstein, P.; Schultz, G.: Leucine synthesis in spinach chloroplasts: partial characterization of 2-isopropylmalate synthase. Biol. Chem. Hoppe-Seyler, **374**, 1105-1108 (1993)
[26] Kohlhaw, G.B.: α-isopropylmalate synthase from yeast. Methods Enzymol., **166**, 414-423 (1988)
[27] Stieglitz, B.I.; Calvo, J.M.: Distribution of the isopropylmalate pathway to leucine among diverse bacteria. J. Bacteriol., **118**, 935-941 (1974)
[28] Casalone, E.; Barberio, C.; Cavalieri, D.; Polsinelli, M.: Identification by functional analysis of the gene encoding a-isopropylmalate synthase II (LEU9) in Saccharomyces cerevisiae. Yeast, **16**, 539-545 (2000)
[29] Chanchaem, W.; Palittapongarnpim, P.: A variable number of tandem repeats result in polymorphic α -isopropylmalate synthase in Mycobacterium tuberculosis. Tuberculosis, **82**, 1-6 (2002)
[30] Junk, D.J.; Mourad, G.S.: Isolation and expression analysis of the isopropylmalate synthase gene family of Arabidopsis thaliana. J. Exp. Bot., **53**, 2453-2454 (2002)

Homocitrate synthase 2.3.3.14

1 Nomenclature

EC number
2.3.3.14

Systematic name
acetyl-CoA:2-oxoglutarate C-acetyltransferase (thioester-hydrolysing, carboxymethyl forming)

Recommended name
homocitrate synthase

Synonyms
2-hydroxybutane-1,2,4-tricarboxylate 2-oxoglutarate-lyase (CoA-acetylating)
EC 4.1.3.21 (formerly)
HCS
acetyl-coenzyme A:2-ketoglutarate C-acetyl transferase
homocitrate synthetase
homocitrate-condensing enzyme
homocondensing enzyme
synthase, homocitrate

CAS registry number
9075-60-9

2 Source Organism

<1> *Escherichia coli* [1]
<2> *Neurospora crassa* (wild type strain STA-4 and lysine-requiring auxotrophic strains 33933 and mutant STL-7 [2]) [2]
<3> *Penicillium chrysogenum* (Wis. 54-1255 [3]) [3, 5, 9, 14, 15, 22, 23]
<4> *Saccharomyces cerevisiae* [4, 7, 8, 16, 19, 20, 23]
<5> *Saccharomycopsis lipolytica* [6]
<6> *Candida pelliculosa* [10, 11]
<7> *Candida maltosa* [12]
<8> *Schizosaccharomyces pombe* [13]
<9> *Yarrowia lipolytica* [17]
<10> *Acetobacter vinelandii* [18]
<11> *Thermus thermophilus* [21]

3 Reaction and Specificity

Catalyzed reaction

acetyl-CoA + H_2O + 2-oxoglutarate = 2-hydroxybutane-1,2,4-tricarboxylate + CoA (<7>, sequential mechanism [12])

Reaction type

condensation

Natural substrates and products

S acetyl-CoA + H_2O + 2-oxoglutarate <1, 3, 4, 7, 8> (<1>, enzymatic formation of homocitric acid, an intermediate in lysine biosynthesis [1]; <4>, first enzyme of the 2-aminoadipic acid pathway for biosynthesis of lysine [8]; <3>, first enzyme of the lysine biosynthetic pathway, feedback regulated by L-lysine. Maximum homocitrate synthase activity in cultures is found at 48 h, coinciding with the phase of high rate of penicillin biosynthesis [9]; <8, 11>, first enzyme of the lysine biosynthetic pathway [13, 21, 23]; <3>, expression of lys1 is partially repressed by high concentrations of lysine in the culture medium, but lysine repression seems to be a weak mechanism of control of the lysine pathway as compared to lysine inhibition of homocitrate synthase [14]; <3>, the enzyme catalyzes the first step of the lysine and penicillin pathway and is highly sensitive to feedback regulation by L-lysine [22]; <3, 4>, the enzyme may play a regulatory function, in addition to its catalytic function, in Saccharomyces cerevisiae but not in Penicillium chrysogenum [23]) (Reversibility: ? <1, 3, 4, 7, 8> [1, 8, 9, 12, 13, 14, 21, 22, 23]) [1, 8, 9, 12, 13, 14, 21, 22, 23]

P 2-hydroxybutane-1,2,4-tricarboxylate + CoA

Substrates and products

S acetyl-CoA + H_2O + 2-oxoadipate <10> (Reversibility: ? <10> [18]) [18]

P 2-hydroxyhexane-1,2,6-tricarboxylate

S acetyl-CoA + H_2O + 2-oxoglutarate <1-11> (Reversibility: ? <1-11> [1-23]) [1-23]

P 2-hydroxybutane-1,2,4-tricarboxylate + CoA <1-11> [1-23]

S acetyl-CoA + H_2O + oxaloacetate <10, 11> (Reversibility: ? <10, 11> [18, 21]) [18, 21]

P 2-hydroxypropane-1,2,3-tricarboxylate + CoA

Inhibitors

1,10-phenanthroline <4> (<4>, 0.01 mM, 91% inhibition [7]) [7]
2,2'-dipyridyl <4> (<4>, 0.01 mM, 82% inhibition [7]) [7]
AMP <3> (<3>, 1 mM, 13-15% inhibition [15]) [15]
Co^{2+} <7> [12]
CoA <3, 5> (<5>, strong competitive inhibitor of acetyl-CoA fixation [6]) [6, 15]
Cu^{2+} <4> (<4>, 0.0001 mM $CuSO_4$, complete inhibition [7]) [7]
DL-4,5-transdehydrolysine <5> (<5>, 5 mM, 90% inhibition [6]) [6]
DL-allo-5-hydroxylysine <5> (<5>, 5 mM, 75% inhibition [6]) [6]

DL-α-aminoadipate <7> [12]
DL-homolysine <6> (<6>, inhibition of enzyme from wild-type strain, enzyme from S-(β-aminoethyl)-L-cysteine resistant mutant strain is not inhibited [10]) [10]
EDTA <3> (<3>, 5 mM, 30% inhibition [15]) [15]
EDTA <5> (<5>, 15 mM [6]) [6]
Hg^{2+} <4, 6, 7> (<4>, 0.01 mM $HgCl_2$, complete inhibition [7]) [7, 10, 12]
L-Arg <4> [8]
L-Lys <3, 4, 5, 6, 7, 11> (<3>, 0.3 mM, 50% inhibition [3]; <4>, 20 mM, 95% inhibition [4]; <5>, 5 mM, complete inhibition [6]; <4>, 10 mM, 80% inhibition. No inhibition by 10 mM D-Lys [7]; <3>, 50 mM, 48% inhibition. Feedback inhibition [9]; <6>, 0.02 mM required for half-maximal inhibition of the enzyme from wild-type strain, enzyme from S-(β-aminoethyl)-L-cysteine resistant mutant strain is not inhibited [10]; <3>, 50% inhibition by 0.053 mM, at 6 mM 2-oxoglutarate [15]; <4>, feedback inhibition of homocitrate synthase by lysine modulates the activation of LYS gene expression by transcriptional activator Lys14p [20]; <3>, no inhibition [5]) [3, 4, 6, 7, 9, 10, 12, 15, 20, 21]
L-norleucine <4> [8]
LL-diaminopimelic acid <5> (<5>, 5 mM, 90% inhibition [6]) [6]
Mg^{2+} <3> (<3>, 10 mM $MgSO_4$, 30% inhibition [15]) [15]
Mn^{2+} <3, 7> (<3>, 0.025 mM, 35% inhibition [15]) [12, 15]
NEM <6> (<6>, 0.1 mM, 90% inhibition [10]) [10]
Na_2HAsO_4 <4> (<4>, 0.01 mM, 67% inhibition [7]) [7]
NaF <4> (<4>, 0.01 mM, 73% inhibition [7]) [7]
PCMB <6> (<6>, 0.1 mM [10]) [10]
S-(β-aminoethyl)-L-cysteine <6> (<6>, 1.1 mM required for half-maximal inhibition of enzyme from wild-type strain, enzyme from S-(β-aminoethyl)-L-cysteine resistant mutant strain is not inhibited [10]) [10]
Zn^{2+} <3> (<3>, 0.025 mM, 60% inhibition [15]) [15]
acetyl-CoA <11> [21]
benzylpenicillin <3> (<3>, 20 mM, partial inhibition is enhanced by lysine [5]) [5]
casein hydrolysate <3> (<3>, about 50% inhibition [5]) [5]
dipicolinic acid <5> (<5>, non-competitive inhibitor of 2-oxoglutarate fixation [6]) [6]
hydroxylysine <4> (<4>, 20 mM, 69% inhibition [4]; <4>, DL-hydroxylysine [8]) [4, 8]
iodoacetic acid <4> (<4>, 0.01 mM, 71% inhibition [7]) [7]
p-hydroxymercuribenzoate <4> (<4>, 0.0001 mM, complete inhibition [7]) [7]
pipecolic acid <5> (<5>, non-competitive inhibition on the fixation of both substrates [6]) [6]
selenalysine <6> (<6>, 7.5 mM, 90% inhibition, half-maximal inhibition at 1.9 mM, competitive inhibition against both acetyl-CoA and 2-oxoglutarate [11]) [11]

thialysine <4, 6, 7> (<4>, 5 mM, complete inhibition [8]; <6>, competitive with acetyl-CoA [11]; <7>, L-thialysine [12]) [8, 11, 12]

Activating compounds

ATP <3> (<3>, required [5]) [5]

Metals, ions

Mg^{2+} <3> (<3>, required [5]; <5>, about 10% stimulation, no absolute requirement [6]; <4>, 0.002 mM $MgCl_2$ has no effect [7]) [5]

Turnover number (min^{-1})

58 <11> (acetyl-CoA) [21]
58 <11> (oxaloacetate) [21]
92 <11> (acetyl-CoA, <11>, reaction with 2-oxoglutarate [21]) [21]

Specific activity (U/mg)

0.657 <10> [18]
Additional information <3, 4, 5> (<5>, rapid method for estimating the activity [6]) [4, 6, 15]

K_m-Value (mM)

0.025 <7> (2-oxoglutarate) [12]
0.028 <11> (acetyl-CoA) [21]
0.032 <11> (acetyl-CoA) [21]
0.033 <7> (acetyl-CoA) [12]
0.044 <11> (2-oxoglutarate) [21]
0.06 <10> (acetyl-CoA, <10>, reaction with 2-oxoglutarate [18]) [18]
0.12 <6> (acetyl-CoA, <6>, enzyme from S-(β-aminoethyl)-L-cysteine resistant mutant strain [10]) [10]
0.2 <6> (acetyl-CoA, <6>, enzyme from wild-type strain [10]) [10]
0.255 <11> (oxaloacetate) [21]
1.24 <10> (2-oxoadipate) [18]
2.2 <3> (2-oxoglutarate) [15]
2.24 <10> (2-oxoglutarate) [18]
2.28 <10> (oxaloacetate) [18]
5.5 <3> (2-oxoglutarate) [5]
10 <6> (2-oxoglutarate, <6>, enzyme from S-(β-aminoethyl)-L-cysteine resistant mutant strain [10]) [10]
11.8 <6> (2-oxoglutarate, <6>, enzyme from wild-type strain [10]) [10]

K_i-Value (mM)

0.008 <3> (L-Lys) [15]
0.0094 <11> (L-Lys) [21]
0.14 <6> (selenalysine, <6>, competitive against 2-oxoglutarate [11]) [11]
0.94 <6> (L-thialysine, <6>, competitive against acetyl-CoA [11]) [11]
1.1 <6> (selenalysine, <6>, competitive against acetyl-CoA [11]) [11]
5 <7> (L-lysine) [12]
5.1 <7> (DL-α-aminoadipate) [12]
15 <7> (L-thialysine) [12]

pH-Optimum

6.9 <3> [5]
7.5-8 <3, 4> [7, 15]
7.5-8.2 <5> [6]
8-8.7 <7> [12]

Temperature optimum (°C)

32 <4> [7]
37 <6> (<6>, wild-type strain [10]) [10]
43 <6> (<6>, enzyme from S-(β-aminoethyl)-L-cysteine resistant mutant strain [10]) [10]
60 <11> [21]

Temperature range (°C)

25-40 <4> (<4>, 25°C: about 60% of maximal activity, 40°C: about 50% of maximal activity [7]) [7]
40-80 <11> (<11>, 40°C: about 70% of maximal activity, 80°C: about 60% of maximal activity [21]) [21]

4 Enzyme Structure

Molecular weight

60000 <11> (<11>, gel filtration [21]) [21]
89000 <10> (<10>, gel filtration [18]) [18]
155000 <3> (<3>, gel filtration [15]) [15]

Subunits

? <9> (<9>, x * 48442, calculation from nucleotide sequence [17]) [17]

5 Isolation/Preparation/Mutation/Application

Source/tissue

mycelium <3> [3, 5, 9]

Localization

cytoplasm <3> (<3>, mainly located in [23]) [23]
cytosol <3> (<3>, main activity located in [15]) [15]
mitochondrion <3> (<3>, 25% of total activity [15]) [15]
nucleus <4> (<4>, two isoenzymes are located in the nucleus [20]) [19, 20, 23]

Purification

<3> [15]
<4> (partial [4]) [4]
<10> (recombinant enzyme [18]) [18]
<11> [21]

Cloning

<3> (overexpression in Penicillium chrysogenum using additional copies of lys1 with its own promoter or under the control of the pcbC promoter in either autonomously replicating or integrative vectors. Tranformants containing 3 to 32 additional copies of the lys1 gene are selected [22]) [14, 22]
<4> (expression in Escherichia coli [19]) [19]
<9> [17]
<10> (expressed at high levels in Escherichia coli [18]) [18]

6 Stability

Temperature stability

70 <11> (<11>, rapid loss of activity above [21]) [21]

Oxidation stability

<10>, loss of approximately 50% of its activity within 2 h upon exposure to air [18]

General stability information

<3>, enzyme is very unstable in crude and protamine sulphate-treated desalted extracts [15]
<3>, freezing and thawing results in 50% loss of activity [15]

Storage stability

<3>, 4-6°C, enzyme in cell-free extract in presence of 20% glycerol, 65% loss of activity after 48 h [15]
<5>, 4°C, stable for at least 1 week [6]
<6>, 0-5°C, 0.1 M potassium phosphate buffer, pH 7.8, containing 20% glycerol and 1 mM dithiothreitol, about 50% loss of activity after 2 weeks [10]

References

[1] Strassman, M.; Ceci, L.N.: Enzymatic formation of homocitric acid, an intermediate in lysine biosynthesis. Biochem. Biophys. Res. Commun., **14**, 262-267 (1964)

[2] Hogg, R.W.; Broquist, H.P.: Homocitrate formation in Neurospora crassa. J. Biol. Chem., **243**, 1839-1845 (1968)

[3] Demain, A.L.; Masurekar, P.S.: Lysine inhibition of in vivo homocitrate biosynthesis in Penicillium chrysogenum. J. Gen. Microbiol., **82**, 143-151 (1974)

[4] Tucci, A.F.; Ceci, L.N.: Homocitrate synthase from yeast. Arch. Biochem. Biophys., **153**, 742-750 (1972)

[5] Masurekar, P.S.; Demain, A.L.: Insensitivity of homocitrate synthase in extracts of Penicillium chrysogenum to feedback inhibition by lysine. Appl. Microbiol., **28**, 265-270 (1974)

[6] Gaillardin, C.M.; Poirier, L.; Heslot, H.: A kinetic study of homocitrate synthetase activity in the yeast Saccharomyces lipolytica. Biochim. Biophys. Acta, **422**, 390-406 (1976)
[7] Gray, G.S.; Bhattacharjee, J.K.: Biosynthesis of lysine in Saccharomyces cerevisiae: properties and spectrophotometric determination of homocitrate synthase activity. Can. J. Microbiol., **22**, 1664-1667 (1976)
[8] Gray, G.S.; Bhattacharjee, J.K.: Biosynthesis of lysine in Saccharomyces cerevisiae: Regulation of homocitrate synthase in analogue-resistant mutants. J. Gen. Microbiol., **97**, 117-120 (1976)
[9] Luengo, J.M.; Revilla, G.; Lopez, M.J.; Villanueva, J.R.; Martin, J.M.: Inhibition and repression of homocitrate synthase by lysine in Penicillium chrysogenum. J. Bacteriol., **144**, 869-876 (1980)
[10] Takenouchi, E.; Tanaka, H.; Soda, K.: Homocitrate synthase of S-(β-aminoethyl)-L-cysteine resistant mutant of Candida pelliculosa. J. Ferment. Technol., **59**, 429-433 (1981)
[11] Shimizu, E.; Yamana, R.; Tanaka, H.; Soda, K.: Effects of a selenium analogue of L-lysine on growth of Candida pelliculosa. Agric. Biol. Chem., **48**, 2871-2872 (1984)
[12] Schmidt, H.; Bode, R.; Lindner, M.; Birnbaum, D.: Lysine biosynthesis in the yeast Candida maltosa: properties of some enzymes and regulation of the biosynthetic pathway. J. Basic Microbiol., **25**, 675-681 (1985)
[13] Ye, Z.H.; Bhattacharjee, J.K.: Lysine biosynthesis pathway and biochemical blocks of lysine auxotrophs of Schizosaccharomyces pombe. J. Bacteriol., **170**, 5968-5970 (1988)
[14] Banuelos, O.; Casqueiro, J.; Fierro, F.; Hijarrubia, M.J.; Gutierrez, S.; Matin, J.F.: Characterization and lysine control of expression of the lys1 gene of Penicillium chrysogenum encoding homocitrate synthase. Gene, **226**, 51-59 (1999)
[15] Jaklitsch, W.M.; Kubicek, C.P.: Homocitrate synthase from Penicillium chrysogenum. Biochem. J., **269**, 247-253 (1990)
[16] Ramos, F.; Verhasselt, P.; Fellers, A.; Peeters, P.; Wach, A.; Dubois, E.; Volckaert, G.: Identification of the gene encoding a homocitrate synthase isoenzyme of Saccharomyces cerevisiae. Yeast, **12**, 1315-1320 (1996)
[17] Perez-Campo, F.M.; Nicaud, J.M.; Gaillardin, C.; Dominguez, A.: Cloning and sequencing of the LYS_1 gene encoding homocitrate synthase in the yeast Yarrowia lipolytica. Yeast, **12**, 1459-1469 (1996)
[18] Zheng, L.; White, H.R.; Dean, D.R.: Purification of the Acobacter vinelandii nifV-encoded homocitrate synthase. J. Bacteriol., **179**, 5963-5966 (1997)
[19] Chen, S.; Brockenbrough, J.S.; Dove, J.E.; Aris, P.: Homocitrate synthase is located in the nucleus in the yeast Saccharomyces cerevisiae. J. Biol. Chem., **272**, 10839-10846 (1997)
[20] Feller, A.; Ramos, F.; Pierard, A.; Dubois, E.: In Saccharomyces cerevisiae, feedback inhibition of homocitrate synthase by lysine modulates the activation of LYS gene expression by Lys14p. Eur. J. Biochem., **261**, 163-170 (1999)

[21] Wulandari, A.P.; Miyazaki, J.; Kobashi, N.; Nishiyama, M.; Hoshino, T.; Yamane, H.: Characterization of bacterial homocitrate synthase involved in lysine biosynthesis. FEBS Lett., **522**, 35-40 (2002)
[22] Banuelos, O.; Casqueiro, J.; Gutierrez, S.; Martin, J.F.: Overexpression of the lys1 gene in Penicillium chrysogenum: homocitrate synthase levels, α-aminoadipic acid pool and penicillin production. Appl. Microbiol. Biotechnol., **54**, 69-77 (2000)
[23] Banuelos, O.; Casqueiro, J.; Steidl, S.; Gutierrez, S.; Brakhage, A.; Martin, J.F.: Subcellular localization of the homocitrate synthase in Penicillium chrysogenum. Mol. Gen. Genet., **266**, 711-719 (2002)

Sulfoacetaldehyde acetyltransferase 2.3.3.15

1 Nomenclature

EC number
2.3.3.15

Systematic name
acetyl-phosphate:sulfite S-acetyltransferase (acyl-phosphate hydrolysing, 2-oxoethyl-forming)

Recommended name
sulfoacetaldehyde acetyltransferase

Synonyms
EC 4.4.1.12 (formerly)
O-acetylase
O-acetyltransferase
Xsc
acetylase
transacetylase

CAS registry number
9012-30-0

2 Source Organism

<1> *Alcaligenes defragrans* [1]
<2> *Bacterium sp.* (taurine-decomposing strain [2-4]) [2-4]
<3> *Pseudomonas aeruginosa* (strain TAU-5 [5,6]) [5, 6]
<4> *Acetinobacter sp.* (strain ICD [7]) [7]

3 Reaction and Specificity

Catalyzed reaction
acetyl phosphate + sulfite = 2-sulfoacetaldehyde + phosphate (<1> requires Mg^{2+}, phosphate, thiamin diphosphate [1])

Reaction type
C-S bond cleavage
transfer of phosphate

Natural substrates and products

S 2-sulfoacetaldehyde + phosphate <1-3> [1, 2, 5]

P acetyl phosphate + sulfite

S Additional information <1-3> (<1-3> involved in taurine catabolism [1-6]) [1-6]

P ?

Substrates and products

S 2-sulfoacetaldehyde + phosphate <1-4> (Reversibility: ? <1-4> [1,2,5,7]) [1, 2, 5, 7]

P acetyl phosphate + sulfite

Inhibitors

EDTA <2> (<2> 5 mM, 60% inhibition [4]) [4]
p-chloromercuribenzoate <2> (<2> 0.1 mM. 100% inhibition [4]) [4]
sulfite <2> (<2> above 1 mM, product inhibition [4]) [4]
thiamine <2> [3]

Cofactors/prosthetic groups

thiamine diphosphate <1-3> [1, 3, 6]

Activating compounds

Additional information <2> (<2> not stimulatory: NADH, NADPH, ADP, ATP, CoA, FAD, pyridoxal phosphate, pyridoxamine phosphate [3]) [3]

Metals, ions

Mg^{2+} <1, 2> (<1,2> required [1,2]) [1, 2]

Specific activity (U/mg)

0.4 <3> [5]

K_m-Value (mM)

5 <2> (2-sulfoacetaldehyde) [4]

pH-Range

6-8.2 <1> [1]
7.5 <2> (<2> assay at [3]) [3, 4]

Temperature optimum (°C)

30 <2> (<2> assay at [3]) [3]

4 Enzyme Structure

Molecular weight

85000 <2> (<2> gel filtration [4]) [4]
220000 <4> (<4> gel zymography [7]) [7]
252000 <1> (<1> gel filtration [1]) [1]

Subunits

tetramer <1> (<1> 4 x 63000-65000, SDS-PAGE, MALDI-TOF MS [1]) [1]

5 Isolation/Preparation/Mutation/Application

Localization
soluble <1, 2> [1, 2]

Purification
<1> [1]
<2> (partial [4]) [4]
<3> (partial [6]) [6]

Renaturation
<2> (thiamine diphosphate and $MgSO_4$ restore enzyme activity of dialyzed preparations [3]) [3]

Cloning
<1> [1]

References

[1] Ruff, J; Denger, K.; Cook, A.M.: Sulphoacetaldehyde acetyltransferase yields acetyl phosphate: purification from Alcaligenes defragrans and gene clusters in taurine degradation. Biochem. J., **369**, 275-285 (2003)
[2] Kondo, H.; Ishimoto, M.: Enzymatic formation of sulfite and acetate from sulfoacetaldehyde, a degradation product of taurine. J. Biochem., **72**, 487-489 (1972)
[3] Kondo, H.; Ishimoto, M.: Requirement for thiamine pyrophosphate and magnesium for sulfoacetaldehyde sulfo-lyase activity. J. Biochem., **76**, 229-231 (1974)
[4] Kondo, H.; Ishimoto, M.: Purification and properties of sulfoactetate sulfo-lyase, a thiamine pyrophosphate-dependent enzyme forming sulfite and acetate. J. Biochem., **78**, 317-325 (1975)
[5] Shimamoto, G.; Berk, R.S.: Taurine catabolism. III. Evidence for the participation of the glyoxylate cycle. Biochim. Biophys. Acta, **632**, 399-407 (1980)
[6] Shimamoto, G.; Berk, R.S.: Taurine catabolism. II. Biochemical and genetic evidence for sulfoacetaldehyde sulfo-lyase involvement. Biochim. Biophys. Acta, **632**, 121-130 (1980)
[7] King, J.E.; Jaouhari, R.; Quinn, J.P.: The role of sulfoacetaldehyde sulfo-lyase in the mineralization of isethionate by an environmental Acinetobacter isolate. Microbiology, **143**, 2339-2343 (1997)

The manufacturer's authorised representative in the EU is Springer
Nature Customer Service Centre GmbH, Europaplatz 3, 69115 Heidelberg,
Germany. If you have any concerns regarding our products, please
contact ProductSafety@springernature.com

Printed and bound by CPI Group (UK) Ltd, Croydon, CR0 4YY
25/09/2026
02222831-0010